Additional Problems with Solutions: A Supplement to

MICROELECTRONIC CIRCUITS

THIRD EDITION

By SEDRA / SMITH

K. C. SMITH

University of Toronto

SAUNDERS COLLEGE PUBLISHING

Harcourt Brace Jovanovich College Publishers

Fort Worth Philadelphia San Diego New York Orlando Austin San Antonio

Toronto Montreal London Sydney Tokyo

Printed in the United States of America.

Sedra/Smith: Problems Book to accompany MICROELECTRONIC CIRCUITS, 3/E

ISBN 0–03–052614–0

4567 021 9876543

CONTENTS

PREFACE

I OVERVIEW

• The Manual Format

This is a collection of problems and solutions with tabulated answers, designed to accompany the Text *"Microelectronics Circuits"*, third edition, by Sedra and Smith, Saunders College Publishing, 1991.

The goal of this **Manual**, captured in the words of the title, is to motivate and assist in the dynamic process of active learning.

The mechanism provided here in **Part I: Problems** is one of a collection of problems keyed to the **Text** in a variety of ways: Most obviously, the problems are grouped according to the sections of the **Text**. Possibly less apparent is their relationship both to segments of the **Text** and to the end-of-chapter problems contained there, about which more will be said shortly. As well, the problems are coded to indicate Complexity (C), Length (L), and Design content (D) with an appended asterisk notation to indicate intensity of the associated attribute.

Solutions provided in **Part II: Solutions** are relatively detailed. While the presentation is in a somewhat compressed format, attention has been given to revealing intermediate analytical and computational steps. As well, additional comments on context interpretation, and additional work are relatively common. Finally, answers are provided in **Part III: Answers** to allow readers to conveniently evaluate their success at problem solving without the inevitable hints that skimming the actual solution might provide.

• An Apology to the User – The Likelyhood of Errors

In a Manual such as this, intended as an aid to the student in a process of active learning, the issue of errors is a very critical one. Obviously, errors embodied in the problem solutions presented here can be very disconcerting to anyone who is less than secure in his knowledge of the subject matter. Thus the reduction of errors has been, and will continue to be, a high priority. It is in the latter sense that your indulgence and help are sought in the conjoined process of error detection and recovery. Certainly I will be most grateful for your help in reporting them.

In this process of error compensation, it is possibly useful to identify the types of errors you will inevitably find. In increasing order of subtlety and criticality, they are.

Typographical errors: There are many types of possible typographical errors which can be broadly characterized as omission, exchange, and replacement, either in word, number, symbol, phrase or sentence constructs. While unnecessarily confusing, they usually have the virtue of being correctable in context. To assist the detection process at its lowest level, solutions are relatively detailed with lots of intermediate calculations, relatively consistent variable naming, and relatively complete use of units for numerical results. Unfortunately, you may possibly find missing solution lines, as well.

Arithmetic Errors: These occur between steps in a computation as a result of calculator misuse or transcription error. They are distinguished from typographical errors by the fact that they propagate. They can be detected only by carefully checking and reproducing the preceeding substitutional and computational steps. Often the integrity of the following solution structure remains, but *not always*. One of the generic methods I use to help ensure structural integrity is an overall test for physical plausibility, or reasonableness, though this is generally not documented. Another, more explicit, attempt to reveal such errors is in the use of frequent *Check* comments which typically use a result in a somewhat global verification process. Incidentally, this is a good approach *for you to use* in your solutions as well!

Conceptual Errors: These are of two kinds, either local or global. The former occur usually as a result of misinterpretation of a symbol, or of the scope of a question. Occasionally you may find a piece of a question that was not answered at all, or answered in a less than complete fashion. The only virtue of this sin is that it is normally detectable. On a far more serious scale will be the occasional occurrence of totally wrong solution methods. These are quite insidious and confusing to a novice, since they can easily be mistaken to be a valid alternative approach. While these are relatively unlikely, they are almost certainly present.

For all of these errors, please accept my apologies. While I have utilized many approaches to minimizing them, the limitations of available time and resources has produced the result you see before you. All that remains to be said, again, is that I beg your indulgence, and look forward to your help in improving the situation.

• Solution Presentation Format

As you will note, the solution format, **Part II: Solutions** in this **Manual** is often less-than-ideal, being basically a run-on string of what would ideally be separated lines. This choice was made in view of the need to reduce the overall size of the **Manual** while making the solution relatively complete, with lots of intermediate steps. Obviously fewer steps in a more structured format would be more readable, and more beautiful, but probably less informative! To help in interpreting the string format, a somewhat-variable attempt at the use of bridging language, sentence structure, and punctuation has been made. For instructive variety, some solutions are presented more elegantly, including more explicit language, both with respect to physical arrangement and description, as well as mathematical structure.

II ADVICE TO THE STUDENT

• Coping with Errors

As noted earlier, I regret that you are likely to find errors in the solutions presented here. My regret concerns the fact that I am distressingly aware that an error of mine can be difficult to separate from a conceptual problem you may have. The only positive thing I can say is that learning to cope with imperfection is "good for the soul". Certainly a lot has been written about the positive effects of moderate stress on mental (and physical) development. Ask any reformed couch potato!

But what can you do? Certainly compare notes with your colleagues. Revel in the possibility that this **Manual** is an ideal candidate for leisure-time conversation, after a hard day in class or study hall. More seriously, it is certain that a minor degree of cross-checking with others can certainly avoid wasted time. Then, and even on your own, if your solution and mine differ, certainly be prepared for a quick check of obvious things - typos, arithmetic, etc. If you do not find the problem quickly, go on to another one, *as a way to test* **yourself**. If you have trouble there as well, suspect *your own need* for more reading and review of the **Text**. Otherwise a bit more work on checking the solutions is appropriate. Bear in mind, that it is regrettable, but true, that there are errors in these **Solutions**. *Feel good about yourself in finding them!* Feel sadness (and compassion) for my failure to do so!

• The Role of Circuit-Related Sketching in Electronics-Problem Solution

The merits of sketching in the solution of problems in Electronics cannot be overemphasized. Properly organized, *sketching* constitutes a highly-efficient information-encoding mechanism, *a language* in which relatively complex issues in electronics design and analysis can be presented and communicated. As well, particularly for those broadly conversant with its idioms and dialects, circuit-related sketching can provide the basis for an enriching aesthetic experience, manifesting a kind of poetry, or "music for the eyes" so to speak. This idea is a very important element in the graphic presentation style seen in the **Text** *"Microelectronics Circuits"*, where a lot of use is made of schematic-circuit and waveform sketches. While the idea is included in the **Text** implicitly, the role of sketching in Laboratory work is made more explicit in the **Manual** *"Laboratory Explorations"*.

Regrettably, here in this **Manual**, *"Trial and Success"*, it has not been possible to properly present anything like a complete view of the potential of sketching as language. There are two reasons, one economic, and one paedogogical.

The paedogogical issue appears first in problem presentations, in the use of circuit sketches in Part I. Thus, there, you see some problems posed almost exclusively in terms of circuit sketches. To better appreciate circuit sketches as language, pause for a moment to reflect on how to present such a problem as one of those, *without a sketch!* For large electronic assemblages, this can be a very daunting problem: For example, for those of you familiar with SPICE as a Circuit Simulator, contrast the sterility of the connection-specification list used there (for example in Appendix C of the **Text**), with the aesthetic elements of the circuit sketch it attempts to describe.

On the other hand, to communicate situational detail using spoken language is also important! Certainly as a student of Electronics, or of engineering in general, you must be able to handle problems presented in spoken-language style. However *one of the best ways of dealing with such problems is to prepare a sketch* of the situation described. Incidentally, for a person proficient in the process of circuit sketching, such a sketch would normally be created incrementally as the text is scanned, then augmented and checked as the text is reread.

In spite of all this, *the economic issue* associated with the creation of well-formed drawings in a published work such as this is a very real one. Regrettably, because of the relatively high costs of production and presentation, there are far fewer sketch-based problems provided to you in this **Manual** than good paedogogy would suggest. In particular, as well, there is a lot of reference to existing figures, say in the **Text**, where reuse of a costly resource is a logical part of a good engineering solution to any (engineering) problem.

More critically, in terms of illustrating the best style for you to emulate, I must emphasize that there are *far too few sketches* used in the **Solutions** part of this Manual. The ones seen usually arise in response to a direct request for a sketch. While this is *paedogogically wrong*, it is *economically necessary*. More concretely, in *your* work in Electronics, normally without these constraints, the very best and most effective style I would recommend is to always try a sketch. "*When in doubt, sketch*", would not be too strong a recommendation to follow. Notice that in the **Text**, an aspect of this idea is embedded in the recurring idea of "working on the diagram" that appears there, for example on page 211. As is illustrated occasionally in the **Solutions** to follow, it is generally *a very good idea* to notate circuit sketches with small calculations, whose role it is to present, memorably, in context, circuit-specific data. The idea of number-coded waveform sketches described on pages 390-391 of Appendix F4 of the **Laboratory Manual**, "*Laboratory Explorations*", is a quite dramatic example of this idea. In a very broad sense, in general, but certainly in the solution of the relatively intricate problems which appear in this **Manual**, first try to capture the specified situation as a sketch. Then, at or near the appropriate node of the circuit, possibly connected by a pointer line or other reference notation, do the calculations that you can, relative, for example, to bias point, signal limits, etc. Use these (possibly approximate) results, then, to guide your more elegant and formal solution, and, as well, to provide a rough check on the plausibility of your final results.

• Solving a Problem – Some General Advice

Read the Problem carefully to see if you understand the general idea it attempts to present. As noted earlier, try to present the situation described in a labelled sketch. The preparation of this sketch may be somewhat iterative – first a rough idea with some labels (left in place on your page), then a refined version added, with complete labelling. Note the idea of progression *without erasure*. As a general rule, don't eliminate earlier work, either by erasure or abandonment, for it represents the path of your progress, the history of the process of your "learning to learn", the shoulders on which your final solution stands, the available evidence of the logical process you can use when reviewing your work, and so on. Perhaps, later, you may want to make your solution more beautiful for final presentation, but this is often not necessary in the engineering workplace, except for very formal reports required by top management. Notice also that in the phrase "left in place", I have attempted to suggest avoiding the scraps of paper, the legendary "back of the envelope", and so on, which are relatively inappropriate in a modern responsible decision-path-traceable engineering-design process.

- In general, it is often a good idea to redraw the circuit (or photocopy it with segmentation and enlargement, if complex), and then do your work while looking at it, and working *on it*, if that is convenient.
- Prepare an informal summary table of the symbolic and numeric values of specified variables and of the values which you must find in your calculations. As suggested in detail in the **Manual** "Laboratory Explorations", organize the solution to your problem by first preparing a tabular format in which you might wish to present the results. Certainly from the point of view of real engineering problem solving, this is a very credible and effective way to both organize your thinking and the ultimate presentation of your work to the boss. Bear in mind, of course, that while all of this is a good idea (else I would not have written it!), it is often difficult to do and/or overkill in a simple situation. Whether you use the idea, or not, depends on your particular situation, in the same sense as does the use of refined sketches. **If it helps, do it!** Notice in general that most of life's problems are amenable to more than one solution style!
- As a generalization of the detailed comments above, always attempt to make the specifications of any problem you face, whether here, now, or later in real life, as *explicit* as you can. That is what the sketches and tables are all about! Set yourself up, as much as you can, for a multisensory input, for the possibility that a rapid review of the situation through, say, a quick glance at a circuit diagram, crystalizes the issue before you, thereby avoiding the forgotten fact, the unnecessary rework, etc.

• Relationship of the Problems Here to the Exercises and Problems in the Text

The problems in this **Manual** are intentionally coupled in a variety of ways to the **Exercises** and **Problems** in the **Text**:

- First, you will see that a fraction of the **Problems** are direct variations of those in the **Text**. By and large, these can be seen to represent several situations: One is of the acknowledged existence of a set of relatively basic, classic problems that bear repeating. Another is where problem variety in some subject is somehow limited. Another is a concern for representing, by example, a general approach to creating numerically-different problems in an area where that is often not straightforward. Another is to provide, in conjunction with the Exercises or Problems, in the **Text**, an opportunity to see the bigger picture as influenced by a particular set of circuit-design parameters, and thereby experience the issue of design variants, by viewing a few sample points in design space.
- Second, a fraction of the **Problems** presented are coupled more subtley to those in the **Text** by being expansions, extensions, or decompositions of them. By *expansion*, I imply the more detailed examination of an interesting aspect of the **Text** problem. By *extension*, I imply the posing of questions which enlarge the domain of analysis, of design, or of application. By *decomposition*, I refer to the reuse of selected parts of a **Text** problem, often over a wider domain of device parameters, loads, frequencies, etc. The enlarged dimensionality implied by the words *expansion* and *extension* is indicative of the fact that the **Problems** presented are often relatively complex. The argument, in support of intended complexity, are many: that real life is complex, that complexity may reinforce in-depth and long-chain thinking, that complexity by added parts implies choice, and, finally, that the existence of **Solutions** as aids tends to bolster a complex situation that could otherwise be quite difficult.

• Some Facts of Interest

This **Manual** contains 645 **Problems** incorporating 90 figures, with **Solutions** embodying 140 figures. Of the 645 problems, more than 168 involve direct design practice.

ACKNOWLEDGEMENTS

I would like to express my particular appreciation to some of those who made this work possible:

- To Laura Fujino, the love of my life, I am indebted for countless hours of discussion on the processes of problem creation and presentation, as well as for the final camera-ready production.
- To Inge Weber for her patience and endurance in preparing the basic text and figures from my less-than-ideal hand-wrought creations.
- To Alexa Barnes, my editor, who quite patiently endured the delay induced by other projects, and who oversaw final production.
- To Professor Peter Boulton, Director of the Computer Systems Research Institute at the University of Toronto, whose facilities and services have been used so intensively in this work.

To these and others more peripherally involved, I am most grateful.

But, for the errors and omissions, you will doubtless find here, I alone am responsible. For them, I must again apologize, and thank you in advance for your tolerance and forebearance in enduring and reporting them.

Kenneth Carless Smith, PhD, FIEEE, PEng
Department of Electrical Engineering
University of Toronto
10 King's College Rd.
Toronto, Ontario, M5S 1A4
Canada

FAX: 416 978 7423

Email: lfujino@csri.toronto.edu

July 1992

PART I
PROBLEMS

pages 1 to 109

CHARACTERIZATION CODE

C Complex

D Design

L Long

Where suffixes * and ** indicate
indicate more and much more
of the preceeding attribute.

Chapter 1

INTRODUCTION TO ELECTRONICS

SECTION 1.1: Signals

1.1 **For the following circuits, identify the signal-source form, whether Thevenin or Norton, and provide, in an organized two-column table, sketches of both standard forms. Where appropriate, reduce the circuit to its single-source, single-impedance form. Be careful with the polarities of voltage and current generators.**

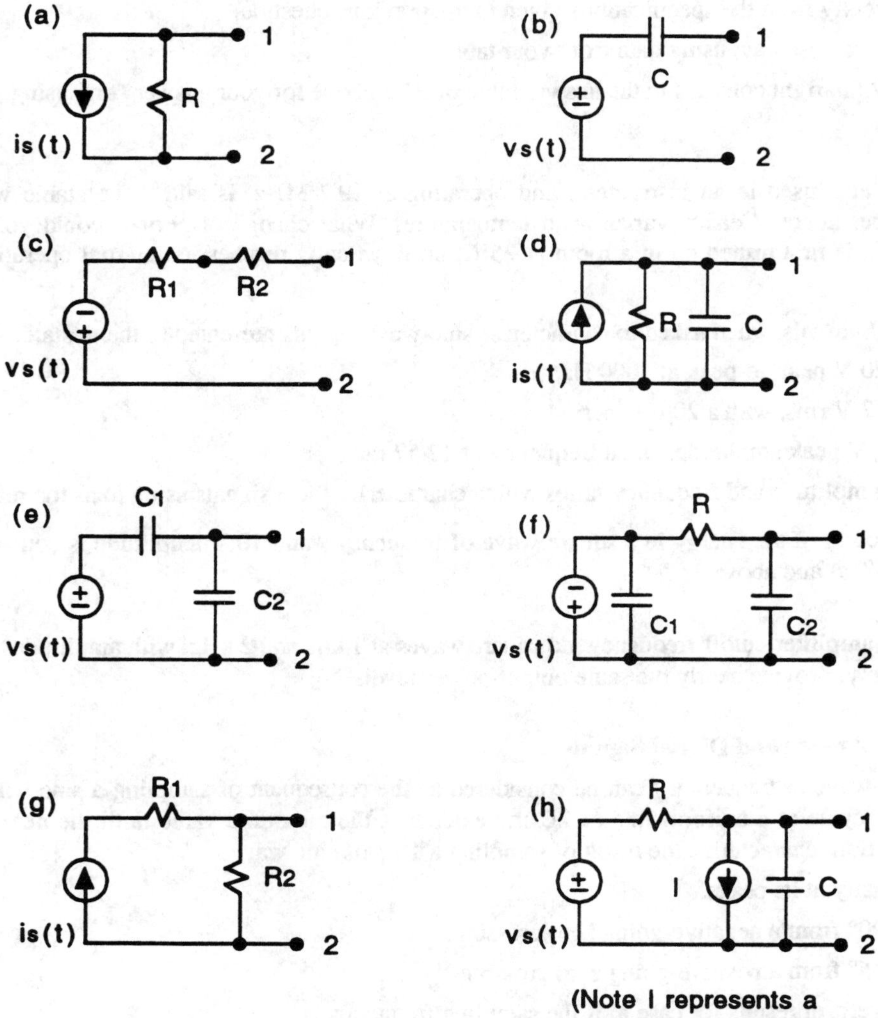

(Note I represents a constant (dc) current)

SECTION 1.2: Frequency Spectrum of Signals

L 1.2 For the following signals whose frequency is expressed either in radians per second or Hertz, find the corresponding value in the alternate form. Provide your answers in a neat five-column format, a line label at the left, Hz next left, rad/s at middle right, and 2 blank columns at the far right.

(a) 60Hz, (b) 754 rad/s, (c) 2513.3 rad/s, (d) 1010 kHz, (e) 97.30 MHz, (f) 1 Hz, (g) 377 rad/s, (h) 1 rad/s, (i) 1 GHz, (j) 400 GHz.

L 1.3 For each part of the previous question find the period of the signal. Express it in seconds in two ways, using 3 significant digits:

a) with one left of the decimal point and with an appropriate power of 10, and

b) using the standard names for subdivisions (seconds(s), milliseconds(ms), microseconds(μs), nanoseconds (ns), picoseconds (ps), femptoseconds (fs)).

Create your answer in two ways:

i) directly from the specifications given in the previous question,

ii) the easiest way, using data from your table.

Use the 2 far-right columns in the answer table of P1.2 above for your answer (first using powers of 10, then names).

1.4 An oscillator used in an instrument, and operating at 10.7 MHz, is said to be stable within 3 parts-per-million per-degree-Celsius variation in temperature. What change of period would you expect from the moment it is first turned on in a room at 25°C, until it finally reaches its internal operating temperature at 50°C?

1.5 Three individuals, when asked to characterize sine-wave signals presented to them, state:

a) 0.20 V peak-to-peak at 1000 Hz,

b) 2.12 V rms, with a 20μsec period,

c) 1.0 V peak amplitude, and a frequency of 12.57 rad/s.

Find the amplitude and frequency ratios which characterize the 3 signals using a) as the reference.

1.6 What fraction of the energy in a square wave of frequency f and 10 V amplitude is contained in harmonics above $9f$? at and above $3f$?

1.7 For what amplifier-cutoff frequency do square waves at 1 khz and 2 kHz, with amplitudes of 1.1V and 1.2V respectively, provide nearly the same output-power levels?

SECTION 1.3: Analog and Digital Signals

1.8 A square wave at frequency f can be considered as the consequent of sampling a sine wave of frequency f twice per cycle at a uniform rate of $2f$, and extending the measured value until the next sample. For this interpretation, characterize the result of sampling a 1V rms sine wave:

a) exactly at its peaks,

b) at 90° from a negative-going zero crossing,

c) at 45° from a positive-going zero crossing.

What waveform results for case a) if the sampling frequency is

i) doubled,

ii) halved?

SECTION 1.4: Amplifiers

L* **1.9** **Measurements made on a set of amplifiers provide the following attributes. Calculate those missing elements needed to characterize each. Each amplifier uses ± 10V supplies with no dc ground connection. Signal connections are with respect to ground, however. Signals are assumed to be sine waves whose peak values are given. Amplifier a) has been completely characterized by way of example.**

	Supply			Input				Output				A_v		A_i		A_p		Eff.
#	I_+ mA	I_- mA	P mW	υ_i mV	i_i µA	R_{in} kΩ	P_{in} µW	υ_O V	i_O mA	R_{load} kΩ	P_{out} mW	ratio V/mV	dB	ratio mA/µA	dB	ratio mW/µW	dB	%
a	3	3	60	1	1	1	.0005	2	20	0.1	20	2	66	20	86	40×10^3	76	33
b		1		20		.01		1		1								
c					10^3	0.1			10		10							10
d			200			.01	10				40	0.2						
e						10		0.5		10						0.1		20

1.10 **An amplifier operating from ± 10V supplies has a linear transfer characteristic passing through (0, 0), but with output saturation at +7V and –9V. If the amplifier gain is 50 V/V, what is the largest sine-wave input having no dc component, that can be applied without clipping?**

1.11 **If for the situation described in P1.10 above, it is desired to have the largest possible unclipped output, and a dc component can be tolerated, what is the rms value of the largest possible sine wave? What is the dc output component? To what dc value must the input be biassed?**

1.12 **An amplifier having a transfer characteristic**

$$\upsilon_O = 8 - 4\,(\upsilon_I - 1)^2$$

with

$$1 \le \upsilon_I \le \upsilon_O + 1\ ,\quad \upsilon_O \ge 0$$

is to operate with a dc output voltage of 4V. For an output signal of ≤1 volt peak amplitude at the input frequency ω, what % second-harmonic distortion results? (HINT: See P 1.11 in the Text)

1.13 Repeat Example 1.2 on page 16 of the Text, for the situation in which

$$\upsilon_O = 5 - 10^{-10}\, e^{40\upsilon_I}$$

for $\upsilon_I > 0$ and $\upsilon_O \geq \upsilon_I$. and the output is biassed at $V_O = +5/2$ volts. Find V_I, $L+$, $L-$, the output sine wave allowed, and the voltage gain A_υ at the bias point.

SECTION 1.5: Circuit Models for Amplifiers

1.14 A voltage amplifier connected to a particular source υ_S has a no-load voltage gain of 100 V/V and a gain of 70 V/V with a 1 kΩ load. What is its output resistance? What is its gain with a 500 Ω load?

1.15 An amplifier, when connected to a 10 kΩ source, has an overall gain (υ_o/υ_s) of 1667 V/V. When a second identical amplifier is connected in parallel to the same source, the corresponding gain for each is found to be 909 V/V. Estimate the input resistance of the amplifiers.

1.16 A voltage amplifier has an open-circuit voltage gain of $A_{\upsilon o}$, an input resistance R_i and an output resistance R_o. Find the condition under which a cascade of n of these amplifiers has the same open-circuit gain as a single amplifier.

D **1.17** A design is required of an amplifier to operate between a 1 MΩ source and a 100 Ω load. You have two amplifiers, each with a gain 10 V/V, but with input and output resistances of A_1 being 1 MΩ and 10 kΩ, and of A_2 being 10 kΩ and 100 Ω respectively. There are two possible ways to connect the two amplifiers between the source and load. Which is best? What is the overall gain? Contrast this with the gain using only one amplifier at a time? If a good fairy granted you one wish – to double (or halve) any one property of either amplifier – is there a best choice to be made? Why?

1.18 A voltage amplifier with a basic gain of 80 dB, has an output resistance of 10 kΩ. What is the voltage gain which results for a load of 1 MΩ, 10 kΩ, 10Ω? What is its equivalent transconductance when operating into a zero-ohm load?

DL* **1.19** This problem is intended to provide you with a basis for insight into Problem 1.15 in the Text.

(a) Evaluate the gain υ_S/υ_O for each of the amplifiers interposed individually between the stated source and load.

(b) >From the process and results of (a), identify where the least losses occur, whether at the source or load, for each amplifier. Use these observations to make 3 lists of amplifiers (put in descending order of merit), as input-stage coupler, output-stage coupler and as provider of gain.

(c) Now consider a design with a pair of amplifiers, picking, as input, an amplifier high on list 1 and reasonable on list 3, and, as output, one high on list 2 and reasonable on list 3.

(d) What is the highest gain you can get from two stages?

(e) Reconsider the process in an attempt to see if you could reach the same conclusion by simply thinking about it, rather than by making explicit lists.

DL* 1.20 You are required to design a two-stage current amplifier to operate between a current source having a 10 kΩ internal resistance and a load of 10 kΩ. Three types of amplifier stage are available:

(1) A low-input-resistance type, with $R_i = 10\ \Omega$, $R_o = 10\ k\Omega$ and $A_{is} = 100A/A$

(2) A high-gain type, with $R_i = 10\ k\Omega$, $R_o = 1\ k\Omega$ and $A_{is} = 1000A/A$

(3) A high-output resistance type, with $R_i = 10\ k\Omega$, $R_o = 100\ k\Omega$ and $A_{is} = 100A/A$.

How many amplifier combinations are there? Rank them by available gain.

D 1.21 Reconsider Problem 1.20 above. Rank the 3 amplifiers on the basis of a *figure of merit* (for current amplifiers) which is $\left(\dfrac{A_{is} \times R_o}{R_i}\right)$. Select the two amplifiers of *lowest rank*, and use only those types to design a two-stage current amplifier of highest-possible gain between a 10 kΩ source and 10 kΩ load. What is the highest available gain?

D 1.22 Reconsider the three amplifiers introduced in Problem 1.20 above as transconductance amplifiers. Restate the specifications of each as a transconductance amplifier. Identify a figure of merit for a transconductance amplifier like that suggested in Problem 1.21 above for a current amplifier. Use this to rank the three as transconductance amplifiers.

1.23 Using the results of Example 1.4 (on page 25 of the Text) for a BJT, characterize its use with E grounded, B as input and C as output, both as a current amplifier and as a transconductance amplifier. Use $r_\pi = 5\ k\Omega$ and $\beta = 200$. What are A_{is} and G_m respectively?

C 1.24 Reconsider the gyrator circuit of Fig. 1.20 (on Page 27 of the Text). For each transconductance amplifier having $g = 1$ mA/V and an input and output resistance, each of 100 kΩ, find the inductance L that results from use of $C = 1\ \mu F$. What equivalent series resistance does it have (due to R_i and R_o)?

SECTION 1.6: Frequency Response of Amplifiers

1.25 In passing through a particular amplifier, an input sine wave of 2 mV peak-to-peak amplitude at 1 kHz emerges with the same waveform shape, an amplitude increased to 2V peak and evidence that is has been delayed by 0.2 ms. For the amplifier transmission, what is the magnitude? What is the phase?

1.26 A direct-coupled amplifier (one whose response extends down to zero frequency) has an upper 3 dB frequency of 100 kHz. What is its bandwidth? When coupled to a signal source using a capacitor, its frequency response is found to deteriorate at low frequencies, the response being reduced by 3 dB at 20 kHz. What is the overall bandwidth of this arrangement?

1.27 Consider the circuits of Fig. 1.23 (on page 31 of the Text). In a particular system application, a new output $V_{out} = V_i - V_o$ is created in each case. What is the type of the corresponding output V_{out} for circuit a)? circuit b)?

1.28 An amplifier, considered to have a high-frequency response which can be characterized as STC, is measured at 3 frequencies, 1 kHz, 10 kHz and 20 kHz, at which the gain magnitude is found to be 11×10^3, 8×10^3, and 4×10^3 V/V, respectively. Estimate the 3 dB frequency and the frequency at which the gain can be expected to drop to 1. At what frequency does a phase lag of 60° or so appear?

CDL 1.29 Consider one stage of the amplifier cascade in Fig. P1.28 (on page 44 of the Text). At what frequency is its response 3 dB down from the midband value? For 2 stages in cascade, what does the 3 dB frequency become? For a modified 2-stage cascade in which one of the resistors is decreased to kR (k≤1), find a process to calculate what the frequency becomes. For what value of k does $f_{3\,dB}$ of the modified 2-stage cascade have a value $\dfrac{0.95}{2\pi RC}$?

L 1.30 A voltage amplifier has the transfer function

$$A_\upsilon = \frac{1000}{\left[1 + j\,\dfrac{f}{10^5}\right]\left[1 + \dfrac{10}{j\,f}\right]}$$

On a Bode magnitude plot, sketch asymtotes representing each of the terms shown. Then sketch the overall (sum) response. What do each of the three terms contribute (in dB) at $f = 1$, 10, 100, 10^4, 10^5 and 10^6 Hz. What is the overall response at the same frequencies? What is the 3 dB bandwidth of the amplifier? Over what frequency range is the phase $0 \pm 6°$?

1.31 A voltage amplifier has the transfer function

$$A_\upsilon = \frac{10^7\,j\,f}{\left[j\,f + 10^5\right]\left[\dfrac{j\,f}{10} + 1\right]}$$

Note that this is not in the most useful standard form. Without converting it explicitly, what are the upper and lower 3 dB frequencies and what is the midband gain (i.e. the gain between the upper and lower cutoffs)? Now reduce A_υ to standard form, and consider the same questions: Do you have a preference for one form over the other?

D 1.32 Consider the transconductance amplifier of Fig. 1.17a) (on page 23 of the Text) driving a load capacitance of C=10pF and driven by a 10 kΩ source, R_s. Find expressions for the gain at low frequencies and the associated upper 3 dB frequency. For one particular amplifying device, namely a BJT, both R_i and R_o are inversely proportional to bias current I, while G_m is directly proportion to it. Typically

$$R_i = \frac{2.5}{I},\quad R_o = \frac{200}{I} \quad \text{and } G_m = 40I$$

Design the circuit bias current so that the resulting upper 3 dB frequency is 1 MHz or more. What is the midband gain A_M that results? Using the expressions you have derived, find the product of gain and bandwidth. What is interesting about it? Use this result to state the gain of an amplifier whose bias is adjusted for a 3 dB frequency of 10 MHz. What current is needed?

1.33 Find the transfer function of the circuit shown: Sketch its magnitude and phase.

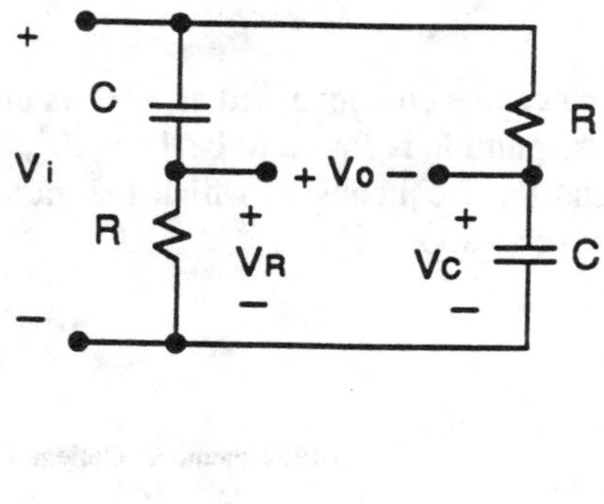

Chapter 2

OPERATIONAL AMPLIFIERS

SECTION 2.1: The OpAmp Terminals

2.1 What is the number of op amps that can be accomodated in an 8-pin IC package? In a 14-pin package? How many unused pins are there in each case?

SECTION 2.2: The Ideal Op Amp

2.2 An otherwise-ideal op amp, known to have a gain of 10^4 V/V, is measured in a circuit to have an output voltage of −3 V. While it would be difficult to measure, what would you expect the voltage from the negative input pin to the positive one to be? If the voltage at the positive pin is known to be +100 mV, what is the voltage you would expect at the negative one?

2.3 For the amplifier described in P2.2 above, connected in the circuit shown in Fig. P2.2 (on page 104 of the Text), what voltage v_I would be required at the input to produce $v_O = 3.5$ V?

SECTION 2.3: Analysis of Circuits Containing Ideal Op Amps – The Inverting Configuration

2.4 An inverting op-amp circuit with the topology of Fig. 2.4 on page 52 of the Text, has $R_1 = 4.7$ kΩ and $R_2 = 47$ kΩ. What closed-loop gain would you expect? In the laboratory, a student accidentally exchanges these two resistors. What gain would you expect him to find?

2.5 The circuit shown in Fig. P2.6 c), (on page 105 of the Text) using an op amp with a gain of 10^4 V/V, is found to have an output voltage of +10 V. What is the voltage required at the inverting input terminal of the op amp for this to occur? What is the current through the grounded 10 kΩ resistor? What is the input voltage, v_I, you would expect? (Hint: First, consider this question assuming that the gain (10^4) is very very high. Then, refine your answer with a calculation in which a very small error correction is made).

D 2.6 Design an op-amp circuit with a gain of −2 V/V, using three 100 kΩ resistors. How many solutions are there? What is the input resistance of each?

D 2.7 Design an inverting op-amp circuit with a gain whose magnitude is 10 V/V using one 220 kΩ resistor and another resistor no greater than 1 MΩ.

DL* 2.8 Design an amplifier with a gain of −20 V/V, an input resistance of −100 kΩ, and no resistor greater than 1 MΩ. (Hint: you need more than 2 resistors!)

2.9 An inverting op-amp circuit is designed to use one 10 kΩ and one 100 kΩ resistor. What are the two possible closed-loop gains you would expect with an ideal op amp? What gains do you get with an op amp whose open-loop gain is only 100 V/V?

2.10 An inverting op-amp circuit designed for a nominal gain of −100V/V uses a very high-frequency amplifier whose open-loop gain is relatively low. What must the amplifier gain be if the closed-loop gain is to lie within 10% of the nominal value? Within 1% of nominal?

2.11 Design an amplifier with a gain of 200 V/V and an input resistance of 100 kΩ using 2 op amps and resistors no larger than 1 MΩ. Share the gain as much as possible between the two amplifiers.

D 2.12 Reconsider P2.11 above if R_{in} must be 2 MΩ. Use a minimum number of resistors.

CD 2.13 Design the circuit of Fig. 2.8 on page 57 of the Text to have an input resistance of 1 MΩ and a gain of −22 V/V using resistors no larger than 1 MΩ. If resistors no smaller than 100 kΩ are available, what do you do?

C 2.14 Consider the circuit of Fig 2.8 of the Text with the grounded end of R_3 connected to input υ_2, and υ_1 connected to R_1. Use the approach in Example 2.2 (on page 57 of the Text) and superposition, to find an expression for υ_O in terms of υ_2 alone, and of υ_1 and υ_2 together.

SECTION 2.4: Other Applications of the Inverting Configuration

2.15 A Miller integrator for which the time constant is 1 ms is driven by a positive step of 1 volt amplitude. What does the output do? At what rate? If the initial output voltage is 10 V, how long does it take for the output to reach 0 V?

2.16 A Miller integrator with a time constant of 10 ms is driven by a 60 Hz sine wave of 0.1 V peak amplitude. Describe the resulting output waveform, in amplitude and phase. Is the output leading or lagging the input?

C 2.17 Consider a differentiator circuit such as that shown in Fig. 2.12 a), on page 62 of the Text, having a 5 ms time constant. For what rate of change of input signal is the output +1 V? An input signal begins to rise from zero volts at t=0 at the rate of 1 V/ms, reaches a value of 20 V, then falls at the same rate to zero volts. Sketch and label the resulting output waveform over an interval of 50 ms.

L 2.18 The circuit of Fig 2.12 a) of the Text is augmented by a resistor r = 100Ω in series with C = 1.0 μF. Resistor R = 10 kΩ. Sketch and label the output if the input is:

a) a positive pulse of 0.1 V amplitude and 10 μs duration,

b) a negative pulse of 50 mV amplitude and 0.1 s duration.

D 2.19 Design a circuit with 3 inputs to provide an output $\upsilon_O = -(\upsilon_1 + 2\,\upsilon_2 + 3\,\upsilon_3)$ using 10 kΩ as the smallest resistor.

CDL 2.20 Design a circuit to combine 3 inputs to form $\upsilon_O = \upsilon_1 + 2\,\upsilon_2 - 3\,\upsilon_3$. Use only inverting amplifiers, with 10 kΩ as the smallest resistor. There is more than one way! Find one which minimizes the total resistance used.

2.21 For the following circuit, find an expression for the output υ_O in terms of υ_1 and υ_2, assuming an ideal op amp.

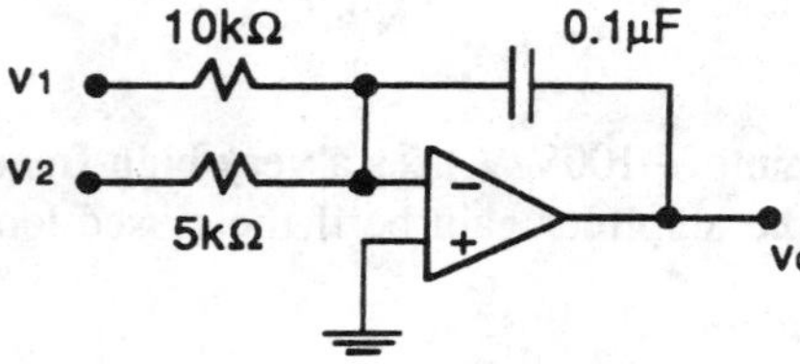

L 2.22 Find the transfer function of the following circuit:

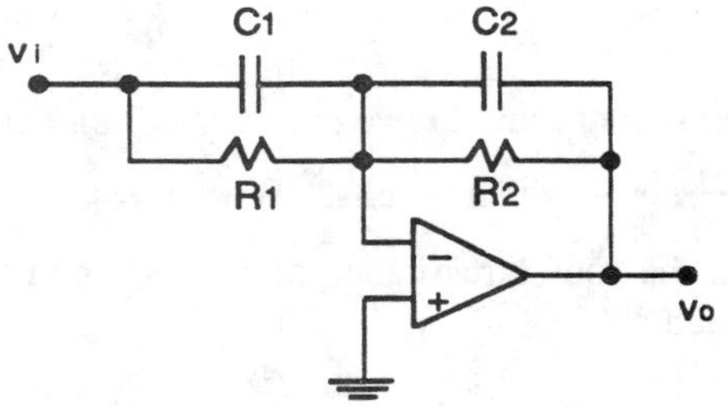

What is the condition for which the output is independent of frequency? Sketch Bode magnitude plots (in rad/s) for 3 cases:

a) $C_2 = 0.1\,C_1 = 0.1\mu F$, $R_2 = 10\,R_1 = 100\text{ k}\Omega$;

b) R_2 is raised to 1 MΩ;

c) R_2 is lowered to 10 kΩ.

SECTION 2.5: The Noninverting Configuration

2.23 A non-inverting op-amp circuit with the topology of Fig 2.14, on page 64 of the Text, has $R_1 = 4.7\text{ k}\Omega$ and $R_2 = 47\text{ k}\Omega$. What closed-loop gain would you expect? In the laboratory, a student accidentally exchanges these two resistors. What gain would you expect her to find?

D 2.24 Design a non-inverting amplifier with a gain of 1.5 V/V using three 1 kΩ resistors. Sketch two solutions.

CL 2.25 Use the circuit idea shown in Fig P2.33 (on page 109 of the Text) to design a circuit whose output is $v_O = v_1 + 2v_2 - 3v_3$, with 10 kΩ as the smallest resistor used. There are several possible ways! Find one.

DL* 2.26 Use the general result outlined in P2.33 on page 109 of the Text for the circuit shown there in Fig P2.33 to create a circuit to provide an output $v_O = 10\,(v_1 - v_2)$. (Hint: Use an additional positive input.) Have you seen this circuit before? What is it called? You may find the latter questions more straightforward after you have read the next Section of the Text.

D 2.27 A designer, needing to provide a unity-gain buffer, considers the use of the circuit topology shown in Fig. 2.17 on page 66 of the Text. However, the amplifier he has available has an open-loop gain of only 10. What closed-loop gain would the simple circuit produce? His boss suggests that he consider the circuit of Fig 2.15 as a solution. As well, she suggests that the smallest resistor used be 10 kΩ. What design would result?

SECTION 2.6: Examples of Op-Amp Circuits

CD 2.28 A designer wishes to use a simple modification of the circuit of Fig 2.18 on page 67 of the Text to implement a centre-zero voltmeter whose scale ends are ±1 volt. The meter movement provided is a 0 to 1 mA unit with a resistance of 50 Ω. Her boss suggests that a solution is possible using a single additional resistor and one of the ±10 V supplies from which the op amp is powered. What is the value of the additional resistor? To what supply is it connected? To what circuit node is the additional resistor connected? What is the required value of R?

CD 2.29 An analog-circuit designer requires a +5 V power source from which to run a small amount of digital logic requiring 20 mA at +5 V. The analog system uses ±15 V supplies which are quite well-regulated (that is stable over time and temperature and reasonably independent of load). Suggest a simple op-amp circuit, using a resistor network operating at 0.5 mA, to do the job. If the op amp requires a bias current of 2 mA from its supplies at no load, what is its total power dissipation when fully loaded at the maximum current required by the logic?

2.30 For a particular difference amplifier using the topology of Fig 2.19 on page 68 of the Text, $R_2 = R_4 = 100$ kΩ and $R_1 = R_3 = 10$ kΩ. What is the gain, $A = \frac{v_O}{v_1 - v_2}$, you would expect? (Be careful of what is asked!).

D 2.31 The difference amplifier described in P2.30 above is connected to two sources, v_{S1} and v_{S2}, each having a 10 kΩ internal resistance. What is the gain $\frac{v_O}{v_{S1} - v_{S2}}$ which results? What must you do to achieve a source-to-output gain of magnitude 10. As well, the source resistance of v_{S2} is found to be only 8 kΩ. What else must you do to achieve true difference action?

L 2.32 Reconsider the difference amplifier analyzed in Example 2.4 on page 68 of the Text, using Fig 2.19 and Fig 2.20, under the condition that resistor R_4 is connected to a 3rd input, v_3. Find the expression corresponding to Equation 2.4 for v_O. Simplify it for the case in which $\frac{R_2}{R_1} = \frac{R_4}{R_3}$.

L 2.33 Consider the circuit shown here which employs an ideal op amp.

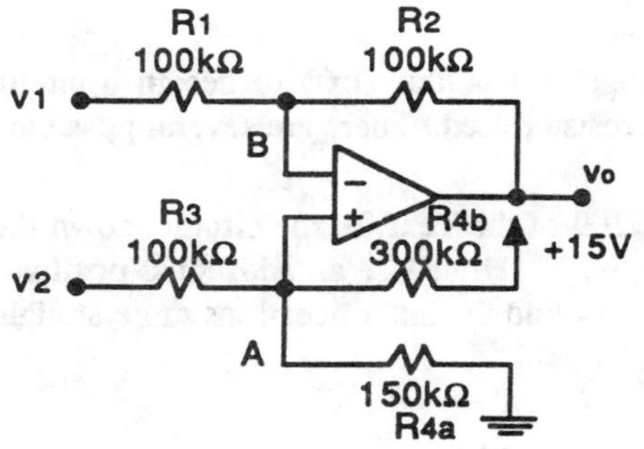

What is the value of v_O for

a) $v_1 = v_2 = 5$ V,

b) $v_1 = v_2 = 0$ V,

c) $v_1 = +3$ V, $v_2 = -2$ V?

Do your analysis from first principles. Afterward, consider the answer to P2.32 above.

D 2.34 Using the circuit of Fig 2.23 on page 71 of the Text, design an instrumentation amplifier with a difference gain of 100 V/V shared equally between the input and output stages. Employ 10 kΩ as the smallest resistor. For your design, what voltages appear on the outputs of A_1 and A_2 for $v_1 = 5.0$ V and $v_2 = 4.9$ V?

2.35 Consider the negative-impedance-converter circuit of Fig 2.24 on page 73 of the Text, in which $R_1 = 10$ kΩ, $R_2 = 10$ kΩ and $R = 11$ kΩ, connected to a signal sourse v_S whose source resistance is 10 kΩ. What voltages result at the positive input, v_I, and the output, v_O, of the op amp for $v_S = +0.5$V? (Hint: Use the expression for R_{in} in the text, then a voltage-divider calculation to find v_I).

CL 2.36 To the circuit shown in Fig 2.26 on page 75 of the Text, with $r = R = 10$ kΩ and $C = 1$μF, a string of 1 volt pulses of constant, but unknown, width is applied from an initial condition of $v_2 = 0$. Sketch the voltage at v_O assuming the pulses are equally spaced. If the output reaches +10 V during the tenth pulse, estimate the pulse width. What are the limits of certainty in your estimate, again assuming all pulses are exactly the same width?

SECTION 2.7: Nonideal Performance of Op Amps

2.37 An internally compensated op amp has f_t of 10 MHz and a dc gain of 10^6 V/V. What is the 3 dB frequency of its open-loop gain? If this amplifier is to be operated at 100 kHz, what gain is available?

2.38 The op amp in P2.37 above is to be used in a closed-loop amplifier having a gain of 20 dB. What corresponding break frequencies would you observe in the inverting and non-inverting versions? For what frequencies is the phase shift of the corresponding amplifier less than 6 degrees?

2.39 The op amp described in P2.37 above is to be used in a system for which low-frequency operation should extend (within 3 dB) to 10 kHz. What is the maximum closed-loop gain available from a single amplifier? >From 2 amplifiers used in cascade? (See the result for 2 amplifiers in cascade developed in P2.53 in the Text, on page 112).

2.40 A measurement of the closed-loop gain of an amplifier shows it to be –25 V/V at 120 kHz and –100 V/V at 5 kHz. Estimate the closed-loop gain at low frequencies and the corresponding 3 dB frequency. What is f_t of the op amp used? (Be careful!)

2.41 An amplifier intended for very-high-frequency operation, yet characterized by a single-pole rolloff, has f_t = 100 MHz and A_0 = 20 V/V. For a design in which the actual (rather than the nominal) closed-loop gain is –10 V/V, what 3 dB frequency results?

SECTION 2.8: The Internal Structure of IC Op Amps

D 2.42 For a particular op amp modelled as in Fig 2.31 of the Text, G_m = 2 mA/V, $R_{i2} = 4\times10^6\Omega$ and μ = 700. Measurements indicate the open-loop gain to be 1.12×10^6 V/V. Estimate R_{01}. Chose a value of C to make the unity-gain frequency equal to 15.9 MHz.

2.43 For a particular op amp, f_t is measured to be 10 MHz and C is known to be 5 pF. What value of G_m must have been used?

SECTION 2.9: Large-Signal Operation of Op Amps

2.44 An op-amp circuit operating from ±10 V supplies has L+ and L– of +8 V and –8.5 V respectively, and a closed-loop gain of –10 V/V. What is the peak-to-peak value of the largest possible input sine wave having zero average, for which the output is not distorted?

2.45 An op amp has a slew rate of 10 V/μsec. What is the highest frequency at which it can reproduce a 6-V peak-to-peak triangle wave at its output?

2.46 An op amp, using a compensation capacitor of 30 pF, has a slew rate of 5 V/μs and f_t = 10 MHz. What are the values of the biasing current and transconductance of the input stage. Verify your values using the result of P2.64 on page 113 of the Text.

2.47 Find an expression for the amplitude of a sine wave for which the small-signal and large-signal (SR-limited) bandwidths are the same. When the small-signal bandwidth is 0.5 MHz and the slew rate is 2 V/μsec, what is the amplitude for which equal bandwidths result?

SECTION 2.10: Common-Mode Rejection

2.48 A differential amplifier has a composite input signal consisting of 2 sine-wave components at different frequencies (60 Hz and 1 kHz) at each of its inputs: Both have a common component of 8 volts peak at 60 Hz. At 1 kHZ, each has a component of 1 mV amplitude, but of 180° relative phase. The output consists of a 0.6 V peak component at 60 Hz and a 60 mV peak component at 1 kHz. Find the difference-mode gain, the common-mode gain, and the CMRR, both as a ratio and in dB.

2.49 An amplifier having a CMRR of 100 dB has a desired difference-mode output of 2 Vpp and a difference mode gain of 200 V/V. For what value of input common-mode signal is the unwanted signal only 1% of the desired output?

2.50 A difference amplifier is constructed to have a differential gain of 10, using an op amp whose CMRR is 120 dB. What is the common-mode gain of the resulting difference amplifier? What is its CMRR?

SECTION 2.11: Input and Output Resistances

2.51 A non-inverting buffer uses an op amp with $A_0 = 10^4$ V/V, $R_{icm} = 10\ M\Omega$, and $R_{id} = 1\ k\Omega$. What is its input resistance at low frequencies?

C 2.52 A power op amp for which $R_{icm} = 50\ M\Omega$, $R_{id} = 10\ k\Omega$, $A_0 = 10^3$ V/V and $f_t = 10^4$ Hz is used in a non-inverting configuration with gain of 10. Characterize its input impedance as a frequency-dependent network. At what frequencies does it show a 45° phase?

2.53 An op-amp circuit with gain −10 V/V uses a CMOS op amp, whose open-loop output resistance is 100 kΩ, whose dc gain is 10^4 V/V, and whose $f_t = 20$ MHz. Characterize the output resistance. At what frequency does its magnitude become 10 times that at dc?

SECTION 2.12: DC Problems

D 2.54 For an amplifier operating with ±4 V saturation limits at a closed-loop gain of −100 V/V, what input offset voltage is required to assure less than 1% reduction in output swing capability due to offset?

D 2.55 An inverting amplifier with gain of −100 V/V and an input resistance of 100 kΩ, uses an op amp with 1 mV offset, a bias current of 30 nA and an offset current of 3 nA. What output offset results with a) a basic uncompensated design b) a bias-current-compensated design? Which offset source dominates in each case? What is the net output offset if it is halved? What compensating resistor do you use?

D 2.56 If the amplifier in P2.55 above is capacitor-coupled at the input, what output offset results in the basic and compensated designs? What compensation resistor should be used?

CD 2.57 Design a direct-coupled inverting op amp with a gain of −100 V/V, the highest possible input resistance, and an output offset ≤0.5 V, using an op amp with 2 mV offset, and bias currents of 1 μA equal to within ±10%. What is R_{in} of your design?

Chapter 3

DIODES

SECTION 3.1: The Ideal Diode

3.1 For the following circuits employing ideal diodes, find the labelled currents, I, and voltages, V, measured with respect to ground.

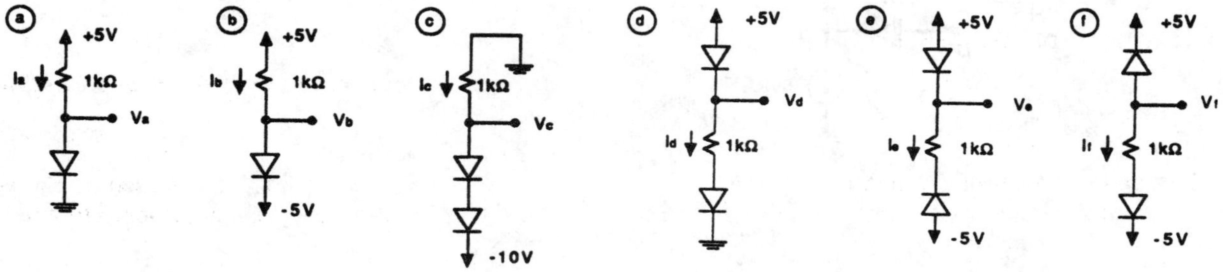

L 3.2 For the following logic gates using ideal diodes:

i) If $V_A = V_E = 5$ V, and $V_B = V_C = V_D = 0$ V, what is the value of V_Y produced?

ii) If logic '1' = 5 V and logic '0' = 0 V, identify the logic function performed.

iii) If logic '1' = 0 V and logic '0' = 5 V, identify the logic function performed.

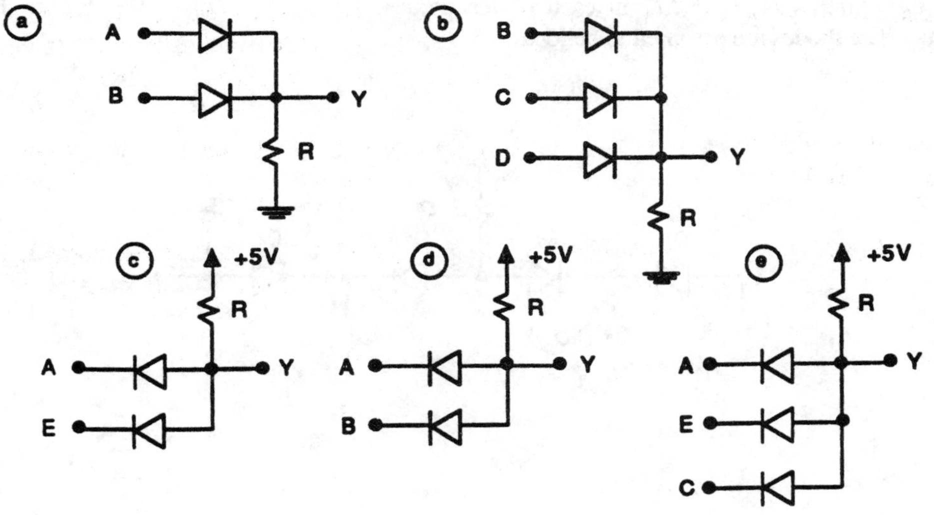

3.3

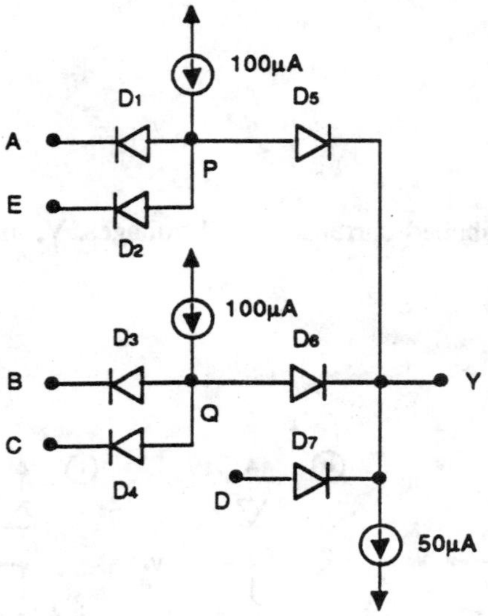

For the conditions stated in P3.2 ii) above, find an expression for the logic function $Y = f(A, B, C, D, E)$? In particular, for the input logic values stated, what is the logic output value?

3.4

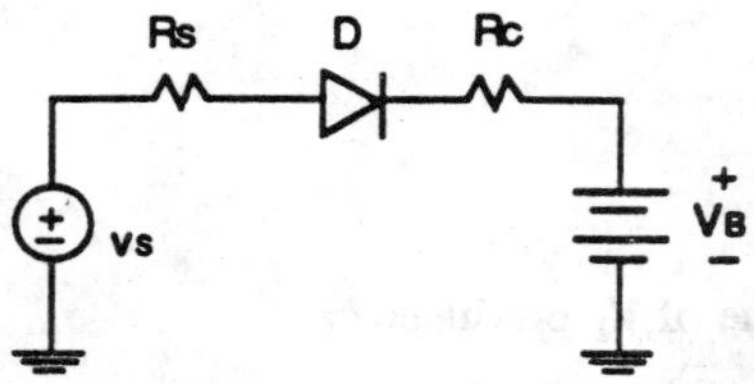

In the battery-charger circuit shown, the sinewave input v_S is 12 V rms, while the battery voltage varies from 12 V to 14 V from the discharged to fully-charged states. $R_S = 10\ \Omega$ is the charging-source resistance, D is an ideal diode and $R_C = 50\ \Omega$ is a current-controlling resistor established by the designer. Sketch and label the diode-current waveform for $V_B = 12$ V. What are its peak and average values? What do the peak and average diode currents become when V_B reaches 14 V?

3.5 Find the currents I_1, I_2, I_3, I_4 in each of the diodes D_1, D_2, D_3, D_4 of the circuit shown. What V_O results? The diodes are assumed to be ideal.

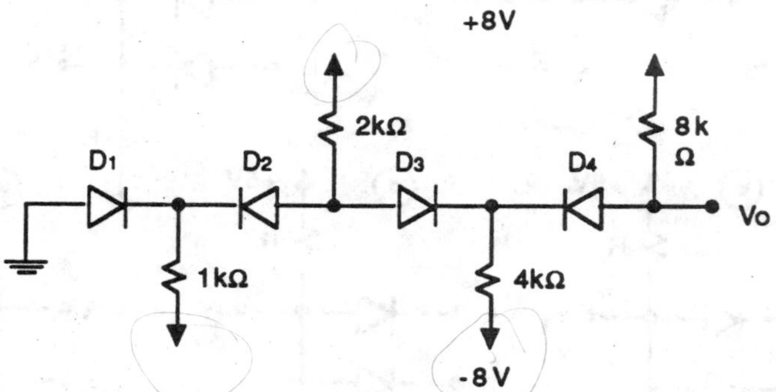

SECTION 3.2: Terminal Characteristics of Junction Diodes

3.6 A small discrete silicon diode (a "100 μA diode") is found to conduct 100 μA at 0.700 V and 1 mA at 0.815 V. Find the values of n and I_S which correspond.

3.7 A diode for which n=1 conducts 0.1 mA at 0.7 V. Find its voltage drop at 1 mA. For what current is its voltage drop equal to 0.815 V?

3.8 A 10-A silicon diode for which n=2 is known to have a forward voltage drop of 0.700 V at 10 A. What is the junction voltage at which it conducts 10 mA? 10 μA?

3.9 A particular "1 mA diode", which at 25° C conducts 1 mA at 0.7 V, is operated at 95° C in a circuit which provides it a constant 100 μA. What does its junction voltage become if $n = 2$?

3.10 For the situation described in P3.9 above, the leakage current at 25° C is 1 nA. What does it become at 95° C? at 100° C?

SECTION 3.3: Analysis of Diode Circuits

L 3.11 A diode described by the exponential characteristic of Fig. 3.12 is connected to a source whose Thevenin-equivalent voltage is V_T and resistance is R_T. Draw load lines and find operating points (V_D, I_D) for:

(a) $V_T = 1$ V, $R_T = 100\ \Omega$

(b) $V_T = 0.9$ V, $R_T = 100\ \Omega$

(c) $V_T = 0.9$ V, $R_T = 90\ \Omega$

Note that the graphical process, while tedious for a single analysis, can be quite effective if a variety of related or similar situations are to be evaluated.

3.12 Determine the diode current I_D and voltage V_D for the circuit in Fig. 3.10 of the text with $V_{DD} = 1.0$ V and $R = 100\ \Omega$. Assume the diode to be the one sketched in Fig. 3.12, having a current of 1 mA at a voltage of 0.7 V, and with a voltage change of 0.1 V per decade of current change.

3.13 Repeat problem P3.12 above utilizing the piecewise-linear model whose parameters are $V_{DO} = 0.65$ V and $r_D = 20\ \Omega$.

3.14 Repeat problem P3.13 above utilizing a piecewise linear model whose parameters are $V_{DO} = 0.70$ V and $r_D = 10\ \Omega$. What do I_D and V_D become if a simple battery of 0.75 V in series with an ideal diode is substituted for the actual junction diode.

L 3.15 In the context of the sequence of problems P3.12, 3.13 and 3.14 preceding, note that the degree of adequacy of a simple model depends on the choice of its parameters in the particular context. To illustrate this dependence, consider Figure 3.10 with V_{DD} reduced to 0.8 V while R remains at 100 Ω. Find the operating point (V_D, I_D) for

(a) the diode characterized by Fig. 3.12, by plotting the load line,

(b) a piecewise-linear diode for which $V_{DO} = 0.65$ V and $r_D = 20\ \Omega$

(c) a piecewise-linear diode for which $V_{DO} = 0.70$ V and $r_D = 10\ \Omega$

(d) a constant-voltage model with $V_D = 0.75$ V.

D 3.16 A series string of 5 diodes is connected through a resistor R to a 10 V supply. For diodes having 0.7 V drop at 1 mA and a 0.1 V/decade characteristic, find R required to establish a total diode-string voltage of 4.0 V.

3.17 In problem P3.16 above if R is reduced to 500 Ω, what does the voltage across the string of 5 diodes become?

3.18 A 1-mA diode having a 0.1 V/decade characteristic operates from a constant-current supply with $V_D = 0.8$ V. If it is shunted by two more identical diodes, what does the voltage drop become?

SECTION 3.4: The Small-Signal Model and its Application

C 3.19 A junction diode for which n=2 operates in a particular circuit with a current that varies over the range 0.1 mA to 10 mA. What is the diode incremental resistance at the extreme values of current? If you were asked to state an "average"resistance, which would be "best" - an arithmetic mean $\left(\frac{r_1 + r_2}{2}\right)$ or a geometric mean $(r_1\ r_2)^{1/2}$? Calculate both, as well as the resistances at 1 mA and at 5.05 mA.

3.20 A diode for which n=2 operates in a circuit for which the current is essentially a constant value of 2 mA. Find the corresponding diode incremental resistance. A second identical diode is used to shunt the first. What does the current in each diode become? What is the incremental resistance of each? What is their parallel combination? What can you conclude about the relation of diode incremented resistance to junction size?

CL 3.21

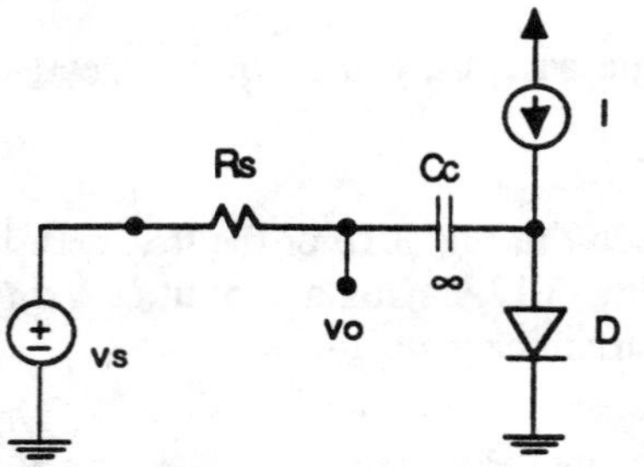

In the circuit shown, v_S is a sine wave of amplitude V and C_C a large capacitor which blocks direct-current, and allows all of I to flow in D. For $R_S = 1\ k\Omega$ and $v_S \leq 10$ mV, find $\frac{v_O}{v_S}$ for I = 10, 1, 0.1 and 0.01 mA. Use $n = 2$.

3.22

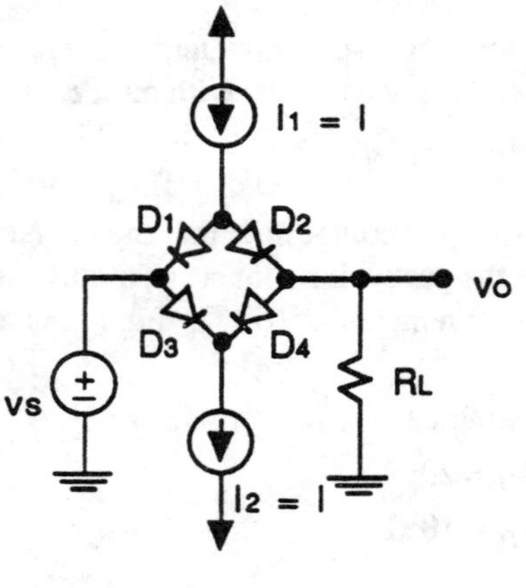

In the circuit shown, v_S is a small signal having a relatively low source resistance. $I_1 = I_2 = I$ is variable. All diodes are identical, with n=2. For $v_O = v_S$ = 0V, how does the current I split among the diodes? In general, what is the relationship amongst the diode currents? Find an expression for the Thevenin equivalent source resistance seen by R_L as a function of I for v_s around zero volts. For $R_L = 10\ k\Omega$, find v_O/v_S for I = 1 mA and 1 µA. For which current is the input signal size most critical?

What is the peak signal for which reasonably linear operation is possible? Note that this circuit, when compared to that in P3.21 above, gives you a preview of a general principal that you will see much more of in this text. It is that an increased use of semiconductors (the diodes and the current sources) reduces the need for other (often much larger) components (such as the capacitor C_C).

3.23 For the design described in problem 3.16 above, of a 4.0 V regulator using five diodes and a 600 Ω resistor with a 10 V supply, the supply is found to vary by ±10%. What output voltage variation results? Use n=2. If, separately, a load of 2 mA is applied to the output, what drop in output voltage would you expect? For both low input voltage and maximum load, what is the lowest output voltage you would find? Express the resulting changes in output voltage (for supply variation alone, load variation alone, and both together) both as absolute and as % changes.

SECTION 3.5: Operation in the Reverse Breakdown Region - Zener Diodes

3.24 A 6.8 V Zener diode specified at 5 mA to have $V_Z = 6.8$ V and $V_Z = 20\ \Omega$ with $I_{Zk} = 0.2$ mA, is operated in a regulator circuit using a 200 Ω resistor and a 9 V supply. Estimate the knee voltage of the Zener. For no load, what is the lowest supply voltage for which the Zener remains in breakdown operation? For the nominal supply voltage, what is the maximum load current for which the Zener remains in breakdown operation? For half this load current, what is the lowest supply voltage for breakdown operation?

3.25 For the situation described in P3.24 above, what are the line regulation and load regulation (as defined in equations 3.25 and 3.26 in the text).

D 3.26 For the situation described in P3.24 above, a modified design is required for the situation in which the supply variation is ±5%, the Zener-diode nominal Zener-voltage variation is ±3%, and the load varies from 2 to 10 mA. Find the value of R for which the minimum zener current is $\geq 2\ I_{Zk}$. For this situation, what are the limits on the output voltage produced. Assume $r_Z = 20\ \Omega$ and $I_{Zk} = 0.2$ mA for all available zeners.

D 3.27 A designer needing a well-regulated 15 V supply in an application where a poorly regulated 24 V source is available, considers the use of a shunt regulator string consisting of two 6.8 V Zeners and two junction diodes. Available Zener diodes are specified to have $V_Z = 6.8$V at 20 mA with $r_Z = 5\ \Omega$. Available junction diodes are 10-mA types modelled at 10 mA by a 0.7 V drop and a 2.5 Ω series resistance. Design a suitable regulator for desired operation with a 15 mA nominal load. What does the output of your design become if the supply is 10% high, the series resistor is 5% low, and the load is accidentally removed? What is the power dissipated in each 6.8 V Zener under the worst combination of these conditions?

SECTION 3.6: Rectifier Circuits

3.28 A half-wave rectifier using diodes for which $V_D = 0.7$ V, is supplied by an 8 V rms sine wave at 60 Hz. What is the peak value of the output voltage for very light loads? For what fraction of a cycle does the diode conduct (first approximately, and then more exactly). What is the average value of the output voltage? What is the peak-inverse voltage across the diode? Now, if the diode resistance is 10 Ω, the source resistance is 50 Ω, and the load resistance is 1 kΩ, what do the peak and average output voltages become?

3.29 In a half-wave rectifier employing an 8 V rms sine-wave supply and driving a 1 kΩ load, a 6.8 V Zener connected with Zener cathode at the output is accidentally substituted for the rectifier diode. Using 6.8 V and 0.7 V drops for diode conduction in the two directions, sketch the output voltage. What is the average value of the output voltage?

C 3.30 In a full-wave rectifier, using a centre-tapped transformer winding whose full-output voltage is 16 V rms, and having a 1 kΩ load, 6.8 V Zener diodes are accidentally installed in place of high-breakdown diodes, but with the same cathode polarities. Sketch the output voltage waveform in the event that the total-winding equivalent resistance is 100 Ω. What peak diode currents flow?

3.31 A transformer secondary winding whose output is a 12 V rms sinusoid at 60 Hz is used to drive a bridge rectifier whose diodes' conduction can be modelled by 0.7 V drops. The load is a 1 kΩ resistor. Sketch the load waveform. What is its peak value? Over what time interval is it zero? What is its average value? What is the *PIV* for each diode?

3.32 A half-wave rectifier employing a 12-V-rms 60-Hz sine-wave source and no *dc* load is filtered using an electrolytic capacitor having a small leakage current. For diodes assumed to have a 0.7 V drop independent of current, characterize the resulting output. What is the *PIV* required of the diode?

D 3.33 To the circuit in P3.32 above, a load which can be modelled as a 1 mA constant current is connected. If an output ripple of 0.4 V pp results, what is the value of the filter capacitor used? For half this ripple, and double the load, what capacitor is necessary? In each case, what average current flows during the diode's conduction interval?

3.34 For both situations described in P3.33 above, but with full-wave rectification, what capacitor values are necessary? What average diode currents flow? What diode *PIV* is required?

D 3.35 A design is required of a full-wave rectifier with capacitor filter to supply 12 volts to a 100 Ω load. A ripple voltage of less than or equal 0.4 V pp is necessary. Diodes are assumed to conduct with 0.7 V drop. Characterize the required 60 Hz transformer secondary, the capacitor and the diodes. For the diodes, provide the required *PIV* and peak-current rating.

C* L* 3.36 Consider a full-wave bridge rectifier operating at 60 Hz from a single transformer-secondary winding having a 1.0 Ω equivalent internal resistance and 20 V pp open-circuit output. The load consists of a 1000 μF capacitor and 200 Ω resistor in parallel. Consider the diodes to have a constant 0.7 V drop during conduction. Hint: To characterize this situation, first consider the ideal case of a zero-impedance source, finding the usual parameters, including the average diode current during conduction. Assuming the diode current to be a triangular pulse, limited by some combination of the sinewave slope and the charging-circuit time constant, find the corresponding average voltage drop in the transformer resistance.

SECTION 3.7: Limiting and Clamping Circuits

3.37

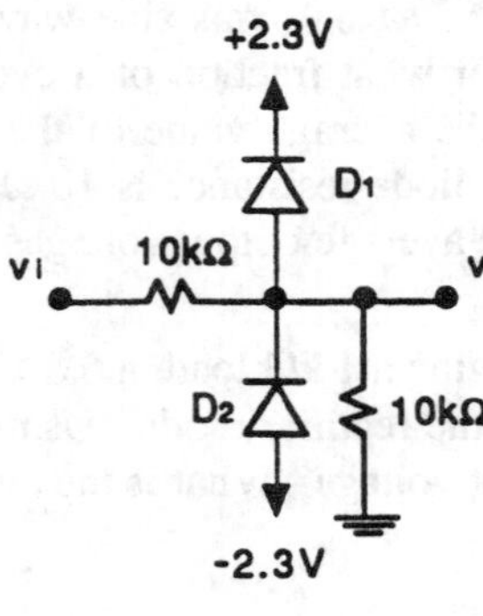

For the following passive symmetric hard limiter, find the upper and lower limiting levels (including a 0.7 V diode drop), the gain *K*, and the upper and lower input threshold levels. What is the input current required at twice the upper threshold value?

CD* 3.38 Convert the circuit in P3.37 above to a soft limiter with $K = \frac{1}{4}$ after limiting begins, by adding two additional components. Add two more components (and additional supplies) for hard limiting at ±5.0 V. Using the new supplies and two more components, create an equivalent circuit having the same hard-and soft-limiting characteristics, but only two power supplies.) (Hint: Your final circuit should employ six resistors, four diodes, and two power supplies.

D 3.39 Design a passive symmetric high-gain pseudo-hard-limiting circuit using four diodes for ±1.4 V limits. Use only a connection to ground (i.e., no additional supplies). This limiter is to be used to create an approximation to a square wave from a sine-wave input. For what peak-to-peak amplitude of the sine wave is the rise and fall time of the pseudo-square wave ≤5% of the wave period. What resistor is required for a peak diode current of about 10mA?

3.40 A simple clamped-capacitor circuit such as that shown in Fig. 3.38 of the text, utilizing a capacitor C and grounded-cathode diode, has a square-wave input with upper and lower levels at 100 V and 10 V respectively. Describe the resulting output for a very high-resistance load to ground and a 0.5 V diode drop at very low currents. What happens as the load resistance R reduces? Describe the output waveform when $RC = 2T$ where T is the period of the input square wave.

3.41 A voltage doubler, consisting of a clamped capacitor and a half-wave rectifier, operates at 20 kHz using two 0.1 μF capacitors. For a sine-wave input signal of 100 V peak what output voltage would you expect at no load? For a peak-to-peak ripple voltage of 5% of the peak voltage, what average output voltage would you expect? What is the load current for which this situation applies?

C* 3.42 Consider the detailed operation of an unloaded half-wave doubler circuit with positive output using equal capacitors C and driven by a 100 V pp square wave. In particular, follow the cycle-by-cycle operation immediately upon turn on with the output voltage equal to zero and the input low. What is the output voltage after the first half cycle (after the input has risen by 100 V and fallen again)? After the first cycle (just after the input rises again)? After the first two cycles? After four cycles? After eight cycles? Sketch the output voltage against time as measured by the number of cycles.

3.43 Continue to think about the operation of the half-wave doubler as suggested in P3.42 above. In particular if the load is a current which discharges a capacitor of value C by 5% in ½ cycle, find the steady-state average output voltage as a function of V_O, the peak-to-peak input (V_O = 100 V here). (Hint: Note a) that the total effective capacitance is 2C for half the cycle and b) that two equal capacitors when joined, share their charge difference equally).

SECTION 3.8: Physical Operation of Diodes - Basic Semiconductor Concepts

3.44 At a particular temperature, the fraction of ionized atoms in a piece of silicon is 10^{-n}. If the material is doped to a level of 1 in 10^m with acceptor atoms, what is the net concentration of holes and electrons in the resulting material?

SECTION 3.9: The *pn* Junction under Open-Circuit Conditions

3.45 For a *pn* junction in which the *n* region is doped at ten times the concentration of the *p* region, in what region is the depletion region largest? By what factor?

SECTION 3.10: The *pn* Junction Under Reverse-Bias Conditions

3.46 For a particular reverse-biased *pn* junction, the terminal current is 10 nA. If the drift current at the operating temperature is 15 nA, what must the voltage-dependent diffusion current be at this particular reverse voltage?

3.47 At a particular operating point of a reverse-biassed *pn* junction, a change of 1 volt produces a transient current increase corresponding to a net charge flow of 0.1 pC. What is the corresponding depletion capacitance of the junction at this operating voltage?

3.48 For a particular junction for which m=1.6, a capacitance C_j of 1.8 pF is measured for a reverse junction voltage of 2 V and 0.2 pF for a voltage of 10 V. What are the corresponding values of V_O and K? What is the capacitance C_j at 0 V?

SECTION 3.11: The *pn* Junction in the Breakdown Region

3.49 A particular *pn* junction for which the breakdown voltage is 120 V, can dissipate 50 mW before its junction temperature becomes high enough to lead to permanent junction failure. What continuous reverse current flow can cause permanent failure? If reverse current flows only 10% of the time at the peaks of a cyclic applied voltage, what peak current can be tolerated?

SECTION 3.12: The *pn* Junction Under Forward-Bias Conditions

3.50 For a junction conducting 1 mA at 700 mV, for which n=2 and a diffusion capacitance of 1pF is associated, what is the value of q_O which applies? For a junction 10 × larger what would q_O be? What is q_M at 1mA?

3.51 Use the relationships given for C_d and q_M to calculate the diffusion capacitance of a junction characterized by n and voltage v at current i.

L 3.52 For a *pn* junction, the following measurements are taken:

$C_j = 0.8$ pF at $v_r = 1$ V reverse bias

$C_j = 0.2$ pF at $v_r = 5$ V reverse bias

$C_j = 0.1$ pF at $v_r = 10$ V reverse bias

$C_T = C_j + C_d = 10$ pf at $v_f = 0.70$ V forward bias for which $i_f = 1$ mA

$r_d = 50\ \Omega$ at $i_f = 1$ mA.

Find K, V_O, m, n, k_C. What is $C_T = C_j + C_d$ at $i_f = 10$ mA?

3.53 For a diode 10 times the junction area of that in P3.52 above, find r_d, C_d, C_j, C_T at $i_f = 1$ mA forward and C_j at 1 volt reverse bias.

Chapter 4

BIPOLAR JUNCTION TRANSISTORS (BJTs)

SECTION 4.2: Operation of the npn Transistor in the Active Mode

4.1 A particular npn BJT operates with the base-emitter junction forward-biassed, the base-to-emitter voltage being 700 mV. For active-mode operation, in what range must the collector-to-base and collector-to-emitter voltages lie? When the transistor is appropriately biassed in the active mode, the collector current is found to be 10 mA. What is the corresponding value of I_S for this transistor if n is assumed to be 1? Under the same conditions, the base current is found to be 100 μA. What is the value of β for this transistor? If measured, what would the emitter current be found to be?

4.2 For the devices and situations described, provide the missing entries in the following table. Line a) is provided by way of example.

Device #	I_C mA	I_B mA	I_E mA	α	β
a	10	.1	10.1	.99	100
b	1				50
c			2	.98	
d		.01		.995	
e			110		10
f		.001			1000

Note that transistors like device f) are constructed with very thin bases in order to achieve high β, but suffer accordingly from reduced breakdown-voltage ratings.

4.3 Consider the first-order large-signal equivalent-circuit models of an npn BJT shown in Fig. 4.5 of the Text, which employ α and β explicitly. Draw these side-by-side to emphasize current flow directly from collector to emitter with base current entering from the left (the final shape can be called a "tilted T", or, in the spirit of livestock branding in the far west of North America, a "Lazy T"). Label all currents and υ_{BE} in each case. What two labels can be applied to each of the two controlled current sources?

4.4 A particular BJT operates in its usual current range with $\upsilon_{BE} = 0.7$ V, $i_C = 1$ mA. What would be its υ_{BE} at $i_C = 0.1$ μA for n=1? Repeat for n=2.

4.5 For a "1 mA transistor", that is one for which $i_C = 1$ mA for $\upsilon_{BE} = 700$ mA, the collector-base reverse current, I_{CBO}, is 0.1 nA at 25°C. Device β is nominally 100. For a transistor operating with its base open-circuited (in which case I_{CBO} constitutes the only source of base current), what collector current flows, at 25°C? At 95°C?

4.6 For a particular BJT fabricated in the style shown in Fig. 4.6 of the Text, the collector-base junction is 100 times larger than the emitter-base junction. If, for this device, normal β = 150, what would you expect it to become if the role of emitter and collector are reversed? That is, estimate β_R.

SECTION 4.3: The pnp Transistor

4.7

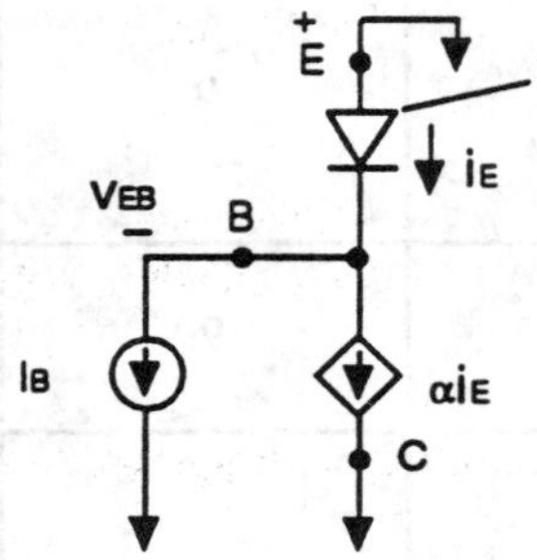

For the accompanying equivalent circuit of a pnp transistor (drawn to emphasize the direction of current flow), $\alpha = 0.975$. For an external current extracted from the base, $I_B = 10$ μA, what collector and emitter currents would you expect? If for this device, $\upsilon_{EB} = 0.70$ V at $I_C = 1$ mA, what value of υ_{EB} would you expect? (Assume that n=1)

4.8 For a particular pnp transistor for which D_B has a scale current of 10^{-13} *A*, and D_E has a scale current of 10^{-11} A, calculate β and i_C for $\upsilon_{BE} = 0.643$ V, for n=1.

SECTION 4.4: Circuit Symbols and Conventions

4.9

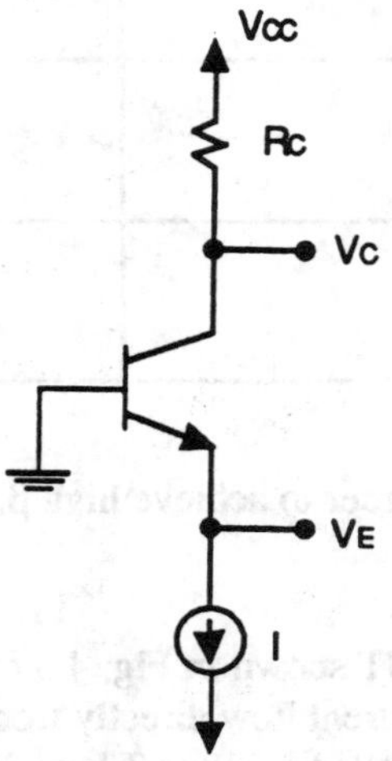

The operation of the BJT in this circuit can be shown to be conveniently independent of device parameters. Of course, it requires the complexity of the current sink I. However, this will be shown in Chapter 6 of the Text to be relatively simply constructed in an integrated-circuit environment. Specifically, for operation of this circuit, consider devices for which υ_{BE} varies from 0.6 to 0.8 V at 1 mA with β variation from 10 to 300.

(a) For $I = 1$ mA what is the range of values expected for emitter current i_E, emitter voltage V_E and collector current i_C?

(b) For $V_{CC} = 10$ V, $R_C = 5$ kΩ, what is the corresponding expected range of values of V_C? Does variation of V_E from device to device matter?

(c) To ensure operation in the active mode $\upsilon_{CB} \geq 0$. What is the largest value of R_C which maintains active-mode operation?

4.10 Reconsider the situation described in P4.9 above, modified to include an additional (signal) current source i_e connected to the emitter. For our purposes here, i_e can be considered to be a current sine wave with peak amplitude of 0.1 mA, that is 1/10 of the emitter-bias current $I = 1$ mA. For this situation, express the collector current i_C in terms of I, i_e and α. What is the largest value of i_C for a) very high β, b) β=10? Under these conditions, what is the largest value you can use for R_C to ensure active-mode operation? Recall that $V_{CC} = +10$ V. For this value of R_C, what is the peak-to-peak value of the signal voltage υ_C at the collector that is produced for a) very high β b) $\beta = 10$?

4.11 Reconsider the situation described in P4.9 above in which I is implemented using a resistor R_E and voltage supply. With a –10 V supply what value of R_E ensures $I_E = 1$ mA for a transistor for which $V_{BE} = 0.7$ V. Select a "pseudo-standard value", one specified to two significant digits, chosen to produce a current on the high side of your calculated value. Now, for V_{BE} varying from 0.6 to 0.8, β from 10 to ∞, and R_E by $\pm 1\%$, what is the largest available collector current? Now chose R_C as large as possible while ensuring active-mode operation. Select a pseudo-standard value (as specified above) on the low side. What is the lowest possible value of υ_C for your chosen R_C varying by $\pm 1\%$?

SECTION 4.5: Graphical Representation of Transistor Characteristics

4.12 A BJT which conducts $i_C = 10$ mA at $V_{BE} = 0.7$ V and 25°C is operated with V_{BE} fixed at 0.62 V. What is the collector current at 0°C, 25°C and 50°C? Assume n=1.

4.13 A BJT operating at a fixed V_{BE} is found to have $i_C = 2.1$ mA at $V_{CE} = 2$ V, and $i_C = 2.19$ mA at $V_{CE} = 9$ V. What is its output resistance r_o at this current level? What value of V_A corresponds? What would its output resistances be at 0.1 mA and 10 mA (approximately)?

4.14 A BJT for which $V_A = 200$ V operates at $V_{CE} = 5$ V at a current of 100 μA. What would its current become (provided breakdown does not occur) if V_{CE} is raised to 50 V?

SECTION 4.6: Analysis of Transistor Circuits at DC

L 4.15 For the following circuits, find node voltages, V_E, V_C, and branch currents I_E, I_C, I_B. Use $V_{BE}=|V_{EB}|=0.7$ V and β=50.

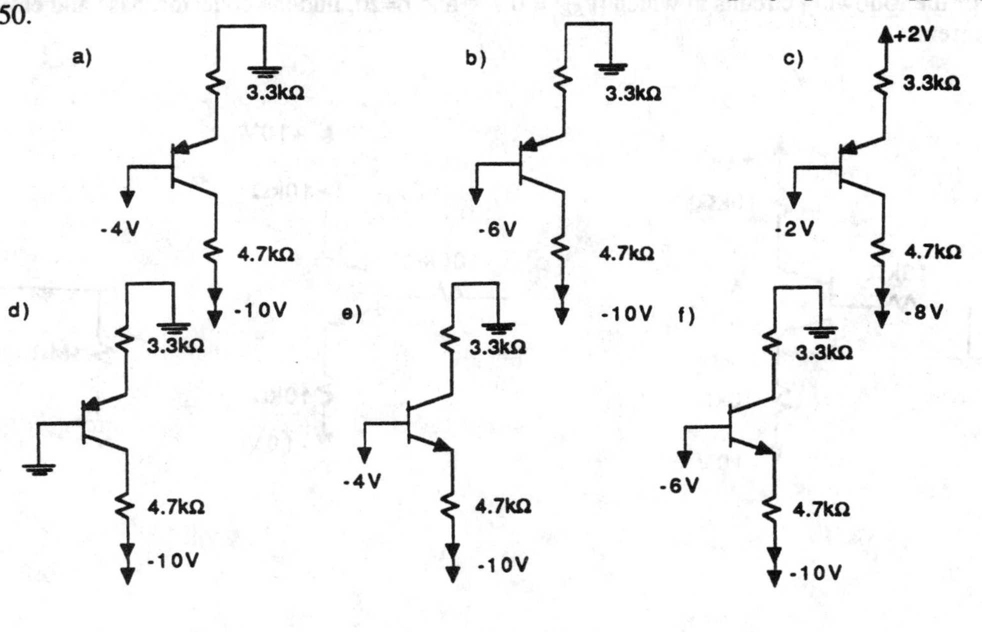

D 4.16 For the circuits shown in P4.15 a), b) above, find emitter and collector resistors (to replace the present ones) such that $I_E = 0.5\text{mA}$ and $V_{BC}=0$ for $\alpha \approx 1$.

4.17 For the following circuits in which $|V_{BE}| = 0.7$ and $\beta=10$, find the collector, emitter and base currents and voltages.

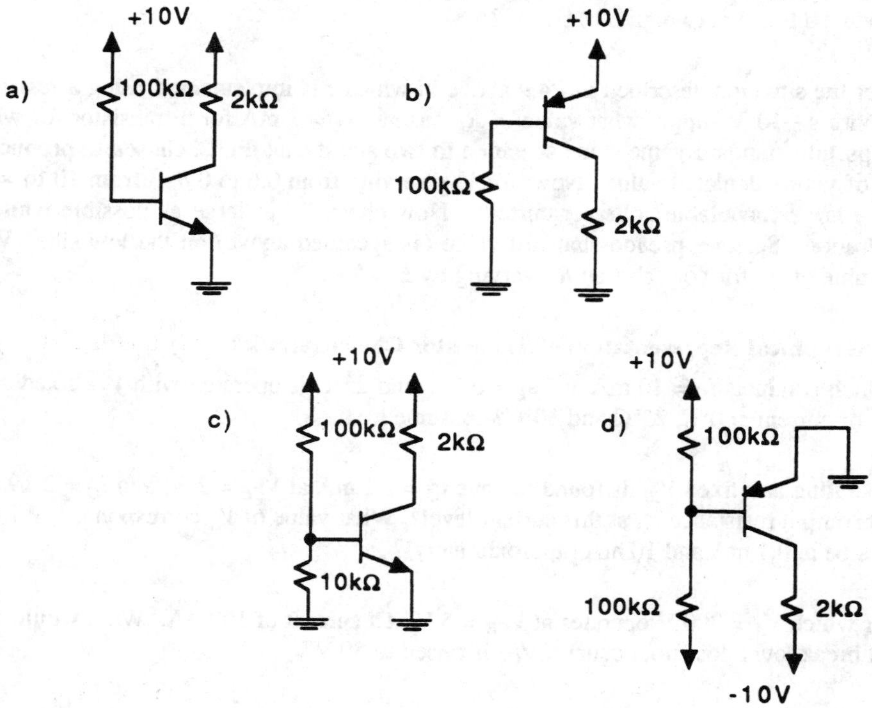

4.18 For the following circuits in which $|V_{BE}| = 0.7$ V and $\beta=20$, find the collector, base and emitter voltages and currents.

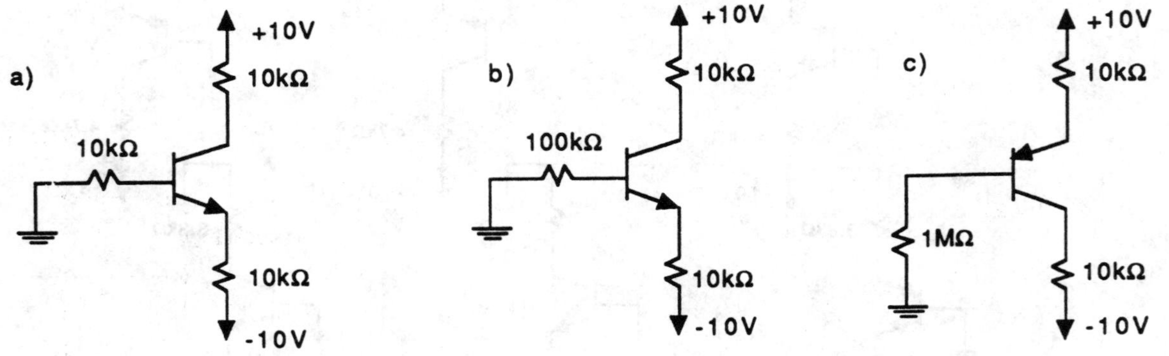

4.19

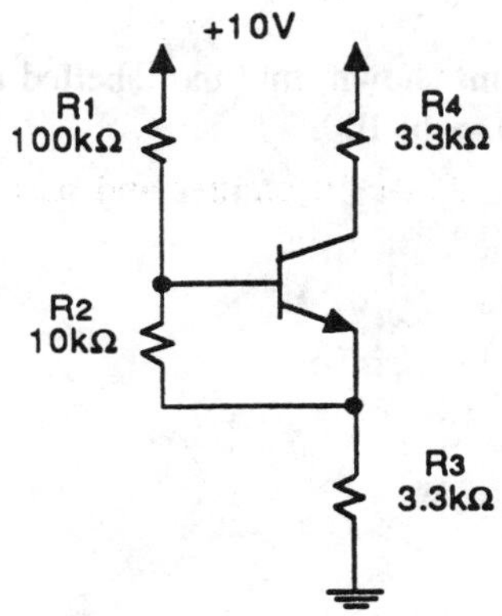

For the circuit shown, find the voltages at the base, emitter and collector for $\beta = \infty$, 100 and 10. Assume $V_{BE} = 0.7$ V.

4.20 For the circuit of P4.19 above, for what value of β does the emitter current reduce to 80% of that for $\beta = \infty$?

4.21

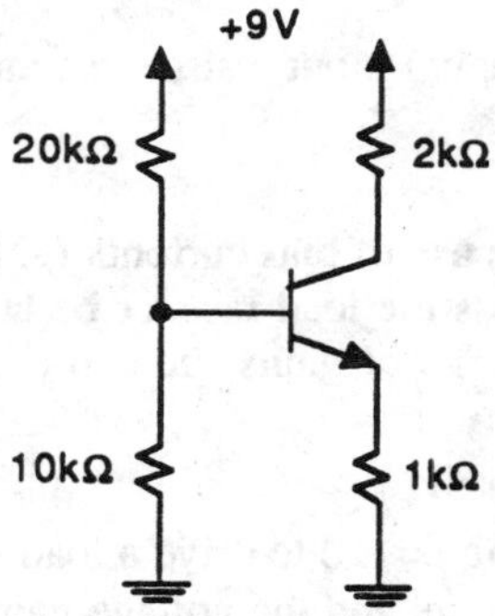

For the circuit shown, find I_E and V_{CE} for $V_{BE} = 0.7$ V and

a) $\beta = \infty$
b) $\beta = 100$
c) $\beta = 10$

L 4.22 For the following circuits find the currents I_C and the voltages V_{CE}. Use β=50 and $V_{BE} = 0.7$ V.

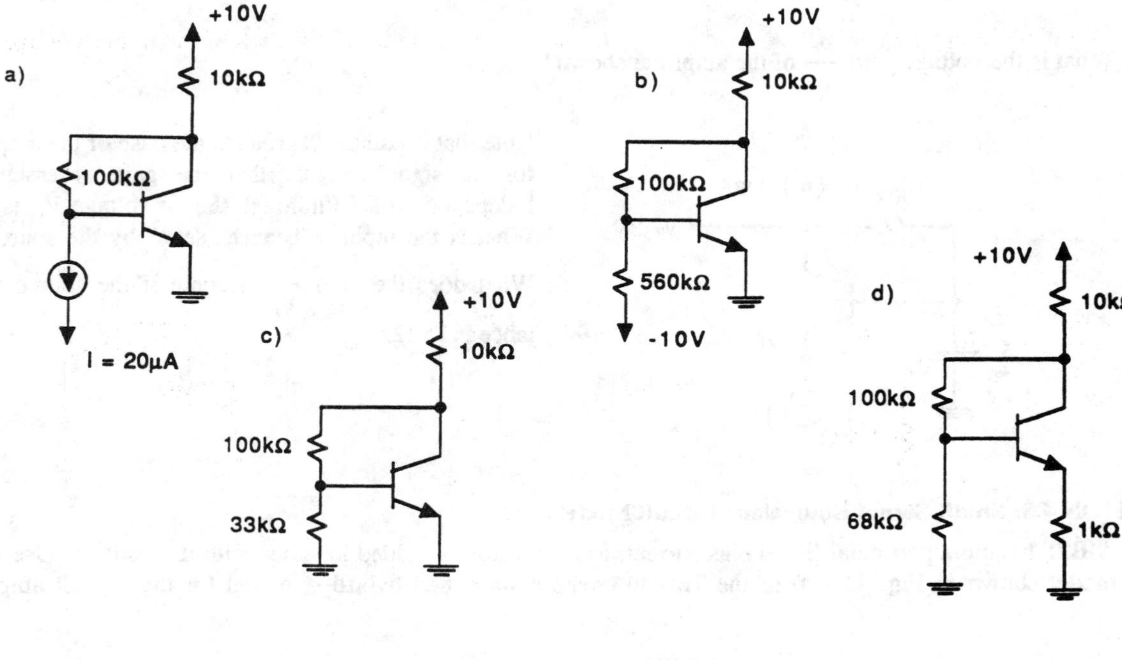

4.23

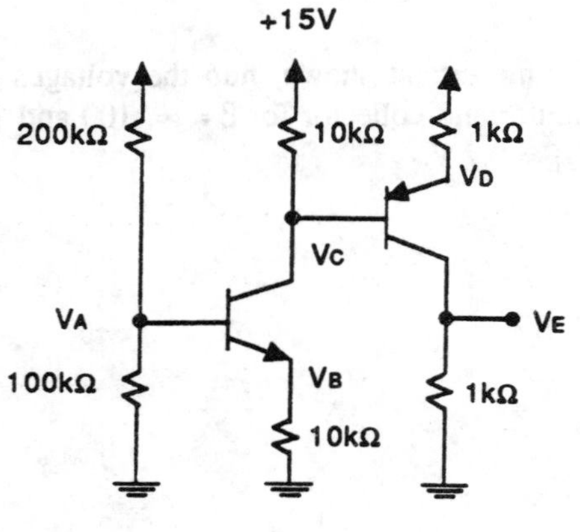

For the circuit shown, find the labelled node voltages when β is a) ∞, b) 100.

SECTION 4.7: The Transistor as an Amplifier

4.24 What values of transconductance apply to BJTs biased at 1 μA, 100 μA, 1 mA, and 100 mA?

4.25 For the current levels listed in P4.24 above, what equivalent small-signal input resistances model operation as seen at the emitter? At the base, for $\beta = 100$?

D 4.26 In the design of a particular amplifier, a young engineer considers the use of bias currents I_E, from 0.1 to 10 mA. Unfortunately, the application requires that the dc voltage across the load resistor be held constant to provide correct biassing of a connected amplifier stage. Find the range of gains she can expect from this gain stage.

4.27 A particular amplifier utilizes a BJT biassed at $I_E = 100$ μA and having β=150 to drive a load of 10 kΩ. For the emitter grounded for signals, what is the input resistance at the base, and the voltage gain from base to collector?

4.28 What is the voltage gain $\frac{v_o}{v_i}$ of the amplifier shown?

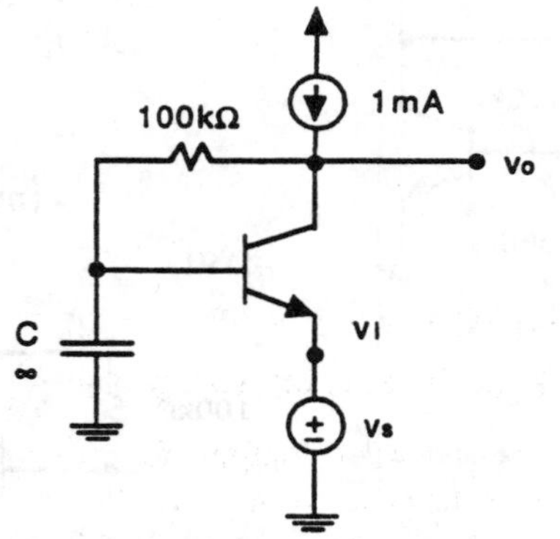

Note that capacitor C grounds the base of the amplifier for *ac* signals. Note that the gain is essentially independent of β (although the *dc* voltage V_O is not). What is the input resistance "seen" by the source v_s. What does the gain $\frac{v_o}{v_s}$ become if the source resistance is 75 Ω.

SECTION 4.8: Small-Signal Equivalent-Circuit Models

4.29 A BJT having a particular β and bias current, has a resistor r_E added in series with the emitter. Use the T model shown in Fig. 4.27 a) of the Text to create a simplified hybrid–π model for the overall amplifier (including r_E). For this model find g_m', r_π' in terms of r_E, g_m and r_π of the basic BJT. What is the equivalent input resistance (r_π') and transconductance (g_m') of the modified amplifier, for β=100, I_C= 1 mA and r_E=3r_e?

L 4.30 An appropriate choice of one of the BJT models of Figs. 4.26 and 4.27 of the Text, often makes the solution of a particular problem somewhat easier. To illustrate, find the gain $\frac{v_o}{v_s}$ for each of the circuits below using the model(s) suggested as Π_{gm}, Π_β, T_{gm}, T_α, corresponding to Fig. 4.26 a) 4.26 b), 4.27 a), 4.27 b) respectively. In each case, assume (for simplicity) that I_E = 1 mA, r_e = 25 Ω, r_π = 2.5 kΩ, $\beta \approx$ 100, α = 0.99 and g_m = 40 mA/V. (Note that biassing is generally not shown in detail).

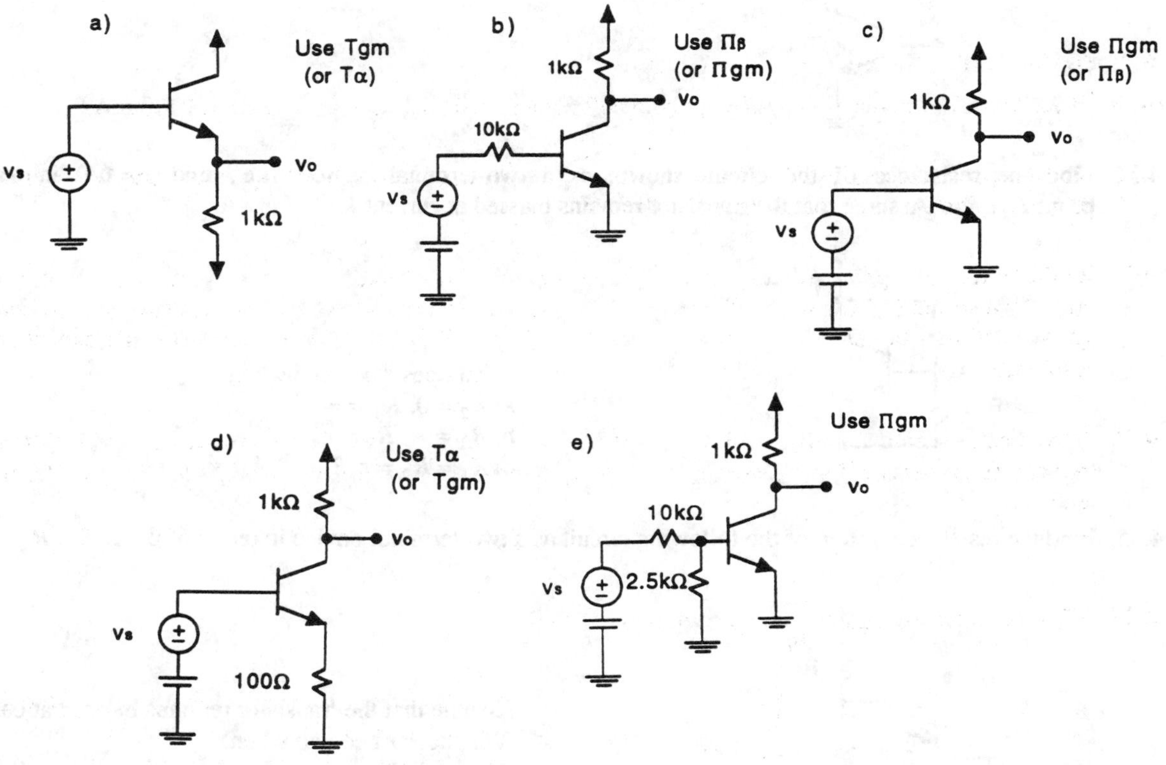

4.31

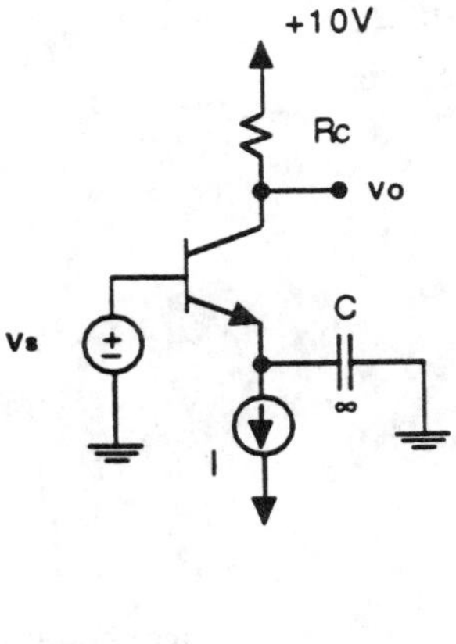

In the circuit shown, I = 1 mA, R_C = 7.5 kΩ and α = 0.99. What is the voltage gain $\frac{v_o}{v_s}$? What is the largest sine-wave output for which the transistor remains in the active region? What is the peak value of the corresponding input?

4.32 For reasonably linear operation of a transistor amplifier, it is customary to limit the base-emitter voltage swing to ±10 mV around the operating point. For a transistor for which n=1, to what fraction of the emitter bias current does this voltage range correspond? For the circuit shown in P4.31 above, find the value of R_C which provides the largest-possible reasonably-linear output while operation remains in the linear region.

4.33

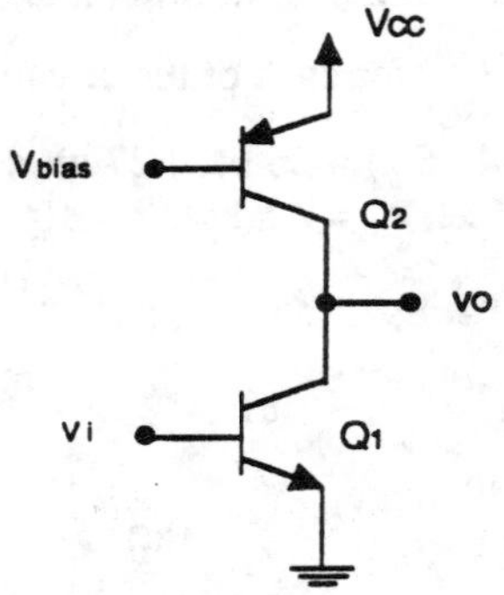

An amplifier employs the components shown together with others that maintain $i_{C1} = i_{C2} = 100\ \mu A$ and $V_O \approx \frac{V_{CC}}{2}$. What is the gain $\frac{v_o}{v_i}$ for $V_A = 200$ V?

C 4.34 Find the resistance of the circuit shown, as a two-terminal device, (i.e., find $r = v/i$) in terms of β, r_e, R_1, R_2. Assume that the transistor remains biassed at current I.

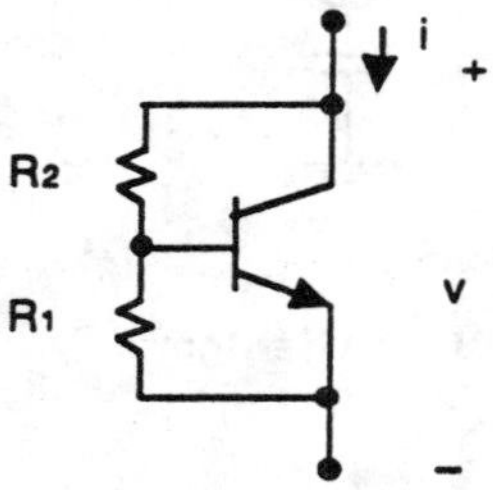

What does r become when:
a) $R_2 = 0$, $R_1 = \infty$
b) $R_1 = \infty$, $R_2 = r_\pi$
c) $R_1 = R_2 = r_\pi$?

4.35 Find the resistance, r=v/i, of the following circuit as a two-terminal device in terms of β, r_e, R_1, R_2.

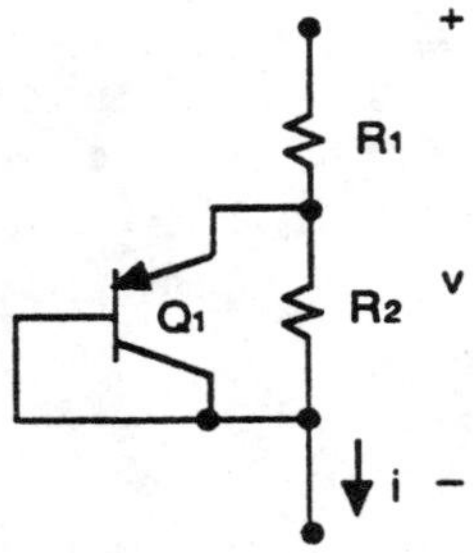

Assume that the transistor remains biassed at current I.
What does r become when:
a) $R_1 = 0$, $R_2 = \infty$
b) $R_1 = r_e$, $R_2 = \infty$
c) $R_1 = 0$, $R_2 = r_\pi$
d) $R_1 = r_e$, $R_2 = r_\pi$

4.36

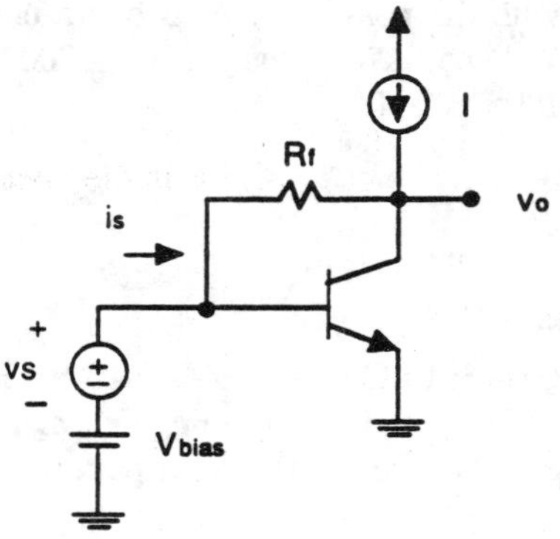

Use the hybrid-π model with g_m and r_o to find the voltage gain, $A_v = \frac{v_o}{v_s}$, and the input resistance, $R_i = \frac{v_s}{i_s}$, of the following circuit, assumed biassed by some external means at $i_C \approx I$. What do these parameters reduce to when $R_f = r_o$. (Hint: Use the fact that the signal at the base must be relatively small).

4.37

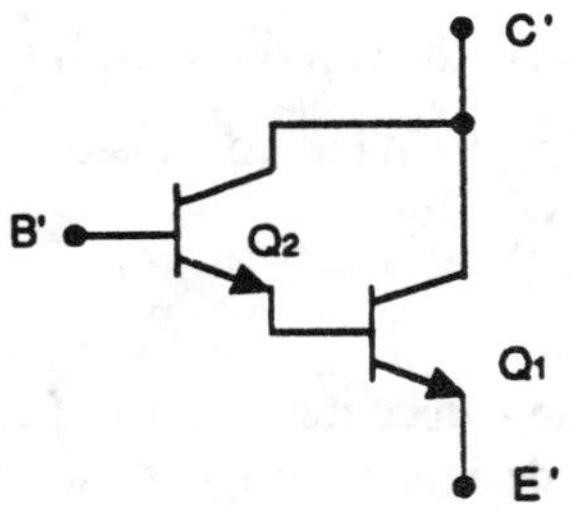

Find the equivalent hybrid-π model of the following circuit (called a Darlington connection) expressing overall values r_π' and g_m' in terms of $r_{\pi 1}$, g_{m1}, $r_{\pi 2}$ and g_{m2}. Now, realizing that the total collector current will flow predominantly in Q_1 and, accordingly, that the parameters of Q_2 will differ from those of Q_1 by a factor of approximately β_1 (for $\beta_1 >> 0$), find a corresponding approximate model.

SECTION 4.9: Graphical Analysis

4.38. Sketch the $i_C - v_{EC}$ characteristics for a pnp transistor having β=200 and V_A=100 V. Sketch the characteristic curves for i_B = 1, 2, 5, 8, 10 μA. Assume for this sketch that $i_C = \beta i_B$ at $v_{EC} = 0$. Sketch the load line for V_{CC} = 10 V and R = 5 kΩ. Operation is defined by a dc bias current of I_B 5μA. Identify the operating point, and estimate its coordinates. For a triangular signal of 3 μA peak superimposed on I_B, find the corresponding signal component of i_C and v_{EC}. For operation in the active region, defined for convenience as $v_{EC} \geq 0$, estimate the maximum peak of an output triangle wave and the corresponding peak signal current. Sketch the output which corresponds to a superimposed base-current triangle wave of 10 μA peak amplitude. Assume for this purpose that v_{EC} can reduce to zero. For what fraction of a cycle is its output clipped?

SECTION 4.10: Biassing the BJT for Discrete-Circuit Design

D 4.39 Consider the one-supply bias scheme in Fig. 4.38 of the Text. For $R_B = R_E$, above what value of β is I_E constant to within 1%?

D 4.40 Using the ⅓ rule ($V_{BB} = V_{CB} = V_{CC}/3$) and $R_B = \beta R_E/10$, provide a design for the circuit of Fig. 4.38 in the Text, in which V_{CC} = 12 V and I_E = 100 mA. For the BJT, β = 50 and V_{BE} = 0.7. Find R_E, R_1, R_2, R_C to the nearest single significant digit. What values of I_E and V_{CE} does your design provide?

D 4.41 For the bias arrangement shown in Fig. 4.39 in the Text, using ±5 V supplies, a design is required for which I_E is fixed to within 5% and a ±1 V signal output range is available for β ≥ 20 and R_C = 1 kΩ.

D 4.42 A design is required of the feedback-bias scheme shown in Fig. 4.40 in the Text which will maintain $V_{CB} \geq 0.5$ V for β ≤ 200, V_{CC} = 5 V and R_C = 3.6 kΩ, with V_{BE}=0.7 V. For β ≥ 50, what is the range of I_E and V_{CB} you achieve?

D 4.43 In the situation described in P4.42 above, a designer faced with the possibility of β being uncontrollably high choses to shunt the base-emitter junction with resistor R_β. What is its value for $\beta_{eq} \leq 200$? Find R_B to meet the other specifications. What ranges of I_E and V_{CB} result for $\beta \geq 50$?

D 4.44 Repeat P4.43 above for $\beta_{eq} \leq 100$.

SECTION 4.11: Basic Single-Stage BJT Amplifier Configurations

4.45 For the circuit in Fig. 4.42 in the text, $R_B = 10\ k\Omega$, $R_E = 1\ k\Omega$, $R_C = 1\ k\Omega$, $V_{CC} = V_{EE} = 10$ V. For the BJT, $\beta = 100$ and $V_A = 100$ V. Find the corresponding values of V_B, V_E, I_C and V_C. Find g_m, r_e, r_π and r_o which correspond. Compare your results with those in Ex. 4.31 (page 247 of the Text). What patterns do you see?

4.46 Consider the amplifier, whose bias design was analyzed in P4.45 above, when connected in the common-emitter configuration with $R_S = R_L = 1\ k\Omega$. Find the values of R_i, G_m, R_o, A_{vo}, A_{is}, A_v, $A_i = \dfrac{R_s + R_i}{R_L} \times A_v$. Compare your results with those in Ex. 4.32 (page 251 of the Text). What conclusions can you reach related to resistance-scaled (or current-scaled) designs?

CD 4.47

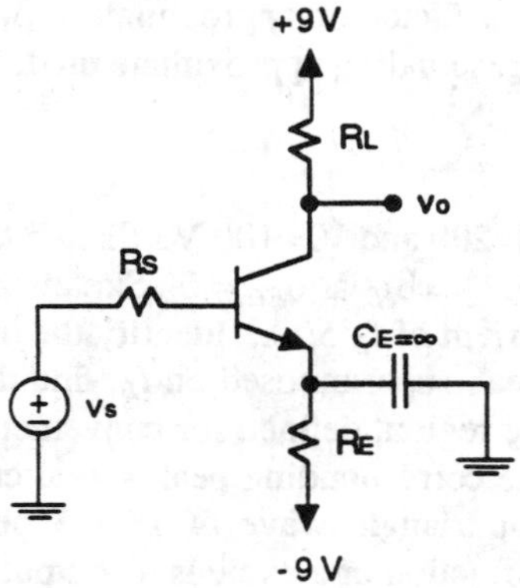

In an attempt to reduce the number of components in a space-critical design, a designer employs the circuit shown, which incorporates the source resistance R_S and load resistor R_L as part of the bias design. In this particular situation, $R_L = 10\ k\Omega$, $R_S = 100\ k\Omega$, $\beta \geq 90$ and $V_A = 100$ V. Design for the highest possible gain and an output-signal swing of 1 Vpp under all bias conditions. What are the extreme values of V_E, V_C and I_C which your design produces? What is the range of voltage gains $\dfrac{v_o}{v_s}$ you expect? What is the range of voltages v_b, and v_s which correspond to ± 1V output?

4.48 A common-emitter amplifier operating between a $10\ k\Omega$ source and a $10\ k\Omega$ load, and with ± 10 V supplies, employs $R_B = R_E = R_C = 10\ k\Omega$. For $V_A = 200$ V and β ranging from 50 to 150, what range of voltage gains $\dfrac{v_o}{v_s}$ results?

4.49 An alternative to the *CE* amplifier described in P4.48 above is considered in which a $100\ \Omega$ part of R_E is left unbypassed. What range of voltage gains $\dfrac{v_o}{v_s}$ results?

4.50

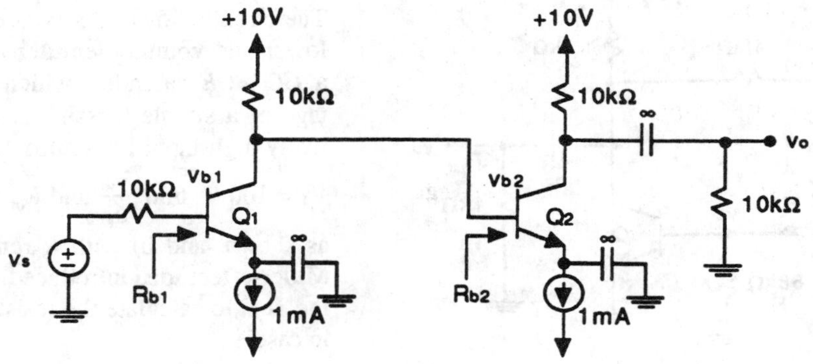

For $\beta = 150$, find $\frac{v_o}{v_{b2}}$, R_{b2}, $\frac{v_{b2}}{v_{b1}}$, R_{b1}, $\frac{v_{b1}}{v_s}$, $\frac{v_o}{v_s}$.

4.51 A common-base amplifier, biassed at an emitter current of 3 mA, employs an unbypassed base resistor of 2 kΩ and $R_C = 3$ kΩ, $R_E = 3$ kΩ and $R_L = 1$ kΩ. For $\beta \geq 150$, what range of input resistances result? What range of voltage gains result from a 100 Ω source?

D **4.52** Provide a design resembling that in Fig. P4.64 on page 291 of the Text which has an input resistance of 10kΩ for $\beta = 50$. What is its voltage gain from the input with a 1 kΩ load?

D **4.53**

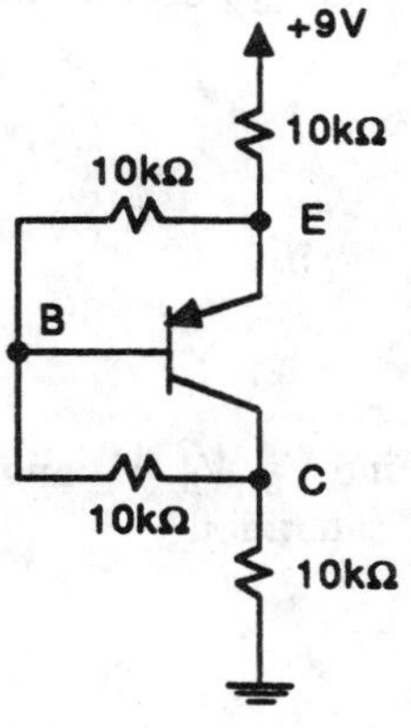

For the circuit shown, evaluate V_E, V_B, V_C and I_C for $\beta = 100$. Show capacitor-coupled connections to a 0 Ω source, a 10 kΩ load and ground to achieve voltage gains of:

a) ≈ + 1 V/V
b) – 1 (Hint: Use an extra resistor)
c) – K where K is large
d) + K where K is large

4.54 For each of the designs created in P4.53 above, calculate the exact gains assuming $\beta = 100$, $V_A = \infty$.

4.55 An emitter follower biassed at 0.1 mA employs a 100 kΩ base resistor and a 50 kΩ emitter resistor. The BJT has $\beta = 50$ and $V_A = 100$ V. When driven by a 20 kΩ source and driving a 2 kΩ load, what is the voltage gain which results?

CL **4.56**

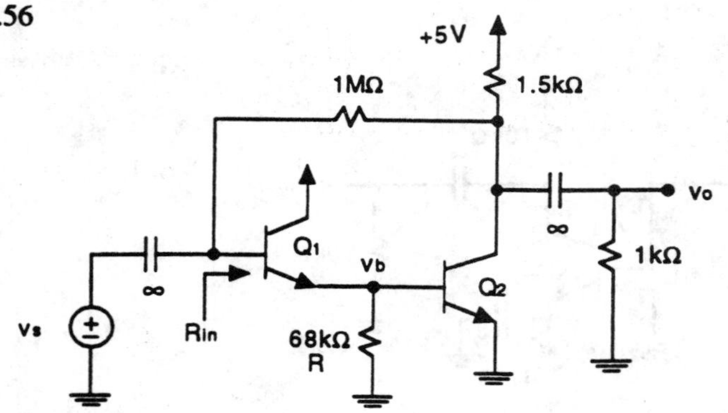

The circuit shown is a combination of a follower and common-emitter amplifier (called a *CC – CE* cascade), which has the advantage of a simple biassing structure and relatively high input resistance. For $\beta = 100$ and $V_A = 100$ V, find $\frac{v_o}{v_s}$ and R_{in} for the circuit a) as shown, and b) with *R* removed. Use the Miller-effect idea introduced in section 7.4 of the Text to calculate the resistance seen by v_s in case a).

SECTION 4.12: The Transistor as a Switch: Cutoff and Saturation

4.57 In the circuit shown in Fig. P4.78 in the Text, the transistor operates with $V_{CE} = 0.2$ V, $V_{BE} = 0.7$ V, and forced β of three. What must the value of R_B be? For $\beta_{forced} \le \beta/2$, what is the largest value R_B can have.

4.58

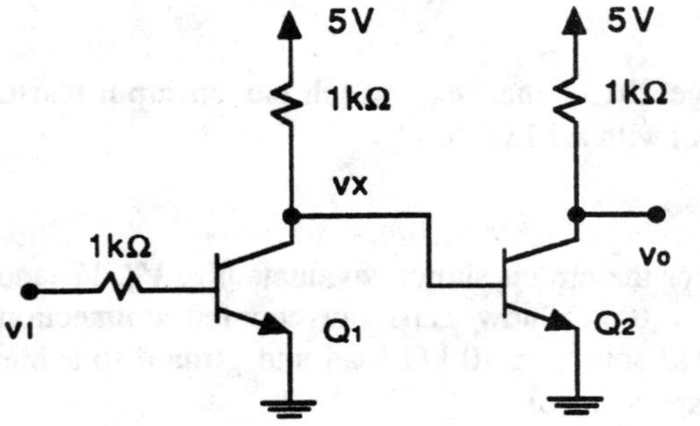

For the circuit shown, characterize the mode of operation of each transistor and the voltages v_X and v_O for v_I equal to
a) 0 V
b) +5 V
Assume $V_{CE\,sat} = 0.2$ V, $V_{BE} = 0.7$ V. At what value of forced β do Q_1 and Q_2 operate when saturated?

L **4.59** In the following circuits, $\beta = 100$, $V_{BE} = 0.7$ V, and $V_{CE\,sat} = 0.2$ V. Find V_E, V_B, V_C and the value of β at which the BJT operates. At what value of *I* does each circuit just leave saturation?

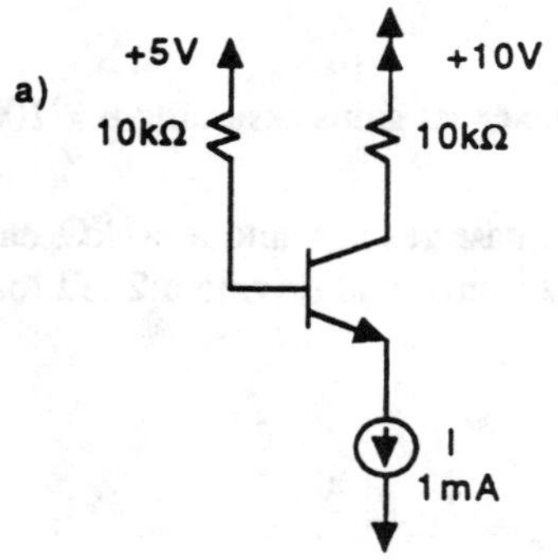

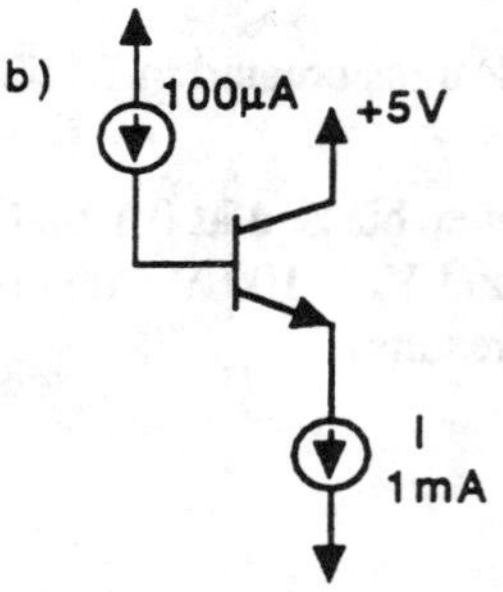

SECTION 4.13: Complete Static Characteristics and Second-Order Effects

4.60 For a particular transistor, for which $V_A = 200$ V, $\beta = 120$, operating in the grounded-base configuration, the collector current is found to increase by 50 nA from its former value of 0.1 mA when the collector voltage is raised by 10 V. Estimate r_o and r_μ. (Hint: Use the result of P4.83 in the Text)

4.61 A particular transistor for which $BV_{CBO} = 50$ V, $BV_{EBO} = 7$ V, $BV_{CEO} = 30$ V is used in the following circuits: In each case, find V_O. Note that X represents an open circuit.

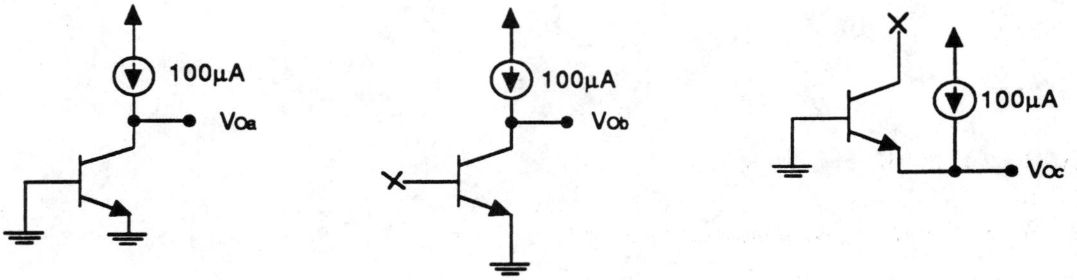

4.62 A particular BJT with grounded emitter and a constant base current of 0.1 mA, is found to have $V_{CE\,sat}$ of 0.2 V at $I_C = 3$ mA and $V_{CE\,sat}$ of 0.1 V at $I_C = 1$ mA. What are the corresponding values of $R_{CE\,sat}$ and $V_{CE\,off}$?

4.63 A BJT operating at a constant collector-to-emitter voltage of 10V is found to have $I_C = 1.20$ mA with $I_B = 11$ μA. When I_B is increased to 12 μA, I_C becomes 1.29 mA. What are the values of h_{FE} and h_{fe} for this transistor in this situation? Estimate V_A.

4.64

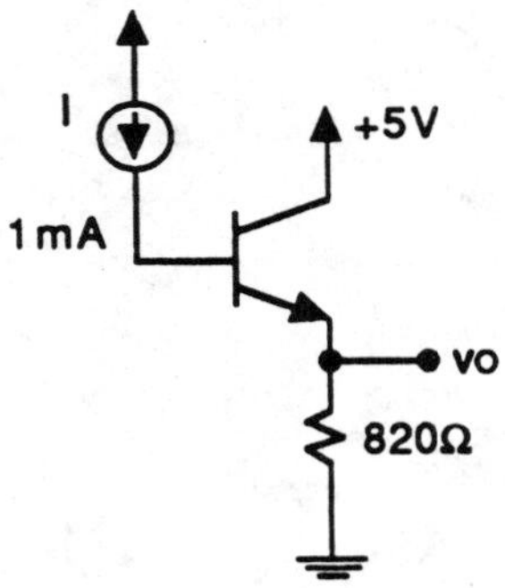

For $V_{CE\,off} = 50$ mV and $R_{CE\,sat} = 50\ \Omega$ for the transistor operated at the base current provided in the circuit shown, what output voltage results? If the base drive were quadrupled, what would you expect V_O to become?

Chapter 5

FIELD-EFFECT TRANSISTORS (FETs)

SECTION 5.1: Structure and Physical Operation of the Enhancement Type MOSFET

5.1 An n-channel enhancement MOS transistor for which $V_t = 1.5$ V is operated with a source voltage of 0 V. For what range of values of υ_{GS} is a channel induced? For $\upsilon_{GS} = 3.0$ V, for what range of values of υ_{DS} is the channel pinched off at the drain end? For what range of values of υ_D does the drain current saturate? For what range of values of υ_D does the transistor operate in the triode mode?

L 5.2 Complete the following table for devices (a) through (g)

#	Channel Type	V_t V	υ_S V	υ_G V	υ_D V	Mode (region)
a	n	1	0	3	2.1	
b	n	2	−2	2	−0.1	
c	p	−2	0	−1	−3	
d	p	−1	2	0	−1	
e		2	−3	0		saturated
f		−2	3	0	−1	
g		−2	3		−3	cutoff

SECTION 5.2: Current-Voltage Characteristics of the Enhancement MOSFET

5.3 The n-channel enhancement MOSFET characteristics shown in Fig. 5.10b of the Text represent the relationship between i_D, υ_{DS} and υ_{GS} for a range of devices for which $K = 0.25 mA/V^2$ with V_t as a parameter. For a particular device for which $V_t = 2$ V, use the data in the following table to locate the point (or points) of operation and, thereby, the missing attribute. Hint: It may help to mark and label the points on the figure in the Text, or a copy of it.

#	v_{GS} V	v_{DS} V	i_D mA
a	8	8	
b		8	9
c	8		9
d	6	6	
e	6	4	
f	6	2	
g		10	16
h	10		12

L 5.4 **The characteristic curves in Fig. 5.10b are even more useful than you have perhaps realized. On a photocopy or other facsimile of the curves, relabel the axes and v_{GS} values to correspond to the following situations:**

a) $K = 0.25\ mA/V^2,\ V_t = 2\ V$

b) $K = 0.25\ mA/V^2,\ V_t = 1\ V$

c) $K = 125\ \mu A/V^2,\ V_t = 2\ V$

d) $K = 0.5\ mA/V^2,\ V_t = 2\ V$

* **e)** $K = 1.0\ mA/V^2$, **with the 5 v_{GS} lines labelled from (and including) the lower axis as 1, 2, 3, 4 and 5 volts. (Hint: The i_D axis remains unchanged)**

** **f)** $K = 0.25\ mA/V^2$, **with the 5 v_{GS} lines labelled from (and including) the lower axis as 2, 3, 4, 5, 6 volts.**

5.5 **A n-channel enhancement MOSFET having $K = 100\ \mu A/V^2$ and $V_t = +1$ V is operated with $V_S = 0$ V and $V_G = 3$ V. For what range of voltages, V_D, on the drain is operation in the triode region? What current flows for $v_{DS} = 2$ V? 1 V? 0.5 V? What is the value of r_{DS} for v_{DS} relatively small? At what value of v_{DS} does it increase beyond its low-voltage value by 1%? 10%?**

DL 5.6 An n-channel enhancement MOSFET having $K = 100\ \mu A/V^2$ and $V_t = +1$ V is to be used for small signals as a linear resistor in the range 1 kΩ to 1 MΩ. What is the corresponding range of values of υ_{GS} required? What are the corresponding ranges of operating current (i_D) and voltage (υ_{DS}) for which the resistance provided is within 10% of its desired value.

5.7 Using equations 5.5 and 5.13, find the values of υ_{DS} for which i_D is 100%, 99%, 90% and 50% of its saturated value. For a device for which $K = 0.25\ mA/V^2$ and $V_t = 2$ V, find the values of υ_{DS} for $\upsilon_{GS} = V_t + |V_t| = 2V_t$

5.8 For a particular MOSFET operating in saturation at $i_D = 2.1$ mA and $\upsilon_{DS} = 3.0$ V, the drain current is found to increase to 2.2 mA when υ_{DS} is raised by 5 V. Find the corresponding output resistance and estimates for the channel-length-modulation factor λ, and the equivalent Early voltage V_A.

5.9 A p-channel MOSFET for which $V_t = -2$ V has a channel width of 100 µm and length of 3 µm. If it is fabricated in a process for which $\mu_n C_{OX} = 20\ \mu A/V^2$ and $\lambda = -.01\ V^{-1}$, estimate the drain current for saturation operation with $\upsilon_{GS} = \upsilon_{DS} = -5$ V. (Hint: note μ_n above, not μ_p)

5.10 A p-channel MOSFET for which the nominal threshold is −2 V (when $\upsilon_{SB} = 0$) operates in a circuit for which normal operation is such that the source voltage takes on any value between 0 V and +5 V. In a process for which $\gamma = 0.6\ V^{1/2}$ and $\phi_f = 0.3$ V, what is the threshold voltage which applies at 5 V? at 0 V? (Hint: Note that the substrate must be connected to a voltage which prevents the substrate-to-channel junction from becoming forward-biassed, i.e. to +5 V)

5.11 For the following circuits employing enhancement MOSFETS, for which $|V_t| = 2$ V and $K = 0.1\ mA/V^2$, find the labelled voltages and currents.

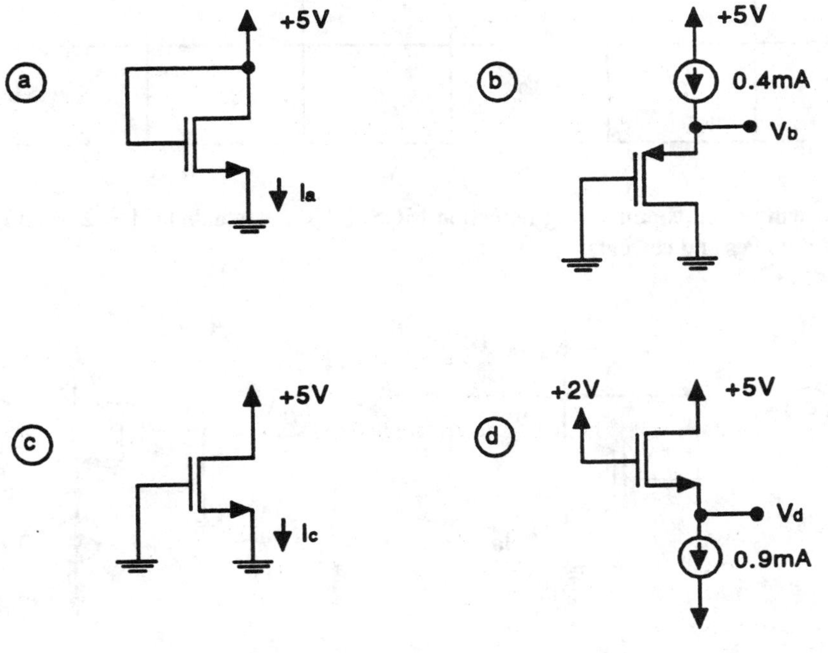

SECTION 5.3: The Depletion-Type MOSFET

L 5.12 A depletion-type n-channel MOSFET for which $K = 1\ mA/V^2$ and $V_t = -4$ V is operated under a variety of conditions as stated partially in the following table: Complete the table by providing the missing entries:

#	υ_S V	υ_G V	υ_{GS} V	υ_D V	υ_{DS} V	Operation Mode	i_D mA
a	0	–4		5			
b	0		–2		3	saturation	
c	0	0		5			
d	0		0		2		
e		0	+1		5		25
f	0	+2		5		triode	
g	0		+2	0			
h		0	+2	0		triode	

5.13 For the following circuits employing depletion MOSFETS, for which $|V_t| = 2$ V and $K = 0.1 mA/V^2$, find the labelled voltages and currents.

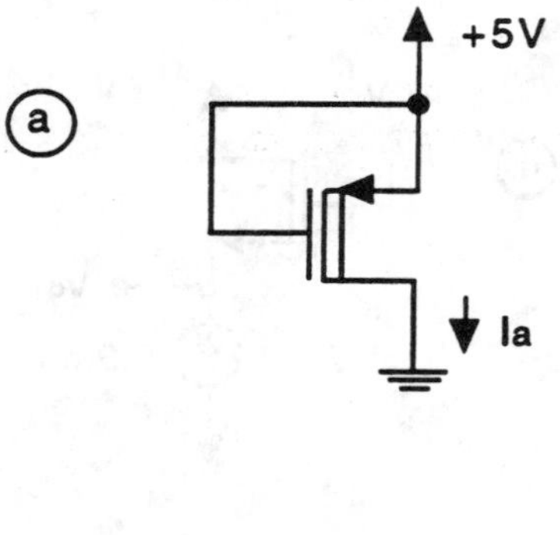

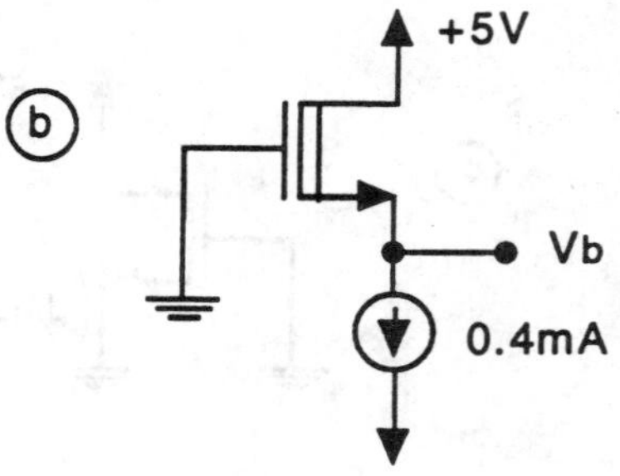

5.13

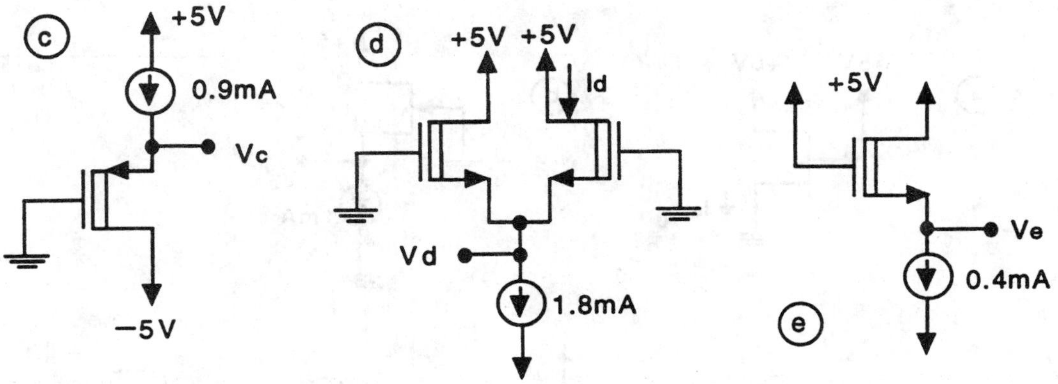

L 5.14

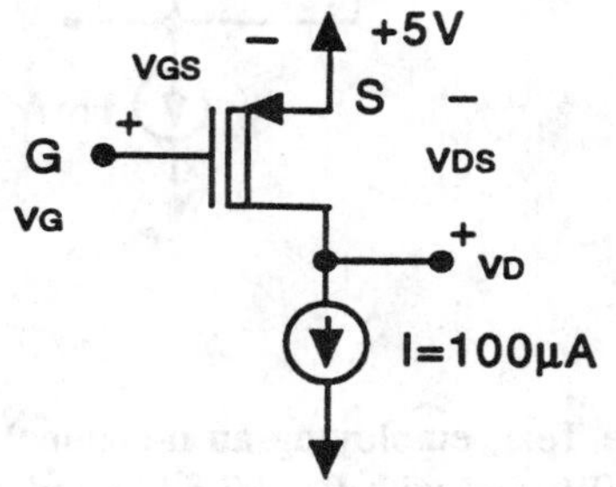

A depletion-type PMOS transistor operates in the following circuit with $v_D = 4.8$ V when $v_G = 5$V, and $v_D = 4.95$ V when $v_G = 0$ V. Find its I_{DSS} and V_t.

SECTION 5.4: The Junction Field-Effect Transistor (JFET)

5.15 An n-channel JFET with $I_{DSS} = 10$ mA and $V_p = -2$ V operates with gate and source grounded and drain connected to a positive voltage V_+. What current flows when $V_+ = +4$ V? +2 V? +1 V? At what value of V_+ does i_D become half its saturation value?

D 5.16 An n-channel JFET for which $I_{DSS} = 10$ mA and $V_p = -2$ V operates with source grounded and drain at +1 V. For what value of gate voltage is the drain current 5 mA? 1 mA?

5.17 An n-channel JFET for which $I_{DSS} = 10$ mA and $V_p = -2$ V operates as a switch with small v_{DS}. What is the series switch resistance for $v_{GS} = 0$ V? −1 V? −2 V?

5.18 An n-channel JFET for which $I_{DSS} = 10$ mA, and $V_p = -2$ V operates at $v_{DS} = 2$ V with $i_D = 5$ mA and at $v_{DS} = 7$ V with $i_D = 5.1$ mA. If v_{GS} is the same in each case, what is its value? What values of r_o, λ and V_A correspond?

L 5.19 For the following JFET circuits using devices for which $I_{DSS} = 4$ mA and $|V_p| = 2$ V, find the labelled voltages and currents.

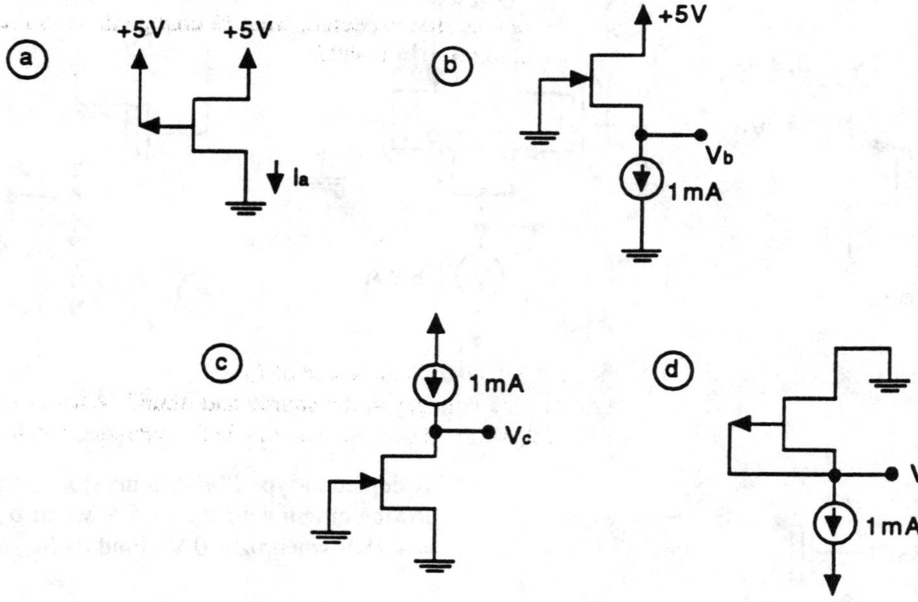

SECTION 5.5: FET circuits at DC

D 5.20 For a circuit whose topology is as shown in Fig. 5.25 of the Text, employing an n-channel enhancement transistor for which $V_t = 1$ V, $\mu_n C_{OX} = 20\ \mu A/V^2$ and $W = 40\ L$, and with $R_D = 7.5$ kΩ and ±5 V supplies, V_D is measured to be +2 V. What is the current I_D? What source voltage would you expect? What is the value of R_S used? Assume $\lambda = 0$.

5.21 A circuit using the topology shown in Fig. 5.25 with $R_D = R_S = 7.5$ kΩ is found to have $V_D = +2$ V. If V_t is known to be 1 V, what value of K applies? For K of half this value, what V_D results? By what factor has it changed to compensate for a 50% change in K?

D 5.22 Design the circuit of Fig. 5.26 of the Text, to obtain $I_D = 0.4$ mA with the transistor specified in P.5.20 above.

D 5.23

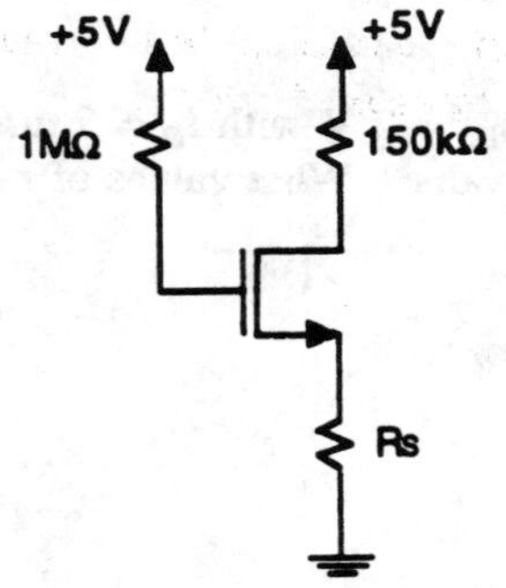

For a transistor for which $V_t = 1$ V and $K = 0.5$ *mA/V* used in the circuit shown, find the value of R_S for which $V_D = +2$ V. Specify it to 1 significant digit. For this value of R_S, what would a more precise measurement of V_D show it to be?

C 5.24 In the following circuit, the transistor employed has nominal values of K and $|V_t|$ of 0.5 mA/V and 1 V respectively.

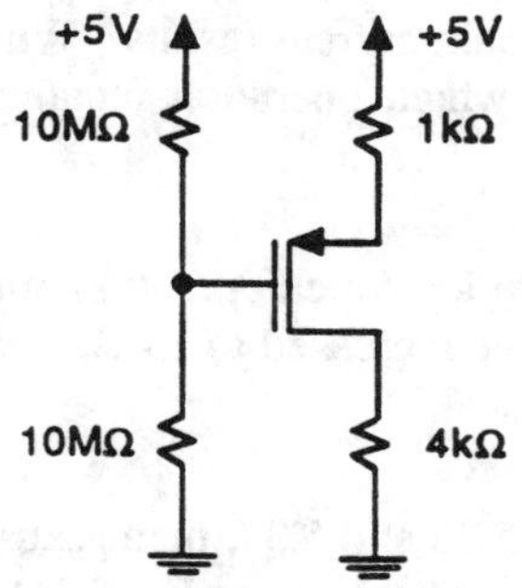

What voltage would you expect at V_D? If the voltage actually measured at V_D is 90% of that expected, what % change in V_t alone or of K alone would account for the result?

5.25 In the circuit shown, $K = 1$ mA/V and $V_t = 2$ V. What is the value of I_{DSS}?

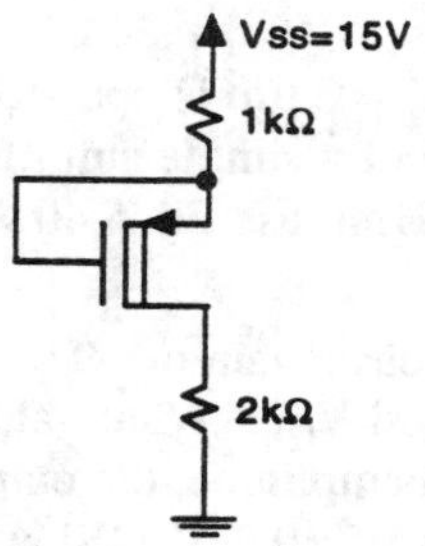

What are the voltages at the source and drain? What is the lowest value of V_{SS} for which the device remains in the saturated mode?

5.26

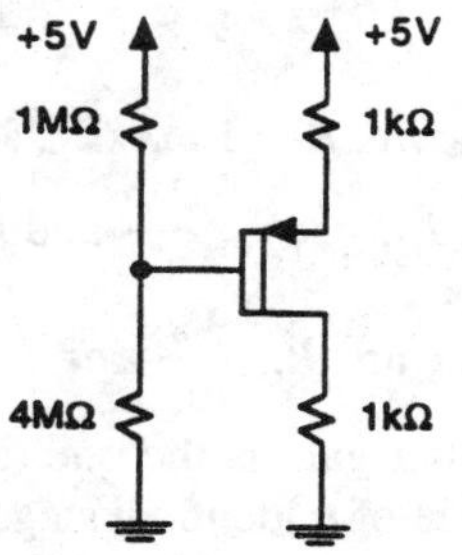

In this circuit, $V_t = 2$ V, $K = 1$ mA/V. What is the corresponding value of V_{DS}?

5.27

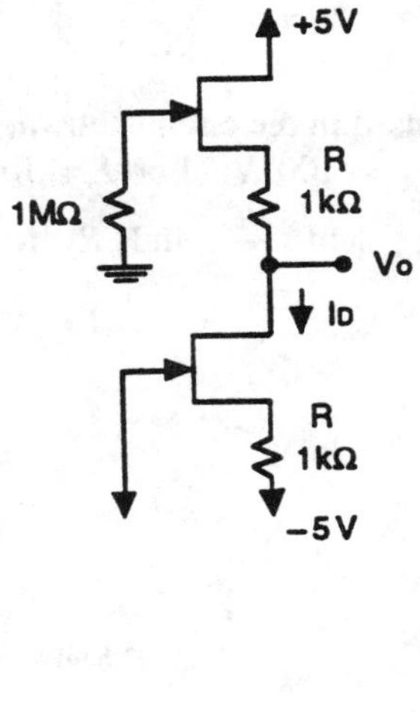

$I_{DSS} = 4$ mA, $V_p = -2$ V. What values of I_D and V_O result? What do they become if both resistors are accidentally replaced by ones of value 2 $k\Omega$?

SECTION 5.6: The FET as an Amplifier

5.28 For the device having $K = 1\ mA/V^2$ and $V_t = 2$ V whose characteristics are shown in Fig 5.34b, draw a load line corresponding to a 12 V supply and a 0.5 kΩ resistor. For $v_{GS} = 5$ V, characterize the operating point. For an *ac* input v_{GS} having a peak value of ±0.5 V, what total variation in drain current results? What is the corresponding average voltage gain? What is the largest sine wave for which operation remains in the saturation mode under these conditions?

5.29 For the situation described in P5.28 above, what is the largest load resistance for which operation remains in the saturation mode for $V_{DD} = 12$ V and $v_{GS} = \pm 0.5$ V. For the load resistance increased to 1 kΩ, what is the ratio of peak voltages at the output produced by a ±0.5 V input signal?

5.30 For the situation described in P5.28 above, find I_D and V_D using Equations 5.36 and 5.37, then g_m using Eq. 5.45. Check your work using Eq. 5.42 for g_m. What is the voltage gain you expect (using Eq. 5.46). Using this result, with a ±0.5 V input what output signal should result? Using Eq. 5.39, what peak output signal voltages would you expect? Compare generally with the results of P5.28 above.

5.31 A p-channel MOSFET, for which $\mu_p C_{OX} = 10\ \mu A/V^2$, $W = 300\ \mu m$, and $L = 3\ \mu m$ is operated at $I_D = 4$ mA. What is the corresponding value of g_m. For what value of R_L is the gain of a simple amplifier equal to −10 V/V? For what peak input signal values is operation reasonably linear. (Hint: use Eq. 5.40 with Eq. 5.36).

5.32 A MOS device operating at a *dc* bias current of 1 mA with a 10 kΩ load has a gain of −9.091 V/V for small signals. When the current is reduced to ¼ mA, the gain reduces to −4.808 V/V. What values of K and V_A apparently prevail? At $I_D = 1$ mA, as a sine wave input signal is raised in amplitude, the output signal peaks are found to change by 10% from their expected value for input peaks of ±0.5 V. What is the value of $(v_{GS} - V_t)$ which apparently applies?

L 5.33

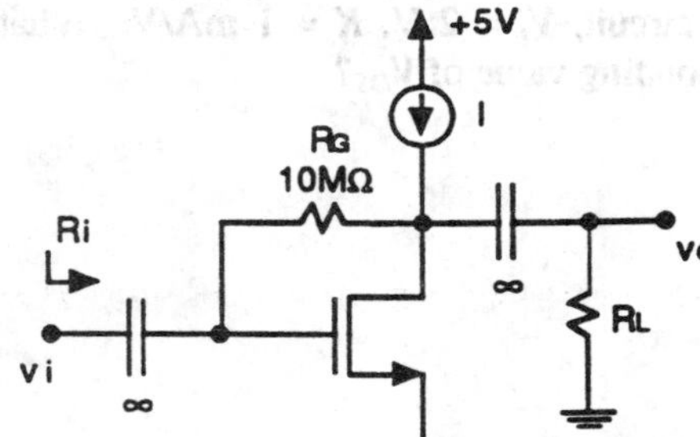

In the circuit shown, $K = 1\ mA/V^2$, $V_t = 1$ V, and $V_A = 50$ V. Find V_D, I_D, g_m, r_o, $\frac{v_o}{v_i}$ and R_i for $R_L = R$. For $I = 1$ mA, what are R_i, $\frac{v_o}{v_i}$ for $R_L = R_G$? r_o?, R_i?. Note that the latter gain is the one to be used for each stage in a cascade of n identical stages.

5.34

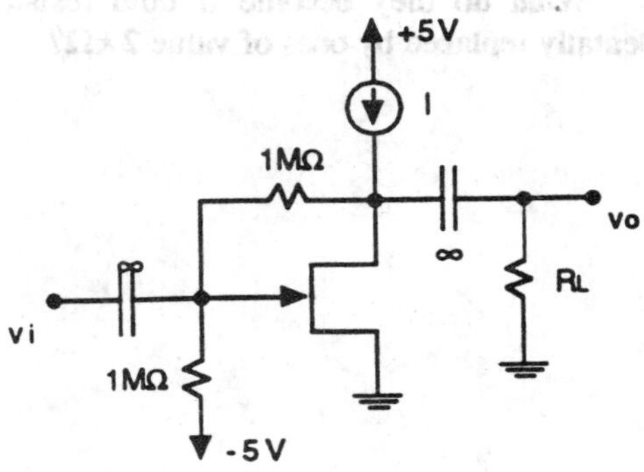

For the FET used in the circuit shown, $V_p = -2$ V, $I_{DSS} = 10$ mA, $V_A = 100$ V. For $I = 10$ mA, what are I_D, V_D, r_o, g_m, and $\frac{v_o}{v_i}$ and R_i for $R_L = \infty$? for $R_L = r_o$?

5.35

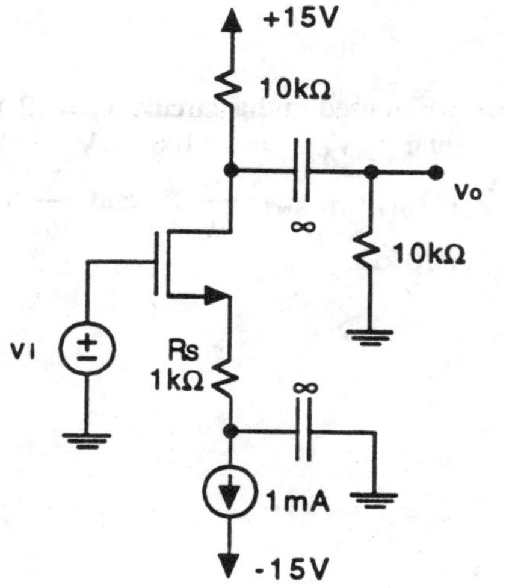

For the situation shown, g_m = 1 mA/V and r_o = 100 kΩ. Find $\frac{v_o}{v_i}$. What does the gain become for R_S = 0? for R_S = 3.76 kΩ?.

SECTION 5.7: Biassing the FET in Discrete Circuits

5.36 A particular n-channel enhancement MOS device for which V_t = 2 V and $K = 0.5\ mA/V^2$ is to be biassed using the circuit of Fig. 5.40a with a 9 V supply. For $R_{G1} = R_{G2}$ = 10MΩ and $R_S = R_D$ = 10 kΩ, find I_D and V_{DS}. For what peak amplitude of output signal will operation remain in the saturation region? What are the corresponding values when a device with V_t = 1 V is substituted in the same circuit?

CD 5.37 A design of a bias circuit using the scheme shown in Fig. 5.40a is required for a family of MOSFETs, for which V_t ranges from 1 to 2 V and K ranges from 0.3 to 0.5 mA/V^2, which provides the largest possible gain using a drain current limited to the range 0.5 to 1 mA. For the largest resistor available being 10 MΩ, what are the values of R_{G1}, R_{G2}, R_S and R_D to be used with a 9 V supply? Arrange that the largest output signal for which operation in saturation is assured, is 0.5V peak.

5.38 A particular n-channel enhancement MOS device for which V_t = 2 V and $K = 0.5\ mA/V^2$ is to be biassed using the circuit of Fig. 5.43 with R_G = 10 MΩ and R_D = 20 kΩ and V_{DD} = 9V. Find I_D and V_{DS}. For what peak amplitude of the output signal will operation remain in the saturation region? What are the corresponding values when a device with V_t = 1 V is substituted?

5.39 The circuit and situation described in P5.38 above is modified by a second resistor, R_{G2} = 10 MΩ shunted from gate to source. Repeat the computations requested there.

D 5.40 Using the circuit of Fig. 5.43, prepare a design for the situation described in P5.37 above, but with a 5 V supply.

D 5.41 Reconsider the situation presented in P5.40 above using the topology of Fig. 5.43, with a resistor R_{G2} added from gate to source to increase the output swing by a factor of 1.5, all other conditions being the same.

D 5.42 For a depletion MOS device for which V_p = –4 V and I_{DSS} = 32 mA, design a bias circuit of the type shown in Fig. 5.40 of the Text for a drain current 8 mA, using a 9 V supply, and the largest possible value of R_D that allows for a drain-signal swing of ±2 V. Find values for R_S, R_D, R_{G1}, R_{G2} using 10 MΩ as the largest available resistor value.

5.43

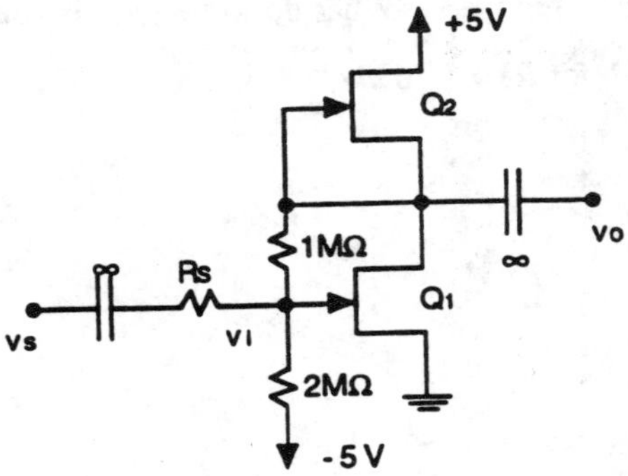

For the FETs used in this circuit, $V_p = -2$ V, $I_{DSS} = 10$ mA, and $V_A = 100$ V. What are I_D, V_{DS1}, r_{01}, r_{02}, g_{m1}, $\frac{v_o}{v_i}$ R_i, and $\frac{v_o}{v_s}$ for $R_S = 0\Omega$? or R_i? or 10 kΩ?

SECTION 5.8: Basic Configurations of Single-Stage Amplifiers

5.44 A common-source amplifier resembling that shown in Fig. 5.45 of the Text employs an FET biassed to have $g_m = 1$ mA/V and $r_o = 50$ kΩ, with $R_D = 10$ kΩ and $R_G = 1$ MΩ to drive a 20 kΩ load. What is the input resistance, the gain with no load, the output resistance, and the loaded gain? Calculate the latter gain two different ways, one direct, and one using a Thevenin-equivalent circuit.

5.45 The circuit described in P5.44 above, using a constant-current source bias, is reconfigured as a common-gate amplifier. What is its input resistance, gain with no load, and gain with load, when driven by a source having a) zero resistance b) a resistance of 1 kΩ?

5.46 A source follower employing a constant-current source supply is measured to have an output resistance of 952 ohms. When its bias current is quadrupled, its output resistance reduces to 455 ohms. Find values for g_m and r_o in the original situation. If this bias current was 1 mA at first, what is V_A for this transistor? At 1 mA bias, for what range of loads is the follower gain ≥ 0.900 V/V?

D **5.47** A source follower, employing a p-channel depletion FET for which $I_{DSS} = 8$ mA, $V_p = +2$ V and $V_A = 40$ V, is biassed at $i_D = I_{DSS}$ by means of a suitably controlled current supply connected to the source. What are the gain and output resistance of the follower for no load? For what value of load does the gain reduce to 0.8 V/V? Suggest a simple biassing-circuit scheme for the follower employing a second identical FET and a ±5 V power supply.

SECTION 5.9: Integrated-Circuit MOS Amplifiers

DL **5.48** Two p-channel transistors, one enhancement and one depletion, are operated as 2-terminal devices. For each, how many connections are there in which current can flow between the two terminals? (Be careful!) In total, how many configurations allow the device to operate in saturation? Sketch them and identify for each the minimum terminal voltage at which current flows in the saturation mode.

L 5.49 For the following circuits, for which $K = 1\ mA/V^2$ and $|V_t| = 2$ V, find the labelled currents and voltages.

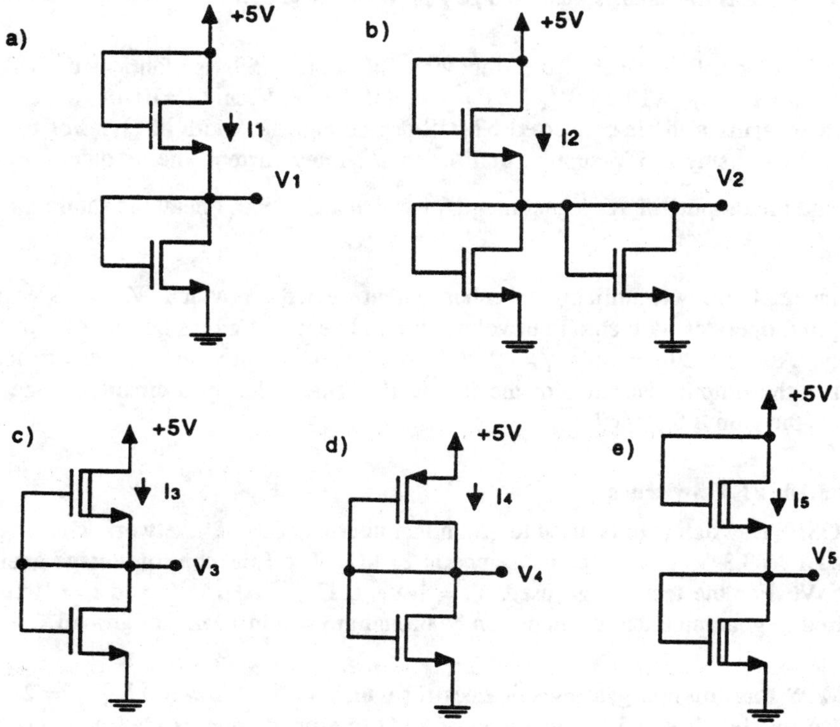

5.50 A particular amplifier employs two enhancement p-channel transistors for which $V_t = -1$ V. For the driver, $K_D = 90\ \mu A/V^2$. For the load, $K_L = 10\ \mu A/V^2$. The power supply is +5 V. Ignoring the body effect, when this amplifier is operated in its linear range, what is the gain, $\frac{v_o}{v_i}$? What is the value of v_I for which $v_O = V_{DD}/2$? What are the upper and lower values of output, and corresponding input voltage, for which Equation 5.78 applies?

5.51 For an NMOS enhancement-load transistor for which $V_{to} = 0.9$V, $2\ \Phi_f = 0.6$V, $\gamma = 0.5V^{1/2}$, $K = 10\mu A/V^2$, connected to a +5 V supply, what is the upper limit of the output-voltage range? (Hint: use Equation 5.25). What are the values of V_t, χ and g_m for outputs near the upper limit, and near 0 V?

5.52 For the load situation described in P5.51 above, employing a driver for which K_D=90 $\mu A/V^2$, what voltage gain results for $V_O = 2.5$ V?

5.53 An NMOS amplifier employs a driver for which $V_t = 1$ V and $K_D = 90\ \mu A/V^2$ with a depletion load for which $V_t = -2$ V and $K_L = 22.5\ \mu A/V^2$. For both devices $V_A = 50$ V. The power supply is +5 V and $\chi = 0.2$. Find the gain for V_O around 2.5 V. What is the range of outputs for which this gain value applies?

5.54 A PMOS mirror employs two enhancement devices for which $V_t = -1$ V, having K values of 10 $\mu A/V^2$ and 90 $\mu A/V^2$, using a 18 μA reference current. For the two possible arrangements, what is the v_{SG} and output current that result?

5.55 A CMOS amplifier using the topology of Fig. 5.59 employs devices for which $|K| = 10\ \mu A/V^2$ and $V_A = 100$ V. What is the small-signal voltage gain which results for $I_{REF} = 25\ \mu A$, $2.5\ \mu A$, or $0.25\ \mu A$?

5.56 A CMOS amplifier using the topology of Fig. 5.59 is fabricated with a process for which $\frac{1}{2}\mu_n C_{OX} = \mu_p C_{OX} = 10\ \mu A/V^2$, $|V_t| = 1$ V, $|V_A| = 50$ V, and $L = 10\ \mu m$, to have $W_n = W_p = 100\ \mu m$. I_{REF} is created using a diode-connected NMOS device half the width of Q_1. For the amplifier biassed to have $V_{GS1} = V_{DS1}$ using a 5 V supply, find the *total* supply current, the *dc* output voltage V_O, the voltage gain $\frac{v_o}{v_i}$, and the output voltage signal range for which all devices operate in saturation.

5.57 A p-channel follower utilizing an enhancement device for which $|V_t| = 0.9$V, $\mu_p C_{OX} W = 1000\ \mu m$ and $L = 10\ \mu m$, operates with an output voltage around zero and a bias current of 1 mA using a +10 V and −5 V supply. As well, $2\Phi_f = 0.6$V, $\gamma = 0.5\ V^{1/2}$. What is the input voltage necessary for an output of zero volts? What is the output resistance of the follower? What is its open-circuit voltage gain? For what range of loads is the gain ≥ 0.5V/V?

SECTION 5.10: FET Switches

D 5.58 A MOSFET switch is to be used to ground an internal node of a network whose open-circuit voltage ranges from 1.1 to 3.3 V and source resistance is 21 kΩ. The FET control voltage available switches between 0 and 5 V. For the technology used, $V_t = 1$ V, $\mu_n C_{OX} = 20\ \mu A/V^2$ and $L = 10\ \mu m$. What switch width is required to guarantee that the node can be brought to within 10 mV of ground?

5.59 A CMOS transmission gate uses devices for which $W_p = 2W_n = 100\ L$, $|V_t| = 2$ V and $\mu_n C_{OX} = 20\ \mu A/V^2$. For control signals of ±5 V and a load of 5 kΩ to ground, what is the fraction of the *ac* input signal lost in the switch, for an input $v_I = v_i + V_I$, for $V_I = -5$ V, 0 V or +5 V?

SECTION 5.11: Gallium-Arsenide (GaAs) Devices - The MESFET

5.60 A GaAs MESFET for which $\beta = 10^{-4} A/V^2$ for each μm of gate width, $\lambda = 0.2\ V^{-1}$ and $V_t = -1.0$ V, and having a width of 100 μm, is operated at $v_{GS} = 0 \pm 0.2$ V, with v_{DS} of about 3 V. Find the range of g_m, r_o and the highest available voltage gain you can expect for such operation.

5.61 The transistor described in P5.60 above is operated with a 3 V supply and a 100 Ω load. What values of v_{DS} result for the inputs stated? What is the corresponding "voltage gain" for a ±0.2 V input signal?

5.62 The amplifier in Example 5.15 and Fig. 5.68 on page 389 of the Text is modified by increasing the width of Q_2 to equal that of Q_1. For the output stabilized at 5 V by some external means, what values of i_{D1}, V_{GS1}, g_{m1}, and small-signal voltage gain result?

5.63 For the situation described in P5.62 above, the output is stabilized at +3 V. What is the value of V_{GS1} required? What are g_{m1} and the gain $\frac{v_o}{v_i}$?

General Problems

5.64

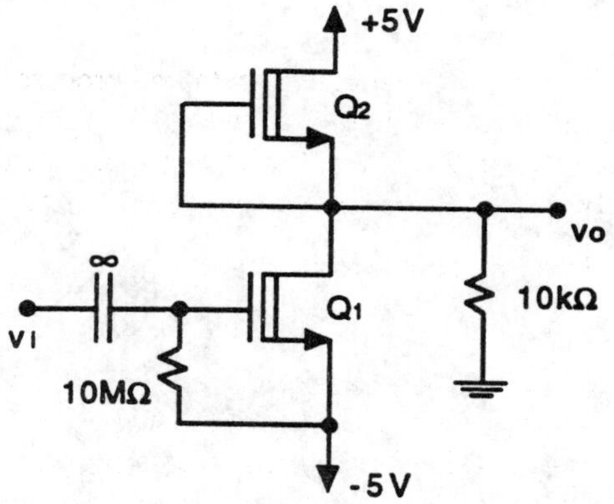

For the circuit shown, using matched devices for which $V_t = -2$ V, $I_{DSS} = 10$ mA and $V_A = 50$V, find V_O, υ_o for $\upsilon_i = 0.1 \sin \omega t$ V.

Chapter 6

DIFFERENTIAL AND MULTISTAGE AMPLIFIERS

SECTION 6.1: The BJT Differential Pair

6.1 For the BJT differential-pair configuration, find the differential signal ($\upsilon_d = \upsilon_{B1} - \upsilon_{B2}$) sufficient to cause

a) $i_{C2} = 99\%\ I$

b) $i_{C1} = 95\%\ I$

c) $i_{C1} = 9.0\ i_{C2}$

L 6.2 For situations related to those shown in Fig. 6.2, some measurements are taken as tabulated below: For all cases, $V_{CC} = +0$ V, $R_C = 4\ k\Omega$, and $I = 2$ mA. For the BJTs assume β is high, $V_{BE} = 0.7$ V, V_{CE} sat = 0.2 V, all essentially independent of the detail of junction-current magnitude. Find the missing values.

Case	υ_{B1} V	υ_{B2} V	$\upsilon_{E1,2}$ V	υ_{C1} V	υ_{C2} V
a	0		−0.7		6
b	2	2		6	
c		1	1.3		
d	−2		0.3	10	
e	1		2.8		3
f	−4	−4			8
g		0	+3.3		
h	1	3.5			

L 6.3 For an npn BJT differential pair using a +10 V supply and collector resistors of 4 kΩ, partial measurements provide results as follows. Find the missing entries, assuming $\alpha = 1$ and $n = 1$.

Case	I mA	υ_{B1} V	υ_{B2} V	υ_E V	υ_{C1} V	υ_{C2} V
a	0.2	0.00		–.700		9.60
b	0.2	0.01	0.00			
c	0.2	0.00	0.05			
d	0.2		0.00	–.675		
e	0.2	–1.00				9.90
f	2.0	–1.00	–1.00			
g	2.0	0.01	0.00			
h	2.0	0.00	0.05			
i	2.0	1.00			3.00	

SECTION 6.2: Small-Signal Operation of the BJT Differential Amplifier

C 6.4 Explore the nature of the small-signal assumption made following Eq. 6.11 in the creation of Eq. 6.12 by including one additional term of the exponential series ($e^x = 1 + x + x^2/2$) in the creation of a higher-order alternative. What is the error made in using Eq. 6.12 for $\frac{\upsilon_d}{2} \leq 10\ mV$? For what value of υ_d is the error made by the linear approximation equal to 10%, 5%, 1%?

6.5 A particular differential amplifier resembling Fig. 6.5, uses $I = 200\ \mu A$, $R_C = 10k\Omega$ and $V_{CC} = +3$ V. What is the differential gain achieved for outputs taken differentially? From one or the other collectors separately? What is the upper limit of common-mode input voltage, for which operation maintains $\upsilon_{CB} \geq$

–0.4 V?

6.6 A differential amplifier resembling that in Fig. 6.5, employs collector resistors of 100 kΩ and a bias source of 200 μA. What is its differential voltage gain for outputs taken differentially? What is its differential input resistance? Transistor $\beta \geq 150$. Emitter resistors are added to double the input resistance. What does the differential voltage gain become?

6.7 A differential amplifier employing 10 kΩ collector resistors, and for which the emitter bias current is 400 μA, uses BJTs for which n=1 and β=200. It is driven differentially by signal sources whose output resistances are 10 kΩ. The emitter-current source has an output resistance of 0.5 MΩ. For outputs taken both differentially and single-endedly, find the differential input resistance, the differential-mode gain from the source, the common-mode input resistance, the common-mode gain, and the CMRR as a ratio and in dB.

6.8 For the situation described in P6.7 above, the collector resistors are mismatched. For outputs taken differentially, find the common-mode gain and CMRR (as a ratio and in dB) for load resistors specified to ±1%, and to ±10%.

C 6.9 For the situation described in P6.7 above, the source resistors are mismatched by 10% and device betas vary by ±10% from their nominal value. For outputs taken differentially, find the nominal differential gain, the worst-case common-mode gain and the corresponding CMRR in dB. Hint: Note that the half-circuit idea does not work directly here; rather current division in a Y-shaped resistor network must be considered.

6.10 For the situation described in P6.9 and P6.7 above, fixed emitter resistors, each of value R_E=9r_e (where r_e is the nominal emitter resistance) are added. What do A_d, A_{cm} and CMRR become for the output taken differentially from matched collector resistors?

SECTION 6.3: Other Non-ideal Characteristics of the Differential Amplifier.

6.11 A BJT differential amplifier operating at a total bias current of 200 μA employs collector resistors that have a ±5% tolerance. What is the worst-case input offset you would expect? If emitter resistors are added, with $R_E = 9r_e$ what input-offset voltage results?

6.12 If in P6.11 above, the added emitter resistors each have a ±5% tolerance, what might the most extreme input offset become? What might be a more realistic estimate of its expected value (Hint: Use the idea in Eq. 6.55). If the collector resistors are now trimmed to have exactly equal values, what does the input offset become?

6.13 Four uncorrelated sources of input offset, to which a differential amplifier is subject, produce essentially equal individual contributions of 2 mV. Estimate the total offset resulting. If the offsets are 0.5, 1, 2, and 4 mV individually, what overall offset might be expected?

6.14 For a BJT differential pair biassed at current I, both a β mismatch of 10% and a source-resistance mismatch of 10% are present. For nominal values of I, β, and R_S of 100 μA, 100, and 100 kΩ, respectively, what worst-case input-voltage offset is possible?

6.15 An npn BJT differential amplifier for which the bias current is 300 μA employs a 15 V supply and 60kΩ resistors. For peak signals of 10 mV across the junctions of each input transistor and $V_{CE\,sat}$ limited to 0.4 V with $V_{BE} = 0.7$ V, what is the most positive usable common-mode input signal?

SECTION 6.4: Biassing in BJT Integrated Circuits

6.16 A diode-connected transistor is operated at a bias current of 100 μA. What is the resistance between its two terminals? If two are connected, a) in parallel, b) in series, to the same biassing source, what do the resistances across the combinations become?

6.17 For what value of β would a simple current mirror have a gain error of 1%? 0.1%?

DL* 6.18 A simple mirror operating at a current of 1 mA is augmented by resistors in series with each emitter across which the nominal voltage drop is 1/10 V_{BE}. For transistors for which $V_{BE} = 0.700$ V at 10 mA and $n = 1$, what resistors would be used (specify to 1 significant digit only). Now one of these resistors is to be laser-trimmed to reduce its value to compensate for a nominal value of β equal to 90. What is its required value? What is the current error at 0.5 mA and at 2 mA with nominal β? At all 3 currents with β = 70?

6.19 A simple current mirror operating at 100 μA employs devices for which β = 150 and $V_A = 150$ V. For what value of output voltage do the two imperfections cancel. Over what output-voltage range is the net error less than 1%?

D 6.20 In the design of a simple current mirror for a particular application, there is a concern for variation of temperature on the output current. Chose a value of R and V_{CC} to provide a nominal current of 100 μA at 25°C, at which $V_{BE} = 0.700$V, with the change at 75°C limited to 5% (Use a junction temperature coefficient of –2 mV/°C).

D 6.21 Given several identical npn transistors and a reference current of 1 mA, sketch the circuit of a multiple-output mirror whose nominal current values are 0.5 mA, 1 mA, and 2 mA. How many transistors do you need? If you also are provided with some matched pnp transistors, sketch an alternative circuit topology assuming only one end of I_{REF} is available. How many transistors do you need? What is the number if both ends of I_{REF} are available (as in Fig. 6.17).

6.22 Repeat the analysis of the circuit of Fig. 6.18, by starting with $i_{B1} = i_{B2} = i$. What does the current gain become with two outputs using transistors Q_{2a} and Q_{2b}?

CDL 6.23 For the circuit of Fig. 6.18, repeat the analysis leading to Eq. 6.67, but maintain all 3 β values separately. For 3 transistors having current gains of β and $\beta(1 \pm k)$, select an optimal placement of each in the circuit. Is there a particular value of k for which your design is particularly good?

CD 6.24 For the Wilson Mirror in Fig. 6.19, follow through the process suggested in P6.23 above for the base-current-compensated mirror.

D 6.25 Design a two-output Widlar current source using a 100 μA reference to provide outputs of both 1 μA and 10 μA using 1mA transistors for which $V_{BE} = 0.700$V at 1mA with $n = 1$.

SECTION 6.5: The BJT Differential Amplifier with Active Loads

6.26 The amplifier shown in Fig. 6.24 uses a bias source of 100 μA with devices for which V_A = 150 V and β = 75. What is its overall transconductance, its open-circuit voltage gain, output resistance and input resistance? What does the voltage gain become when feeding a load equal to the input resistance?

6.27 The differential amplifier in Fig. 6.24, using a 100 μA bias source, is augmented with 500 Ω resistors in the emitters of each of Q_1 through Q_4. For β = 75 and V_A = 150 V, what is the overall transconductance, the output resistance, and the open-circuit voltage gain. (Hint: Note that the common connection between emitter resistors is virtual ground for differential inputs).

6.28

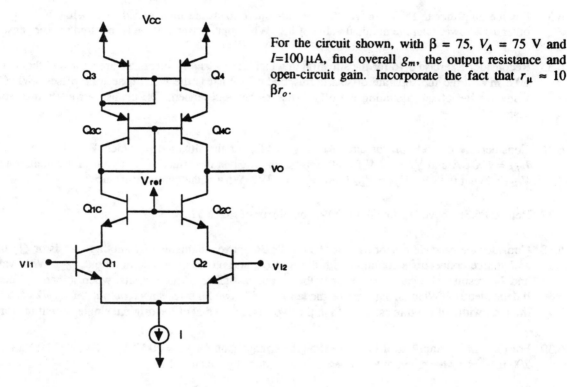

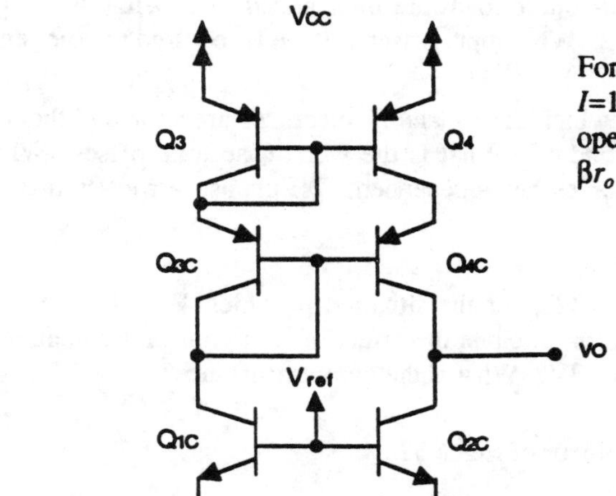

For the circuit shown, with β = 75, V_A = 75 V and I=100 μA, find overall g_m, the output resistance and open-circuit gain. Incorporate the fact that $r_\mu \approx 10\ \beta r_o$.

SECTION 6.6: The JFET Differential Pair

6.29 A pair of JFETs for which I_{DSS} = 2 mA and V_p = –4 V are biassed using a 1 mA current source. For both gates at zero volts, what is the voltage at the common source? For a small differential input υ_d, what is the equivalent g_m of the pair? For what differential input does the current just begin to flow all in 1 transistor? For what range of differential input does the small-signal approximation apply?

6.30 Rework equations 6.90 and 6.91 into a form involving a term $(1 - K\upsilon_{id}^2)^{1/2}$. What is the value of the term for $\dfrac{\upsilon_{id}}{V_p} \Big/ \left[\dfrac{2I}{I_{DSS}}\right]^{1/2}$ = 1, 0.33, 0.1, .033? For what value of $\dfrac{\upsilon_{id}}{V_p} \Big/ \left[\dfrac{2I}{I_{DSS}}\right]^{1/2}$ does the term become 0.90? 0.99?

6.31 The JFET pair in P6.29 above, for which V_A = 50 V, is coupled to a pair of 10 kΩ drain (load) resistors. What voltage gain results for outputs taken differentially? Alternatively, for a load consisting of a current mirror for which V_A is 50 V, what gain results?

SECTION 6.7: MOS Differential Amplifiers

6.32 Consider Fig. 6.30 in the Text. For what values of $\frac{v_{id}}{V_{GS}-V_t}$ does g_m of each device deviate from its value at $v_{id} = 0$ by 10%? 5%? 1%?

6.33 A PMOS differential amplifier utilizing a bias current of $I = 25\ \mu A$ uses devices for which $V_t = 1$ V, $W = 120\ \mu m$, $L = 6\ \mu m$, $\mu_p C_{ox} = 10\ \mu A/V^2$ and $V_A = 50$ V. Find V_{GS}, g_m and the maximum possible voltage gain using current-source loads which are a) ideal, b) have $V_A = 50$ V.

6.34 For the amplifier in P6.33 above, the current-source loads are unbalanced, one being 10% higher and the other 10% lower than the nominal value of $I/2$. What input offset voltage is required to compensate?

6.35 Reconsider Exercise 6.16 for the situation in which the R_D and K tolerances are ±1% and the V_t tolerance is ±0.6 mV. Find the separate offsets corresponding. What is the worst-case total offset. What is a likely value for the offset, assuming the offset sources are independent. (Hint: use a root-sum-of-squares estimate).

6.36 Consider the cascode mirror circuit of Fig. 6.32b, for the situation in which $V_t = 1$ V, $K = 100\ \mu A/V^2$, $I_{REF} = 100\ \mu A$ and $V_A = 20$ V for all transistors. Include the effect of λ in your bias calculations. What is V_{GS1}? What is I_O for $V_{D3} = V_{D4}$? For $V_{D3} = 12$V? What is the output resistance?

6.37 Repeat P6.36 above for the (simple) Wilson mirror of Fig. 6.32c.

6.38 Consider the cascode mirror as described in P6.36 above, augmented by another transistor Q_5 having gate and source connections common with those of Q_3 and providing a second output, I_{02}. What values of I_{01} and I_{02} result? (Be careful!) What is the output resistance of the outputs, when joined? When operated independently? What change can be made to provide two high-resistance outputs of $I_{REF}/2$ each? Compare the total width of the transistors used in the two cases with that of the original single-output mirror.

6.39 For the CMOS amplifier of Fig. 6.34, find the voltage gain for $V_A = 20$ V, $V_t = 1$ V, $K = 100\ \mu A/V^2$, and $I = 200\ \mu A$. For what external load does the gain reduce by a factor of 2?

SECTION 6.8: BiCMOS Amplifiers

6.40 For $I = 10\ \mu A$, find g_m, R_i, r_o and the voltage gain of the CE and CS amplifiers shown in Figs. 6.35a, b. For the BJT, $V_A = 100$ V, $\beta = 100$. For the MOSFET, $V_A = 20$ V, $\mu_n C_{OX} = 20\ \mu A/V^2$, $L = 2\ \mu m$ and $W = 20\ \mu m$.

6.41 For the BiCMOS cascode shown in Fig. 6.35c, using the parameters provided in P6.40 above, find the overall voltage gain $\frac{v_o}{v_i}$, for $I = 10\ \mu A$.

6.42 For the BiCMOS cascode shown in Fig. 6.35c, using the parameters provided in P6.40 above, find the voltage gain $\frac{v_o}{v_i}$, for $I = 100\ \mu A$.

6.43 Consider the BiCMOS double-cascode mirror in Fig. 6.37 for devices as described in P6.40 above, and operating at 10 µA. What does the output resistance become if Q_3 and Q_6 are not used? Recall that $r_\mu \approx 10\,\beta r_o$. What does the output resistance become if Q_6 and Q_3 are retained, but Q_5, Q_2 are eliminated?

SECTION 6.9: GaAS Amplifiers

L **6.44**

For the following GaAs circuits, using devices characterized by the normalized data given in Table 5.2 (see page 389 of the Text) with width in µm as noted near each, find labelled values of *I* and *V*.

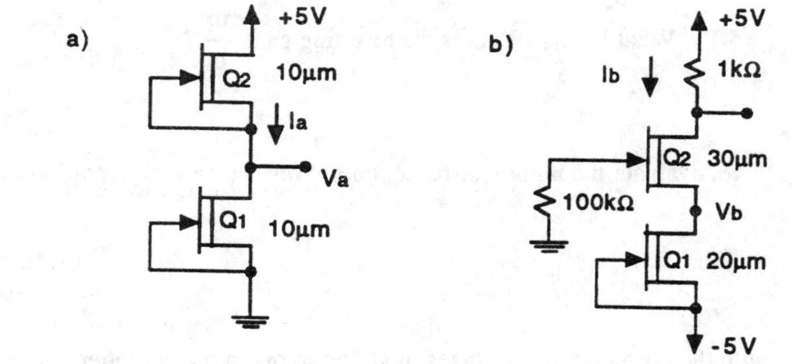

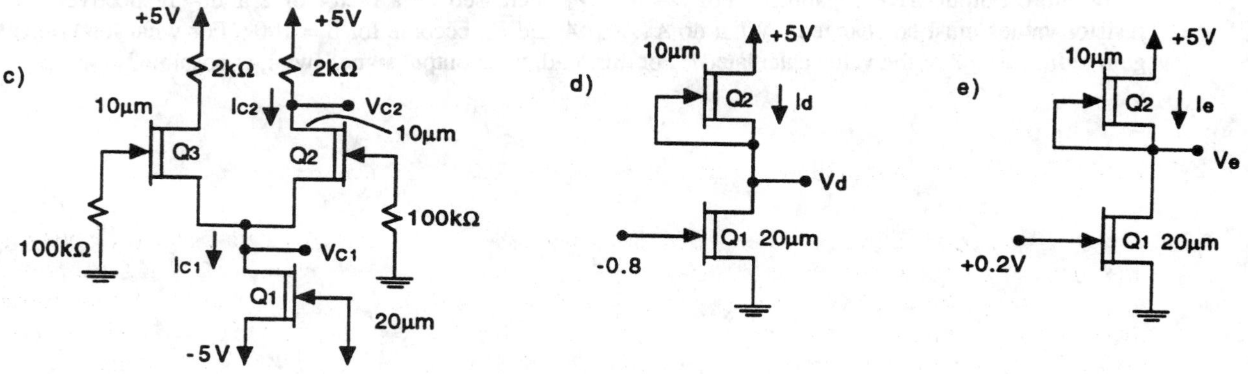

6.45 Repeat Exercise 6.24 for the cascode current source for conditions as stated, except that W_1=5 µm.

L 6.46 The circuit of Fig. 6.40 is extended to a double cascode by adding a transistor Q_3 of width 20 µm whose gate is connected to $V_{BIAS\,2}$, with a change of V_{SS} to 6 V, and V_{BIAS} to $V_{BIAS\,1} = -5.3$ V. For this design provide the data requested in Exercise 6.24 on page 463 of the Text.

D 6.47 Consider the circuit of Fig. 6.41. Using the data from Table 5.2 (on page 389 of the Text) select appropriate values for FET channel and diode widths for operation at *I*=5 mA, with V_{DD}=5 V and $V_A \approx 2$ V. For this design, calculate $\alpha = \dfrac{\upsilon_b}{\upsilon_a}$ from Eq 6.152, and R_o from both Eq. 6.153 and Eq. 6.154.

CDL 6.48 Using Table 5.2 values, design a composite MESFET for nominal operation with $I_{DSS\,eq} = 0.5$ mA, $\upsilon_{DS} = 3$ V and $\upsilon_{DS1} = 0.7$ V. What are W_1 and W_2 required? Using basic relationships for device currents and voltages, what does I_{DSS} become when υ_{DS} is increased to 6 V? Using formal small-signal analysis, what is $R_{o\,eq}$ nominally? How well does this account for the change in I_{DSS} already found?

6.49 Use two of the composite devices created in P6.48 above to implement the amplifier in Fig. 6.43c. For V_{DD} = 6 V, what nominal value of bias V_I is required for $V_O \approx V_{DD}/2$. Using the results of P6.48 above, what is the output resistance of the amplifier? What is the overall gain?

CDL 6.50 For the gain-enhanced MESFET differential amplifier in Fig. 6.45, prepare a design for operation of as many as possible of the devices at a nominal i_D of 0.5 mA and $\upsilon_{GS} = -0.5$ V. What current sources are needed for V_{CM} around 0 V and V_{DD}=5 V? What is V_O? What is the resulting gain $\frac{\upsilon_o}{\upsilon_i}$?

SECTION 6.10: Multistage Amplifiers

6.51 For the multistage amplifier of Fig. 6.46, evaluate the input resistance, output resistance and overall gain for the situations in which

a) $\beta = \infty$

b) $\beta = 50$

L 6.52 Consider the amplifier in Fig. 6.46, with the third and fourth stages modified to operate at a higher current to allow more output-drive capability. For i_{C7} and i_{C8} increased by a factor of 2 and 4 respectively, what resistor values must be changed? What do A_3, A_4, A and R_0 become for $\beta = 100$? For what load does the gain reduce to 0.8 of the value calculated? For this load, what output signal swing is possible?

Chapter 7

FREQUENCY RESPONSE

SECTION 7.1: s-Domain Analysis

7.1 Find the transfer function $T(s) = V_o(s)/V_i(s)$ of the circuit shown: Is this an STC circuit? If so, of what type? For $10C_1 = C_2 = 0.5\mu F$ and $R_1 = 10\ k\Omega$, find the location of the pole(s) and zero(s) and sketch Bode plots for the magnitude and phase responses.

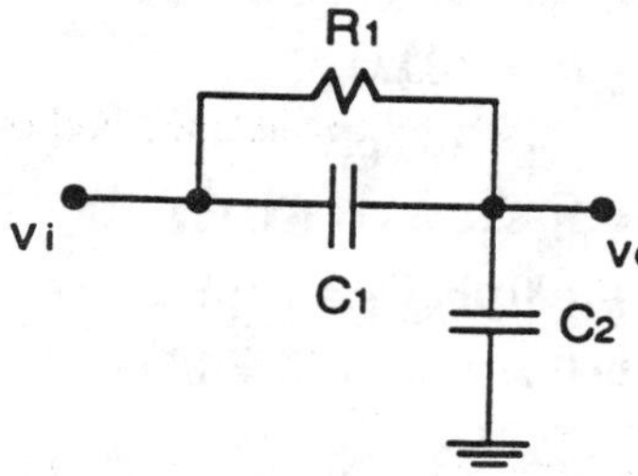

L* 7.2 For the transfer function

$$T(s) = \frac{10^8\ (s)\ (s+10)}{(s+1)\ (s+100)\ (s+10^5)\ (s+10^6)}$$

find

a) the equivalent expression in which the factors are of the form $(1 + \frac{s}{a})$;

b) the gain and phase at very low frequencies and at very high frequencies;

c) the poles and zeros;

d) the greatest-available gain and corresponding phase;

e) the gain at 10^3 rad/s, 10^5 rad/s;

Sketch Bode plots for gain and phase. Note that you have been asked to prepare Bode plots at the very end. Is this appropriate? At what stage in your overall answer would the Bode sketches have been easiest? most useful?

7.3 For the situation presented in P7.2 above, use exact analysis to calculate the amplitude and phase of $T(s)$ at 100 rad/s and 2×10^5 rad/s.

SECTION 7.2: The Amplifier Transfer Function

7.4 For an amplifier whose overall response is characterized as in P7.2 above, find (expressions for) A_M, $F_L(s)$, $F_H(s)$, $A_L(s)$, $A_H(s)$.

7.5 >From a dominant-pole point of view, find approximate transfer functions for $F_L(s)$, $F_H(s)$ and $A(s)$ for an amplifier characterized by the complete transfer function in P7.2 above.

7.6 For the transfer function in P7.2 above, find a value for the lower 3-dB frequency from a dominant-pole viewpoint, using the root-squares approach, and exactly.

7.7 Proceed as in P7.6 above, but for the upper 3-dB frequency.

CD 7.8 An amplifier having the transfer function described in P7.2 above, is augmented by circuitry which causes the addition of a zero and pole at 10^5 and 2×10^6 rad/s respectively. What is the new transfer function? Estimate the new upper 3-dB frequency. Calculate it exactly. Note that the technique demonstrated is called pole-zero cancellation.

CL* 7.9 For the circuit shown, find the upper 3-dB frequency using the method of open-circuit time constants, and exactly, for the conditions that:

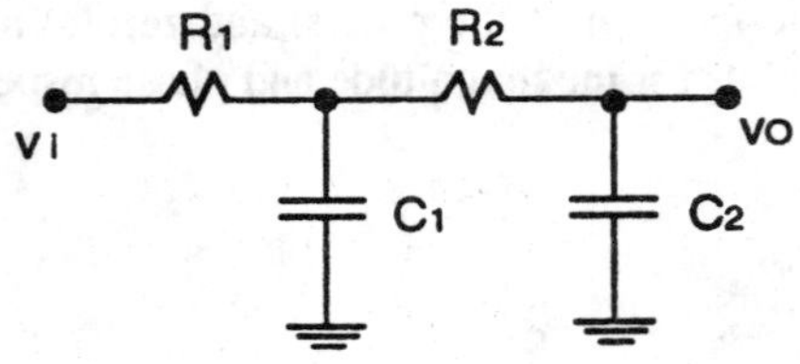

a) $R_1 = R_2 = 10\text{ k}\Omega$, $C_1 = C_2 = 100\text{ pF}$.

b) $R_1 = 10\text{ k}\Omega$, $R_2 = 100\text{ k}\Omega$, $C_1 = 100\text{ pF}$, $C_2 = 10\text{ pF}$.

c) As in b), but with $C_1 = 10\text{ pF}$.

L 7.10 For the circuit shown, find the lower 3-dB frequency using the method of short-circuit time constants, for the conditions that:

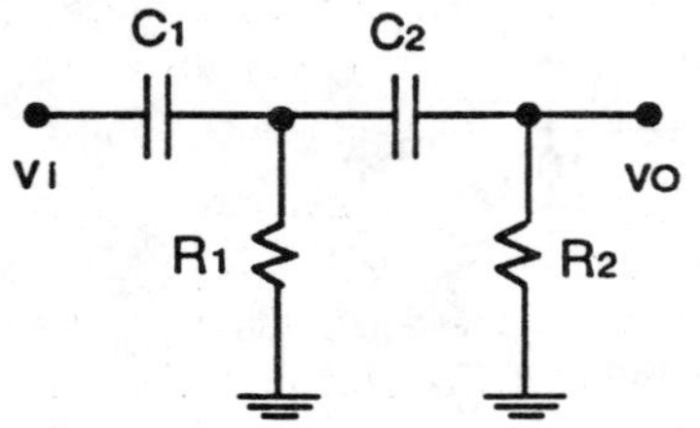

a) $R_1 = R_2 = 10\text{ k}\Omega$ $C_1 = C_2 = 1\ \mu\text{F}$

b) $R_1 = 10\text{k}\Omega$, $R_2 = 100\text{k}\Omega$ $C_1 = 1\ \mu\text{F}$, $C_2 = 0.1\ \mu\text{F}$

c) As in b) but with $C_1 = 0.1\ \mu\text{F}$

SECTION 7.3: Low-Frequency Response of the Common-Source Amplifier

7.11 A MOSFET amplifier using the topology of Fig. 7.9 employing $R = 100\text{k}\Omega$, $R_{G2} = 10\,M\Omega$, $R_{G1} = 22M\Omega$, $R_S = R_D = 10\text{ k}\Omega$, $R_L = 20\text{ k}\Omega$, $C_{C1} = .01\ \mu\text{F}$, $C_{C2} = 0.1\ \mu\text{F}$ and $C_S = 1\ \mu\text{F}$, operates with $g_m = 2$ mA/V. Find the midband gain, 3 poles and a zero.

D 7.12 For the situation described above in P7.11, it is desired to have a single dominant pole at 10 Hz or less, and two coincident ones at about 1 Hz. What values of coupling and bypass capacitors should be used to minimize the total capacitance? Specify the capacitors to 1 significant digit. What poles and zero actually result?

CDL 7.13 The circuit and situation described in P7.11 above is modified by the addition of a resistor r_S in series with C_S. Find expressions for the associated pole, zero and gain corresponding to Equations 7.37, 7.38 and 7.42 respectively. For gain sacrificed by a factor of 2, what do the new pole and zero associated with C_S become?

SECTION 7.4: High-Frequency Response of the Common-Source Amplifier

L 7.14 Find values of the FET unity-gain frequency (for operation in the grounded-source configuration) for

a) A JFET for which $I_{DSS} = 4$ mA, $V_p = -2$V, $C_{gs} = 2$ pF and $C_{gd} = 0.2$ pF, operating at 1 mA.

b) A MOSFET with gate-to-channel capacitance of 0.15 pF, overlap capacitance of 20 fF, gate-to-substrate capacitance of 0.1 pF, having $V_t = 1$ Volt, $K = 100\ \mu A/V^2$, operating at 200 μA.

c) A GaAs MESFET for which g_m = 10 mA/V at relatively high bias currents with $C_{gs} = 0.15$ pF and C_{gd} = 15 fF.

7.15 An n-channel enhancement MOSFET, for which $C_{ox} = 1.0\, fF/\mu m^2$, $\mu_n = .05 m^2/Vs$, $L = 3\ \mu m$, $W = 27\ \mu m$, and $V_t = 0.5$ V, operates with $v_{GS} = v_{DS} = 2.5$ V, and source grounded. The gate overlap is about 0.3 μm. Estimate C_{gs}, C_{gd} and the unity-gain frequency which corresponds.

7.16 A particular FET amplifier is to be operated in one of two grounded-source topologies for which the gain from gate to drain is either –1 or –100 V/V. Its C_{gs} = 200 fF, C_{db} = 100 fF, C_{gd} = 20 fF. Find the equivalent capacitances to ground at the gate and at the drain of each circuit.

7.17 A particular JFET operates in a common-source circuit environment in which g_m = 1 mA/V, r_o = 50 kΩ, C_{gs} = 1 pF, C_{gd} = 0.5 pF, R_S = 100 kΩ, R_{in} = 1MΩ, R_D = 10 kΩ and R_L = 30 kΩ. Find the equivalent input capacitance at the gate, output capacitance at the drain, two poles, and an estimate of the upper 3-dB frequency. What is the highest frequency to which f_H can be raised by lowering R_S? What is the value of R_S which reduces f_H to 90% of that frequency?

7.18 For the situation described initially in P7.17 above, find exact values of the associated poles and zero, and an estimate of f_H. To what do all these change if the signal-source resistance R_S is reduced to 1 kΩ?

C 7.19 A high-performance n-channel MOS device for which V_t = 1 V and f_t = 1 GHz operates at 1 mA and V_{GS} = 2 V in a Common-Source amplifier stage for which the gain is –3 V/V. If C_{gd} is known to be $\le 0.2 C_{gs} \approx C_{db}$, what is the equivalent input capacitance? If driven from a similar amplifier whose output impedance is approximately $(3/g_m) \| 4C_{db}$ what 3-dB frequency would you estimate? Consider this situation in the context of that described in P7.32 of the text.

D 7.20 A MOS amplifier resembling that in Fig. P7.32 of the text uses a resistor R_f connected from the drain to the gate of Q_1 for biassing. If the stage gain is 3, and the output resistance of the source is 10 kΩ, what is the minimum value of R_f to ensure at most a 5% loss in signal at the input?

7.21 An amplifier having a gain of –50 V/V and dominant poles at 50 Hz and 50 MHz is supplied by a negative pulse of 50 mV amplitude and 50 μs duration. Completely characterize the output pulse produced.

SECTION 7.5: The Hybrid-π Equivalent Circuit Model of the BJT

7.22 For a BJT for which V_A = 100 V, β = 100, and the base-spreading resistance is 50 Ω, and operating at 2 mA, estimate all the low-frequency hybrid-π parameters and the maximum open-circuit gain, μ.

C 7.23 A BJT for which n is known to be 1.00, operating with base and collector joined at 10 mA, is found to have an incremental resistance of 2.9 Ω. When a 50 Ω resistor is included in the lead between base and collector, the incremental resistance is found to increase to 3.4 Ω. Estimate values for g_m, r_e, r_π, β and r_x. Briefly, what do you do if it is possible that n is not exactly 1.00?

7.24 For a transistor biased at I_C = 10 mA, the following parameters are measured: h_{ie} = 0.55 kΩ, h_{fe} = 200, h_{re} = 1×10^{-4}, $h_{oe} = 1.2 \times 10^{-4}$ A/V. Determine values for g_m, r_π, r_x, r_μ, r_o and V_A.

7.25 For a particular BJT transistor, for which $C_\mu = 0.5$ pF, operating at 2 mA, with $\beta = 200$ and $f_\beta = 12.7$ MHz, find the corresponding values of unity-gain frequency and C_π. If the bias current is increased to 10 mA, what values of f_β and C_π apply? For the usual situation described, for what range of currents is f_T maintained at the value estimated, as defined by the situation in which $C_\pi \geq C_\mu$?

L 7.26 Complete the table entries below for transistors (a) through (c) under the conditions indicated. Neglect r_x.

Transistor	I_E mA	r_e Ω	g_m mA/V	r_π kΩ	β_o	f_T GHz	C_μ pF	C_π pF	f_β MHz
a		25			200	1	0.5		
b			4	10			0.1		100
c	0.01				150			2	.2

SECTION 7.6: Frequency Response of the Common-Emitter Amplifier

7.27 For a particular BJT *CE* amplifier using the circuit of Fig. 7.26, $R_S = 10$ kΩ, $R_1 \| R_2 = 40$ kΩ, $R_E = 8.2$ kΩ, $R_C = 9.1$ kΩ, $R_L = 10$ kΩ, and $V_{CC} = 5$ V. Under these conditions, I_E is 0.15 mA, at which $\beta = 150$, $r_o = 500$ kΩ, $r_x = 50$ Ω, $f_T = 1$ GHz and $C_\mu = 0.3$ pF. Estimate the midband gain and the upper 3-dB frequency.

7.28 For the situation described in P7.27 above, coupling capacitors of value $C_{C1} = C_{C2} = 1$ µF and $C_E = 10$ µF are used. Calculate the three associated poles and zero, and estimate the lower 3-dB frequency.

D 7.29 For the situation described in P7.27 above, find suitable values for C_{C1}, C_{C2} and C_E so that the dominant low-frequency pole is at 20 Hz, another pole is at 2 Hz, and the third pole and zero coincide.

7.30 For the situation described in P7.27 above, with all coupling and bypass capacitors appropriately large, an additional resistor $R = 350$ Ω is included in series with C_E. What new values of midband gain and upper 3 dB frequency result?

7.31 For the situation described in P7.30 above, the coupling capacitors are $C_{C1} = C_{C2} = 1$ µF, and $C_E = 10$µF. Calculate the three associated poles and zero, and estimate the lower 3 dB frequency.

SECTION 7.7: The Common-Base, Common-Gate and Cascode Configurations

7.32 For a particular BJT Common Base amplifier using the topology of Fig. 7.28, $R_S = 100$ Ω, $R_1 \| R_2 = 40$ kΩ, $R_E = 8.2$ kΩ, $R_C = 9.1$ kΩ and $R_L = 10$ kΩ. For the power supply chosen, $I_E = 0.2$ mA, at which $\beta = 150$, $r_o = 400$ kΩ, $r_x = 50$ Ω, $f_T = 1$ GHz and $C_\mu = 0.3$ pF. Estimate the midband gain and upper 3 dB frequency.

7.33 For the situation described in P7.32 above, capacitors employed are $C_{C1} = 10\ \mu F$, $C_B = 1\ \mu F$ and $C_{C2} = 1\ \mu F$. Find the associated poles and a zero (which occurs when $\frac{1}{C_{BS}} = R_1 \| R_2$) and estimate the lower 3-dB frequency.

7.34 For a particular BJT cascode amplifier using the circuit of Fig. 7.30, $R_S = 10$ kΩ, $R_1 \| R_2 = 15$ kΩ, $R_2 \| R_3 = 15$kΩ, $R_E = 8.2$ kΩ, $R_C = 9.1$ kΩ and $R_L = 10$ kΩ. Under these conditons, I_E is 0.15 mA, at which $\beta = 150$, $r_o = 500$ kΩ, $r_x = 50\ \Omega$, $f_T = 1$ GHz and $C_\mu = 0.3$ pF. Estimate the midband gain and upper 3-dB frequency.

SECTION 7.8: Frequency Response of the Emitter and Source Followers

7.35 For a particular BJT emitter-follower circuit resembling that in Fig. 7.32, $R_S = 10$ kΩ, $R_E = 8.2$ kΩ and $R_L = 10$ kΩ is coupled through a capacitor $C_C = 1\ \mu F$. Under these conditions $I_E = 0.15$ mA, at which $\beta = 150$, $r_o = 500k\Omega$, $r_x = 50\ \Omega$, $f_T = 1$ GHz and $C_\mu = 0.3$ pF. Estimate the midband gain and the upper and lower 3 dB frequencies.

7.36 For a particular FET source-follower circuit operating at a 1 mA bias current, $R_S = 1$ MΩ, $R_L = 10$ kΩ, $g_m = 1$ mA/V, $C_{gs} = C_{gd} = 1$ pF. Estimate the midband gain and upper 3 dB frequency resulting.

SECTION 7.9: The Common-Collector Common-Emitter Cascade

CL 7.37

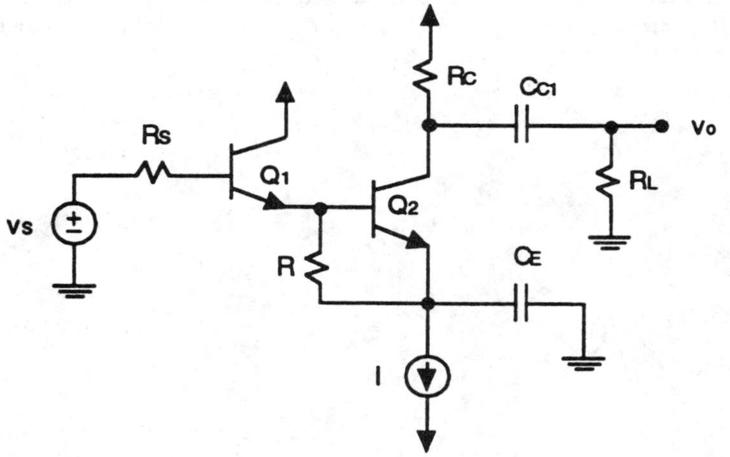

For the circuit shown, $R_S = 100$ kΩ, $R_L = 10$kΩ, $R_C = 9.1$ kΩ, $R_E = 10$ kΩ, $I = 160\ \mu A$, $R = 70$ kΩ, $C_{C1} = 1\ \mu F$, $C_E = 10\ \mu F$. Under these conditions $\beta = 150$, $V_A = 100$ V, $r_x = 50\ \Omega$, $f_T = 1$ GHz and $C_\mu = 0.3$ pF. Estimate the midband gain and the upper and lower 3-dB frequencies. Use $V_{BE} = 0.7$ V. Note that for low-current operation, f_T may reduce, since $C_{\pi\,min} \approx C_\mu$.

L* 7.38 Repeat P7.37 above for the situation in which a) $R = 14$ kΩ, b) ∞.

SECTION 7.10: Frequency Response of the Differential Amplifier

7.39 A BJT differential amplifier operates with a 300 μA emitter-current source and collector resistances of 4 kΩ. For differential input from a source having a total resistance of 10 kΩ, and output to a 10 kΩ load taken differentially, what are values of the gain $\frac{v_O}{v_S}$ and the upper 3-dB frequency? For the BJTs used, $\beta = 150$, $f_T = 1$ GHz and $C_\mu = 0.3$ pF.

D 7.40 The situation described in P7.39 above, is modified so that one BJT base is grounded, with the other connected to a 10 kΩ source. Both collector resistors remain, but the 10 kΩ load is capacitively-coupled to only one of them. There are two possible topologies. Find the midband gain and upper cutoff frequency for each.

7.41 The situation described in P7.40 is further modified so that the 10 kΩ load is taken from the collector of the grounded-base BJT, while the collector of the other BJT is connected directly to the supply. What midband gain and upper cutoff frequency result?

7.42 The situation described in P7.39 above is modified by the addition of resistors equal in value to r_e included in the emitter of each of the BJTs of the differential pair. What midband gain and cutoff frequency result?

C 7.43 For the situation described in P7.39 above, the emitter-current source uses a single transistor for which C_μ = 0.3 pF and V_A = 200 V. Wiring at the common-emitter connection accounts for an additional 0.5 pF. For a common-mode input signal of 5 V peak, what is the corresponding peak voltage on the ends of the load resistor at low frequencies? At what frequency does it reach 1V peak? At what frequency do the input transistors saturate for V_{CC} = 10V?

SECTION 7.11: The Differential Pair as a Wideband Amplifier: The Common-Collector Common-Base Configuration

D 7.44 A wideband amplifier using the topology shown in Fig. 7.42, employs a 300 µA emitter-current source and equivalent load resistance of 2.7 kΩ. Emitter resistances are used to double the usual input resistance presented to the 10 kΩ source. For the BJTs used, β = 150, f_T = 1 GHz and C_μ = 0.3 pF. What midband gain and cutoff frequency results?

Chapter 8

FEEDBACK

SECTION 8.1: The General Feedback Structure

8.1 A feedback amplifier having the structure shown in Fig. 8.1 of the Text, is found to have $x_o = 3.0$ V and $x_f = 0.99$ V when a signal of $x_s = 1.00$ V is applied. What value of x_i results? What must the values of A and β be? What is the open-loop gain? What is the amount of feedback? What is the closed-loop gain? If by accident the connection to the β network were removed, what value would the load voltage tend toward? Is that value likely to be measured? What would happen instead?

8.2 For the circuit in Fig. E8.1 on page 573 of the Text, with $A = 10^2$, find $\frac{R_2}{R_1}$ for a closed-loop voltage gain of 8. What is the corresponding value of β? What is the amount of feedback in decibels? For $V_s = 0.125$ V, find V_o, V_f and V_i. If A increases by 100%, what is the % change in A_f?

SECTION 8.2: Some Properties of Negative Feedback

8.3 An amplifier for which the design was done with $A = 10^3$ and $\beta = 10^{-2}$ is manufactured using an amplifier with half the intended gain. What is the desensitivity factor? What is the sensitivity of closed-loop to open-loop gain $\left(\frac{\partial A_f/A_f}{\partial A/A}\right)$ in dB? What closed-loop gain results?

8.4 An amplifier for which the loop gain is designed to be 89 is suspected to be deteriorating over an extended period of operation at high temperature. The closed-loop gain is found to be 98 rather than the value of 99 measured shortly after original installation. On the assumption that the change has occurred in the active components within the basic amplifier itself, what % deterioration would you expect to find in the open-loop gain if it were possible to measure it directly?

8.5 An amplifier whose open-loop response is characterized by a *dc* gain of 10^4 V/V and a 3 dB rolloff at 10^4 Hz, is connected in a feedback loop for which the overall low-frequency gain is 10^2 V/V. What is the 3 dB rolloff with feedback? What are the values of the Gain-Bandwidth product of the basic amplifier and of the feedback arrangement?

8.6 If in P8.5 above, a manufacturing error reduces the upper 3 dB cutoff to 2×10^3 Hz, what does the closed-loop upper cutoff become? Is this consistent with the sensitivity idea? To appreciate this situation better, find the new gain at 10^4 Hz using a relatively basic calculation. How does the desensitivity factor manifest itself?

8.7 Performance of a basic power amplifier having a signal-to-noise ratio of −3 dB (at the output) is to be improved, using a low-noise preamplifier and feedback, by 40 dB. What is the gain of the preamplifier required? What does the *S/N* ratio at the output become?

8.8 An amplifier exhibiting a non-linear transfer characteristic with a gain of $\geq 10^3$V/V for $v_O \leq 0.1$V, $\geq 10^2$V/V for $0.1\text{V} \leq v_O \leq 1$V, but which limits at $v_O = 1$ V, is connected in a feedback loop with $\beta = 0.01$

V/V. Characterize the transfer characteristic of the closed-loop circuit by finding values of closed-loop gain for the 3 regions of operation, and the values of *input* and *output* voltage which bound them.

SECTION 8.3: The Four Basic Feedback Topologies

8.9 Characterize each of the following amplifiers by feedback type. As well, for each, find β in terms of the labelled components. In all cases, assume that the op amp is ideal.

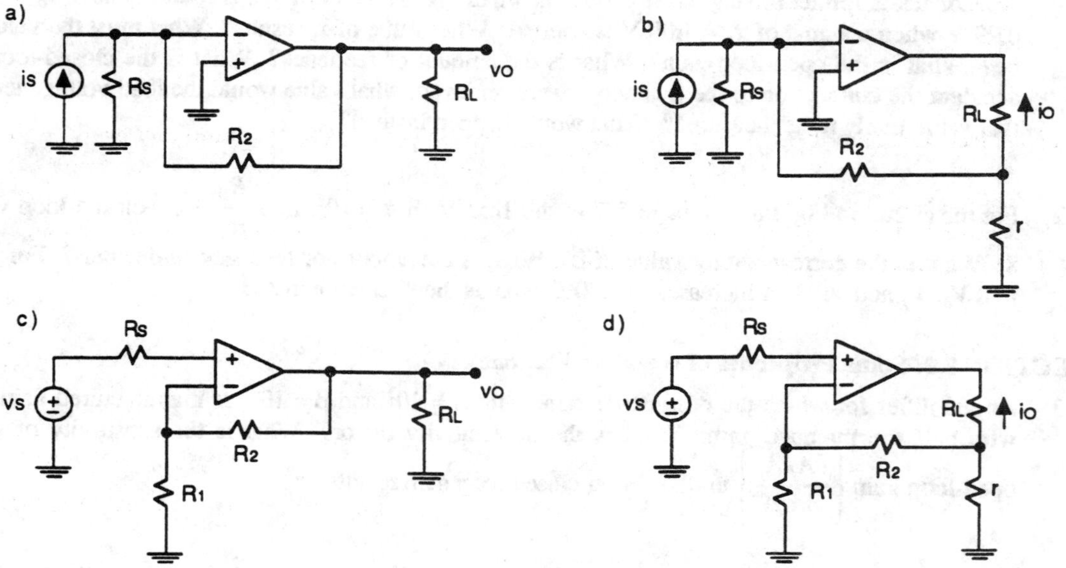

8.10 For each of the circuits below, identify the feedback type and find an expression for β. Assume for the present purposes that g_m for each FET is very high.

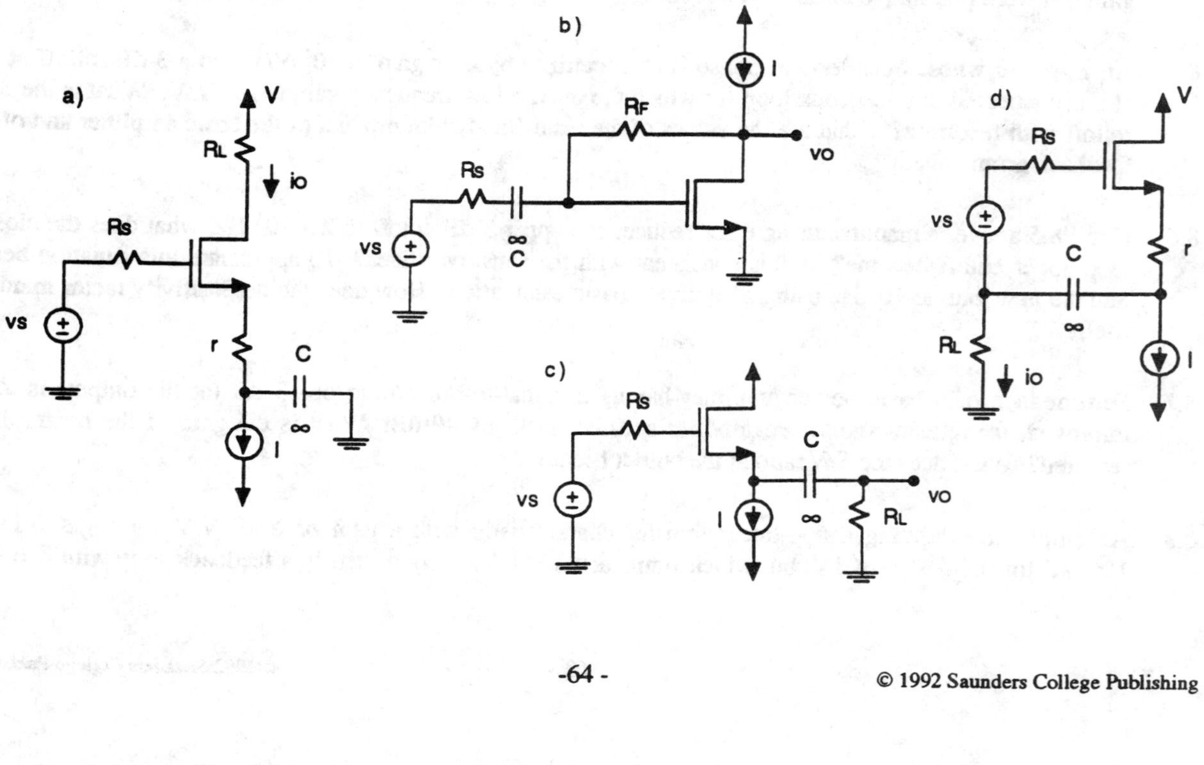

8.11 A series-series feedback circuit representable by Fig. 8.4c in the Text, and which uses an ideal transconductance power amplifier, operates with $V_s = 1$ V, $V_f = 0.1$ V and $I_o = 2$ A. What values of A and β correspond (with correct units noted)?

SECTION 8.4: The Series-Shunt Feedback Amplifier

8.12 A series-shunt feedback amplifier, has an A circuit for which $A = 100$ V/V, $R_i = 10$ kΩ and $R_o = 10$ Ω, and a β circuit with $\beta = 0.1$ V/V, $R_1 = 2$ kΩ and $R_2 = 18$ kΩ. When operating from a zero-impedance source and with no load, what is the overall gain and input and output resistances that result with feedback? If this feedback amplifier is connected between a 0.1 V rms source whose resistance is 10 kΩ and a load of 100 Ω, what does the output voltage become?

8.13 A feedback amplifier connected in the series-shunt topology employs an amplifier having a gain of 900 V/V, an input resistance of 20 kΩ, and an output resistance of 1 kΩ, with a feedback network employing two resistors of 10 kΩ and 190 kΩ at its output and input respectively. The amplifier operates between a 10 kΩ source and a 1 kΩ load. Find A, β, R_{11}, R_{22} and A_f as well as the overall gain and input and output resistances seen by the source and load respectively.

CL 8.14

For the circuit shown, the transistors have $\beta = h_{fe} = 120$. Find $\frac{v_o}{v_s}$, R_{in} and R_{out}. For what values of R_S and R_L (considered separately) does $\frac{v_o}{v_s}$ drop to ½ the value just found. (Hint: be careful!)

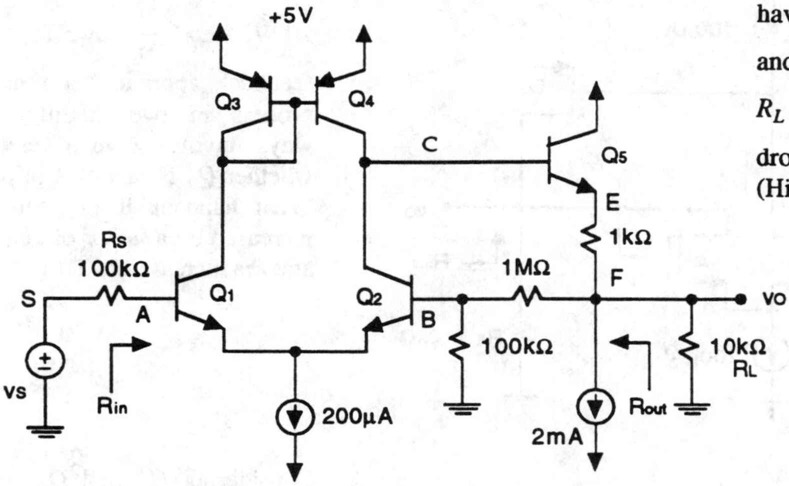

L 8.15

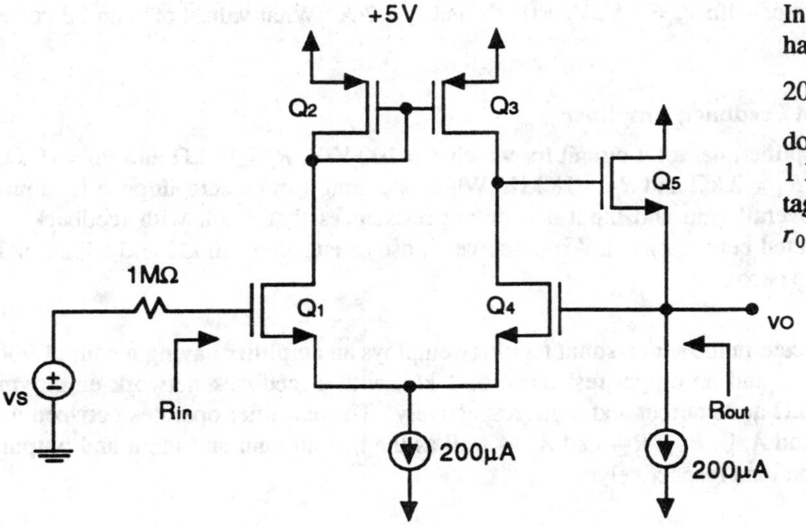

In the circuit shown, all transistors have $V_t = 1$ V, $K = 100\mu A/V^2$, $V_A = 20$ V. Find, $\frac{v_o}{v_s}$, R_{in}, R_{out}. What does the overall gain become with a 1 kΩ load? Estimate the offset voltage. (Hint: consider the effects of r_{01}, r_{02}, r_{03} and r_{04})

L* 8.16

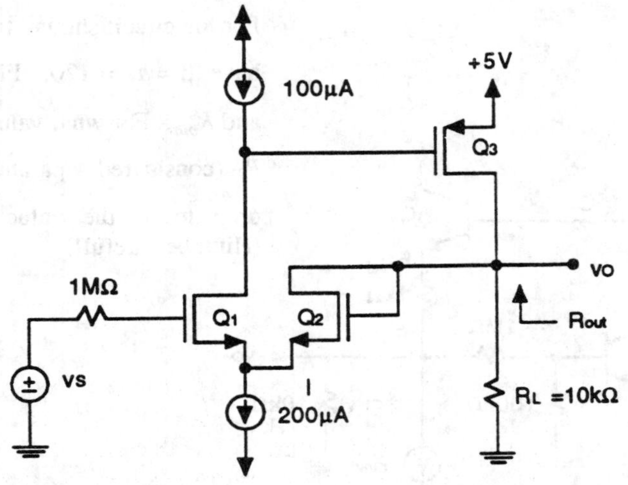

In the circuit shown, all transistors have $V_t = 1$ V, $K = 0.1mA/V^2$, $V_A = 20$ V. Find $\frac{v_o}{v_s}$ and R_{out} using a feedback approach. Formulate the process in two slightly different ways involving interpretations of whether Q_2 is part of A or part of β. What happens if Q_2 and Q_3 are increased by a factor of 10 in width, and I is increased to 1.1 mA?

8.17

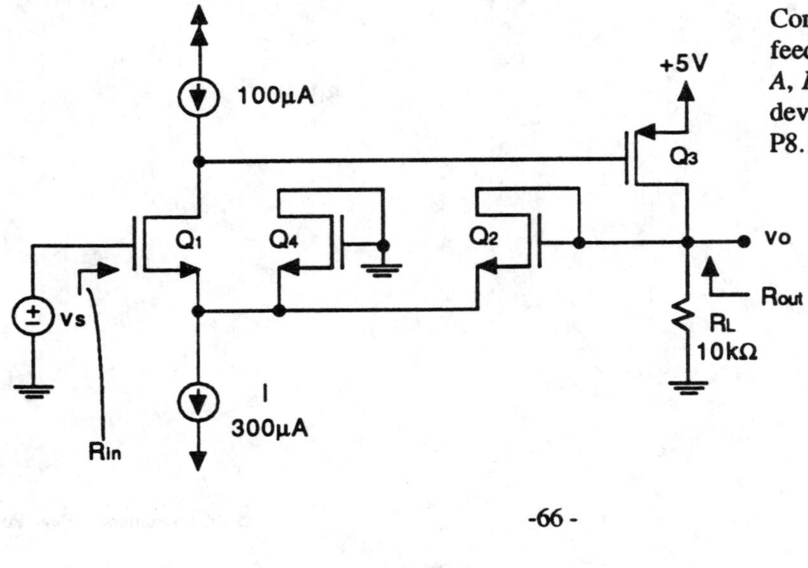

Considering Q_2 and Q_4 to be the feedback network, find A, R_{11}, R_{22}, A_f and R_{out}. Use the device specifications provided in P8.16 above.

SECTION 8.5: The Series-Series Feedback Amplifier

8.18 A feedback amplifier connected in the series-series topology uses a basic amplifier having a gain of 900 V/V, an input resistance of 20 kΩ and an output resistance of 1 kΩ, with a feedback network for which β = 50 V/A, R_{11} = 10 kΩ and R_{22} = 200Ω. The amplifier operates between a 10 kΩ source and a 1 kΩ load. Find A, A_f as well as the resistance R_{in} and R_{out} seen by the source and the load. (Hint: note the specification of A as a voltage amplifier: you must transform it suitably).

8.19 Reconsider the situation described in Example 8.2 and Fig. 8.17 on pages 594 through 597 of the Text, modified to place $R_L = 600\Omega$ between the emitter of Q_3 and the connection to R_{E2} and R_F. The output current is that measured in R_L. Find A, β, A_f and the input and output resistances seen by the source and load respectively.

8.20

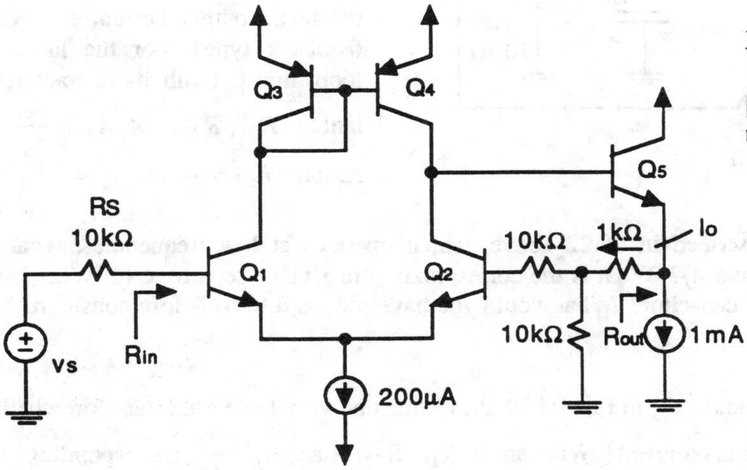

For the circuit shown, $h_{fe} = h_{FE}$ = 100, V_A = 200V. Find β, R_{11}, R_{22}, A, A_f, and the resistances R_{in} and R_{out}.

SECTION 8.6: The Shunt-Shunt and Shunt-Series Feedback Amplifiers

CDL 8.21 A shunt-shunt feedback circuit uses a basic amplifier whose voltage gain is 900 V/V, input resistance is 20 kΩ, and output resistance is 1 kΩ, with a feedback network consisting solely of a 100 kΩ resistor. The amplifier operates between a 10 kΩ source and 1 kΩ load. What is the transresistance of the basic amplifier? What are R_{11}, R_{22}, A, A_f and the input and output resistances presented to the source and load? What does the gain become if the load resistance is halved? What change in R_S (nominally 10 kΩ) is needed to compensate?

L* 8.22

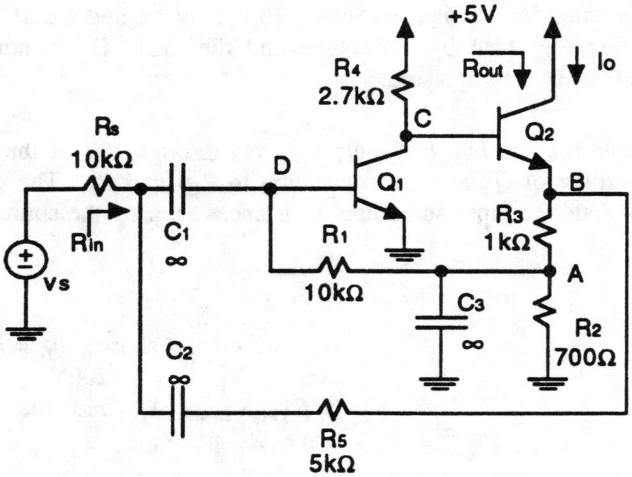

The circuit shown combines two feedback loops: One, involving R_1, R_2 and R_3 operates at *dc*, but because of C_3, not at high frequencies. The other, involving R_3 and R_5 operates at high frequencies, but because of C_2, not at low frequencies. At low frequencies, the feedback is intended to establish the *dc* voltage at *A* at a value which is nearly the voltage V_{BE1}, and thereby the entire bias-current situation at nodes *B* and *C*. What is the feedback type? At high frequencies the object is to create an output current i_o. What is the feedback type? For the latter feedback loop, find β (with its characteristic resistances R_{11}, R_{22}), *A*, $A_f' = \frac{i_o}{v_s}$, R_{in}, R_{out}. Assume $h_{FE} = \infty$, and $h_{fe} = 100$.

8.23 For the shunt-shunt loop described in P8.22 above, which operates at low frequencies, what is β, the corresponding R_{11} and R_{22}, and A_f? What is the corresponding resistance seen by C_3? What is the upper cutoff frequency for a 100 μF capacitor? What would you have judged it to be before considering the feedback situation?

C 8.24 As noted in P8.23 above, capacitor C_3 in Fig. P8.22 above must be very large (and therefore costly). What occurs at high frequencies if it is removed? What are β, R_{11}, R_{22}, *A* and $A_f = \frac{i_o}{v_s}$ corresponding?

D 8.25 Using the results of P8.23, and P8.24 above with C_3 removed, find C_1 and C_2 for poles at 1 Hz and 10Hz respectively.

CDL 8.26 For the circuit of P8.10d above, with *I* such that $g_m = 2$ mA/V, $r_o = 10$ kΩ, $R_S = 100$ kΩ, and $r = 1$ kΩ, find $\frac{i_o}{v_s}$ for a reasonable range of choices of R_L. What is the source resistance associated with i_o? What constrains the values of R_L that can be used?

CL 8.27 Reconsider the situation in P8.23 above with respect to C_3. Consider the effect of moving C_3 to some tap on R_a: That is, split R_1 into two parts R_{1a}, R_{1b} where $R_1 = R_{1a} + R_{1b}$. For what ratio of $\frac{R_{1a}}{R_1}$ is the resistance seen at the tap, a maximum? What is the input resistance seen at the tap at this setting when the low-frequency feedback loop is closed? By what factor can the capacitor C_3 be reduced to maintain the same pole frequency as found in P8.23?

SECTION 8.7: Determining the Loop Gain

C* 8.28 A particular feedback loop intended for relatively high-frequency operation, when opened and terminated, returns a signal of 1.27 V rms when a signal of 20 mV rms at 1 kHz is injected, and 3.1 V rms when a 2 mV rms signal at 10 Hz is injected. What are the loop gains at the two frequencies? Assuming a single capacitor is associated with low-frequency bias stabilization of a direct-coupled amplifier, what do you imagine to be happening? Estimate the lower 3 dB point associated with the loop-gain response. What would you estimate the lower 3 dB response of the closed loop to be? If a third measurement indicates the loop gain to be only about twice as great at 1 Hz as at 10 Hz, estimate a lower bound for the open-loop gain of the basic amplifier. What might the closed-loop gain at 1 kHz be? If the capacitor which stabilizes the bias loop has been identified to be 1 μF, what is the equivalent (open-loop) resistance at the node to which it is attached? (Hint: see P8.29 following).

DL 8.29 A useful circuit, which is also a possible model of the situation alluded to in P8.28 above, is shown here. For an ideal amplifier with $A = 1550$ and $R_1 = 2$ kΩ, $R_2 = 47$ kΩ, find the loop gain at high frequencies and at very low frequencies. For what value of C does the loop gain have a zero at 2.45 Hz? What is the associated pole of the loop gain? What is the corresponding 3 dB frequency of the closed-loop gain? If the capacitor C is made to be 10 μF, what values of R_1 and R_2 must be used to keep the same frequency response?

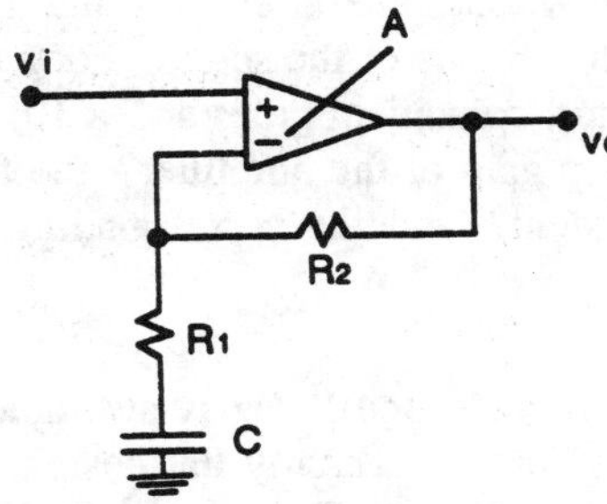

8.30 A non-inverting op-amp circuit for which the two resistors in the β network are 100 Ω and 10 kΩ is measured for loop gain by disconnecting the larger resistor from the output, injecting a 10 mV signal and measuring the returned signal to be 1.2 V. What is the loop gain found? What is the basic op-amp gain?

C 8.31 For a particular situation involving a non-inverting series-shunt feedback amplifier in which a complex unlabelled feedback network is used, measurements at a particular frequency show both the loop gain and the closed-loop gain to be 10. Estimate the (gain) of the basic amplifier and β at this frequency.

8.32 For the circuit shown in P8.15 above, evaluate the loop gain for the conditions stated with a) no load b) a load of 1 kΩ. Using the fact that $\beta = 1$, what closed-loop gain results in each case? Check your results with those obtained in P8.15 using the direct method.

SECTION 8.8: The Stability Problem

8.33 An amplifier with a midband gain of 10^3 and dominant pole at 10^3 rad/s has two other poles coincident at 10^5 rad/s. At what frequency does its total phase shift become 180°? At that frequency, what is its gain magnitude? When incorporated in a feedback loop for which the feedback factor is independent of frequency, for what range of β is the resulting amplifier stable?

8.34 For the situation described in P8.33 above, sketch Nyquist plots for the loop gain $|A\beta|$, for β equal to the critical value, and 20 dB more and less than it.

C 8.35 Through a manufacturing error, an op amp whose *dc* gain is 10^6 V/V and dominant pole is at 10^3 rad/s, acquires a sond non-dominant pole at 10^8 rad/s. When used with a frequency-independent feedback network, what total phase shift can result when $|A\beta| = 1$ for $\beta \le 1$. Can oscillation occur? Unfortunately, in a particular application, a stray capacitance of up to 5 pF to ground is associated with the output of the feedback network. For nominal $\beta = 0.5$, what is the maximum tolerable equivalent output resistance of the β network for which oscillation will **not** begin?

SECTION 8.9: Effect of Feedback on Amplifier Poles

8.36 A *dc* amplifier having a single-pole response with pole frequency at 5×10^3 Hz and unity-gain frequency of 20 MHz is operated in a loop whose feedback factor is 0.125, independent of frequency. Find the low-frequency gain, the 3 dB frequency and the unity-gain frequency of the closed-loop amplifier. By what factor did the pole shift?

D 8.37 Using the amplifier described in P8.36, above, a design is required of an amplifier whose 3 dB frequency is at least 1 MHz. What is the corresponding amount of feedback required? The loop gain? The feedback factor? What low-frequency gain results?

8.38 A two-stage *dc* amplifier having a low-frequency gain of 10^4 K, one pole at 10^5 Hz and a second pole at $10^6/K$, where K is a factor depending on the choice of a resistor in one of the stages. For a particular feedback application using a frequency-independent feedback factor, coincident poles at 5×10^5 Hz are acceptable. What value of K can be used? What is the *dc* open loop gain of the amplifier? the frequency of its second pole? the value of β for which the poles are coincident? and the corresponding low-frequency closed-loop gain?

DL 8.39 A two-pole amplifier with *dc* gain of 10^3 and poles at 10^6 Hz and 2×10^7 Hz is available to a designer interested in exploring bandwidth extension using feedback. For a maximally flat design, what pole frequencies result? What is the corresponding value of Q? of w_o? of the 3 dB frequency? What values of β and closed loop gain are achieved with this design?

CL 8.40 Consider the amplifier of P8.33 above in a feedback loop with frequency-independant β. Find the closed-loop poles as a function of β. Sketch a root-locus diagram. At what frequency and for what β is the complex-pole-pair $Q = 0.707$? What is the corresponding phase margin?

SECTION 8.10: Stability Using Bode Plots

8.41 Use equation 8.48 on page 626 of the Text to explore response peaking as a function of phase margin: For what phase margin is there no peak? For what margin is the peaking factor equal to 2? to 10?

8.42 A particular amplifier with a *dc* gain of 10^4 has one pole at 10^6 Hz and two coincident poles at 10^8 Hz. Provide a sketch of the Bode magnitude and phase plots. Use them to estimate the following: What is the value of frequency-independent β for which the margins are zero? For what value of β is the phase margin 78°? 45°? What are the corresponding closed loop gains? What is the phase margin for $\beta = 3 \times 10^{-2}$?

L 8.43 For the situation described in P8.42 above, express the amplifier transfer function in a form corresponding to Eq. 8.50 on page 627 of the Text. Then use the approach exemplified in Eq. 8.51 to find more precise values for the results requested in P8.42 above.

DL* 8.44 For the parameterized amplifier design described in P8.38 above, and frequency-independent feedback, use the rate-of-closure rule to design an amplifier with 20 dB of feedback at low frequencies and the greatest available bandwidth. What is the available bandwidth? For a closed-loop gain of 10, what is the available bandwidth? For each design, what are the corresponding values of K, *dc* gain, and the second-pole frequency?

SECTION 8.11: Frequency Compensation

D 8.45 For the amplifier described in P8.42 above, consideration is being given to various frequency-compensation ideas. In the case of an *added* dominant pole, what must be its location for a closed-loop gain of 10? of 1? What are the corresponding cutoff frequencies?

8.46 Continue the compensation evaluation begun in P8.45 above, by considering the possibility of lowering the existing dominant pole. To what frequency must it be lowered for a closed-loop gain of 10? of 1? What cutoff frequencies correspond? (Hint: Concerning the double pole, find the frequency at which each pole contributes 22.5° as the frequency to use in the design to give a net margin of 45° or so.)

DL** 8.47 For a particular amplifier in which one pole is at 10^5 Hz and two at 10^7 Hz, one approach to compensation involves lowering the dominant pole to 10^4 Hz. It is realized that two of the poles are controlled by a particular amplifier, for which $C_1 = C_2 = C$, $R_1 = 100\,R_2 = 1$ MΩ and $g_m R_2 = 100$, and across which a Miller capacitor can be installed. What is the required value of C_f? What is the overall effect on the other poles? What is the resulting cutoff frequency? In view of the pole shift, by how much can the dominant pole be raised to maintain the same margins? What new value of C_f is needed? What poles result? What is the resulting cutoff? For comparison, what was the original frequency at which the phase margin was 45°?

Chapter 9

OUTPUT STAGES AND POWER AMPLIFIERS

SECTION 9.1: Classification of Output Stages

9.1 A particular amplifier for which the output-stage bias current is 50 mA is intended to produce single-sine-wave output signals of 1, 10, 100 V rms across a load R_L. In each case, classify the mode of operation that prevails for $R_L = 1$ kΩ, and 0.25 kΩ. If a manufacturing error causes the bias current to reduce to zero, yet the output appears nearly normal for large signals, what mode of operation is apparently possible?

SECTION 9.2: Class-A Output Stage

9.2 The emitter-follower shown in Fig. 9.2 of the Text operates from ± 3 V supplies with R = 1.5 kΩ, using three identical transistors. For $V_{CE\,sat}$ = 0.3 V, what is the largest undistorted sine wave with zero average that can be produced across a 1 kΩ load? a 10 kΩ load? For what range of load resistances is the output symmetrically clipped? For a circuit modification involving connecting a second device in parallel with Q_2, what change occurs?

9.3

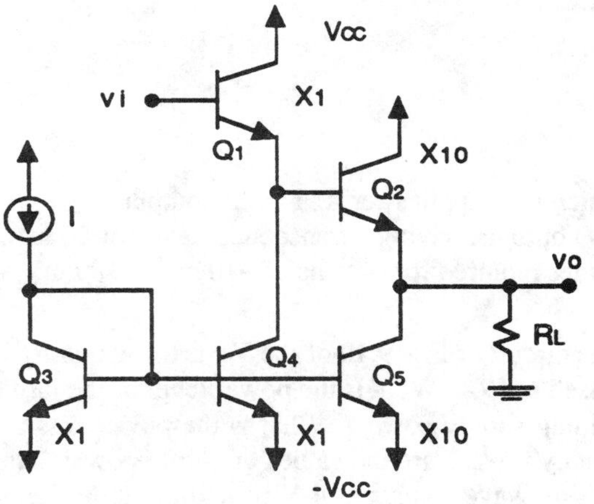

The follower design shown is intended to provide a relatively high input resistance. For V_{CC} = 5 V, V_{EB} = 0.7 V, $V_{CE\,sat}$ = 0.2 V, what is the largest possible unclipped zero-average sine-wave output? For a minimum load resistance of 100 Ω, what is the minimum value of I which maintains $I_{E2} \geq I_{E1}$ for the largest possible undistorted output? Note the transistor sizing indicated by the notation × n.

9.4 The class-A follower shown in Fig. 9.2 of the Text operates from ±9 V supplies with I = 10 mA using identical transistors. Ignoring the power loss in R and Q_3, find the load power, the supply power and the conversion efficiency for:

a) The largest-possible sine-wave output and the smallest-possible load resistance (assuming $V_{CE\,sat}$ = 0.3 V).

b) A sine wave of half the amplitude in a) across a load which is half the resistance of that in a)

c) Repeat a) including the loss in Q_3 and R (note that it is connected to ground!)

d) Repeat b) including the loss in Q_3 and R.

9.5

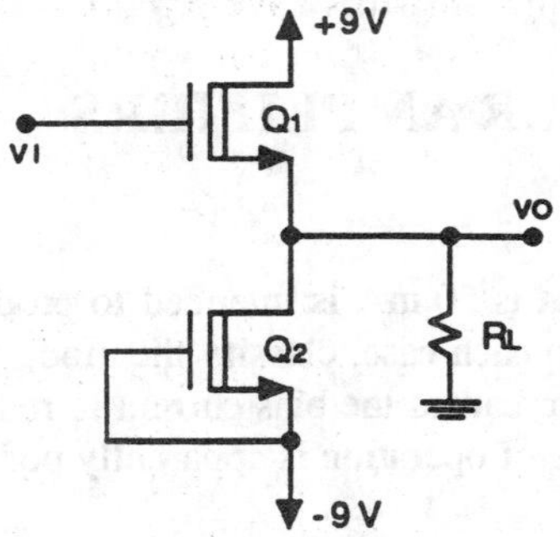

For the *FET* follower shown, V_t = –2 V, I_{DSS} = 10 mA. For υ_O = 0 V, what value of υ_I is measured? For R_L = 1 kΩ, what are the most positive and most negative outputs for which both transistors remain in saturation? What are the corresponding inputs? What are the corresponding values of load power, supply power and conversion efficiency? What is the largest possible relatively undistorted output sine wave, and the corresponding conversion efficiency?

SECTION 9.3: Class-B Output Stage

C 9.6 For the circuit shown, sketch the transfer characteristics for $R_L = \infty$ and R_L = 10 kΩ. For both devices, $|V_t|$ = 1 V and K = 1 mA/V^2. What is the amplitude of the maximum possible output sine wave for which Q_1, Q_2 remain in saturation? What are values for the corresponding input voltage equivalent gain, supply power, load power, and conversion efficiency?

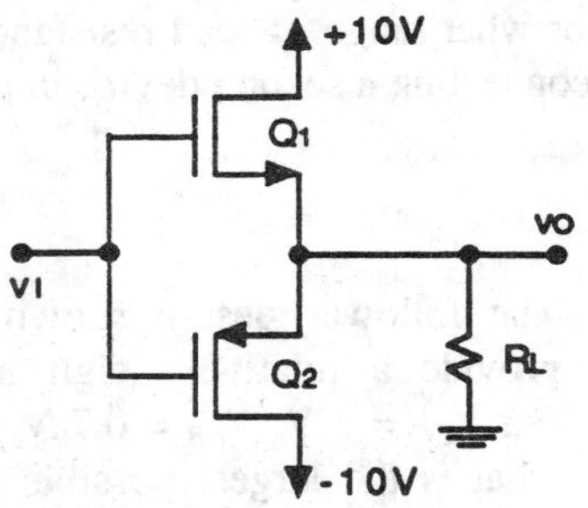

9.7 Consider the circuit of Fig. 9.9 of the Text, in which the supplies for A_o and the output stage are ±3 V, and A_o is a CMOS amplifier with rail-to-rail (i.e. ±3 V) outputs, having a transconductance of 10 mA/V. For R_L = 100Ω, $|V_{BE}|$ = 0.7 V, β = 50, find the input voltages required for outputs of +10 mV, +100 mV and 1 V.

9.8 Consider a particular automotive application of the circuit of Fig. 9.10 of the Text, for which R_L = 16 Ω and $2V_{CC}$ = 12 V nominally, and the base input is biassed at V_{CC}. What is the power level of the largest possible undistorted output signal? What is the corresponding supply power? What is the power dissipation in the two transistors? What is the corresponding efficiency? What are the values of output power, supply power, total device dissipation, and efficiency for output sine-wave signals of 4-V peak amplitudes. For the signal level maintained at the maximum undistorted level found above, but with the supply raised to 14.5 V, what do the total device dissipation and efficiency become?

SECTION 9.4: Class-AB Output Stage

9.9 A BJT class-AB output stage using the structure shown in Fig. 9.11 of the Text employs transistors for which V_{BE} = 0.690 V at 10 mA and n = 1. For a small-signal output resistance at light loads of ≤5 Ω, what quiescent current is needed? What value of V_{BB} should be used? For a 50 Ω load and peak output voltage of 5 V, what is the peak input voltage required? What large-signal gain results? For R_L = 50 Ω, what is the corresponding small-signal gain?

D 9.10 Design the quiescent current of a class-AB MOS output stage so that the incremental voltage gain for υ_I near zero volts, is in excess of 0.99 V/V, for loads larger than 100 Ω. The MOS devices have $|V_t|$ = 1 V and K = 200 mA/V^2. What value of V_{BB} is required?

SECTION 9.5: Biassing the Class AB Circuit

D 9.11 For the situation described in P9.9 above, in which the output transistors have $\beta \geq 30$, design a 2-junction biassing scheme, such as that shown in Fig. 9.14 of the Text, with devices having one-fourth the junction area of those in the output. What is the maximum output current available into a short circuit? For a 50 Ω load, what is the value of the peak output signal for which the current in the bias junctions reduces to 1/10 of the no load value?

DL 9.12 For the situation described in P9.9 above, in which the output transistors have $\beta \geq 30$, design a V_{BE} multiplier, such as that shown in Fig. 9.15 of the Text, with a device for which the junction area and β are the same as those of the output transistors. For a 1-volt positive output across a 50 Ω load, arrange that the bias network current reduces to no less than 20% of its normal value, and that the current in the biassing transistor is no less than half of that. What is the incremental gain for your design with the peak positive output described, and $\beta = 30$?

9.13 For the V_{BE} multiplier embodied in Fig. 9.15 of the Text and operating at total current I, find an expression for the incremental resistance between its terminals in terms of I, R_1, V_{BE} and β, for a multiplication factor $k = 1 + R_2/R_1$, $k \geq 1$. For $k = 2$, $I = 1$ mA, and $\beta \geq 50$, what is the available value of incremental resistance when $R_1 = r_\pi$?

SECTION 9.6: Power BJTs

9.14 The base-emitter junction voltage of a heat-sink-mounted power BJT, measured at a particular test current immediatedly following the application of power, is found to be 630 mV at what is assumed to be $T_J = 30°$. Subsequently, it is found to display a junction voltage of 500 mV when operated for some time at ten times that current and a larger supply voltage at a power level of 45V. Estimate the new junction temperature. If the ambient temperature remains at 30°C, what is the thermal resistance of this device? To maintain operation with a junction temperature of 180°, at the same collector voltage, what collector current would be needed? What voltage drop across the base-to-emitter junction would you expect under these conditions? Use a junction *TC* of –2 mV/°C.

D 9.15 A power transistor, for which the maximum safe junction temperature is believed to be 150°, has a thermal resistance junction-to-case of ≤1.1°C/W. What is the maximum power it can dissipate for a case temperature ≤55°C? For half that power level, to what temperature can the case be allowed to rise? For an ambient temperature of 30°C, what is the thermal resistance of the heat sink needed in each of these situations? For a modular heat sink design for which the rating is 3°C/W for each cm of length, how long a heat sink is needed in each case? If all the thermal resistance measurements cited can be in error by as much as 20%, to what length must the large heat sink be increased to guarantee safe operation?

9.16 A BJT for which $T_{J\,max} = 150°$, and $\Theta_{JC} = 2°C/W$, operates with an electrically-isolating bond to a heat sink for which the resistances are $\Theta_{CS} = 0.5°C/W$, and $\Theta_{SA} = 1°C/W$ for each cm of heat-sink length, respectively. For a 10 cm heat sink what is the maximum power this device can dissipate for ambient temperatures less than 40°C? What is the effect of doubling the heat-sink length? What do you conclude?

9.17 A power BJT operating at a high current density at $I_E = 5$ A is found to have a base current of 0.2 A and a base-input resistance of 0.72 Ω. Estimate a value for the base spreading resistance. Estimate a value for the base-input resistance at $I_E = 3$ A.

SECTION 9.7: Variations on the Class-AB Configuration

D 9.18 A version of the circuit in Fig. 9.24 of the Text, uses *equal* current sources, I, in place of R_1 and R_2. What must their value be for the following situation: the maximum load current is 100 mA, $\beta_{npn} \geq 100$, $\beta_{pnp} \geq 80$; the minimum current in Q_1 and Q_2 must be no less than the maximum base currents of Q_3 or Q_4, nor smaller than 1.5 mA. For the situation in which Q_3 and Q_4 have junction areas 5 times those of Q_1 and Q_2 and conduct 50 mA at 0.7 V, (with $n = 1$ for all), find $R_3 = R_4$ so that the quiescent output current is equal to I calculated above. For what value of R_L is the small-signal gain ≥0.90 for output voltages a) near zero volts b) near +10 volts, c) near −10 volts.

D 9.19 For the output stage using compound devices as shown in Fig. 9.27 of the Text, the output npns are 100 mA devices while the other transistors are 1mA devices. For the npns, $\beta \geq 100$, and for the pnp, $\beta \geq 20$. Find the voltage needed across the V_{BE} multiplier for a standing current, I_Q, of 10 mA in the output, when a) the circuit is as shown, b) the base-emitter junctions of Q_2 and Q_4 are shunted to provide 1 mA currents in each of Q_1 and Q_3. For each situation described, with calculated voltages applied between the bases of Q_1 and Q_3, estimate the total effect of all transistors having a β which is ten times their minimum specified value.

D 9.20 For the circuit of Fig. 9.28 of the Text, in which transistors conduct 1 mA at 0.7 V, with $n = 1$ and $\beta \geq 100$, and $I_{bias} = 1$ mA, find $R_{E1} = R_{E2}$ so that the outgoing short-circuit current is limited to 25 mA. What is the current available without Q_5?

9.21

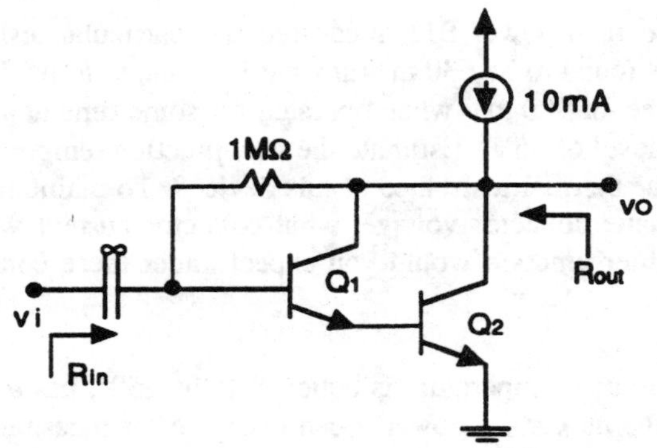

For the circuit shown, with β ranging from 50 to 150, $V_A = 100$ V, and $r_\mu = 10\,\beta r_o$, find the limits of the ranges of values of $g_{m\,eq}$, $\frac{v_o}{v_i}$, R_{in} and R_{out}.

D 9.22

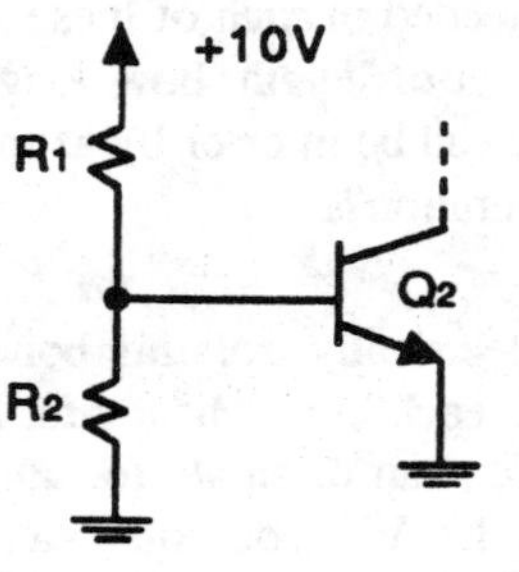

Consider the circuit shown as a simple candidate for use in a thermal-shut-down mechanism. The transistor has V_{BE} of 0.7 V at 100 μA, $\beta = 100$ and a TC of −2 mV/°C at 25°C. Design the circuit so that at 125°C the current in R_1 and Q_2 are each 100 μA. What is the current in Q_2 at 25°C? For what supply voltage (nominally 10 V) would the current at 25°C be double the value found? For operation at 100°C, at what value of supply voltage does the current in Q_2 become 50 μA?

SECTION 9.8: IC Power Amplifiers

D 9.23 Consider the circuit of Fig. 9.30 in the Text. For operation with a supply voltage of 25 V, it is desired to reduce the input bias current to 0.5 μA while maintaining nearly the original gain. Find new values for all the important resistors that must be changed. As a result of your change, what happens to the steady current in Q_{12} and Q_9? Assume $\beta_{npn} = 100$, $\beta_{pnp} = 20$, $|V_{BE}| = 0.7$V.

9.24 For the circuit shown in Fig. 9.30 in the Text, operating from a 27 V supply with all transistors (but Q_7 and Q_9) of the same junction size, and V_A = 100 V, and with β_{npn} = 100 and β_{pnp} = 20, estimate the gain A (as shown in Fig. 9.31) with no external load, and the corresponding 3 dB rolloff. Note that the current level in Q_9 is about 10 times that in Q_{11}.

D 9.25 For the circuit of Fig. 9.33 of the Text, find the value of R_3 for which a load current of 50 mA is shared equally by Q_3 and Q_5. By what factor does the current in Q_3 increase for a total output current of 1A? Q_5 and Q_6 both have an emitter-base voltage of 1.0 V at 1A, while Q_3 and Q_4 are 10 mA units, and Q_1, Q_2 are 1mA units. All show a 0.1 V/decade junction-voltage variation. Use β = 30 for Q_5, Q_6 and 100 for the other devices. What values of R_5 and R_6 ensure a quiescent current of 2mA in Q_3 and Q_4?

D 9.26 Design the bridge amplifier shown in Fig. 9.34of the Text, to provide the largest possible output sine wave from an input sine of 0.1 V peak. The input resistance should be 10 kΩ or more, with the largest available resistor limited to 10 MΩ. The op amps saturate at (at most) 2 V from the supply rails. Use ±12 V supplies.

D 9.27 Modify the connections shown in Fig. 9.34 to create a bridge amplifier with infinite input resistance. Choose values to provide a 20-V peak output from a 1-volt peak input signal.

SECTION 9.9: MOS Power Transistors

9.28 A power MOSFET for which $\mu_n C_{ox} = 30 \mu A/V^2$ has $W = 10^5 \mu m$, $L = 5 \mu m$. For electrons in silicon, $U_{sat} = 5\times10^4$ m/s and $\mu_n = 5\times10^{-2} m^2/Vs$. For V_t = 2 V, find the value of v_{GS} for which velocity saturation begins. What is the corresponding current? What are the currents at twice and half this value? Find g_m at all three values of current.

9.29 For the class-AB amplifier shown in Fig. 9.38 of the Text, operating in the non-saturated-velocity region, K = 200 mA/V^2, $|V_t|$ = 2 V, $|V_{BE}|$ = 0.7V, β is high, and the quiescent currents $I_P = I_N = I_2 = I_3$ = 5 mA. Find the required voltage V_{14} between the bases of Q_1 and Q_4. What is the value of resistor R? If the TC of V_t (at low currents) is –3 mV/°C and of V_{BE} is –2 mV/°C, what fraction of the voltage V_{14} must appear across Q_6? Estimate the voltage gain of the resulting stage for a load of 100 Ω for a) outputs around 0 volts b) around +20 V.

D 9.30

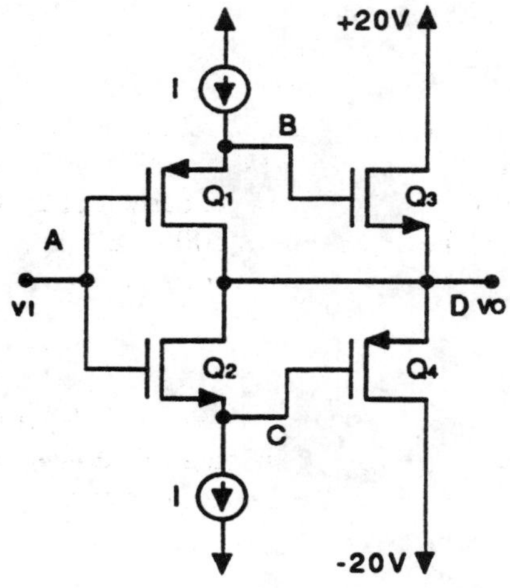

In the circuit shown $K_3 = K_4 = 100\ K_1 = 100 K_2$ = 100 mA/V^2, and $|V_t|$ = 1 volt for all devices. Find I for a quiescent current in Q_3 and Q_4 of 10 mA. What is the gain of the amplifier with a 100 Ω load, for an output voltage of a) 0 V b) +10 V. Find voltages at A, B, C, D for each case. Comment on the mode of operation of each transistor in each case.

Chapter 10

ANALOG INTEGRATED CIRCUITS

SECTION 10.1: The 741 Op-Amp Circuit

D 10.1 In the 741 op-amp circuit of Fig. 10.1 of the Text, transistors Q_{11} and Q_{12} with R_5 establish I_{REF} for the rest of the circuit. For ±15 V supplies and $|V_{BE}| = 0.7$, what value of I_{REF} results? What is I_{REF} for ±5 V supplies? To what value must R_5 be changed to restore I_{REF} to its high-supply value? Specify it to 2 significant digits.

D 10.2 For the situation described in P10.1 above, consider the operation of the Widlar current source involving Q_{10} and Q_{11} with R_4. For both devices having $I_S = 10^{-14}$ A and n=1, calculate the Widlar output current for ±15 V and ±5 V supplies. Choose a new value for R_4 which reestablishes the ±15 V current level in Q_{10} for the ±5 V situation.

CD* 10.3 A designer, attempting to reduce the number of high-value resistors in the 741 design shown in Fig. 10.1, notes the possibility that R_9 can be replaced by another transistor Q_{25} and a resistor R_{12}. Suggest two possible connections and corresponding resistor values to establish a suitable current. Which is best? Why?

C 10.4 Reflect upon the ability of the extended input stage of the 741 Op Amp of Fig. 10.1 (consisting of Q_1, Q_2, Q_3 and Q_4, with Q_5, Q_6, Q_8, Q_9, Q_{10}) to withstand the application of voltages outside the normal operating range. For this purpose, approximate the breakdown voltages as follows: 7 V between the npn base and emitter, and 50 V for both pnp junctions and the npn base-to-collector junction. Specifically, consider a) In+ and In- connected in all possible ways to the ±15 V rails, b) In+ and In- having voltages somewhat outside the supply range. What are the limitations you see?

SECTION 10.2: DC Analysis of the 741

10.5 For the Widlar current source in Fig. 10.2, find the effect on Equation 10.1 of making the base-emitter junction of Q_{10} k times that of Q_{11}. For I_{REF} = 730 μA, R_4 = 5000Ω, k = 0.5, what does I_{C10} become? For what value of R_4 does it become 19 μA again?

10.6 For I_{C10} = 19 μA, and the inputs to the bases of Q_1 and Q_2 both at zero, what voltage on the bases of Q_3 and Q_4 results?

10.7 For the mirror circuit consisting of Q_5, Q_6, Q_7, and assuming all devices conduct approximately the same current with $R_1 = R_2$ = 1kΩ and R_3 = 50kΩ, find the ratio $\frac{I_{C6}}{I_{C3}}$ in terms of β, assumed equal for all devices. What is its value for β = 200?

10.8 For the mirror consisting of Q_5, Q_6, Q_7, find the ratios $\frac{I_{C6}}{I_{C5}}$ for R_1 and R_2 shorted separately. Assume β is high, and operation is at $I_{C5} + I_{C6}$ = 19μA.

10.9 For the extreme situations, described in P10.8 above, what are the corresponding input-offset voltages (between the bases of Q_1, Q_2) which result? Note in general that offsets originating anywhere in the input stage can be compensated by varying R_1 and R_2.

D 10.10 Consider the second stage modified slightly by connecting the lower end of R_9 to the emitter of Q_{17}. What should its value be to maintain the current in Q_{16} at its present value? What must (the new) R_9 be to reduce its current to 4 times that in the base of Q_{17}, for which $\beta = 200$ is assumed?

D 10.11 For the output-stage-biasing scheme shown, what value of R_{10} would be needed to increase the quiescent current in Q_{14}, Q_{20} by 50%? First assume $R_6 = R_7 = 0$. Then consider $R_6 = R_7 = 27\Omega$.

SECTION 10.3: Small-Signal Analysis of the 741 Input Stage

10.12 For a particular application, it is desired to raise the input resistance of a 741-like input stage to 3.6 MΩ for $\beta_n \geq 180$. If a change of input-stage bias current is to be used for this purpose, to what must the current be changed? What is the corresponding value of G_{m1} which results?

10.13 Reconsider the output resistances of Q_4 and Q_6 as calculated in association with Fig. 10.8a, b of the Text. What is the value of R_2 for which $R_{06} = R_{04} = 10.5\text{M}\Omega$? What is the corresponding value of R_{01}? Contrast the open-circuit voltage gains for the original and R_2-modified versions of the input stage.

10.14 In an input-stage design for which the new value of R_1 and R_2, as established for R_2 in P10.13 above, is used, the actual resistors installed are different by 2%. Calculate the corresponding input-offset voltage which results. Contrast that with the result for the original design as calculated in Example 10.1 on page 712 of the Text.

D 10.15 Use the result of Exercise 10.9 on page 714 of the Text to evaluate the required degree of match between R_1 and R_2 to ensure that the CMRR without feedback (that is the ratio of G_{m1} to G_{mcm}) is $\geq$80 dB.

D 10.16 In the calculation of R_0 performed in Exercise 10.10 on page 714 of the Text, its value is seen to be dominated by that of R_{09}. Augment Q_8 and Q_9 with additional resistors which will raise R_{09} to be equal to R_{010}. What resistors are needed? What value of R_0 results? For this situation, find a new value for G_{mcm} as requested in Exercise 10.11, and CMRR as requested in Exercise 10.12.

SECTION 10.4: Small-Signal Analyses of the 741 Second Stage

10.17 Find the input resistance of the second stage in the event that R_3 is replaced by a transistor Q_{25} and 1 kΩ resistor R_8, all connected suitably to the emitter of Q_7.

D 10.18 A designer wishes to capitalize on the change suggested in P10.17 above by lowering R_8 to a value which causes R_{i2} to return to the value of about 4 MΩ. What value of R_8 is appropriate? What value of G_{m2} results? What does R_{017} become? What is the new value of R_{02}? What is the new value of the no-load voltage gain of stage 2? Contrast this with the original value (having the same input resistance)?

SECTION 10.5: Analysis of the 741 Output Stage

10.19 Assuming the existence of some degenerative process which causes the β of Q_{23} to reduce, what is the value for which R_{i3} reduces to 4 times the value of R_{02}? What does the second-stage gain become with $R_L = 2$ kΩ and the conditions assumed just following Equation 10.22 on page 719 of the Text? What does the corresponding output resistance become?

L 10.20 Evaluate the sensitivity of the output-current-limiting scheme involving R_6 to the β of Q_{14} and Q_{15}. In particular, calculate the output current levels at which the excess current available from Q_{13A} (180 μA) is absorbed by Q_5, for the cases in which transistor β is 400, 200 or 100. Recall that $I_{S15} = 10^{-14}$A.

SECTION 10.6: Gain and Frequency Response of the 741

10.21 For the situation described in P10.18 above, calculate the overall gain which results when the amplifier is loaded with $R_L = 2$ kΩ. Find the frequency of the dominant pole associated with C_C = 30 pF. Find the unity-gain frequency f_t. Provide a corresponding Bode plot for this situation. What do you notice about f_t in this and the original cases?

D 10.22 Consider the possibility of a 741-like amplifier in which C_C is available for the customization of frequency response for high-gain applications. What values of C_C would be necessary to maintain phase margins of 45° and of 60° for closed-loop gains of 1000 and of 10^4. In each of the four cases, what upper 3 dB frequencies are possible?

10.23 For the two modified versions of the 741 amplifier alluded to in P10.22 above, for which the phase margin is 45°, calculate the corresponding values of slew rate *SR* and full-power bandwidth f_M, for outputs of ±10 V.

10.24 Reconsider the amplifier shown in Fig. P10.44 on page 759 of the Text, modified to have the bias currents reduced by a factor of 2, to 50 μA and 500 μA respectively, and the junction sizes of Q_6 and Q_7 made 2 times that of Q_5. Assume β = 120 and V_A = 200 V for all devices. What is the classification of the output stage type? What is its standing current? What is the gain with a load of 10kΩ? Calculate the capacitor *C* required for an open-loop 3dB frequency of 1kHz?

SECTION 10.7: CMOS Op Amps

L* 10.25 Consider a CMOS amplifier which uses the topology of Fig. 10.23 of the Text, but with somewhat different device sizing than that in Example 10.2 on page 729 of the Text. In particular, all devices have L = 10μm, the *p* devices have W = 200μm, and the *n* devices are such that $W_6 = 2W_4 = 2W_3$ = 200μm. Generally, V_A = 25 V, $V_{DD} = V_{SS}$ = 5 V, and I_{REF} = 25μA. For all devices, evaluate I_D, $|V_{GS}|$, g_m, and r_o, with V_A ignored in the bias calculations. Present these results in tabular form. Also find A_1, A_2, the dc open-loop gain, the input common-mode range, and the output voltage range. Use $\mu_n C_{ox}$ = 20μA/V^2 = $2\mu_p C_{ox}$, and $|V_t|$= 1V.

L 10.26 Repeat Example 10.2 on page 729 of the Text for I_{REF} = 12μA.

D 10.27 A young designer having forgotten the issue revealed in Eq.10.46 on page 731 of the Text, uses $(W/L)_6 = (W/L)_4$ = 50/10 in a design otherwise the same as considered in Example 10.2. What new value of V_{GS6} is required? Recalculate the value of A_2 and the overall gain which result. What is the input offset voltage produced?

D 10.28 For the modified amplifier described in P10.25 above, find the value of C_C that will result in f_t = 1 MHz. Find the value of *R* that places the transfer-function zero at infinity. Find the frequency of the second pole under the condition that the total capacitance at the output (C_2) is 10 pF. Find the excess phase it introduces at $\omega = \omega_t$. For a 10 pF load, what value of C_C is required to reduce the excess phase shift to 6° at a modified f_t frequency (f_t'). What is the maximum available slew rate for the two cases of excess phase?

SECTION 10.8: Alternative Configurations for CMOS and BiCMOS Op Amps

D 10.29 Consider the mirror used for differential-to-single-ended signal conversion in the cascode first-stage configuration in Fig. 10.25. In this context, consider the relative merit of configuring Q_3, Q_{3C}, Q_4, Q_{4C} as a simple cascode mirror, or as a Wilson mirror. Assuming all transistors identical and characterized by $|V_t|$, K, V_{GS}, and r_o, find the (effect of the) output resistance of each (as seen at the drain of Q_{4C}). Also, find the minimum voltage between $V_{BIAS\,2}$ and $-V_{SS}$ to ensure that all transistors operate in saturation.

10.30 Repeat Exercise 10.28 on page 736 of the Text for the current 2I reduced to 10 μA, but all other conditions the same. What values of R_0 and A_1 result?

D 10.31 A designer inspired by the additional gain provided by the cascode connection as exemplified in Fig. 10.25, contemplates the use of a 3-layer cascode using Q_{1CC}, Q_{4CC} and $V_{BIAS\,3}$ with $Q_{3\,CC}$ and $Q_{4\,CC}$ in a simple cascode connection. Find an expression for the output resistance R_0 now available. For conditions corresponding directly to those in Example 10.28, find the new values of R_0 and A_1. Assuming the values of $V_{BIAS\,2}$ and $V_{BIAS\,3}$ to be optimized, what is the total voltage required from the gates of Q_1, Q_2 to the supply, $-V_{SS}$?

D 10.32 For the circuit in Fig. 10.26 of the Text, and conditions as stated in Exercise 10.29, on page 736, but with $2I = I_B = 10\mu A$, find values of $V_{BIAS\,1}$, $V_{BIAS\,2}$, $V_{BIAS\,3}$ so that Q_6 and Q_7 operate at the edge of saturation. As well, find the output resistance and overall gain in the event that $|V_A| = 25$ V for all devices.

10.33 A folded cascode such as that shown in Fig. 10.26, operates with a bias current of 10 μA flowing in each of the devices Q_1, Q_2, Q_{1C}, Q_{2C} and Q_{3C}, Q_{4C} all having the same value of K and $V_A = 25$ V. For a load capacitance, C_L, of 10 pF, a unity-gain frequency of 1 MHz is found. What is the value of g_m for the critical devices? What is the low-frequency voltage gain of the amplifier? The dominant-pole frequency? The slew rate?

10.34 Repeat Problem P10.59 on page 760 of the Text for the situation in which $I_B = 800\mu A$ and $(W/L)_{1,2} = 600/10$, with $\mu_p\ Cox = 10\mu A/V^2$, and $C_L = 2$ pF. What outut-pole frequency f_t results? For all BJTs of the same size, what is the minimum value of unity-gain frequency of the BJT which ensures that the parasitic pole is at least a factor of 10 higher than the overall f_t.

SECTION 10.9: Data Converters - An Introduction

10.35 A sample-and-hold circuit operating at a 100 kHz rate uses a sampling switch which closes for a 10-nsec interval during each cycle. What is the frequency, f, of the highest-frequency square-wave input for which the output of the S/H provides an adequate representation of the input? Sketch the output of the S/H for input square waves having a frequency f, $0.5f$, $1.1f$ and $2f$. Note in this context that Shannon's sampling theorem states that sampling must be performed at a frequency at least twice that of the highest frequency component to be represented. For a sampling capacitor of 100 pF, what is the maximum value of total source and switch resistance that will normally provide samples which are accurate to within 1%? (Hint: think again of the square-wave situation).

D 10.36 A signal, to be sampled and converted to digital format, has a dynamic range of ±5 V. If it is important to resolve signals and signal changes as small as 0.1 V, what is the number of binary digits (bits) required in the converter? What is the greatest possible resolution that a 10-bit converter would provide?

SECTION 10.10: D/A Converter Circuits

10.37 For the DAC circuit shown in Fig. 10.31, using resistors no smaller than 1 kΩ, what is the largest resistor required to implement an 8-bit converter? (Be careful!). What is the largest value of switch resistance for which the associated error is at most ±½ LSB? Assuming that resistors are adjusted to compensate for the average value of switch resistance, what variability in switch resistance is acceptable if the corresponding error is limited to ±½ LSB? If the two sources of error identified are switch-resistance variation and resistor tolerance, and each is allowed to contribute equally to the overall error, what switch-resistance variation and resistor tolerance are needed for the 8-bit converter described?

10.38 For an R-2R ladder of 8 bits using a 10 V reference, what is the value of R which ensures that the total reference-supply current is 1 mA? What switch resistance would cause an error of ½ LSB if left uncompensated? If compensated by an appropriate change in the 2R resistor, what value of switch resistance, when doubled, produces an error ≤ ½ LSB?

10.39 A D/A circuit modelled after that shown in Fig. 10.33 operates with I_{REF} = 1 mA and R = 0.5 kΩ. If device EB Junction area can be maintained only to within 1%, while other components and parameters are ideal, what is the greatest number of bits possible for which the corresponding error is ± ½ LSB?

SECTION 10.11: A/D Converter Circuits

D 10.40 Sketch the design of a 2-bit flash-derived converter for signals in the range ±1 V using op amps as comparators. How many comparators do you need? What reference voltages would you use? Assuming op amps whose output limiting levels are precisely ±10 V, sketch a converter circuit using 2 op amps, one of which detects the polarity of the input, and, thereby, controls the reference voltage for the second. For the analog input connected to positive input terminals of each op amp, find, for the both the 3-op-amp and 2-op-amp designs, the output codes corresponding to inputs of +0.75 V, +0.25 V, –0.25 V and –0.75 V. Let +10 V correspond to logic '1', and –10 V correspond to logic '0' in both cases.

10.41 The following circuit, whose operation is related to that of the charge-redistribution converter, can be used as a component in a flash converter. With x indicating an NMOS switch operated by non-overlapping positive pulses Φ_A or Φ_B, describe v_O: a) during Φ_A, b) following Φ_A, before Φ_B, c) during Φ_B, d) following Φ_B, for i) $V_A > V_{REF}$, ii) $V_A < V_{REF}$. Assume that the op amp saturates at ±10 V.

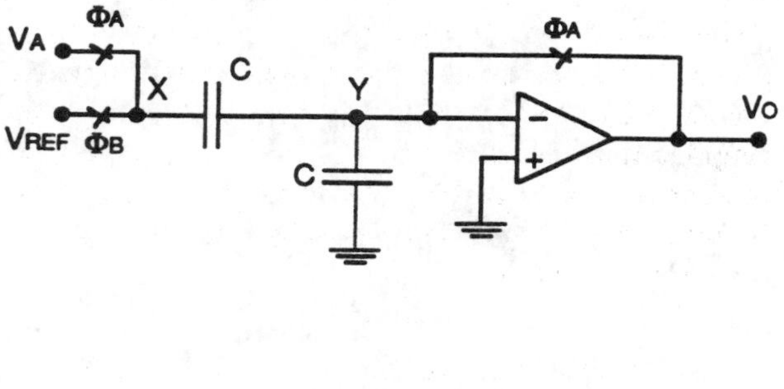

Chapter 11

FILTERS AND TUNED AMPLIFIERS

SECTION 11.1: Filter Transmission, Types, and Specification

L 11.1 The transfer function of a first-order high-pass filter (such as that realized by an *RC* circuit) can be expressed as $T(s) = \frac{s}{s + \omega_o}$ where ω_o is the 3 dB frequency of the filter. Provide a table of values of $|T|$,Φ, G and A at $\omega = \infty$, $2\omega_o$, ω_o, $\omega_o/2$, $\omega_o/5$, $\omega_o/10$, $\omega_o/100$, $\omega_o/1000$.

11.2 A high-pass filter has an equiripple magnitude response and specifications resembling those in Fig. 11.3, but with the passband and stopband exchanged on the frequency axis. Provide a sketch corresponding to this situation. If A_{max} = 1 dB and A_{min} = 50 dB, find $|T|$ at $\omega = \infty$, $\omega = \omega_p$ and $\omega = \omega_s$

11.3 A high-pass filter is required to pass all signals within its passband extending from 4 kHz to ∞, with a transmission variation of at most ±5%. The transmission in the stopband, which extends below 3.2 kHz, should not exceed 0.05% of the maximum passband transmission. Find corresponding values of A_{max}, A_{min}, and the selectivity factor.

11.4 A high-pass filter is specified to have A_{max} = 0.5 dB and A_{min} = 20 dB. It is found that its specifications can be just met with a single-time-constant *RC* circuit with a time constant τ of 10^{-3} *s* and a high-frequency transmission of unity. What are the values of ω_p, ω_s and the selectivity factor which correspond? In a modified design for which A_{min}, A_{max} and ω_p/ω_s are the same, but $f_p = 10^3$ Hz, what value of τ is required? What is the 3 dB frequency of the STC network? What is the resulting attenuation at 100 Hz?

SECTION 11.2: The Filter Transfer Function

11.5 A third-order high-pass filter has transmission zeros at $\omega = 0$ and $\omega = 0.1$ rad/s. Its natural modes are at $s = -1$ and $s = -0.5 \pm 0.8$. The high-frequency gain is unity. Find $T(s)$.

11.6 Find the order N and the form $T(s)$ of a bandpass filter having transmission zeros as follows: one at $\omega = 0$, one at $\omega = 1 \times 10^3$ rad/s, one at 2×10^3 rad/s, one at $6 \times 10^{+3}$ rad/s, one at 12×10^3 rad/s and one at ∞. If this filter has a monotomatically decreasing passband transmission with a peak at the center frequency of 4×10^3 rad/s. and equiripple response in the stop bands, sketch the shape of its $|T|$.

SECTION 11.3: Butterworth and Chebyshev Filters

11.7 Determine the order N of the Butterworth filter for which A_{max} = 0.5 dB, $A_{min} \geq 40$ dB and the selectivity ratio ω_s/ω_p = 1.6. What is the actual value of minimum stopband attenuation realized? If A_{min} is to be exactly 40 dB, to what value can A_{max} be reduced? For A_{max} raised by 0.1 dB, what change in A_{min} results for the same filter order?

D 11.8 Design a Butterworth filter having the following low-pass specifications: f_p = 20 kHz, A_{max} = 1 dB, f_s = 30 kHz and A_{min} = 20 dB. Find *N*, the natural modes, and $T(s)$. What attenuation is provided at 25 kHz? 40 kHz?

11.9 Contrast the attenuation provided by a 3rd-order low-pass Chebyshev filter at $\omega_s = 2\omega_p$ to that provided by a Butterworth filter of the same order. For both, A_{max} = 1 dB. Sketch $|T|$ of both filters on the same axes.

C 11.10 Repeat P11.7 above for a Chebyshev filter.

L 11.11 For P11.7 and P11.10 above, with $\omega_p = 10^3$ rad/s, and a dc gain of 1, find the required natural modes.

SECTION 11.4: First-Order and Second-Order Filter Functions

11.12 Consider the op-amp circuit of Fig. 11.13a driven so as to have a infinite input resistance, a dc voltage gain of 11, and a 3 dB frequency of 10 kHz using 10 kΩ as the lowest resistor value. What is the frequency of the associated zero? What is the gain at high frequencies? For the pole and zero separated by a factor of 100, what must its dc gain become?

D 11.13 Use a combination of the table entries in Fig. 11.13 to design a first-order bandpass filter using a single op amp with 3 dB frequencies at 100 and 1000 Hz. Arrange for a midband gain of –1 V/V and midband input resistance in the vicinity of 10 kΩ. Available capacitors can be specified only to one significant digit.

DL 11.14 Find the transfer function $T(s)$ of the circuit shown: Use it to create a spectrum-shaping network, having a midband gain of –10 V/V, with gain at high and low frequencies of –1 V/V. The 3 dB limits of midband gain should be at 100 and 1000 Hz. The midband input resistance should be in the vicinity of 10 kΩ. Available capacitors can be specified to only one significant digit. Sketch and label a pole-zero plot for this design.

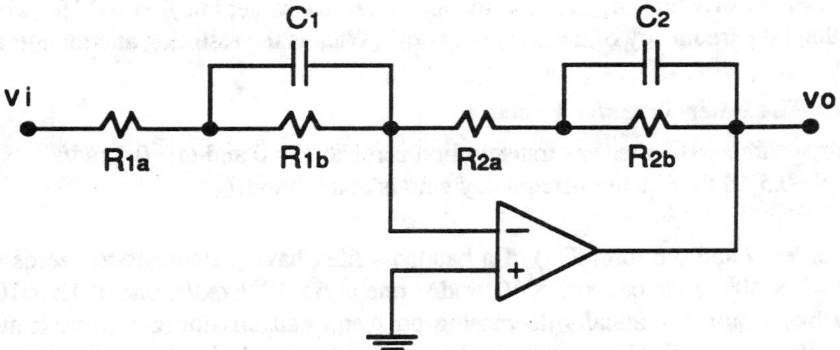

DL 11.15 Use the circuit in P11.14 above, and any insight you may have derived about it, to design a spectrum-shaping network, for which the gain between (3 dB points at) 100 Hz and 1000 Hz is –1 V/V, but the gain at high and low frequencies is –10 V/V. Arrange that the lowest resistance presented to the input source is around 10 kΩ, with the same capacitor-availability restriction as stated in P11.14 above. Provide a labelled pole-zero plot.

11.16 For the op-amp phase-shifter circuit derived from that shown in Fig. 11.14 of the Text by the exchange of R and C, find the transfer function $T(s)$ and the corresponding pole and zero. What phase shift results at $\omega_o = 1/CR$? For an input frequency of 10^4 Hz, and C = 1.59 nF, what values of R are required for phase-shift magnitudes of 6°, 12°, 30°, 60°, 90°, 120°, 150°, 168°, and 174°?

11.17 Use the information in Fig. 11.16c of the Text to obtain the transfer function of a second-order bandpass filter with $\omega_o = 10^3$ rad/s, a maximum gain of 1, and a 3 dB bandwidth of 200 rad/s. What are the frequencies at the stopband edge as characterized by $A_{min} = 20$ dB?

L 11.18 Using the information provided in Fig. 11.16b, find the transfer function for 2 second-order high-pass filters for which $A_{max} = 3$ dB, $\omega_p = 1$ rad/s, and the maximum gain is 1. For one version, the maximum gain occurs at very high frequencies; for the other, the maximum is at relatively low frequencies. Each has the maximum possible Q. For each, what is the lower 3 dB frequency, and A_{min} at $\omega_s = ½$ rad/s?

DL 11.19 Use the result of Exercise 11.15 on page 785 of the Text to find the transfer function of a notch filter that is required to eliminate 60 Hz power-supply hum in an instrumentation system. To accommodate some variability in the power-supply frequency component values, design the filter to provide 20 dB or more attenuation over a band of 1 Hz total width, nominally centred at 60 Hz. The high-frequency transmission is to be unity. What is the 3 dB bandwidth of the notch? At what frequencies is the notch transmission down by 1 dB? by 1%?

SECTION 11.5: The Second-Order LCR Resonator

D 11.20 Using the data in Fig. 11.16c and Equation 11.29, design the LCR resonator of Fig. 11.17b for operation at 1 MHz with a 20 kHz bandwidth, for $R = 10$ kΩ. What is the rms output when driven by a 1 MHz 1 mA-rms current?

D 11.21 The user of an FM radio receiver, discovering that reception is bad due to overload from a nearby broadcast transmitter radiating at 99.9 MHz, considers the use of a "wave trap" employing the notch circuit of Fig. 11.18e. The input resistance of his receiver is 75Ω. Design the trap's LC components so that the loss presented to stations 2 MHz from the interfering one is only 3 dB. If the initial adjustment and stability of the trap components is such that the trap can be off-tuned by as much as 100 kHz, what attenuation would you expect of the offending station's signal?

D 11.22 Using the circuit of Fig. 11.18c, design an RLC high-pass filter having a maximally flat response with a 3 dB frequency of 100 kHz, to operate as part of a circuit for which the loading resistance is 10 kΩ. If coil Q of 50 is available, find suitable values of L and C.

SECTION 11.6: Second-Order Active Filters Based on Inductor Replacement

D 11.23 Design the Inductor-Simulator circuit of Fig. 11.20 of the Text to realize inductances of 10 H and 0.1 H for operation around 1 kHz. Choose a majority of resistors of value 10 kΩ and a capacitor whose impedance is about 10 kΩ at the operating frequency. Available capacitors can be specified to only one significant digit.

DL 11.24 Use the designs prepared in P11.23 above to create a circuit which, when connected to a source having an output resistance of 20 kΩ, provides a bandpass function at 1 kHz. with a) the highest possible Q, b) the least-possible capacitor spread c) equal-valued capacitors (specified to 1 significant digit) and a change of a single resistor but whose value is not less than 1kΩ. What are the Q values which result in each case?

D 11.25 Design a fifth-order Butterworth LP filter having a 3 dB bandwidth of 10^4 Hz, and a dc gain of 10. Use a cascade of the circuits shown in Fig. 11.22a and 11.13a. Assume that capacitors of any required value are available as composed of 2 paralleled parts, each specified to 1 significant digit. Use as many resistors of

33 kΩ as possible. (Your boss has a surplus since all the other employees seem to prefer to use 10 kΩ in their designs!).

CD 11.26 Rework the transfer function of the BP filter given in Table 11.1 on page 799 of the Text, to employ equal capacitors C with two resistors chosen to control Q and ω_o as conveniently as possible, with all the other resistors of some fixed value. For Q ranging from 0.5 to 50, and capacitors specifiable using single digits 1, 2, 3 or 5, what is the range of values, specified by the ratio of maximum to minimum value, required for the resistors selected to tune Q and ω_o.

SECTION 11.7: Second-Order Active Filters Based on the Two-Integrator-Loop Topology

CDL 11.27 Design the KHN circuit of Fig. 11.24a to realize a bandpass filter with center frequency of 10 kHz and 3 dB bandwidth of 500 Hz. Use 1 nF capacitors. Arrange that the input resistance is 100 kΩ, that 100 kΩ resistors predominate, and that no resistor larger than 1 M is used. (Hint: In general, you may have to replace R_3 by a network.) Sketch the complete circuit and specify all components. What center-frequency gain results?

D 11.28 Use the KHN circuit with an output summing amplifier, to design a low-pass notch filter with f_o = 5 kHz, f_n = 7.5 kHz, Q = 10 and a dc gain magnitude of 3. Where possible, use the values found in Exercise 11.22 on page 806 of the Text.

D 11.29 Consider the bandpass filter described in P11.27 above, implemented using the Tow - Thomas biquad. Maintain the same center-frequency gain. Find suitable capacitor and resistor values, (assuming no value restriction).

SECTION 11.8: Single-Amplifier Biquadratic Active Filters

DL 11.30 Design the circuit of Fig. 11.29 to realize a pair of poles for which $\omega_o = 10^5$ rad/s and $Q = 1/\sqrt{2}$. Use the largest possible values of resistance, but with 1 MΩ as the largest available (precisely specifiable) value. Precision capacitors available are specified to one significant digit, either 1, 2 or 5.

D 11.31 Design the Bridged-T-based circuit of Fig. 11.30, to realize a bandpass filter with center frequency of 10 kHz, center-frequency gain magnitude of 1, and 3 dB bandwidth of 500 Hz. Use 1 nF capacitors. What is its input resistance at very-low and very-high frequencies?

CL 11.32 For the bandpass circuit of Fig. 11.30, find an expression for the input impedance as a function of component values and α, and then as a function of center-frequency gain, ω_o and Q. What is the value of input impedance at very low frequencies, very high frequencies, and at the center frequency?

DL 11.33 Design a 7th-order Butterworth low-pass filter with 3 dB bandwidth of 5 kHz and a dc gain of unity, using the cascade connection of 3 Sallen-and-Key biquads (Fig. 11.34c), and a first-order section (Fig. 11.13a). Use a 3.3 nF value for all capacitors.

SECTION 11.9: Sensitivity

11.34 In a particular implementation of the feedback loop of Fig. 11.34a, for which the values of ω_o and Q are given by Equations 11.77 and 11.78 respectively, each of the capacitors C_3 and C_4 is created using the parallel combination of 2 others. In particular, $C_3 = C_{3a} + C_{3b}$ where $C_{3b} = k_3\, C_{3a}$ and $C_4 = C_{4a} + C_{4b}$ where $C_{4b} = k_4\, C_{4a}$ for $0 \le k_3,\ k_4 \le 1$. Find the passive sensitivities of ω_o and Q relative to

C_{3a}, C_{3b}, C_{4a}, C_{4b} in terms of k_3, k_4, and when $k_3 = k_4 = 1$.

L 11.35 The feedback loop of Fig. 11.34, for which ω_o and Q are described by equations 11.77 and 11.78, operates over a range of temperatures for which the resistors, nominally equal, have a resistance which is a function of temperature, T. Thus $R_1 = k_1 R$ and $R_2 = k_2 R$ where $k_1, k_2 = 1 \pm k$, $k << 1$, and where $R = R_o(1 + a(T_o - T))$. Find the sensitivities of ω_o and Q to both k and T. Note that although this is a special situation in which the resistors vary equally around a fixed mean value, the results can be usefully combined with the average resistor sensitivity of –1/2, as needed.

SECTION 11.10: Switched - Capacitor Filters

11.36 For a dc voltage of 1 V applied to the input of the circuit of Fig. 11.35b in which the capacitance C_1 is 0.1 pF, what charge is transferred for each cycle of a 1 MHz clock. What is the average current drawn from the input source? What is the equivalent resistance seen by the input source? For a 2 pF feedback capacitance, what change in output would you expect for each cycle of the input clock? In what direction? For an amplifier saturating at ±10 V, how many cycles does it take for the amplifier output to go from one limit to the other? What is the average slope of the output? What does the slope become for an input of –0.1 V?

D 11.37 Design the circuit of Fig. 11.37b to realize, at the second integrator output, a Butterworth (maximally-flat) low-pass function with $f_{3dB} = 10^5$ Hz, and unity dc gain. Use a clock frequency of $f_c = 1/T_c = 1$ MHz with $C_1 = C_2 = 2$ pF. Find values for C_3, C_4, C_5, C_6 (Note that for a 2nd-order maximally-flat response, $Q = 1/\sqrt{2}$ and $\omega_{3dB} = \omega_o$). Characterize the output of the first integrator: What is its centre frequency? Its 3 dB bandwidth? Its associated maximum gain?

SECTION 11.11: Tuned Amplifiers

11.38 A signal source with internal resistance $R_S = 10\text{k}\Omega$ is connected to the input of a common-emitter BJT amplifier having an emitter resistor R_E. >From base to ground is connected a tuned circuit with $L = 1\mu H$ and $C = 200$ pF. The transistor is biassed at 1 mA, with $\beta = 200$, $C_\pi = 10$ pF and $C_\mu = 1$ pF. The transistor has a load resistance of 5 kΩ. Find ω_o, Q, the 3 dB bandwidth and the center-frequency gain which result for: a) $R_E = r_e$, b) $R_E = 9r_e$. (Hint: Use a Miller calculation for the effect of both C_μ and C_π).

L 11.39 Repeat P11.72 on page 839 of the Text for the situation as described, except that the base is connected to a coil tap located at a fraction k from the grounded end (the one to which the emitter is connected.). Find ω_o, Q, the bandwidth and overall center-frequency gain for: a) $k = 0.5$ and b) $k = 0.1$. Contrast these results with those found in P11.38 above.

D 11.40 A coil having an inductance of 2 μH and intended for operation around 10 MHz, has a Q specified to be 200. What is equivalent series resistance, the equivalent parallel resistance, the value of the resonating capacitor, and the parallel resistance, which, when added, provides a bandwidth of 200 kHz?

D 11.41 A particular LC single-tuned amplifier has a center frequency of 1 MHz and a 3-dB bandwidth of 100 kHz. What is the corresponding Q? How many such amplifiers synchronously-tuned would result in a bandwidth reduced to 50 kHz? For the basic amplifier, what is the 30 dB bandwidth and skirt selectivity? What do these become for the synchronously-tuned cascade?

D 11.42 A two-stage 4th-order stagger-tuned Butterworth design is required with center frequency of 10.7 MHz and an overall bandwidth, B = 250 kHz. Using 3 μH inductors, find C and R for each of the two stages. What is the relative peak gain of each of the two stages?

Chapter 12

SIGNAL GENERATORS AND WAVEFORM - SHAPING CIRCUITS

SECTION 12.1: Basic Principles of Sinusoidal Oscillators

12.1 Consider a sinusoidal oscillator consisting of a non-inverting transconductance amplifier and an RLC tank circuit. The amplifier is characterized by transconductance G_m, input resistance R_{in}, and output resistance R_o. The tank circuit employs a tapped inductor L for which the turns ratio is n (see Fig. 11.42), and whose total series resistance is R_{ls}, with a capacitor C whose (small) loss is represented by a parallel resistance R_{cp}. For the amplifier input connected across the small part of the coil, and the amplifier output connected so as to drive it all, find an expression for the conditions for oscillation, its frequency, and the required relationship amongst parameters.

CD 12.2. For the situation described in P12.1 above, sketch the corresponding circuit topology. Prepare a second sketch to show what must be done if the available transconductance is of an inverting kind. What are the new conditions for oscillation?

12.3 For a situation related to that described in P12.1 above, the connections between the coil and transconductor are exchanged, such that its output drives the tap, while its input is connected across the entire coil. What are the conditions which prevail at the threshold of oscillation?

12.4 For a particular oscillator application, the limiting-amplifier topology shown in Fig. 12.3 of the Text is being considered. The requirement is a design with a maximum gain of 5 which limits at ±2.5 V with a gain of 0.5, and has an input resistance of 10 kΩ. The op amp to be used has an open-loop gain of 1000V/V or more. Supply voltages of ±10 V are available. Assume a diode drop of 0.6 V at the levels of conduction implied. Find appropriate component values. If the op amp, assumed to have a single-pole rolloff, has an f_t of 1 MHz, what is the maximum frequency of operation for which an amplifier phase shift of 2° can be tolerated.

12.5 Reconsider the circuit of P12.8b on page 894 of the Text with R_2 = 10 kΩ, R_1 = 7.5 kΩ and a resistor R_3 = 10 kΩ connected in series with Z_1 and Z_2. Sketch the resulting transfer characteristic under the conditions that V_Z = 6.8 V, V_f = 0.7 V and I_{ZK} is 100 μA. Show what happens if, for a better zener, I_{ZK} reduces to 10 μA.

SECTION 12.2: Op-Amp-RC Oscillators

DL 12.6 Design a basic Wien-Bridge oscillator circuit, using the topology of Fig. 12.4 augmented by a gain-control mechanism consisting of two paralleled anode-to-cathode-connected diodes and a series resistor equal to R_2. The diodes are 1 mA units for which n = 2, and the voltage drop at 1 mA is 0.7 V. Select component values for an output voltage of 2 V peak-to-peak at 10 kHz using 10 nF capacitors. Assume the op amp to have A_o and f_t high enough to be ignored.

C* 12.7 A designer, wishing to use the idea described in P12.6 above at a higher frequency, is concerned with amplifier phase shift. She considers an otherwise-suitable amplifier for which f_t = 1 MHz. What is the highest nominal frequency of operation for which the actual frequency of oscillation is different by 10%.

What must the amplifier closed-loop gain be for operation in this mode?

CL 12.8

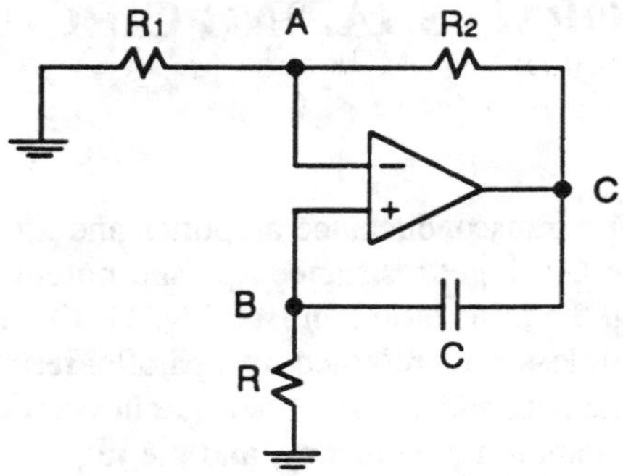

A designer, wishing to employ the phaseshift inherent in a compensated op amp, operated in a closed loop in the vicinity of its 3dB frequency, considers the circuit shown: For an op amp having a relatively large open-loop gain and high-order poles remote from f_t, find expressions for the loop gain and potential conditions for oscillation. For what value of R_2 is the frequency of oscillation $4/RC$? Now for excess phase shift at f_t, what occurs? How could you use this idea to evaluate f_t, and phase margins under various conditions?

CL 12.9 The circuit of Fig. P12.18 of the Text is modified by removing the rightmost resistor R and connecting the lower node of the rightmost capacitor C to the op-amp negative input. Find an expression for the loop gain, the frequency of possible oscillation and the conditions for which it occurs. What is the sensitivity of the frequency of oscillation, and of the critical value of R_f to a deviation of value of *any one of* the resistors R.

12.10 Another possible variant of the phase-shift oscillator of Fig. P12.18 of the Text can be considered in which the right-most RC circuit is removed, and a capacitor C is shunted across R_f. Find an expression for the corresponding loop gain, and the frequency and gain conditions for possible oscillation.

CD 12.11 A phase-shift oscillator is considered which resembles that of Fig. 12.7 of the Text, but uses a positive-gain amplifier. What is the minimum number of RC sections required? What is the corresponding frequency of oscillation and critical value of gain, K?

C* 12.12 For the circuit of Fig. 12.8 of the Text, consider the effect of adding an additional RC section. What are the conditions of oscillation? For the modified and original circuits, find the effect on frequency and critical gain of a simultaneous change in value of all its resistor R. (Hint: Economize your effort by using the technique described in Exercise 12.5 on page 852 of the Text). Note the solution of Ex. 12.5 which emerges in your work.

C*D 12.13 Design the active-filter tuned oscillator shown in Fig. 12.11 of the Text for operation at 10 kHz using 10 nF capacitors, and diodes for which the voltage drop is a 0.7 V at 1 mA. Establish Q to reduce the 3rd-harmonic distortion to less than 1%. What is the output voltage which results?

SECTION 12.3: LC and Crystal Oscillators

C*L** 12.14 For the Colpitts oscilator shown in Fig. P12.21b of the Text, in which both FETs are matched with $I_{DSS} = 4$ mA, $V_p = -2$ V and $V_A = 100$ V, the load resistor is 10 kΩ, and the inductor has a Q of 100 and an inductance $L = 10$ μH at the operating frequency of 1 MHz. Find the values of C_1 and C_2 to ensure oscillation at 1 MHz. With C_2 set 5% lower than the value calculated, and the gate-to-source diode of Q_1 characterized as having a 0.7 V drop at 1 mA and $n = 1$, find the output signal amplitude, assuming the supply voltages to Q_1 and Q_2 are high enough. For supplies of ±3 V, estimate the amplitude of the output signal. (Hint: Use the incremental slope of the characteristic in the upper end of the triode region for a value of v_{DS} slightly less than $|V_p|$).

C*DL* 12.15 A particular quartz crystal is measured to have a series resonance at 2.015 MHz, a parallel resonance at 2.018 MHz, a parallel plate capacitance of 4pF, and a Q of 50×10^3. Find corresponding values of L, C_s, and r. When used in the circuit of Fig. 12.16, C_2 = 10 pF, and C_1 can be as low as 1 pF. For device sizing and V_{DD} such that $g_{mn} = g_{mp}$ = 1 mA/V and V_A is large, with R_f = 1 MΩ, choose values of C_1 and R_1, for which the loop gain can be assumed to be >1. (Hint: Consider the CMOS amplifier as an op amp with finite gain loaded by $R_1 \,||\, R_f$, with gain established by a network consisting of Miller-multiplied R_f reflected back through the C_1/C_2 capacitor ratio, driven by R_1. Using this idea, find a maximum value for R_1).

SECTION 12.4: Bistable Multivibrators

CDL** 12.16 For the CMOS circuit shown, in which all transistors are matched with K = 1 mA/V^2, and $|V_t|$ = 1 V, V_A = 30 V, design a version whose output voltage range extends beyond +4.9V and +0.1V for inputs in the range 0 to 5V, with V_{TL} = 2.0V, and V_{TH} = 3.0V. What is the consequence on the input thresholds of K and V_t varying by ±20%? For amplifier capacitance characterized by 10 pF at each inverter input, and 1 pF at each inverter output, estimate the output rise and fall times. For an input rising and falling at a rate of 1 V/μsec, estimate delay time from the time the source begins to change until regeneration begins.

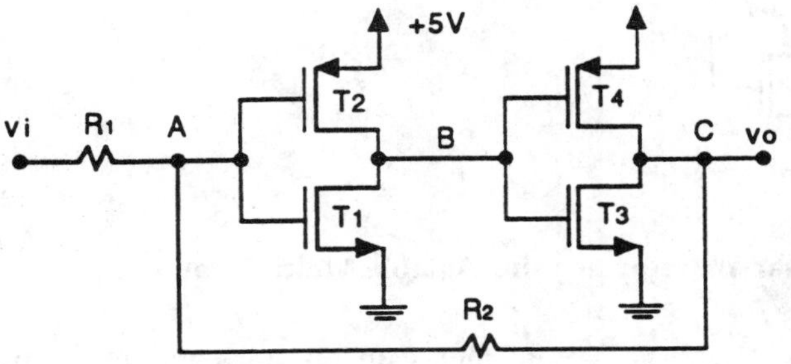

DL 12.17 In the circuit shown, using a high-gain op amp, the output signal is ±13 V or so with ± 15V supplies, and diode drops are 0.7 V for reasonable currents. Design R_1, R_2, and R_3 so that each diode, when conducting, has a current in the range 1 to 4 mA, with as low a current level as possible preferred. What are the input thresholds V_{TL}, V_{TH} for this circuit?

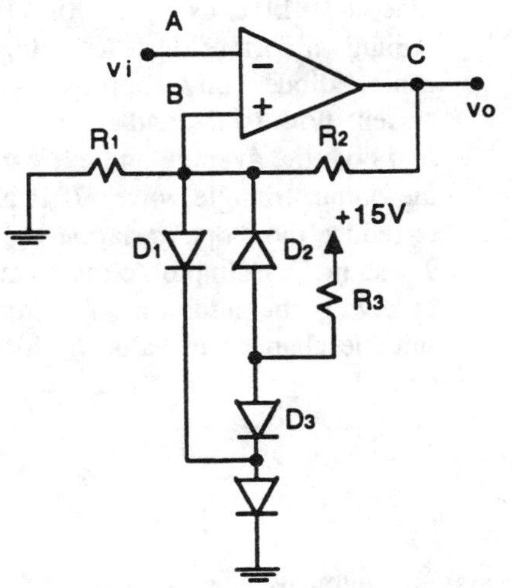

12.18 In the circuit shown, all transistors have $|V_t| = 1$ V and $K_1 = K_2 = K_3 = K_4 = 2K_5 = 2K_6 = 100\mu A/V^2$. Sketch and label the transfer characteristic of the $Q_3 - Q_4$ inverter. Now sketch and label the transfer characteristic υ_o versus υ_i. What are V_{IH} and V_{IL}?

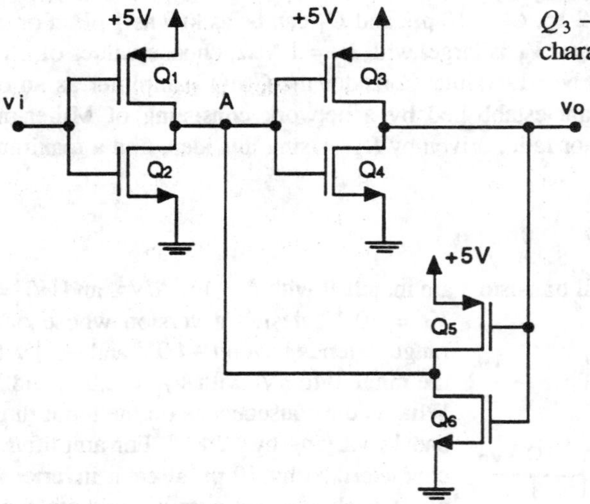

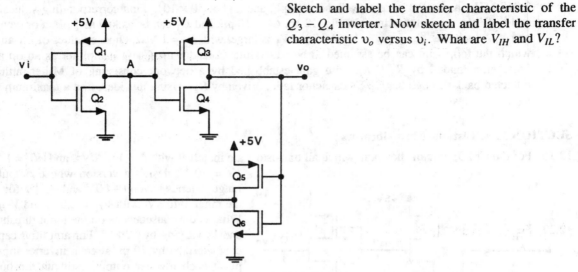

SECTION 12.5: Generation of Square and Triangular Waveforms using Astable Multivibrators

DL 12.19

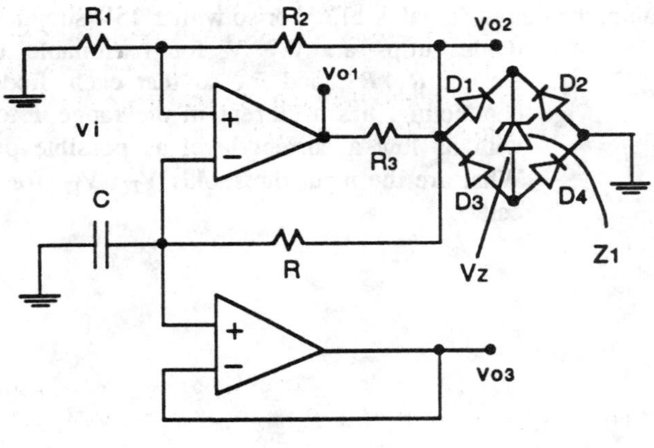

With high-gain op amps for which υ_o of ±13 V is guaranteed, design the circuit shown, using a 6.8V zener diode, to provide a triangle-wave output at υ_3 of ±1 V amplitude at 10 kHz, using a 1000 pF capacitor. Arrange a current of at least 1.0 mA in the zener diode and approximately equal current flow in R_2 and R. For this design, find both the average and extreme slopes of the output triangle wave. If it is important to reduce the slope variation by a factor of 2, show a simple circuit modification involving the addition of 2 resistors and suitable changes in values, which does the job.

CD 12.20 Use the topology of Fig. 12.25 of the Text, implemented using the output-voltage-regulator scheme shown in P12.19 above, to provide square and triangular waves at 10 kHz, each of ±1 V peak amplitude, the square wave being obtained from an appropriate voltage divider using a 1 mA current. Notice that the required bistable is of the non-inverting type. What must its upper and lower thresholds be? Sketch the complete circuit, and select values consistent with specifications provided in P12.19 above.

SECTION 12.6: Generation of a Standardized Pulse – The Monostable Multivibrator

12.21 The circuit shown uses an op amp for which the limiting output voltages are ±10 V. For this design, $R_2 = R_3 = 100\text{k}\Omega$, $R_1 = R_5 = R_4 = 10\text{k}\Omega$, $C_1 = 10$ nF, $C_2 = 1000$ pF. What are the resting voltages at *A, B, C, D*? What is the minimum amplitude of a positive step (*V*) at the input that will trigger the circuit? Sketch waveforms at all nodes for the case in which *V* is a 5 V step. What is the length of the output pulse produced? What is the purpose of R_4? For *V*, a rate-limited rising input of 5 V, what is the slowest rate of rise to ensure triggering? How soon after the output falls can the circuit be retriggered with the expectation of an output pulse of the usual duration?

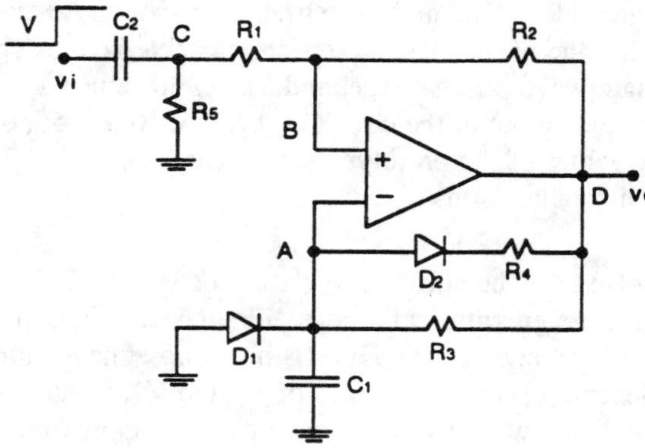

12.22 For the circuit shown, Q_1 through Q_4 have $|V_t| = 1$ V and equal *K*. Correspondingly, their active (amplifying) region is around $V_{DD}/2 = 2.5$ V. Normally *A* is held at 0 V by R_2, *B* is high at 5 V and *C* is low at 0V. When input v_i is applied, *D* rises and *A* rises with it, lowering *B* and causing *C* to rise, replacing v_i through D_2. For $R_1 = R_2 = 10$ kΩ, $C_1 = 10$ nF, sketch the waveforms at all nodes following the rise of v_i. What is the expected maximum duration of v_i? What happens if it stays high longer? In this event, the circuit operates open-loop for the fall of v_o (and does *not* regenerate upon turnoff). If v_i is too long and the gain of each stage is 40 V/V, what output fall time results?

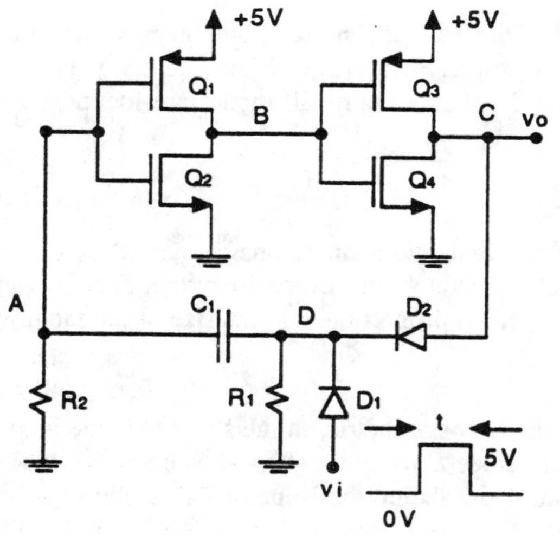

SECTION 12.7: Integrated - Circuit Timers

12.23 With reference to Fig. 12.28, showing the 555 timer operating as a monostable multivibrator with *R* = 10 kΩ and *C* = 10 nF, what is the length of output pulse produced by a negative trigger pulse? What is the maximum length of input pulse which might logically be used? What happens to the flip-flop if the input is longer than this? It is usual in the case of such conflict that the flip flop is so-called "set-dominant" such that v_o (that is, *Q*) remains high (with $\overline{Q}$ low) while *S* is high (held by *V* trigger being low).

12.24 With reference to the situation described in P12.23 above, and for normal operation with a 5 V supply, what is the % change in output pulse length for each 100 mV increase in the saturation voltage of Q_1?

D 12.25 For the astable connection of the 555 timer depicted in Fig. 12.29, using *C* = 10 nF and $R_A = R_B = 10$ kΩ, what is the frequency of the output at v_o? What is the duty cycle? What do these become if R_B is reduced to 1 kΩ? What change must be made to R_A to return to the same frequency? What combination of resistors will provide an output at 10 kHz with a duty cycle of 10%?

SECTION 12.8: Nonlinear Waveform - Shaping Circuits

DL* 12.26 Design a sine-wave shaper using a series resistance R and two shunting diodes connected to ground, one with anode grounded, and the other with cathode grounded. The input is a triangle wave whose amplitude is such that its zero-crossing slope equals that of the sine wave. The diodes are characterized as conducting 1 mA at 0.700 volts, with $n = 2$. Find the triangle-wave peak voltage, and a suitable value for R. Then find the angles Θ ($\Theta = 90°$ at the wave peak) where the output of the circuit is 0.7, 0.65, 0.6, 0.55, 0.5, 0.4, 0.3, 0.2, 0.1 and 0 V. Use these angles to find the values of the prototype sine wave (i.e., $\upsilon_o = 0.7 \sin \Theta$) and the corresponding errors. Present your results in tabular form.

DL* 12.27 Use the results obtained in Example 12.22 of the Text for the square-law shaper of Fig. E12.22 on page 883 of the Text, using ideal diodes, as the basis of a design with real diodes. First consider pseudo-1-mA diodes modelled by an ideal diode, a 600 mV offset and a 100 Ω resistor. Chose new values for R_1, R_2, R_3 and associated voltages V_2 and V_3. Second, using these values of V_2 and V_3, create a revised design using diodes for which the drop is 0.7 V at 1 mA with $n = 2$. In each case, in approximating $i = 0.1\upsilon^2$, arrange that the approximation is perfect at 2, 4, and 8 volts, and calculate the error at 3, 5, 7 and 10 volts.

CD 12.28 Using the idea embodied in Fig. P12.44 on page 900 of the Text and similar component values, prepare a design of a circuit for which the output is $\upsilon_o = \upsilon_i\ \upsilon_2/\upsilon_3$ for υ_1, υ_2, $\upsilon_3 > 0$. Check circuit performance at various combinations of input voltages, say, 0.5, 1, 2, 3 volts. Assume all diodes are identical with 700 mV drop at 1 mA with $n = 2$.

SECTION 12.9: Precision Rectifier Circuits

D 12.29 Combine the circuits of Fig. 12.33 and Fig. 12.34 of the Text into a single one connected both to a common source υ_i and a common load R_o as in 12.30 below. Sketch the composite circuit and its composite transfer characteristic. Since its output is always positive, yet proportioned to the size of the input, what is such a circuit called?

D 12.30 Note that the circuit created in answer to P12.29 above is not symmetric, in the sense that one amplifier is allowed to saturate. Add an additional resistor R_3 and diode D_4 to correct for this, using the idea embodied in the use of D_2 in the more complete sub-circuit. How should the value of R_3 be chosen in relation to R_1, R_2, and R_0?

D 12.31 A designer wishes to create an expanded-scale ac voltmeter in which voltages in the range 100V rms to 140 V rms are displayed. Note that for voltages ≤ 100 V rms, the meter reads 100 V at the left end of the moving-coil-meter scale. Use a combination of the circuit in Fig. 12.35, the idea embodied in Fig. 12.28 on page 887 of the Text, and a 1-mA 50-Ω meter connected in series with $R_4 = 10$ kΩ of the filter. Within the circuit itself, design for signals whose largest value in the normal operating range is ±10 V. Chose R_1 to absorb the very large input voltage involved. Note that ±15 V supplies are available.

12.32

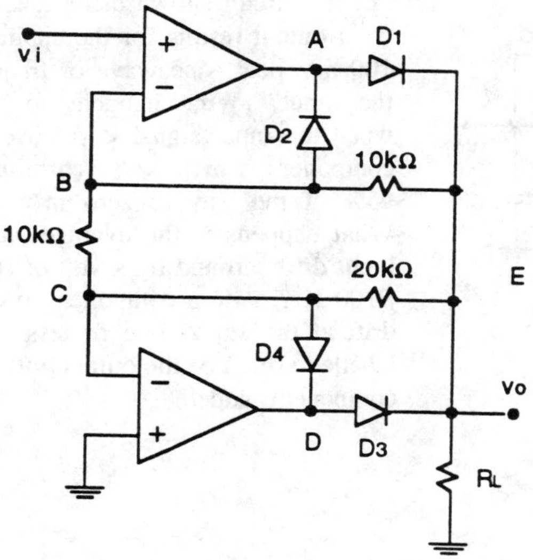

Consider the operation of the circuit shown: What is the output voltage for v_I = +5 V, 0 V, −5 V. What is the input resistance of the circuit? What might the circuit be called? Notice the relationship to the bridge amplifier in Fig. P9.52 on page 697 of the Text.

12.33

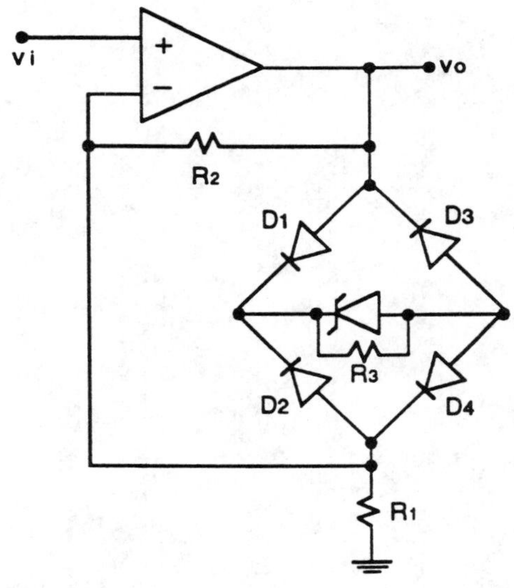

Consider the transfer characteristic of the circuit shown. Sketch and label it for R_1 = 1kΩ, R_2 = 100 kΩ, R_3 = 100 kΩ, with $r_Z = r_D \approx 0$, but V_Z = 6.8 V, V_D = 0.7 V.

DL 12.34

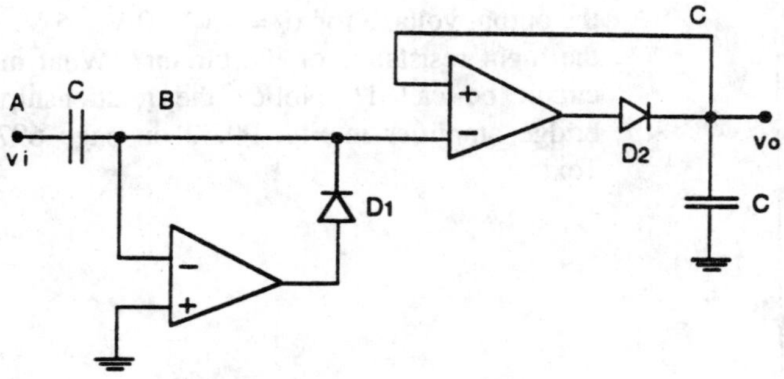

For the circuit shown using ideal op amps, what output results for the application of a 100 mV peak sine wave of frequency f at the input? What happens to the output when the input signal is removed? Add a component which will return the output 95% of the way to zero in a time $10/f$. What happens if the average value of the input drifts around by a volt or so at a rate $f/100$ Hz? Add a component to correct for drift at the stated rate or less. Provide a labelled sketch of the output with these two components added.

Chapter 13

MOS DIGITAL CIRCUITS

SECTION 13.1: Logic Circuits – Some Basic Concepts

13.1 A logic inverter modelled as in Fig. 13.2 in the Text employs a switch for which the offset voltage is 50 mV and the on-resistance is 100Ω. The resistor R_L is 500Ω, and the supply V^+ is 3 V. What are the expected values of the output voltage?

13.2 A particular logic inverter has V_{IL}, V_{IH}, V_{OL} and V_{OH} of 1.2 V, 2.2 V, 0.5 V, and 3.5 V, respectively. What are its low and high noise margins? What is the width of the transition region? What is the average gain of the inverter in its transition region?

13.3 The logic inverter described in P13.1 above, is loaded with a 50 pF capacitor. What are its propagation delays, assumed for simplicity to be measured at the 1.5 V level? For what square-wave input frequency is the output voltage greater than 1.5 V for ¼ of the time? Estimate the total power dissipated at that frequency. What is the corresponding delay-power product?

13.4 The logic inverter described in P13.1 above has a fanout of 3 loads, each of which can be modelled (transiently) as a 0.7 V equivalent-voltage source with 300 ohm resistor. What output voltage levels result? What are the currents flowing to and from each load?

13.5 A particular gate with a 10 pF load, is said to have an average propagation delay of 5 nsec, and a delay-power product of 50 pJ, when operating with an approximately square-wave output at frequency *f*, with 4 V signal swings, and a 5 V supply. If its static power loss is 6 mW with output low, and 1/3 of that with output high, estimate the frequency *f*. What is the delay-power product at frequency *f*/2 with a square-wave output?

13.6 A particular NOR logic gate has $t_{pLH} = 30$ ns and $t_{pHL} = 10$ ns. Five such gates, each with one input low, are connected in a ring. What is the frequency of the oscillation which results? Two additional NOR gates, *A* and *B*, are connected to the ring, gate *A* to the outputs of gates 1 and 3, and gate *B* to the outputs of gates 1 and 4. Sketch the resulting waveforms, paying attention to their relative timing.

SECTION 13.2: NMOS Inverter with Enhancement Load

13.7 For an enhancement-load inverter, V_{OH} is found to be 3.05 V with operation from a 5 V supply. If $V_{t0} = 1$ V, and $2\Phi_f = 0.6$ V, what must γ be?

CL 13.8 For an enhancement-load inverter such as that in Fig. 13.9 of the Text, for which $V_{t0} = 1$ V, $(W/L)_1 = 3.5$, $(W/L)_2 = 1/3.5$, $\mu_n C_{ox} = 20\ \mu A/V^2$, $2\Phi_f = 0.6$ V, $\gamma = 0.5\ V^{1/2}$ and $V_{DD} = 5$ V, find V_{OH}, V_{OL}, V_{OL}, V_{IL}, V_{IH}, NM_H and NM_L, for a) the body effect neglected, and b) the body effect incorporated, with V_{OH} and V_{OL} in non-critical roles as in a), and c) the body effect incorporated, with V_{OH} and V_{OL} as in b). One can view a), b), c) as 3 stages of an iteration involving improvement in the value of V_{OH}, used as an alternative to more-closed-form solutions. Correspondingly, you may note that the formal closed-form approach implied by a) could have been simplified by approximating V_{OL} by 0 in calculating better values of V_{IH} and V_{OL}.

13.9 For the enhancement-load NMOS inverter described in P13.8 above, find the input voltage V_{th} at which $\upsilon_I = \upsilon_O$, with body effect a) neglected, b) included.

13.10 For the inverter described in P13.8 above, driving a 0.1 pF capacitive load, find t_{pLH} and t_{pHL} using the average-current approximations.

13.11 For the situation described in P13.10 above, consideration is being given to making the propagation delays more similar by reducing t_{pLH} by a factor of 2, and/or raising t_{pHL} by a factor of 2. What are the new *W/L* ratios required? What values of V_{OL} would result in each case?

SECTION 13.3: NMOS Inverter with Depletion Load

L 13.12 Consider an NMOS depletion-load inverter whose parameters and dimensions correspond to the enhancement-load inverter described in P13.8 above, namely $V_{tE} = 1$ V, $V_{tDo} = -3$ V, $(W/L)_1 = 7\mu m/2\mu m$, $(W/L)_2 = 2\mu m/7\mu m$, $\mu_n C_{ox} = 20\mu A/V^2$, $2\Phi_f = 0.6$ V, $\gamma = 0.5V^{1/2}$, and $V_{DD} = 5$ V. Find V_{OH}, V_{OL}, V_{IH}, V_{IL}, and the noise margins, taking the body effect into account *only* where necessary.

13.13 For the inverter described in P13.12 above, use the results of P13.22 in the Text to find r_{DS1} and r_{DS2}.

13.14 For the inverter described in P13.12 above, use the results of P13.23 in the Text to find V_{IH} and corresponding υ_o without the body effect. Then, use this υ_o to find V_{tD} with the body effect included, and a correspondingly improved estimate of V_{IH}.

13.15 Use the results of P13.25 in the Text to estimate V_{OL} for the inverter described in P13.12 above. To reduce V_{OL} to half this value, by what factor must the width of Q_1 be increased?

13.16 For the inverter specified in P13.12 above, operating with a 0.1 pF load, use the average-current approximation to find t_{pHL}, t_{pLH} and t_p. Now, use the approximations developed in P13.29 b), c) of the Text to find other estimates for t_{pHL} and t_{pLH}. Use $\alpha = 1 + \frac{\gamma}{V_{t\,DO}} \sqrt{V_{DD}/2}$. With that, estimate *DP* from Eq. 13.29 of the Text.

SECTION 13.4: NMOS Logic Circuits

D 13.17 Repeat P13.30 of the Text for 4-input NOR and NAND gates. As well, use the fact that $\overline{AB} = \overline{A} + \overline{B}$ to find an alternative implementation of the NAND using only a NOR gate with inverters. For all 3 cases, find the dimensions of the input transistors, their area (to which the gate input capacitance is proportional), and the total gate area.

D 13.18 For the circuit of P13.32 of the Text, and device sizes of P13.30 of the Text, find the total device area required. Now sketch an alternative implementation using only NOR gates (as suggested in P13.17 above). Find the total device area required using this approach.

CD 13.19 Combine the circuits of Fig. P13.33 and Fig. P13.34 of the Text to create a buffered bootstrap driver using only enhancement NFETs. To emphasize the buffer aspect, relabel Q_3 of Fig. P.13.34 as Q_5, and add buffer devices Q_3 and Q_4. Use $V_{DD} = 5$ V, $V_t = 1$ V, $\mu_n C_{ox} = 20\mu A/V^2$, $W_1 = 12\mu m$, $W_2 = W_5 = 6\mu m$, $W_3 = 120\mu m$, $W_4 = 240\mu m$, $L_2 = L_5 = 12\mu m$, and $L_1 = L_3 = L_4 = 6\mu m$. For this new circuit, find voltages at the sources of Q_5, Q_2 and Q_4 when υ_I has been high (at 4 V) for some time. What supply current flows?

When υ_I goes low (to 0 V) to what voltages do these nodes rise? (Ignore the body effect). What current is available to begin to charge a load capacitor? When υ_i rises again, what current is available immediately to discharge the load capacitor? To what voltages do the three nodes fall? In all of these calculations, ignore the effects of internal node capacitances.

SECTION 13.5: The CMOS Inverter

13.20 Consider a CMOS inverter for which $\mu_n C_{ox} = 2\mu_p C_{ox} = 20\mu A/V^2$, $(W/L)_n = 8\mu m/2\mu m$, $(W/L)_p$ = $16\mu m/2\mu m$, $V_{tn} = -V_{tp} = 1$ V, and $V_{DD} = 5$ V. What are the resistances from the output node to the supply rails for inputs high (+5 V) and low (0 V). (Hint: See P13.35 of the Text)

13.21 For the inverter described in P13.20 above, with standard inputs of 0 V or V_{DD}, find the maximum currents that can be sourced or absorbed with the output joined to ground or V_{DD} respectively. What do the currents become for an output voltage of $V_{DD}/2$? for an output voltage of 0.1 V_{DD} from either V_{DD} or 0V?

D 13.22 A CMOS inverter for which the device thresholds and K factors are matched, with $|V_t|$ nominally 1 V, operates from a nominal supply voltage of 5 V. For possible variation of both V_t and V_{DD} by ±25%, what ranges of V_{IL}, V_{IH}, V_{OL}, V_{OH} result. What ranges of NM_H and NM_L result between various inverters?

13.23 The CMOS inverter specified in P13.20 above, and having a 0.5 pF load, is switched at a rate of 20 MHz by an input wave whose rise and fall times are each ¼ of the wave period. What is the peak current due to gate self-conduction. Assuming the gate self-current to be triangular in pulse form, flowing for input voltages $\upsilon_I = |V_t|$ to $\upsilon_I = V_{DD} - |V_t|$, find the average supply current due both to self-conduction and capacitor charge. What is the gate power dissipation for this situation, and with the load capacitor removed?

13.24 For the inverter in P13.20 above, loaded with a 0.5 pF capacitor, estimate t_{pHL} and t_{pLH} using Eq. 13.44. What is the corresponding delay-power product for operation in a 5-stage ring counter?

13.25 A CMOS inverter for which $K_p = K_n = 10\mu A/V^2$ and $|V_t| = 0.8$ V, is to be operated in an electronic-watch environment with $V_{DD} = 1.3$ V.

(a) For what range of input voltages does Q_n conduct? Q_p conduct? What happens for $\upsilon_I = V_{DD}/2$?

(b) What range of output voltages is possible? What values of V_{OH} and V_{OL} correspond?

(c) What are the values of V_{IL} and V_{IH} which apply? What currents flow in the transistors for voltages between V_{IL} and V_{IH}?

(d) Sketch the transfer characteristic, under the assumption that the input is a triangle wave extending from V_{OL} to V_{OH}, and that the output is loaded with a very small capacitor to ground.

(e) For the inverter loaded by a 1 pF capacitor, and input driven by an ideal square wave, sketch the output waveform. Estimate the times taken for the output to rise and fall by the amount V_t following a change of state of the input? Estimate the propagation delay, as defined as the time taken for the output to move to the switching threshold of a succeeding gate, following the crossing of its own threshold.

(f) Estimate the frequency of oscillation of a ring of 5 such gates, once an oscillation is initiated. Note that this oscillator is *not* ordinarily self-starting, but must be "kick-started" by an external signal. Otherwise, the ring has a stable state with all voltages around $V_{DD}/2$.

L 13.26 Consider an inverter for which $\mu_n C_{ox} = 2\mu_p C_{ox} = 20\mu A/V^2$, $(W/L)_n = 18\mu m/2\mu m$, $(W/L)_p = 2\mu m/2\mu m$, $V_{tn} = -V_{tp} = -1$ V, and $V_{DD} = 5$ V. What values of V_{OH}, V_{OL}, V_{IL}, V_{IH} and V_{th} apply? For the latter threshold voltage used in defining propagation delay, find t_{pLH} and t_{pHL} with the gate loaded by a 0.5 pF capacitor.

L 13.27 Consider an inverter for which $\mu_n C_{ox} = 2\mu_p C_{ox} = 20\mu A/V^2$, $(W/L)_n = 8\mu m/2\mu m$, $(W/L)_p = 16\mu m/2\mu m$, $V_{tn} = 0.8$ V, $V_{tp} = -1.6$ V, and $V_{DD} = 5$ V. What values of V_{OH}, V_{OL}, V_{IL}, V_{IH}, and V_{th} apply. For the latter threshold used to define propagation delay, find t_{pLH} and t_{pHL}, for this inverter loaded by a 0.5 pF capacitor.

SECTION 13.6: CMOS Gate Circuits

D 13.28 Using the general idea presented in P13.50 of the Text, and assuming the availability of both true and complemented variables (i.e. $A, \bar{A}, B, \bar{B}$ etc.), find a CMOS circuit to implement $Y = (A + B)\, C\, \bar{D}$.

D 13.29 For a technology in which all devices have a channel length of 5μm and in which symmetrical inverter employs $(W/L)_n = 2 = \frac{1}{2}(W/L)_p$, find appropriate device sizes for the circuit in P13.28 above to provide transient switching ability similar to that of the basic inverter.

13.30 For the 2-input NOR circuit of Fig. 13.21a of the Text, in which $\mu_n = 2\mu_p$, $K_n = K$, $V_{tn} = -V_{tp} = 1$ V and $V_{DD} = 5$V, find the value of K_p such that the threshold for a single input active is 2.50 V. What is the corresponding threshold for both inputs tied together?

D 13.31 The circuit shown, illustrates the existence of yet other possibilities for the design of flexible logic functions using CMOS devices.

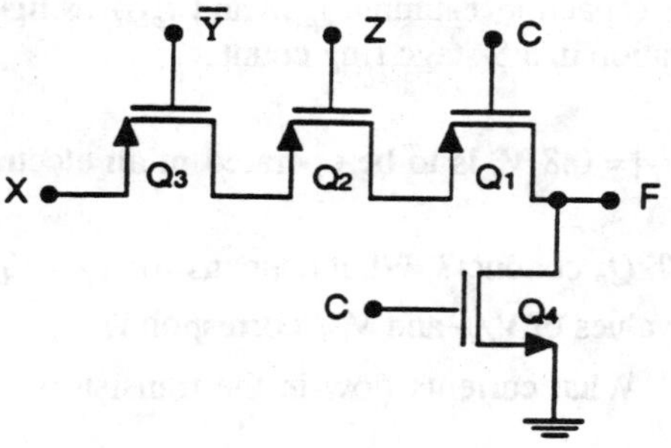

It combines aspects of transmission-gate logic and conventional CMOS logic, with neither in its complete form. In usual applications, the signal C acts as a clock, being normally held high while X, Y and Z change. Finally, C falls while X, Y, C are stable. For signals having 0 V and 5 V levels and using the positive logic convention (where logic '1' is high), prepare a truth table for F. Express F as a function of C, X, Y, Z. In an application in which variable $\bar{X}$ is available, but X is not, suggest a means to provide the required function without adding a (complete) additional inverter.

For dynamic operation of the original circuit equivalent to that of a basic symmetric inverter with $(W/L)_n = 10\mu m/5\mu m$, chose appropriate device widths on the assumption that all device lengths are 5μm.

13.32 Following the direction suggested in P13.57 of the Text, use the data provided to estimate supply current and power dissipation for operation at $V_{DD} = 5$ V and 10 MHz with $C_L = 30$ pF. For your analysis, assume $|V_t| = 1$ V.

13.33 In a buffered-logic family for which minimum overall size and input capacitance is important, the inverter, in its input stage, uses the smallest possible n-channel device of unit area, with a p-channel device of twice its width. In the 2nd and 3rd stages, corresponding devices are each 3 × larger. What is the total area of the buffered inverter? What is the total area of a buffered 4-input NOR whose first stage is made basic-inverter-compatible? What is the input capacitance relative to the inverter? A second design is considered with the same stage-size ratio (i.e. 3×) but using minimum-size input inverters and an intermediate-stage NAND. What is its total area? What is the relative input capacitance presented?

SECTION 13.7: Latches and Flip-Flops

13.34 Two CMOS inverters operating from a 5 V supply, each of which have V_{IL} and V_{IH} of 2.40 V and 2.90 V, respectively, with corresponding outputs of 4.7 V and 0.4 V, are connected as a latch. Approximating the corresponding transfer characteristic of each gate by a straight line between threshold points, sketch the latch open-loop transfer characteristic. At what points are the open-loop input and output voltages equal? What is the loop gain at each?

13.35 Sketch the circuit of a NAND *SR* flip-flop using CMOS, and prepare a truth table whose entries are in terms of stable output voltages available with a 3 V supply, and devices for which $|V_t| < 3/2$ V.

13.36 Consider the *SR* flip-flop of Figure P13.27 of the Text. Let $V_{DD} = 5$ V, $V_t = 1$ V, $V_p = -2$V, with all devices of the same length *L* and devices Q_1 through Q_4 of the same width *W*. What must the width of Q_5 and Q_6 be, so that the *SR* threshold is at 2.5 V? 2.0 V?, 3.0 V? Ignore the body effect.

SECTION 13.8: Multivibrator Circuits

13.37 For the monostable circuit of Fig. 13.31 of the Text, whose waveforms are given in Fig. 13.34 and an expression for *T* given in Exercise 13.24 (on page 954), let $V_{DD} = 5$ V, $V_{th} = 0.6\ V_{DD}$, $R = 22\text{k}\Omega$, $C = 1500$ pF and $R_{on} = 180\Omega$. Find the values of T, ΔV_1, and ΔV_2. By how much does υ_{01} change during the interval *T*? What are the peak sink and source currents of G_1? Use $\upsilon_D = 0.7$ V.

13.38 In a particular CMOS implementation of Fig. 13.31 of the Text, G_2 is a simple inverter and G_1 a NOR, both of which use minimum-sized devices for which $(W/L) = 2$. For this process, $|V_t| = 1$ V, $\mu_n C_{ox} = 2\mu_p C_{ox} = 20\mu A/V^2$, and $V_{DD} = 5$ V. The function of *R* is implemented using a simple current mirror employing two minimum sized p-channel devices and a grounded-source diode-connected n-channel device of 10 times the minimum length. Find the value of *C* for a 10 μs output pulse, accounting for the non-zero value of V_{OL} of G_1 and V_{th} of G_2.

D 13.39 For the astable multivibrator modified as suggested in P13.71 of the Text, provide a design for operation at 1MHz, using a 100 pF capacitor, $V_{DD} = 5$ V, and $V_{th} = 0.44\ V_{DD}$. What values of resistors would you use?

SECTION 13.9: Random-Access Memory (RAM)

13.40 A 1 M-bit memory chip is organized using 4 square arrays, each of which uses a simple NOR decoder for row and column selection. How many inputs would each decoder need? For a bit address consisting (from its left) of array-number bits, row bits and column bits, what are the column-address digits of bit 102,476 on the chip?

13.41 For the NOR address decoder, part of which is shown in Fig. 13.40 of the Text, draw row 13, indicating the connection of its transistors to the first 4 address lines. How many transistors, including the load, are connected to each row line of a 256 K-bit square array?

D*L 13.42 A proposed CMOS static RAM uses cells such as that shown in Fig. 13.41b of the Text, having a cell supply voltage of 5V and word-line selection voltages of 0 at rest, 5 V for reading, and 5 V for writing. Digit-line voltages are precharged to 2.5 V for reading, and 0 and 5 V for writing. The cell itself uses minimum-size devices for which $(W/L) = 2\mu m/3\mu m$ and $|V_t| = 1$ V with $\mu_n C_{ox} = 2.5\mu_p C_{ox} = 25\mu A/V^2$. What is the threshold voltage at the drain of Q_1 or Q_2 at which the cell will change state? What currents supplied to or from the cell, move the cell output voltage half way from its stable state to the threshold? What must the width and length of Q_5 (and Q_6) be to ensure that readout is nondestructive? What currents can be supplied during the writing process by such a device? Is the design viable? Why or why not?

13.43 In a particular dynamic-RAM technology, cell leakage currents can be reduced to 10 fA (1 femp-to-ampere $= 10^{-15}A$). What is the minimum allowable capacitor for a 4 ms refresh interval with a recoverable cell-voltage loss of 1.5 V?

SECTION 13.10: Read-Only Memory (ROM)

D 13.44 Design a bit pattern to be stored in a (14 × 5) ROM which provides the results of division of one two-bit number, *X*, by another, *Y*. The 4-bit word address is to be (x_1, x_0, y_1, y_0). The output is to be (f, q_1, q_0, r_1, r_0) where *F* is the 1-bit overflow (divide-by-0) flag. *Q* is the 2-bit quotient, and *R* is the 2-bit remainder. Give a circuit implementation resembling that in Fig. 13.45, but in which an installed transistor represents a logic 1 internally. (Hint: While this saves ROM transistors, it requires additional inverters). Excluding the input decoder, how many transistors do you need in the heart of the decoder? How many are used in inverters? How many transistors would be required in total, without using the extra inverters? How many transistors would be required for the 4-bit input decoder using the circuit of Fig. 13.40?

SECTION 13.11: Gallium Arsenide Digital Circuits

13.45 Repeat the *dc* analysis of a DCFL NOR gate as sketched in Example 13.4, for the conditions stated there, except with a reduction of the load MESFET width to 5μm. Find V_{OH}, V_{OL}, V_{IL}, V_{IH}, NM_H and NM_L.

13.46 For the situation described in P13.45 above, find the average supply current, the average static power dissipation, the average propagation delay for a 30 fF equivalent load, the dynamic power loss at 2GHz, and the corresponding delay-power product.

L* 13.47 For the *FL* gate of Fig. 13.53 of the Text, whose transfer characteristic is shown in Fig. 13.54 on page 972 of the Text, evaluate the sensitivity of V_{OH}, V_{OL}, V_{IL}, V_{IH}, NM_L, and NM_H to some of the various device parameters. For the nominal design, all $L = 1\mu m$, $W_S = W_L = 20\mu m$, $W_{PD} = 10\mu m$, β (per μm) $= 10^{-4}$ A/V^2, $V_{tD} = -0.9$ V, $\lambda = 0$, $V_{DD} = +3$ V, $V_{SS} = 2$ V. In particular, consider (separate) changes of a) all V_{tD} to −0.8 V, b) all V_{tD} to −1.0 V, c) $W_L = 2W_{PD}$ to 5μm. Note that this design choice will also make propagation delays quite assymetric.

D 13.48 Using the FET-sizing ideas used in the *FL* design of Fig. 13.53, provide a directly-corresponding design for the SDFL NOR in Fig. 13.55b. Using the results for the *FL* design, find V_{OL}, V_{OH}, V_{IL}, V_{IH} and noise margins for your SDFL gate. Note that the design requested, for which the device size ratios correspond directly to those used for Fig. 13.53, will have a problem with fanout. What can be done to increase the fanout from 1 to 4?

Chapter 14

BIPOLAR DIGITAL CIRCUITS

SECTION 14.1: The BJT as a Digital Circuit Element

14.1 Consider P14.2 of the Text for a) $\alpha_F = 0.999$ and b) $\alpha_F = 0.99$. Find i_C and i_E and the voltage drop to ground.

14.2 Consider the possibility that a BJT operating with collector open, and a somewhat-variable base current, can be used as a (constant) voltage regulator for base currents varying from 1 to 4 mA. What value of β_{forced} applies? For a junction-area ratio of 10 to 1, and β_F ranging from 70 to 280, what range of collector-to-emitter voltages result? Note that for a particular transistor, this voltage is relatively constant for varying base current!

14.3

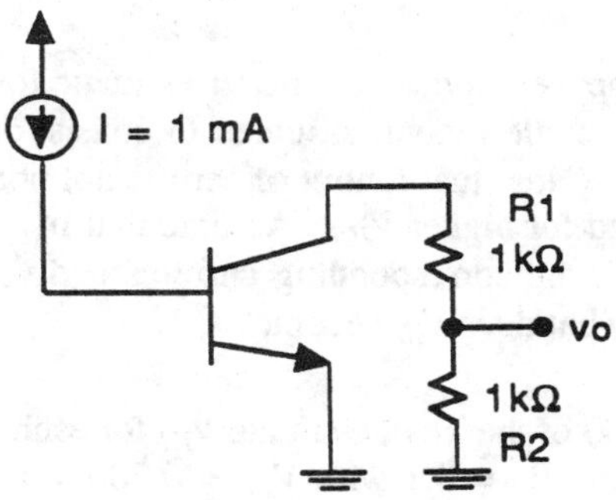

A particular BJT for which $\beta_F = 100$, when operated as suggested in P14.2 above with open collector and 1 mA base current, has a collector-to-emitter voltage of 100 mV. What is the output voltage, v_O, of the circuit shown? For $I = 1$ mA and $R_1 = R_2 = 500\Omega$, what does v_O become? What is the equivalent $R_{CE\,sat}$ of the device itself?

14.4 Considering Table 14.1, on page 994 of the Text, prepare a table for the same transistor but with emitter and collector roles interchanged. What is the limiting value for forced β? Find table entries at 90%, 50%, 20%, 10% and 1% of this value, as well as for $\beta_{forced} = 0$.

14.5 Prepare a table of saturation voltage versus β_{forced} for an npn transistor for which $\alpha_F = 0.995$ and the collector-junction area is 5 times that of the emitter. What is the limiting value of forced β? Tabulate $V_{CE\,sat}$ for β_{forced} values as suggested in P14.4 above. For $I_B = 10$ mA and $I_C = 1$ mA, what voltage exists between collector and emitter with the transistor operated in the normal mode? inverted?

14.6 For a grounded-emitter pnp transistor for which $\beta_F = 200$ and $\beta_R = 2$ operated in a circuit for which $I_B = 1$ mA and $\beta_{forced} = 10$:

a) Calculate and label all currents in the branches of the *EM* model (with diodes etc. reversed from Fig. 14.1b).

b) For $I_S = 10^{-14}$ A, find the voltages across the two junctions and $V_{EC\,sat}$.

c) Verify $V_{EC\,sat}$ using Eq. 14.16.

14.7 A BJT for which the storage time constant is 20 ns and $\beta = 200$, is operated in a circuit for which $I_{C\,sat} = 10$ mA and the base turn-on current, I_{B2}, is 1 mA. Calculate the storage delay under the conditions that base turn-off current, I_{B1}, is a) 0 mA, b) 1 mA, c) 10 mA.

14.8

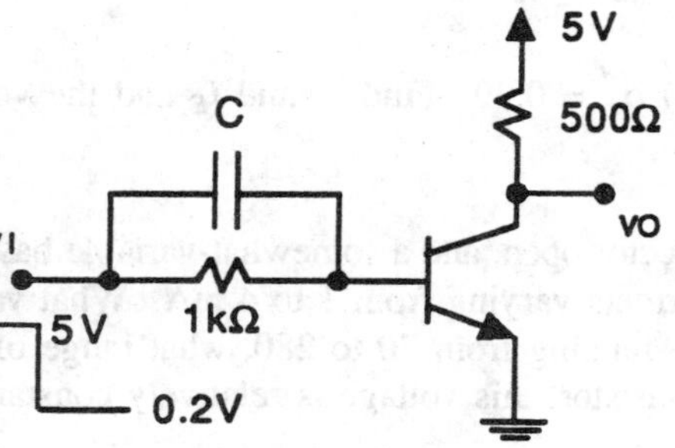

In the following circuit, t_s is measured for various values of C: For $C = 0$, $t_s = 80$ ns; for $C = 8$ pF, $t_s = 30$ ns. Estimate values of τ_s and β if $V_{BE} = 0.7$ V and $V_{CE\,sat} = 0.2$V. What value of C would you chose to reduce t_s to zero? For β twice the present value, what would t_s be with the capacitor you have chosen? What would it have been with $C = 0$?

SECTION 14.2: Early Forms of BJT Digital Circuits

14.9 With reference to Fig. 14.8 and P14.15 on page 1050 of the Text, provide a better estimate for V_{IL}, as that voltage υ_I for which the gain is -1 V/V for a gate with a single fanout, assumed to consist of 450Ω connected to the low-impedance base of a saturated transistor. (Note that fanout of zero is not normal in logic applications, and 2 or more lowers the gain, implying a need for higher V_{IL}). Assume that $\upsilon_{BE} = 0.700$ V at $i_C = 1$ mA and that $\beta = 50$. (Hint: find the required r_π and the corresponding currents and voltages.) For two inputs operating simultaneously (for example joined), what does V_{IL} become?

14.10 For the NAND logic gate shown in Fig. P14.17 on page 1050 of the Text, evaluate V_{IH} for each of the inputs under the condition that for V_{IH}, $\beta_{forced} = \beta_F/2$ and $\beta_F = 50$, $\beta_R = 0.1$ with $V_{BE} = 0.700$ V at $i_E = 1$ mA. (Since α_R is quite small, V_{BE} in saturation can be calculated approximately using the external emitter current).

DL 14.11 Your boss is considering the possibility of raising the input threshold of the DTL circuit of Fig. 14.10. She asks you to help, by calculating, with input 3 high:

a) The base current in Q required to lower the output to $V_{DD}/2$. Use $\beta = 30$.

b) The input threshold voltage V_{th} (at A) corresponding. All junctions have a 0.7 V drop at 1 mA.

c) The maximum current available to Q when A is high

d) The input current flowing from D_1 when A is at 0 V

e) Now, she suggests that you add a diode and change a resistor so that V_{th} is raised by about 0.7 V, while the maximum base drive remains the same. What change do you make? What is the input current needed now when $V_A = 0$? What is the fanout available with this redesign, for which β_{forced} $\leq \beta/2$ is maintained?

D 14.12 Following the general direction suggested by the two-input NAND DTL gate in Fig. 14.10 of the Text, sketch a circuit using only a single transistor to provide the function $Y = \overline{ABC + D + EF}$ using resistor values the same as those in Fig. 14.10. What is the base current which results when D alone is high? When all inputs are high?

SECTION 14.3: Transistor-Transistor Logic (TTL or T²L)

D   14.13 Modify the design of the *IC*-form DTL gate in Fig. 14.11 to double the turnoff current of Q_3. Using the results of Ex. 14.8 on page 1004 of the Text, what does the turnon base current of Q_3 become? What is the absolutely greatest fanout N that ensures that Q_3 is barely saturated? What does V_{OL} become for ¼ that value of fanout? Use $\beta_R = \beta_F/100$, $\beta_F = 50$.

14.14 For the modification and current levels suggested in P14.13 above, and $\tau_s = 10$ ns, what does the storage delay become for $N = 0$? for N equal to ¼ of the maximum fanout? Recalculate these values in the event that β is doubled (from 50 to 100), but with the N values kept the same.

14.15 For the situation described in P14.25 of the Text, what is the minimum value of external load resistance connected to +5 V, that still ensures saturation? For a resistor of twice that value, calculate the 10% to 90% rise and fall times of the output with a 10 pF load capacitor.

D*L 14.16 Consider the circuit of Fig. 14.13 of the Text as the basis of a very-low-voltage BJT logic structure. For V_{BE} = 0.7 V normally, 0.6 V at turnon, and 0.2 V in saturation, with $\beta_F = 40$ and $\beta_R = 0.1$, prepare a design meeting the following specifications: a) the total collector current that can be sustained by Q_3 with $\beta_{forced} = \beta_F/2$ is ≥ 20 mA, b) resistors of a single value, R, are used, c) $V_{CC} = $ V is chosen as small as possible, d) $NM_H \geq 1.5\ NM_L$ with fanout $N = 10$ for which $\beta_{forced} \leq 20$. For what value of N does Q_3 reach the edge of saturation with υ_I high?

D   14.17 Extend the structure of the logic gate of Fig. 14.13 of the Text to provide a logic gate to perform the following function: $y = \overline{AB + CD}$ while maintaining the same input thresholds. Use 6 transistors and 3 resistors in total.

14.18 A manufacturing-process deviation in the production of T²L gates, using the circuit of Fig. 14.23 of the Text, reduces current gain such that $\beta_F = 9$ and $\beta_R = 0.05$. For input high, estimate all node voltages and branch currents, for $V_{BE} = 0.7$ V, and a load of 1kΩ connected to the 5 V supply. What is the largest possible fanout (excluding the 1kΩ load), for which saturation of Q_3 is still possible?

14.19 Repeat the analysis suggested in P14.18 above, with input low (at 0.3 V) and a resistor of 1 kΩ connected from the output to ground.

L*   14.20 For the situation described in P14.18 above, and with input and output joined by a 200Ω resistor, estimate all the node voltages and branch currents. Calculate $V_{CE\ sat}$ for Q_1 relatively precisely using a negative value for β_{forced}.

L*   14.21 Modify your response to P14.20 above for the situations in which a) a load resistor of 200Ω is connected from the output to: a) ground, b) +5 V.

SECTION 14.4: Characteristics of Standard TTL

D   14.22 Using a similar analysis style to that found following Eq. 14.23 on page 1017 of the Text, find the value of R_2 which raises point C of Fig. 14.27b) to 3.0 V. What does the slope of the BC segment become? If R_2 is to be kept at 1kΩ by a desire to maintain the turnoff current level that R_2 provides, what change in R_1 would be needed to raise C to 3.0 V? What change in turnon current to Q_3 does this produce (in absolute value and as a percentage)? What is the effect on gate storage delay? And so you see, once again, the nature of compromise in real design!

14.23 Using the data provided in the answers to Exercise 14.15on page 1018 of the Text, find the noise margins that apply at the interface between two sets of T^2L logic gates, one operating at −55°C and the other at 125°C. Note that there are two pairs of margins depending on the relative temperatures of the driving and driven gates.

14.24 For the circuit of Fig. 14.30 extended to have 4 *OR* inputs, what is the maximum base current provided to Q_3? In a particular circuit with values shown, and no load, with 3 of the 4 *OR* inputs already low, the storage delay is 10 ns. What delay would you expect when all 4 inputs are brought low simultaneously?

14.25 For the tristate gate shown in Fig. 14.20, find the voltage at the tristate input at which the collector current in Q_6 just reaches 1 mA. For all junctions, the voltage drop is 0.700 V at 1 mA, $\beta_F = 50$ and $\beta_R = 0.1$.

14.26 For the tristate gate shown in Fig. 14.20, estimate the possible current flow in the base of Q_3 if the connection from the tristate input is low, but its link to Q_1 is broken. Use $\beta_F = 50$, $V_{BE} = 0.7$ and $V_{CE\,sat} = 0.2$ V.

SECTION 14.5: TTL Families with Improved Performance

L 14.27 A Schottky npn transistor consists of 2 elements: a BJT for which $I_E = 1$ mA at $V_{BE} = 0.75$ V with $n = 1$ and $\beta = 50$, and a Schottky diode for which $I = 1$ mA at 0.5 V with $n = 1$. For emitter grounded, find the base and collector voltages and transistor base and diode currents for input and load currents, respectively, of

a) 1 mA, 0 mA,

b) 1 mA, 1 mA,

c) 1 mA, 10 mA,

d) 10 mA, 10 mA,

e) 10 mA, 1 mA.

L 14.28 For the low-power Schottky TTL gate of Fig. 14.35, find the current that flows in the power supply with a) inputs both high, or b) inputs both low, both for the output i) open-circuited or ii) short-circuited to ground. What is the power dissipated in the gate under all 4 conditions? For a propagation delay of 10 ns average, what is the delay-power product for operation with a 10 pF load at 30 MHz?

SECTION 14.6: Emitter-Coupled Logic (ECL)

L 14.29 Reconsider P14.48 of the Text, for the situation in which V_{IL} and V_{IH} are based on currents ranging from 0.95I to 0.05I.

14.30 Consider Fig. P14.48 on page 1053 of the Text, for the situation in which $V_{BE} = 0.75$ at current $I = 4$ mA. Find R so that $V_{th} = -1.32$ V. What are the values of V_{OH} and V_{OL} that result? Find V_{IH} and V_{IL} for a current splitting in Q_R and the input transistor in the ratio of 1000 to 1. What are the corresponding noise margins?

D 14.31 Modify the circuit of Fig. P14.48 on page 1053 of the Text to create an ECL-to-T^2L converter by connecting the upper supply connections (now grounded) to +5 V, and adding a number of 0.75 V diodes in series with the emitters of Q_2 and Q_3 (to lower the output voltage). Maintain the input threshold at −1.32 V by a suitable choice of $R/2$ with $I = 4$ mA, and corresponding $V_{BE} = 0.75$V. Select the resistors (called R_1) connected to the bases of Q_1 and Q_2, and the number of diodes, N, to meet T^2L worst-case specifications, namely $V_{OL} = 0.5$V and $V_{OH} = 2.7$ V, with symmetrical margins, while keeping R_1 as small as possible.

14.32

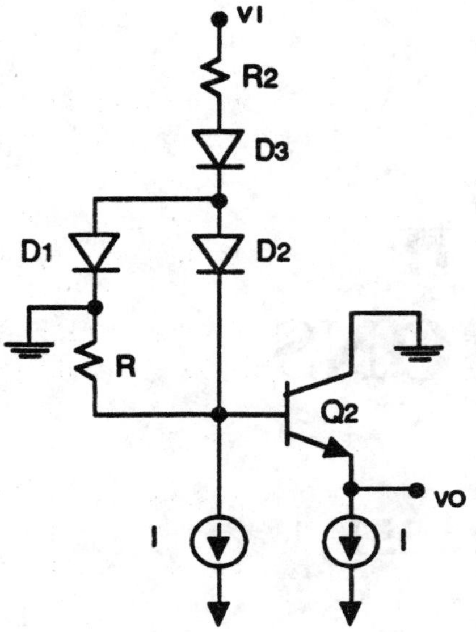

Consider the following circuit as a T^2L-to-ECL converter: Use values of I and R as specified and calculated in P14.30 above. Arrange that the current from the T^2L gate is 8 mA when its output is at the minimum specified value of output, ie $V_{OH} = 2.7$ V. What is the T^2L output current at $V_{OL} \leq 0.5$ V? What are V_{OL} and V_{OH} of the converter circuit? Use 0.75 V for all junctions, when operated at 4 mA.

14.33 For the circuit of Fig. 14.40 of the Text, a manufacturing error reduces the junction size of Q_2 by a factor of 2 and its β to 30. What is the corresponding effect on NM_H?

14.34 For signals whose rise and fall times are 1.2 ns, what length of unterminated gate-to-gate interconnect can be used if a ratio of rise time to return time of 6 to 1 is required. Assume that the signal propagates at 2/3 the speed of light (which is 0.3 mm/ps).

SECTION 14.7: BiCMOS Digital Circuits

L* 14.35 For the circuit of Fig. 14.47 of the Text with $K_1 = \frac{1}{2} K_2, |V_t| = 1$ V, with $R_1 = R_2 = r_{DS1}$ at small V_{DS1} and $V_{GS1} = V_{DD}$, $V_{BE} = 0.7$ V, and $V_{DD} = 5$ V, find V_{th}, V_{OH} and V_{OL} with a) no load, b) a 5 kΩ load to 2.5 V. Assume $\beta = 100$, $V_{BE} = 0.7$ V, and $K_2 = 200\mu A/V^2$.

D 14.36 Following the general direction indicated in Fig. 14.48 of the Text, provide a circuit sketch of a 2-input BiCMOS NOR gate. For Q_1 and Q_2 of the basic inverter having the same W/L ratio, what are the device ratios in an equivalent NOR gate? For one input low, what is the threshold voltage for the other input, the voltage at which two of the 3 active FETs operate in saturation?

PART II
SOLUTIONS

pages 111 to 351

Chapter 1

INTRODUCTION TO ELECTRONICS

SECTION 1.1: Signals

1.1 **Results**: (See "Rough Work/Notes" at the end).

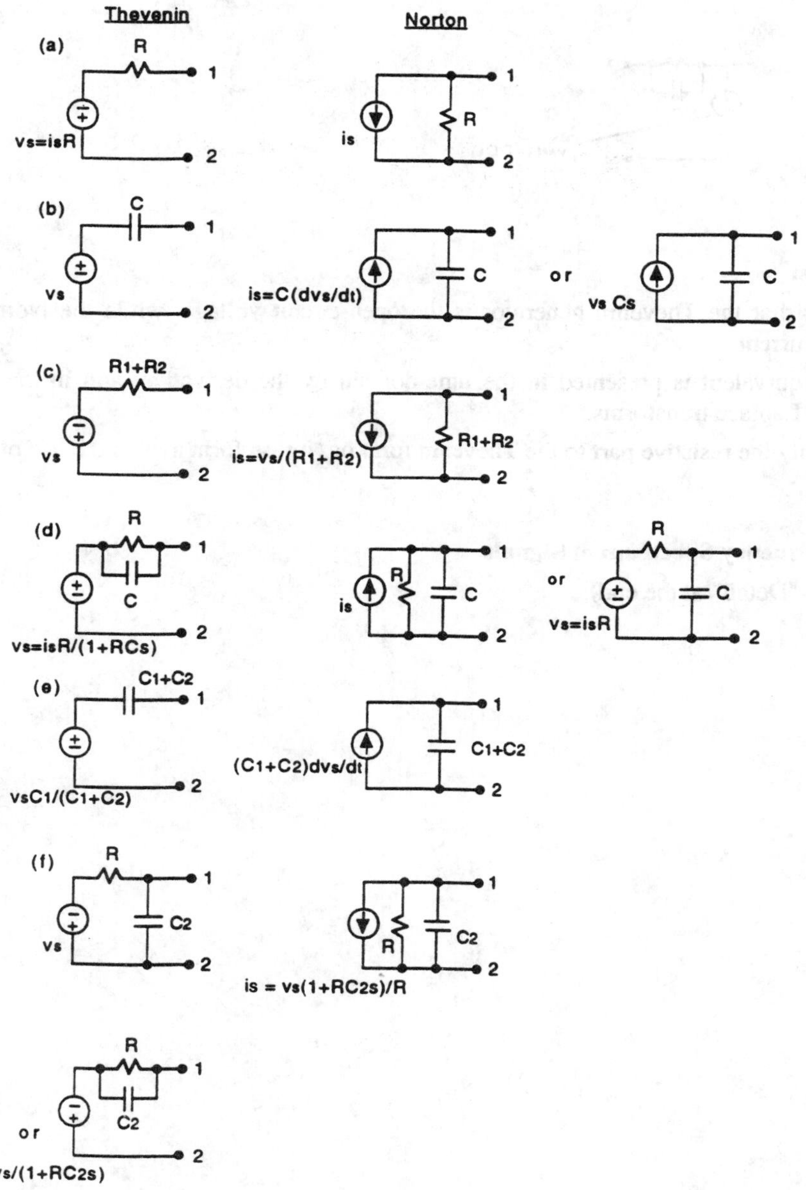

1.1

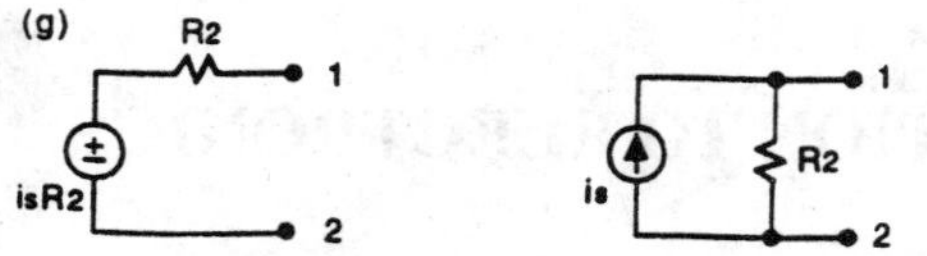

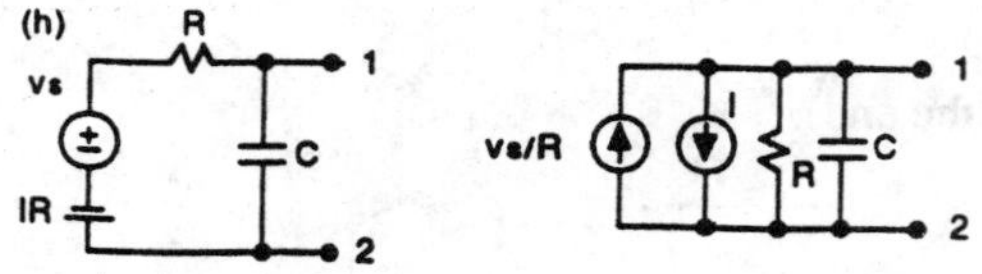

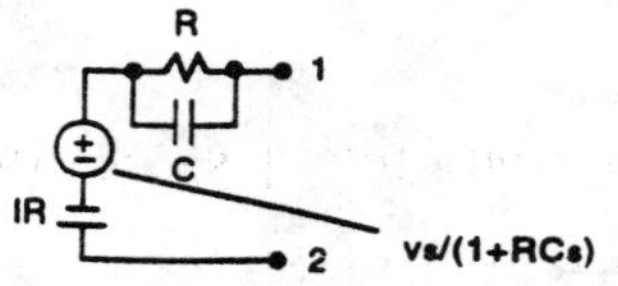

Rough Work/Notes:

a) Simply recall that the Thevenin generator is the open-circuit voltage, while the Norton generator is the short-circuit current:

b) The Norton equivalent is presented in the time-domain by the derivative, and in the complex frequency domain using Laplace transforms.

d) Converting only the resistive part to the Thevenin form or Norton form avoids the use of complex frequency notation.

SECTION 1.2: Frequency Spectrum of Signals

1.2 **Results:** (See "Detail" at the end).

Line Label	Frequency (Hz)	Frequency (rad/s)	Period (s)	Period
a	60	**377**	1.67×10^{-2}	16.7 ms
b	**120.0**	754	8.33×10^{-3}	8.33 ms
c	**400.0**	2513.3	2.50×10^{-3}	2.50 ms
d	1.01×10^{6}	**6.346×10^{6}**	9.90×10^{-7}	990 ns, 0.99μs
e	97.3×10^{6}	**611.4×10^{6}**	1.03×10^{-8}	10.3 ns
f	1	**6.28**	1.00	1.00 s
g	**60.0**	377	1.67×10^{-2}	16.7 ms
h	**0.159**	1	6.29	6.29 s
i	10^{9}	**6.28×10^{9}**	1.00×10^{-9}	1.00 ns
j	4×10^{11}	**25.1×10^{11}**	2.50×10^{-12}	2.50 ps

Detail:

In general, $\omega = 2\pi f$, 2π rad/s ≡ 1 Hz ≡ 6.283 rad/s, and 1 rad/s ≡ 0.159 Hz.
Note that the rightmost columns are the results of P1.3.

1.3 **Results**: These are tabulated in the two rightmost columns of the table above.

Examples:

a) For 60 Hz, period = $\frac{1}{60}$ = **$0.016\dot{6}$ seconds** = 1.66×10^{-2} s = 16.6 or **16.7 ms.**

g) For 377 rad/s, corresponding to $\frac{1}{377}$ = .00265 sec/rad, period = 2π rad = $2\pi \times .00265$ = **.01665** s ≡ **16.7 ms**.

j) For 400 GHz, period = $1/(400 \times 10^9) = .0025 \times 10^{-9} = 2.50 \times 10^{-12}$ = **2.50 ps.**

Conclusion: Clearly, dealing with frequency is easier, either directly from the specification in Hz, or from the tabulated calculation derived from rad/s.

1.4 $\Delta T = 50 - 25 = 25\text{C}°$.

Period of a 10.7 MHz wave = $1/(10.7 \times 10^6) = 93.46$ ns.

Total variation in period expected would be $93.46 \times 10^{-9} \times 3 \times 10^{-6} \times 25 = 7.009 \times 10^{-12} \equiv$ **7.01 ps.**

1.5 a) 0.2 V peak-to-peak ≡ 0.1 V peak ≡ $0.1/\sqrt{2} = 0.0707$ Vrms.

1000 Hz ≡ $2\pi \times 10^3$ rad/s = 6.28×10^3 rad/s, with a period of 1/1000 = 1 ms.

b) 2.12 Vrms, 20μsec period.

Amplitude ratio, $\dfrac{V_b}{V_a} = \dfrac{2.12}{.0707} = 29.98$, or about **30 times.**

Frequency ratio, $\dfrac{f_b}{f_a} = \dfrac{1}{\textit{Period ratio}} = \dfrac{T_a}{T_b} = \dfrac{1\times10^{-3}}{20\times10^{-6}} =$ **50 times.**

c) 1.0 V peak amplitude, 12.57 rad/s frequency.

Amplitude ratio, $\dfrac{V_c}{V_a} = \dfrac{1.00}{0.1} =$ **10 times.**

Frequency ratio, $\dfrac{f_c}{f_a} = \dfrac{12.57}{2\pi \times 10^3} = 2\times10^{-3} = \dfrac{2}{1000} =$ **1/500 times.**

1.6 Fourier series for a square wave of frequency f and 10 V peak amplitude is (from Eq. 1.2 of the Text):

$\upsilon(t) = 4(10)/\pi$ (sin $2\pi ft$ + 1/3 sin $3(2\pi ft)$ + 1/5sin $5(2\pi ft)$ + 1/7sin $7(2\pi ft)$ + 1/9sin $9(2\pi ft)$ + 1/11 sin 11 $(2\pi ft)$+...).

Energy per unit time in a voltage wave $\upsilon(t)$ of duration τ, across a unit load = $\dfrac{1}{\tau}\int_0^\tau \upsilon^2(t)dt$.

In one cycle of the square wave (period $T = 1/f$), associated energies are proportional to:

a) For the square wave: $(10^2)\, T = 100T$ Ws.

b) For the fundamental (first harmonic): $\left[\dfrac{40}{\pi}\dfrac{1}{\sqrt{2}}\right]^2 T = \left[\dfrac{40}{\pi}\right]^2 T/2,$

third harmonic: $\left[\dfrac{1}{3}\right]^2\left[\dfrac{40}{\pi}\right]^2 T/2,$

fifth harmonic: $\left[\dfrac{1}{5}\right]^2\left[\dfrac{40}{\pi}\right]^2 T/2,$

seventh harmonic: $\left[\dfrac{1}{7}\right]^2\left[\dfrac{40}{\pi}\right]^2 T/2,$

ninth harmonic: $\left[\dfrac{1}{9}\right]^2\left[\dfrac{40}{\pi}\right]^2 T/2.$

For the first 9 harmonics: $\left[\frac{40}{\pi}\right]^2 \frac{T}{2}\left[1+\frac{1}{9}+\frac{1}{25}+\frac{1}{49}+\frac{1}{81}\right]$

$$= 1.184 \times \left[\frac{40}{\pi}\right]^2 \frac{T}{2} = 95.95\ T\,\text{Ws}.$$

c) Above the ninth harmonic, total energy is proportional to $100\ T - 95.95\ T = 4.05\ T$, corresponding to $\frac{4.05T}{100T} \times 100 =$ **4.05%** of the total. (See page 152 following Chapter 2, for a graphic view of this).

d) At and above the 3rd harmonic, the total energy is proportional to $100T - \left[\frac{40}{\pi}\right]^2 T/2$.

Of the total, this is: $\dfrac{100T - (\frac{40}{\pi})^2 \frac{T}{2}}{100\,T} \times 100 = \left[100 - \left[\frac{40}{\pi}\right]^2 \frac{1}{2}\right] =$ **18.9%**.

1.7 For a square wave of amplitude V, $\upsilon = \frac{4V}{\pi}$ (sin ωt + 1/3 sin $3\omega t$ + 1/5 sin $5\omega t$ ...).

Assume the pass band includes both fundamentals, i.e., $f > 2$ kHz, and *totally* excludes energy outside the band. For cutoff at $f = 4$ kHz, power levels for unit loads are:

$$P_1 = \left[\frac{4\times 1.1}{\pi}\right]^2 \left[1^2 + (1/3)^2\right] = 1.96 \times 1.111 = 2.18\ \text{W, and}$$

$$P_2 = \left[\frac{4\times 1.2}{\pi}\right]^2 \left[1^2\right] = 2.33\ \text{W, where } P_2 - P_1 = 0.15\ \text{W}.$$

For cutoff at $f = 5+$, between 5 and 6 kHz,

$$P_1 = 1.96\left[1^2 + 1/3^2 + 1/5^2\right] = 2.26\ \text{W},$$

$$P_2 = 2.33\ \text{W, where } P_2 - P_1 = .07\ \text{W}.$$

For cutoff at $f = 8$ kHz,

$$P_1 = 1.96\left[1 + (1/3)^2 + (1/5)^2 + (1/7)^2\right] = 1.96\,(1.172) = 2.29\ \text{W},$$

$$P_2 = 2.33\left[1 + (1/3)^2\right] = 2.33\,(1.11) = 2.59\ \text{W, where } P_2 - P_1 = 0.29\ \text{W}.$$

See that the closest one can get to equal power is for filtering at a frequency **between 5 kHz and 6 kHz,** where the difference is 0.07 W, or $\frac{0.07}{(2.26 + 2.33)/2} \times 100$, or 3% of the average energy level.

SECTION 1.3: Analog and Digital Signals

1.8

1 Vrms corresponds to $\sqrt{2}$ V or 1.414 V peak.

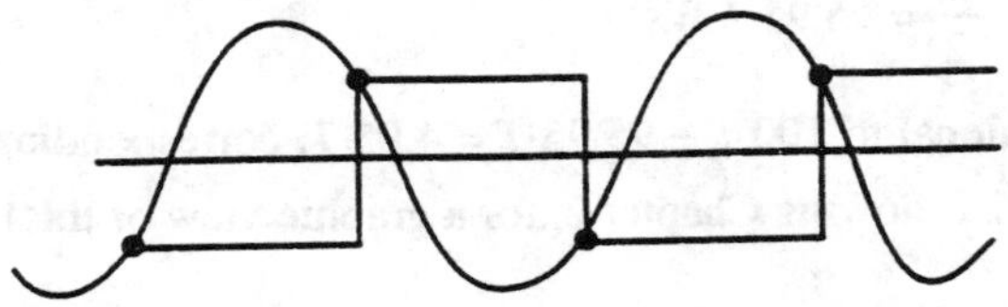

a) For sampling at the peaks, the square wave amplitude would be **1.414 V peak** or 2.818 Vpp.

b) 90° from a negative-going zero crossing is at a negative peak. The next sample is at the positive peak. The result is the same as a).

c) 45° from a positive-going zero crossing, the amplitude is $\sqrt{2}\sin 45° = 1.00$ V. The result would be a square wave of amplitude **1V** and frequency f.

For case a) and sampling at other frequencies:

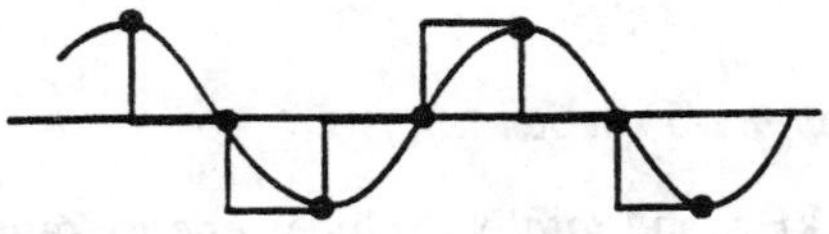

i) For sampling at $2(2f) = 4f$ Hz, the result is a sequence of positive and negative pulses of amplitude $\sqrt{2}$ **V**, width **1/4f**, spaced **1/4f** apart.

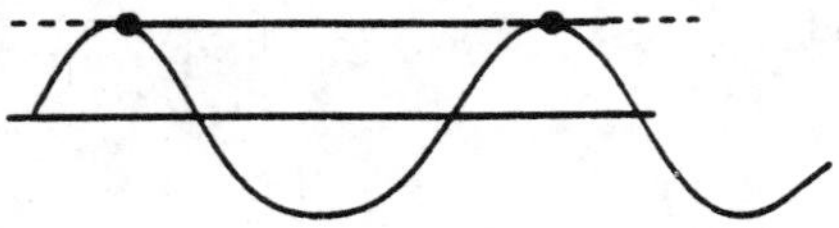

ii) For sampling at $\frac{1}{2}(2f) = f$ Hz, the result is a dc level of $\sqrt{2}$ V = **1.414 V**.

SECTION 1.4: Amplifiers

1.9

b) No dc connection to ground implies $I+ = I- =$ **1 mA**:

$P = 2\,(1\text{ mA} \times 10\text{ V}) = \mathbf{20\ mW}$.

$$i_i = \frac{\upsilon_i}{R_{in}} = \frac{20\times10^{-3}}{.01\times10^{3}} = 2000\mu\text{A} = 2\times10^{3}\mu\text{A} = \mathbf{2\ mA}.$$

$$P_{in} = \frac{20}{\sqrt{2}} \times \frac{2\times10^{3}}{\sqrt{2}}\text{ nW} = \mathbf{20\ \mu W}.$$

$$i_o = \frac{\upsilon_o}{R_L} = 1/1 = \mathbf{1\ mA}.$$

$$P_{out} = \frac{1}{\sqrt{2}} \times \frac{1}{\sqrt{2}} \times 10^{-3} = \mathbf{0.5\ mW}.$$

$$A_\upsilon = \frac{1}{20\times10^{-3}} = 50\text{ V/V} = .05\text{ V/mV} \equiv \mathbf{34\ dB}.$$

$$A_i = \frac{i_o}{i_i} = \frac{1\times10^{-3}}{2\times10^{3}\times10^{-6}} = 0.5\text{ A/A} = 0.5\times10^{-3}\text{ mA/}\mu\text{A} \equiv \mathbf{-6\ dB}.$$

$$A_p = \frac{P_{out}}{P_{in}} = \frac{0.5\times10^{-3}}{20\times10^{-6}} = 25\text{ W/W} = .025\text{ mW/}\mu\text{W} \equiv 10\log 25 = 14\text{ dB, i.e., } \frac{34-6}{2} = \mathbf{14\ dB}$$

$$Eff = \frac{P_{out}}{P} = \frac{0.5 \times 10^{-3}}{20 \times 10^{-3}} = .025 \equiv \mathbf{2.5\%}.$$

	Supply			Input				Output				A_v		A_i		A_p		Eff.
#	I_+ mA	I_- mA	P mW	υ_i mV	i_i µA	R_{in} kΩ	P_{in} µW	υ_O V	i_O mA	R_{load} kΩ	P_{out} mW	ratio V/mV	dB	ratio mA/µA	dB	ratio mW/µW	dB	%
a	3	3	60	1	1	1	.0005	2	20	0.1	20	2	66	20	86	40×10^3	76	33
b	1	1	20	20	2×10^3	.01	20	1	1	1	0.5	0.05	34	0.5×10^{-3}	–6	.025	14	2.5
c	0.05	0.05	1	100	10^3	0.1	50	2	10	0.2	10	0.02	26	0.01	20	0.2	23	10
d	10	10	200	14.1	1.41×10^3	.01	10	2.82	28.2	0.1	40	0.2	46	0.02	26	4	36	20
e	3.1×10^{-3}	3.1×10^{-3}	.063	50	5	10	.125	0.5	.05	10	.013	0.01	20	0.01	20	0.1	20	20

c) $P_{out} = 0.1$ mW, $Eff = 10\% = \frac{P_{out}}{P} \times 100.$

$P = 10\ P_{out} = 10\ (0.1) = \mathbf{1\ mW}.$

$$I = \frac{P}{2(10)} = \frac{1 \times 10^{-3}}{2(10)} = \mathbf{.05\ mA}.$$

$\upsilon_i = i_i\, R_{in} = 10^3 \times 10^{-6} \times 0.1 \times 10^3 = 0.1\ \text{V} = \mathbf{100\ mV}.$

$$P_{in} = \frac{100}{\sqrt{2}} \times 10^{-3} \times \frac{10^3 \times 10^{-6}}{\sqrt{2}} = 5 \times 10^{-5}\ \text{W} = \mathbf{50\ \mu W}.$$

$$p\upsilon_o = \frac{10 \times 10^{-3}}{10/\sqrt{2} \times 10^{-3}} = \sqrt{2}\ \textbf{Vrms} = \mathbf{2\ V\ peak}.$$

$$R_{load} = \frac{2}{10 \times 10^{-3}} = 200\Omega = \mathbf{0.2\ k\Omega}.$$

$$A_\upsilon = \frac{2}{100 \times 10^{-3}} = \frac{2000}{100} = 20\ \text{V/V} = .02\ \text{V/mV} \equiv \mathbf{26\ dB}.$$

$$A_i = \frac{i_o}{i_i} = \frac{10 \times 10^{-3}}{10^3 \times 10^{-6}} = 10\ \text{A/A} = .01\ \text{mA/}\mu\text{A} \equiv \mathbf{20\ dB}.$$

$$A_p = \frac{10 \times 10^{-3}}{50 \times 10^{-6}} = 200\ \text{W/W} = 0.2\ \text{mW/}\mu\text{W} \equiv 10 \log 200 = \mathbf{23\ dB}.$$

d) $I = \dfrac{P}{V_+ + V_-} = \dfrac{200}{2(10)} = \mathbf{10\ mA}.$

$\upsilon_i = (P_{in}\,R_{in})^{1/2} = (10 \times 10^{-6} \times .01 \times 10^3)^{1/2} = 10^{-2}$ V = 10 mV rms = **14.1 mV peak.**

$i_i = \dfrac{\upsilon_i}{R_{in}} = \dfrac{14.1 \times 10^{-3}}{.01 \times 10^3} = 1.41\text{ mA} = \mathbf{1.41 \times 10^3 \mu A}.$

$\upsilon_o = A_\upsilon\,\upsilon_i = 0.2 \times 10^3 \times 14.1 \times 10^{-3} = \mathbf{2.82\ V}.$

$R_L = \dfrac{\upsilon_o^2}{P_{out}} = \dfrac{(2.82/1.41)^2}{40 \times 10^{-3}} = 100\Omega = \mathbf{0.1\ k\Omega}.$

$i_o = \dfrac{\upsilon_o}{R_L} = \dfrac{2.82}{0.1} = \mathbf{28.2\ mA}.$

$A_\upsilon = 0.2 \times 10^3 \equiv 20\log_{10}(200) = \mathbf{46\ dB}.$

$A_i = \dfrac{i_o}{i_i} = \dfrac{28.2 \times 10^{-3}}{1.41 \times 10^3 \times 10^{-6}} = 20\text{ A/A} = 0.02\text{ mA/}\mu\text{A} \equiv \mathbf{26\ dB}.$

$A_p = \dfrac{P_{out}}{P_{in}} = \dfrac{40 \times 10^{-3}}{10 \times 10^{-6}} = 4 \times 10^3\text{ W/W} = 4\text{ mW/}\mu\text{W} \equiv \mathbf{36\ dB}.$

$Eff = \dfrac{P_{out}}{P} = \dfrac{40}{200} \times 100 = \mathbf{20\%}.$

e) $i_o = \dfrac{\upsilon_o}{R_L} = \dfrac{0.5}{10 \times 10^3} = \mathbf{0.05\ mA\ peak}.$

$P_{out} = \upsilon_o i_o = (0.5/\sqrt{2}) \times (.05 \times 10^{-3}/\sqrt{2}) = \mathbf{0.0125\ mW}.$

$P_{in} = \dfrac{P_{out}}{A_p} = \dfrac{.0125 \times 10^{-3}}{0.1 \times 10^3} = \mathbf{0.125 \mu W}.$

$A_p = 0.1 \times 10^3 \equiv 10\log_{10}(100) = \mathbf{20\ dB}.$

$\upsilon_i = (R_{in}P_{in})^{1/2} = (10 \times 10^3 \times .125 \times 10^{-6})^{1/2} = (1.25 \times 10^{-3})^{1/2}$ = .0354 V = 35.4 mV rms = 35.4 × 1.414 = **50 mV peak.**

$i_i = \dfrac{\upsilon_i}{R_{in}} = \dfrac{50\text{mV}}{10 \times 10^3} = \mathbf{5 \mu A}.$

$A_\upsilon = \dfrac{\upsilon_o}{\upsilon_i} = \dfrac{0.5}{50 \times 10^{-3}} = 10\text{ V/V} = 0.01\text{ V/mV} \equiv \mathbf{20\ dB}.$

$A_i = \dfrac{i_o}{i_i} = \dfrac{.05 \times 10^{-3}}{5 \times 10^{-6}} = 10\text{ A/A} = 0.01\text{ mA/}\mu\text{A} \equiv \mathbf{20\ dB}.$

$P = \dfrac{P_o}{Eff} = \dfrac{.0125 \times 10^{-3}}{20/100} = \mathbf{0.0625\ mW}.$

$I = \dfrac{P}{V_+ + V_-} = \dfrac{0.0625 \times 10^{-3}}{2(10)} = 3.125\mu\text{A} = .003125\text{ mA} = \mathbf{3.1 \times 10^{-3}\ mA}.$

1.10 Largest undistorted positive output signal is 7V peak.

∴ undistorted output sine wave can be 7 volt peak.

Corresponding input = $\dfrac{7.0}{50}$ = 140 mV peak.

∴ largest sine wave input (having no dc component) is **140 mV peak**.

1.11 For the largest possible unclipped output, center the output between +7 V and −9 V. The corresponding sine wave has a peak voltage of (7 − −9)/2 = 8 V and an rms of $8/\sqrt{2}$ = **5.66 V**, with an offset of +7 −8 = **−1 V**.

Required dc input offset = −1/50 = **−20 mV**.

Required ac input signal = 8/50 = **+160 mV** peak, or 113 mV rms.

1.12 For $V_O = 4$ V, $V_O = 8 - 4(V_I - 1)^2 = 4$, $-4 = -4(V_I - 1)^2$, $(V_I - 1)^2 = 1$, $V_I - 1 = \pm 1$, and $V_I = 0$ or 2 V, with 0 V forbidden.

Now, for a sine wave input υ_i correctly biassed, $\upsilon_I = 2 + \upsilon_i \cos \omega t$, where,

$$\upsilon_O = 8 - 4(2 + \upsilon_i \cos \omega t - 1)^2 = 8 - 4(1 + \upsilon_i \cos \omega t)^2 = 8 - 4(1 + 2\upsilon_i \cos \omega t + \upsilon_i^2 \cos^2 \omega t)$$

$$= 8 - 4(1 + 2\upsilon_i \cos \omega t + \upsilon_i^2 (1 + \cos 2\omega t)/2)$$

$$= 4 - 8\upsilon_i \cos \omega t - 2\upsilon_i^2 - 2\upsilon_i^2 \cos 2\omega t$$

$$= 4 - 2\upsilon_i^2 - 8\upsilon_i \cos \omega t - 2\upsilon_i^2 \cos 2\omega t.$$

Now, for an output signal (at the input frequency ω) ≤ 1 V peak, $8\upsilon_i \le 1$, $\upsilon_i \le 0.125$ V.

Now, % 2nd harmonic distortion $= \dfrac{2\upsilon_i^2}{8\upsilon_i} \times 100 = \dfrac{2(.125)^2}{8(.125)} \times 100 = \dfrac{.125}{4} \times 100 = \mathbf{3.125\%}$.

1.13 $\upsilon_O = 5 - 10^{-10} e^{40\upsilon_I}$ for $\upsilon_I > 0V$, $\upsilon_O \ge \upsilon_I$.

υ_O is largest when $\upsilon_I = 0$, at which point $\upsilon_O = L+ = 5 - 10^{-10} e^{40(0)} = \mathbf{5\ V}$.

For bias at $V_O = 5/2$, $5/2 = 5 - 10^{-10} e^{40 V_I}$, for which

$e^{40 V_I} = 2.5 \times 10^{10}$, and $V_I = \dfrac{\ln 2.5 \times 10^{10}}{40} = \mathbf{0.598\ V}$.

Now for $L-$: $\upsilon_O = \upsilon_I = 5 - 10^{-10} e^{40\upsilon_I}$.

Solve Iteratively: $\upsilon_I = \dfrac{\ln (5 - \upsilon_I)10^{10}}{40}$ with υ_{I0} = 0.6V. Thus $\upsilon_{I1} = \dfrac{\ln (5 - 0.6)10^{10}}{40} = 0.612$V, and

$\upsilon_{I2} = \dfrac{\ln (5 - 0.612)\, 10^{10}}{40} = 0.613$ V. See convergence: ∴ $L- = \mathbf{0.613\ V}$.

Peak sine-wave allowed is limited by $L-$ to 2.5 − 0.613 = **1.89 V** (peak).

Gain, $A_\upsilon = \left.\dfrac{d\,\upsilon_O}{d\,\upsilon_I}\right|_{V_o = 2.5} = \left.\dfrac{d}{d\,\upsilon_I}\left[5 - 10^{-10} e^{40\upsilon_I}\right]\right|_{V_I = 0.598} = \left.-10^{-10}(40)\, e^{40\upsilon_I}\right|_{V_I = 0.598}$.

i.e., $A_\upsilon = -10^{-10}\left[40\, e^{40(0.598)}\right] = \mathbf{-97.8\ V/V}$.

SECTION 1.5: Circuit Models for Amplifiers

1.14 $A_{\upsilon L} = A_{\upsilon o} \times \dfrac{R_L}{R_L + R_o}$. Thus $\dfrac{R_L + R_o}{R_L} = 1 + \dfrac{R_o}{R_L} = \dfrac{A_{\upsilon o}}{A_{\upsilon L}}$, and $R_o = R_L\left[\dfrac{A_{\upsilon o}}{A_{\upsilon L}} - 1\right] = 1k\left[\dfrac{100}{70} - 1\right] = 0.429$ kΩ.

Use $R_o \approx$ **430 ohms**.

For a 500Ω load, gain = $100 \times \dfrac{500}{500 + 429}$ = **53.8 V/V**.

1.15

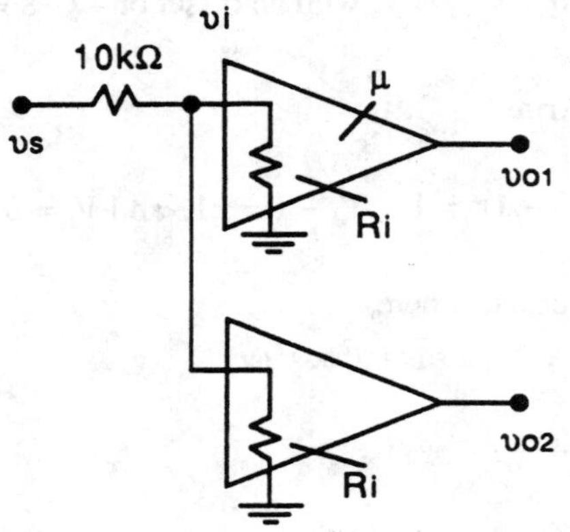

Originally, $\upsilon_o = 1667\ \upsilon_s = \frac{R_i}{R_i + 10k}\ \mu\ \upsilon_s$,

whence $\mu \frac{R_i}{R_i + 10} = 1667$ V/V --- (1).

Now, with a second amplifier connected,

$\mu \frac{R_i/2}{R_i/2 + 10} = 909$V/V --- (2).

$$(1)/(2) \rightarrow \frac{\frac{R_i}{R_i + 10}}{\frac{R_i/2}{R_i/2 + 10}} = \frac{1667}{909} = \frac{R_i + 20}{R_i + 10}.$$

Thus, $1667\ R_i + 16670 = 909R_i + 18180$, and $R_i = \frac{18180 - 16670}{1667 - 909} = 1.99\ k\Omega \approx \mathbf{2.0\ k\Omega}$.

1.16

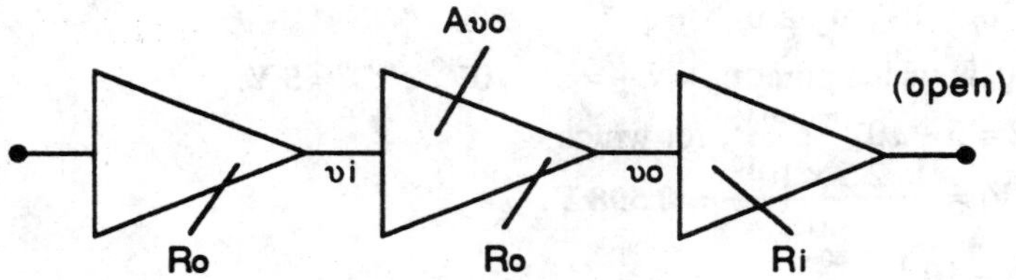

Gain of an internal element in the cascade = $\frac{R_i}{R_i + R_o} \times A_{\upsilon o}$. For the condition stated, this must be 1. Thus,

$\frac{R_i}{R_i + R_o} \times A_{\upsilon o} = 1$, or $\mathbf{A_{\upsilon o} = 1 + \frac{R_o}{R_i}}$.

Particular Cases:

See for $R_o = 0$, $A_{\upsilon o} = 1$;

for $R_o = R_i$, $A_{\upsilon o} = 2$;

for $R_i = \infty$, $A_{\upsilon o} = 1$;

for $A_{\upsilon o} = 11$, $R_o = 10\,R_i$.

1.17

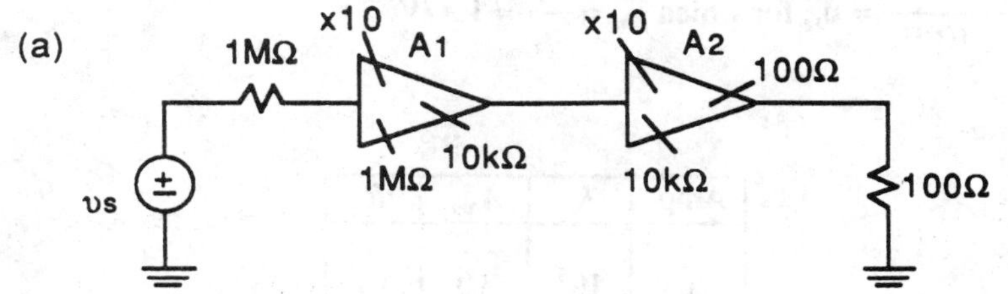

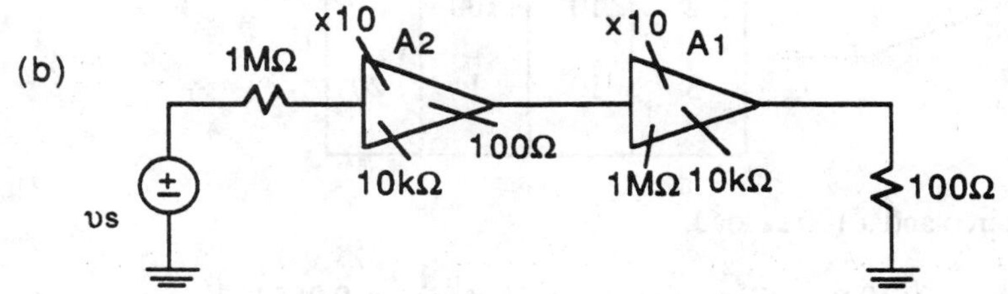

$$\text{Gain } (A_1\,A_2) = \frac{100}{100+100} \times 10 \times \frac{10k}{10k+10k} \times 10 \times \frac{1M}{1M+1M} = 10^2\left[\frac{1}{2^3}\right] = \textbf{12.5 V/V, the best, where}$$

$$\text{Gain } (A_2\,A_1) = \frac{100}{100+10^4} \times 10 \times \frac{10^6}{100+10^6} \times 10 \times \frac{10^4}{10^4+10^6} = 10^2\left[\frac{1}{101}\right]^2 \times (1) \approx \textbf{0.01 V/V.}$$

$$\text{Gain } (A_1) = 10 \times \frac{100}{100+10^4} \times \frac{10^6}{10^6+10^6} = 10\,\frac{1}{101} \times \frac{1}{2} = \textbf{0.05V/V.}$$

$$\text{Gain } (A_2) = 10 \times \frac{100}{100+100} \times \frac{10^4}{10^4+10^6} = 10\left[\frac{1}{2}\right]\left[\frac{1}{101}\right] = \textbf{0.5V/V.}$$

$$\text{Gain (wire)} = 1 \times \frac{100}{100+10^6} \approx \frac{1}{10^4} = \textbf{0.0001 V/V.}$$

Note how important source-to-load matching can be: Only one of 4 possible one-or-two-amplifier combinations does any good; *and* one two-amplifier arrangement is worse than either one-amplifier arrangement!

Choices: **double gain, double input resistance, halve output resistance.**

In general, a change in an input or output resistance by a particular factor provides less than that factor of improvement, due to the comparison process inherent in a voltage divider. Thus **change of gain has the greatest affect** for a given factor. Next choice would be output resistance (reduction by 2) since this implies reduced power loss in many applications.

1.18 80 dB $\equiv 10^{80/20} = 10^4$ V/V:

For 1MΩ load, $A_\upsilon = 10^4 \dfrac{10^6}{10^4+10^6} = \mathbf{0.99 \times 10^4}$ **V/V.**

For 10 kΩ load, $A_\upsilon = 10^4 \dfrac{10^4}{10^4+10^4} = \mathbf{0.5 \times 10^4}$ **V/V.**

For 10 Ω load, $A_\upsilon = 10^4 \dfrac{10}{10^4+10} = \dfrac{1}{1001}\,10^4 = \mathbf{9.99}$ **V/V.**

For a 0 Ω load, $i_o = \dfrac{\upsilon_s \times 10^4}{10^4} = \upsilon_s$, for which $g_m = \dfrac{i_o}{\upsilon_s} = \mathbf{1\ A/V}$.

1.19 Individual Amplifiers:

Amp	R_i	$A_{\upsilon o}$	R_o
1	10^6	10	10^4
2	10^4	100	10^3
3	10^4	1	20

With a $0.5 \times 10^6 \Omega$ source and a 100 Ω load:

a)
$$\left.\frac{\upsilon_s}{\upsilon_o}\right|_1 = \frac{10^6}{0.5 \times 10^6 + 10^6} \times 10 \times \frac{10^2}{10^2 + 10^4} = \frac{1}{1.5} \times 10 \times \frac{1}{101} = \mathbf{0.066\ V/V}.$$
$$\left.\frac{\upsilon_s}{\upsilon_o}\right|_2 = \frac{10^4}{0.5 \times 10^6 + 10^4} \times 100 \times \frac{10^2}{10^2 + 10^3} = \frac{1}{51} \times 100 \times \frac{1}{11} = \mathbf{0.178\ V/V}.$$
$$\left.\frac{\upsilon_s}{\upsilon_o}\right|_3 = \frac{10^4}{0.5 \times 10^6 + 10^4} \times 1 \times \frac{10^2}{20 + 10^2} = \frac{1}{51} \times 1 \times \frac{10}{12} = \mathbf{0.016\ V/V}.$$

b) **One-Stage designs:**

	Loss		
Amp	**Input**	**Output**	**Least Loss**
1	low	high	source
2	med	med	load
3	med	low	load

Ranking **(best first):**

1) As Input Coupler: **1, 2/3, 3/2;**

2) As Output Coupler: **3, 2, 1;**

3) As Provider of Gain: **2, 1, 3.**

c) **Two-Stage designs:**

Input: clearly A_1 is #1 on list 1, and reasonable on list 3;

Output: clearly A_3 is #1 on list 2, but worst on list 3;

But, A_2 is #2 on list 2, *but* #1 on list 3.

Conclude: Try **(A_1 A_2)** and **(A_1 A_3)**.

d) **Highest Gain for a Two-Stage Design:**

For $(A_1\, A_2)$: $\text{Gain} = \dfrac{10^6}{10^6 + 0.5\times10^6}\times 10\times\dfrac{10^4}{10^4+10^4}\times 100\times\dfrac{100}{10^2+10^3} = \dfrac{1}{1.5}\times 10\times\dfrac{1}{2}\times 100\times\dfrac{1}{11}$

$= 30.3$ V/V.

For $(A_1\, A_3)$: $\text{Gain} = \dfrac{10^6}{10^6 + 0.5\times10^6}\times 10\times\dfrac{10^4}{10^4+10^4}\times 1\times\dfrac{100}{100+20} = \dfrac{1}{1.5}\times 10\times\dfrac{1}{2}\times 1\times\dfrac{10}{12}$

$= 2.78$ V/V

Certainly (**A_1 A_2**) seems **best** with an overall gain of **30.3 V/V**.

e) **Reconsidering**:

Certainly, maximizing the gain is a good idea, since coupling is never perfect, (i.e. there is always a loss). Of the highest gain choices, pick the highest input resistance for the input stage and the lowest output resistance for the output stage, i.e., A_1 and A_2 respectively. ∴ chose (**A_1 A_2**).

Try also:

For $(A_2\, A_3)$: $\text{Gain} = \dfrac{10^4}{10^4 + 0.5\times10^6}\times 100\times\dfrac{10^4}{10^4+10^3}\times 1\times\dfrac{100}{100+20} = \dfrac{1}{51}\times 100\times\dfrac{10}{11}\times 1\times\dfrac{10}{12}$

$= 1.49$ V/V.

Now if two amps of the same kind can be used:

For $(A_2\, A_2)$: $\text{Gain} = \dfrac{10^4}{0.5\times10^6 + 10^4}\times 100\times\dfrac{10^4}{10^3+10^4}\times 100\times\dfrac{100}{10^2+10^3} = \dfrac{1}{51}\times 100\times\dfrac{10}{11}\times 100\times$

$\dfrac{1}{11} = 16.2$ V/V.

Note in retrospect, that the only way to possibly better the value of 30.3 V/V is to use $(A_1\, A_1)$, $(A_2\, A_2)$, $(A_2\, A_1)$. See that the loss at the output is too great in the $(A_1\, A_1)$ and $(A_2\, A_1)$ cases.

1.20

Amp	R_i	A_{is}	R_o
1	10	100	10^4
2	10^4	1000	10^3
3	10^4	100	10^5

With 10 kΩ source, and 10 kΩ load:

a) **There are 9 possible amplifier pairs:**

$\text{Gain } 1, 1 = \dfrac{10^4}{10+10^4}\times 100\times\dfrac{10^4}{10+10^4}\times 100\times\dfrac{10^4}{10^4+10^4}$

$$= \frac{1000}{1001} \times 100 \times \frac{1000}{1001} \times 100 \times \frac{1}{2} = 4901 \text{ A/A.}$$

$$\text{Gain 1, 2} = \frac{10^4}{10 + 10^4} \times 100 \times \frac{10^4}{10^4 + 10^4} \times 100 \times \frac{10^3}{10^3 + 10^4}$$

$$= \frac{1000}{1001} \times 100 \times \frac{1}{2} \times 1000 \times \frac{1}{11} = 4541 \text{ A/A.}$$

$$\text{Gain 1, 3} = \frac{10^4}{10 + 10^4} \times 100 \times \frac{10^4}{10^4 + 10^4} \times 100 \times \frac{10^5}{10^5 + 10^4}$$

$$= \frac{1000}{1001} \times 100 \times \frac{1}{2} \times 100 \times \frac{10}{11} = 4541 \text{ A/A.}$$

$$\text{Gain 2, 1} = \frac{10^4}{10^4 + 10^4} \times 10^3 \times \frac{10^3}{10 + 10^3} \times 100 \times \frac{10^4}{10^4 + 10^4}$$

$$= \frac{1}{2} \times 10^3 \times \frac{100}{101} \times 100 \times \frac{1}{2} = 24752 \text{ A/A.}$$

$$\text{Gain 2, 2} = \frac{10^4}{10^4 + 10^4} \times 10^3 \times \frac{10^3}{10^3 + 10^4} \times 10^3 \times \frac{10^3}{10^3 + 10^4}$$

$$= \frac{1}{2} \times 10^3 \times \frac{1}{11} \times 10^3 \times \frac{1}{11} = 4132 \text{ A/A.}$$

$$\text{Gain 2, 3} = \frac{10^4}{10^4 + 10^4} \times 10^3 \times \frac{10^3}{10^3 + 10^4} \times 10^2 \times \frac{10^5}{10^5 + 10^4}$$

$$= \frac{1}{2} \times 10^3 \times \frac{1}{11} \times 10^2 \times \frac{10}{11} = 4132. \text{ A/A.}$$

$$\text{Gain 3, 1} = \frac{10^4}{10^4 + 10^4} \times 100 \times \frac{10^5}{10 + 10^5} \times 100 \times \frac{10^4}{10^4 + 10^4}$$

$$= \frac{1}{2} \times 100 \times 1 \times 100 \times \frac{1}{2} = 2500 \text{ A/A.}$$

$$\text{Gain 3, 2} = \frac{10^4}{10^4 + 10^4} \times 100 \times \frac{10^5}{10^5 + 10^4} \times 10^3 \times \frac{10^3}{10^3 + 10^4}$$

$$= \frac{1}{2} \times 100 \times \frac{10}{11} \times 10^3 \times \frac{1}{11} = 4132 \text{ A/A.}$$

$$\text{Gain 3, 3} = \frac{10^4}{10^4 + 10^4} \times 100 \times \frac{10^5}{10^4 + 10^5} \times 10^2 \times \frac{10^5}{10^4 + 10^5}$$

$$= \frac{1}{2} \times 100 \times \frac{10}{11} \times 10^2 \times \frac{10}{11} = 4132 \text{ A/A.}$$

Summary:

Gain	*Combination*
24752	(2,1)
4901	(1,1)
4541	(1,2), (1,3)

4132	(2,2), (2,3), (3,2), (3,3)
2500	(3,1)

Note the relative superiority of (1,1) is due essentially to its R_i of 10Ω.

1.21 Figure of merit: $A_{is}\ R_o/R_i$

Amp	R_i	A_{is}	R_o	$\frac{A_{is} \times R_o}{R_i}$	Rank
1	10	10^2	10^4	10^5	1
2	10^4	10^3	10^3	10^2	3
3	10^4	10^2	10^5	10^3	2

Lowest ranked are A_2, A_3:

Consider: $G_{2,2}$, $G_{2,3}$, $G_{3,2}$, $G_{3,3}$ with values (from P1.20 above) of 4132, 4132, 4132, 4132, respectively. Thus the highest available gain with the lowest-ranked amplifiers is 4132 A/A.

1.22

Amp	R_i	A_{is}	g_m	R_o	Fig. of Merit (1) $g_m R_o R_i = A_{is} R_o$	Rank (1)	Fig. of Merit (2) $g_m R_o R_i^2 = A_{is} R_o R_i$	Rank (2)
1	10	10^2	10	10^4	10^6	2	10^7	3
2	10^4	10^3	10^{-1}	10^3	10^6	2	10^{10}	2
3	10^4	10^2	10^{-2}	10^5	10^7	1	10^{11}	1

Now, $g_m = \frac{i_o}{v_i} = \frac{i_o}{R_i\, i_i} = \frac{A_{is}}{R_i}$:

For A_1, $g_m = \frac{10^2}{10} = 10$ **A/V.**

For A_2, $g_m = \frac{10^3}{10^4} = 10^{-1}$ A/V.

For A_3, $g_m = \frac{10^2}{10^4} = 10^{-2}$ A/V.

Figure of merit (FM) for a transconductance amplifier is $g_m R_o R_i$. But $g_m R_o R_i = \frac{A_{is}}{R_i} \times R_o \times R_i = A_{is} R_o$. Use as FM1. However, high R_i is obviously very important for the g_m generator. Consider $A_{is} R_o R_i = g_m R_o R_i^2$, as FM2.

1.23 $A_{is} = \frac{i_c}{i_b}\Big|_{v_c=0} = \beta$.

$R_i = \frac{v_{be}}{i_b}\Big|_{v_c=0} = \frac{v_{be}}{v_{be}/r_\pi} = r_\pi$.

$R_o = \frac{v_c}{i_c}\Big|_{i_b=0} = \infty$, as described.

$G_m = \frac{i_c}{v_{be}}$; but $i_c = \beta\, i_b$, and $i_b = \frac{v_{be}}{r_\pi}$.

$\therefore G_m = \frac{\beta\, i_b}{v_{be}} = \frac{\beta\, v_{be}/r_\pi}{v_{be}} = \frac{\beta}{r_\pi}$.

Numerically:

$\therefore$ **$A_{is} = \beta = 200$ mA/mA, and $G_m = \frac{\beta}{r_\pi} = \frac{200}{5 \times 10^3} = 40$ mA/V.**

1.24 Using the result in Example 1.5: $L = \frac{C}{g_1 g_2} = \frac{1 \times 10^{-6}}{1 \times 10^{-3} \times 1 \times 10^{-3}}$ = **1 H**. Now, for the effect of resistance: See the equivalent circuit:

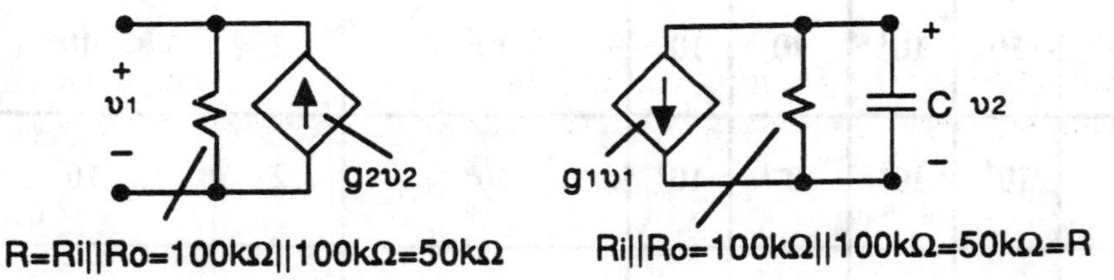

Now $i_1 = \frac{v_1}{R} - g_2 v_2$ --- (1), and $-g_1 v_1 = \frac{v_2}{R} + C\frac{dv_2}{dt}$ --- (2). From (1), $v_2 = -\frac{i_1}{g_2} + \frac{v_1}{g_2 R}$.

From (2), $-g_1 v_1 = \frac{-i_1}{g_2 R} + \frac{v_1}{g_2 R^2} - \frac{C}{g_2}\frac{di_1}{dt} + \frac{C}{g_2 R}\frac{dv_1}{dt}$, or $-g_1 g_2 v_1 = \frac{v_1}{R^2} - \frac{i_1}{R} - C\frac{di_1}{dt} + \frac{C}{R}\frac{dv_1}{dt}$,

or $v_1 = \frac{1}{-g_1 g_2 - \frac{1}{R^2}}\left[-\frac{i_1}{R} - \frac{C\, di_1}{dt} + \frac{C}{R}\frac{dv_1}{dt}\right]$ --- (3).

Consider this to be equivalent to L with r in series, all in parallel with R. Then, for an applied voltage υ_1 and current i_1 into port 1, current in the inductor is $i_1 - \frac{\upsilon_1}{R}$. $\therefore\ \upsilon_1 = L\frac{d(i_1 - \upsilon_1/R)}{dt} + r\,(i_1 - \frac{\upsilon_1}{R})$, or $\upsilon_1 = L\frac{di_1}{dt} - \frac{L}{R}\frac{d\upsilon_1}{dt} + ri_1 - \frac{r}{R}\upsilon_1$, where $\upsilon_1 = \frac{1}{1+r/R}\left[ri_1 + \frac{L\,di_1}{dt} - \frac{L}{R}\frac{d\upsilon_1}{dt}\right]$ --- (4).

Comparing (3), (4), see: $\frac{r}{1+r/R} = \frac{1/R}{g_1 g_2 + 1/R^2}$ --- (5), $\frac{L}{1+r/R} = C\left[\frac{1}{g_1 g_2 + 1/R^2}\right]$ --- (6), $\frac{L}{R}\frac{1}{1+r/R} = \frac{C}{R}\frac{1}{g_1 g_2 + 1/R^2}$ --- (7), the same as (6).

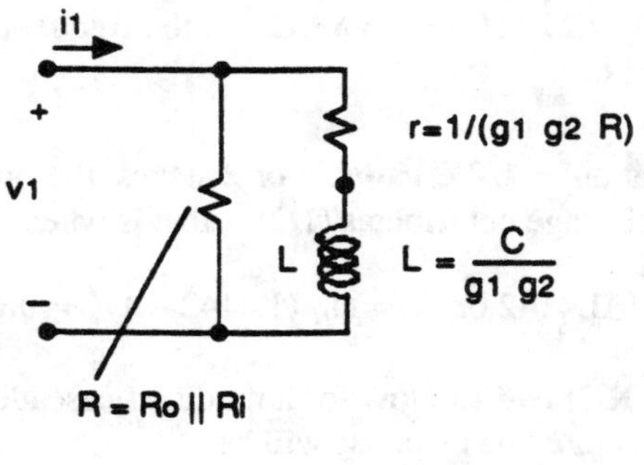

From (5), See $r g_1 g_2 + r/R^2 = 1/R + r/R^2$, or $\mathbf{r} = 1/(g_1 g_2 R) = 1/(10^{-3} \times 10^{-3} \times 50 \times 10^3) = \mathbf{20\Omega}$.

From (6), See $L = C\left[\frac{1}{g_1 g_2 + 1/R^2}\right]\left[1 + \frac{1}{g_1 g_2 R^2}\right]$, and

$$L = C\left[\frac{(g_1 g_2 R^2 + 1)}{(g_1 g_2 R^2 + 1)}\frac{R^2}{g_1 g_2 R^2}\right] = \frac{C}{g_1 g_2},$$

as expected.

SECTION 1.6: Frequency Response of Amplifiers

1.25 $\left|A\right| = \frac{\upsilon_o}{\upsilon_i} = \frac{2(2V)}{2mV}$ = **2000 V/V**. At 1 kHz, period is $\frac{1}{10^3}$ = 1 ms. $\therefore$ Delay of 0.2 ms corresponds to $\frac{0.2}{1} \times 360 = 72°$. Thus, the corresponding phase shift is **72°, lagging**.

1.26 The 3 dB bandwidth = 100 kHz – 0 kHz = **100 kHz**. For capacitor coupling, the bandwidth is 100 kHz – 20 kHz = **80 kHz**.

1.27 See for circuit a) that $\upsilon_{out} = \upsilon_i - \upsilon_o$ is the voltage across R, fed by C, i.e. a **high-pass output**. Correspondingly, for circuit b), $\upsilon_{out} = \upsilon_i - \upsilon_o$ is the voltage across C, a **low-pass output**. In fact, the circuits are really the same, with both output types available: high-pass across R, and low-pass across C.

1.28

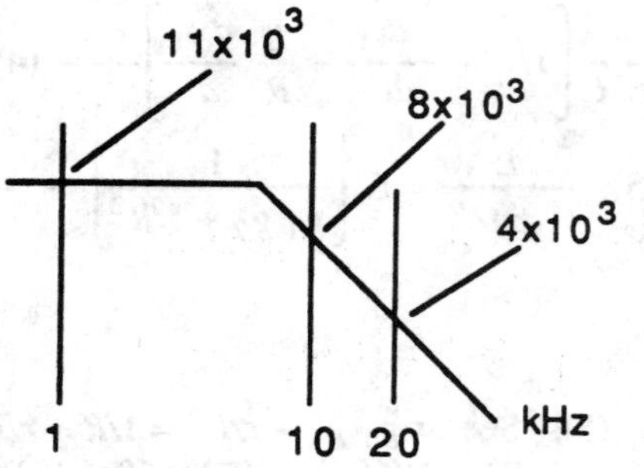

See immediately, that as frequency goes from 10 to 20 kHz, gain drops by a factor of 2 from 8×10^3 to 4×10^3 V/V. Conclude that 10 kHz and 20 kHz are on the –20 dB/decade rolloff, such that $Af = f_t = 20 \times 10^3 \times 4 \times 10^3 = 80 \times 10^6$Hz. Thus f_t, the unity-gain frequency, is **80 MHz**. Now, since at 1 kHz, the gain is only 11/8 larger than at 10× the frequency, one can conclude that the midband gain is likely to be 11×10^3 V/V. Thus the 3 dB frequency is about $\dfrac{80 \times 10^6}{11 \times 10^3}$ = **7.27 kHz** = f_H.
At frequency f, the phase shift is $\tan^{-1} f/f_H$. For $\tan^{-1} f/f_H = 60°$, $f/f_H = \tan 60° = 1.73$.
$\therefore$ $f = 1.73$ (7.27 kHz) = **12.6 kHz** is the frequency where the phase lag is 60°.

1.29 For each stage of the amplifier, the 3 dB frequency is at $\omega_H = 1/RC$ rad/s. For 2 stages, the output reaches $1/\sqrt{2}$ of midband amplitude at the frequency where each stage contributes $(1/2)^{1/4}$, that is when

$$\left[\frac{1}{(1^2 + (\omega/\omega_H)^2)^{1/2}}\right]^2 = \left[\frac{1}{2}\right]^{1/2}, \text{ or } 1 + \left[\frac{\omega}{\omega_H}\right]^2 = 2^{1/2} = 1.4142 \text{ or } \omega = \omega_H\,(1.4142 - 1)^{1/2} = \mathbf{0.644}\ \omega_H.$$

Thus, for 2 stages, the 3 dB frequency becomes **0.644/(RC) rad/s**. Now for a modified cascade, where one stage has $\omega_H = 1/RC$ and the other has $\omega_{H1} = 1/(kRC) = \omega_H/k$, the response will be:

$$T(s) = \frac{K}{(1 + j\,\omega/\omega_H)\,(1 + j\,k\omega/\omega_H)} = \frac{K}{1 + j\,(k+1)\,\dfrac{\omega}{\omega_H} - \dfrac{k\omega^2}{\omega_H^2}}$$

The response is 3 dB down when $\left[1 - \dfrac{k\omega^2}{\omega_H^2}\right]^2 + \left[(k+1)\,\dfrac{\omega}{\omega_H}\right]^2 = 2,$

$$\text{or, } 1 - \frac{2k\omega^2}{\omega_H^2} + k^2\,\frac{\omega^4}{\omega_H^4} + (k+1)^2\,\frac{\omega^2}{\omega_H^2} = 2, \text{ or } k^2\,\frac{\omega^4}{\omega_H^4} + (k^2+1)\,\frac{\omega^2}{\omega_H^2} - 1 = 0, \text{ whence}$$

$$\left[\frac{\omega}{\omega_H}\right]^2 = -\frac{(k^2+1) \pm ((k^2+1)^2 + 4k^2)^{1/2}}{2k^2} = \frac{-(k^2+1) \pm (k^4 + 6k^2 + 1)^{1/2}}{2k^2}$$

$$\text{Thus } \omega = \omega_H \left[\frac{-k^2 - 1 \pm (k^4 + 6k^2 + 1)^{1/2}}{2k^2}\right]^{1/2} \quad ---\ (1).$$

Test: for k = 1: See $\omega = \omega_H\left[\dfrac{-1 - 1 \pm (1 + 6 + 1)^{1/2}}{2}\right]^{1/2} = \omega_H\left[\dfrac{-2 \pm \sqrt{8}}{2}\right]^{1/2} = \omega_H\left[-1 \pm \sqrt{2}\right]^{1/2}$
$= \omega_H(0.414)^{1/2} = 0.644\omega_H$: OK.

Solve by Trial:

For $k = 0.5$: $\omega = \omega_H \left[\dfrac{-.25 - 1 \pm (.0625 + 1.5 + 1)^{1/2}}{2\,(.25)}\right]^{1/2} = \omega_H \left[\dfrac{-1.25 \pm \sqrt{2.5625}}{0.5}\right]^{1/2}$

$= \omega_H \left[(-1.25 \pm 1.60)\,(2)\right]^{1/2} = .837\omega_H$

Now for smaller k, from (1): $\omega = \omega_H \left[-\dfrac{1}{2} - \dfrac{1}{2k^2} \pm \left[\dfrac{1}{4} + \dfrac{1.5}{k^2} + \dfrac{1}{4k^4}\right]^{1/2}\right]^{1/2}$, in general.

For $k = 0.1$: $\omega = \omega_H \left[-0.5 \pm 50 \pm (.25 + 150 + 2500)^{1/2}\right]^{1/2} = \omega_H \left[-50.5 \pm 51.48\right]^{1/2} = 0.99\omega_H$.

For $k = 0.2$: $\omega = \omega_H \left[-0.5 - 12.5 \pm (.25 + 37.5 + 156.25)^{1/2}\right]^{1/2} = 0.964\omega_H$.

For $k = 0.25$: $\omega = \omega_H \left[-0.5 - 8 \pm (.25 + 24 + 64)^{1/2}\right]^{1/2} = 0.946\omega_H$.

Thus, the required value of k is about 0.25.

1.30

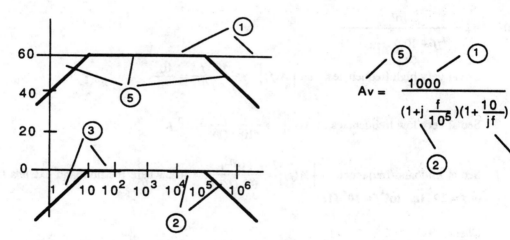

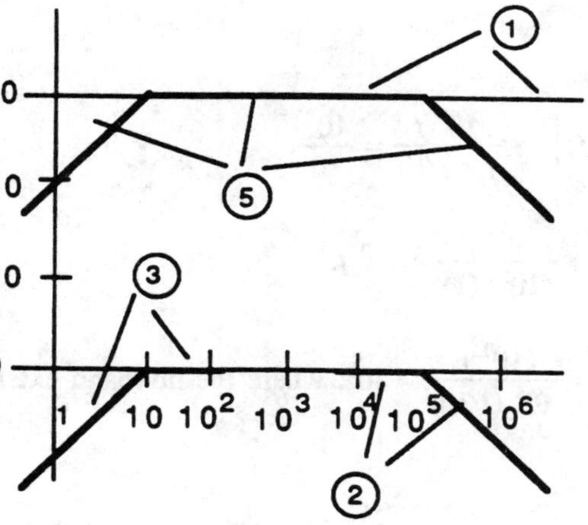

$$A_v = \frac{1000}{\left(1 + j\dfrac{f}{10^5}\right)\left(1 + \dfrac{10}{jf}\right)}$$

(labels on the expression: 5 → numerator, 1 → numerator, 2 → first factor, 3 → second factor)

At $f = 1$: $1000 \equiv 20\log 10^3 = 60$ dB,

$\left|\dfrac{1}{1 + j\dfrac{1}{10^5}}\right| \approx \dfrac{1}{1} \equiv 0$ dB, and $\left|\dfrac{1}{1 + \dfrac{10}{j1}}\right| \approx \dfrac{1}{10} \equiv -20$ dB, for a total of $60 + 0 - 20 =$ **40 dB**.

At $f = 10$: $1000 \equiv 60$ dB,

$\left|\dfrac{1}{1 + j\dfrac{10}{10^5}}\right| \approx 1 \equiv 0$ dB, and $\left|\dfrac{1}{1 + \dfrac{10}{j10}}\right| = \dfrac{1}{\sqrt{2}} \equiv -3$ dB, for a total of $60 + 0 - 3 =$ **57 dB**.

At $f = 100$: $1000 \times \left|\dfrac{1}{1 + j\dfrac{100}{10^5}}\right| \times \left|\dfrac{1}{1 + \dfrac{10}{j100}}\right| \approx 1000 \times 1 \times 1 \equiv 60 + 0 + 0 =$ **60 dB**.

At $f = 10^4$: $1000 \times \left|\dfrac{1}{1 + j\dfrac{10^4}{10^5}}\right| \times \left|\dfrac{1}{1 + \dfrac{10}{j\,10^4}}\right| \approx 1000 \times 1 \times \equiv 60 + 0 + 0 = \mathbf{60\ dB}$.

At $f = 10^5$: $1000 \times \left|\dfrac{1}{1 + j\dfrac{10^5}{10^5}}\right| \times \left|\dfrac{1}{1 + \dfrac{10}{j\,10^5}}\right| \approx 1000 \times 1/2 \times 1 \equiv 60 - 3 + 0 = \mathbf{57\ dB}$.

At $f = 10^6$: $1000 \times \left|\dfrac{1}{1 + j\dfrac{10^6}{10^5}}\right| \times \left|\dfrac{1}{1 + \dfrac{10}{j\,10^6}}\right| \approx 1000 \times \dfrac{1}{10} \times 1 \equiv 60 - 20 + 0 = \mathbf{40\ dB}$.

See, 3 dB bandwidth = 10^5 Hz – 10 Hz = 10^5 Hz.

Phase is 6° at a frequency which is a factor of 10 on the midband side of f_L and f_H.

∴ region for which phase extends from +6° through 0 to –6° is 10 (10 Hz) to $\dfrac{10^5\ \text{Hz}}{10}$, or from 100 hz to 10^4 Hz.

1.31

a) $A_v = \dfrac{10^7\, jf}{(jf + 10^5)\left(\dfrac{jf}{10} + 1\right)}$

See at very high frequencies, that $\left|A(f)\right| \to \dfrac{10^7 f}{f \times f/10} = \dfrac{10^8}{f}$.

See at very low frequencies, $\left|A(f)\right| \to \dfrac{10^7 f}{(10^5)\,(1)} = 10^2 f$.

See at midband frequencies, $\left|A(f)\right| = \dfrac{10^7 f}{10^5\,(f/10)} = 10^3$, where the midband extends from $\dfrac{f}{10} = 1$, or f = **10 Hz**, to f = **10^5 Hz**.

Check:

See, at f = 10 Hz, $A(f) = \dfrac{10^7\, j\,10}{(j\,10 + 10^5)\left(\dfrac{j10}{10} + 1\right)}$, and $\left|A(f)\right| = \dfrac{10^8}{10^5\sqrt{2}} = \dfrac{10^3}{\sqrt{2}}$.

See, at $f = 10^5$ Hz, $A(f) = \dfrac{10^7\, j\,10^5}{(j\,10^5 + 10^5)\left(j\,\dfrac{10^5}{10} + 1\right)} \approx \dfrac{10^{12}}{10^5(j{+}1)\,(10^4)}$, and $\left|A(f)\right| \approx \dfrac{10^3}{\sqrt{2}}$

Thus the midband gain is verified to be 10^3V/V.

b) Now, $A_v = \dfrac{10^7\, jf}{(jf + 10^5)\,(jf/10 + 1)} = \dfrac{10^7/10^5}{\left[\dfrac{jf}{10^5} + 1\right]\left[\dfrac{1}{10} + \dfrac{1}{jf}\right]} = \dfrac{10^3}{\left[1 + \dfrac{jf}{10^5}\right]\left[1 + \dfrac{10}{jf}\right]}$

From this form, we see that the critical frequencies occur when:

$$\frac{f}{10^5} = 1 \rightarrow f = \mathbf{10^5\ Hz}\text{, and } \frac{10}{f} = 1 \rightarrow f = \mathbf{10\ Hz},$$

such that between these frequencies, $A = \dfrac{10^3}{(1 + j\,\varepsilon_1)\,(1 - j\,\varepsilon_2)}$ and $\left| A \right| = \mathbf{10^3\ V/V}$.

Note that the latter approach is more straightforward.

1.32

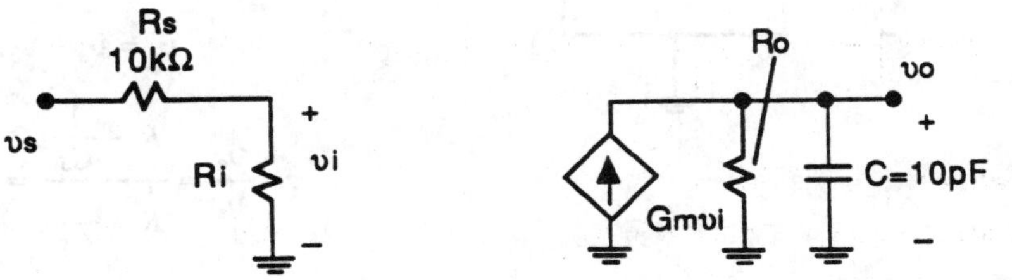

At low frequencies, $A_\upsilon = \dfrac{\upsilon_o}{\upsilon_s} = \dfrac{\mathbf{R_i}}{\mathbf{R_s + R_i}} \times \mathbf{G_m\ R_o = A_M}$. The upper 3 dB frequency, $\omega_H = \mathbf{1/(R_o\ C)}$. Now $R_i = 2.5/I$, $R_o = 200/I$, $G_m = 40\ I$. For $\dfrac{1}{2\pi\, R_o\, C} > 10^6$, i.e. $\dfrac{1}{2\,\pi\, 200/I \times 10 \times 10^{-12}} \geq 10^6$,
$I = 10^6 \times 2\,\pi \times 200 \times 10^{-11} = .0126\,A = 12.6\text{ mA}$.

Now, $A_M = G_m\, R_o\, \dfrac{R_i}{R_s + R_i} = 40 \times 12.6 \times \dfrac{200}{12.6} \times \dfrac{2.5/12.6}{2.5/12.6{+}10}$
$= 40 \times 200 \times .198/(.198 + 10) = \mathbf{155.3\ V/V}$.

Gain-Bandwidth (GB) $= \dfrac{R_i}{R_s + R_i} \times G_m\, R_o \times \dfrac{1}{R_o\, C} = \dfrac{\mathbf{G_m}}{\mathbf{C}} \times \dfrac{\mathbf{R_i}}{\mathbf{R_s + R_i}} = \dfrac{40I}{C}\, \dfrac{(2.5/I)}{R_s + \dfrac{2.5}{I}} = \dfrac{100/C}{R_s + \dfrac{2.5}{I}}$.

See that for large I, $\mathbf{GB} = \dfrac{\mathbf{100}}{\mathbf{R_s\ C}}$, *independent of I*!

Now, Gain $\times 2\pi \times 10^7 = \dfrac{100}{C\, R_s} = \dfrac{100}{10 \times 10^{-12} \times 10^4} = 10^9$, whence Gain $\approx \dfrac{10^9}{2\pi \times 10^7} = \mathbf{15.9\ V/V}$.

See directly (approximately), that the previous design has a gain of 155 and a bandwidth of 1 MHz. Thus this design, with a bandwidth 10 × greater, should have a gain 10 × less. Required current $I = 10^7 \times 2\pi \times 200 \times 10^{-11} = \mathbf{126\ mA}$.

1.33

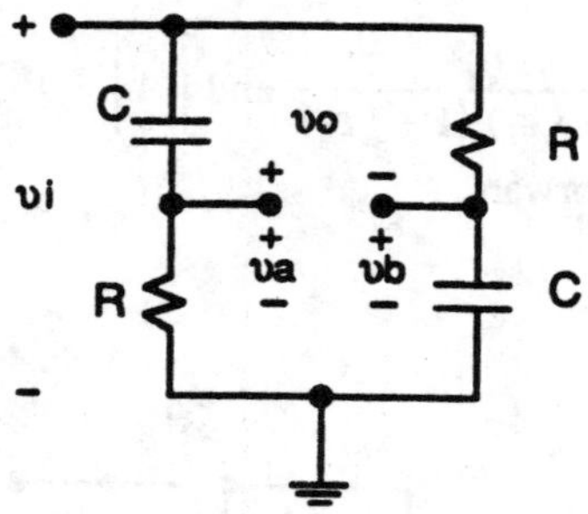

$$\upsilon_a = \frac{R}{R + \frac{1}{Cs}} \times \upsilon_i.$$

$$\upsilon_b = \frac{\frac{1}{Cs}}{R + \frac{1}{Cs}} \upsilon_i.$$

$$\upsilon_o = \upsilon_a - \upsilon_b.$$

$$\therefore \quad \frac{\upsilon_o}{\upsilon_s} = \frac{R - \frac{1}{Cs}}{R + \frac{1}{Cs}} = \frac{\mathrm{R\,Cs - 1}}{\mathrm{R\,Cs + 1}}.$$

At low frequencies , $\frac{\upsilon_o}{\upsilon_s} \to -1$; **Magnitude = 1,** $\Phi = \pm 180°$.

At high frequencies : $\frac{\upsilon_o}{\upsilon_s} \to \frac{RC}{RC} = 1$; **Magnitude = 1,** $\Phi = 0°$.

At $\omega = \frac{1}{RC}$: $\frac{\upsilon_o}{\upsilon_s} = \frac{j-1}{j+1}$, **with magnitude** $\frac{\sqrt{2}}{\sqrt{2}} = 1$, **and** $\Phi = -45° - 45° = -90°$.

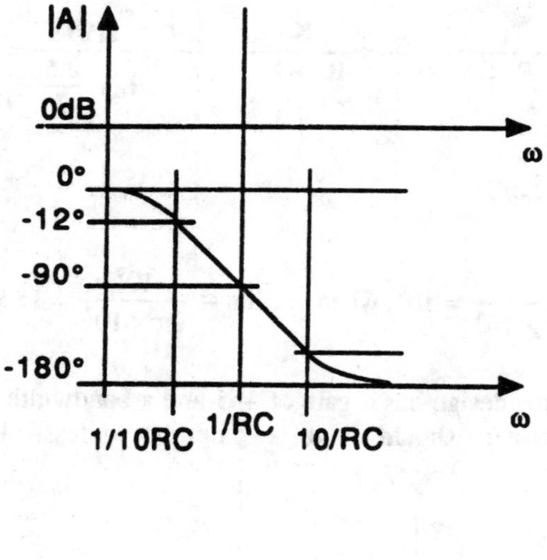

Chapter 2

OPERATIONAL AMPLIFIERS

SECTION 2.1: OpAmp Terminals

2.1 Each op amp has pins for

199z	input :	2	- unique
199z	output:	1	- unique
199z	power :	2	- sharable

Thus an 8-pin package can accommodate: **2 op amps**, using 2(3) + 2 = 8 pins, with **none unused**

Thus a 14-pin package can accommodate: **4 op amps**, using 4(3) + 2 = 14 pins, with **none unused.**

SECTION 2.2: The Ideal Op Amp

2.2 Voltage between input pins, $\upsilon = \upsilon_+ - \upsilon_- = -3V/10^4$ or $\mathbf{-300\ \mu V}$.

In particular, from the negative to the positive input, one would expect $-300\mu V$ or -0.3 mV.

If the positive pin is at +100 mV, the negative would be at 100 − − 0.3 = **100.3 mV**.

2.3 For $\upsilon_o = 3.5$ V: $\upsilon_+ = 3.5\ V/10^4 = 0.35$ mV, across 1 kΩ.

$$\therefore \text{ required } \upsilon_I = \upsilon_+ + \upsilon_{IM} = 0.35 \times 10^{-3} + \left(\frac{0.35 \times 10^{-3}}{1 \times 10^3}\right) 10^6 = .35035\ V \approx \mathbf{0.35\ V}.$$

$$\textit{Check}\text{: Overall gain is } \frac{10^3}{10^3 + 10^6} \times 10^4 \approx \frac{10^3}{10^6} \times 10^4 = 10\ V/V.$$

$\therefore$ 0.35 V at the input produces 3.5 V at the output.

SECTION 2.3: Analysis of Circuits Containing Ideal Op Amps - The Inverting Configuration

2.4

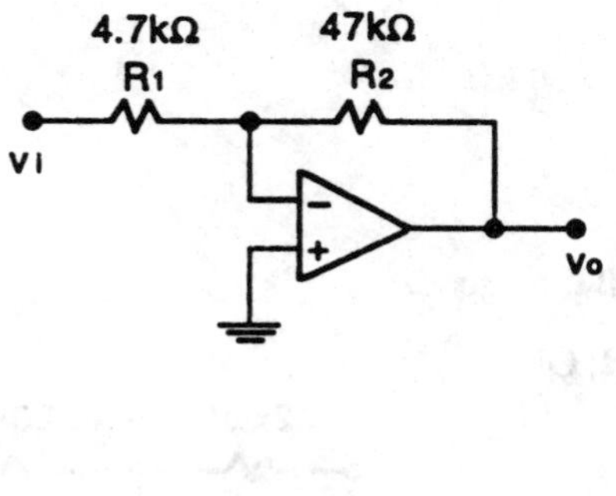

For the desired connection:

$$\frac{\upsilon_o}{\upsilon_i} = -\frac{R_2}{R_1} = -\frac{47}{4.7} = \mathbf{-10\ V/V}.$$

For R_1 and R_2 exchanged:

$$\frac{\upsilon_o}{\upsilon_i} = -\frac{R_2'}{R_1'} = -\frac{4.7}{47} = \mathbf{-0.1\ V/V}.$$

2.5

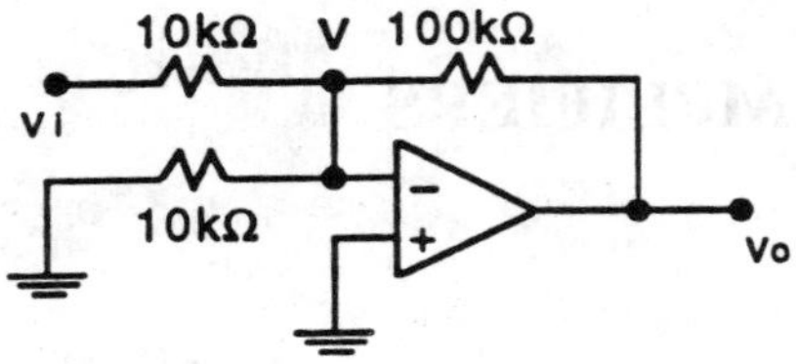

For an ideal op amp: to obtain $\upsilon_o = +10$ V, $\upsilon_- = 0$ V, and the current in the grounded 10kΩ resistor is zero. Thus, $\upsilon_i = -\dfrac{10}{100\text{k}\Omega}\,10\text{k}\Omega =$ **−1 V**.

For gain $= 10^4$: For $\upsilon_o = +10$ V, $\upsilon_- = -10/10^4 = -10^{-3}$ V.

Whence current in 100 kΩ is $(10 - -10^{-3})/10^5 = 10^{-4}$ A, to the υ_- node;

199z current in grounded 10 kΩ is $10^{-3}/10^4 = \mathbf{10^{-7}}$ A, to the υ_- node;

199z current in input 10 kΩ is $(10^{-4} + 10^{-7})$ A $= 10^{-4}$ A, to the input.

$\therefore\ \upsilon_I = -10^{-3} - 10^4\,(10^{-4} + 10^{-7}) = -10^{-3} - 1 - 10^{-3} = -(1 + 2\times 10^{-3}) =$ **−1.002 V**.

2.6 Want Gain of −2 V/V with three 100 kΩ resistors:

There are **2 solutions**:

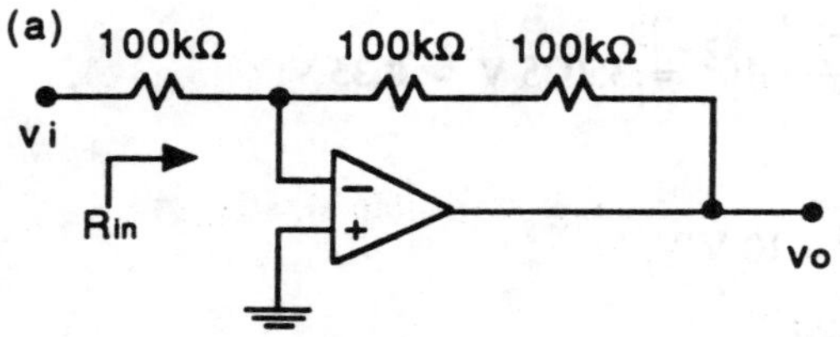

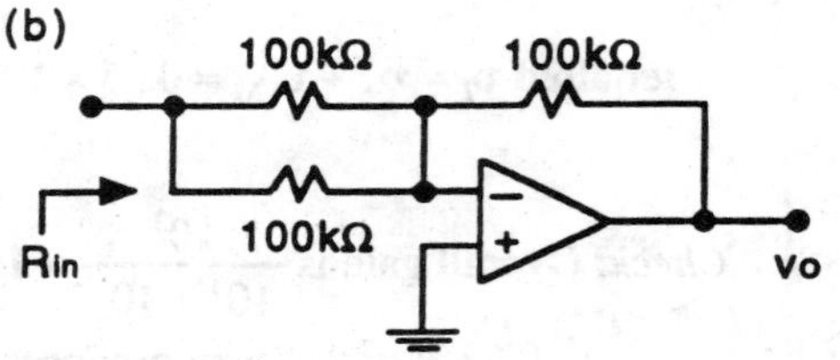

For (a), $\dfrac{\upsilon_o}{\upsilon_i} = -\dfrac{200\text{k}\Omega}{100\text{k}\Omega} = -2$ V/V, and R_{in} = **100 kΩ**.

For (b), $\dfrac{\upsilon_o}{\upsilon_i} = -\dfrac{100\text{k}\Omega}{100\text{k}\Omega/2} = -2$ V/V, and $R_{in} = \dfrac{100\text{k}\Omega}{2}$ = **50 kΩ**.

2.7 There are 2 potential solutions:

a) $R_1 = 220$ kΩ, and $R_2 = 10(220\text{k}\Omega) = 2.2$ MΩ > 1 MΩ; no good.

b) **R_2 = 220 kΩ, and R_1** $= \dfrac{220k\Omega}{10}$ = **22 kΩ** << 1 MΩ; OK.

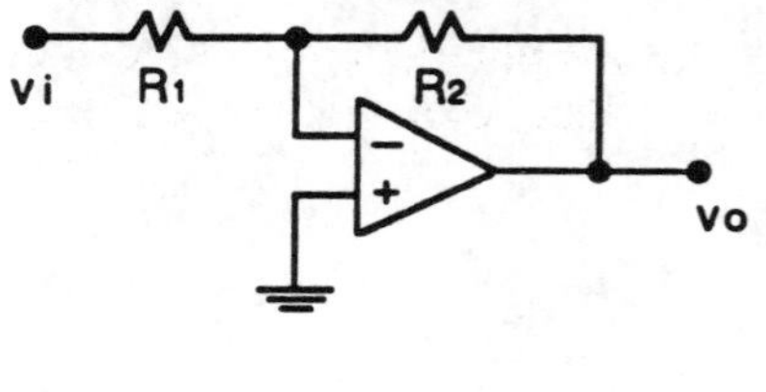

2.8 For an inverting op amp, with $R_{in} = 100\ \text{k}\Omega$, $R_1 = 100\ \text{k}\Omega$.

For a direct design of gain $= -20$ V/V, $R_2 = 20(R_1) = 2\ \text{M}\Omega$; not allowed:

Consider a network for R_2:

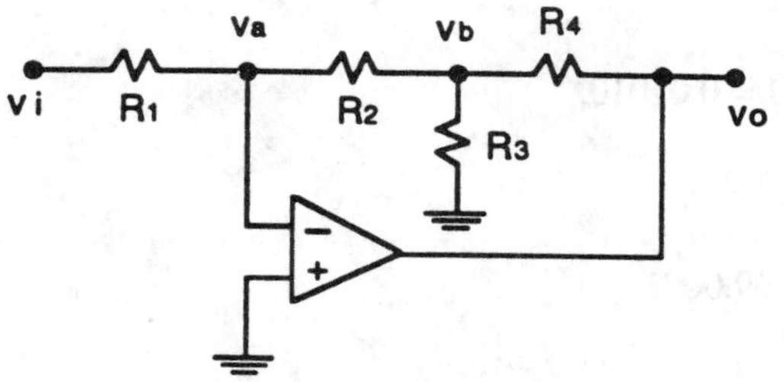

There are many possible designs:

(a) Make $R_1 = R_2 = R_3 =$ **100 kΩ**:

199z $\therefore\ \upsilon_b = -\dfrac{R_2}{R_1}\upsilon_i = -\upsilon_i$

199z and $\upsilon_o = \upsilon_b\left(1 + \dfrac{R_4}{R_2 \parallel R_3}\right)$.

199z $\therefore\ \upsilon_o = -\upsilon_i\left(1 + \dfrac{R_4}{50k}\right) = -20\upsilon_i$.

$\therefore\ 1 + \dfrac{R_4}{50k} = 20 \rightarrow R_4 = (20\text{–}1)\ 50 =$ **950 kΩ**.

(b) Make $R_1 = R_2 = R_4 =$ **100 kΩ**:

See $\upsilon_b = -\upsilon_i$ and, summing currents at υ_b, $\dfrac{\upsilon_o - \upsilon_b}{R_4} = \dfrac{\upsilon_b}{R_3} + \dfrac{\upsilon_b - \upsilon_a}{R_2}$, where $\upsilon_a = 0$.

Thus $\dfrac{\upsilon_o}{R_4} = \upsilon_b\left[\dfrac{1}{R_3} + \dfrac{1}{R_2} + \dfrac{1}{R_4}\right]$, and thus $\dfrac{\upsilon_o}{\upsilon_i} = -R_4\left[\dfrac{1}{R_3} + \dfrac{1}{R_2} + \dfrac{1}{R_4}\right] = -20$ V/V.

199z $\therefore\ \dfrac{100}{R_3} + 1 + 1 = 20$, or $R_3 = \dfrac{100}{20\text{–}2} = \dfrac{100}{18} =$ **5.55 kΩ**.

(c) Make $R_1 = R_3 = R_4 =$ **100kΩ**:

See Thevenin equivalent at υ_b (to the right) is a source $\upsilon_o/2$ with source resistance $R_3 \parallel R_4 = 50$ kΩ. Now, gain to equivalent source ($\upsilon_o/2$) must be 10 V/V:

$\therefore\ R_2 = 10R_1 - 50 = 10(100) - 50 =$ **950 kΩ**.

2.9 Two possible gains:

a) With an ideal op amp:

Gains are $\dfrac{-100}{10} = $ **−10 V/V**, and $\dfrac{-10}{100} = $ **−0.1 V/V**.

b) With an amplifier with gain of $A = 100$ V/V, and $G = \dfrac{-R_2/R_1}{1 + (1 + R_2/R_1)/A}$,

Gains are $G = -\dfrac{100/10}{1 + (1 + 100/10)/100} = \dfrac{-10}{1 + 11/100} =$ **−9.009 V/V**,

199z and $G = -\dfrac{10/100}{1 + (1 + 10/100)/100} = \dfrac{-0.1}{1 + 1.1/100} =$ **−0.0989 V/V**.

See that the error in the high-gain case ≈ 10%, where $G/A \approx 1/10$, 199z and in the low-gain case ≈ 1%, where $G/A \approx 1/100$ (We will see why in Chapter 8).

2.10 From Equation 2.1: $G = \dfrac{-R_2/R_1}{1+(1+R_2/R_1)/A}$. For $\dfrac{R_2}{R_1} = 100 \rightarrow |G| = \dfrac{100}{1+101/A}$.

Now, $|G| \geq 0.9\ (100)$ when $\dfrac{100}{1+101/A} \geq 0.9\ (100)$, or $1 + 101/A \leq \dfrac{1}{0.9} = 1.11$ or $101/A \leq 0.11$.

Whence $A \geq \dfrac{101}{.111} = \mathbf{909\ V/V}$.

Now, $|G| \geq 0.99\ (100)$ when $101/A \leq \dfrac{1}{0.99} - 1 = 0.01010$.

Whence $A \geq \dfrac{101}{.0101} = \mathbf{10^4\ V/V}$.

Check: $G = \dfrac{-100}{1+(1+100)/10^4} = \dfrac{100}{1+101/10^4} = 99.0001$.

2.11

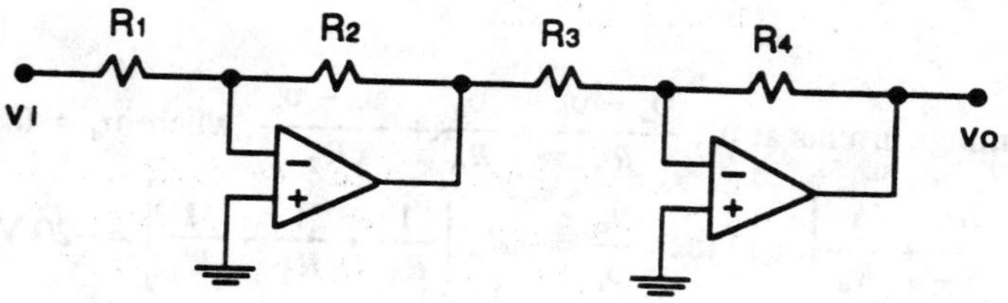

Since $R_{in} = 100\text{k}\Omega$, $R_1 = 100\text{k}\Omega$.
For equal gain/stage,
$G_1, G_2 = \pm\sqrt{200} = -14.14\text{V/V}$,
and $R_2 = 14.14\ (100\text{k}\Omega) = 1.414\text{M}\Omega$: too large.

$\therefore$ Use $R_2 = 1\text{M}\Omega$, for which $G_1 = -10^6/10^5 = -10$ V/V. $\therefore G_2 = -\dfrac{200}{10} = -20\text{V/V}$.

$\therefore$ Use $R_4 = 1\ \text{M}\Omega$, $R_3 = R_4/{-G_2} = 10^6/20 = 50\ \text{k}\Omega$.

Thus, use $\mathbf{R_1 = 100\ k\Omega,\ R_2 = 1\ M\Omega,\ R_3 = 50\ k\Omega,\ R_4 = 1\ M\Omega}$.

2.12 Now for the circuit above, with $R_{in} = 2\text{M}\Omega$, make $R_1 = R_{1a} + R_{1b}$, each $1\text{M}\Omega$, since larger resistors are not available, and $R_2 = 1\ \text{M}\Omega$.

$\therefore G_1 = -1/2$ V/V, and G_2 must be $+200/(-1/2) = -400$ V/V.

Use $R_4 = 1\ \text{M}\Omega$, and $R_3 = 10^6/400 = 2.5\ \text{k}\Omega$.

In summary, use $\mathbf{R_1 = 1\ M\Omega}$ **in series with** $\mathbf{1M\Omega,\ R_2 = 1\ M\Omega,\ R_3 = 2.5\ k\Omega,\ R_4 = 1M\Omega}$.

2.13

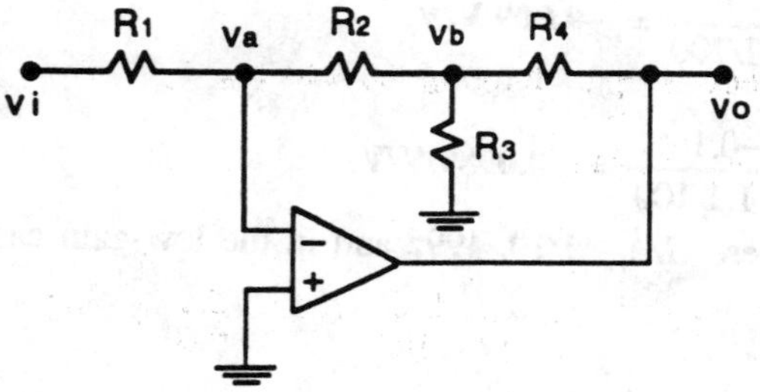

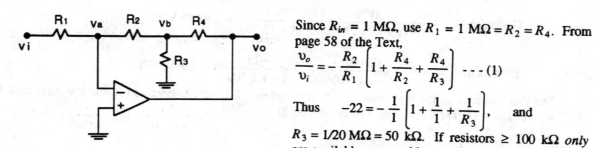

Since $R_{in} = 1\ \text{M}\Omega$, use $R_1 = 1\ \text{M}\Omega = R_2 = R_4$. From page 58 of the Text,

$$\frac{v_o}{v_i} = -\frac{R_2}{R_1}\left[1 + \frac{R_4}{R_2} + \frac{R_4}{R_3}\right] \quad \text{--- (1)}$$

Thus $\quad -22 = -\dfrac{1}{1}\left[1 + \dfrac{1}{1} + \dfrac{1}{R_3}\right]$, and

$R_3 = 1/20\ \text{M}\Omega = 50\ \text{k}\Omega$. If resistors $\geq 100\ \text{k}\Omega$ *only* are available, one could make $R_3 = R_{3a} \parallel R_{3b}$, each of 100 kΩ. *Alternatively*, chose R_3 = 100 kΩ and select a suitable R_4:

$$\therefore\ 22 = \frac{1}{1}\left[1 + \frac{R_4}{1} + \frac{R_4}{0.1}\right], \text{ and } R_4\ (1 + 10) = 22 - 1 = 21\text{k}\Omega.$$

Unfortunately, $R_4 = 21/11 = 1.909$ MΩ is too large! One can see that there are in fact no other choices than using two resistors in parallel for R_3, or in series for R_2 or R_4.

Rewriting (1) above: $\frac{\upsilon_o}{\upsilon_i} = -\left[\frac{R_2}{R_1} + \frac{R_4}{R_1} + \frac{R_2 R_4}{R_1 R_3}\right]$, which, for R_1 = 1 MΩ, becomes $\frac{\upsilon_o}{\upsilon_i} = -\left[R_2 + R_4 + \frac{R_2 R_4}{R_3}\right]$, and for $R_3 = 0.1$ MΩ, then $22 = R_2 + R_4 + 10\, R_2 R_4$.

Now, if either R_2 or R_4 are 1MΩ, say R_2 = 1 MΩ, then $22 = 1 + R_4 + 10R_4$, and $R_4 = 21/11 > 1$ MΩ.

Obviously, using two resistors in series for R_4 is possible also, but not as nice from a practical point of view, since a new circuit node (and connection) must be found. Thus, the first solution (where $R_3 = R_{3a} \parallel R_{3b}$) is the preferred one:

In summary, use **$R_1 = R_2 = R_4$ = 1 MΩ, and R_3 = 100 kΩ || 100 kΩ.**

2.14

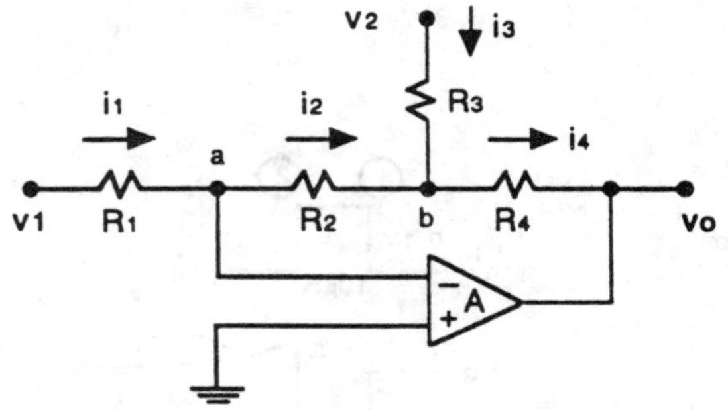

For $\upsilon_1 = 0$: For $A = \infty$, and the feedback working, $\upsilon_a = 0$. Since $i_1 = i_2$, then $\upsilon_b = 0$. Now, for $\upsilon_b = 0$, $i_3 = i_4$, and $\upsilon_o = -(R_4/R_3)\,\upsilon_2$. *For* $\upsilon_2 = 0$: $\upsilon_o = -\upsilon_1\left[\frac{R_2}{R_1}\right]\left[1 + \frac{R_4}{R_2} + \frac{R_4}{R_3}\right]$, as derived in Example 2.2 on page 57 of the Text. Using superposition:

$$\mathbf{\upsilon_o = -\frac{R_2}{R_1}\left[1 + \frac{R_4}{R_2} + \frac{R_4}{R_3}\right]\upsilon_1 - \frac{R_4}{R_3}\upsilon_2.}$$

SECTION 2.4: Other Applications of the Inverting Configuration

2.15 *Using* $\upsilon_o = V_C - \frac{1}{CR}\int_0^t \upsilon_i(t)dt$, see:

For $\upsilon_i = +1$ V, the output $\upsilon_o = V_C - \frac{1}{1\times 10^{-3}}\int_0^t 1\text{dt} = V_C - 10^3\, t$.

That is, the output is a *negative-going ramp* with slope of **1 V/ms** or 1000 V/s, proceeding from V_C = 10 V downward, reaching zero in **10 ms**.

Directly: Following the 1 V step, the input current is $\frac{1}{R}$, charging *C*, causing υ_o to fall at a rate $\frac{V}{T} = \frac{I}{C} = \frac{1}{RC} = \frac{1}{10^{-3}} = 10^3$ V/s, moving 10 V in $\frac{10}{10^3}$ = 10 ms.

2.16 Assuming $V_C = 0$ V, $\upsilon_o = \frac{-1}{CR}\int \upsilon_i\, dt = \frac{1}{CR}\int 0.1 \sin 2\pi\, 60\, t = -\frac{1}{10\times 10^{-3}}\, 0.1\left[\frac{-1}{2\pi\, 60}\right]\cos 2\pi\, 60\, t$

$= 26.5 \times 10^{-3} \cos 2\pi\, 60\, t$

$\therefore$ **the output is a sine wave of 26.5 mV peak, leading the input by 90°**, or, alternatively, is an inverted sine wave, lagging by 90°. Note that the latter idea, that of a lagging inverted output is the most consistent with the STC low-pass view of the Miller integrator.

2.17

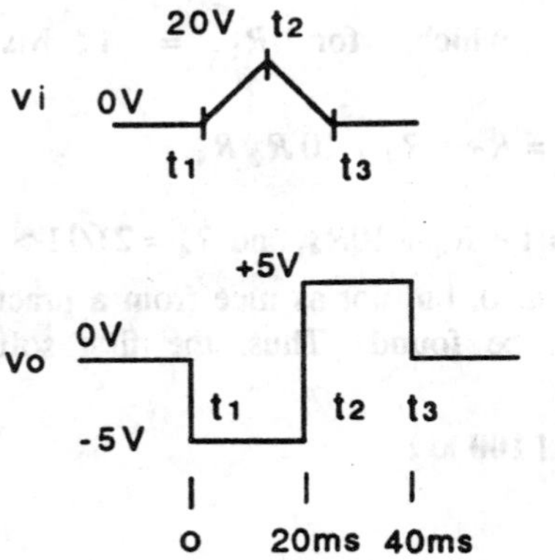

$\upsilon_o = -CR \dfrac{d\upsilon_i}{dt} = -5 \times 10^{-3} \dfrac{d\upsilon_i}{dt} = 1$ V. $\therefore \dfrac{d\upsilon_i}{dt} = -$ 200 V/s. That is, for an output of +1V, the input must **fall at a rate of 200 V/s.**

See $t_1 = 0$,

$t_2 = \dfrac{20-0}{1/10^{-3}} = 20$ ms,

$t_3 = 20 \times 10^{-3} - \dfrac{20}{-1/10^{-3}} = 40$ ms.

For the rise, $\upsilon_o = -5$ ms $\times$ 1 V/ms $= -5$ V.

For the fall, $\upsilon_o = -5$ ms $\times -1$ V/ms $= +5$ V.

2.18

(a)

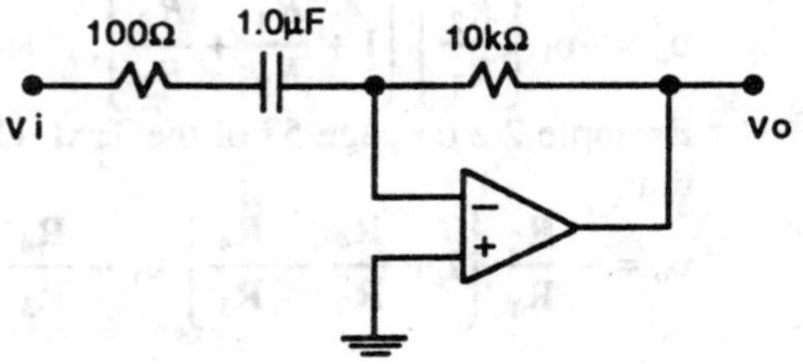

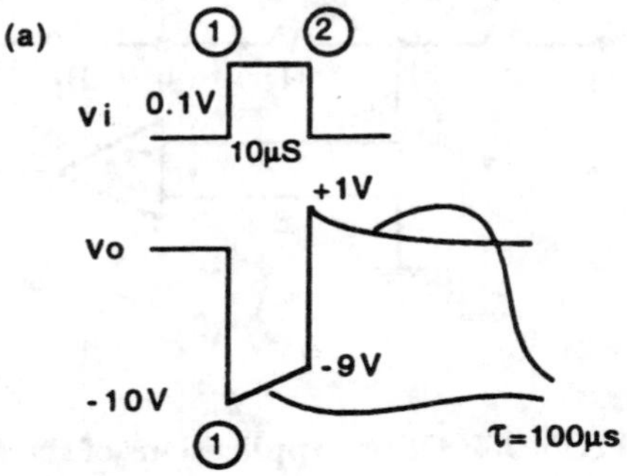

Immediately, upon the input rise, $\upsilon_o = -\left[\dfrac{10k\Omega}{100\Omega}\right] = -10$ V. Time constant: $\tau = 100\Omega \times 1 \times 10^{-6} = 10^{-4}$ s = 100 µs. In 10µs: output rises (almost linearly) to $-10 + \dfrac{10 \times 10^{-6}}{100 \times 10^{-6}} \times 10 = -9$ V. Then, at input fall, output changes by 10 V to +1 V.

(b)

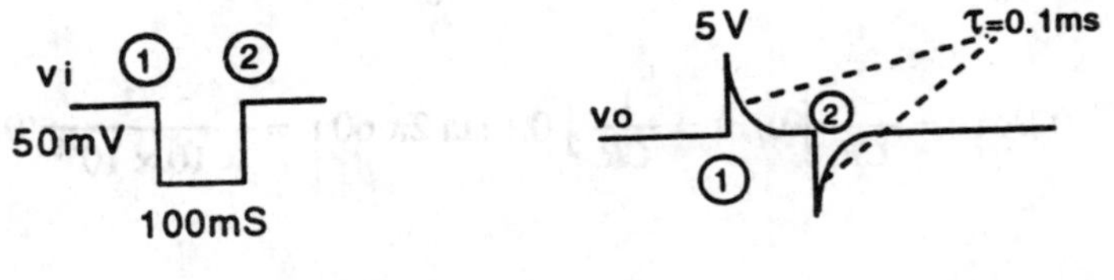

At the falling edge of the input (time t_1) the output rises by 50 mV × 10kΩ/100Ω = 5 V, then begins to fall with a time constant, $\tau = RC = 100 \times 10^{-6} = 10^{-4}$s = 0.1 ms. By the rising edge of the input (100/0.1 = 1000 time constants later), the output has reached 0 V.

2.19

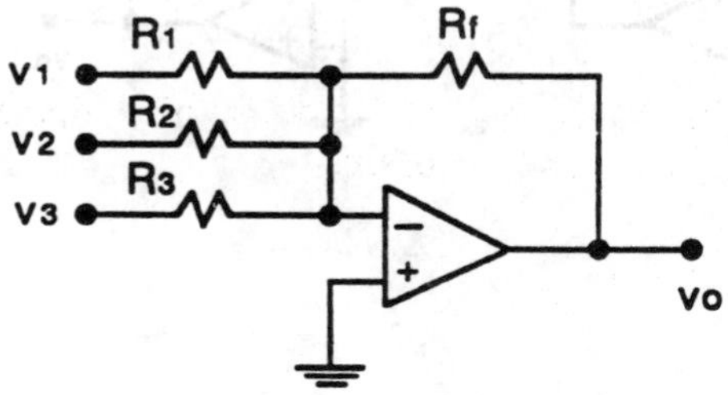

Want, $\upsilon_o = -(\upsilon_1 + 2\upsilon_2 + 3\upsilon_3)$

$$= -3\left[\frac{\upsilon_1}{3} + \frac{\upsilon_2}{3/2} + \frac{\upsilon_3}{1}\right].$$

But, $\upsilon_0 = -R_f\left[\frac{\upsilon_1}{R_1} + \frac{\upsilon_2}{R_2} + \frac{\upsilon_3}{R_3}\right]$.

Thus, make **R_3 = 10 kΩ, R_f = 30 kΩ,**

R_1 = 30 kΩ, and $R_2 = \frac{30}{2}$ = 15 kΩ.

2.20 Want $\upsilon_o = \upsilon_1 + 2\upsilon_2 - 3\upsilon_3$.

See that there are several decompositions:

a) $\upsilon_o = -(-\upsilon_1 - 2\upsilon_2 + 3\upsilon_3)$,

b) $\upsilon_o = -\left\{-(\upsilon_1 + 2\upsilon_2) + 3\upsilon_3\right\}$,

c) $\upsilon_o = -\left\{-(\upsilon_1 + 2\upsilon_2 - 3\upsilon_3)\right\}$, with corresponding circuits:

(a)

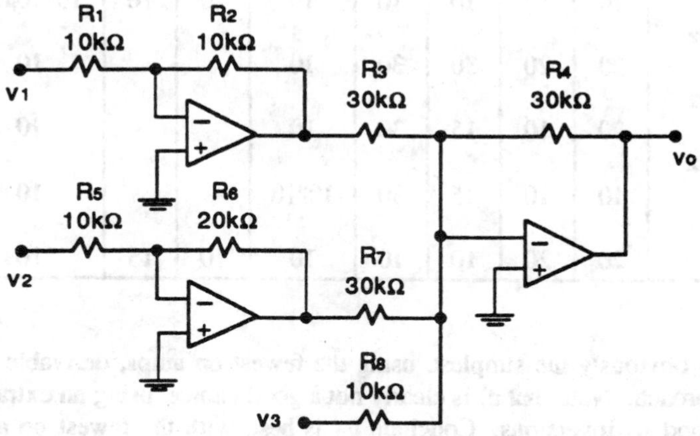

(b)

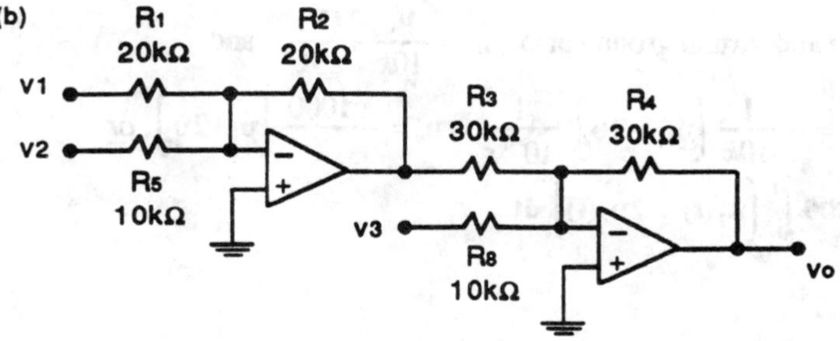

2.20 (Continued)

(c)

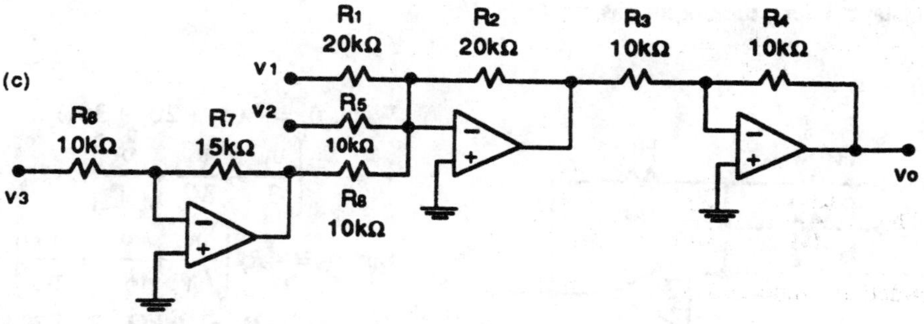

For each circuit, there are many variants of which some are:

199z Version 199z	R_1	R_2	R_3	R_4	R_5	R_6	R_7	R_8	ΣR	# Amp.
199z a_1	10	10	30	30	10	20	30	10	150	3
199z a_2	10	10	30	30	10	10	15	10	125	3
199z a_3	10	10	10	10	10	20	10	10‖10‖10	110	3
199z b_1	20	20	30	30	10			10	120	2
199z b_2	20	10	15	30	10			10	95	2
199z b_3	10	10	15	30	10‖10			10	95	2
199z c_1	20	20	10	10	10	10	15	10	105	3

***Conclusion*:**

Note that b) is obviously the simplest, using the fewest op amps, derivable directly from c) which is the brute force approach. Note that a) is clearly not a good choice, using an extra op amp to separate (unnecessarily) the υ_1 and υ_2 inversions. Conclude b_2 is best, with the fewest op amps, the fewest resistors, the lowest total resistance, and input resistance ≥10 kΩ.

2.21 For an ideal op amp and virtual ground at υ_-, $i_C = \frac{\upsilon_1}{10k} + \frac{\upsilon_2}{5k}$, and

$$\upsilon_o = 0 - \frac{i_C}{Cs} = -\frac{1}{10k}\left[\upsilon_1 + 2\upsilon_2\right]\frac{1}{10^{-7}s} = \upsilon_o = -\frac{1000}{s}\left[\upsilon_1 + 2\upsilon_2\right], \text{ or}$$

$$\upsilon_o = V_o - 1000\int_0^t \left[\upsilon_1(t) + 2\upsilon_2(t)\right] dt.$$

2.22

$$\frac{\upsilon_o(s)}{\upsilon_i(s)} = -\frac{Z_2(s)}{Z_1(s)} = -\frac{R_2 \| \frac{1}{C_2 s}}{R_1 \| \frac{1}{C_1 s}} = -\frac{\frac{R_2/C_2 s}{R_2+1/C_2 s}}{\frac{R_1/C_1 s}{R_1+1/C_1 s}} = -\frac{\frac{R_2}{1+R_2 C_2 s}}{\frac{R_1}{1+R_1 C_{1s}}} .$$

Thus $\frac{\upsilon_o(s)}{\upsilon_i(s)} = -\left[\frac{R_2}{R_1}\right] \times \left[\frac{1+R_1 C_1 s}{1+R_2 C_2 s}\right]$,

independent of frequency if $R_1 C_1 = R_2 C_2$.

(a) For $C_2 = 0.1\ C_1 = 0.1\mu F$, and $R_2 = 10\, R_1 = 10^5 \Omega$:

See $\frac{\upsilon_o}{\upsilon_i} = -10\left[\frac{1+10^4 \times 10^{-6}\, s}{1+10^5 \times 10^{-7}\, s}\right] = -10\left[\frac{1+10^{-2} s}{1+10^{-2} s}\right] =$

– 10V/V, independent of frequency.

(b) For R_2 raised to 1MΩ:

See $\frac{\upsilon_o}{\upsilon_i} = -\frac{10^6}{10^4}\left[\frac{1+10^{-2} s}{1+10^{-1} s}\right] = -100\left[\frac{1+s/100}{1+s/10}\right]$.

(c) For R_2 lowered to 10 kΩ:

See $\frac{\upsilon_o}{\upsilon_i} = -1\left[\frac{1+s/100}{1+s/1000}\right]$.

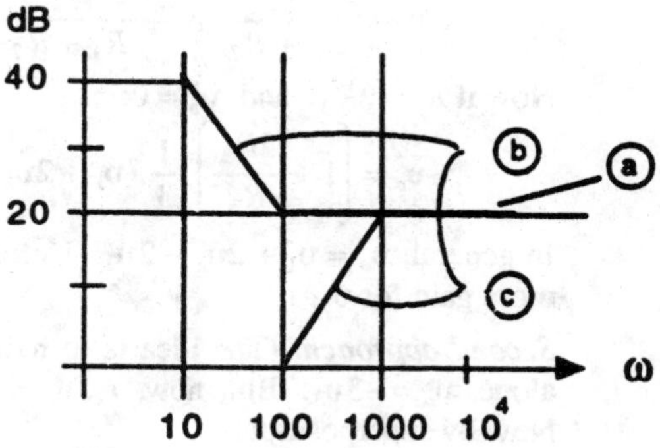

SECTION 2.5: The Non-Inverting Configuration

2.23 Normally: $\frac{\upsilon_o}{\upsilon_i} = 1 + \frac{R_2}{R_1} = 1 + \frac{47}{4.7} = 11$ V/V.

With resistor exchange: $\frac{\upsilon_o}{\upsilon_i} = 1 + \frac{R_2}{R_1} = 1 + \frac{4.7}{47} = 1.10$ V/V.

2.24 Want $\frac{\upsilon_o}{\upsilon_i} = 1.5$ V/V, with three 1 kΩ resistors:

Solutions:

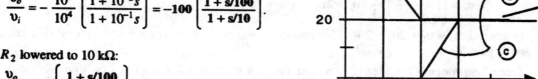

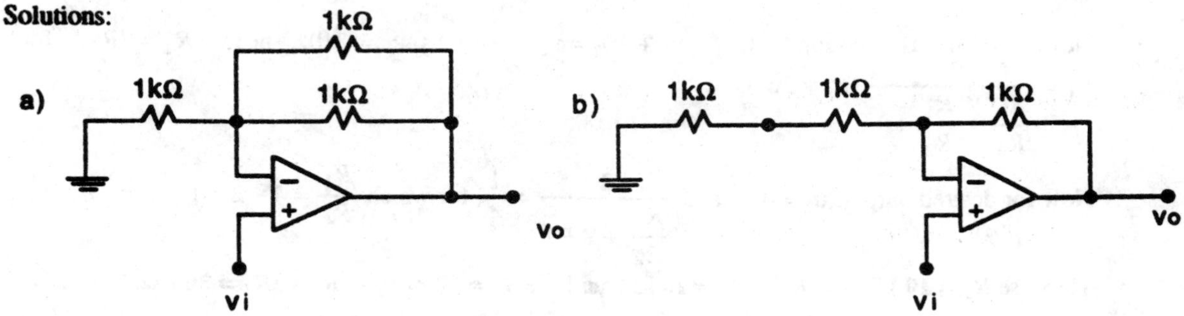

2.25 Want $\upsilon_o = \upsilon_1 + 2\upsilon_2 - 3\upsilon_3$:

Simple approach:

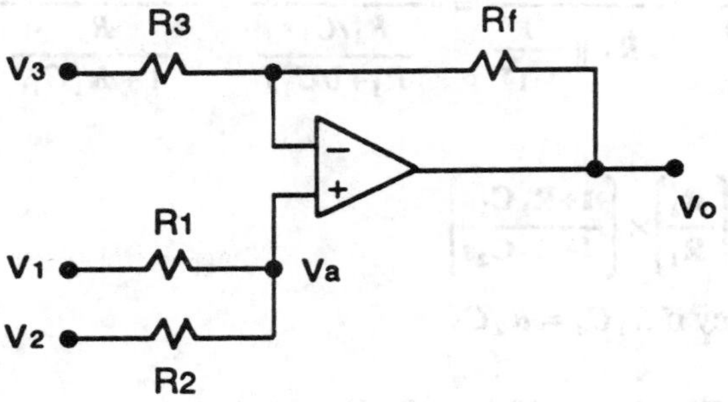

Make $R_2 = R_1/2 \rightarrow R_1 = 2R_2$.

$$\therefore \upsilon_a = \frac{R_2}{R_1 + R_2}\upsilon_1 + \frac{R_1}{R_1 + R_2}\upsilon_2 = \frac{R_2}{3R_2}\upsilon_1 + \frac{2R_2}{3R_2}\upsilon_2 = \frac{1}{3}(\upsilon_1 + 2\upsilon_2).$$

Now if $R_f = 2R_3$, and $\upsilon_3 = 0$:

$$\therefore \upsilon_o = \left[1 + \frac{2R_3}{R_3}\right]\frac{1}{3}(\upsilon_1 + 2\upsilon_2) = (\upsilon_1 + 2\upsilon_2).$$

In general, $\upsilon_o = \upsilon_1 + 2\upsilon_2 - 2\upsilon_3$. Unfortunately, neat and simple, but **not quite right**! We see the need for more gain for υ_3.

Second approach: (The idea is to reduce the gain given to $(\upsilon_1 + 2\upsilon_2)$). Let $R_f = 3R_3$, such that for υ_3 alone, $\upsilon_o = -3\upsilon_3$. But, now, $\upsilon_o/\upsilon_a = 4$. Introduce an additional input $\upsilon_4 = 0$ V connected to υ_a by R_4. Now by superposition,

$$\upsilon_a = \frac{R_2 \parallel R_4}{R_1 + R_2 \parallel R_4}\upsilon_1 + \frac{R_1 \parallel R_4}{R_2 + R_1 \parallel R_4}\upsilon_2 = \frac{\frac{R_2R_4}{R_2+R_4}}{R_1+\frac{R_2R_4}{R_2+R_4}}\upsilon_1 + \frac{\frac{R_1R_4}{R_1+R_4}}{R_2+\frac{R_1R_4}{R_1+R_4}}\upsilon_2,$$

or

$$\upsilon_a = \frac{R_2 R_4}{R_1 R_2 + R_1 R_4 + R_2 R_4}\upsilon_1 + \frac{R_1 R_4}{R_1 R_2 + R_2 R_4 + R_1 R_4}\upsilon_2.$$

Now, if it is required that $\upsilon_1 + 2\upsilon_2 = \upsilon_o$, (for $\upsilon_3 = 0$), need $R_1 = 2R_2$, for which $\upsilon_a = \dfrac{1}{\frac{R_1}{R_4} + \frac{R_1}{R_2} + 1}(\upsilon_1 + 2\upsilon_2)$.

Now for desired output, $\upsilon_o = 4\upsilon_a$, and $\dfrac{1}{\frac{R_1}{R_4} + 2 + 1} = \dfrac{1}{4}$, for which $\dfrac{R_1}{R_4} = 4 - 2 - 1 = 1$, or $R_4 = R_1 = 2R_2$.

Thus use $R_2 =$ **10 kΩ** $\rightarrow R_1 = R_4 =$ **20 kΩ**, and use $R_3 =$ **10 kΩ** $\rightarrow R_f = 3R_3 =$ **30 kΩ**.

2.26

(1)

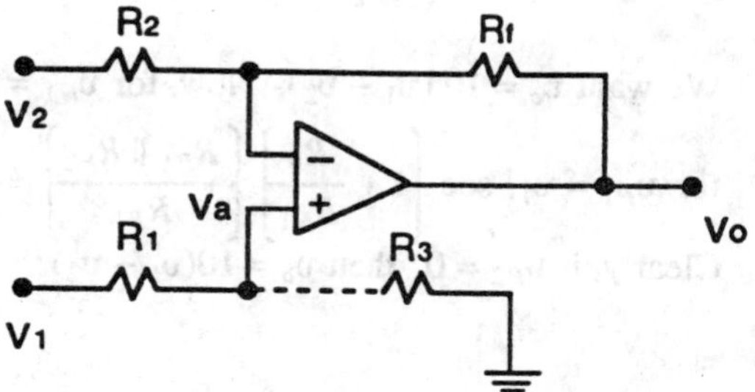

Want $\upsilon_o = 10(\upsilon_1 - \upsilon_2)$. Try: (1)

(2)

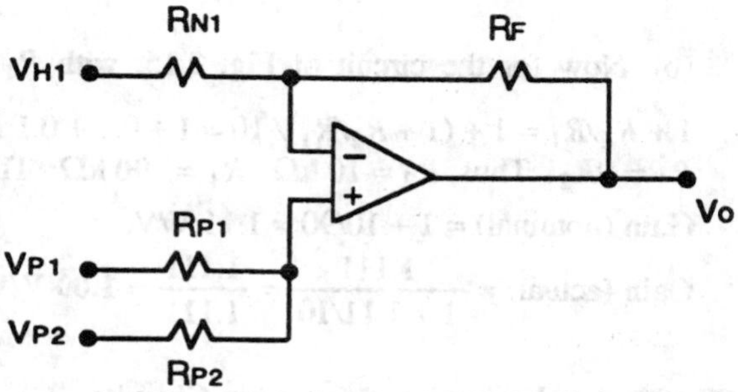

For υ_2 alone, $\upsilon_o = -10\upsilon_2 = -R_f/R_2\ \upsilon_2$. 199z $R_f = 10R_2$

For $R_2 = 10\ k\Omega \rightarrow R_f = 100\ k\Omega$, for which $\upsilon_o = -10\ \upsilon_2$

Now for υ_1 alone, ($R_3 = \infty$), $\upsilon_o = \left[1 + R_f/R_2\right] \upsilon_1 = (1 + 10)\ \upsilon_1 = 11\upsilon_1$

and, together, $\upsilon_o = 11\upsilon_1 - 10\upsilon_2$

We see too much of υ_1: It needs to be attenuated:

Introduce an additional resistor R_3 from υ_a to ground:

$$\therefore\ \upsilon_a = \frac{R_3}{R_1 + R_3}\upsilon_1 = \frac{10}{11}\upsilon_1$$

Now for $R_1 = 10k\Omega$, $R_3/(10 + R_3) = 10/11$, and $11\ R_3 = 100 + 10R_3$,

whence $R_3 = 100k\Omega$, with the result that $\upsilon_o = 11\ \upsilon_1 \times 10/11 - 10\ \upsilon_2 = 10(\upsilon_1 - \upsilon_2)$ as required

In summary: $R_1 = R_2 = 10\ k\Omega$, $R_f = R_3 = 100\ k\Omega$.

This configuration, with $R_f/R_2 = R_3/R_1$, is referred to as a **difference amplifier**.

Alternatively: From the equations supplied in P2.33 on page 109 of the Text: (2)

$$v_o = -\left[\frac{R_F}{R_{N1}}\right] v_{N1} + \left(1 + \frac{R_F}{R_{N1}}\right)\left[v_{P1}\left(\frac{R_{P1} \parallel R_{P2}}{R_{P1}}\right) + v_{P2}\left(\frac{R_{P1} \parallel R_{P2}}{R_{P2}}\right)\right]$$

We want $v_o = 10\,(v_1 - v_2)$. Now, for $v_{N1} = v_2$, see $\frac{R_F}{R_{N1}} = 10$, and for $R_{N1} = 10\text{ k}\Omega$, $R_F = 100\text{ k}\Omega$. Now for $v_{P1} = v_1$, see $\left(1 + \frac{R_F}{R_{N1}}\right)\left(\frac{R_{P1} \parallel R_{P2}}{R_{P1}}\right) = 10 = 11\left(\frac{R_{P2}}{R_{P1} + R_{P2}}\right)$, and for $R_{P1} = 10\text{ k}\Omega$, $R_{P2} = 100\text{ k}\Omega$.
Clearly, if $v_{P2} = 0$, then $v_o = 10(v_1 - v_2)$.

2.27

(a) For the circuit of Fig. 2.17, using the result in Exercise 2.13 on page 67 of the Text,
$G = \frac{1 + R_2/R_1}{1 + (1 + R_2/R_1)/A}$. Now with $R_2 = 0$, $R_1 = \infty$, $A = 10$ V/V, see $G = \frac{1}{1 + (1/10)}$ = **0.909 V/V**.

(b) Now for the circuit of Fig. 2.15, with R_1, R_2 finite, $G = \frac{1 + R_2/R_1}{1 + (1 + R_2/R_1)/10}$. Now for $G = 1.00$, $1 + R_2/R_1 = 1 + (1 + R_2/R_1)/10 = 1 + 0.1 + 0.1\,R_2/R_1$, and $0.9\,R_2/R_1 = 0.1$, $R_2/R_1 = 0.1/0.9$, and $R_1 = 9R_2$. Thus, $R_2 = 10\text{ k}\Omega$, $R_1 =$ **90 kΩ**. Thus
Gain (nominal) = 1 + 10/90 = 1.11 V/V.
Gain (actual) = $\frac{1.111}{1 + 1.11/10} = \frac{1.111}{1.111}$ = **1.00 V/V**, as required.

SECTION 2.6: Examples of Op-Amp Circuits

2.28

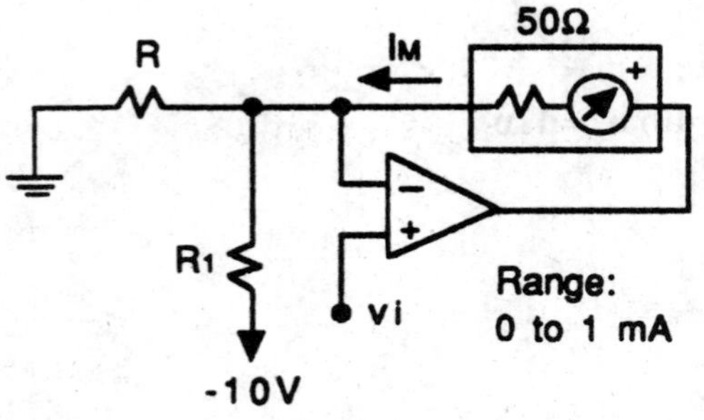

For $v_i = 0$, $v_R = 0$, and the meter is required to read mid-scale. That is, $I_M = 0.5$mA. Thus $R_1 = (0 - -10)/0.5\text{mA} =$ **20kΩ** with the −10V supply. Now, for full-scale reading with $v_i = +1$V, I_M must be 1mA. ∴ $v_i/R + (v_i - -10)/R_1 = 1$mA, and $1/R + (1 + 10)/R_1 = 1$, whence $1/R = 1 - 11/20 = 9/20 = 0.45$, and $R = 1/0.45 =$ **2.22kΩ**. *Check*: for $v_i = -1$ V: $I_M = \frac{-1 - -10}{R_1} - \frac{1}{R} = \frac{9}{20} - \frac{1}{2.22} = 0.45 - 0.45 = 0$ V: OK.

2.29

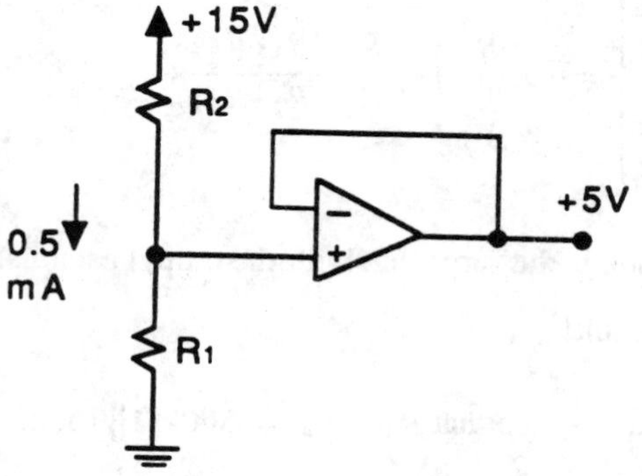

$R_1 = +5V/0.5mA = 10\ k\Omega$; $R_2 = (15-5)/0.5mA = 20\ k\Omega$. Op amp standby dissipation is $2(15V) \times 2mA = 60\ mW$. Op amp dissipation when loaded is $20\ mA\ (15-5) = 200\ mW$. Total op amp dissipation is $200 + 60 = 260\ mW$.

Note also: That the bias network dissipation is $15 \times 0.5\ mA = 7.5\ mW$, and the load power consumption is $5 \times 20\ mA = 100\ mW$.

Thus the total supply power is $60 + 200 + 7.5 + 100 =$ **367.5 mW**.

2.30 From equation 2.4 (or directly by superposition):

$$\upsilon_o = -\frac{R_2}{R_1}\upsilon_1 + \left[1 + \frac{R_2}{R_1}\right]\left[\frac{R_4}{R_3 + R_4}\right]\upsilon_2 = -\frac{R_2}{R_1}\upsilon_1 + \frac{1 + R_2/R_1}{1 + R_3/R_4}\upsilon_2.$$

Hence $\upsilon_o = -\frac{100}{10}\upsilon_i + \frac{1 + 100/10}{1 + 10/100}\upsilon_2 = -\frac{100}{10}\upsilon_1 + \frac{100 + 1000}{100 + 10}\upsilon_2 = 10\,(\upsilon_2 - \upsilon_1)$.

$\therefore$ gain $A = \frac{\upsilon_o}{\upsilon_1 - \upsilon_2} =$ **−10 V/V**: (Note the sign!)

2.31 For this situation, gain $\frac{\upsilon_o}{\upsilon_{s1} - \upsilon_{s2}} = -\frac{100}{10 + 10} =$ **−5 V/V**.

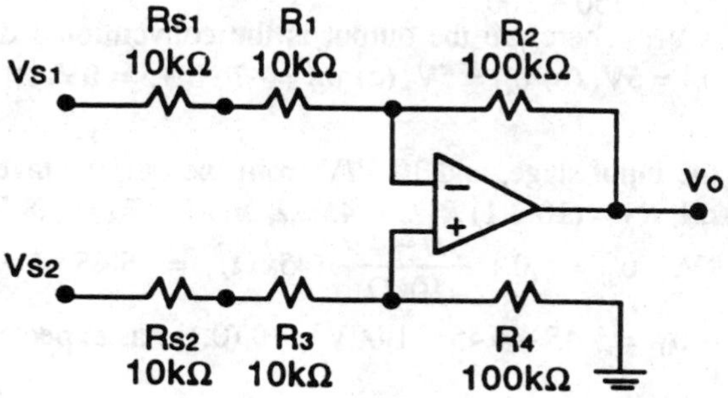

To recover the gain, two possibilities exist:

a) Remove R_1 and R_2 and connect sources directly.

b) Make $R_2 = R_4 = 200k\Omega$.

The latter is probably best for secondary reasons. For example, it reduces (by a factor of 2), the effect of minor variations in R_{S1} and R_{S2}.

For the case in which R_{S2} is 8 kΩ, there are two basic approaches: a) "Pad out" R_3, i.e., add an additional 2kΩ resistor in series with R_{S2} and R_3, or, b) change R_4 to 180kΩ to compensate. (Making $\frac{R_2}{R_{1eq}} = \frac{R_4}{R_{3eq}}$)

2.32 Using superposition, υ_o consists of 3 parts. Thus $\upsilon_o = -\upsilon_1\frac{R_2}{R_1} + \upsilon_2\frac{R_4}{R_3 + R_4}\left[1 + \frac{R_2}{R_1}\right] + \upsilon_3\left[\frac{R_3}{R_3 + R_4}\right]\left[1 + \frac{R_2}{R_1}\right]$, or $\upsilon_o = -\upsilon_1\frac{\mathbf{R_2}}{\mathbf{R_1}} + \upsilon_2\frac{\left[1 + \frac{\mathbf{R_2}}{\mathbf{R_1}}\right]}{\left[1 + \frac{\mathbf{R_3}}{\mathbf{R_4}}\right]} + \upsilon_3\frac{\left[1 + \frac{\mathbf{R_2}}{\mathbf{R_1}}\right]}{\left[1 + \frac{\mathbf{R_4}}{\mathbf{R_3}}\right]}$. Now for $\frac{R_2}{R_1} = \frac{R_4}{R_3}$, see

$$\upsilon_o = -\upsilon_1 \frac{R_2}{R_1} + \upsilon_2 \frac{\left[1+\frac{R_2}{R_1}\right]}{\left[1+\frac{R_1}{R_2}\right]} + \upsilon_3 \frac{\left[1+\frac{R_2}{R_1}\right]}{\left[1+\frac{R_2}{R_1}\right]} = -\upsilon_1 \frac{R_2}{R_1} + \upsilon_2 \frac{R_2}{R_1}\frac{(R_1+R_2)}{(R_1+R_2)} + \upsilon_3,$$

whence $\upsilon_o = \frac{R_2}{R_1}(\upsilon_2 - \upsilon_1) + \upsilon_3$. Note that the operation is the same, but with the output established by υ_3 for $(\upsilon_2 - \upsilon_1) = 0$. This is an interesting and important result!

2.33 *Check* that it is a balanced difference amplifier for υ_1 and υ_2: see that $R_{4b} \| R_{4a}$ = 300kΩ || 150kΩ = 300/3 = 100kΩ = R_2, (OK).

(a) For $\upsilon_1 = \upsilon_2 = 5$ V: See (by superposition), $\upsilon_A = 5\,\frac{100}{100+100} + 15\,\frac{100 \| 150}{300 + 100 \| 150} + 0$, or $\upsilon_A = 5\,(1/2) + 15\,(60/360) = 2.5 + 15/6 = 2.5 + 2.5 = 5\text{V}$. Now, since $\upsilon_A = 5\text{V}$, and $\upsilon_1 = 5\text{V}$, current in $R_1 = 0$, and $\upsilon_O = \mathbf{5V}$.

(b) For $\upsilon_1 = \upsilon_2 = 0\text{V}$, see $\upsilon_A = 15\,\frac{100 \| 150}{300 + 100 \| 150} = 15\left[\frac{60}{360}\right] = \frac{15}{6} = 2.5\text{V}$, and $\upsilon_O = 2.5\,(2) = \mathbf{5V}$.

(c) For $\upsilon_1 = 3\text{V}$, $\upsilon_2 = -2\text{V}$, see $\upsilon_A = -2\left[\frac{1}{2}\right] + 15\left[\frac{1}{6}\right] = -1 + 2.5 = 1.5\text{V}$. Thus $\upsilon_B = 1.5\text{V}$, and

$$\upsilon_O = \upsilon_B - \frac{\upsilon_1 - \upsilon_B}{R_1} \times R_2 = \upsilon_B - \upsilon_1 + \upsilon_B = 2\,(1.5) - 3 = \mathbf{0V}.$$

Alternatively, to use the extended result from P2.32 above, realize that R_{4b} with R_{4a} and the connected supplies have a Thevenin equivalent of $\upsilon_3 = \frac{150}{150+300} \times 15 = 5\text{V}$, and 150k || 300k = 100kΩ. Thus, since $R_1 = R_2 = R_3 = R_4$, $\upsilon_o = \upsilon_2 - \upsilon_1 + \upsilon_3$. Therefore the output is the conventional difference plus the extra (reference) input. Thus (a) $\upsilon_O = \mathbf{5V}$, (b) $\upsilon_O = \mathbf{5V}$, (c) $\upsilon_O = -2 - 3 + 5 = \mathbf{0V}$.

2.34 For an overall gain of 100 V/V, with 10 V/V from the input stage, and 10 V/V from the output stage, see that $1 + 2R_2/R_1 = 10$, and $R_4/R_3 = 10$. For $R_1 = 10\text{k}\Omega$, $R_2 = (10-1)\,R_1/2 = 45\text{k}\Omega$, and for $R_3 = 10\text{k}\Omega$, $R_4 = 100\text{k}\Omega$. Now, for $\upsilon_1 = 5.0\text{V}$, $\upsilon_2 = 4.9\text{V}$, $\upsilon_{01} = 5.0 + \frac{5.0-4.9}{10k\Omega}(45k\Omega) = \mathbf{5.45\ V}$, and $\upsilon_{02} = 4.9 - \frac{5.0-4.9}{10k\Omega}(45k\Omega) = \mathbf{4.45\ V}$. *Check*: $\upsilon_{01} - \upsilon_{02} = 5.45 - 4.45 = 1.00\text{V} = 10\,(0.1)$, as expected!

2.35

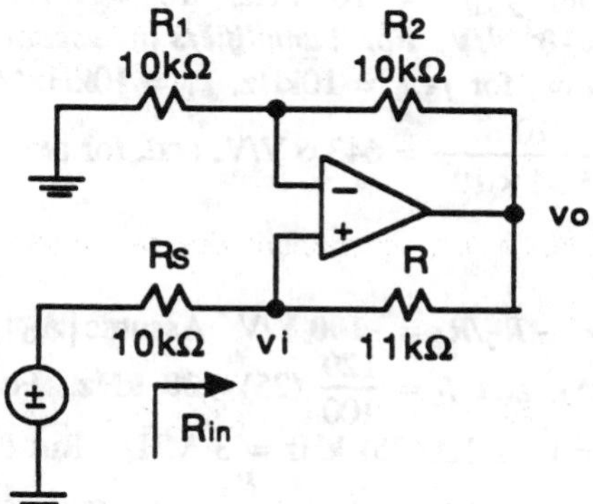

From Example 2.6, $R_{in} = -R\dfrac{R_1}{R_2}$, or $R_{in} = -11\ \text{k}\Omega\left[\dfrac{10}{10}\right] = -11\ \text{k}\Omega$. For the voltage divider with R_S, R_{in}: $\upsilon_I = \dfrac{-11k\Omega}{10k\Omega - 11k\Omega}\upsilon_S = \dfrac{-11}{-1}\upsilon_S = 11\upsilon_S$, and

$\upsilon_O = \left[1 + R_2/R_1\right]\upsilon_I = (1+1)(11)\upsilon_S = 22\upsilon_S$. Now for $\upsilon_S = +0.5$ V, $\upsilon_I = 0.5(11) =$ **5.5 V**, and $\upsilon_O = 2(5.5) =$ **11V**.

2.36

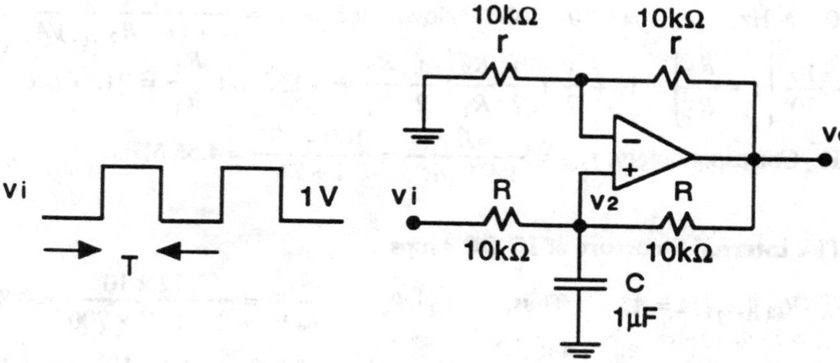

See $\upsilon_2 = \dfrac{1}{CR}\int_0^t \upsilon_i\,dt + V = \dfrac{1}{1\times10^{-6}\times10^4}\int_0^t \upsilon_i\,dt + 0 = 100\int_0^t \upsilon_i\,dt$. Now for each 1 volt pulse: $\Delta\upsilon_2 = 100t\,|_0^T = 100$ T. Now, if 10 V is reached at the end of the tenth pulse: $10\,(100\,T) = 10 - 0 = 10$ V, and $T = \dfrac{10}{100\times10} = 10$ msec. Alternatively, if 10 V is reached at the beginning of the tenth pulse: $9(100\,T) = 10$ V, and $T = \dfrac{10}{9}\times\dfrac{1}{100} = 11.11$ msec. Thus, the pulse width ranges from **10 to 11.11 msec**, and is about **10.55 ms** on the average.

SECTION 2.7: Nonideal Performance of Op Amps

2.37 At and above the cutoff frequency f_{3dB}, $A\times f = f_t = 10^7$. Thus $f_{3dB} = f_t/A_o = 10^7/10^6 =$ **10Hz**. At $f = 100$ kHz, the available gain, $A = f_t/f = 10^7/10^5 =$ **100V/V**.

2.38 Closed-loop gain of 20 dB corresponds to a gain ratio of $10^{20/20} = 10$ V/V. In general, $f_{dB} = f_t/(1 + R_2/R_1)$. For a gain of −10 V/V, $R_2/R_1 = 10$, and $f_{3dB} = 10^7/(1+10) =$ **0.909 MHz**. For a gain of +10 V/V, $1 + R_2/R_1 = 10$, $R_2/R_1 = 9$, and $f_{3dB} = 10^7/(1+9) =$ **1.00 MHz**. The phase shift at the 3 dB frequency is 45°, and 6° at 1/10 the 3 dB frequency. Thus 6° shift is reached at **90.9 kHz** for the inverting amplifier, and at **100 kHz** for the non-inverting, and is less than 6° for all lower frequencies.

2.39 Amplifiers have $A_o = 10^6$ V/V and $f_t = 10^7$ Hz: For maximum bandwidth, use the noninverting form with gain $(1 + R_2/R_1)$. *For a single amplifier*, with f_{3dB} = 10 kHz: $10^4 = 10^7/(1 + R_2/R_1)$. Thus $1 + R_2/R_1 = 10^7/10^4 = 10^3$, and the available gain is **10^3 V/V**. *For 2 amplifiers in cascade*, each with a 3dB frequency at f_1, $f_{3dB} = (\sqrt{2} - 1)^{1/2} f_1 = 0.644 f_1$. Now, for f_{3dB} = 10kHz, f_1 = 10kHz/.644 = 15.54 kHz, with corresponding gain per stage of $\left(1 + \frac{R_2}{R_1}\right) = \frac{10^7}{15.54 \times 10^3} = 643.6$ V/V, and, for two stages in cascade, a gain of $(643.6)^2 = \mathbf{4.14 \times 10^5}$ **V/V**.

2.40 Gain with feedback at low frequencies is likely to be: $-R_2/R_1 = \mathbf{-100}$ **V/V**. Assume $|A_0| >> 100$: Now, for a 3dB frequency f_L < 120 kHz, 100 f_L = 120(25), and $f_L = \frac{120}{100}(25) = \mathbf{30}$ **kHz**. For the closed-loop amplifier, the unity-gain frequency (where $R_2/R_1 = 1$) is 120(25) kHz = 3 MHz. But the unity-gain frequency of the negative-gain amplifier is $\frac{f_t}{(1 + R_2/R_1)}$, which is $\frac{f_t}{2}$ for $\frac{R_2}{R_1} = 1$. Thus $f_t/2$ = 3 MHz, and for the op amp itself, f_t = **6 MHz**.

2.41 f_t = 100 MHz, A_o = 20 V/V: Now, for $\frac{v_o}{v_i} = \frac{-R_2/R_1}{1 + (1 + R_2/R_1)/A_o} = -10$ V/V, $\frac{R_2}{R_1} = 10 + \frac{10}{20}\left(1 + \frac{R_2}{R_1}\right) = 10 + \frac{1}{2} + \frac{1}{2}\frac{R_2}{R_1}$, $\frac{1}{2}\frac{R_2}{R_1} = 10.5$, and $\frac{R_2}{R_1} = 21$. *Check*: $\frac{v_o}{v_i} = \frac{-21}{1 + (22/20)} =$ 10 V/V, OK. Correspondingly, $f_{3dB} = \frac{f_t}{1 + R_2/R_1} = \frac{100 \times 10^6}{1 + 21} = \mathbf{4.55}$ **MHz**.

SECTION 2.8: The Internal Structure of IC OP Amps

2.42 Now, $G_m (R_{i2} \| R_{o1})\mu = A_o$. Thus, $R_{i2} \| R_{o1} = \frac{A_o}{G_m \mu} = \frac{1.12 \times 10^6}{2 \times 10^{-3} \times 700} = 0.8 \times 10^6$, or $R_{01} \| 4 \times 10^6 = 0.8 \times 10^6 = \frac{R_{01} \times 4 \times 10^6}{R_{01} + 4 \times 10^6}$. Thus $4R_{01} = 0.8\, R_{01} + 3.2 \times 10^6$, or $3.2\, R_{01} = 3.2 \times 10^6$, and $R_{01} = \mathbf{1 \times 10^6 \Omega}$. Now, $\omega_t = 2\pi f_t = \frac{G_m}{C}$, or $C = \frac{G_m}{2\pi f_t} = \frac{2 \times 10^{-3}}{2\pi \times 15.9 \times 10^6} = \mathbf{20}$ **pF**.

2.43 Now, $\omega_t = \frac{G_m}{C}$. Thus $G_m = C\,(2\pi f_t) = 5 \times 10^{-12} \times 2\pi \times 10 \times 10^6 = \mathbf{314}$ **μA/V**.

SECTION 2.9: Large-Signal Operation of Op Amps

2.44 The largest possible peak output with zero average is **8 V**. The corresponding input has a peak-to-peak value of 2(8)/10 = **1.6 V**.

2.45 A 6-V pp triangle wave at fHz moves 6 V in $1/2 \times 1/f$ seconds with a slope of $6/(1/2f) = 12f$V/s. Now, if this just matches the slew rate: $12f$ V/s = 10 V/μs, and $f = 10/12 \times 10^6$ Hz = **0.833 MHz**.

2.46 $SR = I_{max}/C$, and $f_t \approx \frac{G_m}{2\pi C}$: Thus, the biassing current, $I_{max} = C(SR) = 30 \times 10^{-12} \times 5$ V/μs, or $I_{max} = 30 \times 10^{-12} \times 5 \times 10^6 = \mathbf{150 \mu A}$. Also, $G_m = 2\pi f_t C = 2\pi \times 10 \times 10^6 \times 30 \times 10^{-12}$ = **1.88 mA/V**. *Check*; Using the result of P2.64 on P113 of the text: $SR = \left(\frac{I_{max}}{G_m}\right)\omega_t = \frac{150 \times 10^{-6}}{1.88 \times 10^{-3}} \times 2\pi \times 10 \times 10^6 = \mathbf{5.01}$ **V/μs**, OK.

2.47 The slew-rate-limited bandwidth of an amplifier with a sinewave peak output V_o is $f_R = \dfrac{SR}{2\,\pi\,V_o}$. Now $f_R = f_b$, when $\mathbf{V_o = SR/(2\,\pi\,f_b)}$. Now, for $SR = 2$ V/μs, and $f_b = 0.5 \times 10^6$, $V_o = 2 \times 10^6/(2\,\pi \times 0.5 \times 10^6) =$ **0.64 V peak**.

SECTION 2.10: Common-Mode Rejection

2.48 The common-mode input (at 60Hz) is 8 V peak, and output is 0.6 V peak. The difference-mode input (at 1 kHz) is 1+1 = 2 mV peak, and output is 60 mV peak. Thus, the Common-mode gain = 0.6V/8V = **0.075 V/V**, and Difference-mode gain = 60mV/2mV = **30 V/V**. Thus the CMRR = 30/.075 = **400** ≡ $20\log_{10} 400$ = **52 dB**.

2.49 CMRR = $10^{100/20} = 10^5$. For a difference-mode output of 2 Vpp, the required common-mode signal output = 1/100 × 2 = 20 mVpp. Thus $\dfrac{200V/V}{20mV/\upsilon_{icm}} = 10^5$, and $\upsilon_{icm} = \dfrac{10^5}{200} \times 20 \times 10^{-3} =$ **10 Vpp**.

2.50

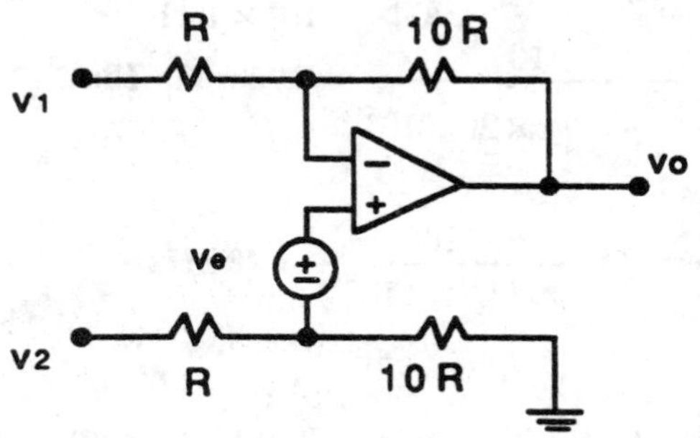

CMRR = 120 dB ≡ 10^6. Now for υ_{icm} at the difference-amplifier input, the common-mode signal at the op amp input is $\dfrac{10}{11}\,\upsilon_{icm}$. The equivalent error at the op amp terminals is $\upsilon_e = ((10/11)\,\upsilon_{icm})/10^6$. This is multiplied by (1 + 10/1) = 11 to appear at the output as $\upsilon_{icm} \times 10^{-5}$. Thus the common-mode gain of the difference amplifier is $\mathbf{10^{-5}}$ **V/V**, and, overall, $CMRR = \dfrac{A_d}{A_{cm}} = \dfrac{A_d}{\upsilon_{ocm}/\upsilon_{icm}} = \dfrac{10}{10^{-5}} = 10^6 \equiv$ **120 dB**, the *same* as that of the basic op amp.

SECTION 2.11: Input and Output Resistances

2.51 For a noninverting buffer, $\beta = R_1/(R_1 + R_2) = 1$. Thus $(1 + A\beta) = 1 + 10^4 \times 1 \approx 10^4$. Now at low frequencies, $R_{in} \approx 2\,R_{icm} \,||\, \left[(1 + A\beta)\,R_{id}\right] = (20 \times 10^6) \,||\, \left[10^4 \times 10^3\right] = (2 \times 10^7) \,||\, 10^7 = 6.67 \times 10^6 =$ **6.7 MΩ**.

2.52 Generally speaking, $R_{in} = 2R_{icm} \,||\, \left[(1 + A\beta)\,R_{id}\right]$. Hence $1 + R_2/R_1 = 10 = (R_1 + R_2)/R_1$, and $\beta = R_1/(R_1 + R_2) = 1/10 = 0.1$, with $A = \dfrac{A_o}{1 + s/(\omega_t/A_o)} \approx \dfrac{\omega_t}{s} = \dfrac{2\pi \times 10^4}{s}$, at high frequencies. Now, at very low frequencies, $R_{in} \approx 2(50 \times 10^6) \,||\, [(1 + 0.1\,(10^3))10^4] \approx 10^8 \,||\, (1.01 \times 10^6) \approx \mathbf{10^6\ \Omega}$. At high frequencies: $Z_{in} = 10^8 \,||\, \left[(1 + \dfrac{2\pi \times 10^4}{s} \times 0.1)\,10^4\right] = 10^8 \,||\, \left[10^4 + \dfrac{2\pi \times 10^7}{s}\right]$.

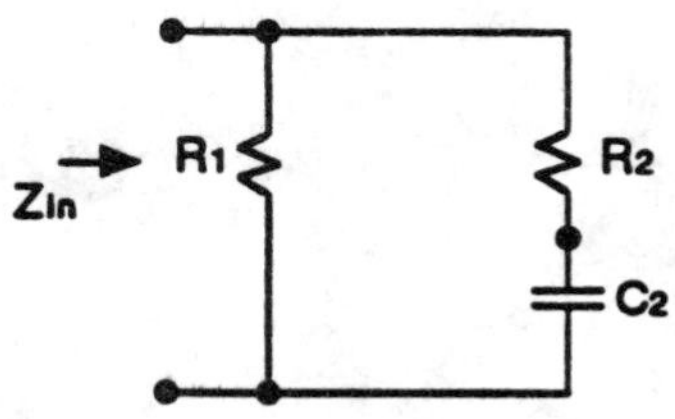

Now, $R_1 = 10^6 =$ **1 MΩ** (the low frequency part including $10^8\Omega$), $R_2 =$ **10 kΩ**, and $C_2 = \dfrac{1}{2\pi \times 10^7} =$ **15.9 nF**. Phase is 45° at $f_1 = \dfrac{1}{2\pi\,C_2\,R_1} = \dfrac{1}{2\pi \times 15.9 \times 10^{-9} \times 10^6} =$ **10 Hz**, and at $f_2 = \dfrac{1}{2\pi\, C_2\, R_2} = \dfrac{1}{2\pi \times 15.9 \times 10^{-9} \times 10^4}$ **= 1kHz.**

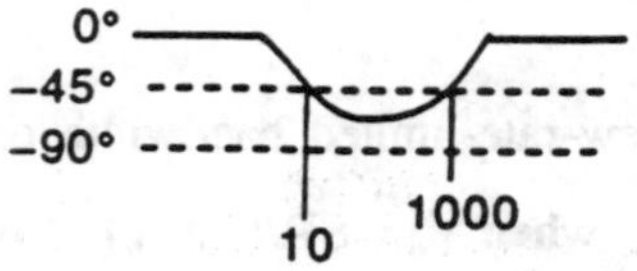

2.53 $R_{out} = \dfrac{R_o}{1 + A\beta}$, where $R_o = 10^5\Omega$, and $\beta = \dfrac{R_1}{R_1 + R_2} = \dfrac{1}{1 + R_2/R_1} = \dfrac{1}{1 + 10} = 0.0909$, with $1 + A\beta = 1 + 10^4 \times \dfrac{1}{11} = 910$. Thus at low frequencies, $R_{out} = \dfrac{100k\Omega}{910} = 110\ \Omega$.

Now for high frequencies, $A \approx \dfrac{\omega_t}{s} = \dfrac{2\pi f_t}{s}$. Thus $Z_{out} = \dfrac{10^5}{1 + \dfrac{2\pi \times 20 \times 10^6}{s} \times \dfrac{1}{11}}$, or

$Y_{out} = 10^{-5} + 2\pi \times 20 \times 10^6 \times \dfrac{10^{-5}}{11} \times \dfrac{1}{s}$.

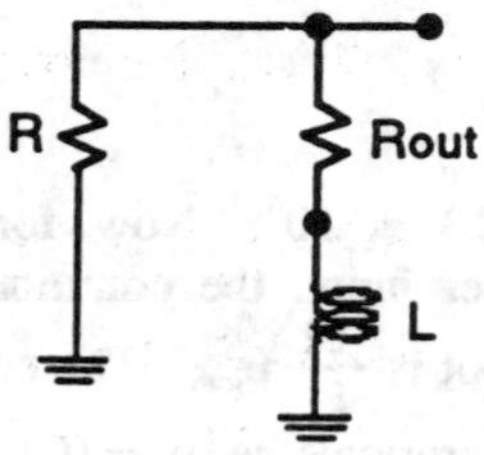

Here $R = 10^5\ \Omega$. Thus $L = \dfrac{11}{2\pi \times 20 \times 10^6 \times 10^{-5}} = 8.75$ mH, and $R_{out} = 110\Omega$ (as above). *Alternatively*: from Fig. 2.42, $R = R_o = \mathbf{10^5\Omega}$, $R_{out} = \dfrac{R_o}{A_0\beta} = \dfrac{10^5}{10^4 \times 1/11} = \mathbf{110\Omega}$, and $L = \dfrac{R_0}{\beta w_t} = \dfrac{10^5}{\frac{1}{11} \times 2\pi \times 20 \times 10^6}$ **= 8.75 mH.** Thus,

$|Z_{out}| \approx \left[(110)^2 + \left[2\pi f\,(8.75) \times 10^{-3}\right]^2\right]^{1/2} = 10(110)$, when $f \approx \dfrac{1100}{2\pi\,(8.75 \times 10^{-3})}$ **= 20 kHz.**

SECTION 2.12: DC Problems

2.54 Nominal 4V peak swing is reduced by less than 4V/100 = 40mV. Since gain = −100 V/V, $R_2/R_1 = 100$ and the gain for the offset voltage is $1 + R_2/R_1 = 101$. Thus, the required offset < $\dfrac{40}{101} \approx$ **0.40 mV.**

2.55 $A_f = -100$; $R_{in} = 100\ \text{k}\Omega = R_1 \rightarrow R_2 = 10\ \text{M}\Omega$.

a) For no compensation: $v_O = (1 + 100)\,(1\text{mV} + 30 \times 10^{-9} \times 100\ \text{k}\,||\,10\ \text{M}) = 101\ (1\ \text{mV} + 3\ \text{mV}) = 404\ \text{mV} \approx$ **0.40 V.**

b) For compensation, using $R_3 = R_1\,||\,R_2 \approx$ **100 kΩ**, only the offset current and offset voltage apply. Thus $v_O = (101)\,(1\text{mV} + 3 \times 10^{-9} \times 100\ \text{k}) = 131.3\ \text{mV} \approx$ **0.13 V.** For case (a), **bias current dominates**; For case (b), **offset voltage dominates**. For each dominant effect halved, the output offset becomes: (a) 101 (1 + 3/2) = **0.25 V**; (b) 101 (1/2 + 0.3) = **0.08 V**.

2.56

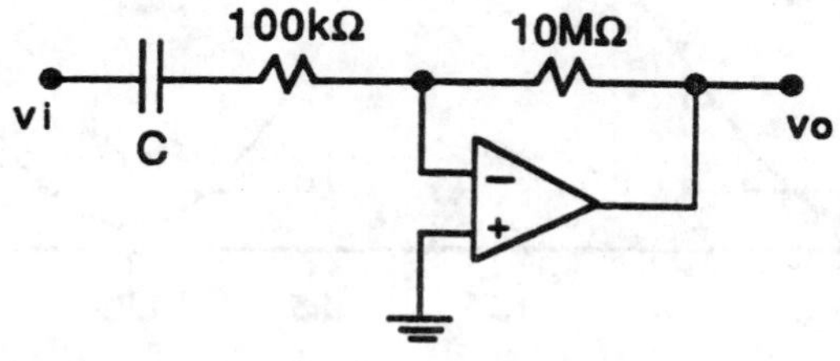

Note that the offset gain = 1V/V:

(a) For no compensation, $v_O = 1 \times 10^{-3} + 30 \times 10^{-9} \times 10 \times 10^{6} = 301\text{ mV} \approx$ **0.3 V**.

(b) For compensation with R_3 = 10 MΩ from the negative input to ground, $v_o = 1\text{ mV} + 3 \times 10^{-9} \times 10 \times 10^{6} = 31\text{ mV} =$ **0.03 V**. Thus, use 10 MΩ to compensate.

2.57

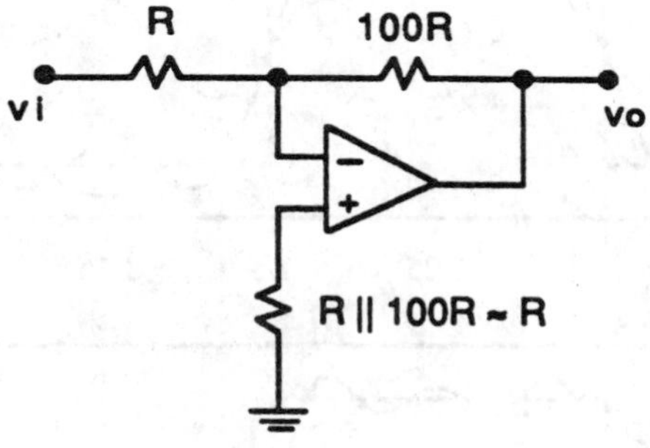

For no offset compensation,

$v_O = 101\,[2\times10^{-3} + (R\,||\,(100\,R))\,1.1 \times 10^{-6}] \leq 0.5\text{V}$. Thus $2 \times 10^{-3} + 1.1 \times 10^{-6}R \leq 4.95 \times 10^{-3}$, and $R = \dfrac{2.95 \times 10^{-3}}{1.1 \times 10^{-6}} = 2.68\text{ k}\Omega$. For compensation, with $R_3 = R$,

$v_O \approx 101\,[2\times10^{-3} + R(2 \times \frac{1}{10} \times 10^{-6})] \leq 0.5\text{V}$. Thus $R = \dfrac{2.95 \times 10^{-3}}{0.2 \times 10^{-6}} = 14.75\text{ k}\Omega$, (use 15 kΩ). For this design, $R_{in} \approx 14.75\text{k}\Omega \approx$ **15 kΩ**.

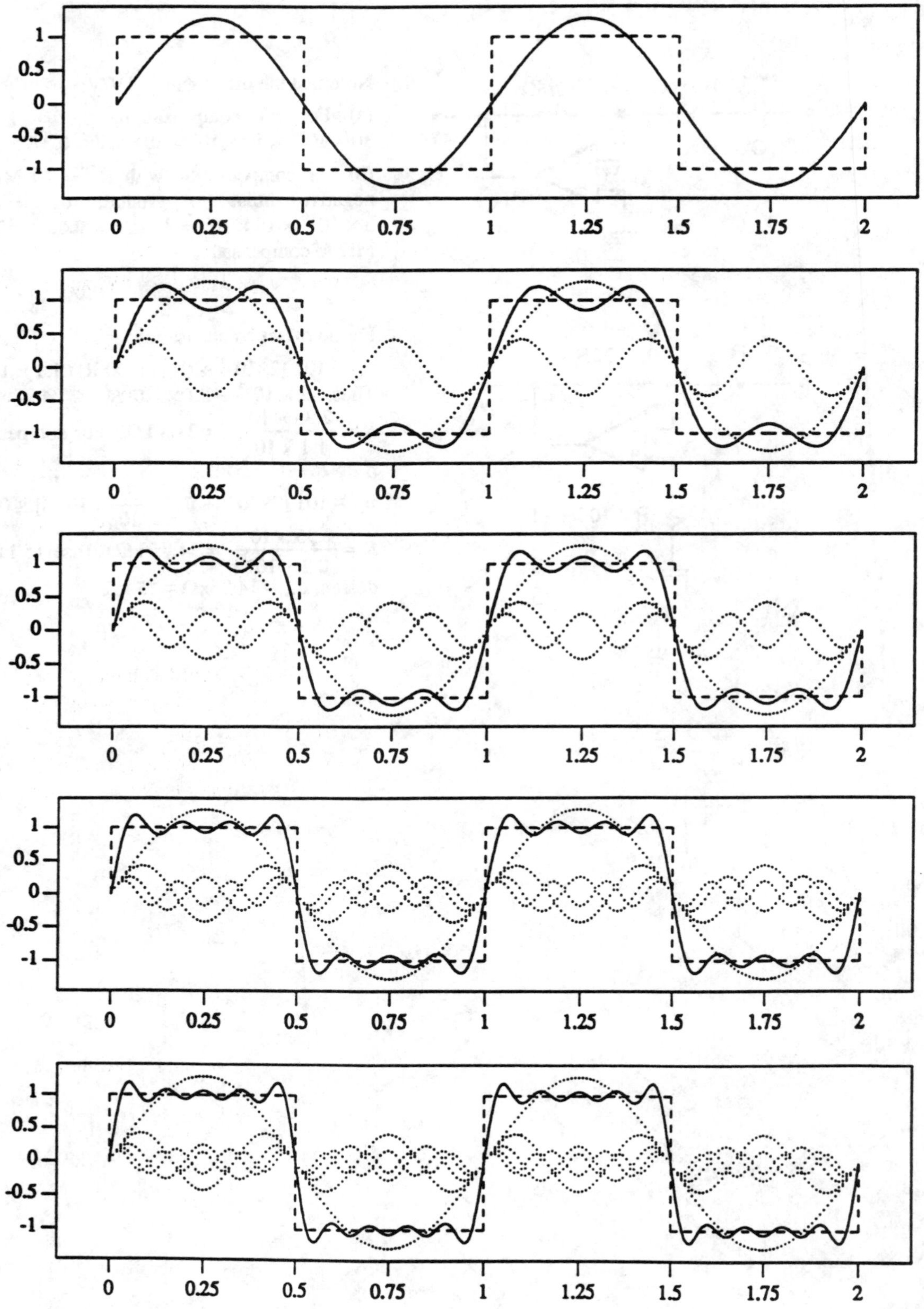
1
0.5
0
-0.5
-1
0
0.25
0.5
0.75
1
1.25
1.5
1.75
2

Chapter 3

DIODES

SECTION 3.1: The Ideal Diode

3.1 Diodes are ideal: Thus the forward voltage drop is 0V and reverse current is 0 mA:

(a) Diode is polarized to conduct by the +5V and 0V connections: Thus V_a = **0V**, and I_a = (5–0)/1kΩ = **5mA**.

(b) As in (a), but the most negative supply is –5V. Thus V_b = –5+0 = **–5V**, and I_b = (+5 – 0 – (–5))/1kΩ = **10mA**.

(c) Diodes both conduct: Thus $V_C = -10 + 0 + 0 =$ **–10V**, and I_C = (0 –0 –0 –(–10))/1kΩ = **10mA**.

(d) Both diodes are polarized to conduct. Thus V_d = +5 –0 = **5V**, and I_d = (+5 –0–0–0)/1kΩ = **5mA**.

(e) Upper diode is polarized to conduct, but lower to cut off. Thus current I_e = **0mA**, and V_e = 5V – 0V = **5V** (with the upper diode conducting the meter current).

(f) Upper diode polarized to cut off, and lower to conduct. Thus current, I_f = **0mA**, and V_f = –5V –0 – 0 = **–5V** (with the lower diode conducting the meter current (assumed very small) through 1kΩ).

3.2 (a)

i) V_A = +5V; upper diode conducts; V_Y = **+5V**.

ii) The output is high if either input is high. For high (+5V) defined as logic '1', the output is 1, if $A = 1$ or $B = 1$; ie, $Y = A + B$. Thus, the function is **logic OR** (in positive logic).

iii) For high (+5V) defined as logic '0', the output is high (logic '0') if either A or B is high (logic '0'). That is $\overline{Y} = \overline{A} + \overline{B}$, or $Y = \overline{\overline{Y}} = \overline{\overline{A} + \overline{B}} = A \bullet B = AB$ (for simplicity). Thus the function is a **logic AND** in negative logic. The AND idea can be verified by noting that the output is low (logic '1') only if A and B are both low.

(b)

i) V_B, V_C and V_D are all 0V. V_Y follows the highest input. Thus V_Y = **0V**.

ii) $Y = A + B + C$, an **OR** (in positive logic).

iii) $Y = A \bullet B \bullet C = ABC$, an **AND** (in negative logic).

(c)

i) See that the output follows the lower of A or E (just as the output followed the upper of A or B in (a)). That is, $V_A = V_E = 5\text{V} \rightarrow V_Y =$ **5V**.

ii) From above, $\overline{Y} = \overline{A} + \overline{B}$, ie $Y = \overline{\overline{Y}} = \overline{\overline{A} + \overline{B}} = A \bullet B = AB$ That is, the function is **AND** or, directly, $Y = A \bullet B = AB$ (since both inputs must be high for the output to be high).

iii) Directly, from logic first principles, or in analogy to the circuit in (a), $Y = A + B$ (in negative logic). That is, the function is **OR**.

(d)

i) $V_A = 5V, V_B = 0V \rightarrow V_Y =$ **0V**.

ii) **AND** (positive logic).

iii) **OR** (negative logic).

(e)

i) $V_A = V_E = 5V, V_C = 0V \rightarrow V_Y =$ **0V**.

ii) **AND**; $Y = A \bullet E \bullet C = AEC$.

iii) **OR** ; $Y = A + E + C$.

3.3 Use positive logic, that is, '1' = 5V, '0' = 0V.
For D_5 open, $P = A \bullet E = AE$.
For D_6 open, $Q = B \bullet C = BC$.
Now, note that the current available from node P or Q (100μA) exceeds that drawn from node Y: Thus if P or Q goes high, Y is also pulled high. Thus $Y = P + Q + D$ and, **Y = AE + BC + D** (in positive logic).
Now, for $V_A = V_E = 5V$; $V_B = V_C = V_D = 0V$; that is $A = E =$ '1' and $B = C = D = 0$, for which $Y = 1 \bullet 1 + 0 \bullet 0 + 0 = 1$. Thus the output is **logic '1'**, or **5V** (for ideal diodes)

3.4 12V rms = 12(1.414) = 16.97V peak
For a 12V battery and an ideal diode, the peak diode current is (16.97 – 12)/(10 + 50) = **82.8mA**
The diode begins to conduct (and ceases to conduct) when 16.97 sinωt = 12V or sinωt = 12/16.97 = .707 or $\omega t = \pi/4 \equiv 45°$.

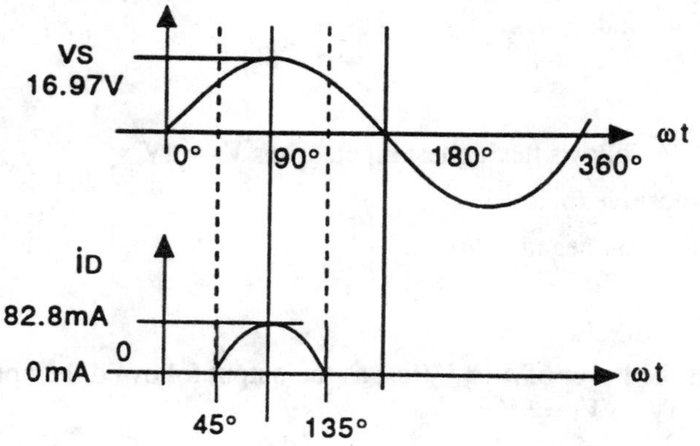

Average value of $i_D = \frac{1}{2\pi}\int_{\pi/4}^{3\pi/4}$ (16.97 sin Θ - 12) dΘ)/60 =

$$\frac{1}{2\pi}\left[-\frac{16.97}{60}\cos\Theta - \frac{12}{60}\Theta\right]_{\pi/4}^{3\pi/4} = \frac{0.282}{2\pi}(0.707 + 0.707) - 0.2\,\frac{3\pi/4 - \pi/4}{2\pi} = (.0635 - .0500)$$

A = **13.5mA**. Now for V_B = 14V: Peak current = (16.97 – 14)/60 = .0495A or **49.5mA** Conduction begins when 16.97 sinωt = 14, or sinωt = 14/16.97 = 0.825, or $\omega t \equiv$ 55.6°.

$$\text{Average current} = \frac{1}{2\pi}\left[-\frac{16.97}{60}\cos\Theta - \frac{14}{60}\Theta\right]_{55.6°}^{180°-55.6°} = -\frac{.282}{2\pi}(-.565 - .565) - 0.233\left(\frac{\frac{124.4 - 55.6}{360}\times 2\pi}{2\pi}\right) = .0507 - .0445 = .0062\text{A or } \mathbf{6.2mA}.$$

3.5

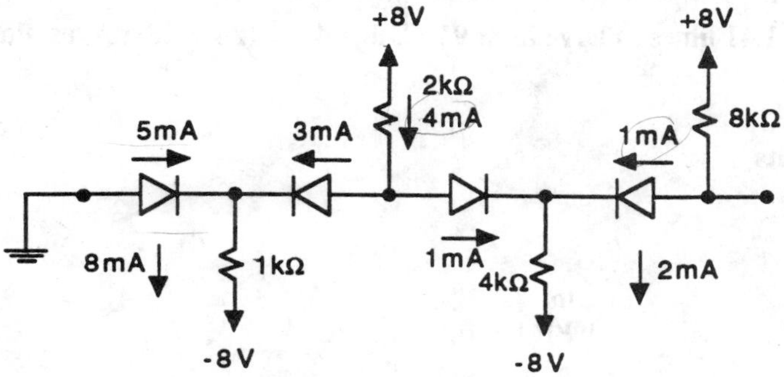

Consider the conducting state of each of the diodes: Assume D_1 conducts; thus its cathode is at 0V. Correspondingly, the current in the 1kΩ resistor, $I_{1K} = (0 - -8)/1k = 8mA$. Assume D_2 conducts; thus its cathode is at 0V, and $I_{2K} = (8 - 0)/2k = 4mA$.

Now, since $I_{1K} > I_{2K}$, then D_1 and D_2 both conduct as assumed. Now, noting that succeeding resistors are progressively larger, assume that both D_3 and D_4 also conduct, and that their anode and cathode voltages are all zero volts. Thus, $I_{4K} = (0 - -8)/4k = 2mA$, and $I_{8K} = (8 - 0)/8k = 1mA$.

Thus, overall: $I_{D4} = I_{8K} =$ **1mA**; $I_{D3} = I_{4K} - I_{D4} = 2 - 1 =$ **1mA**; $I_{D2} = I_{2K} - I_{D3} = 4 - 1 =$ **3mA**; $I_{D1} = I_{1K} - I_{D2} = 8 - 3 =$ **5mA**; Thus diodes are all conducting, and $V_o =$ **0V**.

SECTION 3.2: Terminal Characteristics of Junction Diodes

3.6 Given: $V_1 = 0.700V$ @ $I_1 = 100\mu A$, and $V_2 = 0.815V$ @ $I_2 = 1mA$.

Now $i_D = I_s\, e^{v_D/n\, V_T} \rightarrow v_D = nV_T \ln i_D/I_s = nV_T \ln i_D - nV_T \ln I_S$.

$\therefore\ v_{D1} - v_{D2} = nV_T\, (\ln i_{D1}/i_{D2})$.

Here, $0.700 - 0.815 = n\ V_T \ln 0.1/1 = 0.025n\,(-2.3026)$.

$\therefore\ n = 0.115/(.025(2.303)) = 1.997$ or **2.00**. RE

Now $I_S = i_D\, e^{-v_D/nV_T} = 100 \times 10^{-6}\, e^{-700/(2(25))} = 10^{-4}\, e^{-14} =$ **8.32×10^{-17} A**.

3.7 $i = I_S\, e^{v/nV_T}$, and $v_1 - v_2 = nV_T \ln i_1/i_2$ --- (1).

Thus, $v_1 = v_2 + nV_T \ln i_1/i_2 = 700 + 1(25) \ln 1/0.1 = 700 + 25 \ln 10 = 757.6mV$, or **0.758V.**

Now from (1), $i_1 = i_2\, e^{\frac{v_1 - v_2}{nV_T}} = 0.1\, e^{(815 - 700)/1(25)} =$ **9.95mA.**

3.8 $i = I_S\, e^{v/nV_T} \rightarrow v = nV_T \ln i/I_S$, or $v = v_o + nV_T \ln i/i_o$.

At 10mA, $v = 700 + 2(25) \ln (10 \times 10^{-3}/10) = 355mV$, or **0.355V**.

At 10μA, $v = 700 + 2(25) \ln (10 \times 10^{-6}/10) =$ **9.2mV**.

3.9 At 100μA and 25°C, $v = v_o + nV_T \ln i/i_o = 700 + 2(25) \ln 0.1/1 = 585mV$.

At 100μA and 95°C, $v = 585 + 2.0\,(25 - 95) = 445mV$ or, **0.445V**.

3.10 The leakage doubles for each 10°C rise in temperature.

$\therefore$ at 95°C, it is $2^{(95-25)/10} = 2^7 = 128$ times its value at 25°C, or **128nA,** or 0.128μA.

Now at 100°C, it is $2^{(100-25)/10} = 2^{7.5} = 181$ times its 25°C value, or **181nA**.

Note that it is $2^{(100-95)/10} = 2^{0.5}$, or 1.41 times as large as at 95°C, ie $1.41 \times 128 = 181$nA, as already noted.

SECTION 3.3: Analysis of Diode Circuits

3.11

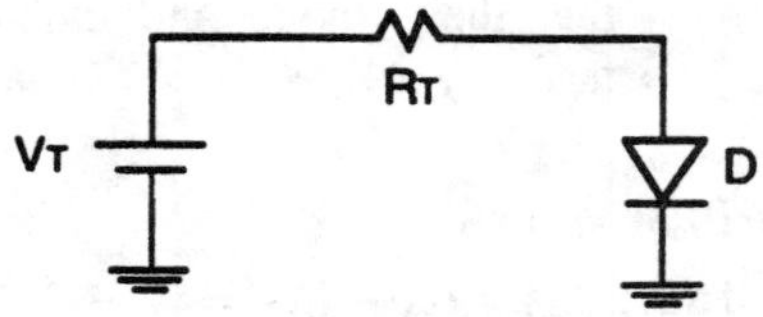

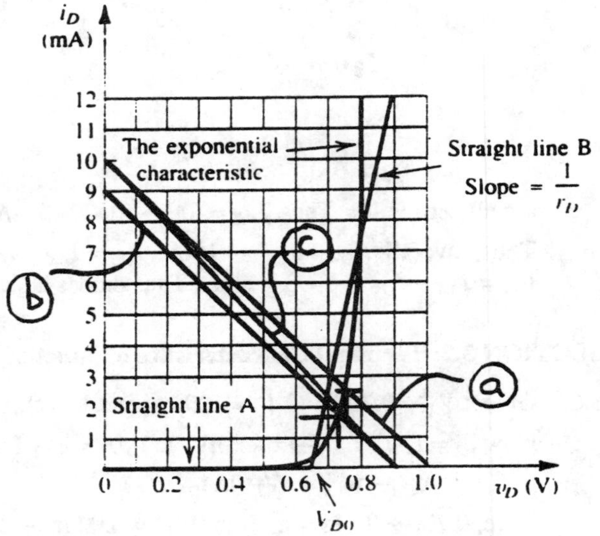

For the Load Lines:

(a) $V_T = 1$V, $R_T = 100\Omega$, $I_T = 1/100 = 10$mA.

(b) $V_T = 0.9$V, $R_T = 100\Omega$, $I_T = 0.9/100 = 9$mA.

(c) $V_T = 0.9$V, $R_T = 90\Omega$, $I_T = 0.9/90 = 10$mA.

From the Graph:

(a) $V_D = \mathbf{0.75V}$, $I_D = \mathbf{2.5mA}$.

Check: $I_D = 10 - V_D/0.1 = 10 - 10V_D = 10 - 10\,(.75) = 2.5$ mA.

(b) $V_D = \mathbf{.73V}$, $I_D = 1.7$mA mA.

Check: $I_D = 9 - V_D/0.1 = 9 - 10V_D$, or $I_D = 9 - 10\,(.73) = 1.7$ mA.

(c) $V_D = \mathbf{.74V}$, $I_D = \mathbf{1.8mA}$

Check: $I_D = 10 - V_D/.09 = 10 - 11.11\,V_D = 10 - 11.11\,(.74) = 1.77$ mA.

3.12 Though the diode is "the one sketched in Fig. 3.12", use the analytical description to provide a comparison with the results in P3.11 a) above.

See, for a diode current i, $V_{DD} = 1.00\text{V} = (100\Omega)\, i + 0.7\text{V} + 0.1 \log i/1$,

or $i = 0.3\text{V}/0.1\text{k} - (0.1/0.1) \log i$, or $i = 3 - \log i$, in mA.

Solve iteratively: Initially with $i = 1$:

$i = 1 \rightarrow i = 3 - \log 1 = 3 - 0 = 3\text{mA}$,

$i = 3 \rightarrow i = 3 - \log 3 = 3 - .48 = 2.52\text{mA}$,

$i = 2.52 \rightarrow i = 3 - \log 2.52 = 2.60\text{mA}$, whence

$I_D = i = 3 - \log 2.60 = \mathbf{2.59mA}$, and $V_D = 0.7 + 0.1 \log 2.59 = \mathbf{0.741V}$.

3.13 Here, for i in mA, $V_{DD} = 1.00 = (100\Omega)\, i + 0.65\text{V} + (20\Omega)\, i$,

or $I_D = i = \dfrac{1.00 - 0.65}{0.1 + 0.02} = \dfrac{0.35V}{0.12k\Omega} = \mathbf{2.92mA}$, and $V_D = 0.65 + 2.92\,(.02) = \mathbf{0.708V}$.

3.14 For $V_{DO} = 0.70\text{V}$, and $r_D = 10\Omega$, $1.00 = 0.1i + 0.70 + 0.01i$, and

$I_D = i = \dfrac{1.00 - 0.70}{0.1 + 0.01} = \dfrac{0.30}{0.11} = \mathbf{2.73mA}$, with $V_D = 0.70 + 2.73\,(0.01) = \mathbf{0.727V}$.

For $V_{DO} = 0.75\text{V}$ and $r_D = 0\Omega$, $I_D = i = \dfrac{1 - 0.75}{0.1 + 0} = \mathbf{2.50mA}$, with $V_D = \mathbf{0.75V}$.

3.15

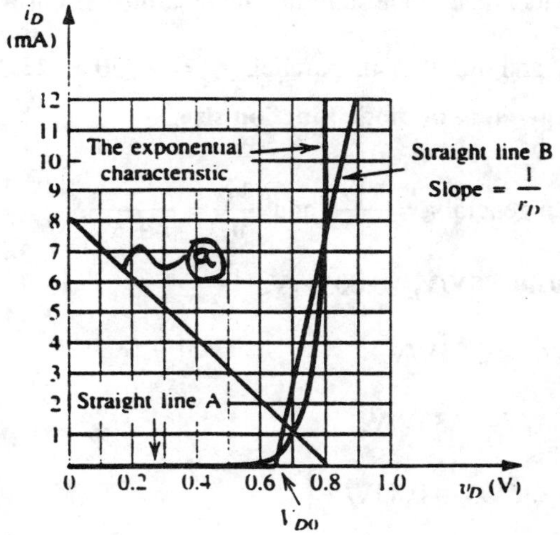

(a) For $V_T = 0.8\text{V}$ and $R_T = 100\Omega$, $I_T = 0.8/100 = \mathbf{8mA}$. From the Figure, see $\mathbf{V_D = 0.7V}$, $\mathbf{I_D = 1mA}$. *Check*: $I_D = 8 - V_D/0.1 = 8 - 10V_D = 8 - 10\,(0.7) = 1$ mA.

(b) $I_D = \dfrac{0.8 - 0.65}{0.1 + 0.02} = \mathbf{1.25mA}$, and $V_D = 0.65 + 0.22\,(1.25) = \mathbf{0.675V}$.

(c) $I_D = \dfrac{0.8 - 0.70}{0.1 + 0.01} = \mathbf{0.91mA}$, and $V_D = 0.70 + 0.01\,(0.91) = \mathbf{0.701V}$.

(d) $I_D = \dfrac{0.8 - 0.75}{0.1} = \mathbf{0.5mA}$, and $V_D = \mathbf{0.75V}$.

3.16 Drop across each diode = 4.0/5 = 0.8V. Thus $0.8\text{V} = 0.7\text{V} + 0.1 \log I_D/1mA$, or $\log I_D = (0.8 - 0.7)/0.1 = 1$, whence $I_D = 10^{\log I_D} = 10^1 = 10$, or $I_D = \mathbf{10mA}$ (as can be seen directly). Now $R = (10 - 4.0)/10\text{mA} = 0.6\text{k}\Omega$, or $\mathbf{600\Omega}$.

3.17 For $R = 500\Omega$ above, with each diode having a drop of υ volts: See $i = \dfrac{10 - 5\upsilon}{0.5} = 20 - 10\upsilon$,

also $\upsilon = 0.7 + 0.1 \log i/1$.

Solve iteratively:

$\upsilon = 0.8\text{V} \rightarrow i = 20 - 8 = 12$, $\upsilon = 0.7 + 0.1 \log 12 = 0.7 + 0.1079 = 0.8079\text{V}$;

$\upsilon = 0.8079\text{V} \rightarrow i = 20 - 8.079 = 11.92$, $\upsilon = 0.7 + 0.1 \log 11.92 = 0.7 + 0.1076 = 0.8076\text{V}$;

$\upsilon = 0.8076\text{V} \rightarrow i = 20 - 8.076 = 11.92$ mA, $\upsilon = 0.7 + 0.1 \log 11.92 = 0.7 + 0.1076 = 0.8076\text{V}$.

Thus the diode-string voltage becomes $5(0.8076) = 4.038\text{V}$, or $\mathbf{4.04V}$.

3.18 After shunting, three identical diodes in parallel share the current equally.
Original current is $i = (1)\ 10^{(0.8-0.7)/0.1} = 10mA$ (as can be seen directly).
Final current in each diode = 10/3 = 3.33mA. Final voltage drop = 0.7 + 0.1 log 3.33/1 = **0.752V**.

SECTION 3.4: The Small-Signal Model and its Application

3.19 In general, $r = nV_T/I_D$. At 0.1mA, $r = \frac{2(25mV)}{0.1mA} = \mathbf{500\Omega}$. At 10mA, $r = \frac{2(25)}{10mA} = 5\Omega$.
Now the "average" resistance would more likely be considered at 1mA (ie $(0.1 \times 10)^{1/2}$), than at (0.1 + 10/2) = 5.05mA. Thus the geometric mean is likely to be most relevant and therefore "best".

See $(r_1 + r_2)/2 = (500 + 5)/2 = \mathbf{252.5\Omega}$, and $(r_1\ r_2)^{1/2} = \sqrt{500 \times 5} = \mathbf{50\Omega}$. Now, at 1mA, $r = \frac{2(25)}{1mA} = \mathbf{50\Omega}$, and at 5.05mA, $r = \frac{2(25)}{5.05mA} = \mathbf{9.9\Omega}$. *Note* the correspondence of the value at 1mA with the geometric mean, as expected. Note that the arithmetic-mean results appear to be less easily interpreted.

3.20 At 2mA: $r = \frac{nV_T}{I_D} = \frac{2(25mV)}{2mA} = \mathbf{25\Omega}$. With a second identical diode shunting the first, the current is shared equally: **1mA** in each. At 1mA: $r = \frac{2(25)}{1} = \mathbf{50\Omega}$, and for two in parallel, $r_{eq} = 50 \| 50 = \mathbf{25\Omega}$. This demonstrates that diode incremental **resistance is independent of diode junction size.**

3.21 For small υ_S, use the small-signal slope resistance, r. In general $r = \frac{nV_T}{I}$ and $\frac{\upsilon_O}{\upsilon_S} = \frac{r}{r + R_S}$.

For I = 10mA: $r = \frac{2(25)}{10} = 5\Omega$; $\upsilon_O/\upsilon_S = \frac{5}{5 + 1000} = \mathbf{0.00498V/V} \approx 0.005V/V$.

For I = 1mA: $r = \frac{2(25)}{1} = 50\Omega$; $\frac{50}{50 + 1000} = \mathbf{0.0476V/V} \approx 0.05V/V$.

0.1mA: $r = \frac{2(25)}{0.1} = 500\Omega$; $\frac{500}{500 + 1000} = \mathbf{0.333V/V} \approx 0.33V/V$.

0.01mA: $r = \frac{2(25)}{.01} = 5000\Omega$; $\frac{5000}{5000 + 1000} = \mathbf{0.833V/V} \approx 0.83V/V$.

3.22 For $\upsilon_O = \upsilon_S = 0$, see that I_1 and I_2 both **split equally** between D_1, D_2 and D_3, D_4 respectively. Thus **all diode currents are equal.**

In general, for $\upsilon_S \neq 0$ (say $+\varepsilon$), the currents in D_1, D_2 redistribute with their sum = I_1. Extra current in D_2 needed to drive R_L high implies that there is less current in D_4 than in the balanced case, the extra part of I_2 flowing in D_3. In general, current needed by R_L for $\upsilon_O \neq 0$ is provided half by that in D_2 increasing and half by that in D_4 decreasing, with corresponding changes in D_1 and D_3, such that load current originates ultimately in υ_S, as Kirchoff's current law would require.

For small υ_S (around zero volts), with all diode currents equal to $i/2$,

$$R_T = \left(r_1 + r_2\right) \| \left(r_3 + r_4\right) = \left(\frac{50}{I/2} + \frac{50}{I/2}\right) \| \left(\frac{50}{I/2} + \frac{50}{I/2}\right) = \frac{200}{I} \| \frac{200}{I} = \frac{\mathbf{100}}{\mathbf{I}}.$$

For I = 10mA, $R_T = \frac{100mV}{10mA} = 10\Omega$, and with $R_L = 10k\Omega$, $\upsilon_O/\upsilon_S = \frac{10k}{10 + 10k} = \mathbf{0.999V/V}$.

For $I = 1\mu A$, $R_T = \frac{100mV}{1\mu A} = 100k\Omega$, and $\upsilon_O/\upsilon_S = \frac{10k}{100k + 10k}$ = **0.0909V/V**.

Concerning Linearity: Signal size (for $R_L = 10k\Omega$) is of little concern at 1mA, since the load current is likely to be a small part of I. But, at low I, the load current may cause the diode current to vary over a wide range. Operation is linear for diode-voltage variation of ±10mV or so: Note that the corresponding current variation is Δi, where $10 = 2(25) \ln \frac{\Delta i}{i}$, or $\ln \frac{\Delta i}{i} = \frac{10}{50} = .20$, or $\Delta i/i = 1.2$, corresponding to a variation of about 20%. Now for a positive output signal, i_{D2} can increase by 20% of its normal current $I/2$, while i_{D4} decreases by the same amount, the output current being the sum of the two changes. Thus for $R_L = 10k\Omega$ and $I = 1\mu A$, υ_O is limited to $2 \times 0.20 \frac{1\mu A}{2} \times 10k\Omega = 2mV$ peak, in which case υ_S is restricted to $\frac{2mV}{.0909}$ = **22mV** peak.

An Alternative View: The largest (positive) output occurs when the drop in D_2 increases by 10mV, and that in D_1 decreases by 10mV. Thus, at the limit $\upsilon_S - \upsilon_O = 2(10) = 20mV$. But for 1μA and a 10kΩ load, $\upsilon_O/\upsilon_S = .09$, or $\upsilon_O = .09\ \upsilon_S$.

$\therefore$ limit is at $\upsilon_S - .09\ \upsilon_S = 20mV$, or υ_S = **22mV**.

3.23 For the regulator, $I_D = \frac{10-4}{600} = 10mA$. For each of 5 diodes $r = \frac{nV_T}{I_D} = \frac{2(25)}{10} = 5\Omega$; Resistance of the total diode string = 5(5) = 25Ω. For ±10% supply variation, expect an output variation of $\pm \frac{25}{25+600} \times (0.1 \times 10) = \pm$ **40mV**, equivalent to $\pm \frac{40mV}{4V} \times 100$ = **±1%**. For a 2mA load increase, the diode current reduces from 10mA to 8mA (a small amount) and the output drops by 5(5Ω) (2mA) = **50mV**, or $\frac{50}{4V} \times 100$ = **−1.25%**. For both effects, assuming approximately linear operation, the combined drop would be −40mV −50mV = **−90mV** or **−2.25%**. The lowest output voltage would be 4.00 − .09 = **3.91V**.

From First Principles: (for a 0.1V/decade current change (which is *not* $n = 2$ precisely!)), V_S = 90% of 10 = 9V. For diode current i and voltage υ, see $\frac{9-5\upsilon}{0.6} = i + 2$ whence $i = 13 - 8.33\ \upsilon$ --- (1), and $\upsilon = 0.7 + 0.1 \log i/1$ --- (2). Thus $\upsilon = 0.7 + 0.1 \log (13 - 8.33\ \upsilon)$.

Iterate: with $\upsilon = 0.8 - .09/5 = .782$ initially. Thus $\upsilon = 0.7 + 0.1 \log (13 - 8.33 (.782)) = 0.7812V$, and $\upsilon = 0.7 + 0.1 \log (13 - 8.33 (.7812)) = 0.78123V$.

Thus the output drop is 5 (0.8000 − .78123) = **93.9mV**.

SECTION 3.5: Operation in the Reverse Breakdown Region – Zener Diodes

3.24 Knee voltage ≈ $6.8V - \frac{20}{1000}(5 - 0.2)$ = **6.70V**. For no load, breakdown is sustained to about **6.7V**. For a 9V supply, and bare breakdown, the load can increase to $\frac{9-6.7}{200\Omega} - 0.2 = 11.5 - 0.2$ = **11.3mA**. For $\frac{11.3}{2}$ = 5.65mA load, the lowest supply voltage for regulation is V, where $\frac{V-6.7}{0.2k} - 5.65 = 0.2$, from which V = 6.7 + 0.2 (5.85) = 7.87V. Thus the lowest supply for regulation with half-maximum load ≈ **7.9V**.

3.25 Line regulation is $\frac{r_2}{R + r_2} = \frac{20}{200 + 20}$ = **.0909V/V = 90.9mV/V**.

Load regulation is $-(r_2 \| R) = -20 \| 200$ = **−18.2Ω** or, **−18.2mV/mA**.

3.26 *Worst case*: V_S low, I_L high, V_Z high, where $\frac{9(.95) - 6.8\,(1.03)}{R} = 10\text{mA} + 2\,(0.2)\text{mA}$, whence $R \le \frac{9\,(.95) - 6.8\,(1.03)}{10 + 0.4} = \frac{8.55 - 7.004}{10.4} = 0.148\text{k}\Omega$. Use **150Ω**.

Now, the lowest output occurs for 10mA load, V_Z low, V_S low, where $I_Z \approx \frac{9\,(.95) - 6.81\,(.97)}{0.148} - 10.0 = \frac{8.55 - 6.596}{0.148} - 10.0 = 3.20\text{mA}$, for which $V_Z \approx 6.8\,(.97) - (5 - 3.2)\,20\Omega = 6.596 - .036 = 6.56\text{V}$.

More precisely: $I_Z = \frac{8.55 - 6.56}{0.148} - 10 = 3.45\text{mA}$, for which $V_Z \approx 6.8\,(.97) - (5 - 3.45)\,20 = 6.596 - .031 = 6.565\text{V} \approx$ **6.57V**.

The highest output occurs for 2mA load, V_Z high, V_S high, where $I_Z \approx \frac{9(1.05) - 6.8\,(1.03)}{0.148} - 2 = \frac{9.45 - 7.004}{0.148} - 2 = 14.52\text{mA}$, for which $V_Z = 6.8\,(1.03) + (14.52 - 5)\,20 = 7.004 + .1904 = 7.194\text{V} \approx$ **7.19V**.

More precisely, $I_Z = \frac{9.45 - 7.194}{0.148} = 15.24\text{mA}$, for which $V_Z = 6.8\,(1.03) + (15.24 - 5.0)\,20 = 7.004 + 0.205 = 7.209$, or **7.21V**.

3.27

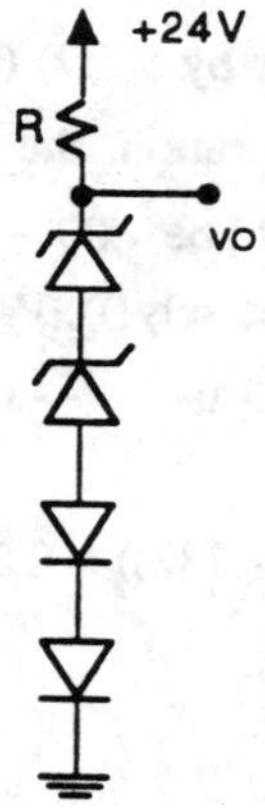

$\Sigma V = 2\,(6.8) + 2\,(0.7) = 15\text{V}$, but since V_Z is specified at 20mA and V_D at 10mA, use an I_Z in between where $2\,(20 - I_Z)\,(5\Omega) = 2\,(I_Z - 10)\,(2.5\Omega)$, or $2\,(20 - I_Z) = I_Z - 10$, $40 - 2I_Z = I_Z - 10$, $3I_Z = 50$, and $I_Z = 16.7\text{mA}$. For a nominal load of 15mA, and nominal 24V supply, $R = \frac{24 - 15}{16.67 + 15} = 0.284\text{k}\Omega$. Use a **270Ω** resistor, as a standard value. For supply 10% high, resistor 5% low, and no load, $I_Z = \frac{24\,(1.1) - V_Z}{(0.270)\,(0.95)} = 102.9 - 3.899V_Z$. Also $V_Z = 15\text{V} + (I_Z - 16.67)\,(2\,(5) + 2\,(2.5)) \times 10^{-3} = 15 + .015\,I_Z - 0.250 = 14.75 + .015\,I_Z$, whence $I_Z = 102.9 - 3.899\,(14.75 + 0.015\,I_Z) = 102.9 - 57.51 - .0585\,I_Z$, and $I_Z = \frac{45.39}{1.0585} = 42.88\text{mA}$, for which $V_Z = 14.75 + .015\,(42.88) =$ **15.39V**. For each zener at 42.88mA, $V_Z = 6.8\text{V} + (42.88 - 20)\,(.005) = 6.911\text{V}$, and $P_D = 6.911 \times 42.88 =$ **296mW**.

SECTION 3.6: Rectifier Circuits

3.28 The peak output is $8\sqrt{2} - 0.7\text{V} = 11.31 - 0.70 =$ **10.6V**: The diode conducts for about **½ cycle**, but more precisely, between the points where $11.31 \sin \Theta = 0.7$, or $\sin \Theta = .0619$, for which $\Theta = 3.55°$ and $180 - 3.55°$, that is, for $\frac{180 - 2\,(3.55)}{360} = 0.48$ or, **48%** of a cycle. Ignoring the diode drop, the average output $= 8\sqrt{2}/\pi =$ **3.60V**. With a constant 0.7V drop for 0.48 of a cycle, the average output is $3.6 - 0.48\,(0.7) =$ **3.26V**. The peak inverse voltage across the diode is peak input $= 8\sqrt{2} =$ **11.3V**. For $R_S = 50\Omega$, $r_D = 10\Omega$ and load of 1kΩ, $\upsilon_O \approx (\upsilon_S - 0.7)\,\frac{1k}{(1 + .01 + .05)k}$ for a half cycle, $\approx 0.94\,\upsilon_S - 0.66$. At the peak, the output is $0.94\,(11.31) - 0.66 =$ **9.97V**, and the average output is $0.94\,(3.26) =$ **3.06V**.

3.29 Peak inputs are $\pm 8\sqrt{2} = \pm 11.31$V. Peak outputs are $11.31 - 0.70 = 10.61$ V, and $-11.31 + 6.80 = -4.51$V.

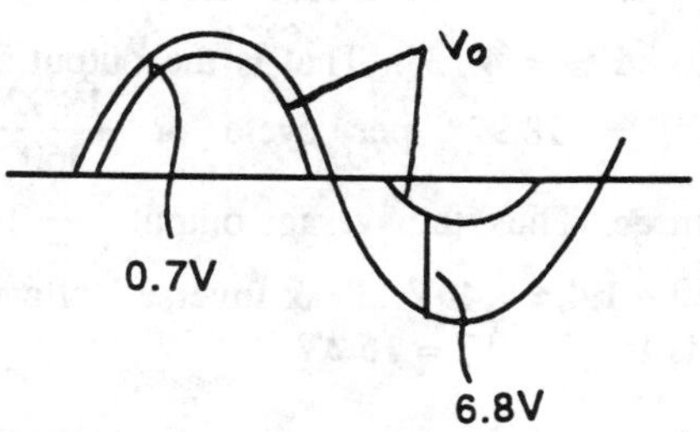

For the positive part (as noted in P3.28), conduction is from 3.6° to 176.4°, or about 48% of the cycle, with average of $((8\sqrt{2}/\pi - 0.7) - 0.7)0.48 = 3.26$V. For reverse conduction: current flows for the parts of the cycle between which $11.31 \sin \Theta = 6.8$V, or $\sin \Theta = \frac{6.8}{11.31} = 0.601$, for which $\Theta = 36.96°$ or 37°, and $180° - 36.96° = 143°$, that is for $(143 - 37)/360 = 29.4\%$ of a cycle. The average (negative) value is $((\frac{11.31}{\pi}) - 0.7)\ 0.294$ or 0.853V. Overall, the average value of the output is $3.26 - 0.85 =$ **2.41V**.

3.30 Full-winding peak transformer output voltage $= 16\sqrt{2} = 22.62$V.

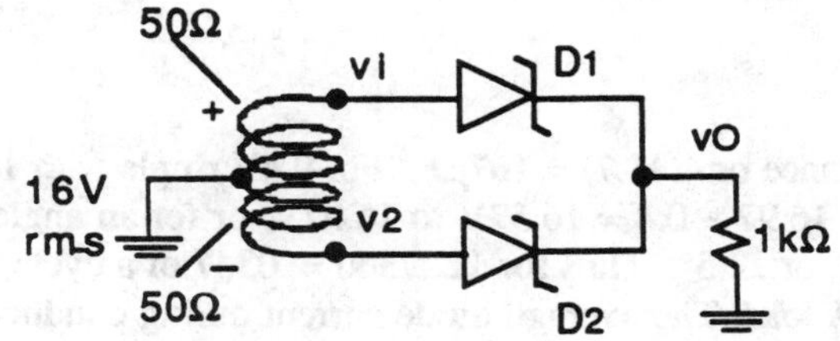

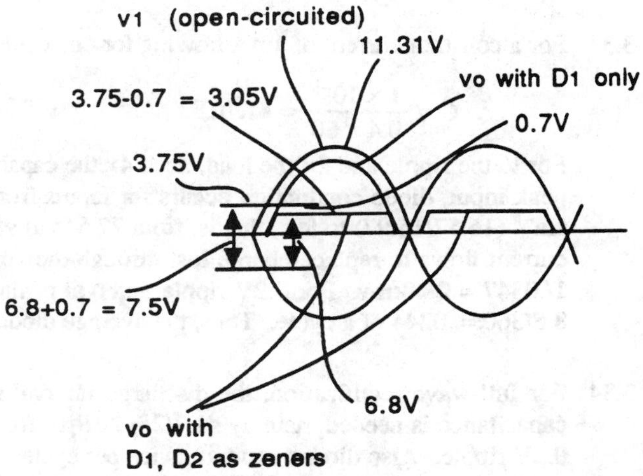

For full-winding voltages $\geq (6.8 + 0.7) = 7.5$V, D_1 conducts in the forward direction (0.7V), while D_2 breaks down (6.8V). At the peak, the shorted-diode current flow is $(22.62 - 7.5)/100\Omega = 151.2$mA with $v_O = 22.62/2 - 0.7 - (100\Omega/2)\ 151.2\text{mA} = (11.31 - 0.7 - 7.56\text{V}) = 3.05$V, (or, see as $7.5/2 - 0.7 = 3.05$V). But current also flows in R_L, $\approx \frac{3.05}{1k} = 3$mA from an equivalent source of $\frac{100}{2} \| \frac{100}{2} = 25\Omega$, to produce an additional drop of $3\text{mA} \times 25\Omega = 0.075$V.

$\therefore\ v_O \approx 3.05 - .075 \approx 2.975$V.

Thus peak value of the output is 2.975V, flat while the sum of the open-circuit winding voltages exceeds 7.5V. Peak diode current flow is $151.2\text{mA} + \frac{3mA}{2} =$ **152.7mA**.

3.31

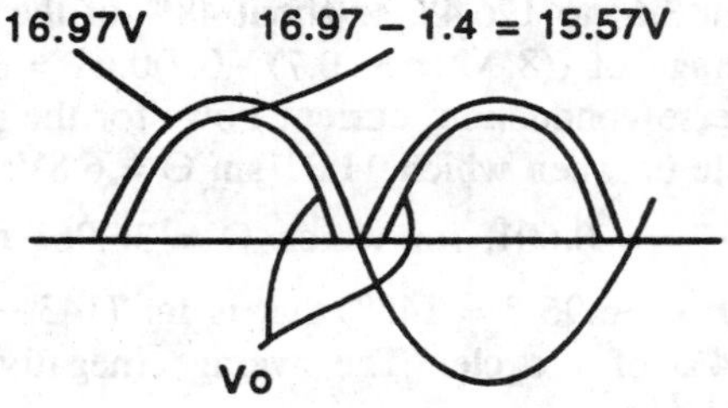

Transformer Peak Voltage = $12\sqrt{2}$ = 16.97V. Load Peak Voltage = 16.97 – 2 (0.7) = **15.57V**. Now, 16.97 sin Θ = 0.7 + 0.7 = 1.4, for which sin Θ = $\frac{1.4}{16.97}$ = .0825, and Θ = 4.73°. That is the output is zero for 4(4.73°) = 18.92° per cycle or $\frac{18.92}{360} \times \frac{1}{60}$ = **0.876msec**. Thus, the average output ≈ $\frac{2}{\pi}$ 16.97 – 1.4 = 10.80 – 1.4 = **9.40V**. Peak Inverse Voltage for each diode is 16.97 – 0.7 = **16.3V**.

3.32 For a 12V sinusoid, peak is $12\sqrt{2}$ = 16.97V. Capacitor charges to the peak less one diode drop (due to the flow of small capacitor leakage currents), that is, to 16.97 – 0.7 = 16.27V. Thus the output is a dc voltage of **16.27V** and the PIV required of the diode is 16.27 + 16.97 = **33.2V**.

3.33 For a constant current of 1mA flowing for one cycle, voltage drop is $V = \frac{IT}{C} = 0.4\text{V} = \frac{1mA \times 1/60}{C}$.

$\therefore\ C = \frac{1 \times 10^{-3}}{0.4 \times 60}$ = **41.7µF**.

For ½ the ripple and 2× the load, need 4× the capacitance or 4(41.7) = **167µF**. For 0.4V ripple with 16.97V peak input, diode conduction occurs for inputs from 16.97 – 0.4 = 16.57V to 16.97V, or for an angle from $\sin^{-1}$ (16.57/16.97) to 90°, that is, from 77.54° to 90°, or 12.5°. Thus for 12.5/360 = .0347 of a cycle, diode current flows to replace charge lost through the 1mA load. The average diode current during conduction is 1/.0347 = **28.8mA**. For 0.2V ripple, interval is $\sin^{-1}$ 16.77/16.97 = 81.2° to 90°, or 8.8°, corresponding to 8.8/360 = .0244 of a cycle. Thus, the average diode current for a 2mA load is 1/.0244 × 2 = **81.8mA**!

3.34 For full-wave rectification, the discharge interval is essentially halved. Thus for the same ripple, half the capacitance is needed, namely 41.7/2 = **20.9µF** for 1mA and 0.4V ripple, and 167/2 = **83.4µF**, for 2mA and 0.2V ripple. Also diodes conduct twice per cycle. Thus the diode average conduction currents are ½ of the originals, ie 28.8/2 = **14.4mA**, and 81.8/2 = **40.9mA** respectively. In each case, diode PIV = 16.27 + 16.97 = **33.24V**.

3.35 12V applied to a 100Ω load implies a 120mA load current. Now $CV = IT$.

Thus $C = \frac{120 \times 10^{-3} \times ½ \times 1/60}{0.4}$ = **2500µF**. Assume negligible transformer resistance: Thus the peak sine wave required is 12 + 0.4/2 + 0.7 = 12.9V and the transformers RMS voltage = $12.9/\sqrt{2}$ = 9.12V per side. Thus the transformer should have an **18.24V rms centre-tapped secondary**.

For diodes: PIV = 12 + 0.2 + 9.12 $\sqrt{2}$ = **25.1V**. Diode current flows from $\sin^{-1}$ (12.9 – 0.4)/12.9 = 75.69° to 90°, or (90 – 75.69)/360 = .0397 of a cycle. Average diode current for each diode = 120/.0397 × ½ = **1.509A**, with the peak current being about twice as high ie (1.509 – 0.12) 2 = **2.78A**!!

3.36

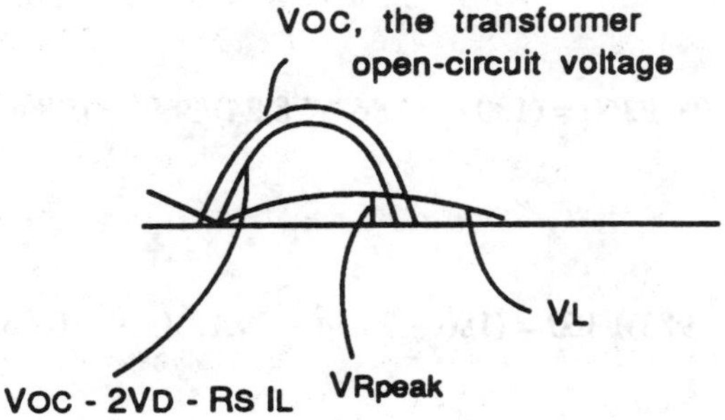

Ideal Case: Peak output = 20 − 0.7 − 0.7 = 18.6V, for which the peak output current = 18.6/200 = 93mA. Ripple voltage is $V = IT/C = \frac{93 \times 10^{-3} \times 1/60 \times 1/2}{1000 \times 10^{-6}}$ = **0.775V**.

Diode conduction is from $\sin^{-1}$ ((20 − .775)/20) = 74.0° to 90°, ie 16° or 16/360 = .0444 of a cycle, with an average diode current = 93/.0444 × ½ = **1.047A**, and an average DC output = 18.6 − 0.775/2 = **18.2V**.

With Source Resistance, R_S, the peak diode current is limited, and the output voltage will drop. As the output voltage drops, the recharging interval will increase, and the ripple will decrease, both because at lower voltages the load current is smaller, and also because during the charging interval(s), the load is supported directly through the diodes.

Now, the average output decreases to 18.2 x where $x \le 1.0$, and the ripple to $0.78x$. Thus the peak diode current (at the sine-wave peak and where (say) the ripple is halfway) is $+.78(x/2)/R_S = (18.6 - 17.8x)/R_S$. Now, assume the current to be triangular in form, flowing for a fraction y of a cycle. Thus average charge from the supply (for two diodes) in one cycle must be that required by the load in the same interval, that is,

$$2(½)(y)\left(\frac{18.6 - 17.8x}{R_S}\right) = \frac{18.2x}{200} \quad \text{--- (1)}$$

Now for y: Diodes begin to conduct at $\sin^{-1} \frac{18.2x - 0.78x/2}{18.6} = \sin^{-1}(0.958x)$ and cease (following the input peak) at $90° + 90° - \sin^{-1} \frac{18.2x + 0.78x/2}{18.6} = 180° - \sin^{-1}(0.999x)$,

where $y = \frac{180 - \sin^{-1} 0.958y - \sin^{-1} 0.999y}{360}$ --- (2)

Now, for $R_S = 1.0\Omega$, (1) → $2(½)y\,\frac{(18.6 - 17.8x)}{1} = \frac{18.2x}{200}$, and $x = 11.0y\,(18.6 - 17.8x)$ --- (3)

Iterate: Try $x = 0.95$ initially:

(2) → $y = (180 - \sin^{-1} .958(.95) - \sin^{-1} .999\,(.95))/360 = (180 - 65.5 - 71.6)/360 = .119$, and

(3) → $x = 11.0\,(.119)\,(18.6 - 17.8\,(.95)) = 2.21$.

Try $x = 1.0$:

(2) → $y = (180 - \sin^{-1} .958 - \sin^{-1} .999)/360 = (180 - 73.3 - 87.4)/360 = 0.0536$,

(3) → $x = 11.0\,(.0536)\,(18.6 - 17.8) = 0.472$.

Try $x = 0.97$:

$y = (180 - \sin^{-1}(.958 \times 0.97) \sin^{-1}(.999)(.97))/360 = (180 - .68.3 - 75.7)/360 = 0.100$,

$x = 11\,(.100)\,(18.6 - 17.8\,(.97) = 1.47$.

Try $x = .99$:

$y = (180 - \sin^{-1}(.958)(.99) - \sin^{-1}(.999 \times .99))/360 = (180 - 71.5 - 81.5)/360 = 0.075$,

$x = 11\,(.075)\,(18.6 - 17.8\,(.99)) = 0.807.$

Try $x = .98$:

$y = (180 - \sin^{-1}(.958)(.98) - \sin^{-1}(.999 \times .98))/360 = (180 - 69.86 - 78.24)/360 = 0.0886,$

$x = 11\,(.0886)\,(18.6 - 17.8\,(.98)) = 1.13.$

Try $x = .983$:

$y = (180 - \sin^{-1}(.958 \times .983) - \sin^{-1}(.999)(.983))/360 = (180 - 70.34 - 79.12)/360 = 0.0848,$

$x = 11\,(.0848)\,(18.6 - 17.8\,(.983)) = 1.028.$

Try $x = .984$:

$y = \left[180 - \sin^{-1}(.958 \times .984) - \sin^{-1}(.999 \times .984)\right]/360 = (180 - 70.50 - 79.43)/360 = 0.0835,$

$x = 11\,(.0835)\,(18.6 - 17.8\,(.984)) = 0.996.$

Try $x = .986$:

$y = \left[180 - \sin^{-1}(.958 \times .986) - \sin^{-1}(.999 \times .984)\right]/360 = (180 - 70.84 - 79.43)/360 = 0.0826,$

$x = 11\,(.0826)\,(18.6 - 17.8\,(.986)) = 0.953.$

Try $x = .985$:

$y = \left[180 - \sin^{-1}(.958 \times .985) - \sin^{-1}(.999 \times .985)\right]/360 = (180 - 70.07 - 79.74)/360 = 0.0821,$

$x = 11\,(.0821)\,(18.6 - 17.8\,(.985)) = 0.964.$

Use $x = 0.985$.

One can conclude that:

a) The iterative process is not a very good one, but the result is probably OK.

b) The output voltage decreases to $18.2x = 18.2\,(.985) =$ **17.9V**, a drop of about 18.2 − 17.9 = **0.3V**, or 0.3/18.2 × 100 = 1.6%, in the transformer resistance of 1 ohm.

SECTION 3.7: Limiting and Clamping Circuits

3.37 The upper limiting level is 2.3 + 0.7 = **3.0V**. The corresponding input threshold level is 3.0 + (3.0/10k) × 10k = **6.0V**. The corresponding lower values are **−3.0V** and **−6.0V** respectively. The gain K (for linear operation) is 10k/(10k + 10k) = **0.5V/V**. At twice the upper threshold, $V_{in} = 2\,(6.0) = 12$V, and the current is (12 − 3)/10k = **0.9mA**.

3.38

(a)

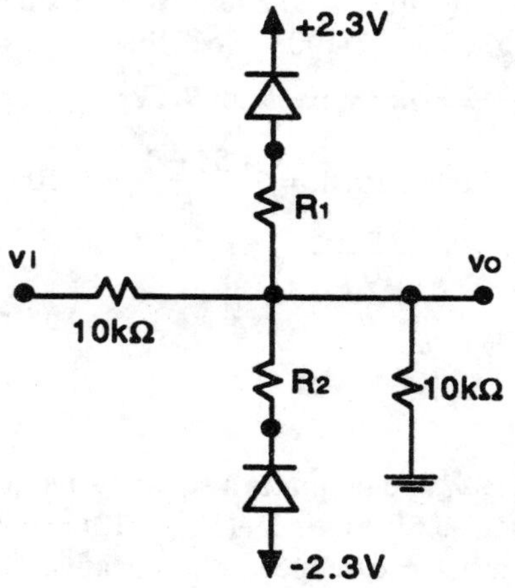

For $v_O \geq 3.0$ V, current begins to flow in R_1. Thus $k = ¼ = \frac{10\|R_1}{10\|R_1 + 10}$.

$\therefore\ 4\frac{10R_1}{10+R_1} = 10 + \frac{10R_1}{10+R_1}$,

$40R_1 = 100 + 10R_1 + 10R_1$,

$20R_1 = 100$,

$R_1 = 5\text{k}\Omega$.

For symmetrical operation, $R_2 = 5\text{k}\Omega$ also. Note that there is also a one-component solution for symmetrical operation:

(b)

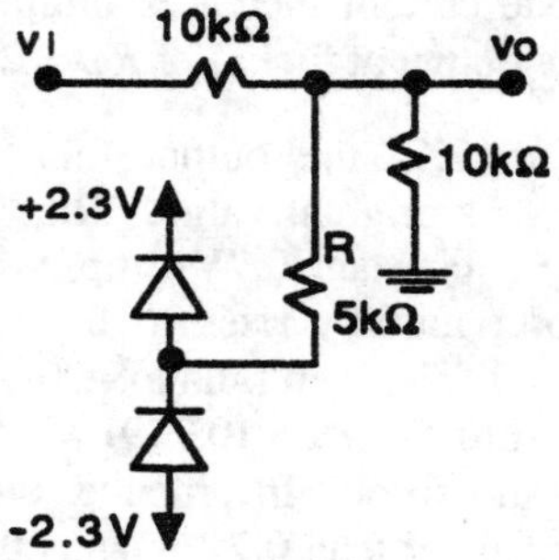

Hard limiting at ±5V can be provided using two additional diodes: D_3, from v_O to +4.3V with anode at v_O, and D_4, from v_O to –4.3V with cathode at v_O.

(c)

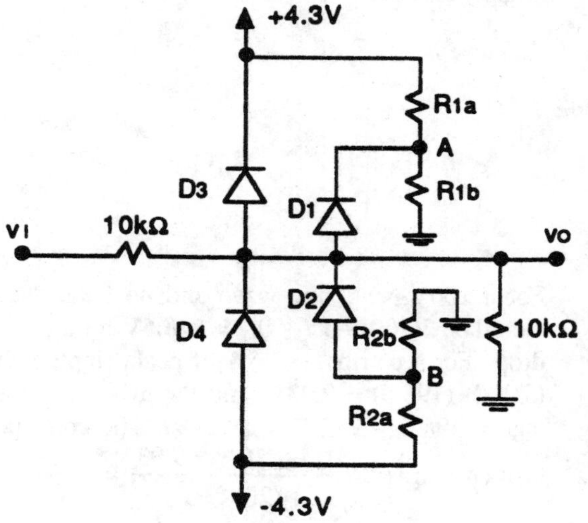

Here $\frac{R_{1a}}{R_{1a}+R_{1b}} \times 4.3 = 2.3$ V and $R_{1a}\|R_{1b} = 5\text{k}\Omega$, that is $\frac{R_{1a}R_{1b}}{R_{1a}+R_{1b}} = 5$, or $\frac{R_{1a}}{R_{1a}+R_{1b}} = \frac{5}{R_{1b}}$. Substitute the last into the first to get:

$\frac{5\,(4.3)}{R_{1b}} = 2.3$, or $R_{1b} = \frac{5\,(4.3)}{2.3} = 9.35\text{k}\Omega$, and $R_{1a} = \frac{5}{9.35}(R_{1a} + 9.35) = 0.535\ R_{1a} + 5 = \frac{5}{1-.535} = 10.75\text{k}\Omega$. Again, a resistor can be saved by replacing R_{1b} and R_{2b} by a single resistor equal to their sum (18.7kΩ) between nodes A and B with no ground connection.

3.39

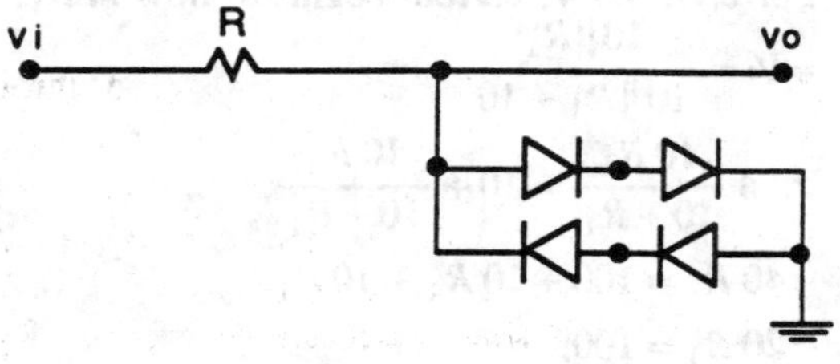

Consider an input $\upsilon_i = A\sin\omega t$. Now, $A \sin \Theta = 1.4$V, where $\Theta/360 = 5/100$. See that $\Theta = 18°$ and $A = \frac{1.4}{\sin 18°}$ = **4.53V peak**, or **9.1Vpp**.

The peak diode current $= \frac{4.53 - 1.4}{R}$ = 10mA. Use a resistor, $R = \frac{4.53 - 1.4}{10}$ = **313Ω**.

3.40

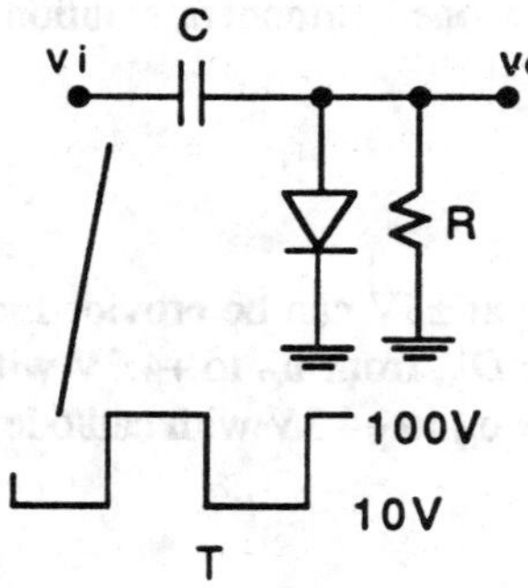

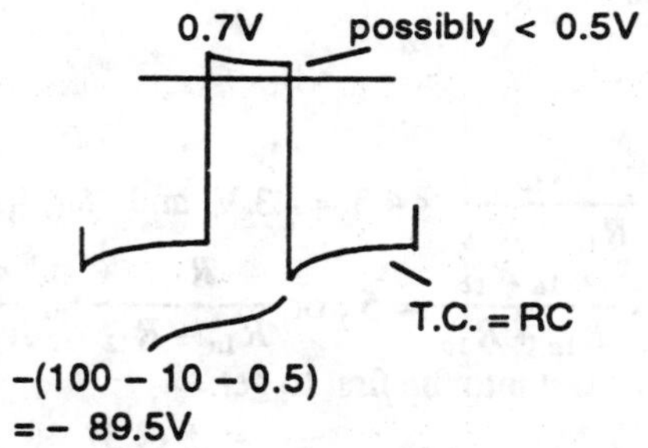

For light load, the output is a square wave of period T going from **+0.5V** to +0.5 –(100 – 10) = **–89.5V**. As the load resistance reduces, the negative side of the waveform is no longer flat at –89.5V, but rather rises toward ground. As well, upon the positive transition, the diode current increases initially, with 0.7V or so positive output at first. For $RC = 2T$, in one half cycle, where $t = T/2$, the output falls to $e^{\frac{-T/2}{2T}} = e^{-1/4}$ = 0.779 of its original value. Thus assuming the diode to have a constant 0.7V drop when conducting, the waveform initially rises to 0.7V, then drops to 0.7 (.779) = 0.55V, then falls to –90 + .55 = –89.5V, then droops (up) to –89.5 (0.779) = –69.7V, then rises to 0.7 V, and so on. In practice, the diode conducts at voltages lower than 0.7 V. Thus the upper level is not simply an exponential, but will fall more rapidly, to slightly less than 0.55V.

3.41

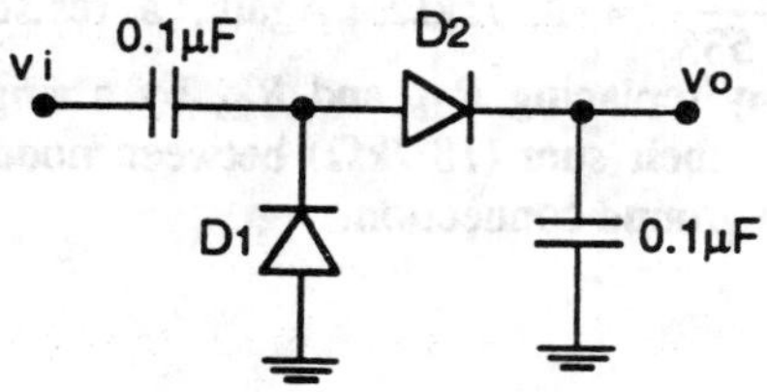

For a 100V-peak sine wave and no load, the output would be 2(100) –0.7 – 0.7 = **198.6V** for a 0.7V diode drop. For a pp ripple of 5% of peak, ripple voltage is (5/100) (198.6) = 9.93V, and the average output voltage = 198.6 – 9.93/2 = **193.6V**. The corresponding load current is $I = \frac{0.1 \times 10^{-6} \times 9.93}{1/(20 \times 10^3)}$ = **19.86mA**.

3.42 Assume ideal diodes having a drop of 0V. Initially, the input capacitor C_1 is charged with 0V on its internal end and input low, while the output capacitor C_2 is discharged. As the input rises by 100V, charge is dumped through the connecting diode D_2 from C_1 into C_2. Since the capacitors are equal and the input voltage change is 100V, the output rises to **50V**.

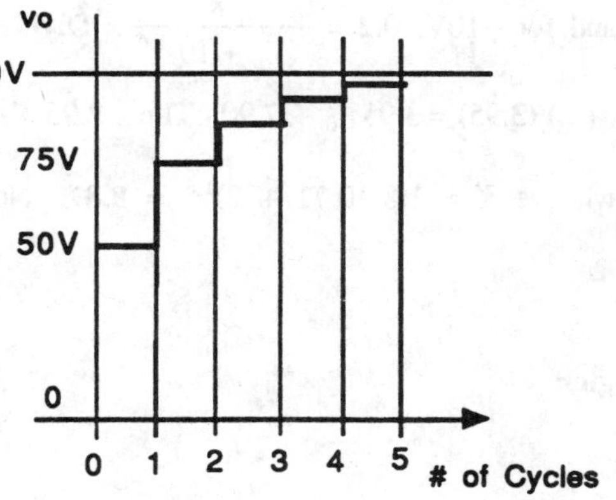

When the input falls, C_1 is recharged through the grounded diode D_1, while the connecting diode D_2 opens, leaving C_2 charged at 50V. Now, when the input rises again, diode D_2 does not conduct until the input rises by 50V. For the remaining 50V change of input, charge is shared equally by C_1 and C_2, while the voltage on C_2 rises by 25V to **75V**. When the input falls, C_1 is recharged. Correspondingly, after 2 cycles, the output becomes $50 + 25 + \frac{25}{2}$ = **87.5V**. After four cycles, the output is $50 + 25 + \frac{25}{2} + \frac{25}{4} + \frac{25}{8}$ = **96.9V**. After eight cycles the output is $50 + 25 + \frac{25}{2} + \frac{25}{4} + \frac{25}{8}\left[1 + \frac{1}{2} + \frac{1}{4} + \frac{1}{8} + \frac{1}{16}\right]$ = **99.8V**.

3.43 Ultimately, an equilibrium is established with the output reaching V_1 at the beginning of a cycle, V_2 half way, and V_3 at the end. Here, $V_2 = V_1\,(1 - \frac{5/100}{2}) = 0.975V$, since two capacitors C supply the load. Also $V_3 = V_2\,(1 - 5/100) = 0.95V_2 = 0.95\,(.975)\,V_1 = 0.92625\,V_1$. But at the start of the next cycle, with $V_1 = (V_O + V_3)/2$, hence $V_1 = (V_O + .92625\,V_1)/2 = V_O/2 + .4631V_1$.

$\therefore\ V_1 = \frac{V_O}{2(1-0.4631)} = \frac{V_O}{2(.5369)} = 0.931V_O,\ V_3 = .9263V_1 = 0.863V_O$, and the average output = $(V_1 + V_3)/2 = (0.931\,V_O + 0.863\,V_O)/2 = 0.897V_O$.

Here, including an approximation of the effect of diode drops, the average output would be about 0.897 (100 – 1.4) = **88.4V**, with ripple = $V_1 - V_3$ = (.931 – .863) 100 = **6.70Vpp**, the output ranging from 88.4 + 6.70/2 = 91.8V, to 88.4 – 6.70/2 = 85.0V.

SECTION 3.8: Physical Operation of Diodes – Basic Semiconductor Concepts

3.44 Acceptor concentration is $N_A = 10^{-m}$. Thus the hole concentration is $p_{po} \approx 10^{-m}$, while the electron concentration is $n_{po} = 10^{-n}$, $n >> m$. *More precisely,* in 10^n atoms, one is ionized at a particular temperature to produce one hole and one electron. As well, the number of acceptor - produced holes is $10^n \times 10^{-m}$. Thus the total number of holes is $1 + 10^n \times 10^{-m}$ in 10^n atoms, where $p_{po} = \frac{1 + 10^n \times 10^{-m}}{10^n} \approx$ **10^{-m}**, while **$n_{po} = 10^{-n}$** directly.

SECTION 3.9: The *pn* Junction under Open-Circuit Conditions

3.45 The depletion region will be larger in the lighter-doped p region. In fact, it will be **10 × larger** there than in the *n* region.

SECTION 3.10: The *pn* Junction Under Reverse-Bias Conditions

3.46 $I_S - I_D = I \rightarrow 15 - I_D = 10$. Thus the diffusion current, I_D = **5nA**.

3.47 $Q = CV \rightarrow C = \dfrac{0.1pC}{1V} = 0.1\text{pF}$. That is, the depletion capacitance is **0.1pF**.

3.48 $C_j = \dfrac{K}{(V_0 - V_D)^m}$. For −2V, $1.8 = \dfrac{K}{(V_0+2)^{1.6}}$, and for −10V, $0.2 = \dfrac{K}{(V_0+10)^{1.6}}$. Divide to get $9 = \left[\dfrac{V_0+10}{V_0+2}\right]^{1.6}$, and $V_0 + 10 = (V_0+2)\,(9^{1/1.6}) = (V_0+2)\,(3.95) = 3.95V_0 + 7.90$. Thus, $2.95\,V_0 = 2.10$, for which V_0 = **0.71V**. Now, $1.8 = \dfrac{K}{(0.71+2)^{1.6}}$, whence $K = 1.8\,(0.71+2)^{1.6}$ = **8.87**. Now, for 0V, $C_j = \dfrac{8.87}{(0.71+0)^{1.6}}$ = **15.3pF**.

SECTION 3.11: The *pn* Junction in the Breakdown Region

3.49 $P_D = V_B\,I_R$. Thus $I_R = \dfrac{P_D}{V_B} = \dfrac{50\times10^{-3}}{120}$ = **0.42mA**.

For breakdown only 10% of the time, a peak current of 0.42/0.1 = **4.2mA** can be tolerated.

SECTION 3.12: The *pn* Junction Under Forward-Bias Conditions

3.50 $q_M = q_O\,(e^{v/nV_T} - 1)$, and $C_d = \left.\dfrac{dq_M}{dv}\right|_{v=V_Q} = \dfrac{q_O}{nV_T}\,e^{v/nV_T}$. Now, for $n = 2$, I_Q = 1mA, and V_Q = 700mV, $1\times10^{-12} = \dfrac{q_O}{25\times10^{-3}}\,e^{700/50}$. Thus $q_O = 25\times10^{-3}\times1\times10^{-12}\,e^{-700/50}$ = **2.08 × 10⁻⁸ pC**. For a junction 10 times larger, q_O would be **20.8 × 10⁻⁸pC**. At 1mA, for the original junction, $q_M = q_O\,(e^{700/50} - 1)$ = **0.025pC**.

3.51 $q_M = q_O\,(e^{v/nV_T} - 1)$, and $C_d = \left.\dfrac{dq_M}{dv}\right|_{v=V_o} = \dfrac{q_O}{nV_T}\,e^{v/nV_T}$. But $i = I_S\,(e^{v/nV_T} - 1) \simeq I_S\,e^{v/nV_T}$. Thus $C_d = \dfrac{q_O}{nV_T}\times\dfrac{i}{I_S} = K(\dfrac{i}{nV_T})$ with K a constant for a given junction.

3.52 $r_d = nV_T/I_A \rightarrow n = (50\Omega\times1mA)/25mV = 2$. Also $C_j = \dfrac{K}{(V_0 - V_A)^m}$. Thus, at −1V, $0.75 = \dfrac{K}{(V_0+1)^m}$ - - - (1). Thus, at −5V, $0.2 = \dfrac{K}{(V_0+5)^m}$ - - - (2). Thus, at −10V, $0.1 = \dfrac{K}{(V_0+10)^m}$ - - - (3). Now (1)/(2) $\rightarrow 0.75/0.2 = \left[\dfrac{V_0+5}{V_0+1}\right]^m = 3.75$ - - - (4), and (2)/(3) $\rightarrow 0.2/0.1 = \left[\dfrac{V_0+10}{V_0+5}\right]^m = 2.0$ - - - (5).

Explore trial solutions of (4), (5).

For $V_O = 0$ V: $\left[\dfrac{0+5}{0+1}\right]^m = 3.75 \rightarrow m = 0.82$,

and $\left[\dfrac{0+10}{0+5}\right]^m = 2.0 \rightarrow m = 1$, $\Delta = 0.18$.

For $V_O = 1.0$: $\left[\dfrac{1+5}{1+1}\right]^m = 3.75 \rightarrow m = 1.21$,

and $\left[\dfrac{1+10}{1+5}\right]^m = 2.0 \rightarrow m = 1.15$, $\Delta = 0.06$.

For $V_O = 2.0$: $\left[\dfrac{2+5}{2+1}\right]^m = 3.75 \rightarrow m = 1.56$,

and $\left[\dfrac{2+10}{2+5}\right]^m = 2.0 \rightarrow m = 1.27$, $\Delta = 0.29$.

See nearer $V_O = 1 \rightarrow$ Try:

$V_O = 0.5$: $\left[\dfrac{0.5+5}{0.5+1}\right]^m = 3.75 = 3.666^m \rightarrow m = 1.02$,

and $\left[\dfrac{0.5+10}{0.5+5}\right]^m = 2.0 = 1.909^m \rightarrow m = 1.07$, $\Delta = 0.05$.

See between 0.5 and 1.0 $\rightarrow$ Try:

$V_O = 0.7$: $\left[\dfrac{0.7+5}{0.7+1}\right]^m = 3.75 = 3.353^m \rightarrow m = 1.095$,

and $\left[\dfrac{0.7+10}{0.7+5}\right]^m = 2.00 = 1.877^m \rightarrow m = 1.10$, $\Delta = 0.005 \simeq 0$.

Conclude n = **2**, m = **1.10**, V_0 = **0.7V** and $K = 0.75\,(0.7 + 1)^{1.1}$ = **1.344**.

Now, at $V_A = 0\text{V}$, $C_{j0} = \dfrac{1.344}{(0.7)^{1.1}} = 2.0\text{pF}$. Now, as $C_T = C_j + C_d = 10\text{pF}$, then $C_d = 10 - 2 = 8\text{pF}$, and $K_c = C_d/I_D = 8\text{pF}/1\text{mA}$ = **8pF/mA**.

Now, at 10mA, $C_T \simeq 10(8) + 2$ = **82pF**.

3.53 For a diode 10× the area of that above, operated at $I_D = 1\text{mA}$, $r_d = \dfrac{nV_T}{I_D} = \dfrac{2 \times 25}{1}$ = **50Ω**, and $C_d = K_c\, I_D = 8 \times 1$ = **8pF**. *But the junction is 10× larger*: Thus $C_{j0} = 2.0\text{pF}(10)$ = **20pF**, and $C_T = 20 + 8$ = **28pF**. *Now at $V_R = 1V$*, with 10× the area, $C_j = \dfrac{K}{(V_0 - -V_R)^m} = \dfrac{1.344 \times 10}{(0.7+1)^{1.1}}$ = **17.2pF**.

Chapter 4

Bipolar Junction Transistors (BJTs)

SECTION 4.2: Operation of the npn Transistor in the Active Mode

4.1 For (ensured) active-mode operation of an npn transistor, $\upsilon_{CB} \geq$ **0V** and $\upsilon_{CE} = \upsilon_{CB} + \upsilon_{BE} \geq$ **700mV**.
In the active mode, $i_C = I_S\, e^{\upsilon_{BE}/V_T}$ (for n =1); $10 \times 10^{-3} = I_S\, e^{700/25}$; or $I_S = 10 \times 10^{-3}\, e^{-28} = \mathbf{6.91 \times 10^{-15}\,A}$.
Also, in the active mode, $\beta = \dfrac{i_C}{i_B} = \dfrac{10mA}{100\mu A} = \dfrac{10}{0.1} = \mathbf{100}$, and $i_E = i_C + i_B$ = 10mA + 0.1mA = **10.1mA**.

4.2 (b) $I_B = \dfrac{I_C}{\beta} = \dfrac{1mA}{50} = \mathbf{20\mu A}$; $I_E = I_C + I_B = 1 + .02 = \mathbf{1.02mA}$; $\alpha = \dfrac{I_C}{I_E} = \dfrac{1.00}{1.02} = \mathbf{0.980}$.

(c) $I_C = \alpha I_E = 0.98\,(2) = \mathbf{1.96mA}$; $I_B = I_E - I_C = 2 - 1.96 = .04mA = \mathbf{40\mu A}$; $\beta = \dfrac{I_C}{I_B} = \dfrac{1.96}{0.04} = \mathbf{49}$.

(d) $I_C = \beta I_B = \dfrac{\alpha}{1-\alpha} I_B = \dfrac{0.995}{1-0.995} \times 0.01 =$ **1.99mA**; $I_E = I_C + I_B$ = 1.99 + .01 = **2.00mA**;
$\beta = \dfrac{\alpha}{1-\alpha} = \dfrac{0.995}{1-.995} = \mathbf{199}$.

(e) $I_C = \alpha I_E = \dfrac{\beta}{\beta+1} I_E = \dfrac{10}{10+1} \times 110 =$ **100mA**; $I_B = \dfrac{I_C}{\beta} = \dfrac{100mA}{10} = \mathbf{10mA}$; $\alpha = \dfrac{I_C}{I_E} = \dfrac{100}{110} = \mathbf{0.909}$.

(f) $I_C = \beta I_B = 1000 \times 0.001 = \mathbf{1mA}$; $I_E = I_C + I_B = 1 + 0.001 = \mathbf{1.001mA}$; $\alpha = \dfrac{I_C}{I_E} = \dfrac{1}{1.001} = \mathbf{0.999}$.

Device #	I_C mA	I_B mA	I_E mA	α	β
a	10	.1	10.1	.99	100
b	1	.02	1.02	0.98	50
c	1.96	.04	2	.98	49
d	1.99	.01	2.00	.995	199
e	100	10	110	0.909	10
f	1	.001	1.001	0.999	1000

4.3

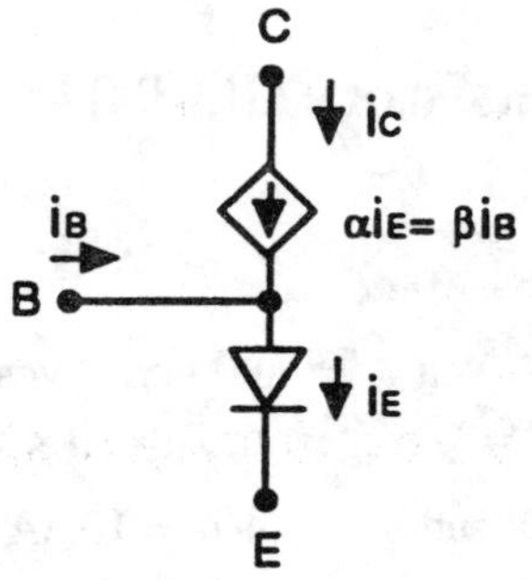

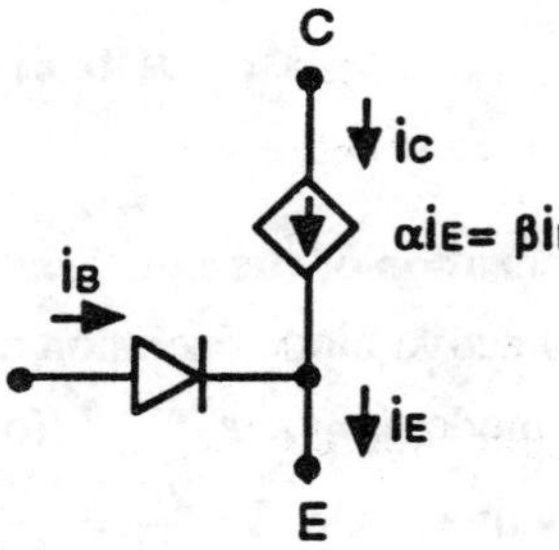

The controlled sources can be labelled $\alpha\, i_E$ or $\beta\, i_B$ where $\alpha\, i_E = \beta\, i_B$

4.4 Generally, $i = I_S\, e^{\upsilon_{BE}/nV_T}$.

For $n = 1$, at 1mA, $10^{-3} = I_S\, e^{700/25}$ --- (1), and at 0.1μA, $10^{-7} = I_S\, e^{\upsilon/25}$ --- (2). Now, dividing, $10^{-3}/10^{-7} = e^{(700-\upsilon)/25} = 10^4$. Taking logarithms, $700 - \upsilon = 25 \ln 10^4 = 230$. Thus, $\upsilon = 700 - 230 = 470$mV, or **0.47V**.

For $n = 2$, see $\upsilon = 700 - 2(25)\ln 10^4 = 700 - 2(230) = 240$mV, or **0.24V**.

4.5 Generally, for the base open, $i_C = \beta\, i_B = 100\, I_{CBO}$.

At 25°C, $I_{CBO} = 0.1$nA, and $i_C = 100\,(0.1 \times 10^{-9})\,A =$ **10nA**.

At 95°C, $I_{CBO} = 0.1 \times 10^{-9} \times 2^{\frac{95-25}{10}} = 10^{-10} \times 2^7 A$, whence $i_C = 100 \times 10^{-10} \times 2^7 =$ **1.28μA**.

4.6 $\alpha_F/\alpha_R = 100$, the relative junction area. Thus $\alpha_R = \dfrac{\alpha_F}{100} = \dfrac{150}{150+1}/100 = .00993$, and $\beta_R = \dfrac{\alpha_R}{1-\alpha_R} = \dfrac{.00993}{1-.00993} = .0092 \cong$ **.01**.

SECTION 4.3: The pnp Transistor

4.7 For $\alpha = 0.975$, $\beta = \dfrac{\alpha}{1-\alpha} = \dfrac{0.975}{1-.975} = 39$. For $I_B = 10\mu A$, $I_C = 39\,(10)\mu A = 3.90\mu A =$ **0.390mA**, and $I_E = 390 + 39 = 429\mu A =$ **0.429mA**. Generally, $\upsilon_{BE2} = \upsilon_{BE1} + nV_T \ln \dfrac{i_{C2}}{i_{C1}}$. Thus at $i_C = 0.390$mA, $\upsilon_{BE} = 700 + 25 \ln \dfrac{0.390}{1.00} =$ **676mV**.

4.8 For D_B, $\dfrac{I_S}{\beta} = 10^{-13}$A, and for D_E, $\dfrac{I_S}{\alpha} = 10^{-11}$A. Thus, $\dfrac{1/\beta}{1/\alpha} = 10^{-2}$, or $\dfrac{\alpha}{\beta} = 10^{-2}$. Correspondingly, $\dfrac{\beta/(\beta+1)}{\beta} = 10^{-2}$, or $\beta + 1 = 100$, and $\beta =$ **99**.

Now, $i_C = I_S\, e^{643/25(1)} = 99 \times 10^{-13}\, e^{25.7} =$ **1.46A!** Note that this is a large transistor, a 14.5-Amp device (where $10^{-11}\, e^{700/25} = 14.5$A).

SECTION 4.4: Circuit Symbols and Conventions

4.9

(a) Here, $i_E = I =$ **1mA**, independent of υ_{BE} or β, V_E ranges from **–0.6V** to **–0.8V**; $i_C = \alpha\, i_E$ ranges from $\frac{10}{10+1}(1) =$ **0.909mA** to $\frac{300}{300+1}(1) =$ **0.997mA**.

(b) $V_C = V_{CC} - R_C\, I_C$, ranges from $10 - 5\,(.909) =$ **5.45V** to $10 - 5\,(.997) =$ **5.02V**.
No, $\mathbf{V_E}$ **variation has no effect on** I_C or V_C.

(c) For $\upsilon_{CB} \geq 0$, at the largest value of i_C, $R_C\, i_C \leq V_{CC} - 0 = 10\text{V}$, whence $R_C \leq \frac{10}{.997} =$ **10.03kΩ** (use **10kΩ**).

4.10 Here, $\mathbf{i_C = \alpha\,(I + i_e)}$. Now, for high β, $\alpha = 1$ and $i_C = 1 + 0.1 =$ **1.1mA**. For $\beta = 10$, $\alpha = \frac{10}{10+1} = 0.909$, and $i_C = (1.1)\,(.909) =$ **1mA**. For high β, the largest allowed R_C is limited to be $R_C = \frac{10-0}{1.1} =$ **9.90kΩ**, and $\upsilon_C = 1(0.1\text{mA})\,(9.09\text{k}\Omega) = 0.909\text{V}$ peak or **1.82 Vpp**. For $\beta = 10$, $\upsilon_C = 0.909\,(0.1\text{mA})\,(9.09\text{k}\Omega) = 0.826\text{V}$ **peak, or 1.65Vpp**.

4.11 For $V_E = 0.7\text{V}$, use $R_E = (0 - 0.7 - -10)/1 =$ **9.3kΩ**. Now the largest i_C occurs for small R_E, small υ_{BE}, and large β. That is, $i_{C\,\max} = \left[\frac{-0.6 - -10}{9.3\,(1-.01)}\right] \times 1 =$ **1.021mA**. For active-mode operation, $R_C \leq \frac{10-0}{1.021} =$ **9.79kΩ**. Use **9.7kΩ**. Now for R_C varying by 1%, the lowest possible value of $\upsilon_C = 10 - (9.7)\,(1.01)\,(1.021) =$ **–2.8mV**. Note that operation is still in the active mode, since $\upsilon_E \equiv -0.70\text{V}$.

SECTION 4.5: Graphical Representation of Transistor Characteristics

4.12 Generally, $TC \approx -2.0\text{mV/°C}$. Now, for 10mA, at 25°C, $V_{BE} = 700\text{mV}$. Thus, for 10mA, at 0°C, $V_{BE} = 700 + (0 - 25)\,(-2.0) = 750\text{mV}$. For 10mA, at 50°C, $V_{BE} = 700 + (50 - 25)\,(-2.0) = 650\text{mV}$. For V_{BE} constant at 620mV, at 25°C, $i = 10e^{(620-700)/25} =$ **0.407mA**; at 0°C, $i = 10\,e^{(620-750)/25} =$ **0.055mA**; at 50°C, $i = 10\,e^{(620-650)/25} =$ **3.012mA**, a nearly 55 to 1 range!!

4.13 Here, $\Delta\, i_C = 2.19 - 2.10 = 0.09\text{mA}$, for $\Delta\, V_{CE} = 9 - 2 = 7\text{V}$. Thus $r_o = \frac{\Delta\,\upsilon}{\Delta\, i} = \frac{7}{.09} =$ **77.8kΩ**, at an average current of $(\,2.19 + 2.10/2\,) = 2.145\text{mA}$. Thus $V_A = 77.8 \times 2.145 =$ **167V**. Correspondingly, at 0.1mA, $r_o = \frac{167}{0.1} = 1.67\text{M}\Omega \equiv$ **1.7MΩ**, and at 10mA, $r_o = \frac{167}{10} = 16.7\text{k}\Omega \equiv$ **17 kΩ**.

4.14 At 100μA, $r_o = \frac{200}{0.1mA} = 2\text{M}\Omega$. For an increase in V_{CE} from 5 to 50V, the increase in current is $\frac{50-5}{2M} =$ 22.5μA. Thus at 50V, the current becomes $100 + 22.5 =$ **122.5μA**.

SECTION 4.6: Analysis of Transistor Circuits at DC

4.15 In general, assume the active mode initially, with a conducting emitter-base junction, then verify that the collector-base junction is *not* conducting. Since $\beta = 50$, $\alpha = \frac{50}{51} = 0.98$.

(a) $V_B = -4\text{V}$; $V_E = -4 + 0.7 =$ **–3.3V**; $I_E = \frac{0 - -3.3}{3.3} =$ **1mA**; $I_C = 0.98(1) =$ **0.98mA**; $I_B = \frac{0.98}{50} =$ **19.6μA**; $V_C = -10 + 4.7\,(0.98) =$ **–5.39V**. See this is OK; operation is in the active mode.

(b) $V_B = -6\text{V}$; $V_E = -6 + 0.7 =$ **–5.3V**; $I_E = \frac{0 - -5.3}{3.3} =$ **1.606mA**; $I_C = 1.606\,(.98) = 1.574\text{mA}$; $I_B = \frac{1.57}{50} = 31.4\mu\text{A}$; $V_C = -10 + 4.7\,(1.574) = -2.60\text{V}$. Since V_C is well above V_B, the transistor is saturated.

In practice, in saturation $\upsilon_{EC} \approx 0.2$V. Thus, $V_C = V_E - 0.2 = -5.3 - 0.2 = $ **−5.5V**, and $I_C = \frac{-5.5 - -10}{4.7}$ = **0.957mA**, with $I_B = I_E - I_C = 1.606 - .957 =$ **0.648mA**.

(c) $V_B = -2$V; $V_E = -2 + 0.7 = -1.3$V; $I_E = \frac{2 - -1.3}{3.3} =$ **1mA**; $I_C =$ **0.98mA**; $I_B = \frac{0.98}{50} =$ **19.6μA**; $V_C = -8 + 4.7\ (0.98) =$ **−3.39V**. See OK, active.

(d) $V_B = 0$V; $V_E =$ **0V**. Thus the transistor is cutoff.
∴ $V_C =$ **−10V**, and $I_B = I_C = I_E =$ **0mA**.

(e) $V_B = -4$V; $V_E = -4 - 0.7 =$ **−4.7V**; $I_E = \frac{-4.7 - -10}{4.7k\Omega} =$ **1.128mA**; $I_C = 1.128\ (0.98) =$ **1.105mA**; $I_B = \frac{1.105}{50} =$ **22.1μA**; $V_C = 0 - 1.105\ (3.3) =$ **−3.67V**. Since $V_C > V_B$, operation is in the active mode, as assumed.

(f) $V_B = -6$V; $V_E = -6 - 0.7 =$ **−6.7V**; $I_E = \frac{-6.7 - -10}{4.7k\Omega} =$ **0.702mA**; $I_C = 0.702\ (0.98) =$ **0.688mA**; $I_B = \frac{0.688}{50} =$ **13.8μA**; $V_C = 0 - 0.688\ (3.3) =$ **−2.27V**. Since $V_C > V_B$, this is active-mode operation, as assumed.

4.16

(a) $V_B = -4$V; $V_E = -4 + 0.7 = -3.3$V; $I_E = \frac{0 - -3.3V}{R_E} = 0.5$mA; Thus, $R_E = \frac{3.3}{0.5} =$ **6.6kΩ**. Now, $I_C = 0.5 \times 1 = 0.5$mA; $V_C = -10 + R_C\ (0.5) = V_B - V_{BC} = -4 - 0 = -4$V; Thus, $R_C = \frac{1}{0.5}\ (-4 + 10) =$ **12kΩ**.

(b) $V_B = -6$V; $V_E = -6 + 0.7 = -5.3$V; $I_E = \frac{0 - -5.3}{R_E} = 0.5$mA; Thus, $R_E = \frac{5.3}{0.5} =$ **10.6kΩ**. Now, $V_C = -10 + R_C\ (0.5) = V_B - V_{BC} = -6 - 0 = -6$V. Thus $R_C = \frac{1}{0.5}\ (-6 + 10) =$ **8kΩ**.

4.17 **Assume active mode, and forward conduction of the base-emitter junction.**

(a) $V_E =$ **0V**; $V_B = 0 + 0.7 =$ **0.7V**; $I_B = \frac{10 - 0.7}{100k\Omega} =$ **93μA**; $I_C = 93\mu A \times 10 =$ **0.930mA**; $I_E = 930 + 93 = 1023\mu A =$ **1.023mA**; $V_C = 10 - 2k\ (0.93) =$ **8.14V**.

(b) $V_E =$ **+10V**; $V_B = +10 - 0.7 =$ **9.3V**; $I_B = \frac{9.3 - 0}{100k} =$ **93μA**; $I_C = 93\ (10) =$ **0.930mA**; $I_E = 11\ (93) =$ **1.023mA**; $V_C = 0.93\ (2) =$ **1.86V**.

(c) $V_E =$ **0V**; $V_B =$ **0.7V**; $I_B = \frac{10 - 0.7}{100k\Omega} - \frac{0.7}{10k\Omega} = 93\mu A - 70\mu A =$ **23μA**; $I_C = (23\mu A)\ 10 =$ **0.230mA**; $I_E = 11\ (23) =$ **0.253mA**; $V_C = 10 - 2\ (0.23) =$ **9.54V**.

(d) See that $V_E =$ **0V**, and that if the transistor conducts with $V_B = -0.7$V, that the current in the upper 100kΩ will exceed that in the lower, providing no net base current. Thus the transistor is cut off, with $V_B = +10 - (10 - -10)\frac{100k\Omega}{100k\Omega + 100k\Omega} =$ **0V**, $V_C =$ **−10V**, and $I_B = I_C = I_E =$ **0mA**.

4.18

(a) Assume active mode and consider the base-to-emitter circuit: Thus $0 - 10k\Omega\ (I_B) - 0.7 - 10k\Omega\ (I_E) = -10$V. But $I_E = (\beta + 1)\ I_B = 21 I_B$. Thus $10\ I_B + 10\ (21\ I_B) = 9.3$V, whence $I_B = \frac{9.3}{220} =$ **42.27μA**. Thus, $V_B = 0 - 42.27\ (10k) =$ **−0.423V**; $V_E = -0.423 - 0.7 =$ **−1.123V**; $I_E = 21\ (42.27) =$ **0.888mA**; $I_C = 20\ (42.27) =$ **0.845mA**; and $V_C = +10 - 10k\ (0.845) =$ **1.55V**.

(b) As before: $I_B = \frac{10V}{100k + 21(10k)}$ = **32.26μA**; V_B = 0 − 100k (32.26) = **−3.23V**; V_E = −3.23 −0.7 = **−3.93V**; I_E = 21 (32.26) = **0.677mA**; I_C = 20 (32.26) = **0.645mA**; V_C = +10 − 10k (0.645) = **3.55V**.

(c) Here, $I_E = \frac{10V - 0V}{10k + 1M/21} = \frac{10}{(10 + 47.6)k}$ = **0.174mA**; V_E = 10 − 10kΩ (0.174mA) = **8.26V**; V_B = 8.26 − 0.70 = **7.56V**; $I_C = \frac{20}{21}$ (0.174) = **0.166mA**; $I_B = \frac{0.174mA}{21}$ = **8.29μA**; V_C = −10 + 0.166 (10) = **−8.34V**.

4.19 For V_{BE} = 0.7V, the current in the base-emitter shunting resistor is 0.7V/10kΩ = 70μA. This flows in the resistor to 10V, creating an equivalent base source of V_{BB} = 10 − 100kΩ (70μA) = 3.0V, with R_{BB} = 100kΩ. Now, for the base-emitter loop and base current I_B, 3V − (100kΩ) I_B − 0.7V − (β + 1) (3.3kΩ) I_B = 0. Thus, $I_B = \frac{2.3}{3.3\,(\beta + 1) + 100}$.

For β = ∞: I_B = **0μA**; V_B = **3V**; V_E = 3 − 0.7 = **2.3V**; $I_E = \frac{2.3}{3.3k\Omega}$ = 0.697mA; V_C = 10 − 3.3 (0.697) = 10 − 2.3 = **7.7V**.

For β = 100: $I_B = \frac{2.3}{3.3\,(101) + 100}$ = 5.35μA; $V_B = 3.0 - 5.35 \times 10^{-6}\,(10^5)$ = **2.465V**; V_E = 2.465 − 0.7 = **1.765V**; I_C = 100 (5.35) = 0.535mA; V_C = 10 − 3.3 (.535) = **8.23V**.

For β = 10: $I_B = \frac{2.3}{3.3\,(11) + 100}$ = 16.87μA; V_B = 3.0 − (16.87) (0.1) = **1.313V**; V_E = 1.313 − 0.7 = **0.613V**; $V_C = 10 - 3.3 \times 10^3\,(16.87 \times 10^{-6})\,(10)$ = **9.44V**.

4.20 From the solution of P4.19, V_{BB} = 3V, R_{BB} = 100kΩ.

For β = ∞: V_E = 3 − 0.7 = 2.3V, with $I_E = \frac{2.3V}{3.3k\Omega}$ = 0.697mA.

For β generally, $I_E = (\beta + 1)\,I_B = \frac{(\beta + 1)\,(2.3)}{3.3\,(\beta + 1) + 100} = \frac{2.3}{3.3 + 100/(\beta + 1)}$.

Thus, $\frac{2.3}{3.3 + 100/(\beta + 1)}$ = 0.8 (0.697) = .5576, or $3.3 + \frac{100}{\beta + 1} = \frac{2.3}{0.5576}$ = 4.125, or 100/(β + 1) = 0.8248, β + 1 = 100/.8248 = 120, whence β = **120**.

4.21 Here, $V_{BB} = \frac{10k}{10k + 20k}$ (9V) = 3V; R_{BB} = 10k||20k = 6.66kΩ.

(a) *For* β = ∞: V_B = 3V; V_E = 3 − 0.7 = 2.3V; $I_E = \frac{2.3V}{1k\Omega}$ = **2.3mA**; V_C = 9 − 2 (2.3) = 4.4V. ∴ $V_{CE} = V_C - V_E$ = 4.4 −2.3 = **2.1V**.

(b) *For* β = 100: $I_E = \frac{3 - 0.7}{1k + 6.6/101}$ = **2.156mA**; V_E = 2.156 (1k) = 2.156V; V_C = 9 − 2k (2.156) × 100/101 = 4.731V; V_{CE} = 4.731 − 2.156 = **+2.575V**.

(c) *For* β = 10: $I_E = \frac{2.3}{1k + 6.6/11}$ = **1.432mA**; V_E = 1.432V; V_C = 9 − 2 (1.432) 10/11 = 6.396V; V_{CE} = 6.396 − 1.432 = **4.964V**.

4.22

(a) Let $I_C = i$. From the supply to ground: 10 = 10k(i + i/β + I) + 100k(I + i/β) + 0.7, or 10 i + 10 i/50 + 100 i/50 = 9.3 − 10I − 100I. Thus 12.2i = 9.3 − 110I −−− (1), or $i = \frac{9.3 - 110\,(.02)}{12.2}$, or $i = \frac{7.1}{12.2}$ = **0.582mA** = I_C, for which $V_{CE} = 10 - 10\,(.02 + \frac{51}{50}\,(0.582))$ = **3.86V**.

(b) For $V_B = 0.7\text{V}$, $I_{560k} = \dfrac{0.7 - -10}{560} = .0191\text{mA}$. From the previous (1) in (a): $i = \dfrac{9.3 - 110\,(.0191)}{12.2} =$ $\mathbf{0.590mA} = I_C$, and $V_{CE} = 10 - 10\,(.0191 + \frac{51}{50}\,(.590)) = \mathbf{3.79V}$.

(c) For $V_B = 0.7 \rightarrow I_{33k} = \dfrac{0.7V}{33k} = .0212\text{mA}$. Thus, $I_C = \dfrac{9.3 - 110\,(.0212)}{12.2} = \mathbf{0.571mA}$, and $V_{CE} = 10 - 10\,(.0212 + \frac{51}{50}\,(0.571)) = \mathbf{3.96V}$.

(d) Let the base voltage be υ and base current be i. $\therefore\ \upsilon = 0.7 + (\beta + 1)\,i\,(1k) = 0.7 + 51\,i$ - - - (1); $\upsilon_C = \upsilon + 100k\,(\upsilon/68k + i) = \upsilon + 1.47\,\upsilon + 100\,i = 2.47\,\upsilon + 100\,i$ - - - (2); Also $\upsilon_C = 10 - 10k\,((\beta + 1)\,i + \upsilon/68k) = 10 - 510\,i - 0.147\,\upsilon$ - - - (3). Now (3) with (2) $\rightarrow 2.47\,\upsilon + 100\,i = 10 - 510\,i - 0.147\,\upsilon$, or $2.617\upsilon + 610\,i = 10$ - - - (4). Then (4) with (1) $\rightarrow 2.617\,(0.7 + 51\,i) + 610\,i = 10$, or $1.832 + 133.5\,i + 610\,i = 10$. Thus, $i = \dfrac{10 - 1.832}{133.5 + 610} = \dfrac{8.18}{743.5} = .011\text{mA}$, whence $\upsilon = 0.7 + 51\,(.011) = 1.261\text{V}$, and $I_C = \beta\,i = 50\,(.011) = \mathbf{0.55mA}$, $V_E = 1k\,(\beta + 1)\,i = 1\,(51)\,(.011) = 0.561\text{V}$, $V_C = 10 - 10k\,(51(.011) + \frac{1.261}{68k}) = 10 - 5.61 - .185 = 4.205\text{V}$. Thus, $V_{CE} = 4.205 - .561 = \mathbf{3.69V}$

4.23

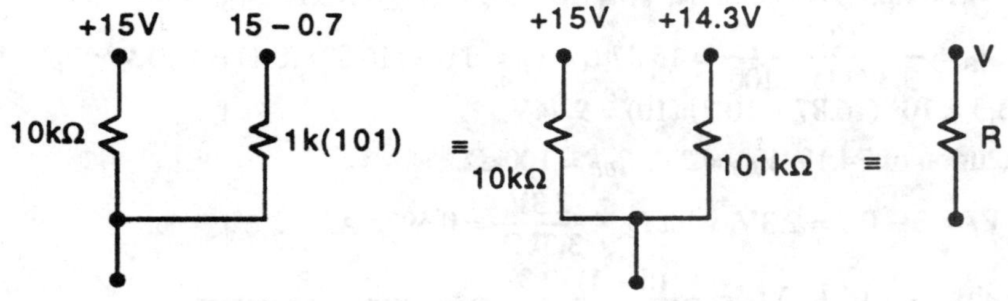

For $\beta = \infty$: $V_A = \dfrac{100}{100 + 200}\,(15) = \mathbf{5V}$; $V_B = 5 - 0.7 = \mathbf{4.3V}$; $V_C = 15 - \dfrac{10k}{10k}\,(4.3) = \mathbf{10.7V}$; $V_D = 10.7 + 0.7 = \mathbf{11.4V}$; $V_E = (\dfrac{15 - 11.4}{1k})\,1k = \mathbf{3.6V}$.

For $\beta = 100$: *At node A*: $V_{AA} = 5\text{V}$, and $R_{AA} = 100\text{k}\Omega \,||\, 200\text{k}\Omega = 66.7\text{k}\Omega$; $\therefore\ I_E = \dfrac{5 - 0.7}{10k\Omega + 66.7/101} = .4034\text{mA}$; $V_B = 1k\,(.4034) = \mathbf{4.03V}$; and $V_A = 4.03 + 0.7 = \mathbf{4.73V}$.

At node C: $V = 15 - \dfrac{10}{10 + 101}\,(0.7) = 14.94$; $R = 10k \,||\, 101k = 9.10k\Omega$. Now, from the collector at node C, $I = 0.4034 \times \dfrac{100}{101} = 0.399\text{mA}$. Thus, $V_C = 14.94 - 0.399\,(9.10) = \mathbf{11.31V}$, and $V_D = 11.37 + 0.7 = \mathbf{12.07V}$.

Thus, $V_E = 0 + 1k\,(\dfrac{15 - 12.07}{1k} \times \dfrac{100}{101}) = \mathbf{2.90V}$.

SECTION 4.7: The Transistor as an Amplifier

4.24 Generally, $g_m = \dfrac{I_C}{V_T}$: For $1\mu\text{A}$, $g_m = \dfrac{10^{-6}}{25 \times 10^{-3}} = 40 \times 10^{-6} = \mathbf{40\mu A/V}$; and for $100\mu\text{A}$, $g_m = \mathbf{4mA/V}$; for 1mA, $g_m = \mathbf{40mA/V}$; for 100mA, $g_m = \mathbf{4A/V}$.

4.25 At the emitter, $r_e = \dfrac{V_T}{I_E} = \dfrac{\alpha\,V_T}{I_C} = \dfrac{\alpha}{g_m} \approx \dfrac{1}{g_m}$. At the base, $r_\pi = \dfrac{V_T}{I_B} = \dfrac{V_T}{I_C/\beta} = \dfrac{\beta}{g_m} = (\beta + 1)r_e$. For $\beta =$

100, α = 0.99. Now, for I_C = 1μA, $r_e = \dfrac{0.99}{40 \times 10^{-6}}$ = 24.75kΩ ≡ 25kΩ, $r_\pi = \dfrac{100}{40 \times 10^{-6}}$ = **2.5MΩ**; for 100μA, $r_e \approx$ **250Ω**, r_π = **25kΩ**; for 1mA, r_e ≡ **25Ω**, r_π = **2.5kΩ**; for 100mA, r_e = **0.25Ω**, r_π = **250Ω**.

4.26 Gain $= -g_m R_L$, where $g_m = \dfrac{\alpha I_E}{V_T} \approx \dfrac{I_E}{V_T}$. The voltage across $R_L = I_C R_L = \alpha I_E R_L$ is a constant, K. Thus $R_L = \dfrac{K}{\alpha I_E}$, for which the $|\text{gain}| = \dfrac{\alpha I_E}{V_T} \times \dfrac{K}{\alpha I_E} = \dfrac{\mathbf{K}}{\mathbf{V_T}}$, a constant! Thus the gain is constant. There is no gain variation possible. The bias current *does not matter!*

4.27 The input resistance at the emitter, $r_e = \dfrac{V_T}{I_E} = \dfrac{25 \times 10^{-3}}{100 \times 10^{-6}}$ = 250Ω. The input resistance at the base, $r_\pi = (\beta + 1) r_e$ = 151 (250) = **37.75kΩ**. The voltage gain, base to collector, is $-\dfrac{\alpha R_L}{r_e} = -\dfrac{150}{151} \times \dfrac{10k\Omega}{250}$ = **–39.73V/V**.

4.28 The collector load is the 100kΩ resistor from collector to (grounded) base. Now, I_E = 1mA, and $r_e = \dfrac{25mV}{1mA}$ = 25Ω. Thus the gain $= -\dfrac{\alpha R_L}{r_e} \approx -\dfrac{100k\Omega}{25\Omega}$ = **–4000V/V**. The resistance "seen" by the source υ_S is r_e = **25Ω**. Now, for R_S = 75Ω, $\dfrac{\upsilon_i}{\upsilon_s} = \dfrac{25}{25 + 75}$ = ¼, and $\dfrac{\upsilon_o}{\upsilon_s} = \dfrac{\upsilon_i}{\upsilon_s} \times \dfrac{\upsilon_o}{\upsilon_i}$ = –4000 × ¼ = **–1000V/V**.

SECTION 4.8: Small-Signal Equivalent-Circuit Models

4.29

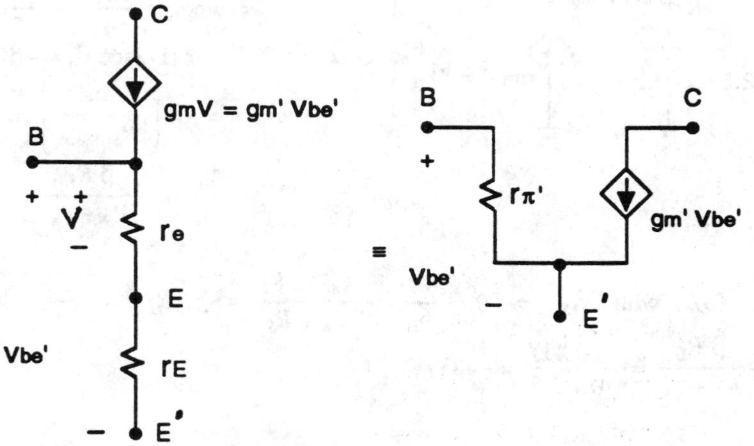

See $\upsilon = \left[\dfrac{r_e}{r_e + r_E}\right] \upsilon_{be}'$, and $g_m' \upsilon_{be}' = g_m \upsilon = g_m \left[\dfrac{r_e}{r_e + r_E}\right] \upsilon_{be}'$. Thus, $\mathbf{g_m' = g_m \dfrac{r_e}{r_e + r_E}}$. Now,
$r_\pi' = \dfrac{\upsilon_{be}'}{i_b} = \upsilon_{be}' / \left[\dfrac{\upsilon_{be}'}{r_e + r_E} - g_m' \upsilon_{be}'\right] = 1/\left[\dfrac{1}{r_e + r_E} - g_m'\right] = 1/\left[\dfrac{1}{r_e + r_E} - g_m \dfrac{r_e}{r_e + r_E}\right] = \dfrac{r_e + r_E}{1 - g_m r_e} = \dfrac{r_e + r_E}{1 - \alpha} = (\beta + 1)(r_e + r_E)$. Now, for I_C = 1mA, r_e = 25Ω, and $r_E = 3r_e$ = 75Ω, $\mathbf{r_\pi' = (\beta + 1)(r_e + r_E)}$ = 101 (25 + 75) = **10.1k Ω**, and $g_m' = \dfrac{g_m r_e}{r_e + r_E} = \dfrac{(\alpha/r_e) r_e}{r_e + r_E} = \dfrac{\alpha}{r_e + r_E}$

$$= \frac{100/101}{25+75} = 9.9 \times 10^{-3}\text{A/A} = \mathbf{9.9mA/V}.$$

4.30

(a)

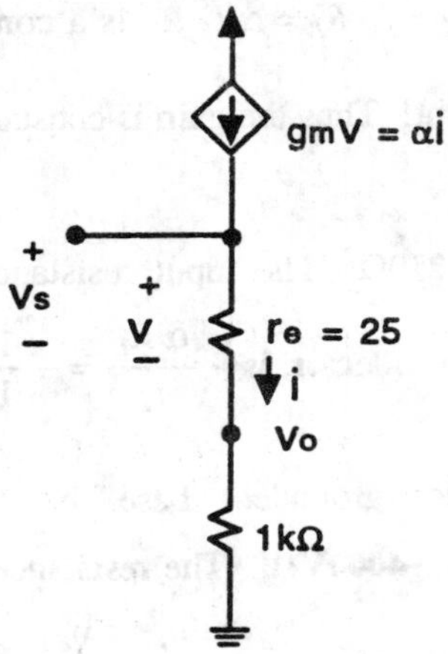

See directly (with either T_{gm} or T_α) that $\upsilon_o = \dfrac{1k\Omega}{25 + 1k\Omega} \times \upsilon_s$, whence

$$\frac{\upsilon_o}{\upsilon_s} = \mathbf{0.976V/V}.$$

(b)

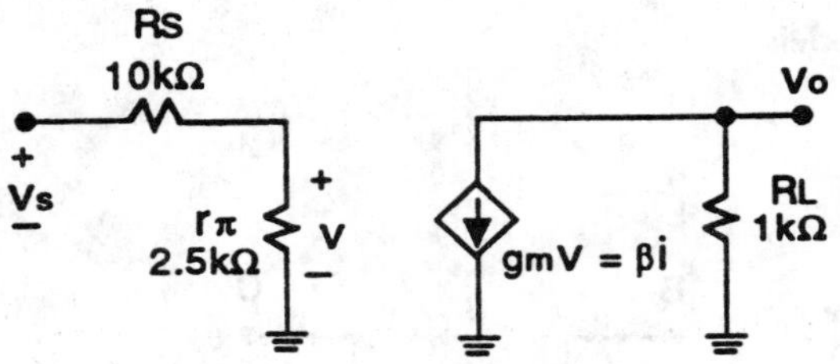

For Π_{gm}, see $\upsilon_o = -g_m\, \upsilon\,(R_i) = -\dfrac{r_\pi}{r_\pi + R_S}(\upsilon_s) \times g_m R_L$, whence $\dfrac{\upsilon_o}{\upsilon_s} = -\dfrac{2.5}{2.5+10} \times 40 \times 1 = \mathbf{-8V/V}$.

For Π_β, see $\upsilon_o = -\beta(i)\, R_L$

$= -\beta\left[\dfrac{\upsilon_S}{R_S + r_\pi}\right] R_L$, whence

$$\frac{\upsilon_o}{\upsilon_s} = -\frac{\beta R_L}{R_S + r_\pi} = -\frac{100(1)}{(10+2.5)} = \mathbf{-8V/V}.$$

(c) As in (b), with $R_S = 0$, $\dfrac{\upsilon_o}{\upsilon_s} = -\dfrac{g_m r_\pi R_L}{r_\pi + R_S} = -g_m R_L = -40 \times 1 = \mathbf{-40V/V}$, or $\dfrac{\upsilon_o}{\upsilon_s} = -\dfrac{\beta R_L}{R_S + r_\pi} = -\dfrac{100(1)}{0+2.5} = \mathbf{-40V/V}$.

(d)

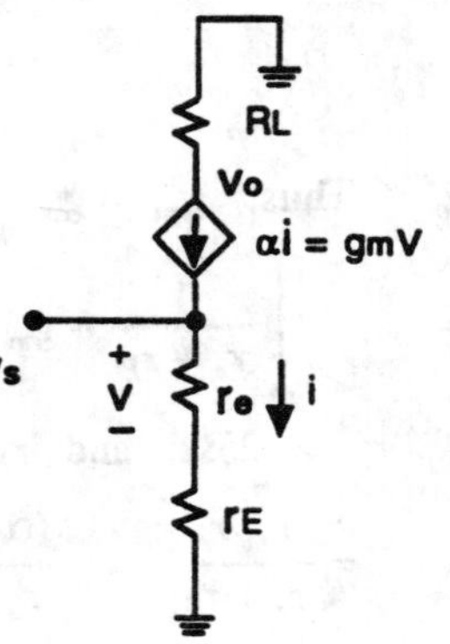

$$\frac{\upsilon_o}{\upsilon_s} = -\frac{\alpha R_L}{r_e + r_E} = -\frac{0.99(1k)}{25+100} = \mathbf{-7.92V/V},$$

or $\dfrac{\upsilon_o}{\upsilon_s} = -g_m R_L \times \dfrac{r_e}{r_e + r_E} = -\dfrac{40 \times 25}{25+100} = \mathbf{-8V/V}$,

with the former (using T_α) being more direct

(e)

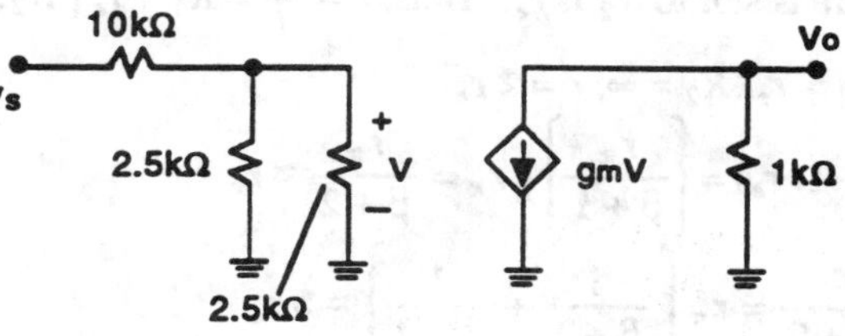

See $\frac{v_o}{v_s} = -g_m R_L \times \frac{R_B \| r_\pi}{R_B \| r_\pi + R_S} = -40 \times 1 \times \frac{2.5/2}{2.5/2 + 10} = \mathbf{-4.44V/V}$.

4.31 Since $I_E = 1\text{mA}$, $r_e - \frac{25mV}{1mA} = 25\Omega$, and $\frac{v_o}{v_s} = -\frac{\alpha R_L}{r_e} = -0.99 \frac{(7.5k\Omega)}{25} = \mathbf{-297V/V}$. For a signal voltage of 0V, $V_O = 10 - 7.5\ (.99)\ (1) = 2.575\text{V}$. For guaranteed active operation, $v_{CE} \geq 0\text{V}$. Thus the largest allowed sinusoid has a peak value of 2.575 V −0V = **2.575V** at the output, and 2.575/297 = **8.67mV peak** at the input.

4.32 Now, $i = I_S e^{v/V_T}$ in general. Thus $\frac{i_1}{i_2} = e^{(v_1 - v_2)/V_T}$, and for $i_2 = I\text{mA}$ initially, $i_1 = Ie^{10/25} = 1.49I\text{mA}$, (or $i_1 = Ie^{-10/25} = 0.670I\text{mA}$). That is, the current **increases by 49%**, or **reduces by 33%**, for ±10mV variation around the operating point. For linear operation over a ±10mV input range, a current of 1.49 I must be tolerated. For $v_{CB} \geq 0$, with $I = 1\text{mA}$, $\frac{10 - 0}{R_C} = 1.49\ I = 1.49\text{mA}$, and $R_C \leq \frac{10}{1.49mA} \leq \mathbf{6.7k\Omega}$.

4.33 For $i_C = 100\mu A$, $r_e = \frac{25mV}{100\mu A} = 250\Omega$, and $r_o = \frac{V_A}{i_C} = \frac{200V}{100\mu A} = 2 \times 10^6\Omega$. Thus, the gain $\frac{v_o}{v_i} = -\frac{\alpha R_L}{r_e} \approx -\frac{2 \times 10^6 \| 2 \times 10^6}{250} = \mathbf{-4000V/V}$.

4.34 For v across the two-terminal device, the voltage across the base-emitter junction is $v_\pi = \frac{R_1 \| r_\pi}{R_1 \| r_\pi + R_2} \times v$.

For this situation, the total current, $i = \frac{v}{R_1 \| r_\pi + R_2} + g_m v_\pi = v \left[\frac{1}{R_1 \| r_\pi + R_2} + \frac{g_m (R_1 \| r_\pi)}{R_1 \| r_\pi + R_2} \right]$

$$= v \left[\frac{1 + \frac{g_m r_\pi R_1}{R_1 + r_\pi}}{R_2 + \frac{r_\pi R_1}{R_1 + r_\pi}} \right] = v \left[\frac{R_1 + r_\pi + \beta R_1}{R_1 R_2 + r_\pi R_2 + r_\pi R_1} \right].$$

Thus, resistance $r = \frac{v}{i} = \frac{R_1 R_2 + r_\pi (R_1 + R_2)}{r_\pi + R_1 (\beta + 1)} = \left[\frac{R_1 R_2}{\beta + 1} + r_e (R_1 + R_2) \right] / (r_e + R_1)$.

(a) For $R_2 = 0$, $R_1 = \infty$, $r = \frac{R_2 + r_\pi (1 + R_2/R_1)}{r_\pi/R_1 + (\beta + 1)} = \frac{0 + r_\pi}{0 + (\beta + 1)} = \frac{r_\pi}{\beta + 1} = \mathbf{r_e}$.

(b) For $R_1 = \infty$, $R_2 = r_\pi$, $r = \frac{r_\pi + r_\pi (1 + 0)}{0 + \beta + 1} = \mathbf{2r_e}$.

(c) For $R_1 = R_2 = r_\pi$, $r = \frac{r_\pi + r_\pi (1 + 1)}{\frac{r_\pi}{r_\pi} + \beta + 1} = \frac{3\, r_\pi}{\beta + 2} \approx \mathbf{3\, r_e}$.

4.35 Resistance presented by the diode-connected transistor to R_2 is r_e. Thus, $r = \frac{v}{i} = R_1 + r_e \| R_2$.

(a) For $R_1 = 0, R_2 = \infty, r = \mathbf{r_e}$. (b) For $R_1 = r_e, R_2 = \infty, r = \mathbf{2\,r_e}$.

(c) For $R_1 = 0, R_2 = r_\pi, r = 0 + r_e \| r_\pi = r_e \| ((\beta + 1)\, r_e) = \left[\frac{r_\pi}{\beta+1}\right] \| r_\pi = \frac{r_\pi}{\beta+2} \approx \mathbf{r_e}$.

(d) For $R_1 = r_e, R_2 = r_\pi, r = r_e + r_e \| r_\pi = r_e + \frac{r_\pi}{\beta+2} = r_\pi \left[\frac{1}{\beta+1} + \frac{1}{\beta+2}\right] \approx \mathbf{2\,r_e}$.

4.36

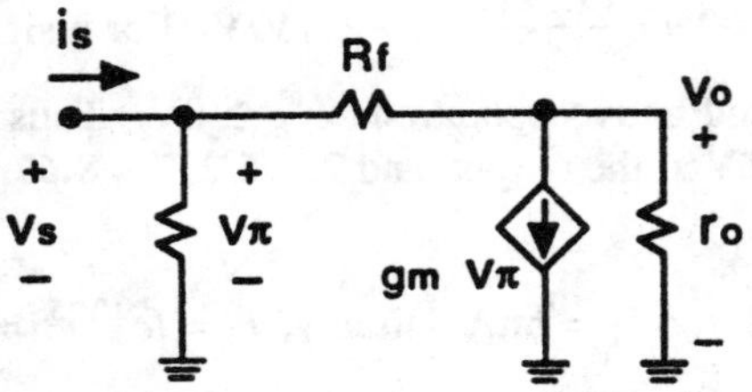

For $v_b = v_s$, small compared to v_o, the gain is $A_v = \frac{v_o}{v_s} = -\mathbf{g_m\,(r_o \| R_f)}$, and the input current $i_s = \frac{v_s}{r_\pi} + v_s \frac{(1 - -g_m\,(r_o \| R_f))}{R_f}$,

whence $R_{in} = \frac{v_s}{i_s} = \frac{1}{\frac{1}{r_\pi} + \frac{1}{R_f/(1 + g_m\,(r_o \| R_f))}}$

$= \mathbf{r_\pi} \| \frac{\mathbf{R_f}}{\mathbf{1 + g_m\,(r_o \| R_f)}}$

Now for $R_f = r_o$, the gain is $A_v = -g_m\,(r_o \| r_o) = -\frac{\mathbf{g_m\,r_o}}{\mathbf{2}}$, and $R_{in} = r_\pi \| \left[\frac{r_o}{1 + g_m\,r_o/2}\right] \approx r_\pi \| \left[\frac{2}{g_m}\right] \approx \mathbf{2r_e}$, that is, very small.

4.37

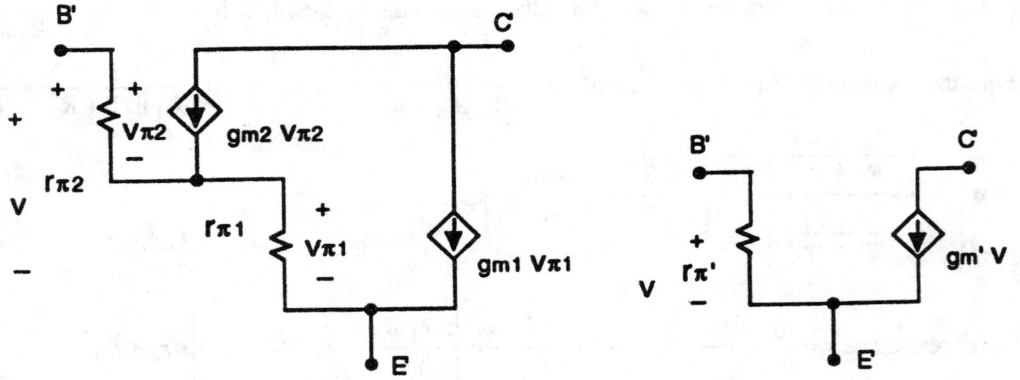

Note that the emitter current of Q_2 is the base current of Q_1, and therefore $r_{e2} = r_{\pi 1}$ and $r_{\pi 2} = (\beta_1 + 1)\, r_{\pi 1}$. Correspondingly, $v_{\pi 1} \approx v_{\pi 2} = v/2$, and $g_{m2} = g_{m1}/(\beta_1 + 1)$. By considering the input base current, see: $r_\pi' = r_{\pi 2} + (\beta_1 + 1)\, r_{\pi 1} = (\beta_1 + 1)\, r_{\pi 1} + (\beta_1 + 1)\, r_{\pi 1} = \mathbf{2\,(\beta_1 + 1)\, r_{\pi 1}}$. By considering the output collector current, see: $g_m' v = g_{m1}\, v_{\pi 1} + g_{m2}\, v_{\pi 2} = g_{m1}\, v/2 + (g_{m1}\, v/2)/(\beta_1 + 1)$ $= (g_{m1}\, v/2)\,[1 + 1/(\beta_1 + 1)] \approx g_{m1}\, v/2$. Thus $g_m' = \mathbf{g_{m1}/2}$.

4.38

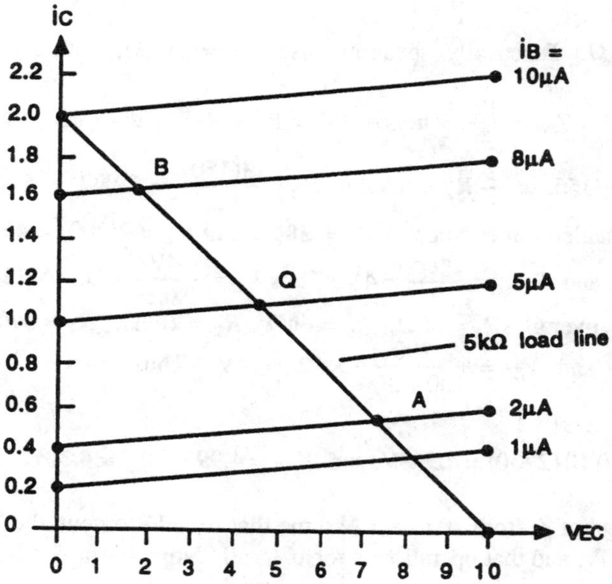

With $\beta = 200$ and $V_A = 100$V, for $i_B = 1, 2, 5, 8$ and 10μA, $i_C = \beta\, i_B = 200, 400, 1000, 1600$ and 2000μA, and $r_o = V_A/i_C$ = 500kΩ, 250kΩ, 100kΩ, 62.5kΩ and 50kΩ, with the current at 10V greater than that at 0V, by $10/r_o$ = 20, 40, 100, 160 and 200μA (ie by 10% (= (10V/100V) × 100)). For V_{CC} = 10V and R_L = 5kΩ, the intercept is $i_C = 10/5k\Omega$ = 2mA.

For the operating point (Q), i_B = 5μA, $V_{EC} \approx$ 5V from the graph, and $i_C \approx$ (5 (200)) (1 + 5/100) ≡ **1050μA**, with υ_{EC} = 10 – 5kΩ (1.05mA) = **4.75V**. *For a ±3μA peak input wave*, operation varies from *Q* to *A* to *B* to *Q* above. *At A*: i_B = 2μA, υ_{EC} ≡ 8V from the graph, with $i_C \approx$ 2 (200) (1 + 8/100)μA = 432μA = 0.432mA, for which υ_{EC} = 10 – 5kΩ (.432) = **7.84V**, and i_C = 2 (200) (1 + 7.84/100) = **0.431mA**. *At B*: i_B = 8μA, υ_{EC} ≡ 2V from the graph, with $i_C \approx$ 8 (200) (1 + 2/100)μA = **1.632mA**, and υ_{EC} = 10 –5 (1.632) = **1.84V**.

Thus the output wave has a positive peak of 4.75 – 1.84 = **2.91V**, at an output current of 1.632 – 1.050 = **0.582mA**, (*Check*: R_L' = 2.91/0.582 = 5kΩ), and a negative peak of 7.84 – 4.75 = **3.09V**, with a current of 1.050 – 0.432 = **0.618mA**, (*Check*: R_L' = 3.09/.618 = 5kΩ).

Note that the positive and negative peaks are different indicating that a (small) signal distortion results.

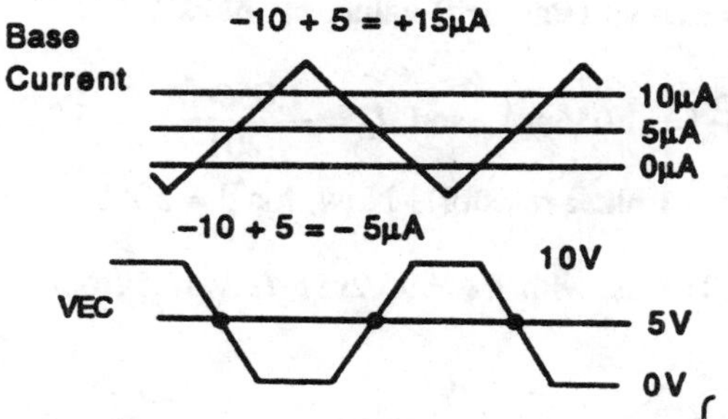

For a ±10μA peak input wave: See from the graph that for base signals of more than +5μA, $i_B \geq$ 10μA and the transistor saturates, and that for signals less than –5μA, $i_B \leq$ 0μA, and the transistor cuts off. For this situation, the output is clipped for **50%** of the cycle.

4.39 From Fig. 4.38b), in general, $I_E = (V_{BB} - V_{BE})/\left[R_E + \frac{R_B}{\beta + 1}\right]$. For $\beta = \infty$, $I_E = \frac{V_{BB} - V_{BE}}{R_E} = I$. For $I_E = .99\, I$, $\frac{R_B}{\beta + 1} = \frac{1}{100} R_E$, or $\beta + 1 = 100\,\frac{R_B}{R_E}$. For $R_E = R_B$, $\beta + 1 = 100$ or $\beta = \mathbf{99}$. That is, for $\beta \geq 99$, I_E is within 1% of its maximum value.

Alternatively, one could interpret the situation to mean ±1% of a nominal value, where the largest occurs for $\beta = \infty$, and the nominal for $\beta = 99$, where $\frac{1}{\beta + 1} = \frac{1}{100}$, and the minimum where $\frac{1}{\beta + 1} = \frac{2}{100}$ for which $\beta = \frac{100}{2} - 1 = \mathbf{49}$.

4.40 See that $V_{BB} = \frac{12}{3}$ = 4V. Generally, $I_E = \frac{V_{BB} - V_{BE}}{R_E + \frac{R_B}{\beta + 1}}$. Here, 100mA = $\frac{4 - 0.7}{R_E + \frac{50}{51} \times \frac{R_E}{10}}$. Thus $R_E = \frac{3.3}{100} / \left[1 + \frac{50}{51} \times \frac{1}{10} \right]$ = .03005kΩ. Practically speaking, use R_E = 30 Ω, with $R_B = \frac{\beta R_E}{10}$ = 50(30)/10 = 150Ω. Now $\frac{R_2}{R_1 + R_2} \times 12 = V_{BB} = \frac{12}{3}$, whence $3R_2 = R_1 + R_2$, or $R_1 = 2R_2$. Now, since $R_1 || R_2$ = 150 = $\frac{R_1 R_2}{R_1 + R_2}$, $\frac{2R_2 (R_2)}{3R_2}$ = 150, or $\frac{2}{3} R_2$ = 150, or $R_2 = \frac{3(150)}{2}$ = 225Ω, and $R_1 = 2R_2$ = 450Ω. For a conservative design, use smaller values, such as R_2 = **200Ω** and R_1 = 2(200) = **400Ω**. Now, for I_E = 100mA, $I_C = \frac{50}{51}(100)$ = 98mA, and $R_C I_C = \frac{V_{CC}}{3}$ = 4V. Thus $R_C = \frac{4V}{98mA}$ = .0408kΩ, for which use R_C = **40Ω**. Now, for the design overall: R_E = **30Ω**, R_C = **40Ω**, R_2 = **200Ω**, R_1 = **400Ω**, where $R_B = 200 || 400 = \frac{200\,(400)}{600}$ = 133.3Ω, and $V_{BB} = \frac{200}{400 + 200} \times 12$ = 4V. Thus, $I_E = \frac{4 - 0.7}{30 + \frac{133.3}{51}}$ = **101.2mA**, $V_{CE} = 12 - 0.1012 \left[\frac{50}{51} \right] 40 - 0.1012\,(30)$ = **4.995V**, and V_{CB} = 4.995 – 0.7 ≅ **4.30V**.

4.41 Assume I_E varies 5% over the entire range of β, from 20 to ∞. Assume that for ±1V output, the base signal is very small. Further, assume that $V_B \approx$ 0V, and that operation is for $v_{CB} \geq 0$. Now, for a –1V output signal and $\beta = \infty$, $\frac{5V - 1V}{1k\Omega} = I_C = I_E$ = 4mA. Since V_E = –0.7V, $R_E = \frac{-0{,}7 - -5}{4}$ = 1.075kΩ. *In practice* one would use R_E = **1.00kΩ**, in which case $V_C \geq 5 - 1k(1) \left[\frac{5 - 0.7}{1} \right]$ = 0.7V, and the collector goes to –0.3V with a 1V signal peak. For R_E = 1.075kΩ, I_E = 0.95 (1mA) = $\frac{0 - -5 - 0.7}{1.075 + \frac{R_B}{20 + 1}}$, whence $R_B = 21 \left[\frac{4.3}{.95} - 1.075 \right]$ = 72.5kΩ. *In practice* use a smaller (standard) value, say **68kΩ**.

4.42 For V_{CB} = 0.5V, and β = 200: $I_E = \frac{5 - 0.7 - 0.5}{3.6k\Omega}$ = 1.056mA, and $I_B = \frac{1.056mA}{201}$ = 5.25μA. Thus, $R_B = \frac{0.5V}{5.25\mu A}$ = 95.2kΩ. (In practice one would use a **100kΩ** resistor). Now, for β = 50, 5 – 3.6 I_E – 95.2 $\frac{I_E}{50 + 1}$ –0.7 = 0, or $I_E = \frac{5 - 0.7}{3.6 + \frac{95.2}{51}}$ = 0.787mA. That is, with R_B = 95.2kΩ, I_E varies from **1.056mA** to **0.787mA** and, V_{CB} from **0.5V** to $95.2 \frac{(0.787)}{51}$ = **1.47V**.

Alternatively, with R_B = 100kΩ: For β = 200: $I_E = \frac{5 - 0.7}{3.6 + \frac{100}{201}}$ = **1.049mA**, and $V_{CB} = 100 \frac{(1.049)}{201}$ = **0.522V**. For β = 50: $I_E = \frac{5 - 0.7}{3.6 + \frac{100}{51}}$ = **0.773mA**, and $V_{CB} = 100 \frac{(0.773)}{511}$ = **1.516V**.

4.43 For this situation, the "base" current is $I_B = \frac{I_E}{\beta + 1} + \frac{V_{BE}}{R_\beta} = \frac{I_E}{\beta_{eq} + 1}$. Now for $\beta = \infty$,and β_{eq} = 200, $R_\beta = \frac{0.7\,(201)}{I_E}$. Now, for V_{CB} = 0.5V, β_{eq} = 200, V_{CC} = 5V, R_C = 3.6kΩ, $I_E = \frac{5 - 0.5 - 0.7}{3..6k\Omega}$ = 1.056mA.

Thus, $R_\beta = \frac{0.7(201)}{1.056} = 133.2\text{k}\Omega$. In practice, use a (smaller) standard value, 130kΩ, or **120kΩ** as it is more commonly available. *With* R_β = 120kΩ, $I_B = \frac{0.7}{120k}$ = 5.83μA, and for $\beta = \infty$ and V_{CB} = 0.5V, $R_B = \frac{0.5}{5.83\mu A}$ = 85.8kΩ. In practice, use a larger standard value, say R_B = **91kΩ**, for which V_{CB} = 91k (5.83μA) = 0.530V, and $I_E = \frac{5 - 0.53 - 0.70}{3.6}$ = 1.047mA. Now for β = 50, base current flows in R_B to produce a voltage drop which combines with a constant voltage drop of V_β = 0.53V in R_B due to R_β. Thus, $I_E = \frac{V_{CC} - V_\beta - V_{BE}}{R_C + R_B/(\beta + 1)} = \frac{5 - 0.53 - 0.7}{3.6 + 91/(51)}$ = 0.700mA, with V_{CB} = 5 –3.6 (0.700) –0.7 = 1.78V. That is, for R_β = 120kΩ, R_B = 91kΩ, and β ≥ 50, I_E varies from **0.700mA** to **1.047mA**, while V_{BC} varies from **1.78V** to **0.530V**.

4.44 For β_{eq} = 100, and using the solution for P4.43, $R_\beta = \frac{0.7(101)}{1.056}$ = 66.95kΩ. Use **68kΩ** as very close (though larger). Now $I_B = \frac{0.70V}{68k}$ = 10.3μA, whence $R_B = \frac{0.50}{10.3}$ = 48.5kΩ. Use **47kΩ** as close (though smaller). For these choices and for $\beta = \infty$, $V_{CB} = \frac{0.70}{68} \times 47$ = **0.484V**, and $I_E = \frac{5 - 0.484 - 0.700}{3.6}$ = **1.06mA**. Now for β = 50, $I_E = \frac{5 - 0.484 - 0.7}{3.6 + \frac{47}{51}}$ = **0.844mA**, and V_{CB} = 5 – 3.6 (.844) –0.7 = **1.26V**.

4.45 Ignore V_A in the bias calculation: Thus $I_E = \frac{0 - 0.7 - -10}{1k + 10k/101}$ = 8.462A, $I_C = \frac{100}{101}$ (8.462) = 8.39mA = **8.4mA**, $V_B = -10k\Omega\,\frac{8.4mA}{100}$ = **–0.84V**, V_E = –0.84 –0.70 = **–1.54V**, V_C = 10 – 8.4mA (1kΩ) = **+1.6V**, in which case, $g_m = \frac{8.4}{25}$ = **336mA/V**, $r_e = \frac{100}{101} \times \frac{1}{336}$ = **2.95Ω**, r_π = 101 (2.95) = **298Ω**, $r_o = \frac{100}{8.4}$ = **11.9kΩ**. In comparison with the results of Ex.4.31, we see for equally-scaled resistors, that the result is characterized by: i) *constant voltages*, ii) *inversely-scaled currents* and iii) *directly-scaled parameters*.

4.46 Using the results of P4.45, $R_i = R_B \parallel r_\pi$ = 10k ∥ .298k = **289Ω**, $G_m = -g_m$ = **–336mA/V**, $R_o = R_C \parallel r_o$ = 1k ∥ 11.9k = **0.922kΩ**, $A_{vo} = G_m R_o$ = –336 (0.922) = **–310V/V**, $A_{is} = G_m R_i$ = –336 (.289) = **–97.1A/A**, $A_v = \frac{R_i}{R_i + R_s} G_m (R_L \parallel R_o)$ = (0.289/(0.289 + 1)) (–336) (1k ∥ 0.922k) = **–36.2V/V**, $A_i = \frac{R_s + R_i}{R_L} A_v = -\frac{1k + .289}{1k} \times 36.2$ = **–46.7A/A**. See for resistance-scaled designs that: a) Parameters scale correspondingly, and b) Gains are constant.

4.47 The need for highest-possible gain for a fixed load implies a large bias current: Thus, for $\beta = \infty$, V_B = 0 and for ±1V swing, and $v_C \geq 0$, V_C = 0 + 1 = 1V and $I_C = I_E = \frac{9 - 1}{10k}$ = 0.8mA, whence $R_E = \frac{0 - 0.7 - -9}{0.8}$ = 10.38kΩ.

Now, if we use R_E = **10kΩ** (as a standard value), we see that v_C falls to 0.7V (for $\beta = \infty$) with v_{CB} = –0.3V, which is often acceptable for linear operation. Otherwise, use R_E = 11kΩ.

For R_E = 10kΩ: *With* $\beta = \infty$: V_E = **–0.7V**, V_C = **+0.7V**, $I_C = \frac{9 - 0.7}{10}$ = **0.83mA**; $r_e = \frac{25mV}{.83}$ = 30.1Ω; $r_\pi = \infty$, $r_o = \frac{100V}{.83}$ = 120.5kΩ, $\frac{v_o}{v_s} = -\frac{10k \parallel 120k}{30.1}$ = **–307V/V**, and for ±1V output, $v_b = v_s = \frac{1}{307}$ = **3.26mV**.

With β = *90*: $V_E = \frac{-9 + 0.7}{100k + 91(10k)} \times$ 100k –0.7 = **–1.52V**, $I_C = \frac{90}{91}\,\frac{9 - 1.52}{10}$ = **0.740mA**, V_C = 9 – 10k

(.740) = **1.60V**, $r_e = \frac{25mV}{0.740}$ = 33.8mA, r_π = 91 (33.8) = 3.074kΩ, $r_o = \frac{100V}{0.740}$ = 135mA, $\frac{\upsilon_o}{\upsilon_s}$ = $-\frac{90\,(10k\,||\,135k)}{3.074k + 10k}$ = **–64.1V/V**. (Alternatively, $\frac{\upsilon_o}{\upsilon_s} = -\frac{3.07}{10 + 3.07} \times \frac{90}{91}\,\frac{(10k\,||\,135k)}{33.8}$ = –64.0V/V). Now for ±1V output, $\upsilon_s = \frac{1}{64.1}$ = **15.6mV**, and $\upsilon_b = \frac{3.07}{10 + 3.07}$ (15.6) = **3.66mV**.

4.48 Approximately, since $R_B = R_C$, I_E is essentially fixed at $\frac{10 - 0.7}{10k}$ = .930mA for reasonable β. Thus, $r_e = \frac{25mV}{0.93mA}$ = 26.9Ω, and $r_o = \frac{200V}{0.93}$ = 215kΩ. Now, for β = 50, r_π = 51 (26.9) = 1.372kΩ, and $\frac{\upsilon_o}{\upsilon_s} = -\frac{1.37k\,||\,10k}{1.37k\,||\,10k + 10k} \times \frac{50}{51} \times \frac{10k\,||\,215k}{26.9} = -0.177 \times 0.98 \times 355 =$ **–37.5V/V**. Now for, β = 150, r_π = 151 (26.9) = 4.062kΩ, and $\frac{\upsilon_o}{\upsilon_s} = -\frac{4.06\,||\,10}{4.06\,||\,10 + 10} \times \frac{150}{151} \times \frac{10k\,||\,215k}{26.9} = -.224 \times .993 \times 355 =$ **–79.0V/V**.

4.49 From P4.48 above, $I_E \approx$ 0.93mA, r_e = 26.9, and r_o = 215kΩ. *For* β = 50, r_{ib} = 51 (26.9 + 100) = 6.47kΩ, and $\frac{\upsilon_o}{\upsilon_s} = -\frac{6.47\,||\,10}{6.47\,||\,10 + 10} \times \frac{50}{51} \times \frac{10k\,||\,215k}{100 + 26.9} = -0.282 \times 0.98 \times 75.3 =$ **–20.8V/V**. *For* β = 150; r_{ib} = 151 (26.9 + 100) = 19.2kΩ, and $\frac{\upsilon_o}{\upsilon_s} = -\frac{19.2\,||\,10}{19.2\,||\,10 + 10} \times .98 \times 75.3 =$ **–29.3V/V**. We see that the design is relatively insensitive to β variation.

4.50 For each transistor, I_E = 1mA, r_e = 25Ω, and r_π = 151 (25) = 3.78kΩ. Thus $\frac{\upsilon_o}{\upsilon_{b2}} = -\frac{150}{151}\,\frac{10k\,||\,10k}{25}$ = **–199V/V**. Now, R_{b2} = **3.78kΩ**, and $\frac{\upsilon_{b2}}{\upsilon_{b1}} = -\frac{150}{151}\,\frac{10k\,||\,3.78k}{25}$ = **–109V/V**. Now R_{b1} = **3.78kΩ**, and $\frac{\upsilon_{b1}}{\upsilon_s} = \frac{3.78k}{10k + 3.78k}$ = **0.274V/V**. Thus $\frac{\upsilon_o}{\upsilon_s}$ = .274 (–109) (–199) = **5934V/V**.

4.51 I_E = 3mA, $r_e = \frac{25mV}{3mA}$ = 8.33Ω. Thus $R_i = R_E\,||\left[r_e + \frac{R_B}{\beta + 1}\right] = 3k\,||\left[8.33 + \frac{2k}{151}\right] = 3k\,||\,21.58$ = **21.4Ω** for β = 150, and **8.3Ω** for β = ∞. Now, the gain from a 100Ω source is: $\frac{\upsilon_o}{\upsilon_s} = \frac{\alpha\,(R_L\,||\,R_C)}{R_S + R_i}$ in general. For β = 150, $\frac{\upsilon_o}{\upsilon_s} = \frac{150}{151}\,\frac{(1k\,||\,3k)}{100 + 21.4}$ = **6.14V/V**. For β = ∞, $\frac{\upsilon_o}{\upsilon_s} = \left[\frac{1k\,||\,3k}{100 + 8.3}\right]$ = **6.92V/V**.

4.52 $R_i = (\beta + 1)\,(r_e + R_E)$ = 10kΩ, $r_e = \frac{25}{0.2}$ = 125Ω, β = 50. Thus 10kΩ = 51 (125 + R_E), whence $R_E = \frac{10^4}{51}$ –125 = **71Ω**. Thus the voltage gain = $-\frac{\beta\,(R_C\,||\,R_L)}{R_i} = -\frac{50\,(10k\,||\,1k)}{10k\Omega}$ = **–4.55V/V**.

4.53 For a base current i and V_{EB} = 0.7V, using KVL: $9 - 10k\Omega\,(101\,i + \frac{0.7}{10k\Omega})$ –0.7, whence $-10k\Omega\,(i + \frac{0.7}{10k\Omega}) - 10k\Omega\,(101\,i + \frac{0.7}{10k\Omega}) = 0$, or 9 – 1010 i – 0.7 – 0.7 – 10 i – 0.7 – 1010 i – 0.7 = 0, 9 – 4 (0.7) = 2030 i, and i = 3.054μA. Thus I_C = 50 (3.054) = **0.153mA**, V_E = 9 – 10kΩ (101 (3.054 × 10^{-3}) + $\frac{0.7}{10k}$) = 9 – 3.084 – 0.7 = **5.216V**, V_B = 5.216 –0.7 = **4.516V**, V_C = 10kΩ (101 (3.054 × 10^{-3}) + $\frac{0.7}{10k\Omega}$) = **3.784V**. *Check*: $V_B - V_C$ = 4.516 – 3.784 = 0.732 ≈ 0.7 + 10kΩ (3.054μA) as required.

For all designs, all couplings are via capacitors:

(a) Source coupled to B; Load to E; (Ground to C)

(b) Source to B; 10kΩ coupled from E to ground (or (better) 10/3kΩ from E); Load to C.

(c) Source to B; Ground to E, Load to C.

(d) Source to E; Ground to B, Load to C.

4.54 For all designs, $I_C = 0.153\text{mA}$, $r_e = \dfrac{25}{.153} = 163.4\Omega$;

(a) Since $\upsilon_e \approx \upsilon_b$, the shunt 10kΩ can be ignored, and $A_\upsilon = \dfrac{10k \| 10k}{0.163 + 10k \| 10k} = \mathbf{0.968V/V}$.

(b)

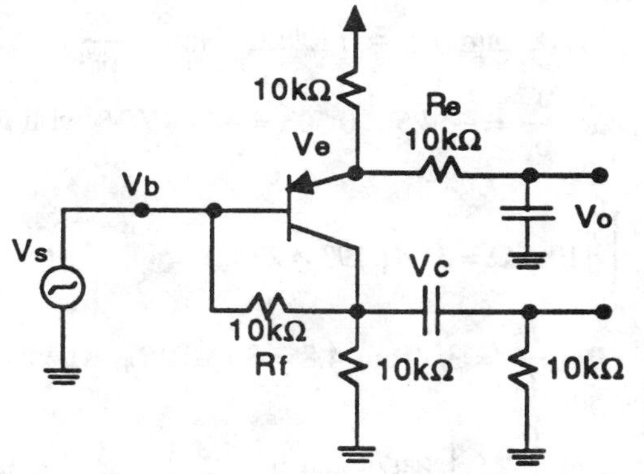

For $\dfrac{\upsilon_o}{\upsilon_b} = -1\text{V/V}$, and $\upsilon_s = \upsilon$, the voltage across R_f is 2υ. Thus $R_{Leq} = R_L \| R_C \| R_f/2 =$ 10k ∥ 10k ∥ 5k = 2.5kΩ, and the gain $= -\alpha \dfrac{R_{Leq}}{R_{Eeq}} = -\dfrac{100}{101} \times \dfrac{2.5k}{10k \| 10k} \approx \mathbf{-0.5V/V}$.

For $R_e = 10/3k\Omega$, Gain $= -\dfrac{100}{101} \times \dfrac{2.5}{10k \| \dfrac{10k}{3}} \approx \mathbf{-1V/V}$.

(c) For K large, signal across R_f is essentially due to the output voltage. Thus $R_{Leq} = R_L \| R_C \| R_f = \dfrac{10k}{3} =$ 3.33kΩ, and the gain $= -\dfrac{3.33k\Omega}{163.4} = \mathbf{-20.4V/V}$.

(d) Base is grounded, and the gain = **+20.4V/V**.

4.55

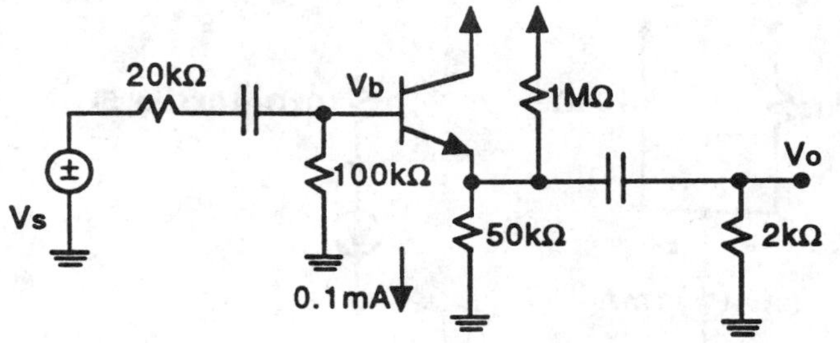

$\beta = 50$; $V_A = 100\text{V}$. Now for $I_C = 0.1\text{mA}$, $r_o = \dfrac{100V}{0.1} = 1\text{M}\Omega$, $r_e = \dfrac{25mV}{0.1mA} = 250\Omega$, R_{Leq} = 2k ∥ 50k ∥ 1M = 1.919kΩ, Gain $\dfrac{\upsilon_o}{\upsilon_b} = \dfrac{1.919k}{250 + 1.919} = 0.885\text{V/V}$, $R_{inb} = 101$ (.25 + 1.919) = 219kΩ, Gain $\dfrac{\upsilon_b}{\upsilon_s} = \dfrac{100 \| 219}{100 \| 219 + 20} = 0.775$. Thus gain $\dfrac{\upsilon_o}{\upsilon_s} =$ $0.775 \times 0.885 = \mathbf{0.685V/V}$.

4.56 *Iterate*: $V_{B2} = 0.7\text{V}$, $V_{B1} = 1.4\text{V}$, $I_{B1} \geq \dfrac{0.7}{68k} \times \dfrac{1}{101} = 0.102\mu\text{A}$. Thus, $V_{C2} \approx 1.4 + 0.1 = 1.5\text{V}$, $I_{C2} \approx \dfrac{5 - 1.5}{1.5} = 2.33\text{mA}$, $I_{B2} = \dfrac{2.33}{101} = 23.1\mu\text{A}$, $I_{B1} \approx \left(\dfrac{0.7}{68k} + 23.1\right)/101 = 0.33\mu\text{A}$. Thus, $V_{C2} = 0.7 + 0.7$

$+ 0.33 \times 10^{-6} \times 1 \times 10^{+6} = 1.73V$. Thus, $I_{C2} = \frac{5 - 1.73}{1.5} = 2.18mA$

(a) See with R = 68kΩ included, $I_{E2} \approx 2.20mA$, $I_{E1} \approx 32\mu A$. Now, $r_{e2} = 25/2.20 = 11.4\Omega$, $r_{\pi 2} \equiv 101 (11.4) = 1148\Omega$, $r_{e1} = 25/.032 = 781\Omega$, $r_{\pi 1} = 781 (101) = 78.9k\Omega$. Thus, $\frac{\upsilon_o}{\upsilon_b} = -\frac{100}{101}\frac{1.5k \| 1k}{11.4} = -52.1V/V$, $\frac{\upsilon_b}{\upsilon_s} = \frac{1.148 \| 68}{0.781 + 1.148 \| 68} = 0.591V/V$, and $\frac{\upsilon_o}{\upsilon_s} = -52.1 \times 0.591 =$ **−30.8V/V**, with $R_{in} = (1.148k \| 68k + 0.78k)\ 101 =$ **193kΩ**.

(b) Now with R = 68kΩ removed, the base current in Q_1 reduces slightly, and the collector of Q_2 lowers by 0.1V or so, with I_{C2} increasing by $\frac{0.1}{1.5} \approx 0.07mA$. Thus $I_{E2} \approx 2.3mA$, $I_{E1} \approx 23\mu A$, with $r_{e2} \approx 10.9\Omega$, $r_{\pi 2} \approx 1.098k\Omega$, $r_{e1} \approx 25/.023 = 1087\Omega$, and $r_{\pi 1} \approx 110k\Omega$. Now, $\frac{\upsilon_o}{\upsilon_b} = -52.1 \times \frac{11.4}{10.9} = -54.5V/V$, $\frac{\upsilon_b}{\upsilon_s} = \frac{1.098}{1.098 + 1.087} = 0.503$, and $\frac{\upsilon_o}{\upsilon_s} = -54.5 \times 0.503 =$ **−27.4V/V**, with $R_{in} = (1.098 + 1.087)\ 101 =$ **221kΩ**.

Thus resistance seen by υ_s for (a) is $\left(\frac{1M\Omega}{1 - -30.8}\right) \| 193k\Omega = 31.4 \| 193 =$ **27kΩ**.

4.57 $I_C = \frac{5 - 0.2}{1k\Omega} = 4.8mA$, and $I_B = \frac{5 - 0.7}{R_B} = \frac{4.3}{R_B}$. But $\frac{I_C}{I_B} = 3$. Thus $4.8 = 3\ (4.3)/R_B$, whence $R_B = \frac{3(4.3)}{4.8} = 2.69k\Omega \approx$ **2.7kΩ**. Now, $\beta_{forced} = \frac{I_C}{I_B} = \frac{\frac{5-0.2}{1k}}{\frac{5-0.7}{R_B}} = R_B\left(\frac{4.8}{4.3}\right) \le \beta/2$, and $R_B \le \frac{4.3}{4.8} \times \frac{\beta}{2} =$ **0.448 βkΩ**.

4.58

(a) $\upsilon_I = 0V \rightarrow$ **Q₁ cutoff, and Q₂ saturated;**

(b) $\upsilon_I = 5V \rightarrow$ **Q₁ saturated, and Q₂ cutoff.** $\beta_{forced} = \frac{I_C}{I_B} = \left(\frac{5 - 0.2}{1k\Omega}\right) / \left(\frac{5 - 0.7}{1k\Omega}\right) = \frac{4.8}{4.3} =$ **1.12.**

4.59 (a) Assume the transistor is saturated. Working on the diagram:

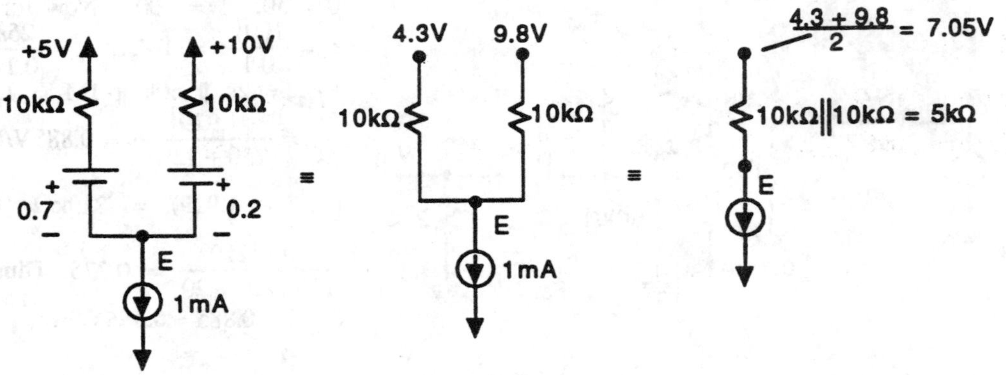

∴ $V_E = 7.05 - 5k\Omega(1mA) =$ **2.05V**, $V_B = 2.05 + 0.7 =$ **2.75V**, $V_C = 2.05 + 0.2 =$ **2.25V**, $I_C = \frac{10 - 2.25}{10k} = .775mA$, $I_B = \frac{5 - 2.75}{10k} = .225mA$, and $\beta_{forced} = \frac{0.775}{.225} =$ **3.44**. For the edge of saturation at $\upsilon = V_E$, $\left[\dfrac{5-0.7-\upsilon}{10k}\right] 100 = \dfrac{10-0.2-\upsilon}{10k}$, for which $430 - 100\,\upsilon = 9.8 - \upsilon$, $99\upsilon = 420.2$, and $\upsilon = \mathbf{4.244V}$, with $I = \dfrac{101}{100}\,\dfrac{(9.8-4.24)}{10k} = \mathbf{0.561mA}$ (at the edge of saturation).

(b) For saturation: $V_C = \mathbf{5V}$, $V_E = 5 - 0.2 = \mathbf{4.8V}$, $V_B = 4.8 + 0.7 = \mathbf{5.5V}$, $I_C = 1 - 0.1 = 0.9\text{mA}$, and $\beta_f = \dfrac{0.9}{0.1} = \mathbf{9}$. For barely linear operation, $I = I_E = (\beta + 1)\, I_B = 101\,(0.1\text{mA}) = \mathbf{10.1mA}$.

4.60 For a grounded-base amplifier, the output resistance r_{ob} is approximately $r_\mu \,||\, \beta r_o$. Here, $r_{ob} = \dfrac{\Delta \upsilon_C}{\Delta i_C} = \dfrac{10}{50 \times 10^{-9}} = 200 \times 10^6\Omega$, with $r_o = \dfrac{V_A}{I_C} = \dfrac{200}{0.1 \times 10^{-3}} = \mathbf{2M\Omega}$. Thus $\dfrac{1}{r_{ob}} = \dfrac{1}{r_\mu} + \dfrac{1}{\beta\, r_o}$, or $\dfrac{1}{200 \times 10^6} = \dfrac{1}{r_\mu} + \dfrac{1}{120 \times 2 \times 10^6}$. That is, for r_μ in MΩ, $\dfrac{1}{r_\mu} = \dfrac{1}{200} - \dfrac{1}{240} = \dfrac{240-200}{200\,(240)}$, or $r_\mu = \dfrac{200}{40}(240) = 1200\text{M}\Omega = \mathbf{1.2G\Omega}$

4.61 $V_{Oa} = BV_{CBS} \approx BV_{CBO} = \mathbf{50V}$; $V_{Ob} = BV_{CEO} = \mathbf{30V}$; $V_{Oc} = BV_{EBO} = \mathbf{7V}$.

4.62 $R_{CE\,sat} = \dfrac{\Delta V_{CE}}{\Delta I_C} = \dfrac{0.2-0.1}{3-1} = \dfrac{0.1V}{2mA} = \mathbf{50\Omega}$. Generally, $V_{CE\,sat} = V_{CE\,off} + R_{CE\,sat}\, I_C$. Thus at $\dfrac{1+3}{2} = 2\text{mA}$, $V_{CE\,sat} = \dfrac{0.1+0.2}{2} = V_{CE\,off} + 0.05(2)$, and $V_{CE\,off} = \dfrac{.3}{2} - .05(2) = .15 - .10 = \mathbf{0.05V}$. Otherwise, we could use values at one of the 1mA or 3mA points, to obtain the same result since the same line is involved in all 3 cases.

4.63 At 1.20mA, $h_{FE} = \dfrac{I_C}{I_B} = \dfrac{1.20mA}{11\mu A} = \mathbf{109}$, and $h_{fe} = \dfrac{\Delta I_C}{\Delta I_B} = \dfrac{(1.29-1.20)mA}{(12-11)\mu A} = \dfrac{.09}{.001} = \mathbf{90}$. If one assigns the increase in *dc* β to the effect of collector voltage on the base width, then at the particular value of $\upsilon_{CE} = 10\text{V}$, $I_C = h_{fe}\, I_B + \left(\dfrac{10}{r_o}\right)$, with $r_o = \dfrac{V_A}{I_C}$. Thus $1.20 = 90\,(11 \times 10^{-3}) + 10\left(\dfrac{1.20}{V_A}\right)$ - - - (1), and also $1.29 = 90\,(12 \times 10^{-3}) + 10\left(\dfrac{1.29}{V_A}\right)$ - - - (2). From (1), $V_A = \dfrac{10\,(1.20)}{(1.20-.99)} = 57.1\text{V}$, and from (2), $V_A = \dfrac{10\,(1.29)}{1.29 - 90\,(12) \times 10^{-3}} = 61.4\text{V}$. Thus $V_A \approx \mathbf{60V}$.

4.64 Assuming saturation, with $I_B = I = 1\text{mA}$, and $V_{CE\,sat} = 0.05 + I_C(0.05)$, we see $\upsilon_o = 5 - V_{CE\,sat}$, and $I_C = \dfrac{5-(0.05 + I_C\,(0.05))}{0.82} - 1.0$. That is, $0.82\, I_C = 5 - .82 - 0.05 - .05\, I_C$, $I_C = 4.75\text{mA}$, and $V_{CE\,sat} = .05 + 4.75\,(0.05) = 0.287\text{V}$, with $\upsilon_O = 5 - .287\text{V} = \mathbf{4.71V}$.

Now for $I = 4(1) = 4\text{mA}$, $I_C = \dfrac{5-(.05 + I_C\,(.05))}{0.82} - 4\text{mA}$, $0.82\, I_C = 5 - .05 - .05\, I_C - 3.28$, $I_C = \dfrac{1.67}{.87} = 1.92\text{mA}$, and $V_{CE\,sat} = .05 + 1.92(.05) = 0.146\text{V}$, with $\upsilon_O = 5 - .146 = \mathbf{4.85V}$.

Chapter 5

FIELD-EFFECT TRANSISTORS (FETs)

SECTION 5.1: Structure and Physical Operation of the Enhancement-Type MOSFET

5.1 *In general*, a channel is induced for $\upsilon_{GS} \geq \upsilon_S + V_t = 0 + 1.5V$. Hence $\upsilon_{GS} \geq$ **1.5V**, here.

In general, the drain end of the channel is pinched off for $\upsilon_{GD} \leq V_t = 1.5V$. Now for $\upsilon_{GS} = 3.0V$ and $V_S = 0$, $\upsilon_G = 3.0V$, and the drain is pinched off for $\upsilon_D \geq 3.0 - 1.5 = 1.5V$. Hence $\upsilon_{DS} \geq$ **1.5V**, here.

In general, saturation occurs for a given υ_{GS}, when $\upsilon_{DS} \geq \upsilon_{GS} - V_t$, or $\upsilon_{DS} \geq 3.0V - 1.5V = 1.5V$, for which $\upsilon_D \geq$ **1.5V**, here (which is, of course, when the drain end of the channel is pinched off).

In general, triode operation occurs for $\upsilon_{DS} \leq \upsilon_{GS} - V_t$, for which $\upsilon_D \leq$ **1.5V**, here.

SECTION 5.2: Current-Voltage Characteristics of the Enhancement MOSFET

5.2

(a) $\upsilon_{DS} = \upsilon_D - \upsilon_S = 2.1 - 0 = 2.1V$; $\upsilon_{GS} = \upsilon_G - \upsilon_S = 3 - 0 = 3V$; $\upsilon_{GS} - V_t = 3 - 1 = 2.0V \leq \upsilon_{DS} \rightarrow \therefore$ **saturated mode.**

(b) $\upsilon_{DS} = \upsilon_D - \upsilon_S = -0.1 - -2 = 1.9V$; $\upsilon_{GS} = \upsilon_G - \upsilon_S = 2 - -2 = 4V$; $\upsilon_{GS} - V_t = 4 - 2 = 2.0 \geq \upsilon_{DS} \rightarrow \therefore$ **triode mode.**

(c) $\upsilon_{SD} = \upsilon_S - \upsilon_D = 0 - -3 = 3V$; $\upsilon_{SG} = \upsilon_S - \upsilon_G = 0 - -1 = 1V < |V_t| = 2V \rightarrow \therefore$ **cutoff mode.**

(d) $\upsilon_{SD} = \upsilon_S - \upsilon_D = 2 - -1 = 3V$; $\upsilon_{SG} = \upsilon_S - \upsilon_G = 2 - 0 = 2V$; $\upsilon_{SG} + V_t = 2 - 1 = 1V \leq \upsilon_{SD} \rightarrow \therefore$ **saturated mode.**

(e) $V_t = 2V \rightarrow$ **n channel**; $\upsilon_{GS} = 0 - -3 = 3V$; Since saturated, $\upsilon_{DS} \geq \upsilon_{GS} - V_t = 3 - 2 = 1V$; $\upsilon_D = \upsilon_{DS} + \upsilon_S \geq 1 - 3 = \mathbf{-2V}$.

(f) $V_t = -2V \rightarrow$ **p channel**; $\upsilon_{SD} = \upsilon_S - \upsilon_D = 3 - -1 = 4V$; $\upsilon_{SG} = \upsilon_S - \upsilon_G = 3 - 0 = 3V$; $\upsilon_{SG} + V_t = 3 - 2 = 1V \leq \upsilon_{SD} \rightarrow \therefore$ **saturated mode.**

(g) $V_t = -2V \rightarrow$ **p channel**; $\upsilon_S = 3V$, $\upsilon_D = -3V$; $\upsilon_S - \upsilon_G \leq -V_t = +2V$ for cutoff; $\therefore -\upsilon_G \leq 2 - \upsilon_S = 2 - 3 = -1V$, and $\upsilon_G \geq$ **1V**.

5.3

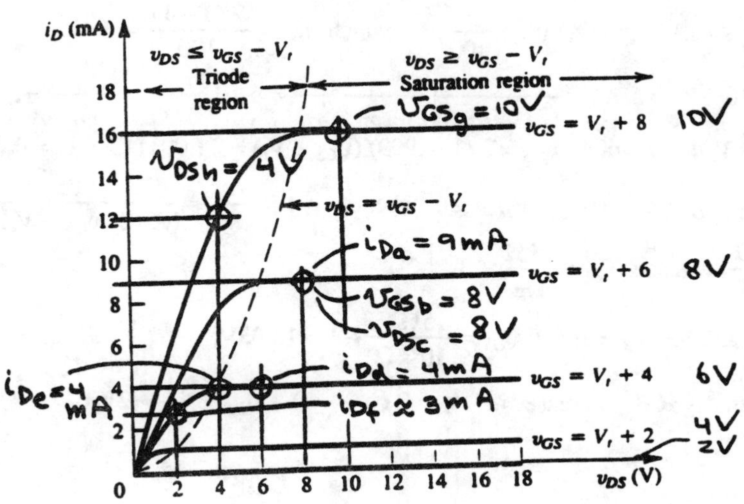

Since $V_t = 2V$, the υ_{GS} are relabelled as shown. See i_{Da} = **9mA**, υ_{GSb} = **8V**, υ_{DSc} = **8V**, i_{Dd} = **4mA**, i_{De} = **4mA**, i_{Df} = **3mA**, υ_{GSg} = **10V**, υ_{DSh} = **4V**.

5.4

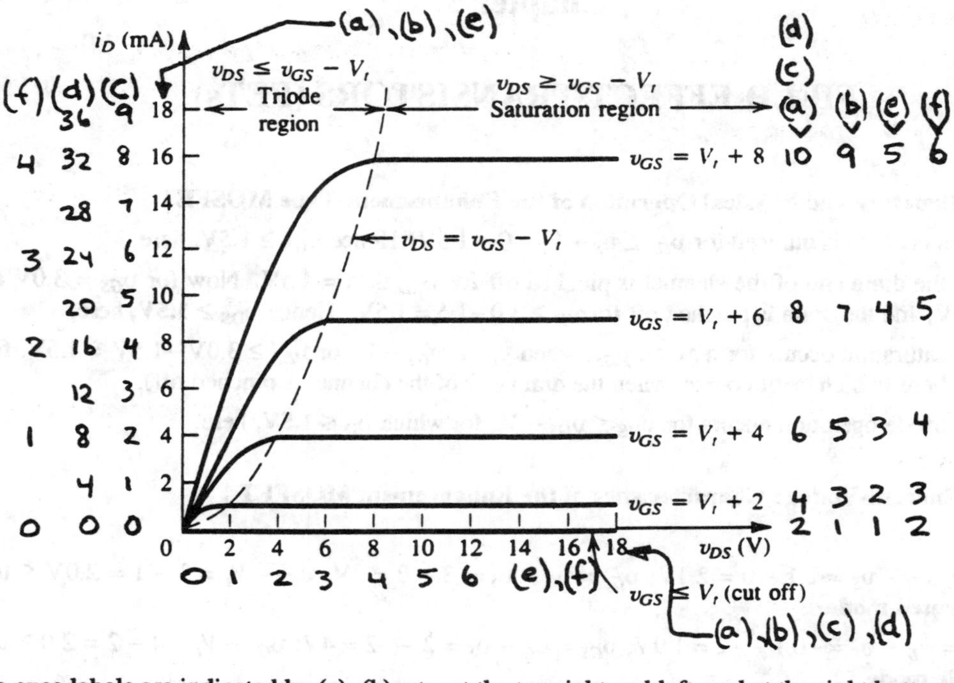

The axes labels are indicated by (a), (b), etc, at the top right and left, and at the right bottom.

5.5 $V_S = 0V$, $V_G = 3V$. Now, for saturation $V_{DS} \ge V_{GS} - V_t$, and for $V_S = 0$, $V_D \ge 3 - 1 = 2V$. Thus the device is in triode operation for $V_D \le 2V$.

In general, in triode mode, $i_D = K\,(2\,(v_{GS} - V_t)\,v_{DS} - v_{DS}^2)$.

For $v_{DS} = 2V$, $i_D = 100 \times 10^{-6}\,(2\,(3-1)\,2 - 2^2) = 100\,(2\,(2)\,(2) - 2^2) = 100\,(8 - 4) = \mathbf{400\mu A}$.

Check, in saturation, $i_D = K\,(v_{GS} - V_t)^2 = 100\,(3-1)^2 = 400\mu A$.

For $v_{DS} = 1V$, $i_D = 100\,(2\,(2)\,1 - 1^2) = \mathbf{300\mu A}$.

For $v_{DS} = 0.5V$, $i_D = 100\,(2\,(2)\,0.5 - 0.5^2) = \mathbf{175\mu A}$.

For v_{DS} very small, $i_D = 2K\,(v_{GS} - V_t)\,v_{DS}$, whence $r_{DS} = \dfrac{v_{DS}}{i_D} = \dfrac{1}{2K\,(v_{GS} - V_t)} = \dfrac{1}{100\,(3-1)} = \mathbf{5k\Omega}$.

Now, r_{DS} increases by 1% when $v_{DS}^2 = (2\,(v_{GS} - V_t)\,v_{DS})\,\dfrac{1}{100}$, for which $v_{DS} = \dfrac{2\,(3-1)}{100} = \mathbf{.04V}$.

Now r_{DS} increases by 10% when $\dfrac{1}{(2\,(v_{GS} - V_t)\,v_{DS} - v_{DS}^2)} = 1.10\,\dfrac{1}{2\,(v_{GS} - V_t)v_{DS}}$, or $2\,(v_{GS} - V_t) - v_{DS} = .909\,(2\,(v_{GS} - V_t))$, for which, $v_{DS} = 2\,(1 - .909)\,(v_{GS} - V_t) = 2\,(.091)\,(3-1) = \mathbf{0.36V}$.

5.6 For triode operation at low v_{DS}, $i_D = K\,(2\,(v_{GS} - V_t)\,v_{DS})$, whence $r_{DS} = \dfrac{v_{DS}}{i_D} = \dfrac{1}{K\,(2\,(v_{GS} - V_t))} = \dfrac{10^6}{100\,(2\,(v_{GS} - 1))}$, or $r_{DS} = \dfrac{5000\Omega}{v_{GS} - 1}$, whence $v_{GS} = \dfrac{5k\Omega}{r_{DS}} + 1$.

For $r_{DS} = 1k\Omega$: $v_{GS} = \dfrac{5}{1} + 1 = \mathbf{6V}$, and for $r_{DS} = 1M\Omega$: $v_{GS} = \dfrac{5k\Omega}{10^3 k\Omega} + 1 = \mathbf{1.005V}$.

For v_{DS} near 0V, and less than a 10% increase in r_{DS}, $2\,(v_{GS} - 1) - v_{DS} \ge \dfrac{1}{1.1}\,(2)(v_{GS} - 1)$, or $v_{DS} \le 2\,(v_{GS} - 1)\,(.0909)$.

For $r_{DS} = 1\text{k}\Omega$, $\upsilon_{GS} = 6\text{V}$, and $\upsilon_{DS} \le 2\,(6-1)\,(.0909) = \mathbf{0.909V}$, for which $i_D = K\,(2\,(\upsilon_{GS} - V_t)\,\upsilon_{DS} - \upsilon_{DS}^2)$ $= 100\,(2\,(6-1)\,.909 - (.090)^2) = 100\,(9.09) - .826) = \mathbf{826\mu A}$.

For $r_{DS} = 1\text{M}\Omega$, $\upsilon_{GS} = 1.005\text{V}$, and $\upsilon_{DS} \le (1.005 - 1.00)\,(.0909) = \mathbf{0.45mV}$, for which $i_D = 100\,(2\,(1.005 - 1)\,(.45 \times 100^{-3}) - (.45 \times 10^{-3})^2) = 100\,(2\,(5 \times .45 \times 10^{-6}) - .45^2 \times 10^{-6}) = 100\,(4.5 - 10^{-6} - .20 \times 10^{-6}) = 430 \times 10^{-6}\mu\text{A} = \mathbf{0.43nA!}$ or, *easier* $= \dfrac{0.45mV}{10^6\Omega} \approx 0.45\text{nA}$.

5.7 For saturation, $i_{DS} = K\,(\upsilon_{GS} - V_t)^2$, and for triode operation, $i_D = K\,((2\,\upsilon_{GS} - V_t)\,\upsilon_{DS} - \upsilon_{DS}^2)$. Now, for i_{Dt} = 100% of i_{Ds}: $K\,(2\,(\upsilon_{GS} - V_t)\,\upsilon_{DS} - \upsilon_{DS}^2) = K\,(\upsilon_{GS} - V_t)^2$, or $\upsilon_{DS}^2 - 2\,(\upsilon_{DS})\,(\upsilon_{GS} - V_t) + (\upsilon_{GS} - V_t)^2 = 0$, whence $\upsilon_{DS} = \dfrac{2\,(\upsilon_{GS} - V_t) \pm \sqrt{2^2\,(\upsilon_{GS} - V_t)^2 - 4\,(\upsilon_{GS} - V_t)^2}}{2}$, for which $\mathbf{\upsilon_{DS} = (\upsilon_{GS} - V_t)}$.

For $i_{Dt} = 0.99\,i_{Ds}$:

$2\,(\upsilon_{GS} - V_t)\,\upsilon_{DS} - \upsilon_{DS}^2 = 0.99\,(\upsilon_{GS} - V_t)^2$, or $\upsilon_{DS}^2 - 2\,(\upsilon_{DS})\,(\upsilon_{GS} - V_t) + 0.99\,(\upsilon_{GS} - V_t)^2 = 0$, whence $\upsilon_{DS} = (\upsilon_{GS} - V_t)\left[1 \pm \sqrt{1 - (.99)^2}\right] = (\upsilon_{GS} - V_t)\,(1 \pm 0.141)$, (where the negative value applies). Thus $\mathbf{\upsilon_{DS} = 0.859\,(\upsilon_{GS} - V_t)}$.

For $i_{Dt} = 0.90\,i_{Ds}$: $\upsilon_{DS} = (\upsilon_{GS} - V_t)\left[1 \pm \sqrt{1 - .9^2}\right]$. Thus $\mathbf{\upsilon_{DS} = 0.564\,(\upsilon_{GS} - V_t)}$.

For $i_{Dt} = 0.50\,i_{Ds}$: $\upsilon_{DS} = (\upsilon_{GS} - V_t)\left[1 \pm \sqrt{1 - .5^2}\right]$. Thus $\mathbf{\upsilon_{DS} = 0.134\,(\upsilon_{GS} - V_t)}$.

Now, for $\upsilon_{GS} = 2\,V_t$ and $V_t = 2\text{V}$, $\upsilon_{DS} = \mathbf{2V, 1.72V, 1.13V}$, and $\mathbf{0.264V}$, respectively.

5.8 $r_o = \dfrac{5V}{(2.2 - 2.1)mA} = \mathbf{50k\Omega}$; $V_A = r_o\,i_D = 50\text{k}\Omega\left(\dfrac{2.1 + 2.2}{2}\right) = \mathbf{107.5V}$; and $\lambda = \dfrac{1}{V_A} = \mathbf{.0093V^{-1}}$.

5.9 Now $K_p = \dfrac{1}{2}\mu_p\,C_{ox}\,\dfrac{W}{L}$. Assuming $\mu_p = \dfrac{1}{2}\,\mu_n$, $K_p = \dfrac{1}{2} \times \dfrac{1}{2} \times 20 \times \dfrac{100}{3} = 166.7\mu\text{A/V}^2$.

For $\upsilon_{GS} = \upsilon_{DS} = -5\text{V}$, $i_D = K\,(\upsilon_{GS} - V_t)^2\,(1 + \lambda\,\upsilon_{DS}) = 166.7\,(-5 - -2)^2\,(1 - .01(-5)) = 166.7\,(3^2)\,(1.05)$ $= \mathbf{1.575mA}$.

5.10 For the substrate connected to $V_B = +5\text{V}$, while the source voltage is varied: $|V_t| = |V_{t0}| + \gamma(\sqrt{2\,\Phi_f + V_{SB}} - \sqrt{2\,\Phi_f}) = +2 + 0.6\,(\sqrt{0.6 + V_{SB}} - \sqrt{0.6})$. For $V_S = +5\text{V}$, $V_{SB} = 0\text{V}$, and $|V_t| = 2\text{V}$, or $V_t = \mathbf{-2V}$. For $V_S = 0\text{V}$, $V_{SB} = 5 - 0 = 5\text{V}$, and $|V_t| = +2 + 0.6\,(\sqrt{0.6 + 5} - \sqrt{0.6}) = 2.955\text{V}$, or $V_t = \mathbf{-2.96V}$.

5.11 (a) $\upsilon_{GS} = \upsilon_{DS} = 5\text{V}$; saturated operation, for which $i_D = K\,(\upsilon_{GS} - V_t)^2 = 0.1\,(5 - 2)^2 = 0.9\text{mA}$. Thus $I_a = \mathbf{0.9mA}$.

(b) $\upsilon_{SG} = \upsilon_{SD}$; saturated operation, for which $0.4 = 0.1\,(\upsilon_{SG} - -2)^2$, or $\upsilon_{SG} + 2 = \pm\sqrt{4} = \pm 2$. Thus $\upsilon_{SG} = 0$, which is not possible, or -4V. Thus $V_b = \mathbf{+4V}$.

(c) $V_{GS} = 0$; cutoff, for which $I_c = \mathbf{0mA}$.

(d) $\upsilon_{GS} = \upsilon$, $\upsilon_{DG} = 3\text{V}$; saturated operation, for which $i_D = 0.9 = 0.1\,(\upsilon - 2)^2$, or $\upsilon - 2 = \sqrt{9} = 3$, or $\upsilon = 5\text{V}$. Thus $V_d = \upsilon_G - \upsilon_{GS} = 2 - 5 = \mathbf{-3V}$.

Section 5.3: The Depletion – Type MOSFET

5.12 For all, $K = 1\text{mA/V}^2$; $V_t = -4\text{V}$ (n-channel depletion mode).

(a) $\upsilon_{GS} = -4 - 0 = \mathbf{-4V}$. $\therefore$ **cutoff**. $\therefore$ $i_D = \mathbf{0mA}$, and $\upsilon_{DS} = 5 - 0 = \mathbf{5V}$.

(b) $\upsilon_G = \upsilon_{GS} + \upsilon_S = -2 + 0 = \mathbf{-2V}$, and $\upsilon_D = \upsilon_S + \upsilon_{DS} = 0 + 3 = \mathbf{3V}$. This implies saturation for which $i_D = (\upsilon_{GS} - V_t)^2 = 1\,(-2 - -4)^2 = \mathbf{4mA}$.

(c) $\upsilon_{GS} = \upsilon_G - \upsilon_S = 0 - 0 = \mathbf{0V}$, and $\upsilon_{DS} = \upsilon_D - \upsilon_S = 5 - 0 = \mathbf{5V}$, for which $\upsilon_{DG} = 5 - 0 = 5\text{V} > |V_t|$. This implies **saturation** operation, for which $i_D = 1\,(0 - -4)^2 = \mathbf{16mA}$.

(d) $\upsilon_G = \upsilon_{GS} + \upsilon_S = 0 + 0 =$ **0V**, and $\upsilon_D = \upsilon_S + \upsilon_{DS} = 0 + 2 =$ **2V**, for which $\upsilon_{DG} = 2 - 0 < |V_t|$. This implies **triode operation**, for which $i_D = K\,(2\,(\upsilon_{GS} - V_t)\,\upsilon_{DS} - \upsilon_{DS}^2) = 1\,(2\,(0 - -4)\,2 - 2^2) =$ **12mA**.

(e) $\upsilon_S = \upsilon_G - \upsilon_{GS} = 0 - 1 =$ **−1V**, and $\upsilon_D = \upsilon_S + \upsilon_{DS} = -1V + 5V =$ **4V**, for which $\upsilon_{DG} = \upsilon_D - \upsilon_G = 4V - 0 = |V_t|$. This implies operation at the **edge of saturation**, for which $i_D = 1\,(1 - -4)^2 =$ **25mA**.

(f) $\upsilon_{GS} = \upsilon_G - \upsilon_S = 2 - 0 =$ **2V**, and $\upsilon_{DS} = \upsilon_D - \upsilon_S = 5 - 0 =$ **5V**, for which $\upsilon_{DG} = \upsilon_D - \upsilon_G = 5 - 2 = 3V < |V_t|$. This implies triode operation, for which, $i_D = K\,(2\,(\upsilon_{GS} - V_t)\,\upsilon_{DS} - \upsilon_{DS}^2) = 1\,[2\,(2 - -4)\,5 - 5^2] = 60 - 25 =$ **35mA**.

(g) $\upsilon_G = \upsilon_{GS} + \upsilon_S = 2 + 0 =$ **2V**, and $\upsilon_{DS} = \upsilon_D - \upsilon_S = 0 - 0 =$ **0V**. See that operation is in the **triode mode**, but with $i_D =$ **0mA**.

(h) $\upsilon_S = \upsilon_{SG} + \upsilon_G = -2 + 0 =$ **−2V**, and $\upsilon_{DS} = \upsilon_D - \upsilon_S = 0 - -2 =$ **2V**. See that $\upsilon_{DG} = \upsilon_D - \upsilon_G = 0 - 0 = 0 < |V_t| \rightarrow$ **triode operation**, for which $i_D = K\left[2\,(\upsilon_{GS} - V_t)\,\upsilon_{DS} - \upsilon_{DS}^2\right] = 1\left[2\,(2 - -4)\,2 - 2^2\right] = 24 - 4 =$ **20mA**.

5.13 *Depletion MOS*: $|V_t| = 2V$, $K = 0.1mA/V^2$.

(a) Saturation with $\upsilon_{GS} = 0$, for which $I_a = i_D = K\,(\upsilon_{GS} - V_t)^2 = 0.1\,(0 - -2)^2 =$ **0.4mA**.

(b) Saturation, for which $0.4 = 0.1\,(\upsilon_{GS} - -2)^2$, and $\upsilon_{GS} + 2 = \pm\sqrt{4} = \pm 2V$, whence $\upsilon_{GS} = -4V$ or 0V. Clearly, 4V is not possible. Thus $V_b = 0 - 0 =$ **0V**.

(c) Saturation, for which $0.9 = 0.1\,(\upsilon_{GS} - -2)^2$, and $\upsilon_{GS} + 2 = \pm\sqrt{9} = \pm 3$, for which $\upsilon_{GS} = -5$ or $+1V \rightarrow$ Clearly −5V is not possible. Thus $\upsilon_{GS} = +1V$, and $V_c = 0 + 1V =$ **+1V**.

(d) Devices are connected symmetrically. $I_d = \frac{1}{2} I = \frac{1.8}{2} =$ **0.9mA**. Also see saturated operation, for which $0.9 = 0.1\,(\upsilon_{GS} - -2)^2$. Thus $\upsilon_{GS} = +1V$ (as in (c)), and $V_d =$ **+1V**.

(e) Here, $\upsilon_{DS} = \upsilon_{GS}$. Thus $\upsilon_{DG} = 0 < |V_t|$ implies triode operation, for which $i_D = 0.4mA = 0.1\left[2\,(\upsilon_{GS} - -2)\,\upsilon_{GS} - \upsilon_{GS}^2\right]$, or $4 = 2\upsilon_{GS}^2 + 4\,\upsilon_{GS} - \upsilon_{GS}^2$, $\upsilon_{GS}^2 + 4\,\upsilon_{GS} - 4 = 0$, whence $\upsilon_{GS} = \frac{-4 \pm \sqrt{4^2 - 4(-4)1}}{2} = \frac{-4 \pm 4\sqrt{2}}{2} = -2 \pm 2\sqrt{2}$. But it must be positive. Thus $= -2 + 2\sqrt{2} = 0.828V$, and $V_e = 5 - .828 =$ **4.172V**.

5.14 Operation is triode mode in both cases: Thus $i_D = K\left[2\,(\upsilon_{GS} - V_t)\,\upsilon_{SD} - \upsilon_{SD}^2\right]$, V_t being negative for a p-channel depletion device, when υ_{SG} is used.

For $\upsilon_D = 4.8V$, $\upsilon_{SD} = 5.0 - 4.8 = +0.2V$, $\upsilon_G = 5.0V$, $\upsilon_{GS} = 5.0 - 5.0 = 0.0V$;

$0.1 = K\left[2\,(0 - V_t)\,(+0.2) - 0.2^2\right]$, or $0.1 = K\left[-0.4V_t - 0.04\right]$ --- (1).

For $\upsilon_D = 4.95$, $\upsilon_{SD} = 5.0 - 4.95 = +.05V$, $\upsilon_G = 0V$, $\upsilon_{SG} = 5 - 0 = +5V$; $0.1 = K\left[2\,(+5 - V_t)\,(+.05) - .05^2\right]$, or $0.1 = K\left[+0.5 + 0.1V_t - .0025\right]$, or $0.1 = K\left[0.4975 + 0.1V_t\right]$ --- (2).

Now (1)/(2) $\rightarrow 1 = \frac{-0.4V_t - 0.04}{+0.4975 + 0.1V_t}$, or $-0.4\,V_t - .04 = +\,0.4975 + 0.1V_t$, or $-0.5V_t = +0.5375$, whence $V_t =$ **−1.075V**. Thus with υ_{GS} as defined, the depletion threshold is 1.075V.

From (1), $0.1 = K\,[\,-0.4V_t\,-.04]$, $K = \frac{0.1}{-0.4(-1.075)-.04} = \frac{0.1}{+.43 - .04} = 0.256mA/V^2$. Now, $i_D = K\,(\upsilon_{GS} - V_t)^2 = I_{DSS}\,(1 - \frac{\upsilon_{GS}}{V_t})^2 = \frac{I_{DSS}}{V_t^2}\,(\upsilon_{GS} - V_t)^2$. Thus $I_{DSS} = K\,V_t^2 = 0.256 \times (1.075)^2 =$ **0.296mA**.

SECTION 5.4: The Junction Field-Effect Transistor (JFET)

5.15 For $V^+ = 4V$, operation, is in saturation and $i_D = I_{DDS}\,(1 - \frac{\upsilon_{GS}}{V_p})^2 = 10\,(1 - \frac{0}{-2}) = \mathbf{10mA}$.

For $V^+ = 2V$, operation is at the edge of saturation and $i_D = \mathbf{10mA}$, also.

For $V^+ = 1V$, operation is in triode mode, and $i_D = \frac{I_{DSS}}{V_p^2}\left[2\,(\upsilon_{GS} - V_p)\,\upsilon_{DS} - (\upsilon_{DS})^2\right] = \frac{10}{2^2}\left[2\,(0 - -2)\,1 - 1^2\right] = \frac{10}{4}\,(4 - 1) = \mathbf{7.5mA}$.

For $i_D = \frac{I_{DSS}}{2} = 5mA = \frac{10}{4}\left[2\,(2)\,\upsilon_{DS} - \upsilon_{DS}^2\right]$, whence $2 = 4\,\upsilon_{DS} - \upsilon_{DS}^2$, $\upsilon_{DS}^2 - 4\,\upsilon_{DS} + 2 = 0$, and $\upsilon_{DS} = \frac{- -4 \pm \sqrt{4^2 - 4(2)}}{2} = \frac{4 \pm 2\sqrt{2}}{2} = 2 \pm \sqrt{2} = 3.414$ or $0.586V$. Clearly $V^+ = \mathbf{0.586V}$.

Check: $i_D = \frac{10}{4}\left[2\,(2)(.586) - .586^2\right] = 5mA$.

5.16 For triode-mode operation: $i_D = \frac{I_{DSS}}{V_p^2}\left[2\,(\upsilon_{GS} - V_p)\,\upsilon_{DS} - \upsilon_{DS}^2\right]$.

Now, for $i_D = 5mA$, $5 = \frac{10}{4}\left[2\,(\upsilon_{GS} + 2)\,1 - 1^2\right]$, whence $2 = 2\,(\upsilon_{GS} + 2) - 1 = 2\,\upsilon_{GS} + 4 - 1$. Thus $\upsilon_{GS} = \frac{2 - 3}{2} = \mathbf{-0.5V}$.

Check: $i_D = \frac{10}{4}\left[2\,(-0.5 - -2)\,1 - 1\right] = 2.5\,[2\,(1.5) - 1] = 5mA$.

Now, for $i_D = 1mA$, $1 = \frac{10}{4}\left[2\,(\upsilon_{GS} + 2)\,1 - 1^2\right]$, or $0.4 = 2\,\upsilon_{GS} + 4 - 1$, whence $\upsilon_{GS} = \frac{0.4 - 3}{2} = \frac{-2.6}{2} = \mathbf{-1.3V}$.

5.17 For triode operation: $i_D = \frac{I_{DSS}}{V_p^2}\left[2\,(\upsilon_{GS} - V_p)\,\upsilon_{DS} - \upsilon_{DS}^2\right]$.

For small υ_{DS}, $i_D = \frac{2I_{DSS}}{V_p^2}\,(\upsilon_{GS} - V_p)\,\upsilon_{DS}$, and $r_{DS} = \frac{\upsilon_{DS}}{i_D} = \frac{1}{\frac{2I_{DSS}}{V_p^2}\,(\upsilon_{GS} - V_p)} = \frac{V_p^2}{2I_{DSS}\,(\upsilon_{GS} - V_p)}$.

Now for $I_{DSS} = 10mA$, $V_p = -2V$: For $\upsilon_{GS} = 0V$, $r_{DS} = \frac{2^2}{2 \times 10\,(0 - -2)} = \mathbf{100\Omega}$. For $\upsilon_{GS} = -1V$, $r_{DS} = \frac{2^2}{2 \times 10\,(-1 - -2)} = \mathbf{200\Omega}$. For $\upsilon_{GS} = -2V$, $r_{DS} = \frac{2^2}{2 \times 10\,(-2 - -2)} = \infty$. (Of course, since at $\upsilon_{GS} = -2V$, the switch is cut off!)

5.18 For $\upsilon_{DS} = 2V$ and $V_p = -2V$, the JFET is just at the edge of saturation, for which $i_D = K\,(\upsilon_{GS} - V_t)^2 = \frac{I_{DSS}}{V_p^2}\,(\upsilon_{GS} - V_p)^2 = I_{DSS}\left(1 - \frac{\upsilon_{GS}}{V_p}\right)^2$, or $5 = 10\left(1 - \frac{\upsilon_{GS}}{-2}\right)^2$, and $1 + \frac{\upsilon_{GS}}{2} = \sqrt{1/2} = .707$. Thus $\upsilon_{GS} = 2\,(.707 - 1) = \mathbf{-0.586V}$.

Check: $i_D = 10\left(1 - \frac{-.586}{-2}\right)^2 = 5.00mA$.

For $\upsilon_{GS} = -0.586V$, $\upsilon_{DS} = 7V$, $5.10 = 5.00\,(1 + \lambda\,(7 - 2))$, or $0.10 = 25\lambda$, whence $\lambda = \frac{0.1}{25} = \mathbf{.004V^{-1}}$. Thus $V_A = \frac{1}{\lambda} = \frac{1}{.004} = \mathbf{250V}$, and r_o (at 5mA) $= \frac{250V}{5mA} = \mathbf{50k\Omega}$.

More painstakingly: In saturation, $i_D = I_{DSS}\left[1 - \frac{\upsilon_{GS}}{V_p}\right]^2 (1 + \lambda\upsilon_{DS})$. Thus, at υ_{DS} = 2V, 5 = 10 $\left[1 - \frac{\upsilon_{GS}}{-2}\right]^2 (1 + \lambda\, 2)$, and at υ_{DS} = 7V, 5.1 = 10 $\left[1 - \frac{\upsilon_{GS}}{-2}\right]^2 (1 + \lambda\, 7)$.

Divide: Thus $\frac{5.1}{5} = \frac{1 + 7\lambda}{1 + 2\lambda} = 1.02$, or $1 + 7\lambda = 1.02 + 2.04\lambda$, whence $\lambda = \frac{.02}{7 - 2.04} = \mathbf{.00403V^{-1}}$.

Thus $5 = 10\left[1 - \frac{\upsilon_{GS}}{-2}\right]^2 (1 + 2\,(.00403))$, or $\left[1 - \frac{\upsilon_{GS}}{-2}\right] = \sqrt{\frac{1}{2(1.00806)}} = .7043$, whence $\upsilon_{GS} = -2$ (.2957) = **0.591V**, for which $V_A = \frac{1}{.00403}$ = **248V**, and $r_o = \left[I_{DSS}\,(1 - \frac{\upsilon_{GS}}{V_p})^2\,\lambda\right]^{-1}$ = $(10\,(.7043)^2\,(.00403))^{-1}$ = **50.02kΩ**.

5.19 Now, $i_D = \frac{I_{DSS}}{V_p^2}\left[2\,(\upsilon_{GS} - V_p)\,(\upsilon_{DS}) - \upsilon_{DS}^2\right]$ in triode mode, or $i_D = I_{DSS}\left[1 - \frac{\upsilon_{GS}}{V_p}\right]^2$ in saturation.

(a) For p-channel; $\upsilon_{GS} = 0$; υ_{GD} = 5V → saturation. Thus $I_a = I_{DSS}$ = **4mA**.

(b) For n-channel; υ_{DG} = 5V → saturation. Thus $i_D = I_{DSS}\left[1 - \frac{\upsilon_{GS}}{V_p}\right]^2$, or $1 = 4\left[1 - \frac{\upsilon_{GS}}{-2}\right]^2$, or $1 - \frac{\upsilon_{GS}}{-2}$ = ± ½, whence υ_{GS} = 2 (± ½ −1) = −1 or −3 (cutoff). Thus $V_b = \upsilon_G - \upsilon_{GS} = 0 - -1 =$ **1V**.

(c) For n-channel; $i_D < I_{DSS}$ → triode. Thus $1 = \frac{4}{2^2}\,(2\,(0 - -2)\,\upsilon_{DS} - \upsilon_{DS}^2)$, or $\upsilon_{DS}^2 - 4\,\upsilon_{DS} + 1 = 0$, whence $\upsilon_{DS} = \frac{4 \pm \sqrt{4^2 = 4(1)}}{2}$ = 0.268V, or very large. Thus V_c = **+0.268V**.

(d) The p-channel device is operating with the gate somewhat forward-biassed in the triode mode. Thus $i_D = \frac{I_{DSS}}{V_p^2}(2\,(\upsilon_{GS} - V_p)\,\upsilon_{DS} - \upsilon_{DS}^2)$, or $1 = \frac{4}{2^2}\,(2\,(V_d - 2)\,(V_d) - (V_d)^2) = 2V_d^2 - 4V_d - V_d^2$, or $V_d^2 - 4V_d - 1 = 0$, whence $V_d = \frac{4 \pm \sqrt{16 - 4(-1)}}{2} = \frac{4 \pm 4.472}{2}$ = **−.236V**.

SECTION 5.5: FET Circuits at DC

5.20 $K_n = K = \frac{1}{2}\,\mu_n\,C_{ox}\,\frac{W}{L} = \frac{1}{2}(20) \times 40 = 0.4\text{mA/}V^2$. Thus $I_D = \frac{V_{DD} - V_D}{R_D} = \frac{5 - 2}{7.5k\Omega}$ = **0.4mA**. In saturation, $i_D = K\,(\upsilon_{GS} - V_t)^2$, or $0.4 = 0.4\,(\upsilon_{GS} - 1)^2$. Thus $\upsilon_{GS} - 1 = 1$, or υ_{GS} = 2V. ∴ $\upsilon_S = \upsilon_G - \upsilon_{GS} = 0 - 2 =$ **−2V**, and $R_S = \frac{\upsilon_S - V_{SS}}{I_D} = \frac{-2 - -5}{0.4} = \frac{3V}{0.4}$ = **7.5kΩ**.

5.21 Now, $I_D = \frac{V_{DD} - V_D}{R_D} = \frac{5 - 2}{7.5}$ = 0.4mA, and $V_S = -5 + 7.5(0.4) = -2$V. Thus V_{GS} = 2V, whence $0.4 = K\,(2 - 1)^2$, and K = **0.4mA/V²**.

Now for $K = \frac{0.4}{2}$ = 0.2mA/V, in the source circuit: $\frac{V_S - -5}{7.5k} = i_D = 0.2\,(-V_S - 1)^2$, or $V_S + 5 = 1.5\,(V_S + 1)^2 = 1.5\,V_S^2 + 3V_S + 1$, or $1.5V_S^2 + 2V_S - 4 = 0$, whence $V_S = \frac{-2 \pm \sqrt{2^2 - 4(-4)\,(1.5)}}{2\,(1.5)}$ $= \frac{-2 \pm \sqrt{4 + 24}}{3} = \frac{-2 \pm 5.29}{3}$ = −2.43 or +1.10 (too small). Thus V_S = −2.43V, and V_D = **+2.43V**, with a corresponding change of $\frac{2.43 - 2.00}{2.00}$ = **21.5%**.

5.22 For Fig. 5.26, $V_{DD} = 10V$, $V_{DS} = V_{GS}$, and operation is in saturation. Thus $V_D = V_{DD} - I_D R$, or $V_D = 10 - 0.4R$. Also $I_D = K(V_{GS} - V_t)^2$, where $K = ½ \times 20 \times 40 = 0.4mA/V^2$, and $V_{GS} = V_D$. Thus $0.4 = 0.4(V_D - 1)^2$, or $V_D - 1 = 1$, or $V_D = 2V$. Thus $R = \frac{10-2}{0.4} = \mathbf{20k\Omega}$.

5.23 For $V_D = 2V$, $I_D = \frac{5-2}{150k} = 20\mu A$. Also $I_D = K(\upsilon_{GS} - V_t)^2 = 0.5 \times 10^{-3}(\upsilon_{GS} - 1)^2 = 20 \times 10^{-6}$, or $\upsilon_{GS} - 1 = (40 \times 10^{-3})^{½} = 0.2$, whence $\upsilon_{GS} = 1.2V$. Now, since $V_G = 5V$, $V_S = 5 - 1.2 = 3.8V$, see $R_S = \frac{3.8}{20\mu A} = 190k\Omega$. Thus, to one significant digit, $R_S = \mathbf{200k\Omega}$.

Now, $I_D = 500(\upsilon_{GS} - 1)^2$ and $I_D = \frac{5 - \upsilon_{GS} - 0}{0.2M\Omega}$, or $\frac{5 - \upsilon_{GS}}{0.2} = 500(\upsilon_{GS} - 1)^2$, or $5 - \upsilon_{GS} = 100\upsilon_{GS}^2 - 200\upsilon_{GS} + 100$, or $100\upsilon_{GS}^2 - 199\upsilon_{GS} + 95 = 0$, or $\upsilon_{GS}^2 - 1.99\upsilon_{GS} + 0.95 = 0$, whence $\upsilon_{GS} = \frac{1.99 \pm \sqrt{1.99^2 - 4(.95)}}{2} = \frac{1.99 \pm \sqrt{.1601}}{2} = \frac{1.99 \pm .4}{2} = 1.195V$, or < 1. Thus, $V_S = 5 - 1.195V = 3.805V$, $I_D = \frac{3.805}{.2} = 19.025\mu A$, and $V_D = 5 - .15 \times 19.025 = \mathbf{2.15V}$.

5.24 $V_G = \frac{10M}{10M + 10M} \times 5V = 2.5V$. Now for $\upsilon_{GS} = \upsilon$, $i_D = \frac{5 - \upsilon - 2.5}{1k} = i$, and $i = K(\upsilon_{GS} - V_t) = 0.5(\upsilon - 1)^2$, $0.5(\upsilon - 1)^2 = \frac{2.5 - \upsilon}{1k}$, $(\upsilon - 1)^2 = 5 - 2\upsilon$, $\upsilon^2 - 2\upsilon + 1 = 5 - 2\upsilon$, $\upsilon^2 = 4$, $\upsilon = 2V$. Thus $V_S = V_G + \upsilon_{GS} = 2.5 + 2.0 = 4.5V$, and $i_D = \frac{5 - 4.5}{1k} = 0.5mA$, $V_D = 0 + 4k(0.5mA) = \mathbf{2V}$.

We actually find $\frac{90}{100}(2V) = 1.8V$, in which case i_D is reduced to 90% or 0.45mA.

For $K = 0.5mA/V^2$ and V_t varying, $5 - 1k\Omega(0.45mA) - \upsilon_{GS} = 2.5V$, whence $\upsilon_{GS} = 5 - 0.45 - 2.5 = 2.05V$. Now, $0.45 = 0.5(2.05 - V_t)^2$, or $V_t - 2.05 = \pm(0.90)^{½} = +.949$, whence $V_t = 2.05 \pm .949 = 1.101V$. That is, V_t could have raised by **10.1%**.

For $V_t = 1V$ and K varying, again $\upsilon_{GS} = 2.05V$, but now, $0.45mA = K(2.05 - 1)^2$, or $K = \frac{0.45}{1.05^2} = 0.408mA/V^2$. Thus *K* could have dropped by $\frac{0.5 - .408}{0.5} = \mathbf{18.4\%}$.

Note that the effect of V_t is essentially direct, a 10% change in current resulting from a 10% change in V_t. However, the change in current is only about 10/18.4 or about 54% of that in *K*, due to negative feedback included in the circuit. (See Chapter 8.)

5.25 For a Depletion Device and $\upsilon_{GS} = 0$, $i_D = I_{DSS} = K(\upsilon_{GS} - V_t)^2 = 1mA/V^2(0-2)^2 = \mathbf{4mA}$. Thus $V_S = 15 - 1k\Omega(4mA) = \mathbf{11V} = V_G$, and $V_D = 0 + 2k\Omega(4mA) = \mathbf{8V}$, whence $\upsilon_{SD} = 11 - 8 = 3V > V_t$. Thus the device operates in saturated mode.

Triode operation begins for $\upsilon_{SD} = V_t = 2V$, in which case $V_{SS} = 8 + 2 + 4 = 14V$, with operation being saturated for $\mathbf{V_{SS} \geq 14V}$.

5.26 See $V_G = \frac{4}{4+1} \times 5 = 4V$. Now $\upsilon_{SG} = 5 - 1k\Omega(i_D) - 4 = 1 - i_D$, or $i_D = 1 - \upsilon_{SG} = 1 + \upsilon_{GS}$. Assuming saturation, $i_D = K(\upsilon_{GS} - V_t)^2 = 1(\upsilon_{GS} - 2)^2$, or $1 + \upsilon_{GS} = (\upsilon_{GS} - 2)^2 = \upsilon_{GS}^2 - 4\upsilon_{GS} + 4$, or $\upsilon_{GS}^2 - 5\upsilon_{GS} + 3 = 0$, whence $\upsilon_{GS} = \frac{5 \pm \sqrt{5^2 - 4(3)}}{2} = \frac{5 \pm 3.61}{2}$, or $\upsilon_{GS} = 0.697V$ (or 4.305V (too large)). Now $V_S = V_G - \upsilon_{GS} = 4 - 0.697 = 3.30V$, and $V_D = 0 + \frac{5 - 3.30}{1} \times 1 = 1.70V$. Thus $V_{DS} = 1.70 - 3.30 = \mathbf{-1.60V}$, and $V_{GD} = 4 - 1.70 = 2.30V > V_t$. Thus, operation is in saturation.

5.27 For the lower device, assumed to be in saturation, $i_D = I_{DSS}\left[1 - \frac{\upsilon_{GS}}{V_p}\right]^2$, and $\upsilon_{GS} = -i_D(1k\Omega) = -i_D$. Thus,

$i_D = 4\left[1 - \frac{-i_D}{-2}\right]^2 = (2 - i_D)^2$, $4 - 4i_D + i_D^2 = i_D$, $i_D^2 - 5i_D + 4 = 0$, whence $i_D = \frac{+5 \pm \sqrt{5^2 - 4(4)}}{2}$ $= \frac{+5 \pm 3}{2} = 1$ or 4mA (not acceptable). $I_D =$ **1mA** and $v_{GS} = -1$V. Now, the upper circuit is the same. Thus, the since the gate is at 0V, source is at +1V and $V_O =$ **0V**.

Now, if *both* resistors are raised to 2kΩ, I_D reduces, but it is the same in both cases, and $V_o = 0$V is retained. Here $v_{GS} = -2i_D$, and $i_D = 4\left[1 - \frac{-2i_D}{-2}\right]^2 = (2 - 2i_D)^2 = 4 - 8i_D + 4i_D^2$, for which $4i_D^2 - 9i_D + 4 = 0$, and $i_D = \frac{9 \pm \sqrt{81 - 4(4)(4)}}{2(4)} = \frac{9 \pm 4.123}{8} =$ **0.61mA**. Now for $I_D = 0.61$mA, $v_{GS} = -1.22$V, but V_O remains at **0V**.

SECTION 5.6: The FET as an Amplifier

5.28

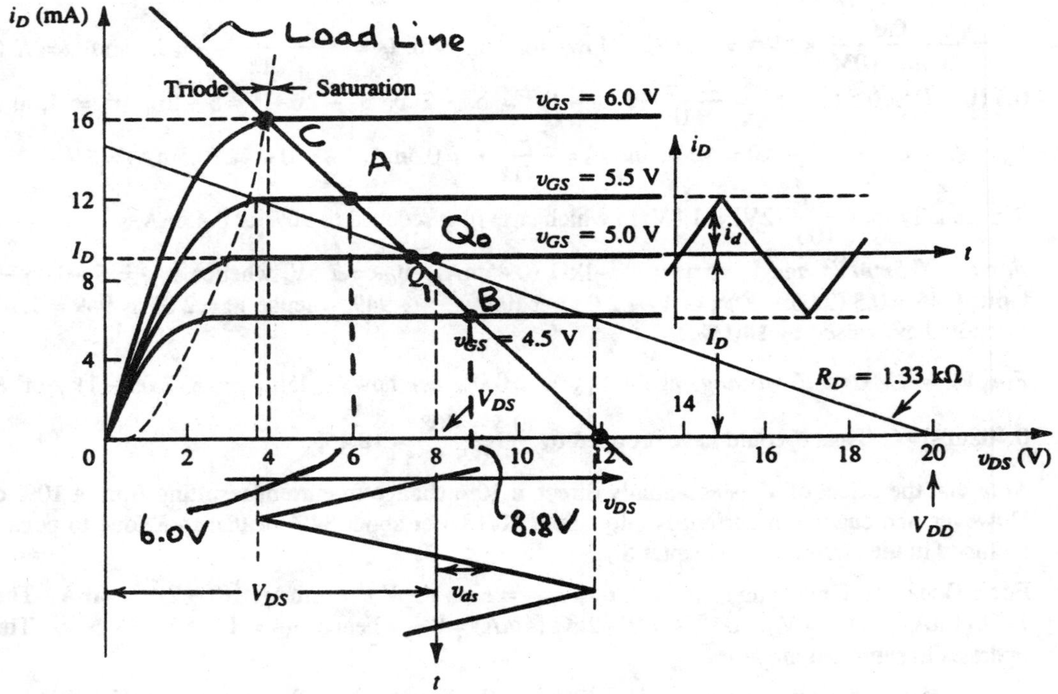

Here, $i_{D\max} = 12\text{V}/0.5\text{k}\Omega = 24\text{mA}$ (from the graph). *Alternatively*, i_D at $v_{DS} = 4\text{V} = (12 - 4)/0.5 = 16\text{mA}$. **Draw the load line which crosses v_{GS} = 5.0V at I_D = 9mA** (using the figure, or the results stated in the Text). For a ±0.5V variation in v_{GS}, i_D increases to 12mA and reduces to 6mA, for a change of **6mA**, with v_{DS} changing from 6.0 V to 8.8V respectively, for a average gain of $\frac{6.0 - 8.8}{0.5 - -0.5} =$ **−2.8V/V**. *Alternatively*, see the average gain is $\frac{-\Delta i_D R_L}{\Delta v_{GS}} = \frac{-(12-6)0.5}{0.5 + 0.5} =$ **−3V/V**, probably a better estimate. Saturation mode extends to point *C*, where $v_{GS} = 6.0$V, $v_{DS} = 4$V and $i_D = 16$mA. Thus the largest input sine wave for saturation operation is 6 − 5 = **1V peak**.

5.29

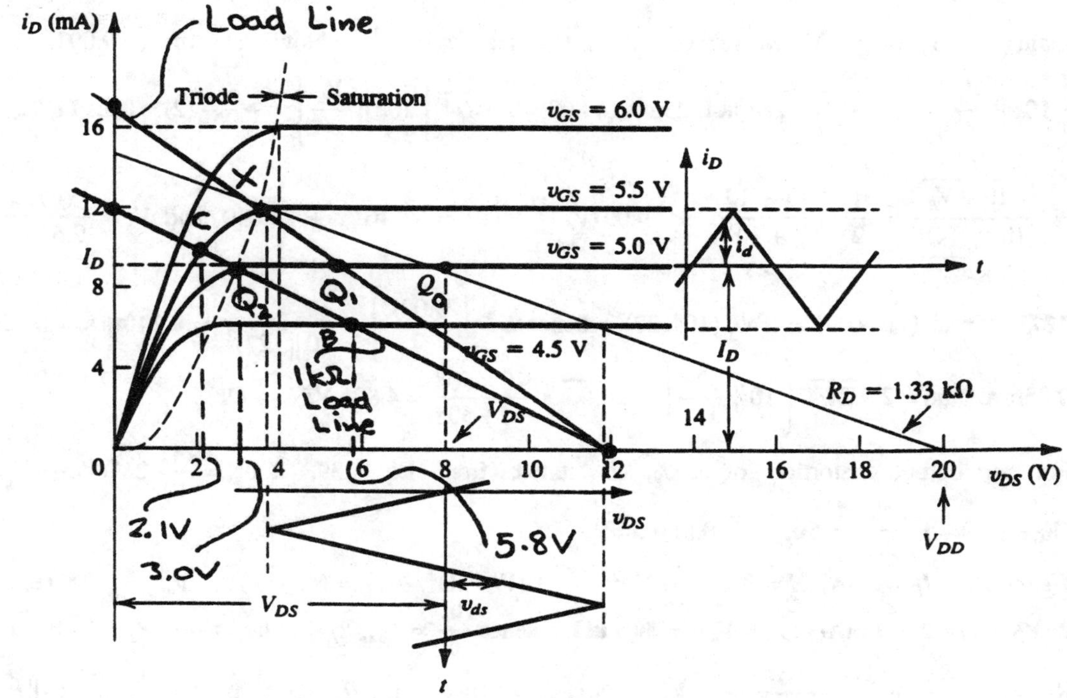

For V_{DD} = 12V, υ_{GS} = ±0.5V, the flattest load line which maintains saturation-mode operation is the one shown (passing through node X). The intercept is at i_D = 17mA, indicating R_L = 12V/17mA = **0.706kΩ.** For R_L = 1kΩ, the i_D intercept is 12V/1kΩ = 12mA, with the load line as shown. Note that for υ_{GS} = 5.0V, the operating point moves to Q_2 at I_D = 9mA and $\upsilon_{DS} \approx$ 3.0V, just at the edge of saturation. For a ±0.5V input, the output voltages range from about 2.1V to 5.8V. Thus the peak ac output voltages are 3.0 – 2.1 = 0.9V and 5.8 – 3.0 = 2.8V, with the ratio being 2.8/0.9 = **3.1 to 1.**

5.30 K = 1mA/V², V_t = 2V, V_{DD} = 12V, R_L = 0.5kΩ, υ_{GS} = 5V ± 0.5V. Thus $i_D = K(\upsilon_{GS} - V_t)^2 = 1(5-2)^2$ = **9mA**, and $\upsilon_D = V_{DD} - R_D\, i_D$ = 12 – 0.5(9) = **7.5V**. See, from Eq. 5.45, and the value of K in the line above, that $g_m = 2\sqrt{K}\sqrt{I_D} = 2\sqrt{1}\sqrt{9}$ = **6mA/V**. From Eq 5.42, $g_m = 2K(V_{GS} - V_t)$ = 2(1)(5 – 2) = **6mA/V**, in correspondence. From Eq 5.46, $\dfrac{\upsilon_d}{\upsilon_{gs}} = -g_m R_D$ = –6mA/V (0.5kΩ) = **–3.0V/V**. For a ±.5V input, expect a ±0.5(–3.0) = **±1.5V** output signal. From Eq 5.39, $i_D = K(V_{GS} - V_t)^2 + 2K(V_{GS} - V_t)\,\upsilon_{gs} + K\upsilon_{gs}^2$. Thus for υ_{gs} = ±0.5V, $i_D = 1(5-2)^2 + 2(1)(5-2)\,0.5 + (1)\,0.5^2$ = 9 + 3 + 0.25 = 12.25mA, for which υ_D = 12 – 0.5 (12.25) = **+5.875V**, and $i_D = 1(5-2)^2 + 2(1)(5-2)(-0.5) + (1)(-0.5)^2$ = 9 – 3 + 0.25 = 6.25mA, for which υ_D = 12 – 0.5 (6.25) = **8.875V**. This is to be contrasted with 7.5 – 1.5 = **6.0V**, and 7.5 + 1.5 = **9.0V**, as calculated from a linearized model.

5.31 $K_p = \frac{1}{2}\mu_p C_{ox}\frac{W}{L} = \frac{1}{2}$ (10) (300/3) = 500µA/V². $g_m = 2\sqrt{K}\sqrt{I_D}$ from Eq 5.45 and the line above, or $g_m = 2\sqrt{0.500}\sqrt{4} = 2\sqrt{2}$ = **2.83mA/V**. Generally, gain = $-g_m R_L$ = –10V/V implies that $R_L = \dfrac{10}{g_m} = \dfrac{10}{2.83}$ = **3.53kΩ**. Operation is reasonably linear for $\upsilon_{gs} \ll 2(V_{GS} - V_t)$, but $I_D = K(V_{GS} - V_t)^2 \rightarrow (V_{GS} - V_t) = \sqrt{I_D/K}$. Thus, linear for $\upsilon_{gs} \ll 2(V_{GS} - V_t) = 2\sqrt{I_D/K}$, that is $\upsilon_{gs} \ll 2\sqrt{4/0.5} = 2\sqrt{8}$ =

5.66V. For 1% nonlinearity, $\upsilon_{gs} \approx$ **0.06V** peak. For 10% nonlinearity $\upsilon_{gs} \approx$ **0.6V** peak.

5.32 Gain $= -g_m (R_L \| r_o)$, where $g_m = 2\sqrt{K I_D}$, and $r_o = \dfrac{V_A}{I_D}$. Now, at 1mA, $9.091 = 2\sqrt{K(1)} \left[10k \| \dfrac{V_A}{1}\right]$ --- (1), and at 0.25mA, $4.808 = 2\sqrt{K/4}\left[10k \| \dfrac{V_A}{1/4}\right]$ --- (2). Thus (1)/(2) $\equiv \dfrac{9.091}{4.808} = \dfrac{2\dfrac{10\,V_A}{10+V_A}}{\dfrac{10\,V_A\,(4)}{10+4\,V_A}} = \dfrac{1}{2}\,\dfrac{4\,V_A+10}{V_A+10}$. $\therefore$ $9.091V_A + 90.91 = 9.616V_A + 24.040$, and $V_A = \dfrac{90.91 - 24.04}{9.616 - 9.091} =$ **127V**. From (1): $9.091 = 2\sqrt{K}(10\|127)$, whence $K = \left[\dfrac{9.091}{2}\right]^2 \dfrac{1}{(10\|127)^2} =$ **0.240mA/V²**. **Check**: At 0.25mA, gain $= 2\sqrt{.24/4}\left[10\| \dfrac{127}{1/4}\right] \approx 2\sqrt{.06} \times \dfrac{10 \times 508}{10+508} = 4.804$ V/V $\rightarrow$ OK.

For an output distortion of 10%, as stated, from Eq 5.39, $K\,\upsilon_{gs}^2 = \dfrac{1}{10}\,2K\,(V_{GS} - V_t)\,\upsilon_{gs}$, or $(V_{GS} - V_t) = \dfrac{10\,\upsilon_{gs}}{2} = 5\upsilon_{gs} = 5\,(0.5) =$ **2.5V**.

5.33 *Generally*: $I_D = I$mA, $I = K(\upsilon_{GS} - V_t)^2 = 1\,(V_D - 1)^2$. $V_D = (\sqrt{I} + 1)$V, $g_m = 2K(\upsilon_{GS} - V_t) = 2\,(V_D - 1) = \mathbf{2\sqrt{I}}$ **mA/V**, $r_o = V_A/I =$ **50/I kΩ**, whence $\dfrac{\upsilon_o}{\upsilon_i} = -g_m\,R_L \| R_G \| r_o \approx -g_m\,R_L = \mathbf{-2R\sqrt{I}}$ **V/V**, and $R_i = \dfrac{R_G}{1 - gain} = \dfrac{\mathbf{10^4}}{\mathbf{1 + 2R\sqrt{I}}}$ kΩ. *For* I = 1mA and $R_L = R_G$: $\upsilon_o/\upsilon_i = -2\sqrt{I}\,(R_G \| R_G \| r_o) = -2 \times (10^4 \| 10^4 \| 50) \approx$ **−100V/V**, $R_i \approx \dfrac{10^4}{1+100} =$ **99kΩ**. For $R_L = r_o$: $\upsilon_o/\upsilon_i = -2\,(10^4 \| 50 \| 50) \approx$ **−50V/V**, $R_i = \dfrac{10^4}{1+50} =$ **196kΩ**. For $R_L = R_i$: $\upsilon_o/\upsilon_i = -2\,(10^4 \| 50 \| R_i) \approx -2\,(50 \| R_i)$, $R_i \approx \dfrac{10^4}{1 + 2\,(50 \| R_i)}$, or $R_i\left[1 + 2\dfrac{(50)R_i}{50 + R_i}\right] = 10^4$, $R_i + \dfrac{100\,R_i^2}{R_i + 50} = 10^4$, $R_i^2 + 50\,R_i + 100\,R_i^2 = 10^4\,R_i + 50 \times 10^4$, $101R_i^2 - 9950\,R_i - 50 \times 10^4 = 0$, $R_i^2 - 98.5\,R_i - 4950 = 0$, whence $R_i = \dfrac{98.5 \pm \sqrt{98.5^2 - 4\,(-4950)}}{2} = \dfrac{98.5 \pm 171.8}{2} =$ **135kΩ**, and $\upsilon_o/\upsilon_i \approx -2\,(50 \| 135) =$ **−73.0V/V**.

5.34 Current in the 1MΩ network can be ignored. Thus $I_D = I =$ **10mA**. Now, $I_D = I_{DSS}\left[1 - \dfrac{\upsilon_{GS}}{V_p}\right]^2$, or $10 = 10\left[1 - \dfrac{\upsilon_{GS}}{V_p}\right]^2 \rightarrow \upsilon_{GS} = 0$V (as could be seen directly). Since $\upsilon_{GS} = 0$, $\upsilon_G = 0$, $\upsilon_D = 0 + \dfrac{0 - -5}{1M} \times 1M =$ **5V**, $r_o = \dfrac{V_A}{I_D} = \dfrac{50V}{10mA} =$ **5kΩ**, $g_m = \dfrac{2\,I_{DSS}}{-V_p}\left[1 - \dfrac{\upsilon_{GS}}{V_p}\right] = \dfrac{2\,(10)}{2}\,(1 - 0) =$ **10mA/V**. For $R_L = \infty$, $\upsilon_o/\upsilon_i = -g_m\,r_o = -10 \times 10^{-3}\,(5 \times 10^3) =$ **−50V/V**. For $R_L = r_o$, $\upsilon_o/\upsilon_i = -g_m\,(r_o \| R_L) = -10 \times \dfrac{5}{2} =$ **−25V/V**. Now R_i = 1MΩ‖(1MΩ/(1 − gain)) in general, or $R_i = 1 \| 1/(1 - -50) =$ **19.2kΩ**, or $1 \| 1/(1 - -25) =$ **37kΩ**, in the two cases.

5.35 Note that while the lower end of r_o is not actually grounded, the signal there is small. Assume it to be zero. For R_S = 1kΩ, gain $\dfrac{\upsilon_o}{\upsilon_i} = -\dfrac{R_C \| R_L \| r_o}{1/g_m + R_S} = -\dfrac{10 \| 10 \| 100}{1/1 + 1} = \dfrac{-4.76}{2} =$ **−2.38V/V**. For R_S = 0Ω, gain $\dfrac{\upsilon_o}{\upsilon_i} = \dfrac{-4.76}{1} =$ **−4.76V/V**. For R_S = 3.76kΩ, gain $\dfrac{\upsilon_o}{\upsilon_i} = \dfrac{-4.76}{1 + 3.76} =$ **−1V/V**.

SECTION 5.7: Biasing the FET in Discrete Circuits

5.36

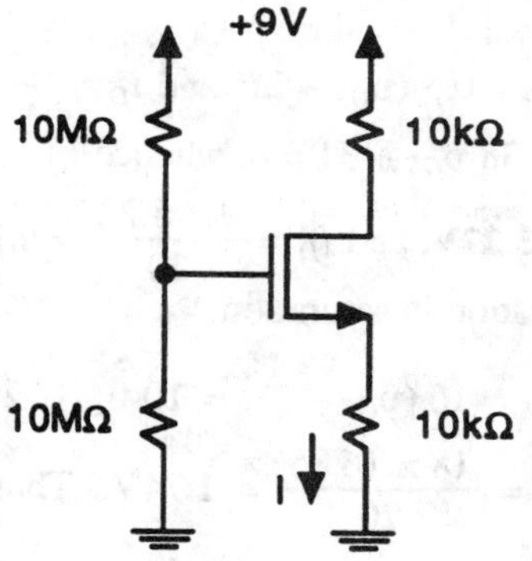

For this device V_t = 2V, K = 0.5mA/V^2. Assume saturation: *For* V_t = 2V, $V_G = 1/2\ (9) = 4.5$V, $I = K(\upsilon_{GS} - V_t)^2 = 0.5\ (\upsilon_{GS} - 2)^2$ --- (1), and $\upsilon_{GS} = 4.5 - 10\ I$ --- (2). Substitute (2) in (1) → $2\ I = (4.5 - 10\ I - 2)^2 = (2.5 - 10\ I)^2 = 6.25 - 50\ I + 100\ I^2$, or $100\ I^2 - 52\ I + 6.25 = 0$, whence $I = \dfrac{52 \pm \sqrt{52^2 - 4\ (100)\ (6.25)}}{2\ (100)}$ = **0.189mA,** (or too large a value). ∴ υ_S = 10k (0.189) = 1.89V, υ_{GS} = 4.5 −1.89 = 2.61V, υ_D = 9 − (10KΩ) (0.189mA) = 9 −1.89 = 7.11V, and υ_{DS} = 7.11 −1.89 = **5.22V** → OK, saturation. Operation remains in saturation until $\upsilon_{GD} \geq V_t$ = 2V, ie, for υ_{GS} = 2.61V, and $\upsilon_{DS} \geq$ 2.61 − 2 = 0.61V. Thus the peak negative-going output signal allowed is 7.11 −(1.89 + 0.61) = **4.61V**. But note that the largest positive-going output signal for cut-off is 1.89V. Now, *For* V_t = 1V, $I = 0.5\ (\upsilon_{GS} - 1)^2$, and $\upsilon_{GS} = 4.5 - 10I$. ∴ $2I = (4.5 - 10\ I - 1)^2 = (3.5 - 10I)^2 = 12.25 - 70I + 100I^2$, and $100I^2 - 72I + 12.25 = 0$.

Thus $I = \dfrac{72 \pm \sqrt{72^2 - 4\ (100)\ (12.25)}}{200}$ = **0.276mA**. ∴ υ_S = 10 (.276) = 2.76V, υ_{GS} = 4.5 − 2.76 = 1.74V, υ_D = 9 − 2.76 = 6.24V, υ_{DS} = 6.24 − 2.76 = **3.48V**. Now, saturation prevails while $\upsilon_{GS} \leq$ 1.0V. Thus the max negative swing is 6.24 − 4.5 + 1 = **2.74V**. The largest positive-going output signal is 9 − 6.24 = 2.76V.

5.37

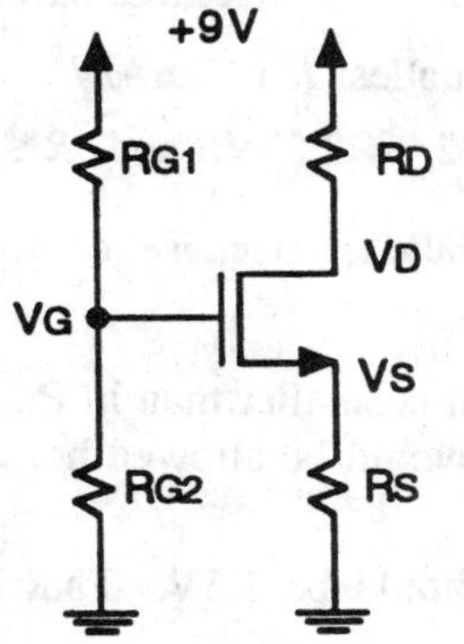

V_t from 1 to 2V, K from 0.3 to 0.5mA/V^2, and I_D from 0.5 to 1mA: Largest current occurs when V_t smallest (1V) and K largest (0.5mA/V^2). Thus $1 = 0.5\ (\upsilon_{GS} - 1)^2$, or $\upsilon_{GS} = \pm\sqrt{2} + 1 = 2.414$V --- (1). Smallest current when V_t largest (2V) and K smallest (0.3mA/V^2). Thus $0.5 = 0.3\ (\upsilon_{GS} - 2)^2$, and $\upsilon_{GS} = \sqrt{1.67} + 2 = 1.29 + 2 = 3.29$V --- (2). From (1), $\dfrac{V_{GG} - 2.414}{R}$ = 1mA, where $R = R_S$. From (2), $\dfrac{V_{GG} - 3.29}{R}$ = 0.5mA. See $V_{GG} - 2.414 = R$, and $2\ V_{GG} - 6.58 = R$. Subtracting, $-V_{GG} + 4.166 = 0 \rightarrow V_{GG}$ = 4.166V. For V_{GG} = 4.166V, R_{G1} = **10M Ω**, $R_{G2} = 4.166/(\dfrac{9 - 4.166}{10})$ = **8.6MΩ** (Use 8.2 MΩ), and $R_S = R$ = 4.166 − 2.414 = **1.75kΩ** (Use 1.8 kΩ).

Check: $\dfrac{4.166 - 3.29}{1.75}$ = 0.507mA. Now for I_D = 1mA, V_t = 1V, and a 0.5V signal, $\upsilon_D > V_{GG} - V_t$. That is, $\upsilon_D > 4.166 - 1.0 = 3.166$, and $R_D = \dfrac{9 - (3.166 + 0.5)}{1mA}$ = 5.33kΩ. (Use **5.1kΩ**).

5.38

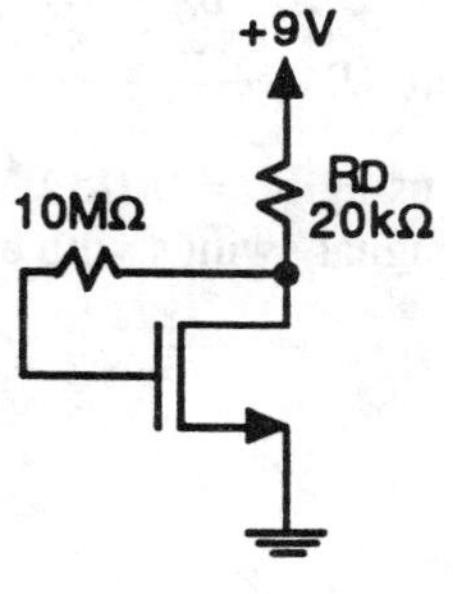

For V_t = 2V, $i_D = K\ (\upsilon_{GS} - V_t)^2$, $= 0.5\ (\upsilon_{GS} - 2)^2$, and $i_D = \dfrac{9 - \upsilon_{GS}}{20k\Omega}$. Thus $9 - \upsilon_{GS} = 20\ (0.5)\ (\upsilon_{GS} - 2)^2$, $9 - \upsilon_{GS} = 10\ \upsilon_{GS}^2 - 40\ \upsilon_{GS} + 40$, $10\ \upsilon_{GS}^2 - 39\ \upsilon_{GS} + 31 = 0$, whence $\upsilon_{GS} = \dfrac{39 \pm \sqrt{39^2 - 4\ (10)\ (31)}}{2\ (10)}$ = 2.79V. ∴ υ_{DS} = **2.79V**, and $I_D = \dfrac{9 - 2.79}{20}$ = **0.311mA**. For negative peak outputs of up to **2Vp**, operation remains in saturation. *For* V_t = 1V, $i_D = 0.5\ (\upsilon_{GS} - 1)^2$, and $i_D = \dfrac{9 - \upsilon_{GS}}{20}$.

$\therefore\ 9 - \upsilon_{GS} = 10\,(\upsilon_{GS} - 1)^2 = 10\,\upsilon_{GS}^2 - 20\,\upsilon_{GS} + 10$, and $10\,\upsilon_{GS}^2 - 19\,\upsilon_{GS} + 1 = 0$, whence $\upsilon_{GS} = \dfrac{19 \pm \sqrt{19^2 - 4\,(10)\,(1)}}{2\,(10)} = \dfrac{19 \pm 17.9}{20} = 1.845\text{V}$. Thus $\upsilon_{DS} = \mathbf{1.85V}$, $I_D = \dfrac{9 - 1.85}{20} = \mathbf{0.358mA}$, with a **1Vp** signal allowed.

5.39 For R_{G2} = 10MΩ from gate to source, $\upsilon_{DS} = 2\,\upsilon_{GS}$. *For* $V_t = 2\text{V}$, $i_D = 0.5\,(\upsilon_{GS} - 2)^2$ and $i_D = \dfrac{9 - 2\,\upsilon_{GS}}{20}$. Thus $9 - 2\,\upsilon_{GS} = 10\,(\upsilon_{GS} - 2)^2 = 10\,\upsilon_{GS}^2 - 40\,\upsilon_{GS} + 40$, and $10\,\upsilon_{GS}^2 - 38\,\upsilon_{GS} + 31 = 0$, whence $\upsilon_{GS} = \dfrac{38 \pm \sqrt{38^2 - 4\,(10)\,(31)}}{2\,(10)} = \dfrac{38 \pm 14.28}{20} = 2.61\text{V}$. Thus $\upsilon_{DS} = \mathbf{5.22V}$, and $I_D = \dfrac{9 - 5.22}{20} = \mathbf{0.189mA}$. For negative peak outputs, 5.22 –2.61 + 2 = **4.61V** is allowed for operation in saturation.

For $V_t = 1\text{V}$, $i_D = 0.5\,(\upsilon_{GS} - 1)^2$ and $i_D = \dfrac{9 - 2\,\upsilon_{GS}}{20}$. Thus $9 - 2\,\upsilon_{GS} = 10\,(\upsilon_{GS} - 1)^2 = 10\,\upsilon_{GS}^2 - 20\,\upsilon_{GS} + 10$, and $10\,\upsilon_{GS}^2 - 18\,\upsilon_{GS} + 1 = 0$, whence $\upsilon_{GS} = \dfrac{18 \pm \sqrt{18^2 - 40}}{20} = \dfrac{18 \pm 16.85}{20} = 1.74\text{V}$. Thus $\upsilon_{DS} = \mathbf{3.48V}$ and $I_D = \dfrac{9 - 3.48}{20} = \mathbf{0.276mA}$, with 3.48 –1.74 + 2 = **3.74V** negative output peaks allowed, while saturated operation prevails.

5.40

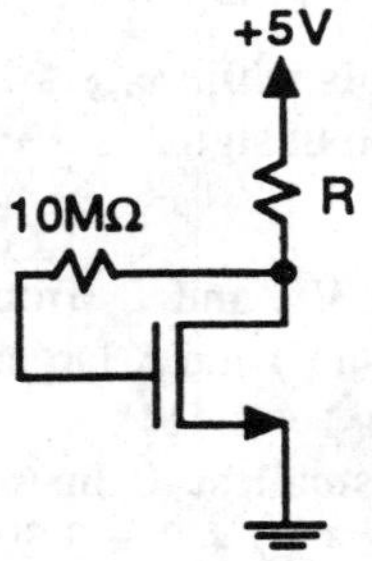

Here, V_t varies from 1 to 2V, K varies from 0.3 to 0.5mA/V^2, I_D varies from 0.5 to 1mA. The largest current (1mA) occurs for the smallest V_t (1V), and largest K (0.5mA/V^2). Thus $1 = 0.5\,(\upsilon_{GS} - 1)^2$. $\therefore\ \upsilon_{GS} = 2.414\text{V}$, and $R \geq \dfrac{5 - 2.414}{1mA} \geq 2.59\text{k}\Omega$. The smallest current (0.5mA) occurs for the largest V_t (2V) and smallest K (0.3mA/V^2). Thus $0.5 = 0.3\,(\upsilon_{GS} - 2)^2$. $\therefore\ \upsilon_{GS} = \sqrt{1.67} + 2 = 3.29\text{V}$, and $R \leq \dfrac{5 - 3.29}{0.5} \leq 3.42\text{k}\Omega$. Use $R = \dfrac{2.59 + 3.42}{2} = \mathbf{3.0k\Omega}$. Because of feedback, the effect of variation is reduced.

The circuit automatically allows a negative signal = V_t > 1V but the gain is smaller than in P 5.37, since R here (3.0kΩ) is less than R_D there (5.2kΩ). Raising R to (say) **3.3kΩ** would be allowed here, and would improve the gain by 10%.

5.41 In P5.40 above, the minimum negative-going signal is 1V. Here it should be 1.5V. That is, we want $\upsilon_{DS} = \upsilon_{GS} + 0.5\text{V}$ for the case in which $V_t = 1\text{V}$.

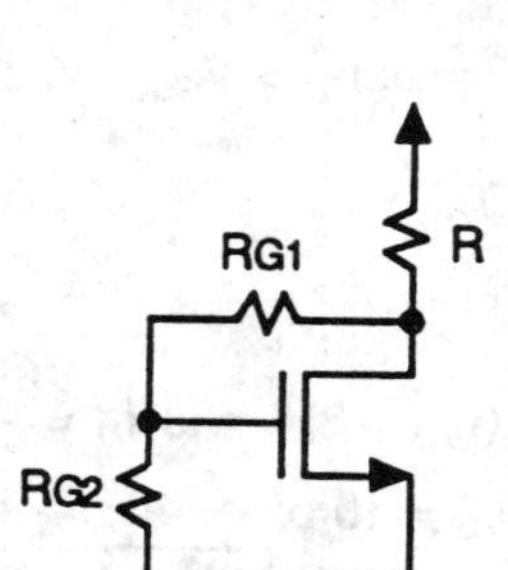

For smallest V_t (and also the largest current), from the results in P5.40, $\upsilon_{GS} = 2.414\text{V}$, and $\upsilon_{DS} = 2.414 + 0.5 = 2.914$, with $R > \dfrac{5 - 2.914}{1mA} = 2.09\text{k}\Omega$. If we use $\mathbf{R_{G2} = 10M\Omega}$, $R_{G1} = \dfrac{10}{2.414}\,(0.5) = 2.07\text{M}\Omega$. Use $\mathbf{R_{G1} = 2.0M\Omega}$. Now for $V_t = 2\text{V}$ (and also the smallest current), from the solution for P5.40 we see, $\upsilon_{GS} = 3.29\text{V}$, and $\upsilon_{DS} = \dfrac{3.29}{10} \times 2.0 + 3.29 = 3.95\text{V}$. Thus $R < \dfrac{5 - 3.95}{0.5} \leq 2.1\text{k}\Omega$.

Notice that a solution *barely exists*, using R = **2.1kΩ**, essentially as a consequence of a demand for large signal swings with a limited supply voltage.

5.42

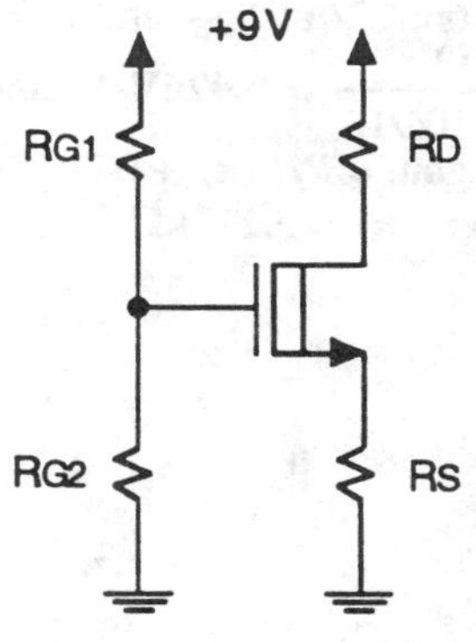

Here, $V_p = -4V$, $I_{DSS} = 32mA$. For $I_D = 8mA$, $8 = 32\,(1 - \frac{\upsilon_{GS}}{V_p})^2$. Thus $(1 - \frac{\upsilon_{GS}}{V_p})^2 = 1/4$, $-\frac{\upsilon_{GS}}{V_p} = \pm 1/2 - 1 = -1.5$ or -0.5, and $\upsilon_{GS} = 0.5V_p = -2V$. Now for a negative swing of 2V to the edge of saturation, $\upsilon_D \geq \upsilon_G + |V_t|$. Now for the largest possible value of R_D, υ_D will be lowest and υ_G lowest. The lowest possible υ_G is 0V, with $R_{G1} = \infty$ and R_{G2} = **10MΩ**, in which case R_S = 2V/8mA = **0.25kΩ**, and $V_D \geq 0 + 4 = 4V$ for lowest swing or 4 + 2 = 6V for no signal, with $R_D \leq 9–6/8$ = **0.375kΩ**. Note that for a 2V positive output swing, υ_D rises to 6 + 2V = 8V, and the transistor is not yet cut off. OK. Note that this design with no biassing supply is relatively sensitive to device variability, all as a result of wanting a large signal swing with a small supply.

5.43 See Q_2 operates with $I_D = I_{DSS}$ = **10mA** as does Q_1 with υ_{GS1} = 0V, and $\upsilon_{DS1} = 0 + 0 - -5/2 \times 1$ = **2.5V**. Thus υ_{DS2} = 5 − 2.5 = 2.5V as well. For each, $r_{01} = r_{02} = r_0 = V_A/I_D$ = 100V/10mA = **10kΩ**. Now $g_m = \frac{2I_{DSS}}{-V_p}\left[1 - \frac{\upsilon_{GS}}{V_p}\right] = \frac{2(10)}{--2}\,(1 - 0)$ = **10mA/V**, and $\upsilon_o/\upsilon_i = -g_m\,(r_{01}\|r_{02}\|R_{G1}) = -10\,(10k\,\|\,10k)$ = **−50V/V**. Correspondingly, $R_i = \left[\frac{10^6}{1 - -50}\right] \| (2k \times 10^6)$ = **19.6kΩ**. For R_S = 0, υ_o/υ_s = **−50V/V**. For $R_S = R_i$ = 19.6kΩ, υ_o/υ_s = −50 (1/2) = **−25V/V**. For R_S = 10kΩ , υ_o/υ_s = 19.6/ (10+19.6) × (−50) = **−33.1V/V**.

SECTION 5.8: Basic Configurations of Single-Stage FET Amplifiers

5.44 R_{in} = **1MΩ**. Gain with no load $= -g_m\,(R_D\|r_o) = -1\,(10k\|50k)$ = **−8.33V/V**. $R_{out} = R_D\|r_o = 10k\|50k$ = **8.33Ω**. Gain with 20kΩ load = −8.33 × 20/ (20+8.33) = **−5.88V/V**. Alternatively, the loaded gain $= -g_m\,(R_D\|r_o\|R_L) = -1\,(10k\|50k\|20k) = -(100/17)$ = **−5.88V/V**.

5.45 $R_{in} = 1/g_m$ = 1/1 = **1kΩ**.

For R_S = 0, Gain with no load $= g_m\,(R_D\|r_o) = 1\,(10k\|50k)$ = **8.33V/V**. Gain with load $= g_m\,(R_D\|r_o\|R_L) = 1\,(10k\|50k\|20k)$ = **5.88V/V**.

For R_S = 1kΩ, Gain with no load = 1/ (1+1) × 8.33 = **4.17V/V**. Alternatively, Gain with no load = $\frac{R_D\|r_o}{1/g_m + R_S} = \frac{10k\|50k}{1/1 + 1}$ = 4.17V/V. Gain with load = 1/ (1+1) × 5.88 = **2.94V/V**.

5.46 $R_{out} = 1/g_m\|r_o$, $r_o \propto 1/i$, $g_m \propto \sqrt{i}$. Originally, $r_o\|1/g_m$ = .952. With four times the current, $(r_o/4)\|1/(\sqrt{4}\,g_m) = (r_o/4)\|(1/(2g_m))$ = 0.455. Now, $\frac{r_o/g_m}{r_o + 1/g_m}$ = .952, $\frac{r_o}{1 + g_m r_o}$ = .952, or $r_o = .952 + .952\,g_m r_o$ – – – (1). Also, $\frac{(r_o/4)(1/2g_m)}{r_o/4 + 1/(2g_m)} = 0.455 = \frac{r_o}{4 + 2g_m r_o}$, or $r_o = 1.82 + .91\,g_m r_o$ – – – (2). From (2), $1.046r_o = 1.904 + .952\,g_m r_o$ – – – (3). Subtract (3) − (1) → $0.046r_o = 0.952$, r_o = **20.62kΩ**. From (1), $20.62 = .952 + .952\,(20.62)\,g_m$, whence g_m = **1.002mA/V**.

If the original current is 1mA, V_A = 1mA × 20.6kΩ = **20.6V**. At 1mA, R_{out} = 952Ω, Gain $= R_L/(R_L + .952) \geq 0.900$ V/V, $R_L \geq 0.9\,R_L + 0.857$, $R_L \geq 0.857/0.1 \geq$ **8.57kΩ**.

5.47

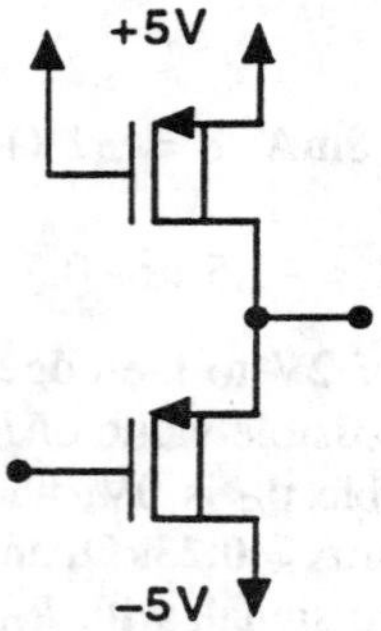

$i_D = I_{DSS}(1 - \upsilon_{GS}/V_p)^2$ = 16mA at υ_{GS} = 0. Thus $g_m = 2\,I_{DSS}(1 - \upsilon_{GS}/V_p) = 2\,(8)\,(1) = 16\text{mA/V}$ at I_{DSS}, and $r_o = 40/16 =$ 2.5kΩ. *For no load,* gain $= \dfrac{2.5k\Omega}{2.5k\Omega + 10^3/16} =$ **0.976V/V**, and $R_{out} =$ 2.5kΩ || 1/16 = **0.061kΩ**. *For load R_L*, gain $= R_L/(R_L + 0.061) \times .976 =$ 0.8. Thus $0.976\,R_L = 0.8\,R_L + 0.0488$, and $R_L =$ **0.277kΩ**.

SECTION 5.9: Integrated-Circuit MOS Amplifiers

5.48

Type	S	G	D	Joined Terminals	Positive End	Saturated Operation Possible
Enhancement	+	−	−	GD	S	Yes
	−	−	+	SG	D	Yes
Depletion	+	+	−	SG	S	Yes
	+	−	−	GD	S	No
	−	+	+	GD	D	Yes
	−	−	+	SG	D	No

There are **2 Enhancement** Configurations, and **4 Depletion** Configurations, with **6 Configurations in total,** for which current flows. Of these, **4 allow saturated operation.**

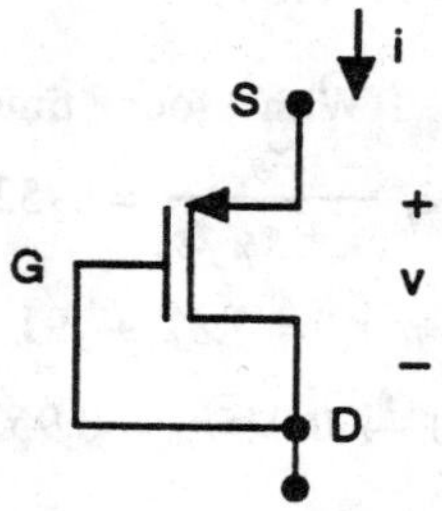

$\upsilon \geq |V_t|$, saturated
$\upsilon < |V_t|$, cutoff

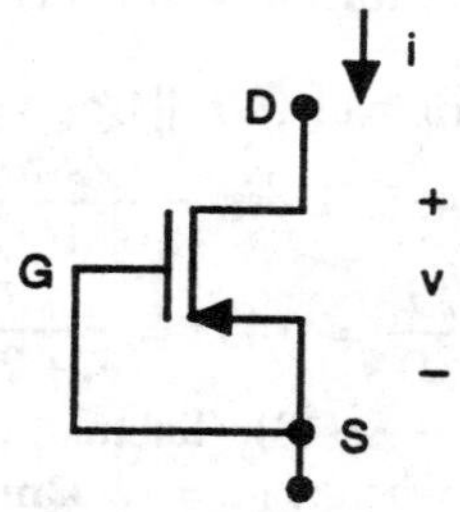

$\upsilon \geq |V_t|$, saturated
$\upsilon < |V_t|$, cutoff

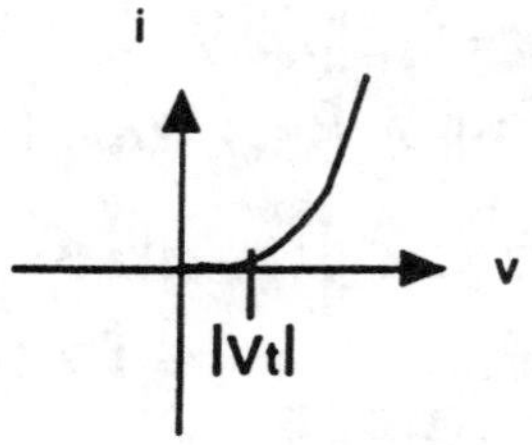

5.48

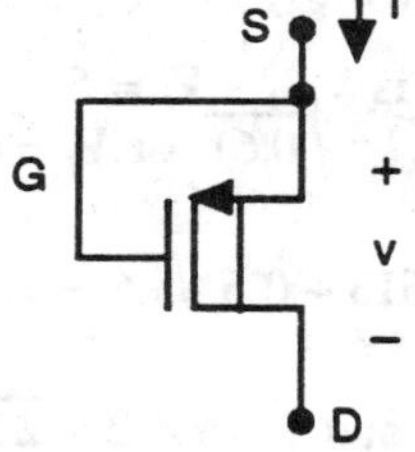

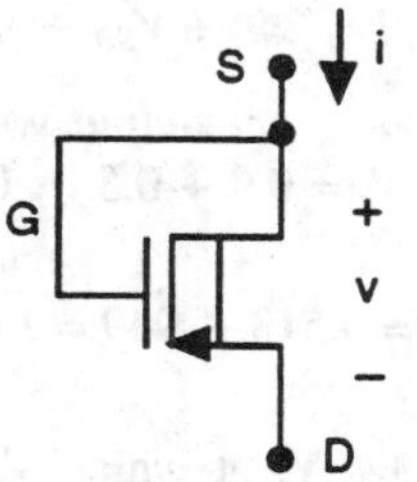

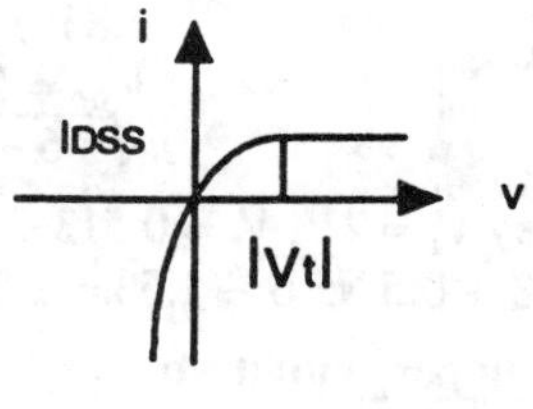

$\upsilon \geq |V_t|$ saturated
$\upsilon < |V_t|$ triode
$\upsilon < 0$ triode

$\upsilon \geq |V_t|$ saturated
$\upsilon < |V_t|$ triode
$\upsilon < 0$ triode

5.49 Assuming no back-bias effect.

(a) Devices identical, $V_1 = 5/2 =$ **2.5V**, $i = K(\upsilon_{GS} - V_t)^2 = 1(2.5-2)^2 = 0.25$mA. Thus $I_1 =$ **0.25mA**.

(b) Lower transistors operate as one with $K = 2\text{mA/V}^2$. Here for $V_2 = \upsilon$, $i = 1(5-\upsilon-2)^2 = 2(\upsilon-2)^2$, $(3-\upsilon)^2 = 2(\upsilon-2)^2$, $3-\upsilon = \pm\sqrt{2}(\upsilon-2) = \pm 1.414\,\upsilon \pm 2.828$. Thus, $3-\upsilon = 1.414\,\upsilon - 2.828$, or $3-\upsilon = -1.414\,\upsilon + 2.828$. Correspondingly $-2.414\,\upsilon = -5.828$, $\upsilon = 2.414$V, for which $i = 2(2.414-2)^2 = 0.343$mA. Thus $V_2 =$ **2.414V**, $I_2 =$ **0.343mA**.

(c) I_3 operates at I_{DSS} if the upper transistor is in saturation, in which case, $i = K(\upsilon_{GS} - V_t)^2 = 1(--2)^2 = 4\text{mA} = I_3$. Also $i = K(\upsilon_{GS} - V_t)^2$, or $4 = 1(\upsilon_{GS} - 2)^2$, $\upsilon_{GS} - 2 = \pm 2$, whence $\upsilon_{GS} =$ **4V** (which is too high to allow the upper transistor to saturate). ∴ upper in triode, lower in saturation. For $\upsilon = V_3$, $i = K(2(0--2)(5-\upsilon) - (5-\upsilon)^2) = K(\upsilon-2)^2$. ∴ $20 - 4\upsilon - 25 + 10\upsilon - \upsilon^2 = \upsilon^2 - 4\upsilon + 4$, $2\upsilon^2 - 10\upsilon + 9 = 0$, $\upsilon = \dfrac{+10 \pm \sqrt{10^2 - 4(2)(9)}}{2(2)} = \dfrac{10 \pm 5.29}{4} = 3.82$V (or an impossibly low value), for which $i = 1(3.82-2)^2 = 3.31$mA. Thus $I_3 =$ **3.31mA** and $V_3 =$ **3.82V**.

(d) From Symmetry, $V_4 =$ **2.5V**, and $I_4 = 1(2.5-2)^2 =$ **0.25mA**.

(e) From Symmetry, $V_5 = 5/2 =$ **2.5V**, and $I_5 = 1(2(2.5--2)(2.5) - 2.5^2) = 5(4.5) - 2.5^2 =$ **16.25mA**.

5.50

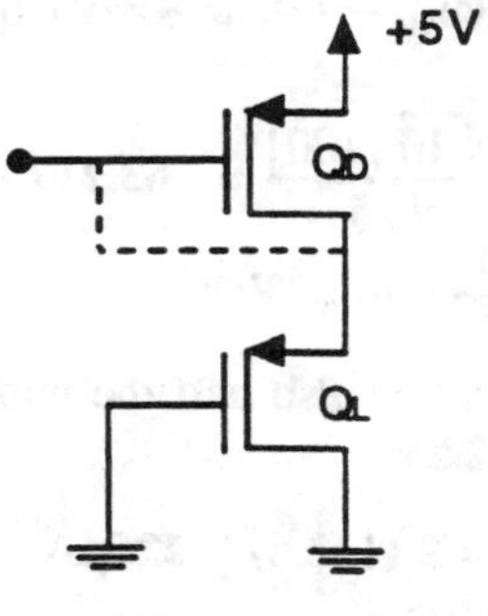

Here $V_t = -1$V, $K_D = 90\mu\text{A/V}^2$, and $K_L = 10\mu\text{A/V}^2$. Ignoring the body effect, and assuming both in saturation, $i = K(\upsilon_{GS} - V_t)^2$. For the driver, $g_{mD} = 2K_D(\upsilon_{GS} - V_t) = 2K_D\sqrt{i/K_D} = 2\sqrt{K_D\, i}$. For the load, $g_{mL} = 2\sqrt{K_L\, i}$. Ignoring r_o, (and back-bias effects), gain $\upsilon_o/\upsilon_i = -g_{mD}\dfrac{1}{g_{mL}} = -\dfrac{2\sqrt{K_D\, i}}{2\sqrt{K_L\, i}} = -\sqrt{K_D/K_L} - \sqrt{90/10} =$ **−3V/V**.

For $\upsilon_O = V_{DD}/2$, $i = K_L(-2.5--1)^2 = 10(1.5^2) = 22.5\mu$A. Now, for Q_D, $22.5 = 90(-(5-\upsilon_I) - -1)^2 = 90(\upsilon_I - 4)^2$, or $\upsilon_I - 4 = \pm\sqrt{22.5/90} = \pm 0.5$, $\upsilon_I = 3.5$ or 4.5 (not possible). Thus $\upsilon_I =$ **3.5V** for $\upsilon_O = V_{DD}/2$. Equation 5.78 applies while devices are both in saturation: ie for υ_O down to $|V_t| =$ **1V**, where $\upsilon_I =$ **+4V** (at cutoff), and υ_O up to $\upsilon_O + 1 = \upsilon_I$.

For Q_D: $i = 90(5 - \upsilon_I - 1)^2$, For Q_L: $i = 10(\upsilon_O - 1)^2$. Thus $90(4-\upsilon_I)^2 = 10(\upsilon_O - 1)^2$, or $3(4-\upsilon_I) = \upsilon_O - 1$, or $\upsilon_O = 13 - 3\upsilon_I$. *Check*: For $\upsilon_I = +4$, $\upsilon_O = 13 - 3(4) = 1$V as found before. Now for $\upsilon_I = \upsilon_O + 1$, $\upsilon_O = 13 - 3(\upsilon_O + 1) = 10 - 3\upsilon_O$. ∴ $4\upsilon_O = 10$V, $\upsilon_O =$ **2.5V**, for which $\upsilon_I = 2.5 + 1 =$ **3.5V**. ***Check***: For υ_O varying from 1V with $\upsilon_I = 4$V, to 2.5V with $\upsilon_I = 3.5$V, that is a 1.5V change results from a 0.5V change.

5.51

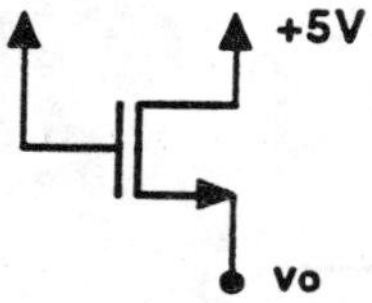

Eq 5.25: $V_t = V_{t0} + \gamma [\sqrt{2\Phi_f + V_{SB}} - \sqrt{2\Phi_f}]$. Eq 5.84: $\chi = \frac{\gamma}{2\sqrt{2\Phi_f + V_{SB}}}$ Eq 5.83: $g_{mb} = \chi\, g_m$. *The upper output voltage limit is* $V_{DD} - V_t = 5 - V_t$ at which $V_{SB} = 5 - V_t$. Thus $V_t = 0.9 + 0.5\,(\sqrt{0.6 + 5 - V_t} - \sqrt{0.6})$, or $V_t = 0.513 + 0.5\sqrt{5.6 - V_t}$.

Iterate: Try $V_t = 2V$, $V_t = 0.513 + 0.5\sqrt{5.6 - 2} = 0.513 + .949 = 1.46$, $V_t = 0.513 + 0.5\sqrt{5.6 - 1.46} = 1.53V$, $V_t = 0.513 + 0.5\sqrt{5.6 - 1.53} = 1.52V$.

Thus the upper output voltage is 5 – 1.52 = **3.48V**, at which V_t = **1.52V**, and $\chi = \gamma/(2\sqrt{2\Phi_f + V_{SB}})$ = $0.5/(2\sqrt{0.6 + 1.52})$ = **0.172**, and g_m = **0**, since i_D is zero at the upper limit. At $\upsilon_O = 0V$, $V_{SB} = 0$ and V_t = **0.9V**. Thus $i = K(\upsilon_{GS} - V_t)^2 = 10\,(5 - 0 - 0.9)^2 = 6.168mA$, for which $g_m = 2K(\upsilon_{GS} - V_t) = 2\,(10)\,(5 - 0 - 0.9)$ = **82µA/V**, and $\chi = \gamma/(2\sqrt{2\Phi_f + V_{SB}}) = 0.5/(2\sqrt{0.6 + 0})$ = **0.323**.

5.52 From P5.51: At $V_O = 2.5V$, $V_{SB} = 2.5V$, $V_t = V_{to} + \gamma(\sqrt{2\Phi_f + V_{SB}} - \sqrt{2\Phi_f}) = 0.9 + 0.5\,(\sqrt{0.6 + 2.5} - \sqrt{0.6}) = 1.39V$. Now, $g_{mL} = 2K(\upsilon_{GS} - V_t) = 2\,(10)\,(5 - 2.5 - 1.39) = 22.2µA/V$, $i = K(\upsilon_{GS} - V_t)^2 = 10\,(5.0 - 2.5 - 1.39)^2 = 12.3µA$, $\chi = \frac{\gamma}{2\sqrt{2\Phi_f + V_{SB}}} = \frac{0.5}{2\sqrt{0.6 + 2.5}} = 0.142$. Thus, $g_{mb} = \chi\, g_{mL} = .142 \times 22.2µA/V = 3.15µA/V$. Also $g_{mD} = 2K\sqrt{i/K} = 2\sqrt{Ki} = 2\sqrt{90 \times 12.3} = 66.54µA/V$ Now the overall voltage gain $= -g_{mD}\,(1/g_{mL} \,||\, 1/g_{mb}) = -66.54\,((1/22.2) \,||\, (1/3.15)) = -66.54µA/V\,(45.0k \,||\, 317k)$ = **– 2.62V/V**, rather than $-\sqrt{90/10} = -3V/V$ that the basic calculation would indicate.

5.53 Assuming V_t includes the back-bias effect, $i_L = K(\upsilon_{GS} - V_t)^2 = 22.5\,(0 - 2.0)^2 = 90µA$, $g_{mL} = 2K(\upsilon_{GS} - V_t) = 2\,(22.5)(2) = 90µA/V$, $g_{mB} = \chi\, g_{mL} = 0.2\,(90) = 18µA/V$, $r_o = V_A/i_L = 50/90µA = 556kΩ$, $g_{mD} = 2\sqrt{K\, i_L} = 2\sqrt{90 \times 90} = 180µA/V$. Thus, gain (around V_D = 2.5V), is $-g_m\,(r_o \,||\, r_o \,||\, 1/g_{mB}) = -180µA/V\,(556k/2 \,||\, 1/18µA/V) = -0.180\,(278k \,||\, 55.6k)$ = **–8.34V/V**.

This again applies reasonably well until the load enters triode operation at $\upsilon_o = V_{DD} - |V_t| = 5 - 2 = 3.0V$, or until the driver enters the triode region at $\upsilon_O = \upsilon$. Now for the lower level, $i_L = 90µA$, that is $90 = 90\,(\upsilon_{GS} - 1)^2 \rightarrow \upsilon_{GS} = 2V$. Now for $\upsilon_{GS} = 2V$, triode operation begins at $\upsilon_{DS} = \upsilon_{GS} - V_t = 2 - 1 = 1V$. In actual fact, as υ_O falls from the middle, that is from 2.5V to 1V, input must rise by $(2.5 - 1)/8.34 \approx 0.2V$. Thus, the output range is from **3.0V** to about **1.2V**.

5.54 There are two possibilities: that 18µA is extracted from the $K = 10µA/V^2$ unit, or the $K = 90µA/V^2$ unit. For $K = 10µA/V^2$, $18 = 10\,(\upsilon_{GS} - 1)^2$, $\upsilon_{SG} = 1 + (18/10)^{1/2}$ = **2.34V**, for which the output current is 9 × 18 = **162µA**, that is, $i = 90\,(2.34 - 1)^2 = 161.6µA$. For $K = 90µA/V^2$, $18 = 90\,(\upsilon_{SG} - 1)^2$, $\upsilon_{SG} = 1 + \sqrt{18/90}$ = **1.45V**, for which the output current is 1/9 × 18 = **2µA**.

5.55 Eq 5.101 indicates $A_\upsilon = -\frac{\sqrt{K}\,|V_A|}{\sqrt{I_{REF}}}$. Let $I = I_{REF}$. For I = 25µA: $A_\upsilon = -\frac{\sqrt{10}\,|100|}{\sqrt{25}}$ = **–63.2V/V**. For I = 2.5µA: $A_\upsilon = -\frac{\sqrt{10}\,(100)}{\sqrt{2.5}}$ = **–200V/V**. For I = 0.25µA: $A_\upsilon = -\frac{\sqrt{10}\,(100)}{\sqrt{0.25}}$ = **–632V/V**.

5.56 For the diode-connected NMOS (call it Q_4) of half the width of Q_1, its width is also half the width of Q_3. Thus, Q_4, Q_3 have the same K and the same V_t. Thus $\upsilon_{SG3} = \upsilon_{SG2} = 5/2 = 2.5V$.

$\therefore I_{REF} = i = 10 \times \frac{100/2}{10}\,(2.5 - 1)^2 = 112.5µA$, with a total supply current = 2 (112.5) = **225µA**. For Q_1, $i = 112.5 = 10 \times 100/10\,(\upsilon_{GS} - 1)^2$. Thus $\upsilon_O = \upsilon_{GS} = 1 + (112.5/100)^{1/2}$ = **2.06V**, and $g_{m1} = 2K(\upsilon_{GS} - V_t) = 2\,(100)\,(2.06 - 1) = 212µA/V$. Thus the gain $\upsilon_o/\upsilon_i = -g_m\,(r_{01} \,||\, r_{02})$, where $r_{01} = r_{02} = r_o = 50/112.5 = 0.444MΩ$. Thus, $\upsilon_o/\upsilon_i = -212 \times 0.444/2$ = **–47.1V/V**. Max positive output = (5 – 2.5 + 1) = **3.5V**. Max negative output = 2.06 – 1 = **1.06V**.

5.57 $K = 1/2\mu_p C_{ox} W/L = (1/2)\,(10)\,(1000/10) = 500µA/V^2$. For $V_O \approx 0V$, $|V_{SB}| = V_{DD} - V_S = +10 - 0 = 10V$.

Now $V_t = V_{t0} + \gamma\left[\sqrt{2\Phi_F + |V_{SB}|} - \sqrt{2\Phi_F}\right] = 0.9 + 0.5\left[\sqrt{0.6+10} - \sqrt{0.6}\right] = 2.14$V. Thus the necessary input voltage = **+2.14V** for an output of zero. Now, $g_m = 2K(\upsilon_{GS} - V_t) = 2\sqrt{KI} = 2\sqrt{0.5\times 1} = 1.414$mA/V. ∴ output resistance of follower = 1/1.414 = **707Ω**. Now $\chi = \gamma/(2\sqrt{2\Phi_F + |V_{SB}|}) = 0.5/(2\sqrt{0.6+10}) =$ 0.077. ∴ load on follower due g_{mb} = 707Ω/.077 = 9.21kΩ. ∴ Gain with no load = 9.21/(.707 + 9.21) = **0.929V/V**. For load R, gain = (9.21 || R)/(.707 + 9.21 || R) ≥ 0.5. Thus 9.21R/(9.21 + R) = 0.5 (0.707 + 9.21R/(9.21 + R)), ∴ 4.60R/(9.21 + R) = 0.3535, 4.60R = 3.26 + .354R, whence R ≥ **0.768kΩ**, for gain ≥ 0.5V/V.

SECTION 5.10: FET Switches

5.58 Now, for 10mV to ground with a 3.3V, 2.1kΩ source, i_D = (3.30 – 0.01)/ 2.1kΩ = 3.29V/21kΩ = 0.157mA. Now in the triode region, $i_D = K\left[2(\upsilon_{GS} - V_t)\upsilon_{DS} - \upsilon_{DS}^2\right] \approx K(2(\upsilon_{GS} - V_t)\upsilon_{DS})$, or 0.157 × 10^{-3} = 1/2 × 20 × 10^{-6} × W/10 × 2 (5 – 1) (10 × 10^{-3}), or 0.157 = 10 × 10^{-6} × 2 (4)W. Thus W = .157/80 × 10^{-6} = **1963μm**, which is quite a large device! Now, if 0.10V were acceptable: $W = \frac{(.30 - .10)/21}{800} \times 10^6 =$ **190μm**.

5.59 K_n = 1/2 (20 × 10^{-6}) (50L/L) = 500μA/V². Assuming $\mu_p = 1/2\mu_n$ with $W_p = 2W_n$, then $K_p = K_n$ = 500μA/V. Now, for operation in the triode mode, $i_D = K\left[2(\upsilon_{GS} - V_t)\upsilon_{DS} - \upsilon_{DS}^2\right] \approx 2K(\upsilon_{GS} - V_t)\upsilon_{DS}$, and $r_{DS} = \upsilon_{DS}/i_D = 1/(2K(\upsilon_{GS} - V_t))$. Now, for V_I = –5V with V_{Gn} = +5V and V_{Gp} = –5V, only the n-channel device conducts with r_{DS} = 1/(2(500 × 10^{-6}) (5 – – 5 – 2)) = 125Ω. Now, with 5kΩ load, ac loss in the switch is 125/(125 + 5000) = .0244 or **2.4%**. Now, for V_I = +5V, with $K_p = K_n$, the result is the same and the loss is **2.4%** in the switch. Now, for V_I = 0V, with V_{Gn} = +5V, V_{Gp} = –5V, both switches conduct (equally), with r_{DS} = 1/(2 (500 × 10^{-6}) (5 – 0 – 2)) = 333.3Ω each. Thus the total switch resistance is 333.3/2 = 167Ω, and the ac switch loss is 167/(167 + 5000) = .0323, or **3.2%**.

SECTION 5.11: Gallium-Arsenide (GaAs) Devices – The MESFET

5.60 Here, from Eq 5.108 and 5.109, $g_m = 2\beta(V_{GS} - V_t)(1 + \lambda V_{DS})$, $r_o \approx 1/(\lambda\beta(V_{GS} - V_t)^2)$, and the highest available gain is $\mu = g_m r_o$.

For υ_{GS} = +0.2V, g_m = 2 (10^{-4}) × 100 (0.2 – –1.0) (1 + 0.2(3)) = 200 × 10^{-4} × 1.2 (1.6) = **38.4mA/V**, $r_o \approx$ 1/(.2 × 100 × 10^{-4} (.2 – – 1.0)²) = **347Ω**, and μ = 38.4 × 10^{-3} × 347 = **13.3V/V**.

For υ_{GS} = –0.2V, g_m = 2 (10^{-4}) × 100 (–0.2 – –1.0) (1 + 0.2(3)) = 20 × 10^{-3} × 0.8 × 1.6 = **25.6mA/V**, r_o = 1/(20 × 10^{-4} (.8)²) = **781Ω**, and μ = 25.6 × 10^{-3} × 781 = **20.0V/V**.

For υ_{GS} = 0V, g_m = 2 (10^{-4}) × 100 (0 – –1) (1.6) = 20 × 10^{-3} (1) (1.6) = **32.0mA/V**, r_o = 1/(20 × 10^{-4} (1.0)²) = **500Ω**, and μ = 32.00 × 500 = **16.0V/V**.

5.61

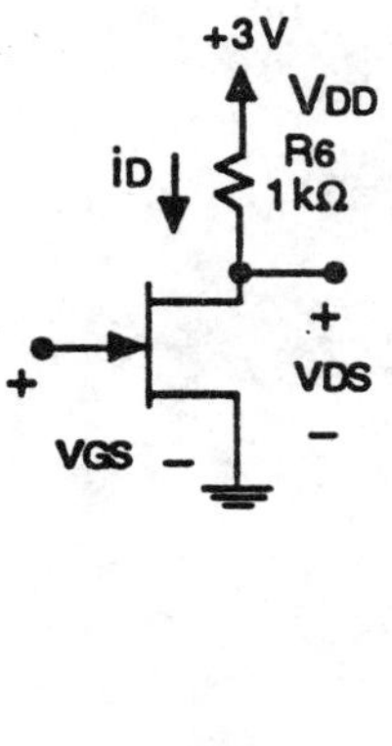

β = 100 × 10^{-4}A/V² = 10mA/V² for a 100μm device. From Eq 5.107, $i_D = \beta(\upsilon_{GS} - V_t)^2(1 + \lambda\upsilon_{DS})$ assuming operation is in saturation, and $\upsilon_{DS} = V_{DD} - i_D R_L$, $i_D = (V_{DD} - \upsilon_{DS})/R_L$. For υ_{GS} = **+0.2V**, (3 – υ_{DS})/0.1 = 10 (0.2 – –1)² (1 + 0.2 υ_{DS}), or 3 – υ_{DS} = 1 (1.2)² (1 + 0.2 υ_{DS}) = 1.44 + 0.288 υ_{DS}, and 1.288 υ_{DS} = 3 – 1.44 = 1.56, υ_{DS} = 1.56/1.288 = **1.211V**. Now this exceeds (0.2 – – 1.0) = 1.2V, OK. For υ_{GS} = **–0.2V**, (3 – υ_{DS})/0.1 = 10 (–0.2 – –1)² (1 + 0.2 υ_{DS}), or 3 – υ_{DS} = 0.64 (1 + 0.2 υ_{DS}) = 0.64 + 0.128υ_{DS}, 1.128υ_{DS} = 2.36, υ_{DS} = 2.36/1.128 = **2.092V**.

For $\upsilon_{GS} = 0$V, $3 - \upsilon_{DS} = 1\,(1)^2\,(1 + 0.2\upsilon_{DS}) = 1 + 0.2\upsilon_{DS}$, $1.2\upsilon_{DS} = 2$, and $\upsilon_{DS} =$ **1.67V**.

Voltage gains: For $\upsilon_{GS} = 0.2$V to -0.2V, "gain" = (1.211 – 2.092)/(0.2 – –0.2) = **–2.2V/V**. For $\upsilon_{GS} = 0.0$V to –0.2V, gain = (1.67 – 2.092)/(0 – – 0.2) = **–2.13V/V**.

5.62 Now, $\beta_1 = \beta_2 = 10\text{mA/V}^2$. Assume that the dc output is stabilized at half the supply voltage. That is, $\upsilon_{DS1} = \upsilon_{DS2} = 5$V. Now, $I_{D2} = \beta_2\,(\upsilon_{GS2} - V_t)^2\,(1 + \lambda\,\upsilon_{DS2}) = 10\,(0 - 1)^2\,(1 + 0.1 \times 5) = 15$mA, and $I_{D1} = I_{D2} =$ **15mA**, with $\upsilon_{GS1} =$ **0V** as well. Thus $g_{m1} = 2\,(10)\,(0 - -1)\,(1 + 0.1 \times 5) =$ **30mA/V**, $r_{01} = 1/(0.1\,(10)\,(0 - -1)^2) = 1\text{k}\Omega$, $r_{02} = 1/(0.1\,(10)\,(0 - -1)^2) = 1\text{k}\Omega$, $A_\upsilon = -30\,(1\text{k}\Omega \,||\, 1\text{k}\Omega) =$ **–15V/V**.

5.63 For $V_O = +3$V, $\upsilon_{DS2} = 10 - 3 = 7$V, $I_{D2} = 10\,(0 - 1)^2\,(1 + 0.1\,(7)) = 17\text{mA} = I_{D1}$. Now for Q_1: $17 = 10\,(\upsilon_{GS1} - -1)^2\,(1 + 0.1(3))$, or $(\upsilon_{GS1} + 1)^2 = 17/(10\,(1 + .3)) = 1.308$, $\upsilon_{GS1} = \pm\,1.144 - 1 = -2.144$ (cutoff) or **+.144V**. Now $g_{m1} = 2 \times (10)\,(.144 - -1)\,(1 + 0.1(3)) =$ **29.7mA/V**, $r_{01} = 1/(0.1\,(10)\,(1.144)^2) = 0.764\text{k}\Omega$, $r_{02} = 1/(0.1\,(10)\,(0 - -1)^2) = 1\text{k}\Omega$. Thus, the gain: $= -\,g_m\,r_{01} \,||\, r_{02} = -29.7\,(1\text{k}\Omega \,||\, 0.764\text{k}\Omega) =$ **–12.9V/V**.

General Problems

5.64 Devices shown are depletion-mode MOS, for which $I_{D1} = I_{DSS}\,(1 - \upsilon_{GS}/V_p)^2 = 10\text{mA} = I_{D2}$. From symmetry, see $\upsilon_{DS2} = \upsilon_{DS1} = 5$V and $V_O =$ **0V**. ∴ current in 10kΩ = 0mA, and bias is as assumed. ∴ $g_{m1} = (2\,I_{DSS}/(-V_p))\,(1 - \upsilon_{GS}/V_p) = 2(10)/2 = 10$mA/V, $r_{01} = V_A/I_D = 50\text{V}/10\text{mA} = 5\text{k}\Omega = r_{02}$, gain $= -\,g_m\,(r_{01} \,||\, r_{02} \,||\, R_L) = -10\,(5\text{k}\Omega \,||\, 5\text{k}\Omega \,||\, 10\text{k}\Omega) = -\,10\,(10\text{k}\Omega/5) = -20$V/V. Thus for $\upsilon_i = 0.1\sin\omega t$, $\upsilon_o =$ **–20 (0.1) sin ωt = –2 sin ωt volts.**

Chapter 6

DIFFERENTIAL AND MULTISTAGE AMPLIFIERS

SECTION 6.1: The BJT Differential Pair

6.1 Eq.6.7, 6.8: $i_{E1} = \dfrac{I}{1+e^{(v_{B2}-v_{B1})/V_T}} = \dfrac{I}{1+e^{-v_d/V_T}}$, $i_{E2} = \dfrac{I}{1+e^{(v_{B1}-v_{B2})/V_T}} = \dfrac{I}{1+e^{v_d/V_T}}$,

(a) $i_{C2} = \alpha\, i_{E2} \approx i_{E2} = 0.99I$, when $\dfrac{1}{1+e^{v_d/V_T}} = 0.99$, or $e^{v_d/V_T} = 1/0.99 - 1 = 0.0101$, or $v_d = V_T \ln .0101$ $= -4.595V_T =$ **115mV**. That is, v_{B1} must be lower than v_{B2} by 115mV.

(b) $i_{C1} = \alpha\, i_{E1} \approx i_{E1} = 0.95I$, when $\dfrac{1}{1+e^{-v_d/V_T}} = 0.95$, or $e^{-v_d/V_T} = 1/0.95 - 1 = .05263$, or $-v_d/V_T = -2.94$, or $v_d = 2.94\ V_T =$ **73.6mV**. That is, v_{B1} must be higher than v_{B2} by 73.6mV.

(c) For $i_{C1} = 9.0\ i_{C2}$ with $i_{C1} + i_{C2} = I$, $I - i_{C2} = 9i_{C2}$, or $i_{C2} = I/10 = 0.1I$, for which $i_{C1} = 0.9I$. Therefore, $\dfrac{I}{1+e^{-v_d/V_T}} = 0.9I$, $e^{-v_d/V_T} = 1/0.9 - 1 = 0.1111$, $-v_d/V_T = -2.197$. That is, $v_d =$ **54.9mV**.

6.2

Case	υ_{B1} V	υ_{B2} V	$\upsilon_{E1,2}$ V	υ_{C1} V	υ_{C2} V
a	0	**0**	–0.7	**6**	6
b	2	2	**1.3**	6	**6**
c	**2.0**	1	1.3	**2**	**10**
d	–2	**1.0**	0.3	10	**2**
e	1	**3.5**	2.8	**10**	3
f	–4	–4	**–4.7**	**4**	8
g	**4.0**	0	+3.3	**3.5**	**10**
h	1	3.5	**2.8**	**10**	**3**

(a) υ_{B1} = 0V, υ_{E12} = –0.7V → υ_{B2} = – 0.7 + 0.7 = **0V** (or lower). υ_{C2} = 6V → υ_{RC2} = 10 – 6 = 4V, and i_{C2} = 4V/4kΩ = 1.0mA. Therefore i_{C1} = 2.0 – 1.0 = 1.0mA, and υ_{C1} = 10 – 4(1) = **6V**. For equal current split, $\upsilon_{B1} = \upsilon_{B2}$ = **0V**.

(b) $\upsilon_{B1} = \upsilon_{B2}$ = 2.0 υ → υ_{E12} = 2.0 – 0.7 = **1.3V**. Now υ_{C1} = 6V → i_{C1} = (10–6)/4 = 1mA, and i_{C2} = 2.0 – 1.0 = 1.0mA, and υ_{C2} = 10 – 4(1) = **6V**.

(c) υ_{E12} = 1.3V. Thus one of υ_{B1}, υ_{B2} = 1.3 + 0.7 = 2.0V. Therefore, υ_{B1} = **2.0V**. Now, since υ_{B2} = 1.0V, i_{C1} = 2.0mA, and υ_{C1} = 10 – 4(2) = **2V**. Also i_{C2} = 0mA, so υ_{C2} = **10V**.

(d) υ_{E12} = 0.3V. Thus one of υ_{B1}, υ_{B2} = 0.3 + 0.7 = 1.0V. Therefore υ_{B2} = **1.0V**, Q_2 conducts 2mA and υ_{C2} = 10 – 2(4) = **2V**, with υ_{C1} = 10V.

(e) υ_{E12} = 2.8V. Thus one of υ_{B1}, υ_{B2} must be 2.8 + 0.7 = 3.5V. Therefore υ_{B2} = **3.5V** and i_{C2} = 2mA. Thus υ_{C2} = 10 – 2(4) = 2V, possibly. Therefore Q_2 is saturated with υ_{C2} = 2.8 + 0.2 = 3.0V, with extra current flowing in the base of Q_2. But Q_1 is cut off and υ_{C1} = **+10V**.

(f) $\upsilon_{B1} = \upsilon_{B2}$ = –4.0 V → $\upsilon_{E1,2}$ = –4.0 – 0.7 = **–4.7V**, υ_{C2} = 8V → i_{C2} = (10–8)/4kΩ = 0.5mA. Thus i_{C1} = 2.0 – 0.5 = 1.5mA, and υ_{C1} = 10 – 1.5(4) = **4V**.

(g) υ_{E12} = 3.3V. Thus one of υ_{B1}, υ_{B2} is at 3.3 + 0.7 = **4.0V**. Thus υ_{B1} = 4.0V, with Q_1 conducting, Q_2 cut off, υ_{C2} = **10V**. For i_{C1} = 2mA, υ_{C1} = 10 – 2(4) = 2V. But υ_{E1} = 3.3V. Thus Q_1 is saturated with $\upsilon_{C1} = 3.3 + 0.2 =$ **3.5V**.

(h) $\upsilon_{E1,2} = 3.5 - 0.7 =$ **2.8V** with Q_1 cut off and Q_2 conducting. Thus $\upsilon_{C1} =$ **+10V**, and υ_{C2} possibly as low as $10 - 2(4) = 2V$. But $\upsilon_{E2} = 2.8V$. Thus $\upsilon_{C2} = 2.8 + 0.2 =$ **3.0V**.

6.3

Case	I mA	υ_{B1} V	υ_{B2} V	υ_E V	υ_{C1} V	υ_{C2} V
a	0.2	0.00	**0.00**	–0.700	**9.60**	9.60
b	0.2	0.01	0.00	**–0.695**	**9.52**	**9.68**
c	0.2	0.00	0.05	**–0.664**	**9.90**	**9.30**
d	0.2	**0.037**	0.00	–9.675	**9.35**	**9.85**
e	0.2	–1.00	**–1.05**	**–1.714**	**9.30**	9.90
f	2.0	–1.00	–1.00	**–1.758**	**6.00**	**6.00**
g	2.0	0.01	0.00	**–0.752**	**5.21**	**6.79**
h	2.0	0.00	0.05	**–0.722**	**9.05**	**2.95**
i	2.0	1.00	**0.951**	**0.228**	3.00	**9.00**

(a) $\upsilon_{C2} = 9.60V \rightarrow i_{C2} = (10.0 - 9.6)/4k = 0.1mA$, that is $i_{C2} = I - 0.1 = 0.2 - 0.1 = 0.1mA \rightarrow \upsilon_{C1} =$ **9.60V**. Since $i_{C1} = i_{C2}$, $\upsilon_{B1} = \upsilon_{B2} =$ **0.00V**. Note that for $i_C = 0.1mA$, $\upsilon_{BE} = 0.700V$.

(b) $\upsilon_d = \upsilon_{B1} - \upsilon_{B2} = .01 - .00 = 10mV$. Thus $i_{E2} = \dfrac{0.2}{1 + e^{10/25}} = \dfrac{0.200}{1 + 1.492} = .0803mA$, and $i_{E1} = 0.200 - .0803 = 0.1197mA$, $\upsilon_{BE1} = 0.700 + 25 \ln (0.1197/0.100) = 0.7045V$. Thus $\upsilon_E = 0.010 - .7045 =$ **–0.695V**, $\upsilon_{C1} = 10 - 4(.1197) =$ **9.52V**, and $\upsilon_{C2} = 10 - 4(.0803) =$ **9.68V**.

(c) $\upsilon_d = \upsilon_{B1} - \upsilon_{B2} = -0.05V$, $i_{E2} = \dfrac{0.2}{1 + e^{-50/25}} = 0.176mA$, $i_{E1} = .200 - .176 = 0.024mA$, $\upsilon_{C1} = 10 - 4(.024) =$ **9.90V**, $\upsilon_{C2} = 10 - 4(.176) =$ **9.30V**, $\upsilon_{BE2} = 0.700 + 25 \ln (0.176/0.1) = .714V$. Thus $\upsilon_E = \upsilon_{B2} - \upsilon_{BE2} = .050 - .714 =$ **–0.664V**.

(d) Assuming $\upsilon_{B1} > 0$ and $i_{E2} < 0.1$mA, $i_{E2} = 0.1\ e^{(675-700)/25} = 0.0368$mA, $i_{E1} = 0.200 - 0.0368 = 0.1632$mA, $\upsilon_{BE1} = 700 + 25\ \ln(0.1632/0.100) = 712.2$mV. Thus $\upsilon_{B1} = 712.2 - 675 = $ **+ 0.037V**, and $\upsilon_{C1} = 10 - 4(.1632) = $ **9.35V**, $\upsilon_{C2} = 10 - 4(.0368) = $ **9.85V**.

(e) $i_{C2} = (10 - 9.90)/4 = 0.025$mA. Thus $i_{C1} = .200 - .025 = 0.175$mA, and $\upsilon_{C1} = 10 - 4\ (.175) = $ **9.30V** See from (c) that $\upsilon_{BE\,2} = 0.664$V, and $\upsilon_{BE\,1} = 0.664\text{V} + .050$V, and since $\upsilon_{B1} = -1.00$V, $\upsilon_{B2} = -1.00 - .050 = $ **−1.05V**, and $\upsilon_E = -1.05 - .664 = $ **−1.714V**.

(f) Inputs equal: current splits equally and $i_{C1} = i_{C2} = 2.0/2 = 1.0$mA, $\upsilon_{C1} = 10 - 4(1) = $ **6.00V**, $\upsilon_{C2} = 10 - 4(1) = $ **6.00V**. Thus $\upsilon_{BE} = 700 + 25\ \ln(1.00/0.1) = 757.6$mV, and $\upsilon_E = -1.00 - .758 = $ **−1.758V**.

(g) $\upsilon_d = \upsilon_{B1} - \upsilon_{B2} = 10$mV. Thus $i_{E2} = \dfrac{I}{1 + e^{\upsilon_D/\upsilon_T}} = \dfrac{2.0}{1 + e^{10/25}} = \dfrac{2.0}{2.492} = 0.803$mA, and $i_{E1} = 2.00 - .803 = 1.197$mA, $\upsilon_{BE\,2} = 700 + 25\ \ln(0.803/0.100) = 752$mV, $\upsilon_E = 0 - .752 = $ **−0.752V**, $\upsilon_{C1} = 10 - 4\ (1.197) = $ **5.21V**, $\upsilon_{C2} = 10 - 4\ (0.803) = $ **6.79V**.

(h) $\upsilon_d = \upsilon_{B1} - \upsilon_{B2} = 0 - .05 = -0.05\text{V} = -50$mV. Now $i_{E2} = \dfrac{2.0}{1 + e^{-50/25}} = 1.762$mA, and $\upsilon_{BE\,2} = 700 + 25\ \ln(1.762/0.1) = 772$mV, that is $\upsilon_E = \upsilon_{B2} - \upsilon_{BE\,2} = 50 - 772 = -\ 722\text{mV} = $ **−0.722V**. Now $i_{E1} = 2.0 - 1.762 = 0.238$mA, and $\upsilon_{C1} = 10 - 4\ (.238) = $ **9.05V**, and $\upsilon_{C2} = 10 - 4\ (1.762) = $ **2.95V**.

(i) $\upsilon_{C1} = 3.00V \rightarrow i_{C1} = (10 - 3)/4 = 1.75$mA. Thus $i_{C2} = 2.00 - 1.75 = 0.25$mA, $\upsilon_{C2} = 10 - 4\ (.25) = $ **9.00V**, $\upsilon_{BE\,1} = 700 + 25\ \ln(1.75/0.1) = 771.6$mV, $\upsilon_{BE\,2} = 700 + 25\ \ln(0.25/0.1) = 722.9$mV. Thus $\upsilon_E = 1.00 - .772 = $ **0.228V**, and $\upsilon_{B2} = .228 + .723 = $ **0.951V**.

SECTION 6.2: Small-Signal Operation of the BJT Differential Amplifier

6.4 Using $e^x \approx 1 + x + x^2/2$ in Eq. 6.11: $i_{C1} = \dfrac{\alpha I\ e^{\upsilon_d/2V_T}}{e^{\upsilon_d/2V_T} + e^{-\upsilon_d/2V_T}} = \alpha I \left[1 + \dfrac{\upsilon_d}{2\,V_T} + \dfrac{\upsilon_d^2}{8\,V_T^2}\right]$

$$\Big/\left[1 + \frac{\upsilon_d}{2\,V_T} + \frac{\upsilon_d^2}{8\,V_T} + 1 - \frac{\upsilon_d}{2\,V_T} + \frac{\upsilon_d^2}{8\,V_T}\right] = \frac{\alpha I}{2}\left[1 + \frac{\upsilon_d}{2\,V_T} + \frac{\upsilon_d^2}{8\,V_T^2}\right] \Big/ \left[1 + \frac{\upsilon_d^2}{8\,V_T^2}\right]$$

$$\approx \frac{\alpha I}{2}\left[1 + \frac{\upsilon_d}{2\,V_T} + \frac{\upsilon_d^2}{8\,V_T^2}\right] \times \left[1 - \frac{\upsilon_d^2}{8\,V_T^2}\right] = \frac{\alpha I}{2}\left[1 + \frac{\upsilon_d}{2\,V_T} + \frac{\upsilon_d^2}{8\,V_T^2} - \frac{\upsilon_d^2}{8\,V_T^2} - \frac{\upsilon_d^3}{16\,V_T^3} - \frac{\upsilon_d^4}{64\,V_T^4}\right]$$

$$= \frac{\alpha I}{2}\left[1 + \frac{\upsilon_d}{2\,V_T} - \frac{\upsilon_d^3}{16\,V_T^3} - \frac{\upsilon_d^4}{64\,V_T^4}\right] = \frac{\alpha I}{2} + \frac{\alpha I}{2\,V_T}\left[\frac{\upsilon_d}{2}\right]\left[1 - \frac{1}{2}\left[\frac{\upsilon_d}{2\,V_T}\right]^2 - \frac{1}{8}\left[\frac{\upsilon_d}{2\,V_T}\right]^3\right] \text{ - - - (1).}$$

Now for $\upsilon_d/2 = 10$mV, $\left[1 - \dfrac{1}{2}\left[\dfrac{\upsilon_d}{2\,V_T}\right]^2 - \dfrac{1}{8}\left[\dfrac{\upsilon_d}{2\,V_T}\right]^3\right] = 1 - \dfrac{1}{2}\left[\dfrac{10}{25}\right]^2 - \dfrac{1}{8}\left[\dfrac{10}{25}\right]^3 = 1 - .08 - .008 = 0.912$.

That is, we see that the higher-order approximation implies a reduction in output current by about **9%** from that derived from the linear one.

Alternatively (and directly) at $\upsilon_d/2 = 10$mV: $i_{C1} = \dfrac{\alpha I}{2}\ \dfrac{2\,e^{10/25}}{e^{10/25} + e^{-10/25}} = \dfrac{\alpha I}{2}\left[\dfrac{2\,(1.492)}{1.492 + 0.670}\right] = 1.380\ \dfrac{\alpha I}{2}$, of which the signal part is $0.380\ \dfrac{\alpha I}{2}$, whereas from Equation 6.12: $i_{C1} = \dfrac{\alpha I}{2}\left[1 + \dfrac{\upsilon_d/2}{V_T}\right] = \dfrac{\alpha I}{2}\left[1 + \dfrac{10}{25}\right] = 1.400\ \dfrac{\alpha I}{2}$ of which the signal part is $0.400\ \dfrac{\alpha I}{2}$. Thus the linear approximation produces a result which is high by .400 −.380/.380 = .053, or about **5%**.

For specified errors, using the result (1) above, but only the term in υ_d^2, we see that the error is about $\frac{1}{2}\left[\frac{\upsilon_d}{2}/V_T\right]^2 = \varepsilon$. Now for $\varepsilon = 10\%$, $\frac{\upsilon_d}{2}/V_T = (2\,(0.1))^{1/2} = .447$, and $\upsilon_d/2 = 0.447\,(25) =$ **11mV**. For 5%, $\upsilon_d/2 = (2\,(.05))^{1/2}\,(25) =$ **7.9mV**. For 1%, $\upsilon_d/2 = (2\,(.01))^{1/2}\,(25) =$ **3.5mV**.

Check with the original (Eq. 6.11): $i_{C1} = \frac{\alpha I}{2}\left[\frac{2\,e^{\upsilon_d/2V_T}}{e^{\upsilon_d/2V_T} + e^{-\upsilon_d/2V_T}}\right]$. For $\upsilon_d/2 = 3.5\text{mV}$,

$i_{C1} = \frac{\alpha I}{2}\left[\frac{2e^{3.5/25}}{e^{3.5/25} + e^{-3.5/25}}\right] = \frac{\alpha I}{2}\,\frac{2\,(1.150)}{1.150 + .869} = 1.139\,\frac{\alpha I}{2}$, whereas, from Eq. 6.12: $i_{C1} = \frac{\alpha I}{2}\,(1 + \upsilon_d/2\,V_T) = \frac{\alpha I}{2}\,91 + 3.5/25) = \frac{\alpha I}{2}\,(1.140)$. Thus the error is $(0.140 - 0.139)/0.139 \equiv 0.7\%$.

6.5 For a differential input of 0V, $i_{C1} = i_{C2} = I/2 = 100\mu A$, and $\upsilon_{C1} = 3 - 0.1\,(10) = 2V = \upsilon_{C2}$. For $I_E = 100\mu A$, $r_e = \frac{V_T}{I_E} = \frac{25mV}{0.1mA} = 250\Omega$. For differential output, $\frac{\upsilon_{od}}{\upsilon_d} = \frac{\alpha\,(10k + 10k)}{250 + 250} \simeq$ **40V/V**. For outputs individually, the gain magnitude is **20V/V**. For $\upsilon_{CB} = -0.4V$ and very small signals, $\upsilon_{C1} = \upsilon_{C2} = 2V$, and $\upsilon_{B1} = \upsilon_{B2} = 2.0 - -0.4 = 2.4V$. That is, the upper limit of the input range is **+2.4V**.

6.6 For $I = 200\mu A$, $r_e = 25mV/100\mu A = 250\Omega$. For differential output, gain is $\alpha\,\frac{R_L}{r_e} \simeq \frac{100k\Omega}{0.25k\Omega} =$ **400V/V**. Differential input resistance is $(\beta + 1)\,(r_e + r_e) = 151\,(.25 + .25) =$ **75.5kΩ**. To double the input resistance, add 0.25kΩ resistors in series with each emitter, at which point the gain (for differential output) is (100k+100k)/(250+250+250+250) = **200V/V**.

6.7

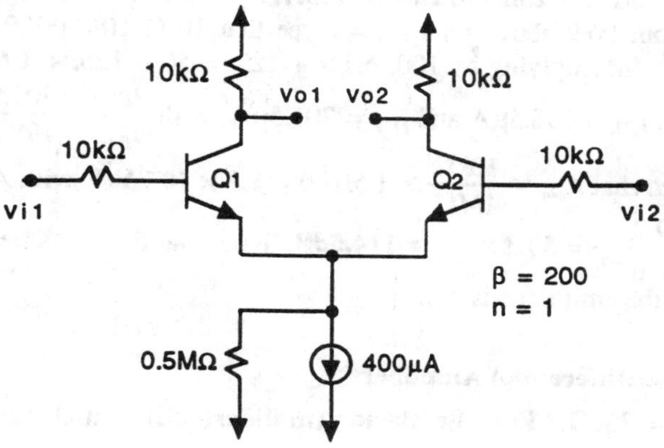

For each transistor, $I_E = 200\mu A$ and $r_e = 25mV/0.2mA = 125\Omega$. Thus the differential input resistance is 2 (201) (.125kΩ) = **50.25kΩ**.

For differential output: Differential gain from bases = 200/201 (10k/.125k) = 79.6V/V. Differential gain from input sources = 50.25/(10+10+50.25) × 79.6 = **56.94V/V**. Common-mode gain = **0V/V**. CMRR = 56.9/0 = ∞, as ratio and in dB. Common-mode input resistance = $(\beta + 1)\,(R) = 201\,(0.5M\Omega) \simeq$ **100MΩ**.

For single-ended output: $|A_{ds}| = \frac{200}{201}\,\frac{10k\Omega}{0.125 + 0.125} \times \frac{50.25}{70.25} = \frac{56.94}{2} =$ **28.5V/V**, $A_{cm} = -\frac{200}{201}\,\frac{10k\Omega}{2\,(0.5M\Omega)} =$ **−.00995V/V**, CMRR = 28.5/.00995 = **2864V/V ≡ 69.1dB**.

6.8 From P6.7 above, for outputs taken differentially $A_d = 56.9V/V$. For A_{cm}: **For matched loads,** it is 10kΩ/1MΩ − 10kΩ/1MΩ = 0V/V. **For ± 1% loads,** it is 10k(1.01)/1M − 10k(.99)/1M = .02(10k)/1M = **2 × 10 $^{-5}$V/V ≡ −94dB**, for which CMRR = 56.9/(2 × 10^{-5}) = **2.85 × 10 ^{6}V/V ≡ 129dB**. **For ± 10% loads,** correspondingly, A_{cm} = **2 × 10^{-4}V/V ≡ −74dB**, and CMRR = **2.85 × 10^{5}V/V ≡ 109dB**

6.9

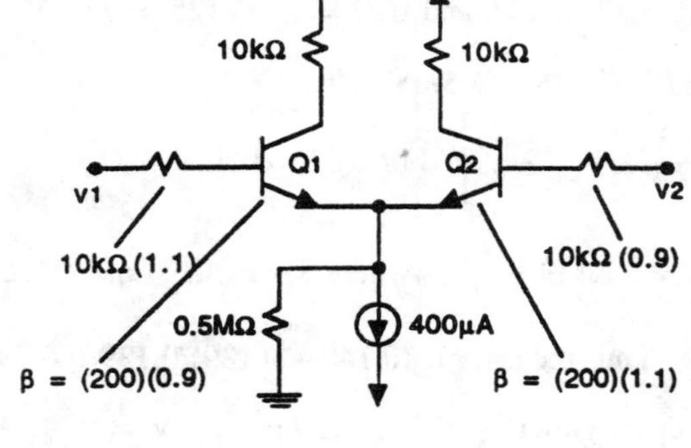

Note that the collector resistors are ideal (that is both are exactly 10kΩ).

For DC Bias: Assume that junction voltages are adequately modelled by $r_{e1}, r_{e2} \approx$ 25mV/200μA = 0.125kΩ nominally. Now for i in Q_1 and $0.400 - i$ in Q_2, $10\,(1.1)\,\frac{i}{200\,(.9)} + 0.125\,i = 0.125\,(0.4 - i) + 10\,(.9)\,\frac{(.4 - i)}{200\,(1.1)}$. Now multiplying by 100: $6.11\,i + 12.5i = 5 - 12.5i + 1.636 - 4.09i$, $35.2i = 6.636$, $i = 0.1885$. That is i_{E1} = 188.5μA, and i_{E2} = 211.5μA.

For signals: Now r_{e1} = 25mV/188.5μA = 132.6Ω, r_{e2} = 25/211.5 = 118.2Ω, $r_{\pi 1}$ = ((200) (.9) +1) 132.6 = 24.0kΩ, $r_{\pi 2}$ = (200 (1.1) +1) 118.2 = 26.1kΩ. Now for common-mode input, base currents split, with 26.1/(24.0 + 26.1) = 0.52 of the input change in the base of Q_1 (and 0.48 in the base of Q_2). Correspondingly, for a total base current i, $i_{C1} = \beta_1\, i_{b1}$ = (200) (0.9) (0.52)i = 93.6i, and $i_{C2} = \beta_2\, i_{b2}$ = (200) (1.1) (.48)i = 105.6i. Thus i_{C2}/i_{C1} = 105.6/93.6 = 1.128. **Note** that the dc current ratio is 211.5/188.5 = 1.122, essentially the same. From either point of view, there will be about a 12% total mismatch in the two output voltages. $A_{cm} \approx \frac{10k\Omega}{2\,(500k\Omega)} \times \frac{12}{100} = \mathbf{12 \times 10^{-4}V/V}$. Since $A_d \approx \frac{10k\Omega + 10k\Omega}{.133 + .118}$ = **79.7V/V**, CMRR = $\frac{79.7}{12 \times 10^{-4}} = \mathbf{66.4 \times 10^3 V/V \equiv 96.4dB}$.

6.10 For I_{bias} = 400μA, $I_{E1} = I_{E2}$ = 200μA nominally, and r_e = 25mV/0.2mA = 125Ω, R_E = 9 (125) = **1.125kΩ**, and $R_E + r_e$ = 1.25kΩ. Using the dc analysis from P6.9 above with $i_{E1} = i$, see that 10 (1.1i)/(200(.9)) + 1.25i = 1.25(0.4 –i) + 10(0.9) (0.4 – i)/(200(1.1)). Multiplying by 100, 6.11i + 125i = 50 – 125i + 1.64 – 4.09i, or 260.2i = 51.64, and i = 0.1985. That is, i_{E1} = 198.5μA and i_{E2} = 201.5μA, with $\frac{i_{E2}}{i_{E1}} = \frac{201.5}{198.5}$ = 1.015. Thus there is about a 1.5% mismatch, such that $A_{cm} \approx \frac{10k\Omega}{1M\Omega} \times 1.5/100 = \mathbf{1.5 \times 10^{-4}V/V}$ with A_d = **79.9V/V**, the same as before, and CMRR = $\frac{79.7}{1.5 \times 10^{-4}} = 53.3 \times 10^4 \equiv$ **114.5dB**. Note that there is a nearly 20dB improvement due to the balancing effect of the emitter resistors.

SECTION 6.3: Other Non-ideal Characteristics of the Differential Amplifier

6.11 For I_{bias} = 200μA, $r_{e1} = r_{e2}$ = 25mV/0.1mA = 250Ω. **For the Basic Amplifier**, differential gain = $\frac{R_C + R_C}{0.25k\Omega + 0.25k\Omega} = 4\,R_C$V/V. Now ±5% variation in R_C produces an output offset of 0.1mA (1.05R_C – 0.95R_C) = 0.1 (0.1)R_C = 0.01R_C. Corresponding input offset (to reduce output to zero) is $V_{OS} = V_O$/gain = $\frac{.01\,R_C}{4\,R_C}$ = **2.5mV**, or from equation 6.49: $|V_{OS}| = V_T \left(\frac{\Delta R_C}{R_C}\right)$ = 25mV (2 (5/100)) = **2.5mV**. **For emitter resistors of $R_E = 9r_e$:** Here, the differential gain = $\frac{R_C}{0.25\,(1 + 9)} = 0.4R_C$. Now to compensate an output offset of .01R_C, we need $V_{OS} = \frac{.01R_C}{0.4R_C}$ = **25mV**.

6.12 From P6.11 above, to compensate for ±5% R_C variation, one needs a 2.5mV input offset with no emitter resistors. This involves an increase in one of the collector currents to 105μA and decrease in the other to

95μA. Now with $R_E = 9\,r_e$ = 9 (250) = 2250Ω nominally, but actually ranging from 0.95(2250) = 2.1375kΩ, to 1.05(2250) = 2.3625kΩ, equivalent offset can reach .105(2.3625) –.095(2.1375) = .2481 – .2031 = 45mV. Total maximum offset is approximately 2.5 + 45 = **47.5mV**.

For uncorrelated variation: For nominal R_E (from P6.11 above), acquire 25mV due to R_C variation. For nominal R_C and varying R_E, to achieve 100μA in each transistor, we need an offset of 0.1(2.3625) – 0.1(2.1375) = 22.5mV. (See the worst case is again 22.5 + 25 = 47.5mV, as an alternative approach). **For no correlation,** $V_{OS} = (22.5^2 + 25^2)^{1/2}$ = **33.6mV**. **For collector resistors trimmed**: Collector currents will be both 100μA and, as above, V_{OS} = **22.5mV**.

6.13 **For equal 2mV offsets:** $V_{OS} = (2^2 + 2^2 + 2^2 + 2^2)^{1/2} = (16)^{1/2}$ = **4mV**.

For unequal offsets: $V_{OS} = (0.5^2 + 1^2 + 2^2 + 4^2)^{1/2} = (.25 + 1 + 4 + 16)^{1/2} = (21.25)^{1/2}$ = **4.61mV**.

6.14 For the offset totally compensated, the collector currents in both transistors will be 100/2 = 50μA. Assume β_1 is 5% high and β_2 is 5% low, while R_{S1} is 5% low and R_{S2} is 5% high. Thus the total offset is $I_{B1}R_{S1} - I_{B2}R_{S2}$ = 50/(105 + 1) × 100 (.95) – 50 /(95 + 1) × 100 (1.05) or $\frac{50\times10^{-6}\times100\times10^3}{100}\left(\frac{.95}{1.06} - \frac{1.05}{.96}\right) = 50\times10^{-3}$ (.896 – 1.094) = -50×10^{-3} (.198) = –9.9mV. Thus the offset can be as large as **9.9mV**.

6.15 For each transistor, r_e = 25mV/150μA = .1667kΩ. For v_{be} = 10mV, v_c = 60kΩ/.1667kΩ × 10mV = 3.6V. Thus, the lowest collector voltage is 15V – 150μA × 60kΩ –3.6V, or 15 – 9 – 3.6 = 2.4V. For bare saturation, the base voltage can exceed this by 0.7 – 0.4 = 0.3V. Thus, the highest usable common-mode input is 2.4 + 0.3 = **2.7V**.

SECTION 6.4: Biassing in BJT Integrated Circuits

6.16 As seen from the emitter, I_E = 100μA and r_e = 25mV/100μA = 250 Ω. Thus the resistance between the terminals is 250Ω.

For two in parallel, the current divides (say equally) with each r_e = 25mV/50μA = 500Ω. The parallel resistance is then 500Ω || 500Ω = **250Ω**, as before. One can see this directly since the junctions are bigger, but the current is the same.

For two in series, the current in each is the same; the resistance of each is the same, and the total resistance is 250 + 250 = **500Ω**.

6.17 From Eq.6.63, $\frac{I_O}{I_{REF}} = \frac{1}{1 + 2/\beta}$. For 1% error, $\frac{1}{1+2/\beta}$ = 0.99, 1 = 0.99 + 1.98/β, β = 1.98/.01 = **198**. For 0.1% error, $\frac{1}{1+2/\beta}$ = 0.999, $\beta = \frac{1.998}{1 - .999}$ = **1998**.

6.18 At 1mA, V_{BE} = 700 + 25 ln (1mA/10mA) = 642.4mV.

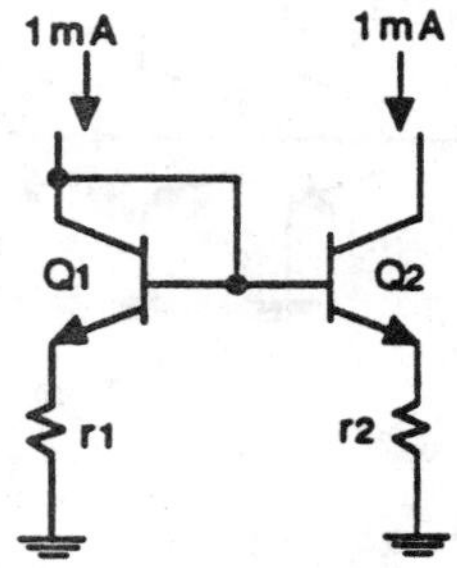

Required r = (642.4mV)(0.1)/1mA = 64.2Ω. Use r = **60Ω**. **Now for β = 90, I_R = 1mA** and I_O = 1mA, I_{B2} = 1/90, I_{E1} = (1 – 1/90) = 0.9889, and I_{C1} = 0.9780mA, I_{E2} = 91/90(1) = 1.0111, and I_{C2} = 1.0000mA. V_{BE1} = 700 + 25 ln (0.9780/10) = 641.9mV, V_{BE2} = 700 + 25 ln (1/10) = 642.4mV, V_{r1} = 60 (.9889) = 59.3mV, V_{r2} = 60 (1.0111) = 60.67mV, $V_{BE1} + V_{r1}$ = 641.9 + 59.3 = 701.2mV. Thus r_2 must be adjusted to $\frac{701.2 - 642.4}{1.0111}$ = **58.15Ω**. **Now for β = 90, and at 0.5mA**: $I_{B2} \approx$ 0.5/90, and $I_{E1} \approx$ 0.5 – 0.5/90 = 0.5 (.9889) = .4944mA, and I_{C1} = 90/91 (.494) = .489mA, and $V_{BE1} + V_{r1}$ = 700 + 25 ln (0.489/10) + .4944 (60) =

654.2mV. Assume $I_{C2} \approx 0.5$mA $\rightarrow V_{BE2} = 700 + 25 \ln (0.5/10) = 625.1$mV. Thus $I_{E2} \approx \dfrac{654.2 - 625.1}{58.15} =$ 1.5007mA, and $I_{C2} = 90/91\ (.509) = 0.4952$mA. Gain = .4952/.5000 = **0.990A/A.**

Now at 2.0mA: $I_{B2} \approx 2/90$, $I_{E1} = 2 - 2/90 = 2\ (.9889) = 1.9778$mA, and $I_{C1} = 90/91\ (1.9778) = 1.9561$mA, and $V_{BE1} + V_{r1} = 700 + 25 \ln (1.9561/10) + 1.9778\ (60) = 777.9$mV. Assume $I_{C2} \approx 2$mA $\rightarrow V_{BE2} = 700 +$ $25 \ln (2/10) = 659.8$mV, $I_{E2} = \dfrac{777.9 - 659.8}{58.15} = 2.032$mA, $I_{C2} = 90/91\ (2.04) = 2.009$mA. Gain = 2.009/2.00 = **1.005A/A.**

Now for $\beta = 70$, and at 0.5mA: $I_{B2} \approx 0.5/70$, $I_{E1} \approx 0.5 - 0.5/70 = 0.5\ (.9857) = .4929$mA, $I_{C1} = 70/71$ (.4929) = .4859mA, $V_{BE1} + V_{r1} = 700 + 25 \ln (0.4859/10) + .4929\ (60) = 654.0$mV. Assume $I_{C2} \approx 0.5$mA $\rightarrow V_{BE2} = 700 + 25 \ln (0.5/10) = 625.1$mV, $I_{E2} = \dfrac{654.0 - 625.1}{58.15} = .4964$mA, $I_{C2} = 90/91\ (.4964) =$ .4909mA, gain = .4909/.5000 = **0.982A/A.**

And at 1.0mA; $I_{B2} \approx 1.0/70$, $I_{E1} \approx 1.0 - 1.0/70 = 1.0\ (.9857) = .9857$mA, $I_{C1} = 70/71\ (.9857) = .9718$mA, $V_{BE1} + V_{r1} = 700 + 25 \ln (0.9718/10) + 0.9857\ (60) = 700.9$mV. Assume $I_{C2} \approx 1.0 \rightarrow V_{BE2} = 700 + 25 \ln$ $(1/10) = 642.43$mV. Thus $I_{E2} = \dfrac{700.9 - 642.4}{58.15} = 1.005$mA, $I_{C2} = 70/71\ (1.005) = .991$, gain = .991/1.00 = **.991A/A.**

And at 2.0mA: $I_{B2} \approx 2.0/70$, $I_{E1} = 2.0\ (.9857) = 1.9714$mA, $I_{C1} = 1.9714\ (70/71) = 1.944$, $V_{BE1} + V_{r1} =$ $700 + 25 \ln (1.944/10) + 1.9714\ (60) = 777.3$. Assume $I_{C2} \approx 2.0$mA $\rightarrow V_{BE2} = 700 + 25 \ln (2/10) =$ 659.8mV. Thus $I_{E2} = \dfrac{777.3 - 659.8}{58.15} = 2.021$mA, $I_{C2} = 70/71\ (2.021) = 1.993$mA, gain = 1.993/2.0 = **.996A/A.**

6.19 $\dfrac{I_O}{I_R} = \dfrac{1}{1 + 2/\beta}$, $I_O = 100\mu A\ \dfrac{1}{1 + 2/150} = 98.684\mu$A. Thus I_O is low by 100 – 98.684 = 1.316μA, for which r_O = 150V/100μA = 1.5MΩ. To compensate, $\dfrac{V_{out}}{1.5 \times 10^6} = 1.316 \times 10^{-6}$, or $V_{out} = 1.5\ (1.316) =$ **1.974V**.

For a net error of <1%, output current must range from 99μA to 101μA.

At 99μA, r_o contributes 99 – 98.684 = 0.316μA, for which $V_{out} = .316 \times 1.5 =$ **.474V**.

At 101μA, r_o contributes 101 – 98.684 = 2.316μa, for which $V_{out} = 2.316 \times 1.5 =$ **3.474V**.

6.20 For a change from 25°C to 75°C, V_{BE} drops by (75 – 25)2 = 100mV. For 100mV to be a 5% change, the drop in R must be 100mV/.05 = 2V. That is, V_{CC} = **2.7V**, and R = 2/100μA = **20kΩ**.

6.21

(a)

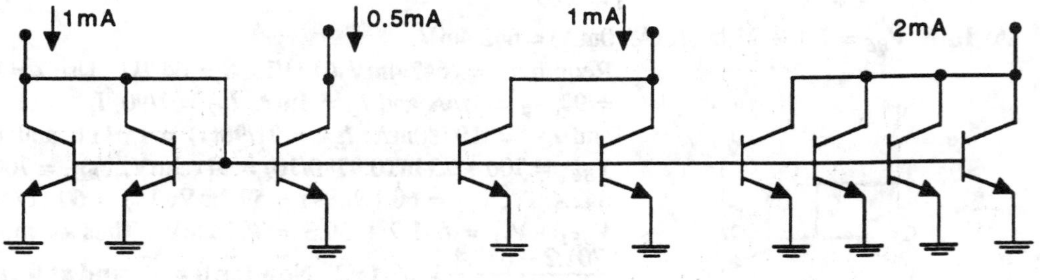

Need **9 BJTs**.

(b)

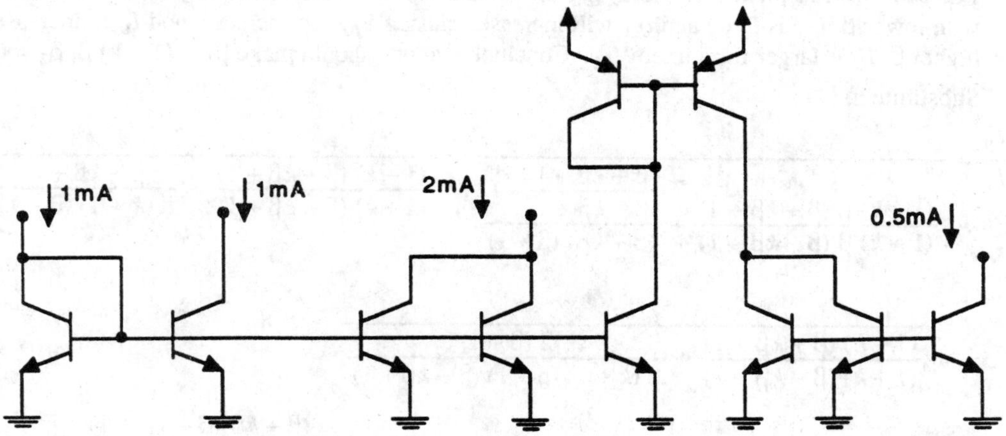

Need **10 BJTs.**

(c) For both ends of I_R available, T_A is not needed. Require **9 BJTs.**

6.22

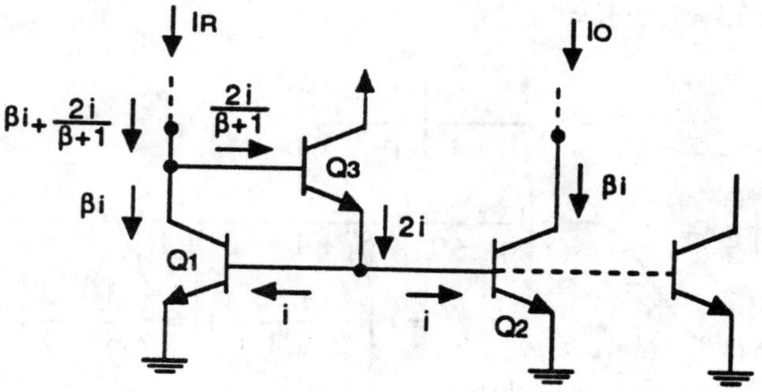

$$\frac{I_O}{I_R} = \frac{\beta i}{\beta i + \frac{2i}{\beta+1}} = \frac{\beta}{\beta + \frac{2}{\beta+1}}, \text{ or}$$

$$\frac{I_O}{I_R} = \frac{1}{1 + \frac{2}{\beta^2+\beta}}, \text{ as noted before.}$$

For two outputs, I_O: $I_R = \beta i + \frac{3i}{\beta+1}$, and $\frac{I_O}{I_R} = \frac{1}{1+\frac{3}{\beta^2+\beta}}$

6.23

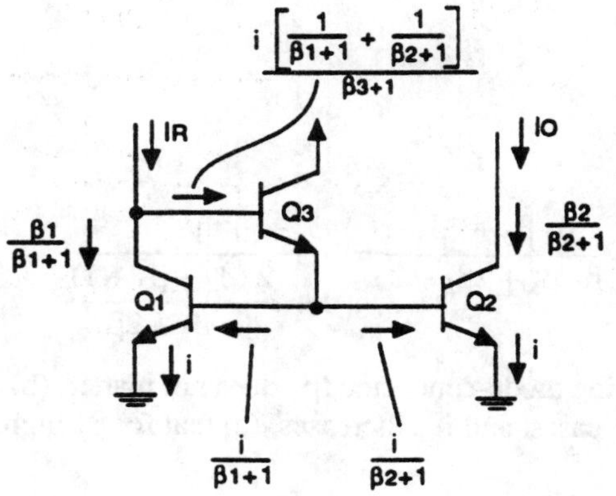

$$\frac{I_O}{I_R} = \frac{\frac{\beta_2}{\beta_2+1}\, i}{\frac{\beta_1}{\beta_1+1}\, i + i \left(\frac{1/(\beta_1+1) + 1/(\beta_2+1)}{\beta_3+1}\right)}$$

$$= \frac{1}{\frac{\beta_1(\beta_2+1)}{\beta_2(\beta_1+1)} + \frac{\frac{\beta_2+1}{\beta_1+1}+1}{\beta_2\,(\beta_3+1)}} \quad ---(1)$$

For the optimal location: See a) Q_3 provides a second-order effect, and any β would be OK. b) For Q_1 with lower β, i_{C1} is fixed and i_{E1} will increase, causing V_{BE1} to increase, and I_O to increase. c) For Q_2 with higher β, I_O is larger for a fixed V_{BE}. Conclude that one should make $\beta_1 = (1-\mathbf{k})\,\beta$, $\beta_2 = (1+\mathbf{k})\,\beta$, $\beta_3 = \beta$. Substitute in (1)

$$\frac{I_O}{I_R} = \frac{1}{\dfrac{(1-k)\,\beta\,(\beta+k\beta+1)}{(1+k)\,\beta\,(\beta-k\beta+1)} + \dfrac{\dfrac{\beta+k\beta+1}{\beta-k\beta+1}+1}{(\beta+k\beta)\,(\beta+1)}} = \frac{1}{\dfrac{(1-k)\,(\beta+k\beta+1)}{(1+k)\,(\beta-k\beta+1)} + \dfrac{\beta+k\beta+1+\beta-k\beta+1}{\beta\,(k+1)\,(\beta+1)\,(\beta-k\beta+1)}}$$

$$= \frac{1}{\dfrac{(1-k)\,(\beta+k\beta+1)}{(1+k)\,(\beta-k\beta+1)} + \dfrac{2\,(\beta+1)}{\beta\,(k+1)\,(\beta+1)\,(\beta-k\beta+1)}}$$

$$= \frac{1}{\dfrac{(1-k)\,(\beta+k\beta+1)}{(1+k)\,(\beta-k\beta+1)} + \dfrac{2}{\beta\,(1+k)\,(\beta-k\beta+1)}} = \frac{(\beta+k\beta)(\beta-k\beta+1)}{(\beta-k\beta)(\beta+k\beta+1)+2}$$

$$= \frac{\beta^2-k\beta^2+\beta+k\beta^2-k^2\,\beta^2+k\,\beta}{\beta^2+k\beta^2+\beta-k\beta^2-k^2\,\beta^2-k\,\beta+2} = \frac{\beta^2+\beta+k\beta-k^2\beta^2}{\beta^2+\beta-k\beta-k^2\beta^2+2}.$$

This becomes one, if $k\beta = 2 - k\beta$, or $k\beta = 1$, or $\mathbf{k} = 1/\beta$!

6.24

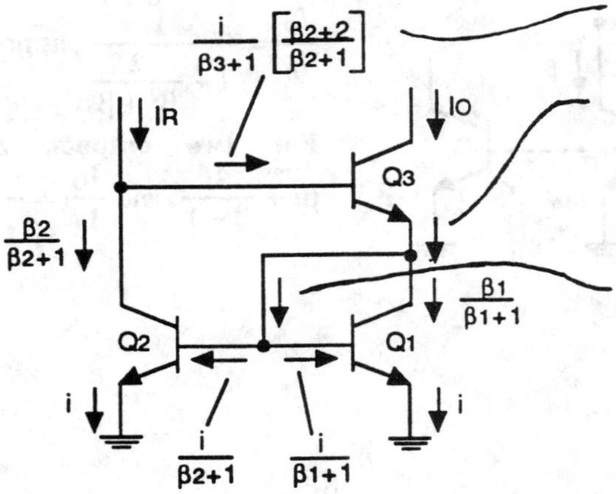

$$\frac{i}{\beta_3+1}\left(\frac{\beta_2+2}{\beta_2+1}\right)$$

$$\left[\frac{\beta_1}{\beta_1+1}+\frac{1}{\beta_1+1}+\frac{1}{\beta_2+1}\right] i = i\left[\frac{1}{\beta_2+1}+1\right] = \left[\frac{\beta_2+2}{\beta_2+1}\right] i$$

$$i\left[\frac{1}{\beta_2+1}+\frac{1}{\beta_1+1}\right]$$

$$\frac{\mathbf{I_O}}{\mathbf{I_R}} = \frac{\dfrac{\beta_3}{\beta_3+1}\left(\dfrac{\beta_2+2}{\beta_2+1}\right)}{\dfrac{1}{\beta_3+1}\left[\dfrac{\beta_2+2}{\beta_2+1}\right]+\dfrac{\beta_2}{\beta_2+1}} = \frac{\beta_3\,(\beta_2+2)}{\beta_2+2+\beta_2\,(\beta_3+1)} = \frac{\beta_2\,\beta_3+2\beta_3}{\beta_2\,\beta_3+2\beta_2+2} = \frac{1}{1+\dfrac{2\,(\beta_2-\beta_3+1)}{\beta_2\,\beta_3+2\beta_3}}$$

For optimal placement of transistors: See (a) that Q_1, being diode-connected, β_1 does not matter, (b) that for Q_2 with low beta, i_{C2} being fixed, i_{E2} increases, V_{BE} increases, and i_{E1} increases, (c) that for β_3 high, I_O

increases. Thus use $\beta_1 = \beta$, $\beta_2 = (1 - \mathbf{k})\,\beta$, $\beta_3 = (1 + \mathbf{k})\,\beta$. Thus $\dfrac{I_O}{I_R} = \dfrac{\beta_2\,\beta_3 + 2\beta_3}{\beta_2\,\beta_3 + 2\beta_2 + 2} = \dfrac{(1+k)\,(1-k)\,\beta^2 + 2\,(1+k)\,\beta}{(1+k)\,(1-k)\,\beta^2 + 2\,(1-k)\,\beta + 2} = \dfrac{\beta^2 - k^2\beta^2 + 2k\beta + 2\beta}{\beta^2 - k^2\,\beta^2 - 2k\beta + 2\beta + 2}$. This is unity if $2k\beta = -2k\beta + 2$, or $k\beta = ½$, or $\mathbf{k} = \dfrac{\mathbf{1}}{\mathbf{2\beta}}$.

6.25 Assume very large β: For the 100μA reference: $V_{BE} = 700 + 25\ \ln\,(0.1/1) = 642.4$mV. For the 1μA output: $V_{BE} = 700 + 25\ \ln\,(10^{-3}/1) = 527.3$mV. For the 10μA output: $V_{BE} = 700 + 25\ \ln\,(10^{-2}/1) = 584.9$mV. Thus, for 1μA output, $R_E = \dfrac{(642.4 - 527.3) \times 10^{-3}}{10^{-6}} =$ **115kΩ**. For 10μA output, $R_E = \dfrac{(642.4 - 584.9)\ 10^{-3}}{10^{-5}} =$ **5.75kΩ**

SECTION 6.5: The BJT Differential Amplifier with Active Loads

6.26 Each transistor conducts 50μA, for which r_o = 150/50μA = 3MΩ, and r_e = 25mV/50μA = 500Ω, G_m = 2/(500+500) = **2mA/V**, $A_\upsilon = -\dfrac{2}{500 + 500}\,(3M\Omega\,||\,3M\Omega) = -2 \times 10^{-3} \times \dfrac{3}{2} \times 10^6 =$ **−3000V/V**. $R_o = 3M\Omega\,||\,3M\Omega$ = **1.5MΩ**, R_{in} = (75 + 1) (500 + 500) = **76kΩ**. For a 76kΩ load, $A_\upsilon = -2 \times 10^{-3}\,(1.5M\Omega\,||\,76k\Omega) =$ **−144.7V/V**.

6.27 For each of the transistors, with an emitter resistor $R_E = r_e$, the output resistance increases to $R_o = r_o\,(1 + g_m\,R_E{}') \approx 3 \times 10^6\,(1 + 500/(500)) = 6$MΩ. Thus, the output resistance is approximately 6MΩ || 6MΩ = 3MΩ. The overall transconductance is 2/(500+500+500+500) = **1mA/V**. The open-circuit voltage gain is − 1mA/V × 3MΩ = **−3000V/V**.

6.28 I = 100μA → Collector current for all transistors is 50μA, for which r_o = 75V/50μA = 1.5MΩ, and r_e = 25mV/50μA = 500Ω, and r_π = (75 + 1) (.50) = 38kΩ. For the cascode transistors, $R_o = r_o\,(1 + g_m\,r_\pi) = r_o\,(\beta + 1) = 1.5 \times 10^6\,(76) = 114$MΩ. Also $r_\mu \approx 10\,(75)\,(1.5 \times 10^6) = 1125$MΩ. Thus the output resistance (in MΩ) is 114 || 114 || 1125 || 1125 = 114/2 || 1125/2 = **51.8MΩ**. Overall, $g_m = G_m \approx$ 2/(500+500) = **2mA/V**, A_υ = −2mA/V × 51.8MΩ = **103.5 × 10³V/V**.

SECTION 6.6: The JFET Differential Pair

6.29 For I_B = 1mA, I_D = 0.5mA for each half. Thus $i_D = 0.5 = 2\,(\,1 - \dfrac{\upsilon_{GS}}{V_P})^2$, $1 - \dfrac{\upsilon_{GS}}{V_P} = \sqrt{¼} = ½$, $V_{GS} = \dfrac{V_P}{2} =$ −2V. Thus for $\upsilon_{G1} = \upsilon_{G2}$ = 0V, υ_S = **+2V**. For each, $g_m = \dfrac{I_{DSS}(2)}{-V_P}\,(1 - \dfrac{\upsilon_{GS}}{V_P}) = \dfrac{2\,(2 \times 10^{-3})}{-4}\,(1 - \dfrac{-2}{-4}) =$ 0.5mA/V. For the pair, and input υ_d, $g_m = \dfrac{1}{1/0.5 + 1/0.5} =$ **.25mA/V**. For all the current in one transistor (say Q_1), $1 = 2\,(1 - \dfrac{\upsilon_{GS}}{V_P})^2$ or $\upsilon_{GS\,1} = V_P\,(1 - \sqrt{½}) = 0.293\,(-4) = -1.17$V, and $\upsilon_{GS\,2} = -4.00$V. Thus the differential signal for cutoff is 4.00 − 1.17 = **2.83V**. For the small-signal approximation, $|\dfrac{\upsilon_{id}}{V_P}| << \sqrt{\dfrac{2I}{I_{DSS}}} = \sqrt{\dfrac{2(1)}{2}} = 1$. Thus, for a factor-of-10 margin, $\upsilon_{id} \approx \dfrac{1}{10}\,(1)\,|V_p| =$ **0.4V**.

6.30 From Eq. 6.90:

$$i_{D1} = \frac{I}{2} + \upsilon_{id}\,\frac{I}{-2V_p}\left[\frac{2\,I_{DSS}}{I} - \left(\frac{\upsilon_{id}}{V_p}\right)^2\left(\frac{I_{DSS}}{I}\right)^2\right]^{½} = \frac{I}{2}\left[1 + \frac{\upsilon_{id}}{-V_p}\left(\frac{2I_{DSS}}{I}\right)^{½}\left[1 - \upsilon_{id}^2\left(\frac{I_{DSS}}{2IV_p^2}\right)\right]^{½}\right].$$

The required term is $\mathrm{T} = \left(1 - \left(\dfrac{\upsilon_{id}}{V_p}\right)^2\left(\dfrac{I_{DSS}}{2I}\right)\right)^{½} = \left(1 - \left(\dfrac{\upsilon_{id}}{V_p}\Big/\left(\dfrac{2I}{I_{DSS}}\right)^{½}\right)^2\right)^{½} = (1 - (\,)^2\,)^{½}$. Now, for ( ) = 1, $T = (1 - 1^2)^{½} =$ **0**, For ( ) = 0.33, $T = (1 - 0.33^2)^{½} =$ **0.944**; For ( ) = 0.1 , $T = (1 - 0.1^2)^{½} =$ **.995**;

For ( ) = 0.033, $T = (1 - (.033)^2)^{1/2}$ = **0.9995**. Now, for $T = 0.9$, ( ) = $\frac{\upsilon_{id}}{V_p}\left[\frac{I_{DSS}}{2I}\right]^{1/2} = (1 - 0.9^2)^{1/2} = (0.19)^{1/2}$ = **0.436**. Now, for $T = 0.99$, ( ) = $\frac{\upsilon_{id}}{V_p}\left[\frac{I_{DSS}}{2I}\right]^{1/2} = \sqrt{1 - .99^2} = \sqrt{.0199}$ = **0.141**. Thus for Eq. 6.93 of the Text, << implies that ( ) = 1/7 would produce only a 1% distortion, while ( ) = 1/2 causes only about 10% distortion.

6.31 For each FET, i_D = 0.5mA, for which $r_o = \frac{V_A}{i_D} = \frac{50}{0.5}$ = 100kΩ, and $g_m = \frac{2\,I_{DSS}}{-V_p}\sqrt{\frac{I/2}{I_{DSS}}} = \frac{\sqrt{2I\,I_{DSS}}}{-V_p}$ = $(2(1)2)^{1/2}/(-(-4))$ = 0.5mA/V. For two 10kΩ load resistors, the differential output gain is $\frac{10k\Omega + 10k\Omega}{1/g_m + 1/g_m} = \frac{20}{2+2}$ = 5V/V, if r_o is ignored, or $\frac{2\,(10k\Omega \,||\, 100k\Omega)}{4k\Omega}$ = **4.55V/V**, if it is included. For a current-mirror load, the net load resistance is $r_o \,||\, r_o$ = 100kΩ/2 = 50kΩ, and gain is $\frac{2\,(50k\Omega)}{2k\Omega + 2k\Omega}$ = **25V/V**.

SECTION 6.7: MOS Differential Amplifiers

6.32 Here, $g_m = \frac{I}{V_{GS} - V_t}$ for each transistor – – – (1), with a maximum at $i_{D1} = i_{D2} = I/2 = K\,(V_{GS} - V_t)^2$ – – – (2), where $V_{GS} - V_t = \left[\frac{KI}{2}\right]^{1/2}$, and $g_{m\,\max} = \frac{I}{(KI/2)^{1/2}} = \sqrt{2I/(K)}$, or $g_{m\,\max}$ = 2K $(V_{GS} - V_t)$, (from (1) with (2)). In general, $i_D = K\,(\upsilon_{GS} - V_t)^2$, and $g_m = 2K\,(\upsilon_{GS} - V_t) = 2K\,(V_{GS} \pm \frac{\upsilon_{id}}{2} - V_t)$. For a 10% g_m drop, $2K\,(V_{GS} \pm \frac{\upsilon_{id}}{2} - V_t) = 0.9\,(2K)\,(V_{GS} - V_t)$, when $V_{GS} \pm \frac{\upsilon_{id}}{2} - V_t = 0.9\,V_{GS} - 0.9V_t$, or $0.1\,(V_{GS} - V_t) = \pm\frac{\upsilon_{id}}{2}$, or $\frac{\upsilon_{id}}{V_{GS} - V_t}$ = ± **0.2**. For 5%: $\frac{\upsilon_{id}}{V_{GS} - V_t}$ = ± 2(.05) = ± **0.1**. For 1%: $\frac{\upsilon_{id}}{V_{GS} - V_T}$ = ± 2(.01) = **±0.02**.

6.33 $i_D = K\,(\upsilon_{GS} - V_t)^2$, $K = \frac{1}{2}\,\mu_p\,C_{ox}\,W/L = \frac{1}{2} \times 10 \times 120/6$ = 100μA/V². For equal current division, 25/2 = 100 $(\upsilon_{GS} - 1)^2$, $\upsilon_{GS} - 1 = \pm(1/8)^{1/2}$, and υ_{GS} = 1.354V. Thus V_{GS} = **1.35V**. Now, $g_m = 2K\,(\upsilon_{GS} - V_t)$ = 2 (100) (1.354 – 1) = **70.8μA/V**, and $r_o = \frac{V_A}{I_D} = \frac{50}{25/2}$ = 4MΩ. Maximum gain will occur for outputs taken differentially: (a) For ideal loads: $A_\upsilon = \frac{2\,(4 \times 10^6)}{2\,(1/70.8 \times 10^{-6})}$ = **283.2V/V**, (b) For loads with V_A = 50V: $A_\upsilon = \frac{2\,(4 \times 10^6 \,||\, 4 \times 10^6)}{2\,(1/70.8 \times 10^{-6})}$ = 283.2/2 = **141.6V/V**.

6.34 **With balanced loads,** $i_D = \frac{25}{2} = K\,(\upsilon_{GS} - V_t)^2$, with υ_{GS} = 1.354V (from P6.33 above). Now for i_D = 0.9 (25/2) = 100 $(\upsilon_{GS} - 1)^2$, $\upsilon_{GS} = (\frac{0.9}{8})^{1/2} + 1$ = 1.3354V. For i_D = 1.1 (25/2) = 100 $(\upsilon_{GS} - 1)^2$, $\upsilon_{GS} = (\frac{1.1}{8})^{1/2} + 1$ = 1.371V. Thus the input offset = 1.371 – 1.335 = **36mV**.

6.35 From Exercise 6.15, K = 200μA/V², V_{GS} = 1.25V, g_m = 100μA/V, V_t = 1V. For an R_D mismatch of ± 1%, $V_{OS} = \left[\frac{V_{GS} - V_t}{2}\right]\frac{\Delta R_D}{R_D} = \frac{1.25 - 1}{2} \times \frac{1}{100}$ = **±1.25mV**. For a K mismatch of ±1%, V_{OS} = $\left[\frac{V_{GS} - V_t}{2}\right]\frac{\Delta K}{K} = \frac{1.25 - 1}{2} \times \frac{1}{100}$ = **±1.25mV**. For V_t tolerance of ±0.6mV, V_{OS} = **±0.6mV**. Thus the worstcase offset is 1.25 + 1.25 + 0.6 = **3.1mV**. The likely offset = $\sqrt{1.25^2 + 1.25^2 + 0.6^2}$ = **1.87mV**.

6.36 Here, $I_O \approx I_{REF} = 100\mu A$.

Roughly: $i_D = K(\upsilon_{GS} - V_t)^2$, $100 = 100(\upsilon_{GS} - 1)^2 \rightarrow \upsilon_{GS} = 2V$. Now, for Q_1, Q_4, $\upsilon_{DS} \approx 2V$ and $(1 + \frac{\upsilon_{DS}}{V_A}) = 1 + \frac{2}{20} = 1.1$. This could be ignored, but let us include it in a more basic calculation:

More precisely: For Q_1, Q_4, $\upsilon_{DS} = \upsilon_{GS} = \upsilon$, and $100 = 100(\upsilon - 1)^2(1 + \upsilon/20)$, or $1 = (\upsilon^2 - 2\upsilon + 1)(1 + .05\upsilon)$. Thus $\upsilon^2 - 2\upsilon + 1 + .05\upsilon^3 - 0.1\upsilon^2 + .05\upsilon = 1$ or $.05\upsilon^3 + 0.9\upsilon^2 - 1.95\upsilon = 0$, or $.05\upsilon^2 + 0.9\upsilon - 1.95 = 0$, $\upsilon = (-0.9 \pm \sqrt{0.9^2 + 4(1.95)(.05)})/(2(.05)) = (-.9 \pm 1.095)/0.1 = 1.954V$. Thus V_{GS1} = **1.954V**. Now for $V_O = V_{D3} = V_{D4}$, see I_O = **100μA**, from symmetry arguments. For each device, $r_o \approx \frac{V_A}{I_D} = \frac{20}{100\mu A} = 200k\Omega$, and $g_m = 2K(\upsilon_{GS} - V_t) \approx 2(100 \times 10^{-6})(2 - 1) = 200\mu A/V$. For the whole mirror, $R_{out} \approx g_{m3} r_{o3} r_{o2} = 200 \times 10^{-6} \times 200 \times 10^3 \times 200 \times 10^3 = 8 \times 10^6 \Omega$ = **8MΩ**. Thus, for $V_O = 12V$, with the standard output being at 2 + 2 = 4V, the extra current = (12 – 4)/8MΩ = 8/8 = 1μA. Thus I_O = **101μA** for $V_O = +12V$.

6.37 Assume $I_O \approx I_{REF} = 100\mu A$. From results of P6.36 above, $V_{GS} \approx 2.0V$. Thus $V_{D1} \approx 2.0 + 2.0 = 4V$. Correspondingly, for Q_1, with $\upsilon_{GS} = \upsilon \approx 2.0V$ and $\upsilon_{DS} \approx 2V$, $100\mu A = 100\mu A/V^2 (\upsilon - 1)^2 (1 + \frac{2\upsilon}{20})$, $1 = (\upsilon^2 - 2\upsilon + 1)(1 + 0.1\upsilon) = \upsilon^2 - 2\upsilon + 1 + 0.1\upsilon^3 - 0.2\upsilon^2 + 0.1\upsilon$, or $+0.1\upsilon^3 + 0.8\upsilon^2 - 1.9\upsilon = 0$, $0.1\upsilon^2 + .8\upsilon - 1.9 = 0$, $\upsilon = \frac{-0.8 \pm \sqrt{.8^2 + 4(1.9)(.1)}}{2(.1)} = 1.916V$. Thus, $\upsilon_{GS1} \approx 1.916V$ = **1.92V**. Now for Q_2, $i_D = 100(1.916 - 1)^2(1 + \frac{1.916}{20})$ = **91.9μA**. Now, for Q_3 with $V_O \approx 2(1.92) = 3.84V$, $I_O = 91.9\mu A$. From P6.75 of the Text, $R_o \approx (g_m r_o) r_o \approx$ **8M Ω** using the results of P6.36 above. Thus for $V_O = 12V$, $I_O = 91.9 + (12 - 3.84)/8$ = **92.9μA**.

6.38

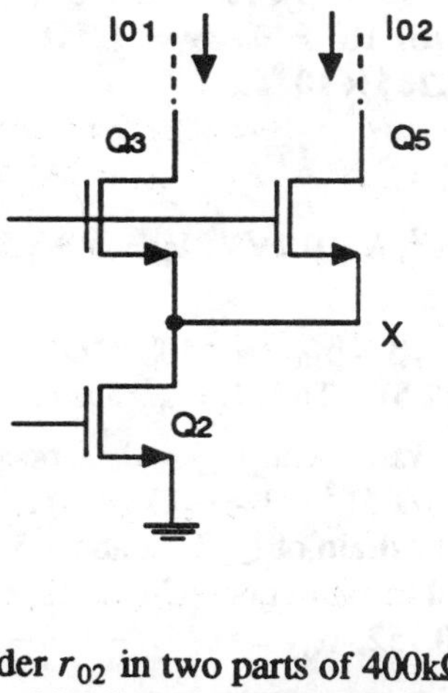

Generally speaking, see that currents in Q_3 and Q_5 will be the same provided the output voltages are the same. Further, the current in each will slightly exceed 100/2 μA since the voltage at node X will rise slightly due to doubling of the equivalent K of Q_3, Q_5. Since the current in each of Q_3, Q_5 is only slightly more than half what it was, each output resistance will be twice as large. In particular, with outputs joined, the output resistance will be only slightly smaller than before: **Now to check these ideas:** $i_{D2} \approx 100\mu A \approx (100 + 100)(\upsilon_{GS} - 1)^2$. Thus $\upsilon_{GS} = 1 \pm \sqrt{1/2} = 1.707V$, rather than 2.0V previously. Thus node X will rise about 0.3V, and i_{D3} will increase by 0.3/200kΩ = 1.5μA. The current in Q_3, Q_5 will be about **50.8μA**, for which gm = $2K(\upsilon_{GS} - V_t) = 2(100 \times 10^{-6})(1.707 - 1) = 141\mu A/V$. Now for outputs joined, $r_{035} = 20/(100 + 1.5) = 197k\Omega$.

Consider r_{02} in two parts of 400kΩ each, with $R_{03} = R_{05} = 197k \times 400k \times 141\mu A/V = 11.1M\Omega$. Together, $R_{035} = 11.1/2$ = **5.6MΩ**. When outputs operate independently, the output resistance decreases a lot since node X is grounded via $1/g_m$ of the other output. From Eq. 6.135 (full version) $R_{03} = r_{03} + 1/g_{m5} + g_{m3} r_{03} 1/g_{m5} \approx 2 r_{03}$ = **800kΩ**. An improved circuit would split both Q_3 and Q_2 into two parts, with each pair of half-size transistors driven from the gate of Q_4, Q_1 respectively. The total device width needed would be same as in the original design, that is **4W** for each original transistor of width W. The version with only Q_3 duplicated uses a total width of **5W** and has poorer performance!

6.39 For $I_B = 200\mu A$, I_D for each transistor = 100μA, $r_o = V_A/I_D = 20/100\mu A = 200k\Omega$, $i_D = K(\upsilon_{GS} - V_t)^2$, $100 \times 10^{-6} = 100 \times 10^{-6}(\upsilon_{GS} - 1)^2 \rightarrow \upsilon_{GS} = 2V$, $g_m = 2K(\upsilon_{GS} - V_t) = 2(100)(2 - 1) = 200\mu A/V$. Gain

$A_\upsilon = -\dfrac{2\,(200k \,\|\, 200k)}{1/0.2 + 1/0.2} = \mathbf{-20V/V}$. Gain reduces by a factor of two for a load of **100kΩ**.

SECTION 6.8: BiCMOS Amplifiers

6.40

(a) Here, $i_C = 10\mu A$, $g_m = \dfrac{10 \times 10^{-6}}{25 \times 10^{-3}} = \mathbf{400\mu A/V}$, $r_o = 100V/10\mu A = \mathbf{10M\Omega}$, $R_i = \beta/g_m = 100/400 \times 10^{-6}$ $= \mathbf{250k\Omega}$, $A_\upsilon = -400 \times 10^{-6} \times 10 \times 10^6 = \mathbf{-4000V/V}$.

b) For $i_D = 10\mu A = ½ \times 20 \times 20/2\,(\upsilon_{GS} - V_t)^2 = 100\,(\upsilon_{GS} - V_t)^2$. Thus $\upsilon_{GS} - V_t = (10/100)^{½} = 0.316$, $g_m = 2K\,(\upsilon_{GS} - V_t) = 2\,(100 \times 10^{-6})\,0.316 = \mathbf{63.2\mu A/V}$, $r_o = 20/10\mu A = \mathbf{2M\Omega}$, $R_i = \infty$, $A_\upsilon = -63.2 \times 10^{-6} \times 2 \times 10^6 = \mathbf{-126.4V/V}$, that is much, much less!

6.41 For $I = 10\mu A$, $g_{m2} = 400\mu A/V$, $r_{\pi 2} = \dfrac{\beta_2}{g_{m2}} = \dfrac{100}{400 \times 10^{-6}} = 250k\Omega$, $r_{02} = 10M\Omega$, $g_{m1} = 63.2\mu A/V$, $r_{01} = 2M\Omega$. From Eq. 6.135, $R_{out} \simeq g_{m2}\, r_{02}\, (r_{01} \,\|\, r_{\pi 2}) = 400 \times 10^{-6} \times 10 \times 10^6 \times (2 \times 10^6 \,\|\, .25 \times 10^6) = 888 \times 10^6\Omega$. Thus $\dfrac{\upsilon_o}{\upsilon_i} = -63.2 \times 10^{-6} \times \dfrac{100}{101} \times .888 \times 10^9 = \mathbf{-55.6 \times 10^3 V/V}$.

6.42 For $I = 100\mu A$, see by scaling from P6.41 above, $g_{m2} = 4mA/V$, $r_{\pi 2} = 25k\Omega$, $r_{02} = 1M\Omega$, $r_{01} = 200k\Omega$, and $\upsilon_{GS1} - V_t = (100/100)^{½} = 1$. Thus $g_{m1} = 2\,(100 \times 10^{-6})\,1 = 200\mu A/V$, $R_{out} \simeq 4 \times 10^{-3} \times 10^6 \times (200 \times 10^3 \,\|\, 25 \times 10^3) = 88.8 \times 10^6$, $\upsilon_o/\upsilon_i = -200 \times 10^{-6} \times 100/101 \times 88.8 \times 10^6 = \mathbf{-17.6 \times 10^3 V/V}$.

6.43 From P6.40 at 10μA: For the BJTs, $g_m = 400\mu A/V$, $r_\pi = 250k\Omega$, $r_o = 10M\Omega$, $r_\mu = 10\,(100)10^7 = 10 \times 10^9\Omega$. For the MOS, $g_m = 63.2\mu A/V$, $r_o = 2M\Omega$.

For the circuit as shown, $R_{out} \simeq 2 \times 10^6 \times 63.2 \times 10^{-6} \times (10 \times 10^6 \times 400 \times 10^{-6} \times 250 \times 10^3) \,\|\, (10 \times 10^9) = 126.4 \times (4000 \times 250 \times 10^3) \,\|\, 10^{10} = 126.4 \times .909 \times 10^9 = \mathbf{115 \times 10^9\Omega}$. *With* Q_3, Q_6 *not used,* $R_{out} \simeq 10^{10} \,\|\, (10^7 \times 400 \times 10^{-6} \times (.25 \times 10^6) \,\|\, (10 \times 10^6)) = (10 \,\|\, .976)\, 10^9 = \mathbf{0.889 \times 10^9\Omega}$. *With* Q_5, Q_2 *eliminated,* $R_{out} \simeq 2 \times 10^6 \times 63.2 \times 10^{-6} \times (10 \times 10^6 \,\|\, 10 \times 10^9) = \mathbf{1.264 \times 10^9\Omega}$.

SECTION 6.9: GaAs Amplifiers

6.44 For a 1μm long, 1μm wide *GaAs* device, $V_t = -1.0V$, $\beta_1 = 100\mu A/V^2$, $\lambda = 0.1V^{-1}$, $V_A = 10V$, $I_S = 10^{-15}A$, $n = 1.1$.

(a) Symmetry would indicate that $V_a = 5/2 = \mathbf{2.5V}$, $\beta = \beta_{10} = 10\,(100) = 1mA/V^2$, $i_D = \beta\,(\upsilon_{GS} - V_t)^2\,(1 + \lambda\,\upsilon_{DS}) = 1 \times 10^{-3}\,(0 - -1)^2\,(1 + 0.1\,(2.5))$. Thus $I_a = \mathbf{1.25mA}$.

(b) Assume operation is in saturation. Lower transistor (Q_1) operates with $\upsilon_{GS} = 0$. Upper one (Q_2) is 1.5 × larger: Thus $1.5\,(\upsilon_{GS2} - -1)^2 = (0 - -1)^2$, or $\upsilon_{GS2} \simeq (1/1.5)^{½} - 1 = -0.18V$. Thus, $i_{D1} = 20 \times 100 \times 10^{-6}\,(0 - 1)^2\,(1 + 0.1\,(5 - .18)) \simeq 2.964mA$. See that the drain of Q_2 is at about 5 – 1kΩ (2.96) = 2.04V, while the gate is at 0V. See that operation is indeed in saturation. *Check*: $2.964 = 30 \times 100 \times 10^{-6}\,(\upsilon_{GS} - -1)^2\,(1 + 0.1\,(2.96 - -0.18))\,(\upsilon_{GS} + 1)^2 = 0.752$, $\upsilon_{GS} = 0.867 - 1 = -0.132V$ OK. ∴ $V_b \simeq \mathbf{-0.13V}$, and $I_b \simeq \mathbf{2.95mA}$.

(c) See Q_1 operates at $\upsilon_{GS} = 0$ with $V_{c1} \simeq 0V$. $I_{c1} \simeq 20 \times 100 \times 10^{-6}\,(0 - -1)^2\,(1 + 0.1(5)) = \mathbf{3.0mA}$. Then $I_{S3} = I_{S2} \simeq 3.0/2 = \mathbf{1.5mA} = I_{c2}$, and $V_{C2} \simeq 5 - 2(1.5) = \mathbf{2V}$. Thus, $1.5 = 10\,(100 \times 10^{-6})\,(\upsilon_{GS} - -1)^2\,(1 + 0.1(2))$, or $(\upsilon_{GS} + 1)^2 = 1.5/(1(1.2)) = 1.25$, $\upsilon_{GS} = 1.12 - 1 = 0.12V$. Thus $V_{c1} \simeq \mathbf{-0.12V}$. *Check*: $I_{c2} = 1\,(0.12 + 1)^2\,(1 + 0.1(2)) = 1.505mA$.

(d) See that Q_1 is near cutoff, though it is larger. Assume V_d is near +5V, say at 5V, in which case $i_{D1} = 2 \times 100 \times 10^{-6}\,(-0.8 - -1)^2)\,(1 + .1\,(5 - 0)) = 0.12mA$. Now $i_{D2} = \beta\left[2\,(\upsilon_{GS} - V_t)\,\upsilon_{DS} - \upsilon_{DS}^2\right](1 + \lambda\,\upsilon_{DS})$. Let $\upsilon_{DS} = \upsilon$, which is small, such that the λ term can be ignored. Thus $0.12 \times 10^{-3} = 10 \times$

100×10^{-6} $[2\ (0 - -\ 1)\ \upsilon - \upsilon^2]$, or $0.12 = (2\upsilon - \upsilon^2)$, $\upsilon^2 - 2\ \upsilon + 0.12 = 0$, and $\upsilon = \dfrac{--2 \pm \sqrt{4 - 4\,(.12)}}{2} = 0.062\text{V}$. Thus $V_d = 5 - .062 =$ **4.94V** and $I_d =$ **0.12mA**.

(e) See that Q_1 is turned on with V_e near 0V. For Q_2, $i_D = 10 \times 100 \times 10^{-6}\ (0 - -\ 1)^2\ (1 + 0.1\ (5)) =$ 1.5mA. For Q_2, $i_D = 20 \times 100 \times 10^{-6}\ (0.2 - -\ 1)^2\ (1 + 0.1\ (0)) = 2.88\text{mA}$. Thus, Q_2 is in triode mode with $\upsilon_{DS} = \upsilon$, assumed small and ignored. Correspondingly, $1.5\text{mA} = 2.0\text{mA}\ [2\ (0.2 - -\ 1)\ \upsilon - \upsilon^2]$, and $\upsilon^2 - 2.4\upsilon + .75 = 0$, $\upsilon = \dfrac{2.4 \pm \sqrt{2.4^2 - 4(.75)}}{2} = 0.37\text{V}$. Thus $V_e =$ **0.37V**, and $I_e =$ **1.5mA**.

6.45 Here, $\beta_1 = 100 \times 10^{-6} \times 5 = 0.5\text{mA/V}^2$:

(a) $i_{D1} = 0.5\ (0 - -\ 1)^2\ (1 + 0.1\ (1)) = 0.55\text{mA}$, for $\upsilon_{DS1} = |V_t|$. For $V_O = -3\text{V}$, $V_{SS} = -5\text{V}$, $V_{S2} = -4\text{V}$, $i_{D2} = 0.55 \times 10^{-3} = 20\ (100) \times 10^{-6}\ (\upsilon_{GS} - -\ 1)^2\ (1 - 0.1\ (-3 - -4))$. Thus $(\upsilon_{GS} + 1)^2 = .25$, $\upsilon_{GS} = \pm .5 - 1 = -0.5\text{V}$. $\therefore V_{bias} = -4 - 0.5 =$ **−4.5V**.

(b) Lowest $V_O = -4V + |V_t| =$ **−3V**.

(c) For $V_O = -3\text{V}$, r_o of Q_2 does not matter, the current being established by Q_1 at a value $I_O =$ **0.55mA**.

(d) From Eq. 5.109: $r_o = \dfrac{1}{\lambda\,\beta\,(V_{GS} - V_t)^2}$. Thus $r_{o1} = \dfrac{1}{0.1 \times 0.5\ (0 - -1)^2} = 20\text{k}\Omega$, and $r_{o2} = \dfrac{1}{0.1 \times 2.0\ (-0.5 - -1)^2} = 20\text{k}\Omega$. From Eq 5.108: $g_m = 2\beta\ (\upsilon_{GS} - V_t)\ (1 + \lambda\ \upsilon_{DS})$. Thus $g_{m2} = 2\ (2)\ (-0.5 - -1)\ (1 + 0.1\ (1)) = 2.2\text{mA/V}$. Thus $R_{out} = g_{m2}\ r_{o2}\ r_{o1} = 2.2 \times 10^{-3} \times 20 \times 10^3 \times 20 \times 10^3$ = **880kΩ**.

(e) Now for the output raised from −3V to +1V, ie by 4V, $\Delta I = \dfrac{4V}{880 \times 10^3} =$ **4.5μA**.

6.46

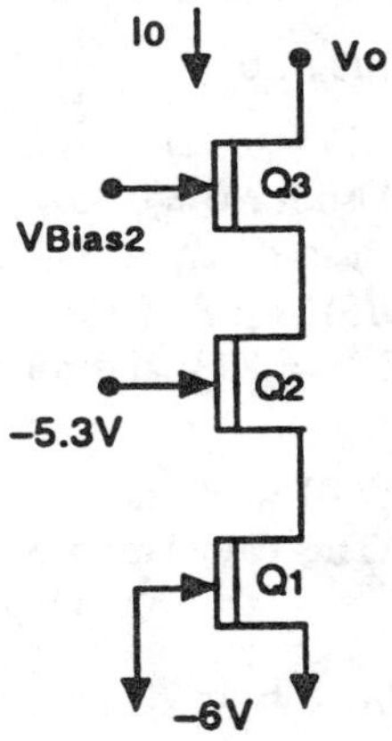

Add Q_3 with $W_3 = 20\mu\text{m}$, $\beta_3 = 2\text{mA/V}^2$, $\beta_2 = 2\text{mA/V}^2$, $\beta_1 = 1\text{mA/V}^2$, $V_t = -1\text{V}$, $\lambda = 0.1\text{V}^{-1}$,

(a) Use the results of Ex. 6.24. Since Q_2, Q_3 are the same size, $V_{bias\,1} = -5.3\text{V}$ permits $\upsilon_{DS1} = 1\text{V}$. Thus $V_{bias\,2} = V_{bias\,1} + 1.0 =$ **−4.3V**. *Check*: Now, for $\upsilon_{DS1} = 1\text{V}$, $i_{D1} = 1\text{mA/V}^2\ (0 - -\ 1)^2\ (1 + 0.1(1)) = 1.1\text{mA}$. Now for Q_2, $1.1\text{mA} = 2\text{mA/V}^2\ (\upsilon_{GS} - -\ 1)^2\ (1 + 0.1(1))$, or $1 = 2\ (\upsilon_{GS} + 1)^2$, and $\upsilon_{GS} = \sqrt{1/2} - 1 = .707 - 1 \simeq -0.3\text{V}$. OK.

(b) Now, Q_3 remains in saturation for $\upsilon_{DS} \geq \upsilon_{GS} - V_t$ or $\upsilon_{DS} > -0.3 - -1 = 0.7\text{V}$. Thus, the output can be as low as $-4.3 + 1.0 =$ **−3.3V**.

(c) Now as Q_3 is operating just as Q_2, V_O can go as low as $-4.3 + 0.3 + 1 = -3.0\text{V}$, at which point $I_O =$ **1.1mA**.

(d) $r_{01} = \dfrac{1}{0.1 \times 1 \times (0 - -1)^2} = 10\text{k}\Omega$, $r_{02} = \dfrac{1}{0.1 \times 2 \times (-0.3 - -1)^2} = 10.2\text{k}\Omega = r_{03}$, $g_{m2} = 2\ (2)\ (-0.3 + 1) = 2.8\text{mA/V} = g_{m3}$. Thus, $R_{02} = g_{m2}\ r_{02}\ r_{01} = 2.8 \times 10.2 \times 10.0 = 285.6\text{k}\Omega$, and $R_{03} = g_{m3}\ r_{03}\ R_{02} = 2.8 \times 10.2 \times 285.6 =$ **8.16MΩ**

(e) For a 4V change in output voltage, $\Delta I = \dfrac{4}{8.16 \times 10^6} =$ **0.49μA**.

6.47 See Q_1 and Q_3 have the same width. Thus for $V_{DD} = 5\text{V}$ and $V_A = 2\text{V}$, $\upsilon_{DS3} = \upsilon_{DS1} = (5-2)/2 = 1.5\text{V}$, and $V_E = V_B = 5 - 1.5 = 3.5\text{V}$. See $5 \times 10^{-3} = W \times 100 \times 10^{-6}\ (0 - -1)^2\ (1 + 0.1\ (1.5))$. Thus $W_1 = W_3 =$ **43.5μm**, and $W_2 = 43.5/2 =$ **21.7μm**. Note that the diodes are sized to give 0.75V drop each for $I/2 = 2.5\text{mA}$. Now, $r_o = \dfrac{1}{\lambda\,\beta\,(\upsilon_{GS} - V_t)^2}$, $g_m = 2\beta\ (\upsilon_{GS} - V_t)\ (1 + \lambda\ \upsilon_{DS})$. Thus $r_{01} = \dfrac{1}{0.1\ (4.35)\ (0 + 1)^2} =$

$2.3k\Omega = r_{03}$, and $r_{02} = \dfrac{1}{0.1\,(2.17)\,(0+1)^2} = 4.61k\Omega$. Also $g_{m1} = 2\,(4.35)\,(0+1)\,(1+0.1\,(1.5)) = 10mA/V$ $= g_{m3}$, and $g_{m2} = 2\,(2.17)\,(0+1)\,(1+0.1\,(5-2)) = 5.64mA/V$. Now, from Eq. 6.152: $\alpha = \dfrac{\upsilon_b}{\upsilon_a} =$

$$\frac{10\times 2.30\times\dfrac{5.64\times 4.61}{5.64\times 4.61+1}+\dfrac{2.30}{2.30}}{10\times 2.30+\dfrac{2.30}{2.30}+1} = \frac{23.15}{25} = \mathbf{0.926V/V}.$$

From Eq. 6.153: $R_o = \dfrac{r_{01}}{1-\alpha} = \dfrac{2.3}{1-.926} =$ **31.1kΩ**. From Eq. 6.154: $R_o = r_{01}\,(g_{m3})\,\dfrac{r_{03}}{2} = 2.3\times 10\times\dfrac{2.3}{2} =$ **26.5kΩ**.

6.48 $I_{DSS\,eq} = 0.5mA$, $\upsilon_{DS} = 3V$, $\upsilon_{DS1} = 0.7V$. Now operate with $\upsilon_{GS1} = 0V$.

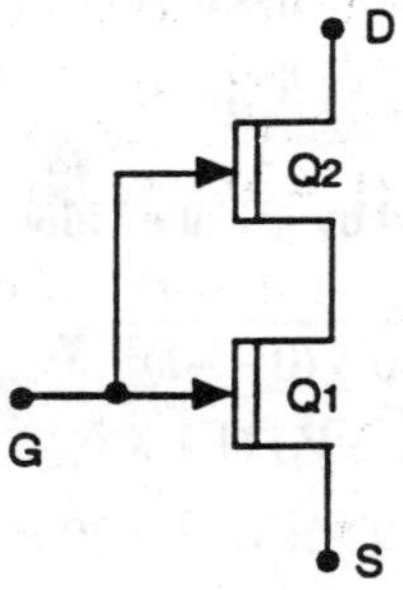

Thus $\upsilon_{GS2} = -0.7V$, with $\upsilon_{DS2} = 3-0.7 = 2.3V$, $0.5 = W_1\,(0.1)\,(0-1)^2 \times (1+0.1\,(0.7))$, or $W_1 = \dfrac{0.5}{0.1}\,\dfrac{1}{1^2}\,\dfrac{1}{1.07} = \mathbf{4.67\mu m}$. Also $0.5 = W_2\,(0.1)\,(-0.7--1)^2\,(1+0.1\,(2.3))$, or $W_2 = \dfrac{0.5}{0.1}\times\dfrac{1}{(0.3)^2}\times\dfrac{1}{1.23} =$ **45.2μm**. For $\upsilon_{DS} = 6V$, the current I_{DSS} will increase, with $\upsilon_{DS1} = \upsilon$ increasing as well. Thus $i = 0.467\,(0--1)^2\,(1+0.1\,\upsilon)$, and $i = 4.52\,(-\upsilon+1)^2\,(1+0.1\,(6-\upsilon))$, $(1+0.1\upsilon) = 9.68\,(\upsilon^2-2\upsilon+1)\,(1.6-0.1\upsilon)$, $1+0.1\upsilon = 15.5\,\upsilon^2-31\upsilon+15.5-.968\,\upsilon^3+1.94\,\upsilon^2-.968\,\upsilon$, $0.968\upsilon^3-17.34\upsilon^2+32.07\upsilon-14.5 = 0$, $\upsilon^3-18\upsilon^2+33.13\,\upsilon-14.98 = 0$.

Solve this cubic iteratively: $\upsilon = \dfrac{18\upsilon^2+14.98-\upsilon^3}{33.13} = .543\upsilon^2+.452-.030\upsilon^3$. Now, for $\upsilon = 0.7$: $\upsilon = .543\,(.7)^2+.452-.03\,(.7)^3 = .266+.452-.01 = .708$; for $\upsilon = 0.71$: $\upsilon = .543\,(.71)^2+.452-.030\,(.71)^3 = .274+.452-.011 = .715$. See that the process does not converge: *Thus reformulate it*: $\upsilon^2 = \dfrac{\upsilon^3+33.13\upsilon-14.98}{18}$, $\upsilon = (.056\upsilon^3+1.84\upsilon-.832)^{1/2}$. Now, for $\upsilon = 0.7$: $\upsilon = (.056\,(.7^3)+1.84\,(.7)-.832)^{1/2} = (.0192+1.288-.832)^{1/2} = (.4752)^{1/2} = .689$. See that now it will diverge for lower values: Also for $\upsilon = 0.8$: $\upsilon = (.056\,(.8^3)+1.84\,(.8)-.832)^{1/2} = (.027+1.472-.832)^{1/2} = .818$. Again we see that it also diverges, but that a solution lies between 0.7 and 0.8, but nearer to 0.7. Now, for $\upsilon = 0.75$: $\upsilon = (.056\,(.75)^3+1.84\,(.75)-.832)^{1/2} = (.024+1.38-.832)^{1/2} = 0.756$; for $\upsilon = 0.74$: $\upsilon = (.056\,(.74)^3+1.84\,(.74)-.832)^{1/2} = .743$; and for $\upsilon = 0.73$: $\upsilon = (.022+1.343-.832)^{1/2} = .7302$.

Now conclude $\upsilon = \upsilon_{DS1} = 0.73V$, for which Q_1, $i_D = 0.467\,(1^2)\,(1+.1\,(0.73)) = .501mA$, and for Q_2, $i_D = 4.52\,(-.73--1)^2\,(1+0.1\,(6-0.73)) = 4.52\,(.0729)\,(1.527) = .503mA$. This calculation is very sensitive due to squared term. Use $i_D =$ **.502mA**.

Small-Signal analysis: $r_o = \dfrac{1}{\lambda\,\beta\,(\upsilon_{GS}-V_t)^2}$, $g_m = 2\,\beta\,(\upsilon_{GS}-V_t)\,(1+\lambda\,\upsilon_{DS})$. Here $R_{out} = r_{02}\,g_{m2}\,r_{01}$, for which $r_{02} = \dfrac{1}{0.1\,(45.2\times 0.1\times 10^{-3})\,(-0.7--1)^2} = 24.6k\Omega$, $r_{01} = \dfrac{1}{0.1\,(.467\times 10^{-3})\,(0+1)^2} = 21.4k\Omega$, $g_{m2} = 2\,(4.52\times 10^{-3})\,(-0.7--1)\,(1+0.1\,(3-0.7)) = 3.34mA/V$. Thus $R_{out} = 24.6\times 10^3\times 21.4\times 10^3\times 3.34\times 10^{-3} =$ **1.76MΩ**. Now, the expected current increase as υ_{DS} rises from 3V to 6V is $(6\text{–}3)/(1.76\times 10^6) = 1.7\mu A$. The earlier estimate of change from .500mA to .502mA by 2μA is quite consistent. The small signal scheme is certainly more straightforward!

6.49 From the results of P6.48 above, for $V_{DD} = 6V$ and $V_O \approx 6/2 = 3V$, the upper composite would conduct 0.5mA, while the lower one would be biased at $V_I =$ **0V** to correspond. For each composite, the output resistance is 1.76MΩ. For the amplifier, the output resistance is 1.76/2 = **.88MΩ**, and the gain is $-g_m\,R_{out}$. Here, $g_m = g_{m1} = 2\beta_1\,(\upsilon_{GS1}-V_t) = 2(0.467\times 10^{-3})(0--1) =$ 0.934mA/V, and the gain is $-0.934\times 10^{-3}\times -0.88\times 10^6 =$ **−822V/V**.

6.50 Though it is not required explicitly by the specification, make all devices the same size. For $V_{cm} \approx 0V$, $V_{DD} = 5V$, and Q_1 and Q_3 matched, the voltages across each will be the same, and $\upsilon_{DS1} = \upsilon_{DS3} = \frac{5-0-0.5}{2} \approx 2.25V$. Thus $0.5 \times 10^{-3} = W_1\ (0.1 \times 10^{-3})\ (-0.5 - -1)^2\ (1 + 0.1\ (2.25))$, or $W_1 = \frac{0.5}{0.1} \times \frac{1}{(0.5)^2} \times \frac{1}{1.225} = \mathbf{16.3\mu m} = W_2 = W_3$. Use $I = \mathbf{1.0mA}$: Now for 0.5mA in Q_1, the voltage at the sources of Q_1 and Q_2 is 0.5V, and that at the source of Q_3 is 0.5 + 2.75 = 2.75V, with 2.75 − 0.5 = 2.25V at its gate. Thus $\upsilon_{DS2} \approx 2.25 - 0.5 = 1.75V$. Now, $0.5 \times 10^{-3} \approx 16.3\ (0.1 \times 10^{-3})\ (\upsilon_{GS} - -1)^2\ (1 + 0.1\ (1.75))$. Thus $\upsilon_{GS2} = \pm\left[\frac{0.5}{1.63} \times \frac{1}{1.175}\right]^{1/2} - 1 = .5109 - 1 = -0.489V$. There would be an **(11mV offset)** to maintain V_O at 0.5 + 1.75 = **2.25V**. Note that since the drain voltages of Q_1 and Q_2 are different, as required by Q_3 operating at $\upsilon_{GS3} \neq 0$, while Q_1 and Q_2 are matched and share a source connection, there must be an input offset voltage. Otherwise, for gate inputs connected, and the drain supply exactly half that of the source bias supply, the output will rise until Q_3 enters the triode mode and $\upsilon_{GS3} = 0$. For the offset acceptable, and nominal operation with $i_D \approx 0.5mA$ and $\upsilon_{DS} \approx 2.25V$, and $\upsilon_{GS} = -0.5V$, $g_m = 2\beta\ (\upsilon_{GS} - V_t)\ (1 + \lambda\ \upsilon_{DS}) = 2\ (0.1 \times 16.3 \times 10^{-3})\ (-0.5 - -1)\ (1 + 0.1\ (2.25)) = \mathbf{2.00mA/V}$. $r_o = \frac{1}{\lambda\beta\ (\upsilon_{GS} - V_t)^2} = \frac{1}{0.1\ (1.63 \times 10^{-3})\ (-0.5 + 1)^2} = 24.5k\Omega$. Now from Eq. 6.159 of the Text: $\frac{\upsilon_o}{\upsilon_i} = \frac{-g_{m1}\ r_{01}}{\left[\frac{g_{m1}\ r_{01} + 1}{g_{m2}\ r_{02} + 1} - \frac{g_{m3}\ r_{03}}{g_{m3}\ r_{03} + 1}\right]} \approx \frac{-2.00 \times 24.5}{\frac{2 \times 24.5 + 1}{2 \times 24.5 + 1} - \frac{2 \times 24.5}{2 \times 24.5 + 1}} \approx -\frac{49}{\frac{50}{50} - \frac{49}{50}} = -49\ (50) = \mathbf{2450V/V}$.

Alternatively, for $W_1 = W_2$, but W_3 smaller, so $\upsilon_{DS1} = \upsilon_{DS2} = 2.0V$, and $\upsilon_{GS3} = 0V$ with $\upsilon_{GS1} = \upsilon_{GS2} = -0.5V$, all operating at 0.5mA nominally. Find: $0.5 \times 10^{-3} = W_1\ (0.1 \times 10^{-3})\ (-0.5 + 1)^2\ (1 + 0.1\ (2.0))$, $W_1 = W_2 = \frac{0.5}{0.1} \times \frac{1}{(0.5)^2} \times \frac{1}{1.2} = \mathbf{16.7\mu m}$. Now for W_3: $0.5 \times 10^{-3} = W_3\ (0.1 \times 10^{-3}\ (0 + 1)^2\ (1 + 0.1\ (5 - 2))$, or $W_3 = \frac{0.5}{0.1} \times \frac{1}{1^2} \times \frac{1}{1.3} = \mathbf{3.85\mu m}$. Now $g_{m3} = 2\ (0.1)\ (3.85)\ (0 - -1)\ (1 + 0.1\ (3)) = 1mA/V$, $r_{03} = \frac{1}{0.1\ (0.1)\ (3.85)\ (1^2)} = 25.97k\Omega$, and $g_{m1} = g_{m2} = 2\ (1.67)\ (-0.5 - -1)\ (1 + 0.1)\ (2)) = 2.004mA/V$, with $r_{01} = r_{02} = \frac{1}{0.1\ (1.67)\ (-0.5 + 1)^2} = 23.95k\Omega$. Thus the gain is $\frac{\upsilon_o}{\upsilon_i} = -\frac{2.00\ (24.0)}{\frac{2.00\ (24.0) + 1}{2.00\ (24.0) + 1} - \frac{1\ (26)}{1\ (26 + 1)}} = -\frac{48}{\frac{49}{49} - \frac{26}{27}} = \mathbf{-1296V/V}$.

SECTION 6.10: Multistage Amplifiers

6.51 **DC bias for $V_{BE} = 0.7V$**, $\beta = \infty$, and ±15V supplies, $I_{E9} = \frac{15 - 0.7}{28.6} = 0.5mA$. Thus $I_{C3} = 0.5mA$, $I_{C1} = I_{C2} = 0.25mA \rightarrow r_e = 100\Omega$, $I_{C6} = 2.0mA$, $I_{C4} = I_{C5} = 1.0mA \rightarrow r_e = 25\Omega$. $V_{B7} = 15 - 3(1) = 12V$, $V_{E7} = 12 + 0.7 = 12.7V$, $I_{E7} = \frac{15 - 12.7}{2.3} = 1mA \rightarrow r_e = 25\Omega$. $V_{B8} = -15 + 1\ (15.7) = 0.7V$, $V_O = 0V$, $I_{E8} = \frac{0 - 15}{3k} = 5mA \rightarrow r_e = 5\Omega$.

AC for $\beta = \infty$: $R_{in} = \infty$, $R_{out} = 3k\Omega || 5\Omega = 5\Omega$, Gain $= \frac{20k + 20k}{100 + 100} \times \frac{3k}{25 + 25} \times \frac{15.7k}{2.3k + .025k} \times \frac{3k}{3k + .005k}$ $= 200 \times 60 \times 6.75 \times .998 = \mathbf{80.9 \times 10^3 V/V}$.

AC for $\beta = 50$: $R_{in} = 51\ (100 + 100) = \mathbf{10.2k\Omega}$, $R_{out} = 3k\Omega || 5 + \frac{15.7k}{51} = 3k\Omega || .313k\Omega = \mathbf{283\Omega}$

For gain: Second-stage R_{i2} = 51 (25 + 25) = 2.55kΩ, third-stage R_{i3} = 51 (2.3 + .015) = 117.6kΩ, fourth-stage R_{i4} = 3k (51) = 153kΩ. Thus the gain = $\frac{40k\,||\,2.55k}{0.2k} \times \frac{50}{51} \times \frac{3k\,||\,117.6k}{(.025k)\times 2} \times \frac{50}{51} \times \frac{15.7k\,||\,153k}{2.325k} \times \frac{50}{51} \times \frac{3k}{3.005k} = \frac{2.397}{0.2} \times \frac{50}{51} \times \frac{2.925}{.05} \times \frac{50}{51} \times \frac{14.24}{2.325} \times \frac{50}{51} \times \frac{3}{3.005}$ = **4040V/V**.

6.52 To raise i_{C7} from 1mA to 2mA, and i_{C8} from 5mA to 20mA, reduce R_4 to 2.3kΩ/2 = **1.15kΩ**, R_5 to 15.7/2 = **7.85kΩ**, R_6 to 3/4 = **0.75kΩ**. Now R_{i4} = 101 (25/20 + 750) = 75.85kΩ,
$A_3 \approx -\frac{R_5\,||\,R_{i4}}{r_{e7}+R_4} = -\frac{7.85k\,||\,75.85k}{12.5+1150}$ = **−6.12V/V**, $A_4 \approx \frac{R_6}{r_{e8}+R_6} = \frac{750}{\frac{25}{50}+750}$ = **0.998V/V**. Now, $A = A_1 A_2 A_3 A_4$ = 22.4 × (−59.2) × (−6.12) × 0.998 = **8099V/V**, and $R_o = R_6\,||\,(r_{e8} + R_5/(\beta+1))$ = 750 || (1.25 + 7850/101) = 750 || 78.97 = **71.4Ω**. For load R_L, loss is $\frac{R_L}{R_L+71.4}$ = 0.8. Thus $R_L = 0.8R_L + 57.12$, $0.2R_L$ = 57.12, R_L = **286Ω**. For a 286Ω load, the upper swing is limited by Q_7 saturating. Assume 0V between the emitter and collector of Q_7, and look down from the collector of Q_7. See an equivalent load resistance at the emitter of Q_8 of 286Ω || 750Ω = 207Ω connected to a supply of (.286/(.286+.75)) (−15) = −4.14V. At its base, see a resistor of 101 (207) = 20.9kΩ to a supply of −4.14 + 0.7 = −3.44V. The equivalent circuit is as shown:

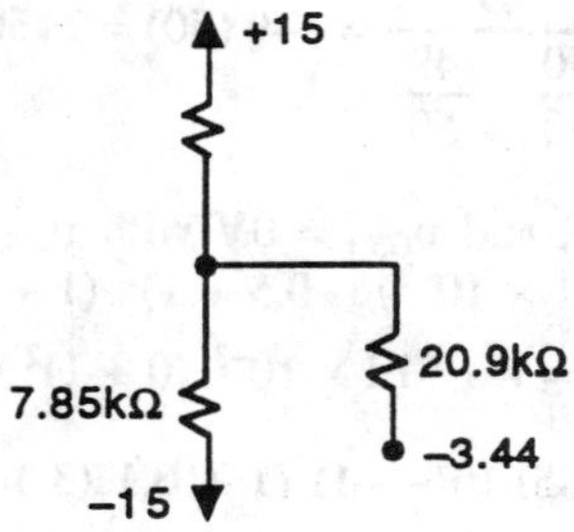

Looking down from the collector of Q_7, see 20.9 || 7.85 = 5.71kΩ to −3.44 + (20.9/(20.9+7.85)) (−15+3.44) = −11.84V.

Thus, $V_A = +15 \frac{-1.16k}{1.16+5.71}$ (15 − − 11.84) = 10.47V (for which Q_5 does not saturate). The maximum positive output is 10.47 − 0.7 = **+9.77V**. For negative output, with Q_8 cutoff, the output is (286/(286+750)) (−15) = **−4.14V**. Thus with a 286Ω load, the output can swing from **9.8V** to **−4.1V**.

Chapter 7

FREQUENCY RESPONSE

SECTION 7.1: s-Domain Analysis

7.1 Using the voltage-divider rule: $T(s) = \frac{V_o(s)}{V_i(s)} = \frac{Z_{shunt}}{Z_{shunt} + Z_{series}}$. Here $Z_{series} = \frac{1}{C_1 s} || R_1 = \frac{R_1/C_1 s}{R_1 + 1/C_1 s} = \frac{R_1}{1 + R_1 C_1 s}$. $\therefore T(s) = \frac{V_o(s)}{V_1(s)} = \frac{1/C_2 s}{1/C_2 s + R_1/(1 + R_1 C_1 s)} = \frac{1}{1 + R_1 C_2 s/(1 + R_1 C_1 s)}$, or $\mathbf{T(s) = \frac{1 + R_1 C_1 s}{1 + R_1 (C_1 + C_2) s}}$. That is, we see a single-time-constant response with a time constant $\tau = R_1 (C_1 + C_2)$, where the resistor R_1 "sees" $C_1 + C_2$, the parallel capacitance of C_1 and C_2 when the source is shorted.

Type of STC:

a) See that the circuit passes dc directly via R_1, with the signal reduced at high frequencies. Thus it is a low-pass (LP) STC circuit.

b) See for $s = 0$, $V_o(s)/V_i(s) = 1$, and for $s = \infty$, $V_o(s)/V_i(s) = C_1/(C_1 + C_2) < 1$.

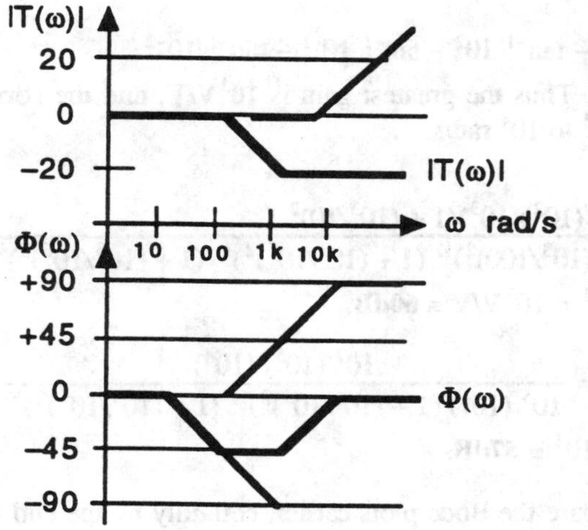

Now, for $R_1 = 10^4\Omega$, $C_1 = (0.5/10)\mu F$, and $C_2 = 0.5\mu F$. There is a pole at $\omega_p = 1/(R_1 (C_1 + C_2)) = 1/(10^4(.05+0.5)10^{-6}) =$ **182 rad/s**, and a zero at $\omega_z = 1/(R_1 C_1) = 1/(10^4(.05)10^{-6}) =$ **2000 rad/s**. From the Figure, or directly, see $T(\omega) = \frac{1 + j\omega R_1 C_1}{1 + j\omega R_1 (C_1 + C_2)}$. Now, $|T(\omega)| = \left[\frac{1 + (\omega R_1 C_1)^2}{1 + (\omega R_1 (C_1 + C_2))^2}\right]^{1/2}$. Thus, $|T(0)| = 1 \equiv 0dB, |T(\infty)| = \frac{C_1}{C_1 + C_2} = \frac{0.5/10}{0.5/10 + 0.5} = \frac{1/10}{1/10 + 1} = \frac{0.1}{0.1 + 1} = \frac{1}{11} \equiv -20.8\text{dB}$.

$\Phi(\omega) = \tan^{-1} \omega R_1 C_1 - \tan^{-1} \omega R_1 (C_1 + C_2)$. $\Phi(0) = 0 - 0 = 0°$, $\Phi(\infty) = 90 - 90 = 0°$. $\Phi(\omega_p) = \tan^{-1} \frac{R_1 C_1}{R_1 (C_1 + C_2)} - \tan^{-1} 1$ $= \tan^{-1}\left[\frac{0.5/10}{0.5/10 + 0.5}\right] - \tan^{-1} 1 = \tan^{-1} 1/11 - \tan^{-1} 1 = 5.19° - 45° = -39.8°$. $\Phi(\omega_z) = \tan^{-1} 1 - \tan^{-1}\left[\frac{R_1 (C_1 + C_2)}{R_1 C_1}\right] = \tan^{-1} 1 - \tan^{-1} 11 = 45° - 84.8° = -39.8°$.

For $\omega_m = (\omega_p \omega_z)^{1/2} = (182 \times 2000)^{1/2} = 603$ rad/s, $\Phi(\omega_m) = \tan^{-1} \frac{603}{2000} - \tan^{-1} \frac{603}{182} = 16.8° - 73.2°$ $= -56.4°$.

7.2

(a) $$T(s) = \frac{10^{14}\,(s)\,(s+10)}{(s+1)\,(s+100)\,(s+10^5)\,(s+10^6)}$$
$$= \frac{10^{14}\,(10)\,(s)\,(1+s/10)}{(100)\,(10^5)\,(10^6)\,(1+s)\,(1+s/100)\,(1+s/10^5)\,(1+s/10^6)}$$
or $$\mathbf{T(s)} = \frac{10^2\,s\,(1+s/10)}{(1+s)\,(1+s/100)\,(1+s/10^5)\,(1+s/10^6)}.$$

(b) As $s \to 0$, $|T(0)| = \frac{10^2\,(0)\,(1)}{(1)\,(1)\,(1)\,(1)} = \mathbf{0V/V}$, $\Phi(0) = \tan^{-1} 1/0 + \tan^{-1} 0/1 - 4\tan^{-1} 0/1 = \mathbf{90°}$.

As $s \to \infty$, $|T(\infty)| = \frac{(\infty)^2}{(\infty)^4} = \frac{1}{\infty^2} = \mathbf{0V/V}$, $\Phi(\infty) = 2\tan^{-1}(\infty) - 4\tan^{-1}(\infty) = -2\tan^{-1}(\infty) =$ **−180°**.

(c) Poles at $s =$ **−1, −100, -10^5, -10^6 rad/s.** Zeros at $s =$ **0, −10, ∞, ∞ rad/s.**

(d) As the frequency rises, the zeros increase the gain while the poles reduce it. We see that the gain increases from 0 to the first pole at 1 rad/s, then begins to rise again at 10 rad/s until 100 rad/s where it flattens, beginning to fall again at 10^5 rad/s, and more at 10^6 rad/s. Thus, we see that the gain is greatest from 100 rad/s to 10^5 rad/s. Thus at $\omega = 10^3$ rad/s, $|T(10^3)| = \frac{10^2\,(10^3)\,(10^2)}{10^3\,(10)\,(1)\,(1)} = 10^3$,

$\Phi(10^3) = 90 + \tan^{-1} 10^2 - \tan^{-1} 10^3 - \tan^{-1} 10 - \tan^{-1} 10^{-2} - \tan^{-1} 10^{-3}$

$\simeq 90 + 90 - 90 - 84 - 0 - 0 =$ **6°**. Also at $\omega = 10^4$ rad/s, $|T(10^4)| = \frac{10^2\,(10^4)\,(10^3)}{10^4\,(10^2)\,(1.1)\,(1)}$
$= 0.9\times10^3$, and

$\Phi(10^4) \simeq 90 + \tan^{-1} 10^3 - \tan^{-1} 10^4 - \tan^{-1} 10^2 - \tan^{-1} 10^{-1} - \tan^{-1} 10^{-2}$

$\simeq 90 + 90 - 90 - 90 - 0 - 0 =$ **0°**. Thus the greatest gain is **10^3V/V**, and the corresponding phase is **0°**, occuring from about 10^3 to 10^4 rad/s.

(e) Gain at 10^3 rad/s,

$$T(\omega = 10^3) = \frac{(10^2)(10^3)(1 + (10^3/10)^2)^{1/2}}{(1 + (10^3/1)^2)^{1/2}\,(1 + (10^3/100)^2)^{1/2}\,(1 + (10^3/10^5)^2)^{1/2}\,(1 + (10^3/10^6)^2)^{1/2}}$$
$$\simeq \frac{10^2\,(10^3)\,(10^2)}{10^3\,(10)\,(1)\,(1)} = 10^{2+3+2-3-1} = \mathbf{10^3 V/V \equiv 60dB}.$$

At 10^5 rad/s, $$T(\omega = 10^5) = \frac{10^2\,(10^5)\,(10^4)}{10^5\,(10^3)\,(1 + (10^5/10^5)^2)^{1/2}\,(1 + (10^5/10^6)^2)^{1/2}}$$
$$= \frac{10^2\,(10^5)\,(10^4)}{10^5\,(10^3)\,(1.414)\,(1)} = 0.707 \times 10^3 \mathbf{\equiv 57dB}.$$

It certainly would have been better to prepare the Bode plots earlier, certainly by the end of part (a). It is actually possible to sketch the pole and zero locations immediately, and to sketch the *shape* of the magnitude plot. However the absolute magnitude requires a calculation like the conversion done in part (a) (see P7.4 following). Probably easiest after (c), most useful before (d), and usefully possible before (b).

(See the Bode Plot on the following page.)

7.2 **(Continued)**

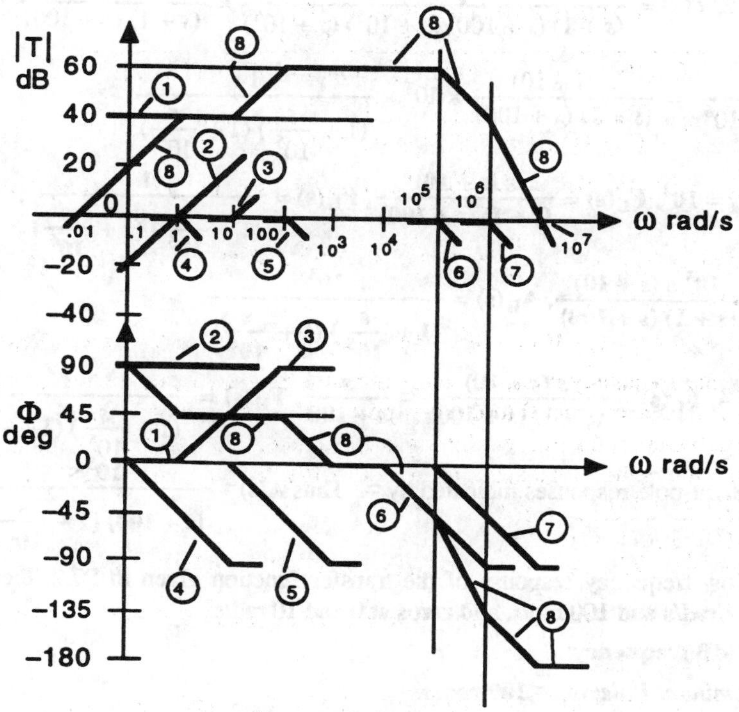

7.3 From P7.2 above, $T(s) = \dfrac{10^2\, s\,(1 + s/10)}{(1 + s)\,(1 + s/100)\,(1 + s/10^5)\,(1 + s/10^6)}$.

For $\omega = 100$ rad/s, $|T(\omega)| = \dfrac{10^2\;10^2\;(1 + (100/10)^2)^{1/2}}{(1 + 100^2)^{1/2}\,(1 + (100/100)^2)^{1/2}\,(1 + (100/10^5)^2)^{1/2}\,(1 + (100/10^6)^2)^{1/2}}$

$= \dfrac{10^2 \times 10^2 \times 10^1}{10^2\,(2^{1/2})\,(1)\,(1)} = 0.707 \times 10^{2+2+1-2} = 0.707 \times 10^3 \equiv$ **57dB**. $\Phi(\omega) = 90 + \tan^{-1} 10 - \tan^{-1} 100 - \tan^{-1} 1 - \tan^{-1} 10^{-3} - \tan^{-1} 10^{-4} = 90 + 84.3 - 89.4 - 45 - 0.1 - 0 =$ **39.8°**.

For $\omega = 2 \times 10^5$ rad/s,

$$| T(\omega) | = \frac{10^2 \times 2 \times 10^5 \,(1 + (\frac{2 \times 10^5}{10})^2)^{1/2}}{(1 + (2 \times 10^5)^2)^{1/2}\,(1 + (\frac{2 \times 10^5}{100})^2)^{1/2}\,(1 + (\frac{2 \times 10^5}{10^5})^2)^{1/2}\,(1 + (\frac{2 \times 10^5}{10^6})^2)^{1/2}}$$

$$= \frac{10^2 \times 2 \times 10^5\,(2 \times 10^4)}{2 \times 10^5\,(2 \times 10^3)\,(1 + 4)^{1/2}\,(1 + 0.2^2)^{1/2}} = \frac{4 \times 10^{11}}{4\,(\sqrt{5})\,(\sqrt{1.04}) \times 10^8} = \frac{10^3}{2.236 \times 1.02}$$

$= 0.4385 \times 10^3 \equiv$ **52.8dB**.

$\Phi(\omega) = 90 + \tan^{-1}(2 \times 10^4) - \tan^{-1}(2 \times 10^5) - \tan^{-1}(2 \times 10^3) - \tan^{-1}(2) - \tan^{-1}(0.2) = 90 + 90 - 90 - 90 - 63.4 - 11.3 =$ **−74.7°**.

SECTION 7.2: The Amplifier Transfer Function

7.4 From P7.2, $T(s) = \dfrac{10^8\, s\,(s+10)}{(s+1)\,(s+100)\,(s+10^5)\,(s+10^6)} = \dfrac{s\,(s+10)}{(s+1)\,(s+100)} \times \dfrac{1}{(1+\frac{s}{10^5})\,(1+\frac{s}{10^6})}$

$\times \dfrac{10^8}{10^5 \times 10^6} = \dfrac{s\,(s+10)}{(s+1)\,(s+100)} \times 10^3 \times \dfrac{1}{(1+\frac{s}{10^5})\,(1+\frac{s}{10^6})}$.

Thus, $\mathbf{A_M = 10^3}$, $\mathbf{F_L(s) = \dfrac{s\,(s+10)}{(s+1)\,(s+100)}}$, $\mathbf{F_H(s) = \dfrac{1}{(1+\frac{s}{10^5})\,(1+\frac{s}{10^6})}}$,

$\mathbf{A_L(s) = \dfrac{10^3\, s\,(s+10)}{(s+1)\,(s+100)}}$, $\mathbf{A_H(s) = \dfrac{10^3}{(1+\frac{s}{10^5})\,(1+\frac{s}{10^6})}}$.

7.5 From P7.4, $\mathbf{F_L(s)} = \dfrac{s\,(s+10)}{(s+1)\,(s+100)} \approx \mathbf{\dfrac{s}{s+100}}$, $\mathbf{F_H(s)} = \dfrac{1}{(1+\frac{s}{10^5})\,(1+\frac{s}{10^6})} \approx \mathbf{\dfrac{1}{1+\frac{s}{10^5}}}$, with the dominant-pole responses indicated by $\approx$. Thus $\mathbf{A(s) \approx \dfrac{10^3\, s}{(s+100)\,(1+\frac{s}{10^5})}}$.

7.6 For the low-frequency response of the transfer function given in P7.2, there are 2 low-frequency poles, at 1 rad/s and 100 rad/s, and zeros at 0 and 10 rad/s.

For the 3dB frequency:

(a) *Dominant Pole*: $\omega_L \approx$ **100 rad/s.**

(b) *Root Squares*: $\omega_L = (100^2 + 1^2 - 2\,(10^2)\,)^{½} = 10^2\,(1 + 10^{-4} - 2 \times 10^{-2})^{½} =$ **99 rad/s.**

(c) Exactly: $\dfrac{(\omega^2 + 10^2)\,(\omega^2)}{(\omega^2 + 1^2)\,(\omega^2 + 100^2)} = \dfrac{1}{2}$ --- (1),

$2\omega^4 + 2(10)^2\,\omega^2 = \omega^4 + 100^2\,\omega^2 + \omega^2 + 100^2$,

$\omega^4 - \omega^2\,(9801) - 10000 = 0$, $\omega^2 = [\,- -9801 \pm \sqrt{9801^2 - 4(-10000))}\,]/2 = [9801 \pm 9803]/2$, that is $\omega^2 = 9802$, or $\omega_L =$ **99.00 rad/s.** Note above in (1) that the zero at zero frequency *must* be included in the calculation. Why?

7.7 For the upper 3dB frequency of the transfer function in P7.2, two poles, at 10^5 and 10^6 rad/s.

For the 3dB cutoff:

(a) *Dominant pole*: $\omega_H =$ **10^5 rad/s.**

(b) *Root squares*: $\omega_H = 1/\sqrt{\dfrac{1}{(10^5)^2} + \dfrac{1}{(10^6)^2}} = 1/\left[\dfrac{1}{10^5}\,(1 + \dfrac{1}{10^2})^{½}\right] = \dfrac{10^5}{(1 + 1/100)^{½}}$, or

$\omega_H =$ **0.995×10^5 rad/s.**

(c) *Exactly*: $F_H(s) = \dfrac{1}{(1 + s/10^5)\,(1 + s/10^6)}$, $T^2\,(\omega) = \dfrac{1}{(1 + (\omega/10^5)^2)\,(1 + (\omega/10^6)^2)} = 1/2$,

$1 + \omega^2\,(1/10^{10} + 1/10^{12}) + \omega^4/10^{22} = 2$, $\omega^4 + \omega^2\,(10^{12} + 10^{10}) - 10^{22} = 0$,

$\omega^4 + 101 \times 10^{10}\,\omega^2 - 10^{22} = 0$,

$\omega^2 = \dfrac{-101 \times 10^{10} \pm \sqrt{101^2 \times 10^{20} + 4 \times 10^{22}}}{2}$,

$\omega^2 = \dfrac{-101 \times 10^{10} \pm 102.96 \times 10^{10}}{2} = .9805 \times 10^{10}$ and $\omega_H =$ **$.990 \times 10^5$ rad/s.**

7.8 Old $T(s) = \dfrac{s(s+10)}{(s+1)(s+100)} \times 10^3 \times \dfrac{1}{(1+s/10^5)(1+s/10^6)}$.

Modified $T(s) = \dfrac{s(s+10)}{(s+1)(s+100)} \times 10^3 \times \dfrac{1}{(1+s/10^5)(1+s/10^6)} \times \dfrac{(1+s/10^5)}{(1+s/(2\times10^6))}$.

Note the *pole-zero cancellation* with the result that T(s) = $\dfrac{\mathbf{10^3\, s\,(s+10)}}{\mathbf{(s+1)(s+100)(1+s/10^6)(1+s/(2\times10^6))}}$, and $\omega_H \simeq 1/\sqrt{(1/(1\times10^6))^2 + (1/(2\times10^6))^2}$ = $10^6\,(1+1/4)^{-1/2} = \mathbf{0.894\times10^6}$ **rad/s.**

Exactly, $\dfrac{1}{(1+(\omega/10^6)^2)(1+(\omega/(2\times10^6))^2} = 1/2,\ 1+\dfrac{\omega^2}{10^{12}}(1+\dfrac{1}{4})+\dfrac{\omega^4}{2(10^{24})} = 2,$

$\omega^4 + 2 \times 10^{12}\ \omega^2 - 2(10^{24}) = 0,\ \therefore\ \omega^2 = \dfrac{-2\times10^{12} \pm \sqrt{4\times10^{24} - 4(-2\times10^{24})}}{2}$ = $\dfrac{10^{12}(-2\pm3.46)}{2} = 0.73\times10^{12}$. Thus $\omega_H = \mathbf{0.856\times10^6}$ **rad/s.**

7.9

(1) *Using open-circuit time constants*: For C_1, $R_{eq} = R_1$, and $\tau_1 = R_1 C_1$. For C_2, $R_{eq} = R_1 + R_2$, and $\tau_2 = C_2(R_1+R_2)$.

(a) For $R_1 = R_2 = 10k\Omega$, $C_1 = C_2$ = 100pF, $\tau_1 = 10^4\times100\times10^{-12} = 10^{-6}$s, $\tau_2 = 100\times10^{-12}\ (10^4+10^4) = 2\times10^{-6}$s, $\omega_H = \dfrac{1}{\tau_1+\tau_2} = \dfrac{1}{1\times10^{-6}+2\times10^{-6}}$ $= 1/3\times10^{-6} = 333\times10^3$ rad/s, or = $\mathbf{0.333\times10^6}$**rad/s.** Alternatively, the sum of squares approach yields $\omega_H = 1/(1^2+2^2)^{1/2}\times10^6 = \mathbf{0.447\times10^6}$**rad/s.**

(b) For R_1 = 10kΩ, R_2 = 100kΩ, C_1 = 100pF, C_2 = 10pF, $\tau_1 = 10^4\times100\times10^{-12} = 10^{-6}$s, $\tau_2 = 10\times10^{-12}\ (10\times10^4+1\times10^4) = 10\times10^{-8}\ (11) = 1.1\times10^{-6}$s, $\omega_H = \dfrac{1}{(1+1.1)\,10^{-6}} = \mathbf{0.476\times10^6}$ **rad/s.** Alternatively, the sum of squares approach yields $\omega_H = 1/(1^2+1.1^2)^{1/2}\times10^6 = \mathbf{0.673\times10^6}$ **rad/s.**

(c) R_1 = 10kΩ, R_2 = 100kΩ, $C_1 = C_2$ = 10pF, $\tau_1 = 10^4\times10\times10^{-12} = 0.1\times10^{-6}$s, $\tau_2 = 10\times10^{-12}\times(10^5+10^4) = 10\times10^{-12}\times10^4\ (11) = 1.1\times10^{-6}$s, $\omega_H = 1/((0.1+1.1)\ 10^{-6})$ = $\mathbf{0.833\times10^6}$ **rad/s.** Alternatively, using the sum of squares, ω_H = $1/(0.1^2+1.1^2)^{1/2}\times10^6 = \mathbf{.905\times10^6}$ **rad/s.**

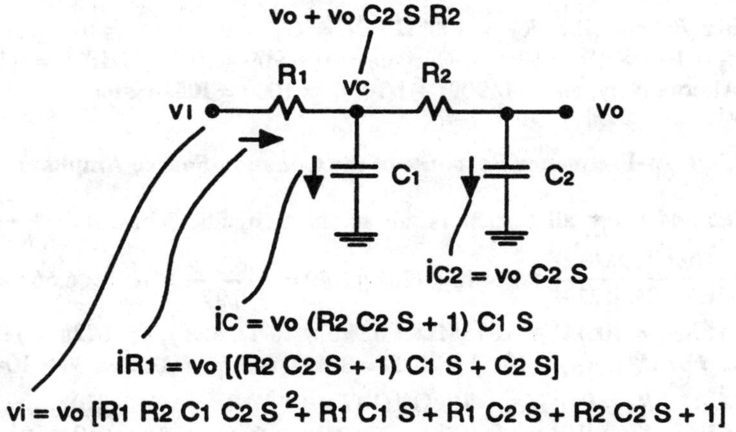

(2) *Exactly*: $T(s) = \dfrac{\upsilon_o}{\upsilon_i} = \dfrac{1}{s^2 R_1 R_2 C_1 C_2 + s(R_1 C_1 + R_1 C_2 + R_2 C_2) + 1}$,

$T(j\omega) = \dfrac{1}{j\omega(R_1 C_1 + R_1 C_2 + R_2 C_2) + (1 - \omega^2 R_1 R_2 C_1 C_2)}$. Now response is 3dB down when $\omega^2 (R_1 C_1 + R_1 C_2 + R_2 C_2)^2 + (1 - \omega^2 R_1 R_2 C_1 C_2)^2 = 2$. Now, for particular cases:

(a) $R_1 = R_2 = 10k\Omega$, $C_1 = C_2 = 100pF$. Now, $\omega^2 (3 \times 10^4 \times 10^{-10})^2 + (1 - \omega^2 \times 10^4 \times 10^4 \times 10^{-10} \times 10^{-10})^2 - 2 = 0$, $9 \times 10^{-12} \omega^2 + (1 - \omega^2 10^{-12})^2 - 2 = 0$, $9\omega^2 10^{-12} + 1 - 2\omega^2 10^{-12} + \omega^4 10^{-24} - 2 = 0$, $\omega^4 + \omega^2 (7 \times 10^{12}) - 10^{24} = 0$, $\omega^2 = \dfrac{-7 \times 10^{12} \pm 10^{12}\sqrt{7^2 - -4}}{2} = 10^{12}\dfrac{(-7 \pm 7.28)}{2} = .14 \times 10^{12}$, and $\omega_H = \mathbf{0.374 \times 10^6}$ **rad/s.**

(b) $R_1 = 10k\Omega$, $R_2 = 100k\Omega$, $C_1 = 100pF$, $C_2 = 10pF$. Now, $\omega^2 (10^4 \times 10^{-10} + 10^4 \times 10^{-11} + 10^5 \times 10^{-11})^2 + (1 - \omega^2 \times 10^4 \times 10^5 \times 10^{-10} \times 10^{-11})^2 - 2 = 0$, or $\omega^2 (10^{-6} + 10^{-7} + 10^{-6})^2 + (1 - \omega^2 10^{-12})^2 - 2 = 0$, or $\omega^2 (2.1 \times 10^{-6})^2 + (1 - 10^{-12}\omega^2)^2 - 2 = 0$, $4.41 \times 10^{-12}\omega^2 + 1 - 2 \times 10^{-12}\omega^2 + \omega^4 10^{-24} - 2 = 0$,

$\omega^4 10^{-24} + 2.41 \times 10^{-12}\omega^2 - 1 = 0$, $\omega^4 + 2.41 \times 10^{12}\omega^2 - 10^{24} = 0$,

$\omega^2 = 10^{12}(-2.41 \pm \sqrt{2.41^2 - -4})/2 = 10^{12}(-2.41 \pm 3.132)/2 = 0.309 \times 10^{12}$.

Thus $\omega_H = \mathbf{0.601 \times 10^6}$ **rad/s.**

(c) $R_1 = 10k\Omega$, $R_2 = 100k\Omega$, $C_1 = 10pF$, $C_2 = 10pF$. Now, $\omega^2 (10^4 \times 10^{-11} + 10^4 \times 10^{-11} + 10^5 \times 10^{-11})^2 + (1 - \omega^2 \times 10^4 \times 10^5 \times 10^{-11} \times 10^{-11})^2 - 2 = 0$. Thus $\omega^2 (1.2 \times 10^{-6})^2 + (1 - \omega^2 (0.1) 10^{-12})^2 - 2 = 0$, $1.44 \times 10^{-12}\omega^2 + 1 - 2\omega^2 (0.1) 10^{-12} + \omega^4 (.01) 10^{-24} - 2 = 0$, $\omega^4 + 1.24 \times 10^{+14} - 10^{+26} = 0$,

$\omega^2 = (-1.24 \times 10^{14} \pm \sqrt{1.24^2 \times 10^{28} - 4(-10^{26})})/2 = 10^{14}(-1.24 \pm \sqrt{1.24^2 + .04})/2 = 0.00801 \times 10^{14} = 0.801 \times 10^{12}$. Thus $\omega_H = \mathbf{0.895 \times 10^6}$ **rad/s.**

7.10 Using short-circuit time constants: For C_1, $R_{eq} = R_1 \| R_2$, and $\tau_1 = C_1 R_1 R_2 / (R_1 + R_2)$. For C_2, $R_{eq} = R_2$, and $\tau_2 = C_2 R_2$.

(a) For $R_1 = R_2 = 10k\Omega$, $C_1 = C_2 = 1\mu F$, $\tau_1 = 1 \times 10^{-6} \times 10^4/2 = 0.5 \times 10^{-2}s$, and $\omega_1 = 1/0.5 \times 10^{-2} = 200$ rad/s; $\tau_2 = 10^{-6} \times 10^4 = 10^{-2}s$, and $\omega_2 = 100$ rad/s. Thus $\omega_L \approx 100 + 200 =$ **300 rad/s**. Using the sum of squares idea, $\omega_L = (100^2 + 200^2)^{1/2} =$ **224 rad/s.**

(b) For $R_1 = 10k\Omega$, $R_2 = 100k\Omega$, $C_1 = 1\mu F$, $C_2 = 0.1\mu F$, $\tau_1 = 10^{-6} \times 10k\Omega \| 100k\Omega = 0.909 \times 10^{-2}s$, $\tau_2 = 10^5 \times 10^{-7} = 10^{-2}s$. Thus $\omega_L = (1/0.909 + 1/1) \times 1/10^{-2} =$ **210 rad/s**. Alternatively, $\omega_L = ((1/.909)^2 + (1/1)^2)^{1/2} \times 100 =$ **147 rad/s.**

(c) For $R_1 = 10k\Omega$, $R_2 = 100k\Omega$, $C_1 = C_2 = 0.1\mu F$, $\tau_1 = 10^{-7} \times .909 \times 10^{+4} = 0.909 \times 10^{-3}s$, $\tau_2 = 10^{-7} \times 10^5 = 10^{-2}s$. Thus $\omega_L = 1/(0.909 \times 10^{-3}) + 1/10^{-2} = 1100 + 100 =$ **1200 rad/s.** Alternatively, $\omega_L = (1/.909^2 + 1/10^2)^{1/2} \times 1000 =$ **1054 rad/s.**

SECTION 7.3: Low-Frequency Response of the Common-Source Amplifier

7.11 *For Midband Gain*: all capacitors are ac short circuits. Thus, $A_\upsilon = \dfrac{-R_{G1} \| R_{G2}}{R + R_{G1} \| R_{G2}} \times g_m R_L \| R_D = -\dfrac{10M \| 22M}{100k + 10M \| 22M} \times 2 \times 10^{-3} (20k \| 10k) = -\dfrac{6.875}{6.975} \times 10^{-3} \times 6.667 \times 10^3 =$ **– 13.14 V/V.**

For C_{C1}: $R_{C1} = 100k\Omega + 6.875M\Omega = 6.975 \times 10^6\Omega$, and $f_{p1} = 1/(2\pi \times 6.975 \times 10^6 \times 0.01 \times 10^{-6}) =$ **2.28Hz**. *For* C_{C2}: $R_{C2} = 10k\Omega + 20k\Omega = 30k\Omega$, and $f_{p2} = 1/(2\pi \times 30 \times 10^3 \times 0.1 \times 10^{-6}) =$ **53.1Hz.**

For C_S: $R_S' = R_S \| 1/g_m = 10k\Omega \| (1/(2 \times 10^{-3})) = 0.5k \| 10k = 476\Omega$, and $f_{pS} = 1/(2\pi \times .476 \times 10^3 \times 1 \times 10^{-6}) =$ **334.3Hz**, with $f_{zS} = 1/(2\pi \times 10 \times 10^3 \times 1 \times 10^{-6}) =$ **15.9Hz.**

7.12 Use the largest capacitor at the source, where the resistance level is least. Now, $C_S = 1/(2\pi \times .476 \times 10^3 \times 10) = 33.4\mu F$. Use C_S = **30μF**. Now, $C_{C1} = 1/(2\pi \times 6.975 \times 10^6 \times 1)$ = 0.0236μF. Use C_{C1} = **0.02μF**. Now, $C_{C2} = 1/(2\pi \times 30 \times 10^3 \times 1) = 5.3\mu F$. Use C_{C2} = **5μF**.

Actual critical frequencies are: $f_{pS} = 1/(2\pi \times .476 \times 10^3 \times 30 \times 10^{-6})$ = **11.1Hz**, $f_{p1} = 1/(2\pi \times 6.975 \times 10^6 \times .02 \times 10^{-6})$ = **1.14Hz**, $f_{p2} = 1/(2\pi \times 30 \times 10^3 \times 5 \times 10^{-6})$ = **1.06Hz**, $f_{zS} = 1/(2\pi \times 10 \times 10^3 \times 30 \times 10^{-6})$ = **0.53Hz**.

7.13 Modify Equations 7.35 through 7.36 for the addition of r_S in series with C_S: $I_d(s) = I_S(s) = \dfrac{V_g(s)}{1/g_m + Z_S} = g_m V_g(s) \dfrac{Y_S}{g_m + Y_S}$. Here $Y_S = \dfrac{1}{Z_S} = \dfrac{1}{R_S} + \dfrac{1}{1/(s C_S + r_S)}$

$$= \frac{1}{R_S} + \frac{s C_S}{1 + s r_S C_S} = \frac{s C_S (R_S + r_S) + 1}{R_S (s C_S r_S + 1)}.$$ Thus $$i_d(s) = g_m V_g(s) \frac{\dfrac{s C_S (R_S + r_S) + 1}{R_S (s C_S r_S + 1)}}{g_m + \dfrac{s C_S (R_S + r_S) + 1}{R_S (s C_S r_S + 1)}}$$

$$= g_m V_g(s) \frac{s C_S (R_S + r_S) + 1}{s C_S (g_m R_S r_S + R_S + r_S) + R_S g_m + 1} = V_g(s) \frac{(R_S + r_S)}{g_m R_S r_S + R_S + r_S}$$

$$\times \frac{s + \dfrac{1}{C_S (R_S + r_S)}}{s + \dfrac{1 + R_S g_m}{C_S (g_m R_S r_S + R_S + r_S)}}.$$ See that there is a zero at $\omega_z = 1/(C_S (R_S + r_S))$, and a pole at

$$\omega_p = \frac{1}{C_S \left[\dfrac{R_S + r_S (1 + g_m R_S)}{1 + g_m R_S}\right]},$$ where the resistor associated with C_S at the pole frequency is

$$r_S + 1/g_m \| R_S = r_S + \frac{R_S/g_m}{1/g_m + R_S} = r_S + \frac{R_S}{1 + g_m R_S} = \frac{r_S (1 + g_m R_S) + R_S}{1 + g_m R_S},$$ as noted. Now the

equivalent **transconductance** is $\dfrac{1}{1/g_m + R_S \| r_S} = \dfrac{g_m}{1 + \dfrac{g_m R_S r_S}{R_S + r_S}} = \dfrac{\mathbf{g_m (R_S + r_S)}}{\mathbf{g_m R_S r_S + R_S + r_S}}$, as noted.

Now for the situation in P7.11, with g_m = 2mA/V, R_S = 10kΩ, the gain is reduced by a factor of two, when $\dfrac{R_S + r_S}{g_m R_S r_S + R_S + r_S} = 1/2$, or $1 + \dfrac{g_m R_S r_S}{R_S + r_S} = 2$, or $g_m R_S r_s = R_S + r_S$, or $r_S = \dfrac{R_S}{g_m R_S - 1} = \dfrac{10 \times 10^3}{2 \times 10^{-3} \times 10 \times 10^3 - 1} = \dfrac{10 \times 10^3}{19}$ = **526Ω**. *For the new pole*: $R_p = r_S + (1/g_m) \| R_S$ = $526 + (1/(2 \times 10^{-3})) \| (10 \times 10^3) = 526 + 476$ = **1000Ω**. This should have been obvious since the gain was to have been reduced by a factor of 2, and $1/g_m$ = 500Ω. Thus $f_{PS} = 1/(2\pi \times 1000 \times 1 \times 10^{-6})$ = **159Hz**, and $f_{ZS} = 1/(2\pi \times 1 \times 10^{-6} (10k\Omega + .526k\Omega))$ = **15.1Hz**.

SECTION 7.4: High-Frequency Response of the Common-Source Amplifier

7.14 From Equation 7.46, $f_T = g_m / [\, 2\pi (C_{gs} + C_{gd})\,]$. Generally, $i_D = I_{DSS}(1 - \dfrac{v_{GS}}{V_p})^2$ - - - (1).

(a) In (1): $1 = 4\,(1 - \dfrac{v_{GS}}{-2})^2 \rightarrow 1 + \dfrac{v_{GS}}{2} = (\tfrac{1}{4})^{1/2}$ = 1/2. Thus $\dfrac{v_{GS}}{2} = -1/2$, $v_{GS} = -1$. Now $g_m = \dfrac{2I_{DSS}}{-V_p}(1 - \dfrac{v_{GS}}{V_p}) = \dfrac{2(4)}{--2}(1/2)$ = 2mA/V. Thus $f_T = \dfrac{2\times 10^{-3}}{2\pi(2 + 0.2)\times 10^{-12}}$ = **144.7MHz.**

(b) $i_D = K(\upsilon_{GS} - V_t)^2$, $200 = 100(\upsilon_{GS} - 1)^2 \rightarrow (\upsilon_{GS} - 1) = \sqrt{2}$, $g_m = 2K(\upsilon_{GS} - V_t) = 2(100 \times 10^{-6})\sqrt{2} = 283\mu A/V$, $C_{gs} = 0.15 \times 10^{-12} + 20 \times 10^{-15} + 0.1 \times 10^{-12} = 0.27pF$, and $C_{gd} = 20 \times 10^{-15}F = 0.02pF$. Thus $f_T = \dfrac{283 \times 10^{-6}}{2\pi(.27 + .02) \times 10^{-12}} = \mathbf{155.3MHz}$.

(c) $f_T = \dfrac{10 \times 10^{-3}}{2\pi(.15 + .015) \times 10^{-12}} = \mathbf{9.65GHz}$.

7.15 $K = 1/2\mu\, C_{ox}\, W/L = 1/2\,(.05 \times 10^{12}) \times 1 \times 10^{-15} \times 27/3 = .225 \times 10^{-3} A/V^2 = 225\mu A/V^2$; $g_m = 2K(\upsilon_{GS} - V_t) = 2(225 \times 10^{-6})(2.5 - 0.5) = 900\mu A/V$; $C_{gd} = L_d\, W\, C_{ox} = .3 \times 27 \times 1 \times 10^{-15} =$ **8.1fF**. $C_{gs} = 2/3\, WL\, C_{ox} + L_d\, W\, C_{ox} = 2/3 \times 27 \times 3 \times 1 \times 10^{-15} + 8.1 \times 10^{-15} = 54 + 8.1 =$ **62.1fF**. Thus $f_T = \dfrac{g_m}{2\pi(C_{gs} + C_{gd})} = \dfrac{900 \times 10^{-6}}{2\pi(62.1 + 8.1) \times 10^{-15}} = \mathbf{2.04GHz}$.

7.16 $C_g = C_{gs} + C_{gd}(1 - \text{gain}) = 200 + 20(1 - -1) =$ **240fF**, for gain of $-1V/V$, or $= 200 + 20(1 - -100) =$ **2220fF**, for gain of $-100V/V$. $C_d = C_{db} + C_{gd}(1 - 1/\text{gain}) = 100 + 20(1 - - 1/1) =$ **140fF**, for gain of $-1V/V$, or $= 100 + 20(1 - - 1/100) =$ **120.2fF**, for gain of $-100V/V$.

7.17 Gain, gate to source, is $-g_m(r_o \| R_D \| R_L) = -1 \times 10^{-3}(50 \| 10 \| 30) \times 10^3 = -6.52V/V$. $C_g = 1pF + 0.5pF(1 - -6.52) =$ **4.76pF**, $C_d = 0.5(1 - - 1/6.52) =$ **.577pF**. Input pole: $f_{pg} = 1/(2\pi(4.76 \times 10^{-12})(100k\Omega \| 1M\Omega)) =$ **.368MHz**. Output pole: $f_{pd} = 1/(2\pi \times .577 \times 10^{-12}(50 \| 10 \| 30) \times 10^3) =$ **42.3MHz**. Upper 3dB frequency $f_{H1} = f_{pg} =$ **0.37MHz**. For R_S reduced to zero, the output pole dominates, and $f_{H2} = f_{pd} =$ **42.3MHz**. Now for $f_{H3} = 0.9(42.3) = 38MHz$, with R_S non-zero and using $f_{pg}' = f_g$. Now $\left[\dfrac{1}{f_g^2} + \dfrac{1}{f_{pd}^2}\right]^{1/2} = \dfrac{1}{38.0}$, $\dfrac{1}{f_g^2} + \dfrac{1}{42.3^2} = \dfrac{1}{38.0^2}$, $\dfrac{1}{f_g} = (.000693 - .000559)^{1/2}$. Thus $f_g =$ **86.4MHz**. Now $f_g = 1/(2\pi(4.76 \times 10^{-12})R_S)$, whence $R_S = 1/(2\pi(4.76 \times 10^{-12})(86.35 \times 10^6)) =$ **387Ω**

7.18 Using the results of P7.17 and Equations 7.59 and 7.60, $\omega_{P1} = 1/\left[C_{gs} + C_{gd}(1 + g_m R_L') + C_{gd}(R_L'/R')\right] R'$

$$= \frac{1}{[1.0 + 0.5(1 + 6.52) + 0.5(6.52/90.9)]\, 90.9 \times 10^3 \times 10^{-12}} = \frac{1}{[4.76 + .036] \times 90.9 \times 10^{-9}}$$

$=$ 2M rad/s, or $f_{p1} =$ **0.365MHz**. $\omega_{p2} = \dfrac{C_{gs} + C_{gd}(1 + g_m R_L') + C_{gd}(R_L'/R')}{C_{gs}\, C_{gd}\, R_L'}$

$= \dfrac{4.796 \times 10^{-12}}{1 \times 10^{-12} \times 0.5 \times 10^{-12} \times 6.52 \times 10^3} = 1.47Grad/s$, or $f_{p2} =$ **234MHz**. $f_z = \dfrac{g_m}{2\pi C_{gd}} = \dfrac{1 \times 10^{-3}}{(2\pi(0.5 \times 10^{-12}))} =$ **318MHz**, for which $f_H \approx f_{p1} =$ **0.365MHz**. Now, for $R_S = 1k\Omega$, $f_z =$ **318MHz**, $f_{p1} = 1/(2\pi(4.76 + 0.5 \times (6.52/1))\, 10^{-12} \times 10^{3)} =$ **19.8MHz**, $f_{p2} = \dfrac{(4.76 + 3.26) \times 10^{-12}}{2\pi(1 \times 0.5 \times 10^{-24} \times 6.52 \times 10^3)}$ $=$ **392MHz**, for which $f_H \approx$ **19.8MHz**.

7.19 $i_D = K(\upsilon_{GS} - V_t)^2 \rightarrow 1 = K(2 - 1)^2 = K$; $g_m = 2K(\upsilon_{GS} - V_t) = 2(1)(2 - 1) = 2mA/V$. Thus $f_T = \dfrac{g_m}{2\pi(C_{gs} + C_{gd})} \rightarrow 10^9 = \dfrac{2 \times 10^{-3}}{2\pi(C_{gs} + C_{gd})}$. Thus $C_{gs} + C_{gd} = \dfrac{2 \times 10^{-3}}{2\pi(10^9)} = 0.318pF$. Now, if $C_{gd} = 0.2\, C_{gs} = C_{db}$, then $1.2\, C_{gs} = .318pF$, $C_{gs} = 0.265pF$, $C_{gd} = C_{db} = .053pF$. Thus $C_{in} = .265 + .053(1 - - 3) =$ **0.477pF**. For input source, resistance $3/g_m = 3/(2 \times 10^{-3}) = 1.5k\Omega$, and capacitance $= 4(.053pF) = .212pF$, $f_H \approx \dfrac{1}{2\pi(.212 \times 10^{-12})(1.5 \times 10^{-3})} =$ **500MHz**. Problem P7.32 of the Text provides the topology for which this high performance is possible.

7.20 $R_{in} = \dfrac{R_f}{1 - -gain} = \dfrac{R_f}{1 + 3} = \dfrac{R_f}{4}$, $\dfrac{R_{in}}{R_{in} + 10k\Omega} = 0.95 = \dfrac{R_f/4}{R_f/4 + 10}$. Thus $R_f/4 = .95\, R_f/4 + 9.5$, $R_f =$

4(9.5)/.05 = **760kΩ**! See R_f must be very high, even in such a low-impedance circuit.

7.21 The output pulse is *positive* with amplitude = 50 (50 × 10^{-3}) = **2.5V** and duration of 50μs. Its transition times are $\frac{2.2}{2\pi f_H} = \frac{2.2}{2\pi \times 50 \times 10^6}$ = **7ns**. Its sag (or droop) = $\frac{\Delta V}{V} = \frac{t_p}{\tau_L} = 2\pi f_L t_p = 2\pi \times 50 \times 50 \times 10^{-6}$ = 0.0157, or **1.6%**.

SECTION 7.5: The Hybrid-π Equivalent Circuit Model of the BJT

7.22 $r_o = V_A/I_C$ = 100/2mA = **50kΩ**, $g_m = I_C/V_T$ = 2mA/25mV = **80mA/V**, $r_\pi = \beta/g_m$ = 100/80 × 10^{-3} = **1.25kΩ**, r_x = **50Ω**, $r_\mu \simeq 10\,\beta\, r_o$ = 10 (100) 50kΩ = **50MΩ**, $\mu = g_m r_o$ = 80 × 10^{-3} × 50 × 10^3 = **4000V/V**.

7.23 For r_x = 0, the measured resistance $r = r_e$. For n = 1, $r_e = \frac{V_T}{I_E} = \frac{25}{10}$ = **2.5Ω**. More generally, $r = r_e + \frac{r_x}{\beta + 1}$ = 2.9. Thus $\frac{r_x}{\beta+1}$ = 2.9 –2.5 = 0.4Ω, r_x = 0.4(β + 1). For the extra 50Ω resistor, 3.4 = 2.5 + $\frac{r_x + 50}{\beta + 1}$. Thus $\frac{r_x + 50}{\beta + 1}$ = 0.9, r_x = (0.9) (β + 1) –50, but also: r_x = (0.4) (β + 1). Subtract: 0 = 0.5 (β + 1) –50, β + 1 = 50/0.5 = 100, β = **99**. Thus r_x = 0.4 (100) = **40Ω**, $g_m = \alpha/r_e = \frac{99}{100}$/25 = **39.6mA/V**, r_π = (β + 1) r_e = 100 (2.5) = **250Ω**. For $n \neq$ 1.0, $r_e = n\, V_T/I_E$ = 2.5n. To find n, add another measurement with another resistor in the base, say 100Ω, to get a 3rd equation.

7.24 $g_m = I_C/V_T$ = 10/25 = **400mA/V**, $r_\pi = h_{fe}/g_m$ = 200/(400 × 10^{-3}) = **0.5kΩ**, $r_x = h_{ie} - r_\pi$ = 0.55 – 0.5 = **50Ω**, $r_\mu = r_\pi/h_{re}$ = 0.5 × 10^3/(1 × 10^{-4}) = **5MΩ**, $r_o = (h_{oe} - h_{fe}/r_\mu)^{-1}$ = (1.2 × 10^{-4} – 200/(5 × 10^6))$^{-1}$ = (1.6 × 10^{-4})$^{-1}$ = **6.25kΩ**, $V_A = I_C\, r_o$ = 6.25 × 10^3 × 10 × 10^{-3} = **62.5V**.

7.25 $f_\beta = 1/(2\pi\,(C_\pi + C_\mu)\, r_{pi})$. For I_C = 2mA, g_m = 2mA/25mV = 80mA/V, $r_\pi = \beta/g_m$ = 200/80 = 2.5kΩ, ∴ 12.7 × 10^6 = 1/(2π (C_π + 0.5) × 10^{-12} × 2.5 × 10^3).

∴ C_π = 5.01 –0.5 = **4.51pF**, and $f_T = \beta_o\, \omega_\beta$ = 200 (12.7 × 10^6) = **2.54GHz**. For I_C = 10mA, g_m = 10/2 × 80 = 400mA/V, r_π = 2/10 × 2.5 = 0.5kΩ, C_π = 10/2 × 4.51 = **22.55pF**, f_β = 1/(2π (22.55 + 0.5) × 10^{-12} × 0.5 × 10^3) = **13.8MHz**. For $C_\pi = C_\mu$ = 0.5pF, $\frac{4.51pF}{2mA} = \frac{0.5pF}{I_C}$. Thus $I_C = \frac{0.5 \times 2}{4.51}$ = **0.222mA**. That is, f_T is maintained at 2.5GHz for currents > 0.22mA or so.

7.26

Transistor	I_E mA	r_e Ω	g_m mA/V	r_π kΩ	β_o	f_T GHz	C_μ pF	C_π pF	f_β MHz
a	**1.0**	25	**39.8**	**5.025**	200	1	0.5	**5.83**	**5**
b	**0.1025**	**244**	4	10	**40**	**4**	0.1	**0.059**	100
c	0.01	**2500**	**0.397**	**378**	150	**0.03**	**0.1**	2	0.2

(a) $r_e = \frac{V_T}{I_E} \rightarrow I_E = \frac{V_T}{r_e} = \frac{25\times10^{-3}}{25}$ = **1mA**. $g_m = \frac{\alpha}{r_e} = \frac{200/201}{r_e}$ = **39.8mA/V**. $r_\pi = (\beta+1)\, r_e$ = 201 (25) = **5.025kΩ**. $C_\pi = \frac{1}{2\pi\,(r_\pi/\beta)\, f_T} - C_\mu = \frac{200}{2\pi\,(5.025\times10^3)\times1\times10^9} - 0.5\times10^{-12}$ = **5.83pF**. $f_\beta = \frac{f_T}{\beta} = \frac{1\times10^9}{200}$ = **5MHz.**

(b) $g_m\, r_\pi = \beta$. Thus $\beta = 4\times10^{-3}\times10\times10^3$ = **40**. $r_e = \frac{r_\pi}{\beta+1} = \frac{10\times10^3}{41}$ = **244Ω**. $I_E = \frac{V_T}{r_e} = \frac{25\times10^{-3}}{244}$ = **0.1025mA**. I_C = 40/41 (.1025) = **0.1mA**. $f_T = \beta f_\beta$ = 40 (100) = **4GHz**. $C_\pi = 1/(2\pi f_\beta\, r_\pi) - C_\mu = 1/(2\pi\times100\times10^6\times10\times10^3) - 0.1\times10^{-12} = (0.159-0.1)10^{-12}$ = **0.059pF**.

(c) $r_e = \frac{V_T}{I_E} = \frac{25mV}{.01mA}$ = **2500Ω**. $g_m = \frac{\alpha}{r_e} = \frac{150}{151}\times\frac{1}{2500}$ = **0.397mA/V**. $r_\pi = (\beta+1)\, r_e$ = 151 (2.5k) = **377.5kΩ**. $f_T = \beta f_\beta = 1.0\times0.2\times10^6$ = 30MHz = **0.03GHz**. $(C_\pi + C_\mu) = \frac{1}{2\pi f_\beta\, r_\pi} = \frac{1}{2\pi\times.2\times10^6\times378\times10^3}$ = 2.1pF. Thus C_μ = 2.1 – 2 = **0.1pF**.

SECTION 7.6: Frequency Response of the Common-Emitter Amplifier

7.27 $r_e = \frac{25\times10^{-3}}{0.15\times10^{-3}} = 166.7\Omega$, $r_\pi = 166.7\times151 = 25.17k\Omega$, $A_M = -\frac{25.17\,\|\,40}{25.17\,\|\,40 + 10}\times\frac{9.1\,\|\,10\,\|\,500}{166.7} = -\frac{15.45}{25.45}\times\frac{4.93\times10^3}{166.7} = -0.607\times25.6 = -17.95 \approx$ **–18V/V**. Now, $C_\pi + C_\mu = \frac{g_m}{2\pi f_T} = \frac{\frac{150}{151}\times\frac{1}{166.7}}{2\pi\times10^9}$ = 0.948pF, C_π = .948 –.30 = .648pF.

Input Pole: $C_T = 0.648 + 0.3\,(1 - -25.6) = 8.63pF$, $R_T = 25.17k\,\|\,40k\Omega\,\|\,10k\Omega + 50 = 6.12k\Omega$,

$f_{p1} = \frac{1}{2\pi\times8.63\times10^{-12}\times6.12\times10^3}$ = 3.00MHz.

Output Pole: $C_T = 0.3pF$, $R_T = 9.1k\Omega\,\|\,10k\Omega\,\|\,500k\Omega = 4.93k\Omega$,

$f_{p2} = \frac{1}{2\pi\times0.3\times10^{-12}\times4.93\times10^3}$ = 108MHz. ∴ The upper 3dB frequency is $f_H \approx$ **3.0MHz**.

7.28 *For* C_{C1}: $R_{C1} = 10k\Omega + 40k\Omega\,\|\,25.2k\Omega = 10k\Omega + 15.5k\Omega = 25.5k\Omega$,

$f_{p1} = \frac{1}{2\pi\times25.5\times10^3\times1\times10^{-6}}$ = **6.24Hz.**

For C_{C2}: $R_{C2} = (9.1\,\|\,500k\Omega + 10k\Omega) = 18.94k\Omega$, $f_{p2} = \frac{1}{2\pi\times18.94\times10^3\times1\times10^{-6}}$ = **8.4Hz.**

For C_E: $R_{CE} = \left[167 + \frac{50 + 40k\,\|\,10k}{151}\right]\|\,8.2k = 220.3\,\|\,8.2k = 214\Omega$, $f_{pE} = \frac{1}{2\pi\times214\times10\times10^{-6}}$ = **74.4Hz**, $f_z = \frac{1}{2\pi\times8.2\times10^3\times10\times10^{-6}}$ = **1.94Hz.**

for F_L: $f_L = \left[6.24^2 + 8.4^2 + 74.4^2 - 2\,(1.94)^2\right]^{1/2}$ = (39+71+5535–7.5)$^{1/2}$ = **75.1Hz.**

7.29 Use data from P7.28 above, where for C_{C1}, R_{C1} = 25.5kΩ; for C_{C2}, R_{C2} = 18.9kΩ; for C_E, R_{CE} = 0.214kΩ. Now for f_{pE} = 20Hz, $C_E = \frac{1}{2\pi\,(.214\times10^3)\times20}$ = **37.2μF**, for which

$f_z = \dfrac{1}{2\pi \times 8.2 \times 10^3 \times 37.2 \times 10^{-6}}$ = **0.521Hz**. Now, for f_{p1} = 0.521Hz, $C_{C1} = \dfrac{1}{2\pi \times 25.5 \times 10^3 \times .521}$ = **12.0μF**. Now, for f_{p2} = 2Hz, $C_{C2} = \dfrac{1}{2\pi \times 18.9 \times 10^3 \times 2}$ = **4.2μF**. **Alternatively**, for f_{p2} = .521Hz, $C_{C2} = \dfrac{1}{2\pi \times 18.9 \times 10^3 \times .521}$ = **16.2μF**, and for f_{p1} = 2Hz, $C_{C1} = \dfrac{1}{2\pi \times 25.5 \times 10^3 \times 2}$ = **3.12μF**. This is better, although is takes a larger total capacitance, since it makes the pole - zero cancellation independent of β.

7.30 For R = 350 in series with C_E, using data from P7.27 above, the total equivalent emitter resistance becomes 166.7 + 350||8.2k = (.167 + .336) kΩ = 0.503kΩ, and at the base, $R_{in} = 40k\Omega \| (151$ (.503kΩ)) = 40||75.95 = 26.2kΩ. Thus $A_M = -\dfrac{26.2}{26.2+10} \times \dfrac{(9.1\|10\|500)k\Omega}{.503k\Omega} = -\dfrac{26.2}{32.4} \times \dfrac{4.72}{.503}$ = −0.724 × 9.38 = **−6.79V/V**.

Now, from P7.27: C_π = 0.65pF, C_μ = 0.3pF.

Input Pole: C_T = 0.65 (1 − (.336/.503)) + 0.3 (1 − −9.38) = .216 + 3.114 = 3.33pF, $R_T = 26.2k\Omega \| 10k\Omega$ = 7.24kΩ, $f_{p1} = \dfrac{1}{2\pi \times 3.33 \times 10^{-12} \times 7.24 \times 10^3}$ = **6.60MHz**.

Output Pole: As before, f_{p2} = **108MHz**. Thus f_{p1} dominates, and $f_H \simeq$ **6.6MHz**.

7.31 From P7.30: *For* C_{C1}: R_{C1} = 10kΩ + 26.2kΩ = 36.2kΩ, $f_{p1} = \dfrac{1}{2\pi\,(36.2 \times 10^3) \times 1 \times 10^{-6}}$ = **4.40Hz**. *For* C_{C2}: $R_{C2} = 10k\Omega + (9.1\|500)\,k\Omega$ = 18.9kΩ, $f_{p2} = \dfrac{1}{2\pi\,(18.9 \times 10^3) \times 1 \times 10^{-6}}$ = **8.42Hz**. *For* C_E: $R_{CE} = 350 + \left[167 + \dfrac{50 + 40k\|10k}{151}\right] \| 8.2k$ = 350 + 214 = 564Ω, $f_{pE} = \dfrac{1}{2\pi \times 564 \times 10 \times 10^{-6}}$ = **28.2Hz**. $f_Z = \dfrac{1}{2\pi\,(8.2 + .35) \times 10^3 \times 10 \times 10^{-6}}$ = **1.86Hz**. *For* f_L: f_L = $(4.40^2 + 8.42^2 + 28.2^2 - 2\,(1.86)^2\,)^{1/2} = (19.4 + 70.9 + 795.2 - 6.9)^{1/2}$ = **29.6Hz**.

SECTION 7.7: The Common-Base, Common-Gate and Cascode Configurations

7.32 r_e = 25/0.2 = 125Ω, $f_T = \dfrac{\alpha/r_e}{2\pi\,(C_\pi + C_\mu)} \rightarrow C_\pi = \dfrac{\alpha/r_e}{2\pi\, f_T} - C_\mu = \dfrac{\frac{150}{151} \times \frac{1}{125}}{2\pi \times 1 \times 10^9}$ −0.3 = 1.26 −.3 = .96pF. $A_M = \alpha \times \dfrac{r_e \| R_E}{r_e \| R_E + R_S} \times \dfrac{R_C \| R_L \| r_o}{r_e}$ = (.993) $\dfrac{.125\|8.2}{.125\|8.2 + .1} \times \dfrac{9.1\|10\|400}{.125}$ = (.993) $\dfrac{.1231}{.2231} \times \dfrac{4.71}{.125}$ = 20.65V/V ≃ **20.7V/V**. $f_{p1} = \dfrac{1}{2\pi\, C_\pi\,(r_e \| R_E \| R_S)}$ = $\dfrac{1}{2\pi\,(.96 \times 10^{-12})\,(.125\|8.2\|.100) \times 10^3}$ = 3GHz. $f_{p2} = \dfrac{1}{2\pi\, C_\mu\,(R_C \| R_L \| r_o)}$ = $\dfrac{1}{2_\pi\,(0.3 \times 10^{-12})\,(4.71 \times 10^3)}$ = 112.6MHz. Thus $f_H \simeq$ **113MHz**, with A_M = **20.7V/V**.

7.33 *For* C_{C1}: $R_T = 100 + 125\|8.2k\Omega$ = 100 + 123 = 223Ω, $f_{p1} = \dfrac{1}{2\pi\,(223) \times 10 \times 10^{-6}}$ = **71.4Hz**.

$R_T = (9.1\|400 + 10)$ kΩ = 18.9kΩ, $f_{p2} = \dfrac{1}{2\pi\,(18.9 \times 10^3)\,(1 \times 10^{-6})}$ = **8.42Hz**.

For C_B: $R_T = R_1 \| R_2 \| [(\beta + 1)\,(r_e + R_S \| R_E)] = 40 \times 10^3 \| [(151)\,(125 + 8.2k\Omega \| 100)]$ = $40k\Omega \| 33.8k\Omega$ = 18.3kΩ, $f_{pB} = \dfrac{1}{2\pi \times 18.3 \times 10^3 \times 1 \times 10^{-6}}$ = **8.70Hz**.

$f_z = \dfrac{1}{2\pi \times 40 \times 10^3 \times 1 \times 10^{-6}} = \mathbf{3.98Hz.}$

Now, $f_L \approx (71.4^2 + 8.42^2 + 8.70^2 - 2\,(3.98)^2\,)^{1/2} = (5098 + 71 + 76 - 32)^{1/2} = \mathbf{72.2Hz.}$

7.34 $I_E = .15\text{mA}$, $r_e = 25/.15 = 167\Omega$. Thus $C_\pi = \dfrac{\alpha/r_e}{2\pi\, f_T} - C_\mu = \dfrac{\frac{150}{151} \times \frac{1}{167}}{2\pi \times 10^9} - 0.3 = .946 - .3 = 0.65\text{pF}.$

At the input: $C_T = C_\pi + C_\mu \;\; (1 \;-\; -\;1) \;=\; .65 \;+\; .3\;(2) \;=\; 1.25\text{pF},$ $R_T = 15k\Omega \,||\, 10k\Omega \,||\, (151(167)) = 15 \,||\, 10 \,||\, 25.2 = 4.85k\Omega$, $f_{p1} = \dfrac{1}{2\pi\,(4.85 \times 10^3)\,(1.25 \times 10^{-12})} =$ **26.3MHz.**

At the emitter of Q_2: $C_T \approx C_\pi + C_\mu\,(1 -- 1/1) = .63 + 2\,(.3) = 1.25\text{pF}$, $R_T = 167\Omega$,

$f_{p2} = \dfrac{1}{2\pi\,(167)\,(1.25 \times 10^{-12})} = \mathbf{953MHz.}$

At the collector: $C_T \approx C_\mu = 0.3\text{pF}$, $R_T = 9.1k\Omega \,||\, 10k\Omega = 4.76k\Omega$.

$f_{p3} = \dfrac{1}{2\pi \times 4.76 \times 10^3 \times 0.3 \times 10^{-12}} = \mathbf{111MHz.}$ Thus $f_H \approx \left[\dfrac{1}{26.3^2} + \dfrac{1}{111^2} + \dfrac{1}{953^2}\right]^{-1/2} =$ **25.6MHz.**

Midband gain $A_M = -\dfrac{15k\,||\,(151 \times 167)}{15k\,||\,(151 \times 167) + 10k} \times \left[\dfrac{150}{151}\right]^2 \times \dfrac{9.1k\Omega \,||\, 10k\Omega}{.167k\Omega} = -\dfrac{15\,||\,25.2}{15\,||\,25.2 + 10} \times \left[\dfrac{150}{151}\right]^2 \times \dfrac{9.1\,||\,10}{.167} = -\dfrac{9.40}{19.40} \times \left[\dfrac{150}{151}\right]^2 \times \dfrac{4.76}{.167} = \mathbf{-13.6V/V.}$

SECTION 7.8: Frequency Response of the Emitter and Source Followers

7.35 $r_e = \dfrac{V_T}{I_E} = \dfrac{25mV}{.15mA} = 167\Omega$, $g_m = \dfrac{150}{151} \times \dfrac{1}{167} = 5.95\text{mA/V}$, $r_\pi = 151\;(167) = 25.2k\Omega$, $C_\pi = \dfrac{\frac{150}{151} \times \frac{1}{167}}{2\pi \times 10^9} - 0.3 = 0.947 - .3 = 0.65\text{pF}.$

See $A_M = \dfrac{8.2\,||\,10}{8.2\,||\,10 + .167} \times \dfrac{151\,(.167 + 8.2\,||\,10)}{151\,(.167 + 8.2\,||\,10) + 10} = \dfrac{4.5}{4.667} \times \dfrac{151\,(4.667)}{151\,(4.667) + 10} = \mathbf{0.963V/V.}$

For f_H: $C_T = C_\mu + C_\pi\,(1 - (4.5/4.667)) = 0.3 + 0.65\,(1 - .964) = 0.323\text{pF}$,

$R_T = 10k\Omega \,||\, (151\,(.167 + 8.2\,||\,10)k\Omega) = 10\,k\Omega \,||\, 705k\Omega = 9.86k\Omega$,

$f_H \approx f_{p1} = \dfrac{1}{2\pi \times 9.86 \times 10^3 \times .323 \times 10^{-12}} = \mathbf{50.0MHz.}$

For f_L: $C_{C1} = 1\mu\text{F}$, $R_{C1} = 10k\Omega + 8.2k\Omega \,||\, (.167 + \dfrac{10k\Omega}{151}) = 10k\Omega + 8.2k\Omega \,||\, .233k\Omega = 10.23k\Omega$, $f_L = \dfrac{1}{2\pi \times 1 \times 10^{-6} \times 10.23 \times 10^3} = \mathbf{15.6Hz.}$

7.36 $r_s = 1/g_m = 1k\Omega$, $A_m = \dfrac{10k\Omega}{(10 + 1)\,k\Omega} = \mathbf{0.909V/V.}$

For f_H: $C_T = 1 + \dfrac{1}{1 + 1 \times 10} = 1.09\text{pF}$, $R_T = 100k\Omega$, $f_H = \dfrac{1}{2\pi \times 10^5 \times 1.09 \times 10^{-12}} = \mathbf{1.46MHz.}$

SECTION 7.9: The Common-Collector Common-Emitter Cascade

7.37 Since $V_{BE} \approx 0.70\text{V}$, $I_R \approx 0.70/70\text{k}\Omega = 10\mu\text{A}$. Thus $I_{E2} \approx 160 - 10 = 150\mu\text{A}$, $r_{e2} = \dfrac{V_T}{150\mu A} = 167\Omega$, $g_{m2} = \dfrac{150}{151} \times \dfrac{1}{167} = 5.95\text{mA/V}$, $r_{\pi 2} = 151\,(167) = 25.2\text{k}\Omega$, $I_{B2} = \dfrac{150}{151} \approx 1\mu\text{A}$. Now $I_{E1} = 1 + 10 = 11\mu\text{A}$, $r_{e1} = \dfrac{25mV}{11\mu A} = 2.27\text{k}\Omega$, $r_{\pi 1} = 151\,(2.27) = 343\text{k}\Omega$.

For Q_2: $C_\pi = \dfrac{5.95 \times 10^{-3}}{2\pi \times 10^9} - 0.3 = 0.65\text{pF}$. For Q_1: $C_\pi = 0.3\text{pF}$. For A_M: $R_{in2} = 70k\Omega \| 25.2k\Omega = 18.5\text{k}\Omega$. $A_M = \dfrac{9.1 \| 10}{.167} \times \dfrac{150}{151} \times \dfrac{18.5}{18.5 + 2.37} \times \dfrac{(18.5 + 2.27)\,151}{(18.5 + 2.27)\,151 + 100}$

$= -\dfrac{4.76}{0.167} \times .993 \times \dfrac{18.5}{20.77} \times \dfrac{(20.77)\,151}{3136 + 100} = \mathbf{-24.4V/V}$.

For f_H: At the base of Q_2: $C_T = .65 + 0.3\,(1 + 4.76 \times 5.95) = 9.45\text{pF}$, $R_T = 70k\Omega \| 25.2k\Omega \| (2.27k\Omega + \dfrac{100k\Omega}{151}) = 70 \| 25.2 \| 2.93 = 18.5 \| 2.93 = 2.53\text{k}\Omega$, $f_{p1} = \dfrac{1}{2\pi \times 2.53 \times 10^3 \times 9.45 \times 10^{-12}} = 6.66\text{MHz}$.

At the collector of Q_2: $f_{p2} = \dfrac{1}{2\pi \times 4.76 \times 10^3 \times 0.3 \times 10^{-12}} = 111\text{MHz}$.

At the base of Q_1: $C_T \approx 0.3 + 0.3\,(1 - (18.5/18.5 + 2.27)) = .333\text{pF}$, $R_T = 100k\Omega \| (151\,(2.27 + 18.5)) = 100\text{k}\Omega \| 3.1\text{M}\Omega = 96.9\text{k}\Omega$, $f_{p3} = \dfrac{1}{2\pi \times 96.9 \times 10^3 \times .333 \times 10^{-12}} = 4.93\text{MHz}$.

$\therefore\ f_H \approx \left[\dfrac{1}{4.93^2} + \dfrac{1}{6.66^2} + \dfrac{1}{111^2}\right]^{-1/2} = \mathbf{3.96MHz}$.

For f_L: For C_{C1} $R_{C1} = 10\text{k} + 9.1\text{k} = 19.1\text{k}\Omega$, $f_{p1} = \dfrac{1}{2\pi \times 1 \times 10^{-6} \times 19.1 \times 10^3} = 8.33\text{Hz}$.

For C_E: $R_{CE} = 167 \| 70k + \dfrac{70k \| (2.27k + 100k/151)}{151} = .167 \| 70 + .019 = 186\Omega$,

$f_{p2} = \dfrac{1}{2\pi \times 186 \times 10 \times 10^{-6}} = 85.6\text{Hz}$, and $f_z = 0\text{Hz}$.

$\therefore\ f_L = (8.33^2 + 85.6^2)^{1/2} = \mathbf{86Hz}$.

7.38 **a) For R = 14kΩ:** $I_R = \dfrac{.700}{14} = 50\mu\text{A}$, $I_{E2} = 160 - 50 = 110\mu\text{A}$, $r_{e2} = \dfrac{25}{110} = 227\Omega$, $g_{m2} = \dfrac{150}{151} \times \dfrac{1}{227} = 4.38\text{mA/V}$, $r_{\pi 2} = 34.3\text{k}\Omega$, $I_{B2} = \dfrac{110}{151} = 0.73$, $I_{E1} = .73 + 50 = 50.7\mu\text{A}$, $r_{e1} = \dfrac{25}{50.7} = 493\Omega$, $g_{m1} = \dfrac{150}{151} \times \dfrac{1}{493} = 2.01\text{mA/V}$, $r_{\pi 2} = 151\,(.493) = 74.5\text{k}\Omega$.

For Q_2: $C_\pi = \dfrac{4.38 \times 10^{-3}}{2\pi \times 10^9} - .3 = 0.40\text{pF}$.

For Q_1: $C_\pi = \dfrac{2.01}{2\pi \times 10^9} - .3 \rightarrow$ Use 0.30pF.

For A_M: $R_{in2} = 14k\Omega \| 34.3k\Omega = 9.94\text{k}\Omega$, $A_M = -\dfrac{4.76}{.227} \times .993 \times \dfrac{9.94}{9.94 + .493} \times \dfrac{(9.94 + .493)\,151}{(10.4)\,151 + 100}$

$= -\dfrac{4.76}{.227} \times .993 \times \dfrac{9.94}{10.4} \times \dfrac{1570.4}{1670.4} = \mathbf{-18.7V/V}$.

For f_H: *At the base of Q_2*: $C_T = 0.40 + 0.3\,(1 + 4.76 \times 4.38) = 6.95\text{pF}$, $R_T = 14 \| 34.3 \| (.493 + \dfrac{100}{151}) = 14 \| 34.3 \| 1.16 = 9.94 \| 1.16 = 0.963\text{k}\Omega$,

$f_{p1} = \dfrac{1}{2\pi \times .963 \times 10^3 \times 6.95 \times 10^{-12}} = 23.8\text{MHz}$. Also $f_{p2} = 111\text{MHz}$.

At the base of Q_1: $C_T \simeq 0.3 + 0.3\,(1 - \dfrac{9.94}{9.94 + .493}) = 0.314\text{pF}$, $R_T = 100 \| (151\,(.493 + 9.94)) = 100k\Omega \| 1.58M\Omega = 94.0\text{k}\Omega$, $f_{p3} = \dfrac{1}{2\pi \times 94.0 \times 10^3 \times .314 \times 10^{-12}} = 5.39\text{MHz}$.

Thus $f_H = \left[\dfrac{1}{5.39^2} + \dfrac{1}{23.8^2}\right]^{-1/2}$ **= 5.26 × 10⁶Hz.**

For f$_L$: f_{p1} = 8.33Hz. Now, *For* C_E: $R_{CE} = .227k \| 14 + \dfrac{(.493k + \frac{100}{151}) \| 14k}{151} = 0.234\text{k}\Omega$, $f_{p2} = \dfrac{1}{2\pi \times 234 \times 10 \times 10^{-6}} = 68.0\text{Hz}$. Thus $f_L = (68^2 + 8.33^2)^{1/2}$ **= 68.5Hz.**

b) **For R** $= \infty$: $I_R = 0$, $I_{E2} = 160\mu\text{A}$, $r_{e2} = \dfrac{25}{160} = .156\text{k}\Omega$, $g_{m2} = \dfrac{150}{151} \times \dfrac{1}{156} = 6.37\text{mA/V}$, $r_{\pi 2} = 151\,(.156) = 23.6\text{k}\Omega$, $I_{B2} = \dfrac{160}{150} = 1.07\mu\text{A}$, $I_{E1} = 1.07\mu\text{A}$, $r_{e1} = \dfrac{25 \times 10^{-3}}{1.07 \times 10^{-6}} = 23.4\text{k}\Omega$, $r_{\pi 1} = 151\,(23.4) = 3.5\text{M}\Omega$. *For* Q_2: $C_{\pi 2} = \dfrac{6.37 \times 10^{-3}}{2\pi \times 10^{-9}} - 0.3\times10^{-12} = .714\text{pF}$. *For* Q_1: $C_{\pi 1} = 0.3\text{pF}$.

For A_M: $R_{in2} = r_{\pi 2} = 23.6\text{k}\Omega$, $A_M = \dfrac{-4.76}{.156} \times .993 \times \dfrac{23.6}{23.6 + 23.4} \times \dfrac{(23.6 + 23.4)\,151}{(23.6 + 23.4)\,151 + 100}$
$= \dfrac{-4.76}{.156} \times .993 \times \dfrac{23.1}{47.0} \times \dfrac{7097}{7197}$ **= −15.0V/V.**

For f$_H$: *At the base of* Q_2: $C_T = 0.714 + 0.3\,(1 + 6.37 \times 4.76) = 10.1\text{pF}$, $R_T = 23.6 \| (23.4 + 100/151) = 11.92\text{k}\Omega$, $f_{p1} = 1/(2\pi \times 11.92 \times 10^3 \times 10.1 \times 10^{-12}) = 1.32\text{MHz}$, $f_{p2} = 111\text{MHz}$.

At the base of Q_1: $C_T = 0.3 + 0.3\,(1 - \dfrac{23.6}{23.6 + 23.4}) = 0.449\text{pF}$, $R_T = 100\text{k}\Omega \| (151\,(23.6 + 23.4)) = 99\text{k}\Omega$, $f_{p3} = 1/(2\pi \times 99 \times 10^3 \times 0.449 \times 10^{-12}) = 3.58\text{MHz}$. Thus $f_H = \left[\dfrac{1}{1.32^2} + \dfrac{1}{3.58^2}\right]^{-1/2} =$ **1.24MHz.**

For f$_L$: f_{p1} = 8.33Hz. Now *For* C_E: $R_{CE} = .156 + \dfrac{(23.4 + (100/151))}{151} = .315\text{k}\Omega$, $f_{p2} = 1/(2\pi \times 315 \times 10 \times 10^{-6}) = 50.5\text{Hz}$. Thus $f_L = (50.5^2 + 8.33^2)^{1/2}$ **= 51.2Hz.**

Summary (also using P7.37)

R(kΩ)	A_M(V/V)	f_L(Hz)	f_H(MHz)
14	−18.7	68.5	5.26
70	−24.4	86.0	3.96
∞	−15.0	51.2	1.24

See that R is important in improving gain and bandwidth. The worst idea is $R = \infty$. Perhaps a good design would be at $R = \sqrt{14 \times 70} = 31\text{k}\Omega$ *(check for interest).*

SECTION 7.10: Frequency Response of the Differential Amplifier

7.39 For each, $I_E = 300/2 = 150\mu\text{A}$, $r_e = 166.7\Omega$, $g_m = 150/151 \times 1/166.7 = 5.96\text{mA/V}$, $r_\pi = 166.7 \times 151 = 25.2\text{k}\Omega$, $C_\pi = \dfrac{5.96 \times 10^{-3}}{2\pi \times 10^9} - 0.3 \times 10^{-12} = .65\text{pF}$.

For Gain: $\frac{\upsilon_o}{\upsilon_s} = \frac{\frac{150}{151}((4+4)||10)}{2\,(.1667)} \times \frac{2\,(25.2)}{10 + 2\,(25.2)}$ = **11.05V/V**. *For f_H*: (From Eq.7.110), $f_H = \frac{1}{2\pi\,(\frac{10}{2}||25.2) \times 10^3 (0.65 + 0.3\,(1 + 5.96 \times 4||(\frac{10}{2}) \times 10^{-12}} = \frac{1}{2\pi\,(4.172)\,4.92 \times 10^{-9}}$ = **7.75MHz.**

7.40 As in P7.39: $r_e = 166.7\Omega$, $g_m = 5.96$mA/V, $r_\pi = 25.2$kΩ, $C_\pi = 0.65$pF, $C_\mu = 0.3$pF.

For R_L connected to the collector of the input transistor: Gain $= -\frac{150}{151}\frac{(4||10)}{2\,(.1667)} \times \frac{2\,(.1667)\,151}{2\,(.1667)\,151 + 10} = -\frac{150}{151}\frac{(2.86)}{.333} \times \frac{50.3}{60.3}$ = **−7.12V/V.**

Gain from base to collector of the input transistor $= -\frac{150}{151}\frac{2.86}{0.333} = -8.53$V/V, (from which there is a Miller-multiplied C_μ, and a voltage divider with two C_π in series). Thus, $C_T = 0.65/2 + 0.3\,(1 - -8.53) = 3.18$pF. $R_T = 10k\Omega||(2\,(.1667)\,151 = 10k\Omega||50.3k\Omega = 8.34$kΩ, $f_H \approx 1/(2\pi \times 8.34 \times 10^3 \times 3.18 \times 10^{-12})$ = **6MHz.**

For R_L connected to the collector of the grounded-base amplifier: Gain = **+7.12V/V**. Gain from base to collector of the input transistor is − (150/151) (4)/0.333 = −11.92V/V. At the input base $C_T = 0.65/2 + 0.3\,(1 - -11.92) = 4.20$pF, $R_T = 8.34$kΩ, $f_H \approx 1/(2\pi \times 8.34 \times 10^3 \times 4.2 \times 10^{-12})$ = **4.54MHz.**

7.41 Parameters as in P7.40: Gain = **+7.12V/V**, $C_T = 0.65/2 + 0.3 = .625$pF, $R_T = 8.34$kΩ, $f_{p1} = 1/(2\pi \times 8.34 \times 10^3 \times .625 \times 10^{-12}) = 30.5$MHz. Now check the output pole, where $C_T = 0.3$pF, $R_T = 4k\,||\,10k = 2.857$kΩ, $f_{p2} = 1/(2\pi \times 2.857 \times 10^3 \times 0.3 \times 10^{-12}) = 186$MHz. Thus, $f_H \approx$ **29.3MHz.**

7.42 From P7.39, $r_e = 166.7\Omega$, $g_m = 5.96$mA/V, $r_\pi = 25.2$kΩ, $C_\pi = 0.65$pF, $C_\mu = 0.3$pF. With r_e added in each emitter lead, and load taken differentially, $A_M = \frac{\frac{150}{151}((4+4)||10)}{4\,(.1667)} \times \frac{4\,(25.2)}{10 + 4\,(25.2)} = \frac{4.44 \times 0.993}{4\,(.1667)} \times \frac{100.8}{110.8}$ = **6.02 V/V.**

At each input, gain from base to collector of $Q_1 = \frac{1}{2} \times \frac{4.44 \times .993}{2\,(.1667)} = -6.68$V/V, that is $C_T \approx 0.65/2 + 0.3\,(1 - -6.68) = 2.63$pF, $R_T = (10/2)||(2(25.2)) = 4.55$kΩ, where $f_H = 1/(2\pi \times 4.55 \times 10^3 \times 2.63 \times 10^{-12})$ = **13.3MHz.**

7.43 For the current source, $r_o = 200\text{V}/300\mu\text{A} = 667$kΩ, $C_T = 0.3 + 0.5 = 0.8$pF. Thus $V_{out} = \frac{4 \times 10^3 \times \frac{150}{151}}{2 \times 667 \times 10^3} \times 5\text{V}$ = **14.9 mV**. For 1V peak on the load ends, $\frac{4k\Omega}{2\,Z_C} \times 5\text{V} = 1\text{V} \rightarrow Z_C = 10$kΩ, and $f = 1/(2\pi(0.8) \times 10^{-12} \times 10 \times 10^3)$ = **20MHz.**

For saturation: Quiescent collector voltage $= V_{CC} - I/2\,(4\text{k}) = 10 - .15 \times 4 = 9.4$V. For 5V peak on the bases, saturation begins for a collector voltage of 5.0V, for which the peak load signal is 9.4 − 5V = 4.4V. This occurs at a frequency of (4.4/1) × 20 = **88MHz.**

SECTION 7.11: The Differential Pair as a Wideband Amplifier: The Common-Collector Common-Base Configuration

7.44 $I_E = 150\mu$A, $r_e = 25/150 = 166.7\Omega$, Add $r_E = r_e =$ **166.7Ω** to double the input resistance. Now, $C_\pi = \frac{150/151 \times 1/166.7}{2\pi \times 10^9} - 0.3\text{pF} = .65\text{pF}$, $A_M = \frac{2.7 \times 10^3 \times \frac{150}{151}}{4\,(166.7)} \times \frac{4 \times 166.7 \times 151}{4 \times 166.7 \times 151 + 10} = \frac{268 \times 10^3}{667}$

$\times \dfrac{\mathbf{100.7k\Omega}}{\mathbf{110.7k\Omega}} = \mathbf{3.66V/V}$.

For f_H: $C_T = 0.65/4 + 0.3 = .4625\text{pF}$, $R_T = (4\ (166.7)\ (151)) \,||\, 10k = 100.7k \,||\, 10k = 9.10\text{k}\Omega$, $f_H = 1/(2\pi \times 9.10 \times 10^3 \times 0.4625 \times 10^{-12}) = \mathbf{37.8MHz}$.

Chapter 8

FEEDBACK

SECTION 8.1: The General Feedback Structure

8.1 From Fig. 8.1, see $x_i = x_s - x_f = 1.00 - 0.99 =$ **0.01V**. Thus $A = x_o/x_i = 3.00/.01 =$ **300V/V**. Thus $\beta = x_f/x_o = 0.99/3.0 =$ **0.33V/V**. The open-loop gain is $A =$ **300V/V**. The amount of feedback is $(1 + A\beta) = 1 + 300\,(.33) =$ **100**. The closed-loop gain is $A_f = x_o/x_s = A/(1 + A\,\beta)$, where, directly, $A_f = 3.0/1.00 =$ **3.0V/V**, and, indirectly, is $300/(1 + 300\,(.33)) = 300/100 =$ **3.0V/V**. For the β network disconnected, $\upsilon_i = \upsilon_s$, and υ_o tends toward $A\,\upsilon_s = 300\,(1.0) =$ **300V**. This value would *not* be measured, since much before it is reached, **the output would** typically **limit** (or saturate).

8.2 See from $A_f = A/(1 + A\,\beta)$, that $8 = 10^2/(1 + 10^2\,\beta)$, or $1 + 10^2\,\beta = 10^2/8$, whence $\beta = (10^2/8 - 1)/10^2 =$ **0.115**. Now $\beta = R_1/(R_1 + R_2) = 0.115$, or $(R_1 + R_2)/R_1 = 1/0.115 = 8.696 = 1 + R_2/R_1$. Thus $R_2/R_1 = 8.696 - 1 =$ **7.696**. Amount of feedback is $1 + A\,\beta = A/A_f = 10^2/8 =$ **12.5V/V**. $\equiv 20\log_{10} 12.5 =$ **22dB**. For $V_s = 0.125$V, $V_o = A_f\,V_s = 8\,(.125) =$ **1V**, $V_f = \beta\,V_o = 1 \times .115 =$ **0.115V**, $V_i = V_o/A = 1/10^2 =$ **0.01V**, or $V_i = V_s - V_f = 0.125 - 0.115 =$ **0.01V**, as expected. For A increasing by 100%, A becomes $2(10^2) = 200$, and $A_f = A/(1 + A\,\beta)$ becomes $200/(1 + 200\,(0.115)) = 8.33$, rather than the former value of 8. That is, A_f increases by $(8.33 - 8)/8 \times 100 =$ **4.12%**.

SECTION 8.2: Some Properties of Negative Feedback

8.3 For the original design, $A_f = A/(1 + A\,\beta) = 10^3/(1 + 10^3 \times 10^{-2}) = 90.9$. For the fabricated design, $A_f = 0.5 \times 10^3/(1 + 0.5 \times 10^3 \times 10^{-2}) =$ **83.3** results. Now, the desensitivity factor, $1 + A\,\beta$, as designed, was $1 + 10^3 \times 10^{-2} =$ **11**, and, as fabricated, is $1 + 0.5\,(10^3)\,(10^{-2}) =$ **6**. Thus for (small) changes around the original design, one would expect the original 50% reduction in A to result in a $50/11 = 4.5\%$ reduction in A_f. Using the changed desensitivity factor, a $50/6 = 8.33\%$ change would be expected. Now, conceptually (for small changes), the sensitivity, $\dfrac{\partial A_f/A_f}{\partial A/A} = \dfrac{1}{1 + A\,\beta} = \dfrac{1}{11} = .091 \equiv$ **−20.8dB**. Actually, $\dfrac{\partial A_f/A_f}{\partial A/A} \approx \dfrac{\Delta A_f/A_f}{\Delta A/A} = \dfrac{(90.9 - 83.3)/90.9}{(10^3 - 0.5\,(10^3))/10^3} = \dfrac{7.6/90.9}{500/1000} = 0.167 \equiv$ **16.7%** The resulting manufactured closed-loop gain is (as calculated above) **83.3**.

8.4 We know that $A\beta = 89$, and $A_f = A/(1 + A\beta)$. Thus, ideally, $99 = A/(1 + 89)$, for which $A = 99\,(90) = 8910$, and $\beta = 89/8910$. After a time, $A_f = 98 = A/(1 + A\,(89/8910))$, or $A = 98 + 98\,(A)\,(89)/8910 = 98 + .9789A$. $A = 98/(1 - 0.9789) = 4645$, lower by more than a factor of 2. *Check*: $A_f = \dfrac{4645}{1 + 4645\,(89/8910)} = 98.000$. Thus gain A has reduced by $(8910 - 4645)/8910 \times 100 =$ **47.9%**.

8.5 For the closed loop, $1 + A\beta = A/A_f = 10^4/10^2 = 100$, and $\beta = (100 - 1)/10^4 = 99 \times 10^{-4}$. Thus the closed-loop 3dB frequency is $(10^4)\,100 =$ **10^6Hz**. For the basic amplifier, $GB = 10^4 \times 10^4 =$ **10^8Hz**. For the feedback arrangement, $GB = 100 \times 10^6 =$ **10^8Hz**, the same! (as expected!)

8.6 For f_{3dB} of A reduced to 2×10^3Hz, the 3dB closed-loop frequency reduces to $2 \times 10^3 \times 10^2 =$ **2×10^5Hz**. On the surface, it appears that the desensitivity idea is not working, since the percentage change of the open-loop and closed-loop 3dB frequencies are the same. For the original amplifier at 10^4Hz, the open-loop gain is down by 3dB, to $10^4/1.414 =$ **$.707 \times 10^4$**, a percentage reduction of about 30%. Correspondingly, the percentage reduction in closed-loop gain $= 30/1 + A\beta = 30/100 = .3\%$, from 100 to $100\,(1 - .3/100) \approx 99.7$. For the manufactured amplifier, for which $f_{3dB} = 2 \times 10^3$Hz, the gain at 10^4Hz is $A\,10_4 = \dfrac{A_m}{(1^2 + (f/f_H)^2\,)^{½}} =$

$$\frac{10^4}{(1+(10^4/(2\times10^3))^2)^{1/2}} = \frac{10^4}{(1+25)^{1/2}} = 0.196\times10^4. \text{ Now } A_f = \frac{A}{1+A\,(\beta)} = \frac{.196\times10^4}{1+.196\times10^4\,(99\times10^{-4})} =$$

96.06. This corresponds to a drop of about 4% in gain for a change of f_{3dB} (from 10^4Hz to 2×10^3Hz) of 80% in frequency, an improvement of about 20 times. Note that at 10^4Hz, $(1 + A\beta)$ is 26, correspondingly. Thus the desensitivity factor is still at work, maintaining the gain for frequencies above the cutoff.

8.7 With a low-noise preamplifier of gain A_2, $S/N = (V_s/V_n)\,A_2$. For an improvement of 40dB, $A_2 = 10^{40/20}$ = **100V/V**. New $S/N = -3\text{dB} + 40\text{dB} =$ **37dB**.

8.8

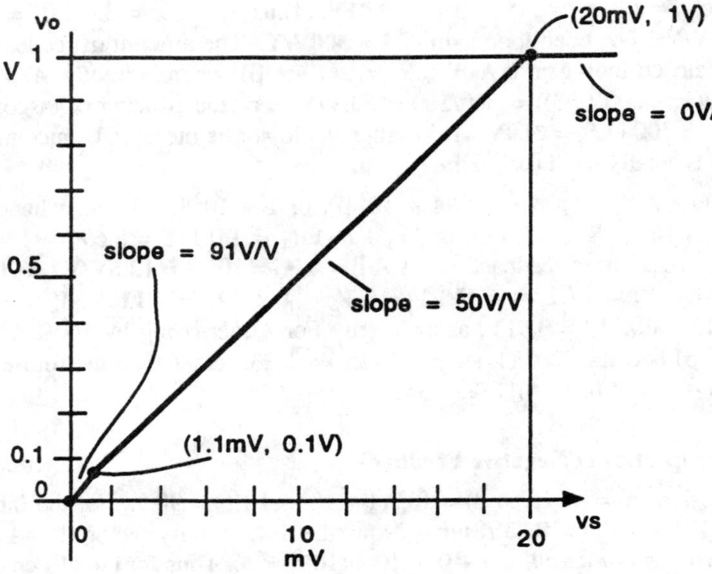

For $\upsilon_o \le 0.1$V, and $A = 10^3$V/V, $A_f = \dfrac{10^3}{1+10^3\,(.01)} = 90.9$V/V. Now for $\upsilon_o = 0.1$V, $\upsilon_s = 0.1/90.9 \equiv$ **1.1mV**, and $\upsilon_i = 0.1/10^3 = 0.1$mV. For $0.1 < \upsilon_o < 1.0$, and $A = 10^2$V/V, $A_f = \dfrac{10^2}{1+10^2\,(.01)} =$ **50V/V**. Now for $\upsilon_o = 1.0$V, $\upsilon_s = 1.0/50 \equiv$ **20mV**, and $\upsilon_i = 1.0/10^3 = 1$mV. For $\upsilon_i > 1$mV, υ_o limits at 1V, and $A = 0$, for which $A_f = 0$, as well.

SECTION 8.3: The Four Basic Feedback Topologies

8.9 For the circuits shown, the feedback type is:

(a) **Shunt-Shunt**: $\beta = \dfrac{I_f}{V_o} = -\dfrac{V_o/R_2}{V_o} = -\dfrac{\mathbf{1}}{\mathbf{R_2}}$.

(b) **Shunt-SeriesK**: Now, assuming $R_2 >> r$, $\beta = \dfrac{I_f}{I_o} \approx -\dfrac{I_o\,r/R_2}{I_o} = -\dfrac{\mathbf{r}}{\mathbf{R_2}}$.

(c) **Series-Shunt**: $\beta = \dfrac{V_f}{V_o} = \dfrac{R_1}{R_1+R_2} \times \dfrac{V_o}{V_o} = \dfrac{\mathbf{R_1}}{\mathbf{R_1+R_2}}$.

(d) **Series-Series**: Now, assuming $(R_1+R_2) >> r$, $\beta = \dfrac{V_f}{I_o} \approx \dfrac{R_1}{R_1+R_2} \times \dfrac{I_o\,r}{I_o} = \dfrac{\mathbf{r\,R_1}}{\mathbf{R_1+R_2}}$.

For (b), assuming $R_2 \simeq r$, $I_f = \dfrac{-r}{r+R_2} \times I_o$, and $\beta = \dfrac{I_f}{I_o} = \dfrac{\mathbf{-r}}{\mathbf{r+R_2}}$.

For (d), assuming $(R_1+R_2) \approx r$, $V_r = (r \,||\, (R_1+R_2))\, I_o = \dfrac{r\,(R_1+R_2)}{r+R_1+R_2} \times I_o$, and

$$\beta = \frac{V_f}{I_o} = \frac{\dfrac{R_1}{R_1+R_2} \times V_r}{I_o} = \frac{R_1}{R_1+R_2} \times \frac{r(R_1+R_2)}{r+R_1+R_2} \times \frac{I_o}{I_o} = \frac{\mathbf{r\,R_1}}{\mathbf{r+R_1+R_2}}.$$

8.10 Assuming g_m is very high, the basic gain A is high, and $r_s = 1/g_m \approx 0$.

(a) **Series-Series**: $\upsilon_f = r\, i_o$, and $\beta = \upsilon_f/i_o = \mathbf{r}$.

(b) **Shunt-Shunt**: $i_f = -\upsilon_o/R_F$, and $\beta = i_f/\upsilon_o = \mathbf{-1/R_F}$.

(c) **Series-Shunt**: $\upsilon_f = \upsilon_o$, and $\beta = \upsilon_f/\upsilon_o = \mathbf{1}$.

(d) **Series-Series**: $\upsilon_f = i_o\, r$, and $\beta = \upsilon_f/i_o = \mathbf{r}$.

8.11 Here, $V_i = V_s - V_f$ = 1V − 0.1V = 0.9V. Thus $A = I_o/V_i$ = 2A/0.9V = 2.22A/V = **2.22S**, and $\beta = V_f/I_o$ = 0.1V/2A = .05V/A = .05Ω = **50mΩ**.

SECTION 8.4: The Series-Shunt Feedback Amplifier

8.12 For the β network, β = 2/(2 + 18) = 0.1, as suggested. Also, R_{11} = 2kΩ || 18kΩ = 1.8kΩ, and R_{22} = 2kΩ + 18kΩ = 20kΩ. For the A circuit, there are losses at the input and output. *For a zero impedance source and no load:* Thus

$$A = \frac{R_{ia}}{R_{ia}+R_S+R_{11}} \times A_a \times \frac{R_L||R_{22}}{R_L||R_{22}+R_{oa}} = \frac{10}{10+0+1.8} \times 100 \times \frac{\infty||20}{\infty||20+.01} = 84.7\text{V/V}$$

Now for A = 84.7V/V, β = 0.1V/V, $A\beta$ = 84.7 × 0.1 = 8.47, and $1 + A\beta$ = 9.47. Thus $A_f = A/(1+A\beta)$ = 100/9.47 = **10.6V/V**.

Now, $R_i = R_S + R_{ia} + R_{11}$ = 0 + 10kΩ + 1.8kΩ = 11.8kΩ, and $R_o = R_L \,||\, R_{oa} \,||\, R_{22}$ = ∞ || 10Ω || 20kΩ = 10Ω. Thus with feedback, $R_{if} = R_i\,(1+A\beta)$ = 11.8 (9.47) = **11.2kΩ**, and, $R_{of} = R_o/(1+A\beta)$ = 10/9.47 = **1.06Ω**. *For a 0.1V rms, 10kΩ source and 100Ω load,* $\upsilon_{out} = 0.1 \times \dfrac{112}{10+112}$ × 10.6 × 100/(100 + 1/.06) = **0.963V** rms.

8.13 *For the feedback network*: β = 10/(1290 + 10) = **0.05V/V**, R_{11} = 10kΩ || 190kΩ = **9.5kΩ**, and R_{22} = 10kΩ + 190kΩ = **200kΩ**.

For the A circuit: $A = \dfrac{R_{ia}}{R_{ia}+R_S+R_{11}} \times A_a \times \dfrac{R_L||R_{22}}{R_L||R_{22}+R_{oa}} = \dfrac{20}{20+10+9.5} \times 900 \times \dfrac{1||200}{1||200+1}$ = **227V/V**. $R_i = R_{ia} + R_S + R_{11}$ = 20 + 10 + 9.5 = 39.5kΩ $R_o = R_{oa} \,||\, R_L \,||\, R_{22}$ = 1 || 1 || 200 = 0.499kΩ.

Overall, $A_f = \dfrac{A}{1+\beta A} = \dfrac{227}{1+227\times.05} = \dfrac{227}{12.35}$ = **18.4V/V**, $R_{if} = R_i\,(1+A\beta)$ = 39.5 × 12.35 = 488kΩ, $R_{of} = R_o/(1+A\beta)$ = 0.499/12.35 = 40.4Ω. Resistance seen by the source is $R_{if} - R_S$ = 488 − 10 = **478kΩ**. Resistance seen by the load is $R_{of} \,||\, (-R_L)$ = 40.4Ω || (−1kΩ) = $\dfrac{.0404\times 1}{1-.0404}$ = .0421kΩ = **42.1Ω**. Overall gain, $V_o/V_s = A_f$ = **18.4V/V**.

8.14 *For the β circuit* (between nodes F and B): β = 100kΩ/(1MΩ + 100kΩ) = 0.0909, R_{11} = 100kΩ || 1MΩ = 90.9kΩ, R_{22} = 100kΩ + 1MΩ = 1.1MΩ.

For the A circuit (between nodes A, B to F): For Q_1, Q_2, I_E = 100μA. Thus r_e = 25/0.1 = 250Ω, r_π = 250 (120 + 1) = 30.25kΩ. Thus R_{id} = 2 (30.25) = 60.5kΩ. For Q_5, assuming $V_o \approx$ 0V, I_E = 2mA, and r_e = 25/2 = 12.5Ω, with r_π = 12.5 (121) = 1.51kΩ. Input resistance to the right at node C is R_{ic} = (120) [12.5 + 1kΩ + 10kΩ || 1.1MΩ] = 1.32MΩ. Now ignoring device r_o, the gain from S to F, is $A = \dfrac{60.5}{100+90.9+60.5} \times \dfrac{120}{121}$

$\times 2 \times \dfrac{1.32 \times 10^6}{250+250} \times \dfrac{10k\Omega \| 1.1M\Omega}{10k\Omega \| 1.1M\Omega + 1k\Omega + 12.5\Omega}$ or A = 1143V/V, with R_i = (100 + 90.9 + 60.5)kΩ = 251.4kΩ, and R_o = 10kΩ || 1.1MΩ || (R_{Q5} + 1kΩ). Now, since r_o of the devices has been ignored, the resistance driving node C is infinite, and thus the output resistance at the emitter of Q_5 is also infinite (however strange that may be). {Note that if V_A = 200V, $r_{02} \simeq r_{01} = \dfrac{200}{100(10^{-6})}$ = 2MΩ, and r_{oc} = 1MΩ, and $R_{Q5} \simeq 10^6/121$ = 8.3kΩ, which with 1kΩ, will reduce R_o by about a factor of 2, not a large effect}. Ignoring the latter effect, R_o = 10kΩ || 1.1MΩ = 9.91kΩ. Now, A = 1143V/V, β = 0.0909, $A\beta$ = 103.9 and $1 + A\beta$ = 104.9. Thus, $\dfrac{\upsilon_o}{\upsilon_s} = \dfrac{A}{1+A\beta} = \dfrac{1143}{104.9}$ = **10.9V/V**. Now, $R_{if} = R_i\,(1 + A\beta)$ = (251.4kΩ) (103.9) = 26.1MΩ, and $R_{in} = R_{if} - R_S$ = 26.1 – .1 = **26MΩ**. Now, $R_{of} = R_o/(1 + A\beta)$ = 9.91kΩ/103.9 = 95.4Ω, and $R_{out} = R_{of} \| (-R_L)$ = 95.4 || (–10kΩ) = **96.3Ω**. For source resistance and load resistance considered separately in the reduction of υ_o/υ_s to ½, the required R_S = **26MΩ** and the required R_L = **96.3Ω**. Clearly in practice, reduction of gain through loading will predominate.

8.15 Assume $\upsilon_o \simeq 0$ and that current splits equally between Q_1, Q_4. Now for Q_1, Q_2, Q_3, Q_4, i_D = (200μA)/2 = 100μA = $K\,(\upsilon_{GS} - V_t)^2$. Thus, $\upsilon_{GS} - 1 = (100/100)^{½}$ = 1, υ_{GS} = 2V, $g_{m1} = 2K\,(\upsilon_{GS} - V_t) = 2\,(100 \times 10^{-6})\,(2-1)$ = 200μA/V, $r_{01} = V_A/I_D = 20/100 \times 10^{-6}$ = 200kΩ.

For Q_5, i_D = 200μA, $(\upsilon_{GS} - 1) = (200/100)^{½}$ = 1.414, υ_{GS} = 2.414V, $g_{m5} = 2\,(100 \times 10^{-6})\,(1.414)$ = 282μA/V, r_{o5} = 20V/200μA = 100kΩ. For the β circuit, β = 1.0, R_{11} = 0Ω, $R_{22} = \infty$. For the A circuit, $R_i = R_S + R_{ia} + R_{11}$ = 1MΩ + ∞ + 0 = ∞, $R_o = r_{o5} \| \dfrac{1}{g_{m5}} = 10^5 \| \dfrac{1}{282 \times 10^{-6}}$ = 100kΩ || 3.55kΩ = 3.36kΩ,

$$A \simeq \frac{2\,(r_{03} \| r_{04})}{1/g_{m1} + 1/g_{m4}} \times \frac{r_{05}}{r_{05} + 1/g_{m5}} = \frac{2\,(200k\Omega/2)}{2\,(1/200 \times 10^{-6})} = \frac{100k\Omega}{100k\Omega + 3.55k\Omega} = 19.3\text{V/V}.$$

Thus $A\beta$ = 19.3 × 1 = 19.3, and $1 + A\beta$ = 20.3. Thus $A_f = \dfrac{A}{1+A\beta}$ = 19.3/20.3 = **0.951V/V** = υ_s/υ_o, $R_{of} = R_o/(1 + A\beta)$ = 3.36/20.3 = 166Ω, that is R_{out} = **166Ω**. Now $R_{if} = \infty\,(20.3) = \infty$, that is R_{in} = ∞Ω. Now for **1kΩ load**, the overall gain, υ_o/υ_s = 1kΩ/(166 + 1kΩ) × .951 = **0.816V/V**.

Concerning offset: For V_o = 0V, V_{G5} = 2.41V, V_{G2} = 5 – 2 = 3V, $V_{S1} = V_{S2}$ = 0 – 2 = –2V, that is current in $r_{01} = \dfrac{3 - -2}{200k\Omega}$ = 25μA, in $r_{02} = \dfrac{5-3}{200}$ = 10μA, in $r_{03} = \dfrac{5-2.41}{200k\Omega}$ = 12.95μA, in $r_{04} = \dfrac{2.41 - -2}{200k\Omega}$ = 22μA. Thus, assuming an ideal mirror, the net current offset at the gate of Q_5 is 25 – 10 + 12.95 – 22 = 5.95μA. Input offset to compensate is $V_{os} = I_{os}/g_m$. Here $G_m = \dfrac{2}{r_{s1} + r_{s2}} = \dfrac{2}{1/g_{m1} + 1/g_{m2}} = \dfrac{1}{1/g_{m1}} = g_{m1}$ = 200μA/V. Thus V_{os} = 5.95μA/200μA/V = **29.8mV**.

8.16 See 100μA in the drain of Q_1 forces i_{D1} = 100μA. Since I = 200μA, thus i_{D2} = 200 – 100 = 100μA. Since $i_{D1} = i_{D2}$, for υ_s = 0, then υ_o = 0, and i_{R_L} = 0. Thus i_{D3} = 100μA also. Now 100μA = 100μA $(\upsilon_{GS} - 1)^2$, whence υ_{GS} = 2V. Also $g_m = 2\text{K}\,(\upsilon_{GS} - V_t)$ = 200μA/V, $r_s = 1/g_m$ = 5kΩ, and r_o = 20/100μA = 200kΩ.

For Q_2 seen as part of A, the feedback is a wire for which β = 1, R_{11} = 0Ω, and R_{22} = ∞Ω. Now, the A circuit consists, at the input, of υ_s connected via 1MΩ to the gate of Q_1 and the gate of Q_2 grounded (through R_{11} = 0Ω). At the output, R_L only is connected to υ_o. Thus $A = \dfrac{200k\Omega}{5k\Omega + 5k\Omega} \times \dfrac{10k\Omega \| 200k\Omega}{5k\Omega}$ = 38.1V/V, $R_i = \infty$, and $R_o = r_{03} \| R_L$ = 200kΩ || 10kΩ = 9.52kΩ. Thus $A\beta$ = 38.1 × 1 = 38.1, and $1 + A\beta$ = 39.1, and $A_f = \dfrac{A}{1+A\beta} = \dfrac{38.1}{39.1}$ = **0.974V/V**, $R_{if} = R_i\,(1 + A\beta) = \infty$, $R_{of} = R_o/(1 + A\beta)$ = 9.52/39.1 = 243Ω, that is R_{out} = 243 || (–10kΩ) = **249Ω**, and υ_o/υ_s = **0.974V/V**.

Now for Q_2 seen as part of β, the feedback is a resistor r_{s2} = 5kΩ, for which β = 1, and R_{11} = 5kΩ, and R_{12} = ∞Ω. Now the A circuit, at the input, includes R_{11} = 5kΩ to ground from the source of Q_1, and otherwise is as before. Thus A = 38.1V/V, and β = 1, as before, with the **same results**.

Now for Q_2 *and* Q_3, both $10\times$ *wider with* $I = 1.1$mA: $i_{D1} = 100\mu$A as before, while $i_{D2} = i_{D3} = 1$mA. But K_2, K_3 are each 1mA/V, and $\upsilon_{GS1} = \upsilon_{GS2} = \upsilon_{GS3}$ as before, but $g_{m2} = g_{m3} = 10\ (200\mu A/V) = 2mA/V$, $r_s = 1/g_m = 0.5k\Omega$ and $r_{03} = 20/1mA = 20k\Omega$. Now using the first idea (ie Q_2 part of A), $A = \dfrac{200k\Omega}{5k\Omega + 0.5k\Omega} \times \dfrac{10k\Omega \| 20k\Omega}{0.5k\Omega} = 484.8V/V$, $A\beta = 484.8$, $A\beta + 1 = 485.8$, $R_o = 10k\Omega \| 20k\Omega = 6.66k\Omega$. Thus $\dfrac{\upsilon_o}{\upsilon_s} = A_f = \dfrac{A}{1 + A\beta} = \dfrac{484.8}{485.8} =$ **0.998V/V**, $R_{of} = 6.66k\Omega/484.8 = 13.75\Omega$, $R_{out} = 13.75 \| (-10k\Omega) =$ **13.8Ω**. See a great improvement in performance as a unity-gain buffer.

8.17 Following the idea in 8.16 above, see $\upsilon_o \approx 0V$ and $i_{D1} = i_{D4} = i_{D2} = i_{D3} = 100\mu A$, for which $g_{m1} = g_{m2} = g_{m3} = g_{m4} = 200\mu A$, $r_s = 5k\Omega$, and $r_o = 200k\Omega$ for each. Here, consider the A circuit to consist of Q_1 with source grounded through R_{11}, and Q_3 loaded by R_{22} and R_L. Correspondingly, the β circuit consists of Q_2 and Q_4 with $r_{s2} = r_{s4} = 5k\Omega$, where $\beta = 5/(5 + 5) = ½$, $R_{11} = 5k\Omega \| 5k\Omega =$ **2.5kΩ**, $R_{22} = 5k\Omega + 5k\Omega =$ **10kΩ**. Thus, $A = \dfrac{200k\Omega}{5k\Omega + 2.5k\Omega} \times \dfrac{10k\Omega \| 10k\Omega \| 200k\Omega}{5k\Omega} =$ **26.0V/V**. Now, $R_o = 10k\Omega \| 10k\Omega \| 200k\Omega = 4.88k\Omega$, $A\beta = 26.0 \times ½ = 13.0$, $A\beta + 1 = 14.0$, $A_f = A/(A\beta + 1) = 26/14 =$ **1.857V/V**, $R_{of} = 4.88k\Omega/14 = 0.349k\Omega$, $R_{out} = 0.349 \| (-10k\Omega) =$ **362Ω**.

SECTION 8.5: The Series-Series Feedback Amplifier

8.18 For the A circuit: At the input, R_S, R_{ia}, and R_{11} are in series such that $R_i = 10k\Omega + 20k\Omega + 10k\Omega = 40k\Omega$. At the output, R_{oa}, R_L, and R_{22} are in series such that $R_o = 1k\Omega + 1k\Omega + 0.2k\Omega = 2.2k\Omega$.

Now $A = \dfrac{i_o}{\upsilon_s} = \dfrac{R_{ia}}{R_i} \times \dfrac{A_\upsilon}{R_o} = \dfrac{20}{40} \times \dfrac{900}{2.2k\Omega} =$ **204.5mA/V**. Now $\beta = 50V/A = 0.05V/mA$, and $A\beta = 204.5 \times .05 = 10.23$, and $A\beta + 1 = 11.2$, $A_f = 204.5/11.2 =$ **18.3mA/V**. $R_{if} = 40k\Omega \times 11.2 = 448k\Omega$, $R_{in} = R_{if} - R_S = 448 - 10 =$ **438kΩ**, $R_{of} = 2.2k\Omega \times 11.2 = 24.64k\Omega$, $R_{out} = R_{of} - R_L = 24.64 - 1 =$ **23.6kΩ**.

8.19 Using the results of Example 8.2 as much as possible: As before $\beta = V_f'/I_o' = 11.9\Omega$, with I_o' now in the emitter of Q_3. $A = I_o'/V_i' = 20.51/.99 =$ **20.7A/V**, with I_o' corrected for being in the emitter of Q_3. $R_i = 13.65k\Omega$, $R_o = R_{22} + R_L + R_{03}$. Here $R_{22} = R_{E2} \| (R_F + R_{E1}) = 100 \| (640 + 100) =$ **88.1Ω**. $R_L = 600\Omega$, and $R_{03} = r_{e3} + R_{C2}/(\beta + 1) = 6.25 + 5k\Omega/101 =$ **55.8Ω**. Thus $R_o = 88.1 + 600 + 55.8 =$ **744Ω**. Now $A = 20.7A/V$, $\beta =$ **11.9Ω**, $A\beta = 20.7 \times 11.9 = 246.3$, $A\beta + 1 = 247.3$, $A_f = A/(1{+}A\beta) = 20.7/247.3 =$ **83.7mA/V**. Thus $R_{if} = 13.65 \times 247.5 =$ **3.38MΩ** $= R_{in}$ (since $R_S = 0$), $R_{of} = (744)\ (247.3) =$ **184kΩ**, and $R_{out} = 184k\Omega - 0.6k\Omega =$ **183kΩ**, as seen by R_L.

8.20 *For* Q_1, Q_2, Q_3, Q_4, $i_E \approx (200\mu A)/2 = 100\mu A$ for balanced operation. Thus $r_e = 25mV/0.1mA = 250\Omega$, $r_\pi = 101\ (250) = 25.25k\Omega$, and $r_o = 200V/0.1mA = 2M\Omega$.

For Q_5, $i_E = 1mA$. Thus $r_e = 25\Omega$, $r_\pi = 101\ (25) = 2.525k\Omega$ and $r_o = 200V/1mA = 200k\Omega$. Now, consider the 10kΩ at the base of Q_2 (included to compensate for the dc drop in R_S) to be part of the A circuit. Thus the β circuit consists only of the 10Ω resistor, and $\beta = \upsilon_f/i_o =$ **10Ω**, for which $R_{11} =$ **10Ω** and $R_{22} =$ **10Ω**.

For the A circuit: For the input series connection, $R_i = 10k\Omega + 2(25.25k\Omega) + 10k\Omega + 10\Omega = 70.5k\Omega$, and for the output series connection, $R_o = R_{22} + R_L + (r_e + (1/(\beta + 1))\ (r_{04} \| r_{02})) \| r_{05} = 10 + 10^3 + (25 + \dfrac{2 \times 10^6}{2(101)}) \| (200 \times 10^{3}) = 10.5k\Omega$. Now, $A = \dfrac{I_{E2}}{V_i} \approx \dfrac{2\ (25.25k\Omega)}{2\ (25.25k\Omega) + 10k\Omega + 10k\Omega + 10} \times$

$$\frac{2\dfrac{100}{101}\,(2M\Omega \| 2M\Omega \| (101\ (1k\Omega + 10\Omega + 25\Omega))\ )}{2\ (250)} \times \frac{1}{(1k\Omega + 10\Omega + 25\Omega)}$$

$= 0.716 \times 376.2 \times .966 \times 10^{-3} = 0.260A/V =$ **260mA/V**.

$A\beta = .260A/V \times 10V/A = 2.60$, $A\beta + 1 = 3.60$, $A_f = \dfrac{I_o}{V_s} = \dfrac{A}{1{+}A\beta} = \dfrac{0.260}{3.60} = .0722A/V$ or **72.2mA/V**, $R_{of} = R_o\ (1 + A\beta) = 10.5k\Omega \times 3.6 = 37.8k\Omega$, whence $R_{out} = R_{of} - R_L = 37.8 - 1 =$ **36.8kΩ**. Now

$R_{if} = R_i\ (1 + A\beta) = 70.5\ (3.6) = 254\text{k}\Omega$, whence $R_{in} = R_{if} - R_S = 254 - 10 =$ **244kΩ**.

SECTION 8.6: The Shunt-Shunt and Shunt-Series Feedback Amplifiers

8.21 *For the basic amplifier*: $R_m = \dfrac{\upsilon_o}{i_i} = \dfrac{A_\upsilon\,\upsilon_i}{\upsilon_i/R_i} = A_\upsilon\, R_i = 900\text{V/V} \times 20\text{k}\Omega =$ **18MΩ**. For the purposes of shunt-shunt analysis, convert the input into a current source $I_S = V_S/10\text{k}\Omega$ with a shunt $R_S = 10\text{k}\Omega$.

For the β circuit: $\beta = \dfrac{I_f'}{V_o'} = -\dfrac{1}{R_f} = -\dfrac{1}{100 \times 10^3} =$ **-10^{-5}A/V**, $R_{11} =$ **100kΩ**, and $R_{22} =$ **100kΩ**.

For the A circuit: The input network consists of $10\text{k}\Omega \| 100\text{k}\Omega \| 20\text{k}\Omega = 6.25\text{k}\Omega$. The gain is $R_m = 18\text{M}\Omega$ (for input current flowing into the 20kΩ input resistor). The output network consists of a series resistor of $R_{oa} = 1\text{k}\Omega$ and a shunt load of $R_{22} \| R_L = 100\text{k}\Omega \| 1\text{k}\Omega$. Now $A = \dfrac{V_o'}{I_i'} = -\dfrac{10k \| 100k}{10k \| 100k + 20k} \times 18 \times 10^6 \times \dfrac{1k \| 100k}{1k + 1k \| 100k}$ V/A $= -\dfrac{9.09}{9.09 + 20} \times 18 \times 10^6 \times \dfrac{0.99}{1 + 0.99} = -2.80 \times 10^6\text{V/A} = -2800\text{V/mA} =$ **−2.8V/μA**, $A\beta = -2.8 \times 10^6 \times -10^{-5} = 28$, $1 + A\beta = 29$, $A_f = \dfrac{V_o}{I_s} = \dfrac{-A}{1 + \beta A} = \dfrac{-2.8 \times 10^6}{29} =$ **−96.6V/mA**. For A, $R_i = 6.25\text{k}\Omega$, $R_{if} = 6.25\text{k}\Omega/29 = 216\Omega$, $R_{in} = 216 \| (-10\text{k}\Omega) =$ **221Ω**, $R_o = 1\text{k}\Omega \| 100\text{k}\Omega \| 1\text{k}\Omega = 498\Omega$, $R_{of} = 498/29 = 17.16\Omega$, $R_{out} = 17.16 \| (-1\text{k}\Omega) =$ **17.5Ω**.

Now, for a load of $1k/2 = 500\Omega$, the gain reduces to $500/(500 + 17.5) \times (-96.6\text{V/mA}) =$ **−93.3V/mA**. To compensate at the input, seen as a fixed current source, we require that the source resistance increase from 10kΩ in order that more of the available input current enters the amplifier. For a source I_S, R_S and $R_{in} = 221\Omega$, we want the same output voltage for the original and new loads. That is, $\dfrac{R_S}{221 + R_S} I_S \times 93.3 = \dfrac{10k\Omega}{221 + 10k\Omega} \times 96.6$, or $R_S = (221 + R_S)\ (1.013)$, or $R_S = 221/(-.013) = -17\text{k}\Omega$. Thus it is not possible for normal input circuits to compensate if I_S is fixed. Alternatively, if the input voltage is fixed at $V_S = I_S\ (10\text{k}\Omega)$, then we may lower R_S (from 10kΩ) so that the output is the same for the original and new loads. That is, $\dfrac{221}{221 + R_S} \times I_S\ (10k) \times \dfrac{93.3}{221} = \dfrac{221}{221 + 10k} \times I_S\ (10k) \times \dfrac{96.6}{221}$, or $\dfrac{93.3}{221 + R_S} = \dfrac{96.6}{221 + 10k}$, $221 + 10\text{k}\Omega = (221 + R_S)\ 1.035$, whence $R_S = 10.221/1.035 - 221 =$ **9.65kΩ**.

8.22 At low frequencies, with the R_1, R_2, R_3 loop viewed as defining the voltage at node A, operation is as a voltage regulator, with the reference voltage being V_{BE1}. Here, R_1 acts as a resistive-wire connection to the input with a voltage comparison being made across the base-emitter of Q_1. Thus the feedback is of the **series-shunt variety**. As a result of it, $V_A \approx 0.7\text{V}$, $I_{R2} = 0.7/700 = 1\text{mA} = I_{E2}$, $V_B = 0.7 + 1(1) = 1.7\text{V}$, $V_C = 1.7 + 0.7 = 2.4\text{V}$, $I_{C1} \approx (5 - 2.4)/2.7\text{k}\Omega \approx 1\text{mA}$, with $r_{e1} = r_{e2} = 25\text{mV}/1\text{mA} = 25\Omega$, and $r_{\pi 1} = r_{\pi 2} \approx 2.5\text{k}\Omega$.

At high frequencies, feedback is of the **shunt-series variety**, with I_o as output and $I_s = V_s/R_S$ as input. For this, the β network consists of R_3 and R_5 with $\beta = \dfrac{I_f'}{I_o'} = \dfrac{-R_3}{R_3 + R_5} = \dfrac{-1}{1 + 5} =$ **−0.166A/A**, $R_{11} = R_5 + R_3 =$ **6kΩ**, $R_{22} = R_5 \| R_3 = (5 \times 1)/(5 + 1) =$ **0.833kΩ**.

The A circuit consists at the input, of $R_S \| R_{11} \| r_{\pi 1} = 10\text{k}\Omega \| 6\text{k}\Omega \| 2.5\text{k}\Omega$ at the base of Q_1 fed by I_i and I_o' emerging from the emitter of Q_2 and connected to R_{22} to ground. The output resistance associated with I_o' is $R_{02} = r_{e2} + R_4/(\beta + 1) = 25 + 2.7\text{k}\Omega/101 = 51.7\Omega$. Now, using a current-divider approach with device β, $A = \dfrac{I_o'}{I_i'} = -\dfrac{10 \| 6}{10 \| 6 + 2.5} \times 100 \times \dfrac{2.7}{2.7 + 101\ (.025 + .833)} \times 101 = -3.75/6.25 \times 100 \times 2.7/89.4 \times 101 =$ **−183A/A**. That is, $A\beta = -183 \times (-{,}166) = 30.5$, $1 + A\beta = 31.5$, $A_f = \dfrac{I_o}{I_s} \dfrac{-183}{31.5} = 5.81\text{A/A}$. Now $A_f' = \dfrac{I_o}{V_S} = \dfrac{I_o}{I_S\, R_S} = 5.81/10^4 =$ **0.581mA/V**. Now $R_i = R_3 \| R_{11} \| r_{\pi 1} = 10\text{k}\Omega \| 6\text{k}\Omega \| 2.5\text{k}\Omega = 1.5\text{k}\Omega$.

Thus, R_{if} = 1.5kΩ/31.5 = 47.6Ω, $R_{in} = 47.6 || (-10k\Omega) \approx$ **50Ω**. Also $R_o = R_{22} + R_{02}$ = 833.3 + 51.7 = 885Ω, and R_{of} = 885 (31.5) = 27.9kΩ. This is the resistance seen (for example) by a low-resistance load inserted between node B and the emitter of Q_2. Since V_A is assumed large and $r_{02} = \infty$, then R_{out} is also **infinite** (independent of the feedback detail). But, even if r_{02} were very low, R_{out} would still be extremely high (see Eq. 6.78 on page 437 of the Text), since R_{of} is quite high.

8.23 For the dc loop, a series-shunt configuration, β = **−1.0** (via R_1) with R_{11} = **10kΩ**, and $R_{22} = \infty\Omega$. For A, $R_i = r_{\pi 1} \approx 2.5k\Omega$, $R_o = R_3 + r_{e2} + \dfrac{R_4}{\beta + 1}$, or $R_o = (700\Omega) || (1k\Omega + 25\Omega + \dfrac{2.7k\Omega}{101})$ = 700 || 1051.7 = 420Ω, and the output of the A circuit consists of a 700Ω load. Thus $A = -\dfrac{100}{101} \times \dfrac{2.7k\Omega || (101\,(25\Omega + 1.7k\Omega))}{25\Omega} \times \dfrac{700}{1725}$ = −43.4V/V. Thus, $A\beta$ = 43.3, $A\beta + 1$ = 44.4, $A_f = \dfrac{A}{1 + A\beta} = \dfrac{-43.4}{44.4}$ = **−0.977V/V**. At node A, R_o = 420Ω, $R_{of} = \dfrac{420}{44.4}$ = 9.46Ω. Thus the resistance seen by C_3 is **9.46Ω**! For a 100μF capacitor, $f_H = \dfrac{1}{2_\pi \times 9.46 \times 100 \times 10^{-6}}$ = **168Hz**. Note that before your exposure to the effects of feedback, you may have considered the resistance seen by R_3 to be (roughly) $700 || 1k\Omega || 10k\Omega \approx 0.4k\Omega$, with $f_H = \dfrac{1}{2\pi \times 0.4 \times 1^{-3} \times 100 \times 10^{-6}}$ = **3.98Hz**!

8.24 When C_5 is removed, we end up with a complex feedback network, consisting of R_3, R_2, R_1, R_5 in a shunt-series feedback loop. Refer to P8.22 above for the basic calculations. For β, I_o' sees two paths to I_f', one through R_5 and one through R_3. The resistance in the R_3 path is 1kΩ + 0.7kΩ || 10kΩ = 1.654kΩ. $I_f' = -I_o'\left(\dfrac{1.654}{5 + 1.654} + \dfrac{5}{5 + 1.654} \times \dfrac{0.7}{10 + 0.7}\right)$, or $I_f/I_o' = \beta$ = −(.2486 + .0492) = **−0.298A/A** and $R_{11} = ((R_5 + R_3) || R_1) + R_2$ = (5 + 1) || 10 + .7 = **4.45kΩ**, R_{22} = 1.654kΩ || 5kΩ = **1.24kΩ**. Using current ratios directly, $A = \dfrac{I_o'}{I_i'} = -\dfrac{10||4.45}{10||4.45 + 2.5} \times 100 \times \dfrac{2.7}{2.7 + 101\,(.025 + 1.24)} \times 101 = -\dfrac{3.08}{5.58} \times 100 \times \dfrac{2.7}{130.4} \times 101$ = **−115.4A/A**. Now, $A\beta$ = −115.4 × (−.298) = 34.4, $A\beta + 1$ = 35.4, $A_f = \dfrac{I_o}{I_s} = -\dfrac{115.4}{34.4}$ = 3.35, $A_f' = \dfrac{I_o}{V_S} = \dfrac{3.35}{10k\Omega}$ = **0.335mA/V**.

8.25 Using results from P8.23: *For* C_1, $R_{inD} \approx (r_{\pi 1} || R_1)(A\beta + 1) = (2.5k\Omega || 10k\Omega) \times 43.4$ = 86.8kΩ, $R_{source} \approx R_S || R_5$ = 10k || 5k = 3.33kΩ. Now, for 1Hz cutoff, $C_1 \approx \dfrac{1}{2\pi \times 1 \times (3.3 + 86.8) \times 10^3}$ = **1.77μF**. *For* C_2, $R_T \approx R_5 + R_S || R_{inD}$ = 5k + 10k || 86.8k = 14kΩ. For 10Hz cutoff, $C_2 = \dfrac{1}{2\pi \times 10\,(14 \times 10^3)}$ = **1.14μF**.

8.26 For Fig. P8.10 d) in P8.10 above, with $g_m = 2\text{mA/V}$, $r_o = 10\text{k}\Omega$, $R_S = 100\text{k}\Omega$, $r = 1\text{k}\Omega$, for $R_L = R \geq 0$, find I_o/V_S and R_{out} (facing R_L). This is a series-series feedback circuit, with the output current monitored by r and a corresponding voltage fed to the source of the transistor. Thus $\beta = V_f'/I_o' = (I_o'\, r)/I_o' = r = 1\text{k}\Omega$, for which $R_{11} = r = 1\text{k}\Omega$, and $R_{22} = r = 1\text{k}\Omega$. The A circuit is as shown, where:

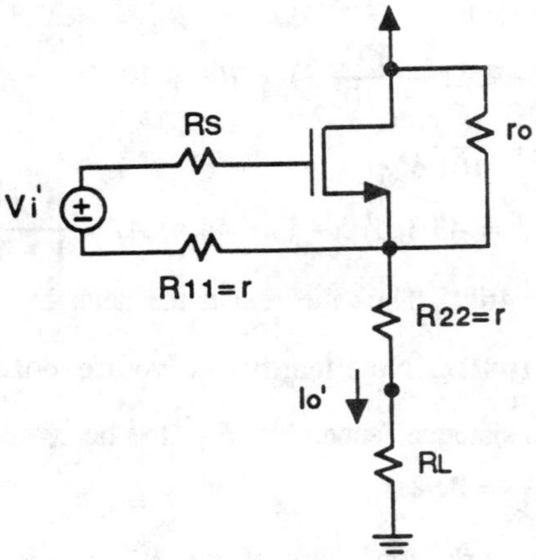

$A = \dfrac{I_o'}{V_i'} = \dfrac{g_m\, V_i'}{V_i'} \times \dfrac{r_o}{r_o + R_L + r} = 2\text{mA/V} \times (10/11 + R_L)$ where $R = R_L$ in kΩ and $R_i = \infty$.

For $R_L = 0$, $A = 2\,\dfrac{10}{11} = 1.82\text{mA/V}$. For $R_L = 10\text{k}\Omega$, $A = 2\,\dfrac{10}{11+10} = 0.95\text{mA/V}$. For $R_L = 100\text{k}\Omega$, $A = 2\,\dfrac{10}{11+100} = 0.09\text{mA/V}$. Generally, $\beta = \dfrac{V_f'}{I_o} = r = 1\text{k}\Omega$, Thus $A\beta = 1.82 \times = 1.82$ to 0.95 to 0.09 for R_L from 0 to 10kΩ to 100kΩ respectively, and $1 + A\beta = 2.82$, to 1.95, to 1.09 correspondingly, with $A_f = \dfrac{I_o}{V_S} = \dfrac{1.82}{2.82} = .645\text{mA/V}$ for $R_L = 0$, or $\dfrac{0.95}{1.95} = .487\text{mA/V}$ for $R_L = 10\text{k}\Omega$, or $\dfrac{0.09}{1.09} = .082\text{mA/V}$ for $R_L = 100\text{k}\Omega$.

See the gain is 3dB down from the 0-ohm value, when $\dfrac{2\,\dfrac{10}{11+R_L}}{1 + \dfrac{2\,(10)}{11+R_L}} = \dfrac{1}{\sqrt{2}}\,(.645) = .456$. This is of the form $2x = .456 + .456\,(2x)$. Thus $\dfrac{10}{11+R_L} = \dfrac{.456}{2\,(1 - .456)} = 0.419$, for which $R_L = \dfrac{10}{.419} - 11 = \mathbf{12.85k\Omega}$. It is a further 3dB down (ie 6dB altogether) at $R_L = 31\text{k}\Omega$. For the output resistance of A, see that since the transistor gate and source are joined through the (floating) signal source, the MOS output resistance is just r_o. Thus $R_o = R_L + r + r_o = 11 + R$. Now $1 + A\beta = 1 + 1\,(\dfrac{20}{11+R}) = \dfrac{R+31}{11+R}$, and $R_{of} = R_o\,(1 + A\beta) = (11+R)\,(\dfrac{R+31}{11+R}) = (R + 31)\text{k}\Omega$. Now the resistance seen by R_L is $R_{out} = R_{of} - R = R + 31 - R = \mathbf{31k\Omega}$ (!). The limit on the value of R_L used depends on the degree to which transconductance constancy is required, and on the size of output voltage that can be tolerated while maintaining linear operation. Certainly for loads from 0kΩ to 12.9kΩ, the transconductance varies only by 3dB, and by another 3dB for R_L up to 31kΩ.

8.27 Note the solution of P8.23. See that $R_1 = 10\text{k}\Omega = R_{1a} + R_{1b}$ with a tap at x. Let $R_{1a} = R$. Now at the tap, without feedback, the resistance is $R_{ox} \simeq (R_{1a} + r_{\pi 1})\,||\,(R_{1b} + R_2\,||\,(R_3 + r_{e2} + \dfrac{R_4}{\beta+1}))$, or $R_{ox} = (R + 2.5)\,||\,(10 - R + 0.7\,||\,(1 + 0.05)) = (R + 2.5)\,||\,(10.42 - R) = \dfrac{(2.5+R)\,(10.42-R)}{12.92}$. Now, $\dfrac{\partial R_{ox}}{\partial R} = 0$, when $(2.5 + R)\,(-1) + (10.42 - R)\,1 = 0$, $1042 + 2.5 - 2R = 0$, or $R = 12.92/2 = 6.5\text{k}\Omega$, for which $\dfrac{R_{1a}}{R} = \mathbf{0.65}$, $R_{ox} = \dfrac{(2.5+6.5)\,(10.4-6.5)}{12.9} = 2.72\text{k}\Omega$, and $R_{of} = \dfrac{R_{ox}}{1 + A\beta} = \dfrac{2.72}{44.4} = \mathbf{61.3\Omega}$, with the low-frequency loop closed. For this situation, for 100µF, $f_H = \dfrac{1}{2\pi \times 61.3 \times 100 \times 10^{-6}} = \mathbf{26Hz}$. For $f_H = 168\text{Hz}$, as in P8.23, $C_3 = 100\mu\text{F} \times 26/168 = \mathbf{15.5\mu F}$, or **15.5%** of the (100µF) capacitor needed with R_1 untapped.

SECTION 8.7: Determining the Loop Gain

8.28 At 1kHz, $A\beta = \frac{1.27}{20 \times 10^{-3}}$ = **63.5V/V**. At 10Hz, $A\beta = \frac{3.1}{2 \times 10^{-3}}$ = **1550V/V**. Assume the amplifier to be direct-coupled, and that the feedback network employs a single capacitor to ground which limits the loop gain at higher frequencies by maintaining β lower there. Further, assume that the loop-gain frequency response around 1kHz is relatively flat, and (separately) that at 10Hz, it falls as frequency rises. Since a single-pole response is postulated, it must fall (proportional to $1/f$) at 20dB/decade, from 1550V/V at 10Hz to reach 63.5V/V at 10 × 1550/63.5 = 244Hz. Thus the corner (3dB) frequency for the feedback network seems to be at **244Hz**. Now, as frequency is lowered below 244Hz, β increases and $A\beta$ increases, reaching 1550 at 10Hz and only 2 × 1550 = 3100V/V at 1Hz. Thus there must be an associated pole of β at about 10Hz/2 = 5Hz with the loop gain at dc being about 3100V/V. Since the loop gain $A\beta$ is 3100 at dc, and 63.5 at 1kHz, while β is frequency-dependent, it is likely that at 1kHz, β is at most 63.5/3100 = 0.0205, and it may even be less if feedback is not unity at dc. Now assuming the worst case, that is that A is at least **3100**: Then at 1kHz, $A_f = \frac{A}{1 + A\beta} = \frac{3100}{1 + 63.5}$ = **48V/V**. For the closed loop, the lower 3dB frequency is a 3dB frequency of the β network, occuring in particular where β is 3dB high, at which the closed-loop gain is essentially 3dB down (for $A\beta >> 1$). For a closed-loop pole at 244Hz with C = 1μF, $R_{eq} = \frac{1}{2\pi \times 244 \times 1 \times 10^{-6}}$ = **650Ω**.

8.29 At high frequencies, $\beta = \frac{R_1}{R_1 + R_2} = \frac{2}{2 + 47}$ = 0.0408V/V, and $A\beta$ = 0.0408 (1550) = **63.3V/V**. At very low frequencies, β = 1V/V, and $A\beta$ = 1 (1550) = **1550V/V**. Now β has a zero when the magnitude of the reactance of C equals the resistance of R_1, for $C = \frac{1}{2\pi \times 2 \times 10^3 \times 2.45}$ = **32.5μF**. The associated pole is at $f_p = \frac{1}{2\pi\,(R_1 + R_2)\,C} = \frac{1}{2\pi\,(49k\Omega) \times 32.5 \times 10^{-6}}$ = **0.100Hz**. See at high frequencies, that the gain will be $\frac{1550}{1 + 1550\,(.0408)}$ = 24.1V/V, and at low frequencies is essentially 1.00V/V. See that f_L is the frequency at which β begins to rise, ie at the frequency at which β has a zero, ie at 2.45Hz, ie f_L = **2.45Hz**. Generally speaking, $A_f = \frac{A}{1 + A\beta} \approx \frac{1}{\beta}$. Thus a pole of the closed-loop response will occur at a zero of β. For the capacitor reduced from 32.5μF to 10μF, resistors must be raised by 32.5/10 = 3.25 times, to R_1 = 2 (3.25) = 6.5kΩ (use **6.8kΩ**), and R_2 = 47 (3.25) = 152.8kΩ (use **150kΩ**).

8.30 Loop gain = 1.2V/10mV = **120V/V**. From the β network, $\beta = \frac{100\Omega}{100\Omega + 10k\Omega}$ = 0.0099. Thus A = 120/.0099 = **1.21 × 10⁴V/V**, is the basic op-amp gain.

8.31 $|A\beta\,(f)| = 10, |A_f\,(f)| = 10$. Now $A_f = \frac{A}{1 + A\beta}$. Assume that at a particular frequency there is no phase shift in A, but that $\beta = a + jb$, and $A_f = x + jy$. Thus $x + j\,y = \frac{A}{1 + A\,(a + j\,b)} = \frac{1}{a + 1/A + j\,b} = \frac{a + 1/A - j\,b}{(a + 1/A)^2 + b^2}$. Thus $x = \frac{a + 1/A}{(a + 1/A)^2 + b^2}$, $y = \frac{-b}{(a + 1/A)^2 + b^2}$. Now $x^2 + y^2 = 10^2$, and $A^2\,(a^2 + b^2)$ = 10^2. Combining: $\frac{(a + 1/A)^2 + b^2}{((a + 1/A)^2 + b^2)^2} = 100 = A^2\,(a^2 + b^2)$, whence $a^2 + b^2 \approx \frac{1}{1 + A^2\,(a^2 + b^2)}$, (say). $A^2\,(a^2 + b^2)^2 + (a^2 + b^2) - 1 = 0$, $a^2 + b^2 = \frac{-1 \pm \sqrt{1 - 4(-1)\,A^2}}{2A^2}$, or $A^2\,(a^2 + b^2) = \frac{-1 \pm \sqrt{1 + 4A^2}}{2}$. Thus $\frac{-1 \pm \sqrt{1 + 4A^2}}{2} = 100$, $-1 \pm \sqrt{1 + 4A^2} = 200$, $1 + 4A^2 = (201)^2$, $4A^2 = 40400$, $A^2 = 10100$, A = 100.5. Now $A^2\,(a^2 + b^2) = 10^2$, $a^2 + b^2 = \frac{0.10}{(100.5)^2}$. Thus $|\beta| = (a^2 + b^2)^{1/2}$ = 10/100.5 = .0995.

Correspondingly, at this frequency, A = **100.5V/V** and $|\beta|$ = **0.0995**.

8.32 In Fig. P8.15 here, open the loop at the gate of Q_4 and inject a signal υ. Now, for Q_1, Q_2, Q_3, Q_4, i_D = 100μA, that is $i_D = K(\upsilon_{GS} - V_t)^2$, or $100 = 100(\upsilon_{GS} - 1)^2$, whence υ_{GS} = 2V, for which $g_m = 2K(\upsilon_{GS} - V_t) = 2(100)(1)$ = 200μA/V, $r_s = 1/g_m$ = 5kΩ and r_o = 20/100μA = 200kΩ. Also for Q_5, $200 = 100(\upsilon_{GS} - V_t)^2$, and $(\upsilon_{GS} - V_t) = \sqrt{2}$ = 1.414V, whence g_m = 2 (100) 1.44 = 282μA/V, $r_s = 1/g_m$ = 3.546kΩ, and r_o = 20/200μA = 100kΩ. Now, *for no load*, $A\beta = -\dfrac{\upsilon_o}{\upsilon} = \dfrac{2\,(200k \,||\, 200k)}{2\,(5k)} \times \dfrac{100k}{100k + 3.55k} = 20 \times$ 0.966 = **19.3V/V**. Since $\beta = 1$, $A_f = \dfrac{A}{1 + A\beta} = \dfrac{19.3/1}{1 + 19.3}$ = **0.951V/V**, the same as originally found. Now, *for* 1kΩ *load*, $A\beta = \dfrac{2\,(200k \,||\, 200k)}{2\,(5k)} \times \dfrac{100k \,||\, 1k}{100k \,||\, 1k + 3.55k} = 20 \times \dfrac{.99}{.99 + 3.55}$ = **4.36V/V**, and with $\beta = 1$, $A_f = \dfrac{4.36/1}{1 + 4.36}$ = **0.813V/V**, where the earlier calculation yields A_f = 0.816V/V.

SECTION 8.8: The Stability Problem

8.33 At frequency ω, $\Phi = \tan^{-1} \omega/10^3 + 2\tan^{-1} \omega/10^5 = 180°$. Now at $\omega = 10^5$rad/s: $\Phi = \tan^{-1}100 + 2\tan^{-1}1$ = 89.4 + 2 (45) = 179.4°. Thus the phase shift is 180° at (slightly above) **10^5 rad/s**.

$$|A(10^5)| = \frac{10^3}{\left[\left[1 + \left(10^5/10^3\right)^2\right]\left[1 + \left(10^5/10^5\right)^2\right]\left[1 + \left(10^5/10^5\right)^2\right]\right]^{½}} = \frac{10^3}{10^2(\sqrt{2})(\sqrt{2})} = \frac{10}{2} = \mathbf{5V/V}.$$

Now for $\beta < 1/5$, $A\beta < 1/5 \times 5 = 1$, when Φ = 180 °, with no margins. For a 20dB gain margin, $\beta = 1/10 \times 1/50 =$ **0.02 or less**. For $\beta = 0.02$, $A\beta = 1$, when $A = 50$. Now $20\log_{10} 50$ = 34dB. Since the midband gain is 60dB, the gain drop is 60 – 34 = 26dB, implying a 3dB frequency of 26/20 or 1.3 decades above 10^3rad/s, that is at $10^3 \times 10^{1.3} = 2 \times 10^4$rad/s.

Check: At 2×10^4rad/s, $\dfrac{10^3}{\left[1 + \left(\dfrac{2\times10^4}{10^3}\right)^2\right]^{½}\left[1 + \left(\dfrac{2\times10^4}{10^5}\right)^2\right]^{½+½}} = \dfrac{10^3}{(20.02)(1.04)} = 48$, for which Φ $= \tan^{-1}20 + 2\tan^{-1} 0.2 = 87.14 + 2(11.3) = 110°$, for a phase margin of 70°.

8.34

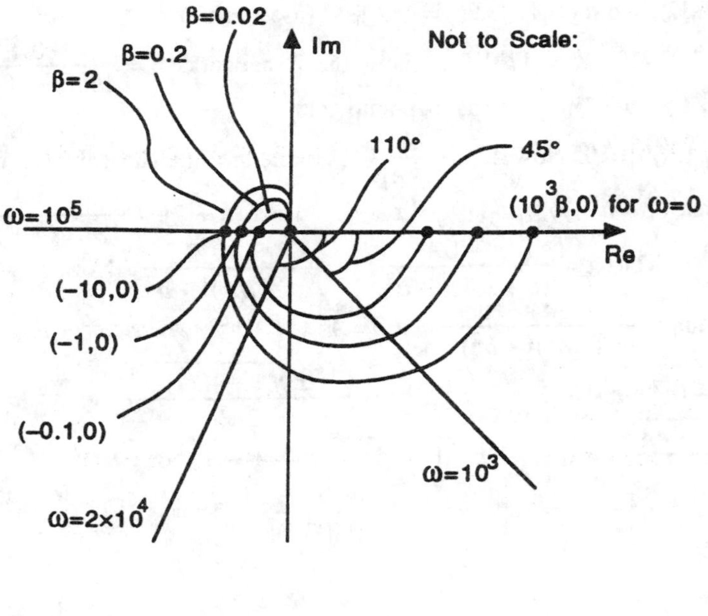

From P8.33: $A = 10^3$, $\omega_{p1} = 10^3$rad/s, $\omega_{p2} = \omega_{p3} = 10^5$rad/s, $\beta_{critical} = 0.2$. Now $A\beta(\omega)$ $= \dfrac{10^3\,\beta}{(1 + j\omega/10^3)(1 + j\omega/10^5)(1 + j\omega/10^5)}$. Now from the solution of P8.33: Know at 10^5rad/s, $|A\beta| = 5\beta$, $\Phi(A\beta) = 180°$, and at 10^3rad/s, $|A\beta| = 0.7 \times 10^3\,\beta$, $\Phi(A\beta) = 45°$, and at 2×10^4rad/s, $|A\beta| = 48\beta$, $\Phi(A\beta) = 110°$. The plot is for $\beta = 0.2$, 2, and 0.02. Note that for 3 poles, the maximum shift is 270°.

8.35 For $\beta = 1.0$, $|A\beta| = 1$ at $\omega = 10^3 \times 10^6 = 10^9$rad/s, at which the total phase shift is $\tan^{-1}(10^9/10^3) + \tan^{-1}(10^9/10^8) = 90° + 84.3° =$ **174.3°**. Thus, **oscillation does not occur**, although there is only 5.7° phase margin. Now for $\beta = 0.5$, $A = 2$, when $A\beta = 1$, whence $A = \dfrac{10^6}{\left[1 + \left(\omega/10^3\right)^2\right]^{½}\left[1 + \left(\omega/10^8\right)^2\right]^{½}} = 2$, for ω just above 10^8rad/s. At 10^8rad/s, $A = \dfrac{10^6}{10^5\,(1+1)^{½}} = 7.07$. At 3×10^8rad/s, $A = \dfrac{10^6}{3 \times 10^5\,(1 + 3^2)^{½}} \simeq 1$. At 2×10^8rad/s, $A = \dfrac{10^6}{2 \times 10^5\,(1 + 2^2)^{½}} = 2.24$. At 2.1×10^8rad/s, $A = \dfrac{10^6}{2.1 \times 10^5\,(1 + 2.1^2)^{½}} = 2.05$. Now at 2.1×10^8rad/s, $\Phi = 90° + \tan^{-1}\dfrac{2.1 \times 10^8}{10^8} = 90° + 64.5° = 154.5°$, for which the phase margin is $180° - 154.5° = 25.5°$. Now with an input capacitance, the β network provides a 3rd pole at ω_p. Oscillation can begin if $\tan^{-1}\left(\dfrac{2.1 \times 10^8}{\omega_p}\right) = 25.5°$, or $\dfrac{2.1 \times 10^8}{\omega_p} = 0.477$, or $\omega_p = \dfrac{2.1 \times 10^8}{0.477} = 4.4 \times 10^8$rad/s. Now for $C =$ 5pf, $\omega = \dfrac{1}{RC}$, $R = \dfrac{1}{\omega C} = \dfrac{1}{4.4 \times 10^8 \times 5 \times 10^{-12}} =$ **454Ω**. Thus for the β network consisting of two equal resistors R, $R/2 < 454$, or $R < 910\Omega$. To ensure stability, even smaller resistors would be needed, say 100Ω.

SECTION 8.9: Effect of Feedback on Amplifier Poles

8.36 The low-frequency amplifier gain, $A = \dfrac{f_t}{f_p} = \dfrac{20 \times 10^6}{5 \times 10^3} = 4 \times 10^3$V/V. With feedback, $A_f = \dfrac{A}{1 + A\beta} = \dfrac{4 \times 10^3}{1 + 4 \times 10^3 \times .125} = \dfrac{4000}{501} =$ **7.98V/V**. $f_{3dB} = f_p(1 + A\beta) = 5 \times 10^3 \times 501 =$ **2.5MHz**. The unity-gain frequency $= 2.5 \times 7.98 = 19.95$MHz, or **20MHz**. Thus the pole is shifted by the amount of the feedback factor, that is by about 500.

8.37 From P8.36 above, for $A = 4000$V/V and $f_p = 5$kHz. To achieve $f_{pf} = 10^6$Hz, the amount of feedback, $1 + A\beta = \dfrac{10^6}{5 \times 10^3} =$ **200**. Correspondingly, the loop gain is $A\beta =$ **199**, and the feedback factor, $\beta = 199/4000 = 0.04975$. Use $\beta =$ **0.05**, for which the low-frequency gain is $A_f = \dfrac{4000}{1 + 4000 \times 0.5} =$ **19.9V/V**.

8.38 $A(s) = \dfrac{10^4 K}{(1 + s/10^5)(1 + s/(10^6/K))}$. For loop closure with a feedback factor β, the poles are: $s = -1/2\,(\omega_{p1} + \omega_{p2}) \pm 1/2\,[(\omega_{p1} + \omega_{p2})^2 - 4(1 + A_o\beta)\,\omega_{p1}\,\omega_{p2}]^{½}$ See that the poles are coincident at 5×10^5 when $5 \times 10^5 = 1/2\,(10^5 + 10^6/K)$, $10 = 1 + 10/K$, or $10/K = 9$, $K = 10/9 =$ **1.11**. The poles are coincident when $(\omega_{p1} + \omega_{p2})^2 - 4(1 + A\beta)\,\omega_{p1}\,\omega_{p2} = 0$, or $(10^5 + 10^6/K)^2 = 4(1 + A\beta)\,10^5 \times 10^6/K$, or $1 + A\beta = \dfrac{(1 + 10/K)^2}{4\,(10/K)} = \dfrac{(1+9)^2}{4\,(9)} = \dfrac{10^2}{4\,(9)} = 2.778$. Thus $A\beta = 1.778$, and $\beta =$ **1.778 × 10⁻⁴**. Thus the dc open-loop gain of the amplifier is $10^4 K =$ **1.11 × 10⁴V/V**, with poles of **10⁵Hz** and **9 × 10⁵Hz**. The Low-frequency closed-loop gain is $A_f = \dfrac{A}{1 + \beta A} = \dfrac{1.11 \times 10^4}{2.778} =$ **4 × 10³V/V**.

8.39 For a maximally-flat design, $Q =$ **0.707**, where $Q = \dfrac{((1 + A\beta)\,\omega_{p1}\,\omega_{p2})^{½}}{(\omega_{p1} + \omega_{p2})}$ --- (1) and $\omega_{p1} + \omega_{p2} = \omega_o/Q$, with poles at $-1/2\ (\omega_{p1}+\omega_{p2}) \pm 1/2\ ((\omega_{p1}+\omega_{p2})^2 - 4\ \omega_{p1}\ \omega_{p2}\ (1+A_o\beta))^{1/2}$. That is, poles are at $\frac{\omega_{p1}+\omega_{p2}}{2} \pm 1/2\ ((\omega_{p1}+\omega_{p2})^2\ (1-4\,Q^2))^{1/2}$, or $\frac{\omega_{p1}+\omega_{p2}}{2}\ (1 \pm j\ (4\,Q^2-1)^{1/2})$, or $\frac{10^6+20\times 10^6}{2} \times (1 \pm j\ (4\,(.707)^2 - 1)^{1/2})$ = **10.5×10^6 (1 ± j) Hz** (as seen directly from Fig. 8.32 above and the text following), and $\omega_o = Q\ (\omega_{p1}+\omega_{p2}) = .707\ (21 \times 10^6) = 14.85 \times 10^6$Hz = **14.85MHz** (that is $\sqrt{2} \times 10.5$MHz). Now from the characteristic equation (Eq. 8.37), $s^2 + s\ (\omega_{p1}+\omega_{p2}) + (1+A_{o\beta})\ \omega_{p1}\ \omega_{p2} = 0$, and (1) above, $s^2 + s\ (\omega_{p1}+\omega_{p2}) + Q^2\ (\omega_{p1}+\omega_{p2})^2$ is the denominator of the transfer characteristic, which for $s = j\omega$, is $(-\omega^2 + j\omega\ (21 \times 10^6) + (21 \times 10^6)^2/2)$. Now at the 3 dB frequency (*omega*$_c$ $(\omega_c\ (21 \times 10^6))^{2)} + ((21 \times 10^6)^2/2 - \omega_c^2)^2 = 1/2$. Normalizing to 21×10^6Hz, $\omega_c^2 + (1/2 - \omega_c^2)^2 = 1/2$, or $\omega_c^2 + 1/4 + (-\omega_c^2) + \omega_c^4 = 1/2$. Thus $\omega_c^4 = 1/4$, $\omega_c^2 = 1/2$, $\omega_c = 0.707$ (normalized), and, in general, $\omega_c = 0.707 \times 21 \times 10^6$ = **14.85MHz** (that is, ω_o). Now, from (1), $(1+A_o\beta) = \frac{(Q\ (\omega_{p1}+\omega_{p2}))^2}{\omega_{p1}\ \omega_{p2}} = \frac{(0.707\ (21 \times 10^6))^2}{(1 \times 10^6)\ (20 \times 10^6)} = \frac{2\ (21)\ (21)}{1 \times 20} = 44.1$. Thus $A_o\beta = 43.1$, $\beta = 43.1/10^3$ = **0.0431**, and $A_f = \frac{A}{1+A\beta} = 10^3/44.1$ = **22.7V/V**.

8.40 $A_o = 10^3$, $\omega_{p1} = 10^3$rad/s, $\omega_{p2} = \omega_{p3} = 10^5$rad/s, and $A(s) = \frac{10^3}{(1+s/10^3)\ (1+s/10^5)^2}$.

$$\text{Thus } A_f(s) = \frac{\dfrac{10^3}{(1+s/10^3)\ (1+s/10^5)^2}}{1+\dfrac{\beta\ 10^3}{(1+s/10^3)\ (1+s/10^5)^2}} = \frac{10^3}{(1+s/10^3)\ (1+s/10^5)^2 + \beta\ 10^3}.$$

Denominator is $(1+s/10^3)\ (1+2s/10^5+s^2/10^{10}) + \beta\ 10^3 = 1 + 2s/10^5 + s^2/10^{10} + s/10^3 + 2s^2/10^8 + s^3/10^{13} + \beta\ 10^3 = (\beta\ 10^3 + 1) + s\ (2/10^5 + 1/10^3) + s^2\ (1/10^{10} + 2/10^8) + s^3/10^{13}$. Normalize to $s/10^5$, that is $10^3\ \beta + 1 + s\ (102) + s^2\ (201) + 100\ s^3$. Now divide by 100 and set equal to 0: $s^3 + 2.01\ s^2 + 1.02s + 10\ \beta + .01 = 0$ – – – (1), which is in the form $(s+a)\ (s+b-jc)\ (s+b+j\,c) = 0 = (s+a)\ (s^2 + sb + jsc + bs + b^2 + jbc - jcs - jbc + c^2) = (s+a)\ (s^2 + 2\,sb + b^2 + c^2) = (s^3 + 2s^2\,b + sb^2 + sc^2 + as^2 + 2abs + ab^2 + ac^2) = s^3 + s^2\ (a+2b) + s\ (b^2+c^2+2ab) + a\ (b^2+c^2)$ – – – (2). Now, see $a + 2b = 2.01$, or $a = 2.01 - 2b$ – – – (3), $b^2 + c^2 + 2ab = 1.02$ – – – (4), $ab^2 + ac^2 = 10\beta + .01 = x$ – – – (5). Now, (4) + (3) $\rightarrow b^2 + c^2 + 2b\ (2.01 - 2b) = 1.02$, $b^2 + c^2 + 4.02b - 4b^2 = 1.02$, $c^2 + 4.02b - 3b^2 = 1.02$ – – (6).

Special Cases: (i) Two poles are coincident when $c = 0$

(3) $\rightarrow a = 2.01 - 2b$ – – – (7),

(4) $\rightarrow b^2 + 2ab = 1.02$ – – – (8),

(5) $\rightarrow ab^2 = x$ – – – (9).

(9) $\rightarrow$ (8) $b^2 + 2x/b = 1.02 \rightarrow b^3 + 2x = 1.02b$ – – – (10)

(9) $\rightarrow$ (8) $x/a + 2ab = 1.02$, $x + 2a^2b = 1.02a$

(9) $\rightarrow$ (7) $b^2\ (2.01 - 2b) = x$, $2.01b^2 - 2b^3 = x \rightarrow b^3 = 2.01/2\ b^2 - x/2$ – – – (11)

(10) + (11) $\rightarrow 2.01/2\ b^2 - x/2 + 2x = 1.02b$, $2.01b^2 - 2.04b + 3x = 0$

$b = 2.04 \pm \frac{\sqrt{2.04^2 - 4\ (+3x)2.01}}{2\ (2.10)} = 0.507 \pm \sqrt{.258 - 1.49}$. The two are identical when $0.258 = 1.49x = 1.49\ (10\beta + 0.01)$, or $\beta = \frac{0.173 - 0.01}{10}$ = **0.0163**, at which $b = 0.507$ and $a = 2.01 - 2\ (.507) = 0.996$.

Denormalizing, the poles are at about **-1×10^5, -0.5×10^5**, and **-0.5×10^5 rad/s**.

(ii) Now the $j\omega$ axis is reached when $b = 0$, for which:

(3) $\rightarrow a = 2.01$, and

(4) $\rightarrow c^2 = 1.02$, $c = 1.01$,

where from (5), $x = 10\beta + .01 = ac^2 = 2.01 \times 1.02$, whence $\beta = \dfrac{2.05 - .01}{10} = \mathbf{.204}$. In this case, the poles are approximately at $\mathbf{-2 \times 10^5}$ **rad/s** and $\mathbf{\pm\, j\, 10^5}$ **rad/s**.

(iii) Now, $Q = .707$ for the complex pole pair implies that $b = c$, (4) $\rightarrow 2b^2 + 2ab = 1.02$, (5) $\rightarrow 2ab^2 = x$, (6) $\rightarrow b^2 + 4.02b - 3b^2 = 1.02$, $2b^2 - 4.02b + 1.02 = 0$,

$b = \dfrac{4.02 \pm \sqrt{4.02^2 - 4(2)(1.02)}}{2\,(2)} = \dfrac{4.02 \pm 2.83}{4} = 1.71$ or 0.298. Now, from (3), for $a = 2.01 - 2b$ positive, $b = 0.298$ for which $a = 2.01 - 2\,(.298) = 1.41$, and $x = ab^2 + ac^2 = 2ab^2 = 2\,(1.41)\,(.298)^2 = .250$. Thus $x = 10\beta + .01 = 0.250$, and $\beta = \dfrac{.240}{10} \approx \mathbf{0.024}$. Accounting for the initial normalization, the pole locations for $Q = 0.707$ are at $\mathbf{-\,1.41 \times 10^5}$ **rad/s, and at** $\mathbf{(-0.298 \pm j\, 0.298) \times 10^5}$ **rad/s.**

Now for $Q = 0.707$, for which $\beta = 0.024$, using the normalized frequency. $w = \omega/10^5$ rad/s, see

$$T(w) = \frac{10^3}{(1 + j100w)\,(1 + jw)^2 + 24} = \frac{10^3}{(1 + j100w)\,(1 - w^2 + 2jw) + 24}$$

$$= \frac{10^3}{1 - w^2 + 2jw + j100w - j100w^3 - 200w^2 + 24} = \frac{10^3}{25 - 201w^2 + j\,(102w - 100w^3)} \text{ - - - (12).}$$

Thus $|T| = \dfrac{10^3}{((25 - 201w^2)^2 + (102w - 100w^3)^2)^{1/2}}$. Now at the closed-loop unity-gain frequency, $|T| = 1$.

Thus (squaring), $10^6 = (25 - 201w^2)^2 + (102w - 100w^3)^2 = 625 - 10050w^2 + 40401w^4 + 10404w^2 - 20400w^4 + 10000w^6$, whence $w^6 + 2w^4 + .035w^2 - 100.06 = 0$. Solve $w^6 + 2w^4 - 100 = 0$ by trial. For $w = 2$, $2^6 + 2 \times 2^4 - 100 = 64 + 32 - 100 = -4$. For $w = 2.02$, $(2.02)6 + 2 \times (2.02)4 - 100 = 67.9 + 33.3 - 100 = 1.2$. Use $w = 2$ as an approximate solution. Now for $w = 2$, from (12), $\Phi(w) = -\tan^{-1} \dfrac{102(2) - 100(2^3)}{25 - 201(2^2)} = -\tan^{-1} \dfrac{204 - 800}{25 - 804} = -\tan^{-1} \dfrac{596}{779} = -37.4°$. Thus the phase margin for $\beta = 0.024$, with $Q = 0.707$, occurring at about 2×10^5 rad/s, is about $180 - 37.4$ or **143°**.

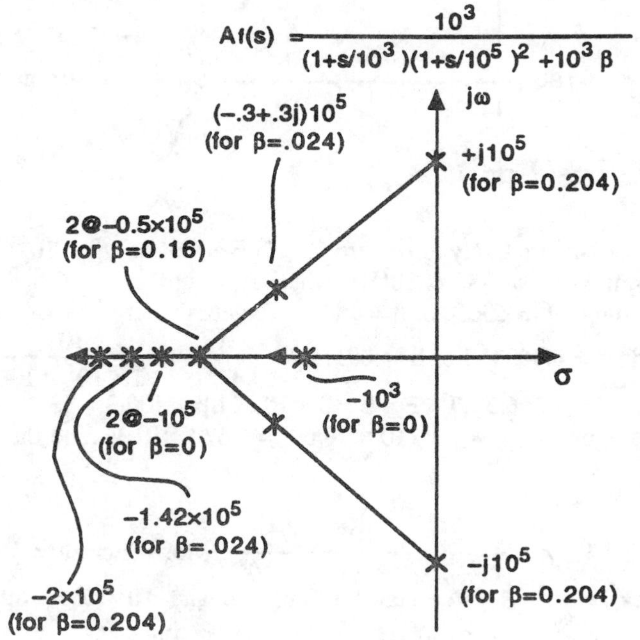

SECTION 8.10: Stability Using Bode Plots

8.41 From Eq.8.48: At the unity-loop-gain frequency, $A_f(j\omega) = \dfrac{1/\beta\, e^{-j\Theta}}{1 + e^{-j\Theta}}$, and $|A_f(j\omega)| = \left|\dfrac{1/\beta}{|1 + e^{-j\Theta}|}\right| =$
$\dfrac{1/\beta}{|1 + \cos\Theta - j\sin\Theta|} = \dfrac{1/\beta}{((1 + \cos\Theta)^2 + \sin^2\Theta)^{1/2}} = \dfrac{1/\beta}{(1 + \cos^2\Theta + 2\cos\Theta + \sin^2\Theta)^{1/2}} = \dfrac{1/\beta}{(2(1 + \cos\Theta))^{1/2}}$.
Now as noted on page 626 of the Text, for a margin of 45°, Θ = 180–45 = 135°, and $|A_f| = \dfrac{1/\beta}{(2(1 + \cos 135°))^{1/2}} = 1.307/\beta$.
There is no peak when $(2(1 + \cos x))^{1/2} = 1$, or $1 + \cos x = 1/2$, $\cos x = -1/2$, $x = 120°$, and phase margin = 180 – 120 = **60°**, (and, of course, greater). *For a peaking factor of 2*: $2 = 1/(2 + 2\cos\Theta)^{1/2}$, or $2 + 2\cos\Theta = 0.25$, $\cos\Theta = -1.75/2 = -0.875$, or $\Theta = 151°$, for which the margin is **29°**. *For a peaking factor of 10*: $2 + 2\cos\Theta = 1/100$; $\cos\Theta = -1.99/2 = -0.995$, $\Theta = 174.3°$, for which the margin is **5.7°**.

8.42

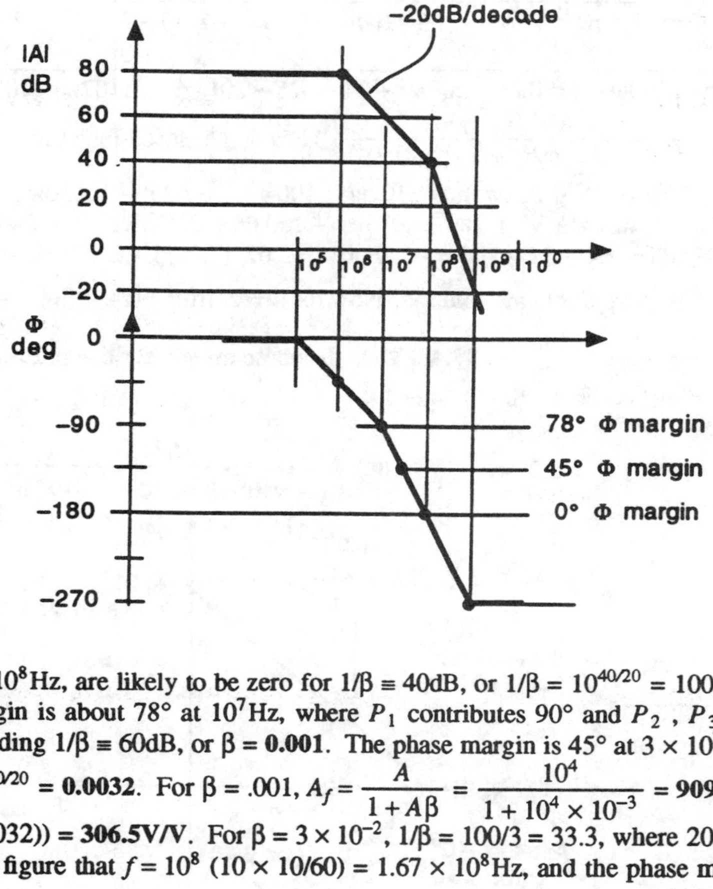

See that margins at 10^8 Hz, are likely to be zero for $1/\beta \equiv 40$dB, or $1/\beta = 10^{40/20} = 100$, or $\beta =$ **0.01**. Moreover, the phase margin is about 78° at 10^7Hz, where P_1 contributes 90° and P_2, P_3 each 6° to the total shift. The corresponding $1/\beta \equiv 60$dB, or $\beta =$ **0.001**. The phase margin is 45° at 3×10^7Hz, for which $1/\beta \equiv$ 50dB, and $\beta = 1/10^{50/20} =$ **0.0032**. For $\beta = .001$, $A_f = \dfrac{A}{1 + A\beta} = \dfrac{10^4}{1 + 10^4 \times 10^{-3}} =$ **909V/V**. For $\beta = .0032$, $A_f = 10^4/(1 + 10^4(.0032)) =$ **306.5V/V**. For $\beta = 3 \times 10^{-2}$, $1/\beta = 100/3 = 33.3$, where $20 \log 33.3 = 30.5\text{dB} \approx$ 30dB. See from the figure that $f = 10^8\ (10 \times 10/60) = 1.67 \times 10^8$Hz, and the phase margin = – 1/6 (90) = **–15°**.

8.43 For the situation in P8.42, $A = \dfrac{10^4}{(1 + j\,f/10^6)(1 + j\,f/10^8)^2}$, for which $\Phi = -\tan^{-1} f/10^6 - 2\tan^{-1} f/10^8$.
For $\Phi = -180°$, check $f = 10^8$. See $\Phi = -\tan^{-1} 10^8/10^6 - 2\tan^{-1} 10^8/10^8 = -89.43 - 2(45) = 179.4°$.
For $f = 1.1 \times 10^8$ $\Phi = -\tan^{-1} 110 - 2\tan^{-1} 1.1 = -89.5 - 2(47.7) = 185°$.

For $f = 1.01 \times 10^8$, $\Phi = -\tan^{-1} 101 - 2\tan^{-1} 1.01 = -89.4 - 2(45.29) = 180°$.
Thus margins are zero at $f = \mathbf{1.01 \times 10^8}$ **Hz**.

Now, at $f = 3 \times 10^7$, $\Phi = -\tan^{-1} \dfrac{3 \times 10^7}{10^6} - 2\tan^{-1} \dfrac{3 \times 10^7}{10^8} = -88.1 - 2(16.7) = 121.5$ for a margin of 180 – 121.5 = 58.5° (rather than 45°).

For $f = 4 \times 10^7$Hz, $\Phi = -\tan^{-1} 40 - 2\tan^{-1} .4 = -88.6 - 2(21.8) = -132.2°$, for a margin of 180° –132.2° = 47.8°.

For $f = 4.2 \times 10^7$Hz, $\Phi = -\tan^{-1} 42 - 2\tan^{-1} .42 = 88.6 - 2(22.8) = 134.2°$, for a margin of 180° –134.2° = 45.8°.

Thus margins are 45° at $f \approx \mathbf{4.3 \times 10^7}$ **Hz**.

Now at $f = 10^7$, the phase margin is $180 - \tan^{-1} \dfrac{10^7}{10^6} - 2\tan^{-1} \dfrac{10^7}{10^8} = 180 - 84.3 - 11.4 = 84.3°$ (not 78° as suggested). For $f = 1.2 \times 10^7$, the phase margin is 180 – 85.2 – 13.7 = 81.1°. For $f = 1.4 \times 10^7$, the phase margin is 180 – 85.9 – 15.9 = 78.2°. Thus the margin is 78° at $f \approx \mathbf{1.4 \times 10^7}$ **Hz**.

Now, at $f = 1.4 \times 10^7$Hz, $|A| = \dfrac{10^4}{(1 + 14^2)^{1/2}(1 + .14^2)} = \dfrac{10^4}{14.04\,(1.02)} = 698$V/V. Thus β can be 1/698 = **.00143** (where the Φ margin = 78°).

Now at $f = 4.3 \times 10^7$Hz, $|A| = \dfrac{10^4}{(1 + 43^2)^{1/2}(1 + .43^2)} = \dfrac{10^4}{(43.01)\,(1.185)} = 196.2$V/V. Thus β = 1/196.2 = **0.0051** (where the phase margin is 45°).

8.44 The available amplifier has a gain of 10^4K with poles at 10^5Hz and 10^6/K Hz.

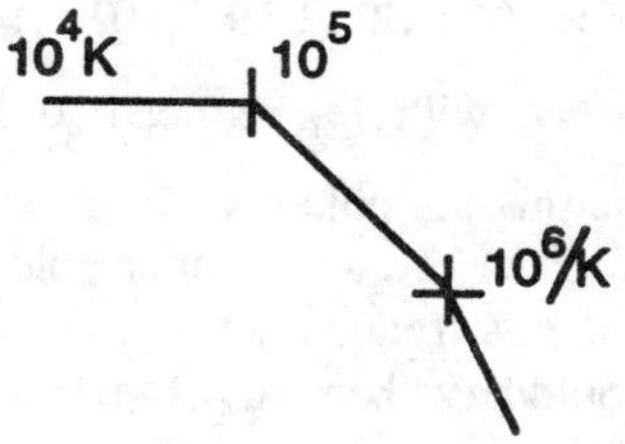

(A) For 20dB of feedback, $10^{20/20} = 1 + A\beta$. Thus $1 + 10^4K(\beta) = 10$, $\beta = 9 \times 10^{-4}/K$. From the rate-of-closure rule, the $1/\beta$ and A lines should intersect at 10^6/K Hz, where $A \approx 10^4K \times \dfrac{10^5}{10^6/K} = 10^3K^2$, and where $A\beta = 1$, or $\beta = \dfrac{1 \times 10^{-3}}{K^2}$. Now $\beta = \dfrac{1 \times 10^{-3}}{K^2} = \dfrac{9 \times 10^{-4}}{K}$, for $K = \dfrac{1 \times 10^{-3}}{9 \times 10^{-4}}$ = **1.11**, for which the bandwidth is $10^6/K$ = **0.9MHz**, and the low-frequency gain is $A_f = \dfrac{A}{1 + A\beta} = \dfrac{10^4K}{1 + 10^4K(9 \times 10^{-4}/K)} = \dfrac{10^4\,(0.9)}{10}$ = **900V/V**.

The second pole is at $10^6/K = 10^6/1.11$ = **0.9MHz**. Now from Eq. 8.36, the closed-loop poles are at $-1/2(\omega_{p1} + \omega_{p2}) \pm \sqrt{(\omega_{p1} + \omega_{p2})^2 - 4(1 + A_o\beta)\,\omega_{p1}\,\omega_{p2}}$.
Here $A_o = 0.9 \times 10^4$V/V, $\beta = 1 \times 10^{-3}$V/V, $\omega_{p1} = 10^5$Hz, $\omega_{p2} = 9 \times 10^5$Hz. Thus the closed-loop poles are at $-\dfrac{(1 + 9) \times 10^5}{2} \pm \sqrt{(\dfrac{(1 + 9)}{2} \times 10^5)^2 - (1 + 0.9 \times 10^4 \times 1 \times 10^{-3})(10^5)(9 \times 10^5)}$, or $-5 \times 10^5 \pm [(5 \times 10^5)^2 - 10 \times 9 \times 10^{10}]^{1/2}$, or $-5 \times 10^5 \pm 10^5(25 - 90)^{1/2}$, or $-10^5(-5 \pm 8.06j)$.

(B) Now, for $A_f = \dfrac{A}{1 + \beta A} = \dfrac{10^4K}{1 + \beta\,10^4K} = 10$, $10^3K = 1 + \beta\,10^4K$, $\beta = \dfrac{10^3K - 1}{10^4K} = 0.1 - 10^{-4}/K$. Now, using the rate-of-closure rule, the intersection of the $1/\beta$ and A lines will be at $10^6/K$ Hz, where $A = 10^4K \times \dfrac{10^5}{10^6/K} = 10^3K^2$. Now, at $10^6/K$ Hz, $A\beta = 1$. Thus $10^3K^2(0.1 - 10^{-4}/K) = 1$, $100K^2 - 0.1K - 1 = 0$, for which $K \approx 0.10$. That is, K = **0.10**, dc gain = **10V/V**, $\beta = 0.1 - 10^{-4}/0.1 = .099$, $A = 10^4K = 10^3$, and bandwidth = $10^6/0.1$ = **10MHz**. The second pole is at $10^6/K = 10^6/0.1$ = **10MHz**.

SECTION 8.11: Frequency Compensation

8.45 $A = 10^4$, $f_{p1} = 10^6$Hz, $f_{p2} = f_{p3} = 10^8$Hz. For an added dominant pole at f_{p0}, the second pole would be at 10^6Hz. Now for $A_f = 10$ (and A very high), $\beta \simeq 1/A_f = 0.1$. Thus $1/\beta \equiv 20$dB, while $A \equiv 80$dB. Thus the dominant pole must drop the response by $80 - 20 = 60$dB, using three decades. $\therefore f_{p0} = f_{p1}/10^3 = 10^6/10^3 =$ **10^3Hz**. Similarly for $A_f = 1$, f_{p0} = **10^2Hz**. In both cases the closed-loop 3dB frequency would be **10^6 Hz**.

8.46 For the existing pole lowered, from $f_{p1} = 10^6$Hz to f_{p1}', consider the effective second pole at f where $\tan^{-1} f/10^8 = 22.5°$, or $f = 0.414 \times 10^8$Hz. Now for a gain of 10, must lower f_{p1} by a factor of 1000 below 0.414×10^8Hz to f_{p1}' = **4.14×10^4Hz**. Now for a gain of one, similarly need f_{p1}' = **4.14×10^3Hz**. In each case, $f_{3dB} \simeq$ **40MHz**.

8.47 Now $f_{p1} = \dfrac{1}{2\pi\, C_1 R_1} = \dfrac{1}{2\pi \times C \times 10^6}$, and $f_{p2} = \dfrac{1}{2\pi\, C_2 R_2} = \dfrac{1}{2\pi \times C \times 10^4}$, a frequency 100 times larger. That is, $\dfrac{1}{2\pi\, C \times 10^6} = 10^5$, or $C = \dfrac{1}{2\pi \times 10^5 \times 10^6}$ = **1.59pF**.

Now, for Miller compensation, $f_{p1}' \simeq \dfrac{1}{2\pi\,(C_1 R_1 + C_2 R_2 + C_f(g_m R_1 R_2 + R_1 + R_2))}$, or

$10^4 \simeq \dfrac{1}{10 \times 10^{-12} \times 10^6 + 10 \times 10^{-12} \times 10^4 + 6.28\, C_f\,(100 \times 10^6 + 10^6 + 10^4)}$, or

$10^4 \simeq \dfrac{1}{10^{-5} + 10^{-7} + 6.8\, C_f\,(1.01 \times 10^8)}$, or $10^{-1} + 10^{-3} + 6.87\, C_f\,(10^{12}) = 1$. Thus $C_f = \dfrac{1}{6.87 \times 10^{12}}$ (1 – .1 – .001), or C_f = **0.146pF**. Check (From Eq.8.58): $C_f \simeq \dfrac{1}{2\pi\, g_m R_2 R_1 f_{p1}} = \dfrac{1}{2\pi \times 10^2 \times 10^6 \times 10^4} =$ **0.159pF**.

Now $f_{p2}' = \dfrac{g_m C_f}{2\pi\,(C_1 C_2 + C_f(C_1 + C_2))} = \dfrac{100/10^4 \times .146 \times 10^{-12}}{2\pi\,((1.59 \times 10^{-12})^2 + 0.146 \times 10^{-12}\,(2)\,(1.59 \times 10^{-12}))}$ $= \dfrac{0.146}{1.59 \times 10^{-9} + 0.292 \times 10^{-9}}$ = **77.6MHz**, and f_{p3} remains at **10MHz**, with f_{p1} reduced to 10^4Hz, or 10kHz. Thus the closed-loop cutoff frequency raises to **10MHz**. Note that the pole split lowered the dominant pole by a factor of 10 and raised the upper pole by a factor of 8 or so. Had the upper pole remained double at 10MHz, and assuming a double pole behaves as a single pole at a frequency for which each contributes 45°/2, that is at $10^7 \tan^{-1}(22.5°) = .414 \times 10^7$Hz, the cutoff would have been at 4.1MHz. Thus pole splitting allows the same phase shift at 10MHz as formerly at 4.14MHz. Thus it seems that the dominant pole could be raised by the factor 10/4.14 = 2.4, to 2.4×10^4Hz for roughly the same margins. For this situation, $f_{p1}' \simeq \dfrac{1}{2\pi\, g_m R_2 C_f R_1}$, and $C_f = \dfrac{1}{2_\pi \times 100 \times 2.4 \times 10^4 \times 10^6}$ = **0.066pF**, for which $f_{p2}' =$ $\dfrac{g_m C_f}{C_1 C_2 + C_f(C_1 + C_2)} = \dfrac{100/10^4 \times 0.066 \times 10^{-12}}{2\pi\,(1.59 \times 10^{-12})\,(1.59 \times 10^{-12}) + 1.59 \times 2 \times 10^{-12}\,(0.066 \times 10^{-12})} =$ $\dfrac{0.066}{1.59 \times 10^{-9} + 0.066 \times 10^{-9}}$ = **39.8MHz**. Thus the poles are at 24kHz, 10MHz and 39.8MHz. Again, the closed-loop cutoff frequency will be at about **10MHz**, whereas the original frequency for which the phase margin is 45° would have been about **4.1MHz**.

Chapter 9

OUTPUT STAGES AND POWER AMPLIFIERS

SECTION 9.1: Classification of Output Stages

9.1 Peak voltages applied are: 1.414V, 14.14V, and 141.4V. Peak load currents for a 1kΩ load are: 1.4mA, 14mA, 141mA. With a 50mA bias current, corresponding operating modes are **A, A, AB**, respectively. For a load of 0.25kΩ, the peak load currents are 5.7mA, 57mA, and 566mA, with operation in modes **A, AB, AB** respectively. For an nearly-normal large output at zero bias current, **class B** operation is apparently possible.

SECTION 9.2: Class-A Output Stage

9.2 $I_{\max} = \dfrac{0 - -3 - 0.7}{1.5k\Omega} = 1.53\text{mA}$. For a 1kΩ load, this will support a negative output peak of −1.53V, and for 10kΩ, a peak of −15.3V. In the latter case, saturation will occur earlier at −3 + 0.3 = −2.7V. For positive inputs, the positive peak is 3.0 − 0.3 = 2.7V, independent of load. Thus for a 1kΩ load, the largest sine wave is **1.53V peak**, and for a 10kΩ load, it is **2.7V peak**. For a negative output at −2.7V with $I = 1.53\text{mA}$, $R \geq 2.7\text{V}/1.53\text{mA} =$ **1.76kΩ**. For a second device connected in parallel with Q_2, I doubles to 3.06mA, and load resistances down to 1.76/2 = **0.88kΩ** can be accommodated with a −2.7V peak signal.

9.3 The output signal is voltage-limited by the saturation of Q_1 to $\upsilon_o = V_{CC} - 0.2 - 0.7 = V_{CC} - 0.9 = 5 - 0.9 =$ 4.1V peak, or current-limited to $\upsilon_O = 10\ IR_L$. Thus for large enough I, the largest possible zero-average unclipped output is **4.1V peak**. For $I_{E2} \geq I_{E1}$, there is $10I - I = 9I$ available to the load. For a 4.1V peak, $9I \geq 4.1/100\Omega$, $I \geq 4.56\text{mA}$. Thus the minimum I required is **4.56mA**.

9.4 (a) The largest-possible sine-wave output is 9 − 0.3 = 8.7V peak. The smallest-possible load resistance is 8.7V/10mA = 870Ω. ∴ Load power $= \dfrac{(8.7/\sqrt{2})^2}{0.870} =$ **43.5mW**. Supply power = 2 (9) × 10 = **180mW**. Conversion efficiency = 43.5/180 × 100 = **24.2%**.

(b) For a signal of 8.7/2 = 4.35V across a load of 870/2 = 435Ω, the load power $= \dfrac{(4.35/\sqrt{2})^2}{0.435} =$ **21.75mW**. Supply power = 2 (9) (10) = **180mW**. Conversion Efficiency = 21.75/180 × 100 = **12.1%**.

(c) The loss in Q_3 and R is 9 V × 10mA = 90mW, the supply power = 2 (9) 10 + 9 (10) = **270mW**. For (a), efficiency = 43.5/270 × 100 = **16.1%**.

(d) For (b), efficiency = 21.75/270 × 100 = **8.06%**.

9.5 For matched FETs, no load, and $\upsilon_O = 0$, $I_{D2} = I_{DSS} = 10\text{mA} = I_{D1}$ and $\upsilon_{GS1} = 0\text{V}$. Thus $\upsilon_I =$ **0V**. For operation in saturation, $\upsilon_{DG} \geq |V_p| = 2\text{V}$. Thus the negative limit of υ_O is −9 + 2 = **−7V**, for which $I_L =$ 7V/1kΩ = 7mA, and $I_{D1} = 10 - 7 = 3\text{mA}$. Generally, $i_D = I_{DSS}\,(1 - \dfrac{\upsilon_{GS}}{V_p})^2$. ∴ $3 = 10\,(1 - \dfrac{\upsilon_{GS}}{-2})^2$, $1 + \dfrac{\upsilon_{GS}}{2} = (\dfrac{3}{10})^{1/2} = .548$, $\upsilon_{GS} = 2\,(.548 - 1) = -0.905\text{V}$. Thus the corresponding input is −7 −.90 = **−7.9V**.

For positive outputs, the input limit for saturation is $\upsilon_I = 9 - 2 =$ **7V**, for which $\upsilon_O = \upsilon$. ∴ $10 + \dfrac{\upsilon}{1k\Omega} = 10\,(1 - \dfrac{7-\upsilon}{-2})^2 = 10\,(1 + 3.5 - \upsilon/2)^2$, or $10 + \upsilon = 10\,(4.5 - \upsilon/2)^2 = 202.5 - 45\upsilon + 2.5\upsilon^2$. Thus, $2.5\upsilon^2 - 46\upsilon + 192.5 = 0$, $\upsilon = \dfrac{46 \pm \sqrt{46^2 - 4(2.5)\,(192.5)}}{2\,(2.5)} = \dfrac{46 \pm 13.8}{5} =$ **6.43V** $= \upsilon_O$. *Check*: $i_{D1} = 10\,(1 - \dfrac{7 - 6.43}{-2})^2 = 16.5\text{mA}$. Compare with 6.43/1kΩ + 10 = 16.4. OK.

Now for a fixed (dc) output signal of $\upsilon_O = 6.43$V: Load power = $\frac{(6.43)^2}{1k\Omega}$ = **41.3mW**. Supply power = 9 (10 + 6.43) + 9 (10) = 147.9 + 90 = **238mW**. Efficiency = 41.3/238 × 100 = **17.4%**.

Now for a dc output of $\upsilon_O = -7.0$V. Load power = $7^2/1k\Omega$ = **49mW**. Supply power = 9 (10 − 7) + 9 (10) = 27 + 90 = **117mW**. Efficiency = 49/117 × 100 = **41.9%**

Largest-possible relatively-undistorted sine wave output is **6.43V peak**, for which Load power = $\frac{(6.43/\sqrt{2}^2)}{1k\Omega}$ = **20.7mW**. Supply power = 9(10 + 6.43/π + 10) = 198.5mW, where 6.43/ π is the average value of the 6.43V half-sine current pulse. Efficiency = 20.7/198.5 × 100 = **10.4%**.

SECTION 9.3: Class-B Output Stage

9.6 *For* $R_L = \infty$: $i_D = 0$ and $\upsilon_{GS} = V_t$ as υ_I varies. For $|\upsilon_I| \le 1$V, $\upsilon_O = 0$. For $1 \le |\upsilon_I| \le 11$V, $|\upsilon_O| = |\upsilon_I| - 1$. For $|\upsilon_I| \ge 11$V, $|\upsilon_O| = 10$V. *For* $R_L = 10k\Omega$: For υ_O = 1V, $i_D = 1/10^4$ = 100μA = 1mA $(\upsilon_{GS} - 1)^2$, $\upsilon_{GS} = \pm (1/10)^{½} + 1$ = 1.32V, υ_I = 2.32V. For υ_O = 9V, $i_D = 9/10^4 = 1 \times 10^{-3}$ $(\upsilon_{GS} - 1)^2$, $\upsilon_{GS} = \pm (.9)^{½} + 1$ = 1.95V, υ_I = 10.95V ≈ 11V. For υ_O = 10V, $i_D = 10/10^4$ = 1mA = 1mA (2 $(\upsilon_{GS} - 1)$ υ_{DS}−υ_{DS}2). Say υ_{DS} = 0.1. Thus 1 ≈ 2 $(\upsilon_{GS}$ − 1) (0.1), υ_{GS} = 1/(2(0.1)) + 1 = 6V, ∴ υ_I = 16V. For υ_O = 9.5V, i_D ≈ 1mA = 1mA (2 $(\upsilon_{GS}$ − 1) 0.5 − 0.5^2), υ_{GS} − 1 = 1 + .25, υ_{GS} = 2.25V, υ_I = 2.25 + 9.5 = 11.75. For Q_1, Q_2 in saturation, the largest possible sine wave output is **9V peak** or **18Vpp**. The corresponding input voltage is **20Vpp** for no load, and **22Vpp** for 10kΩ load. Equivalent gain is 18/20 = **0.9V/V**, or 18/22 = **0.82V/V** respectively. Supply power is **0mW** for no load, and 10 (9/√ 2)/10k = **6.36mW**, for 10kΩ load. Load power is **0mW**, or $\frac{(9/\sqrt{2})^2}{10k\Omega}$ = **4.05mV** respectively. Efficiency is ∞ (for no load), or 4.05/6.36 × 100 = **63.7%** for a 10kΩ load.

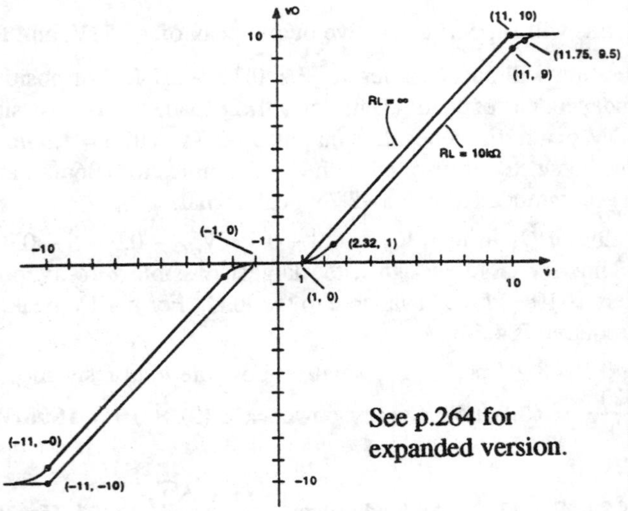

See p.264 for expanded version.

9.7 *For* υ_O = +10mV, υ_{base} = 0.710V, $i_L = 10 \times 10^{-3}/100$ = 0.1mA, i_{base} = 0.1/50 = 2μA, for which the amplifier input voltage is $\upsilon_{in} = \frac{2\mu A}{10mA/V} = \frac{2 \times 10^{-6}}{10 \times 10^{-3}}$ = 0.2mV. Thus υ_I = 10 + 0.2 = **+10.2mV**.

For υ_O = +100mV, υ_{base} = 0.800V, i_L = 100 − 10^{-3}/100 = 1mA, $i_{base} = \frac{10^{-3}}{50}$ = 20μA, $\upsilon_{in} = \frac{20 \times 10^{-6}}{10 \times 10^{-3}}$ = 2mV. Thus υ_I = **+102mV**.

For υ_O = +1V, υ_{base} = 1.7V, i_L = 1/100 = 10mA, i_{base} = 10/50 = 200μA, $\upsilon_{in} = \frac{200 \times 10^{-6}}{10 \times 10^{-3}}$ = 20mV. Thus υ_I = **+1.02V**.

9.8 Assuming $\upsilon_{CE\,sat}$ = 0V, the largest possible undistorted output is **6V peak** or 6/√2 = **4.24V rms**. Corresponding output power = $(6/\sqrt{2})^2/16$ = **1.125W**. Current from the supply is a half sine wave of 6/16 = .375A peak, whose *average value* is .375/π = 0.119A. That is supply power = 12 (0.119) = **1.43W**. Efficiency = 1.125/1.43 × 100 = **78.7%**. Power loss in both transistors is 1.43 −1.125 = **.305W**. Power loss in each transistor is the same = 0.305/2 = **0.153W**.

For 4V peak output: Output power = $(4/\sqrt{2})^2/16$ = **0.5W**. Supply power = (12 × 4/16 × 1/π) = **0.95W**. Total device dissipation = 0.95 –0.5 = **0.45W**. Efficiency = 0.5/.95 × 100 = **52.6%**.

For a +14.5V Supply, and 6V peak output: Output power = **1.125W**. Supply power = 14.5 (.119) = **1.73W**. Device dissipation = 1.73 –1.125 = **0.60W**. Efficiency = 1.125/1.73 × 100 = **65%**.

SECTION 9.4: Class-AB Output Stage

9.9 For $r_{out} \le 5\Omega$, $r_{eN} || r_{eP} = 5\Omega = r_e/2$. Thus $r_e = 10\Omega$, and I_E = 25mV/10Ω = **2.5mA**, the quiescent current, for which $V_{BB}/2$ = 690 + 25 ln (2.5/10) = 655mV, and V_{BB} = 2 (655) = **1.31V**. For 5V peak output and 50Ω load, I_L = 5/50 = 100mA, υ_{BE} = 690 + 25 ln 100/10 = 748mV. Thus υ_I = 5.00 + .748 – .655 = **5.09V**. Large-signal gain = 5.00/5.09 = **0.982V/V**. For small changes around 0V and a 50Ω load, gain = 50/(5 + 50) = **0.91V/V**. For small changes around +5V, and a 50Ω load, r_e = 25mV/100mA = 0.25Ω, gain = 50/(50 + .25) = **0.995V/V**.

9.10 For each device, biased at current I, $I = 200\,(\upsilon_{GS} - 1)^2$, and $g_m = 2\,(200)\,(\upsilon_{GS} - 1)$ mA/V with $r_s = 1/g_m$. For a 100Ω load, gain = $\dfrac{100}{100 + r_s/2}$ = 0.99, 100 = 99 + .495r_s, r_s = 1/.495Ω, or $1/r_s$ = .495 A/V = 495mA/V. Thus 2 (200) (υ_{GS} –1) = 495, υ_{GS} – 1 = 495/400 = 1.237, υ_{GS} = 2.237V, V_{BB} = 2 (2.237) = **4.48V**.

SECTION 9.5: Biassing the Class AB Circuit

9.11 For each junction, V_j = 0.675V, to maintain an output quiescent current of 2.5mA for which I_B = 2.5/31 = 81μA. Correspondingly, the quiescent current of the biassing junctions is 2.5/4 = 625μA. Thus I = 625 + 81 = 706μA. Now for a short-circuit output, the maximum available current = 706μA (30 + 1) = **21.9mA**. For junction bias reduced to 0.1 (625μA), available base current = 706 – 0.1 (625) = 643.5μA, for which i_E = 31 (643.5) = 19.9mA, and υ_O = 50 × 19.9 = **1.00V** across 50Ω. Thus the peak output for 50Ω is **1.00V**.

9.12 I_Q = 2.5mA for which υ_{BE} = 675mV and i_B = 81μA. For a 1V positive output across 50Ω, i_L = 1V/50 = 20mA, for which i_B = 20/31 = 645μA. Now for a normal bias network current I, $(I - 645)/I \times 100 = 20$, $I - 645 = 0.2I$, $0.8I = 645$, I = 806μA. Thus the resistor-network current level = 1/2 (20%) 806 = 80.6μA. Here V_{BB} = 2 (0.675) = 1.35V, and $R_1 + R_2$ = 1.35/80.6μA = 16.75kΩ. Now the normal current in the bias transistor is 806 – 80.6 – 81 = 644μA, for which V_{BE1} = 690 + 25 ln (0.644/10) = 621mV, and I_{B1} = 644/30 = 21.5μA. That is R_1 = 621/80.6 – 21.5 = 10.5kΩ. Use **10kΩ**, for which I_{R2} = 621/10 + 21.5 = 83.6μA, R_2 = 1.350 – 0.621/83.6 = **8.7kΩ**. In practice, use 8.2kΩ with a variable (or fixed) series resistor. Now for υ_O = 1V, i_{EN} = 20mA, i_{BN} = 645μA, and i_{E1} = 806 – 645 – 80.6 = 80.4μA. That is, r_{e1} = 25mV/80.4μA = 311Ω, $r_{\pi 1}$ = 31 (311) = 9.64kΩ, and for the multiplier with input υ, current i is $\dfrac{\upsilon}{8.7k + 10k || 9.64k} \times \left[1 + 31\left[\dfrac{10}{10 + 9.64}\right]\right] = \dfrac{\upsilon\,(16.78)}{13.6}$, or $r_{eq} = \upsilon/i$ = 16.78/13.6 = 1.23kΩ. Now, at the peak output of 1V, r_{eN} = 25mV/20mA = 1.25Ω, and $r_{\pi N}$ = 31 (1.25Ω) = 38.75Ω. Gain at the peak = $\dfrac{50}{50 + 1.25} \times \dfrac{31\,(1.25 + 50)}{31\,(1.25 + 50) + 1.23k\Omega}$ = 0.976 × 0.564 = **0.55V/V**.

For signals around 0V, with a 50Ω load, R_0 = 5Ω as before, but i_{E1} rises to 644μA for which r_{e1} = 25/.644 = 38.8Ω, and r_{eq} reduces to about 38.8/311 or 0.12 of the previous, or 0.12 (1.23kΩ) or about 150Ω. Thus the gain ≈ $\dfrac{50}{50 + 5} \times \dfrac{(50 + 5)31}{(50 + 5)31 + 150}$ = 0.91 × 0.92 = **0.84V/V**.

9.13 From the sequel to Eq. 9.33 of the Text, neglecting β for the bias situation, $k = 1 + R_2/R_1 = (R_1 + R_2)/R_1$. Now, for a junction voltage υ_{BE}, the emitter current of the multiplying transistor = $I - \dfrac{k\,V_{BE}}{R_1 + R_2}$, where I_E =

$I - V_{BE}/R_1$. Thus $r_e = \dfrac{V_T}{I_E} = \dfrac{V_T}{I - V_{BE}/R_1}$. Now for a small incremental voltage υ applied across the multiplier, a current i_B flows in the R_1, R_2 network, where $i_B = \dfrac{\upsilon}{R_2 + R_1 \| r_\pi} = \dfrac{(R_1 + r_\pi)\upsilon}{R_1\, r_\pi + R_2\,(R_1 + r_\pi)}$, and i_C in the transistor collector is $i_C = \dfrac{R_1 \| r_\pi}{R_2 + R_1 \| r_\pi} \times \upsilon \times \dfrac{\beta}{\beta + 1} \times \dfrac{1}{r_e} = \dfrac{R_1\, r_\pi}{R_1\, r_\pi + R_2\,(R_1 + r_\pi)} \times \dfrac{\beta \upsilon}{r_\pi} = \dfrac{\beta R_1\, \upsilon}{R_1\, r_\pi + R_2\,(R_1 + r_\pi)}$. Thus $i = i_B + i_C = \left[\dfrac{\beta R_1 + R_1 + r_\pi}{R_1\, r_\pi + R_2\,(R_1 + r_\pi)} \right] \upsilon$, $r_{eq} = \dfrac{\upsilon}{i} = \dfrac{R_1\, r_\pi + R_2\,(R_1 + r_\pi)}{(\beta + 1)R_1 + r_\pi} = \dfrac{(\beta + 1)r_e\, R_1 + R_1\, R_2 + (\beta + 1) r_e\, R_2}{(\beta + 1)R_1 + (\beta + 1)r_e} = \dfrac{\mathbf{r_e\,(R_1 + R_2) + R_1\, R_2/(\beta + 1)}}{\mathbf{R_1 + r_e}}$, with $r_e = V_T/(I - V_{BE}/R_1)$, and $k = (R_1 + R_2)/R_1$. Now, $k = 1 + R_2/R_1$, or $R_2/R_1 = k - 1$, or $R_2 = (k - 1)\, R_1$, and $R_1 + R_2 = k\, R_1$. Thus, $\mathbf{r_{eq}} = \dfrac{\mathbf{r_e\,(k\,R_1) + R_1^2\,(k - 1)/(\beta + 1)}}{\mathbf{R_1 + r_e}}$, with $r_e = V_T/(I - V_{BE}/R_1)$.

Now for $k = 2$, $I = 1\text{mA}$, $\beta \geq 50$, $R_1 = r_\pi = (\beta + 1)r_e$, $r_e = \dfrac{25}{1 - 700/R_1}\ \Omega$, with R_1 in ohms, $r_{eq} = \dfrac{2\, r_e\, R_1 + R_1^2/51}{R_1 + r_e} = \dfrac{2\, r_e\,(51)\, r_e + (51\, r_e)^2/51}{51\, r_e + r_e} = r_e \left\{ \dfrac{102 + 51}{52} \right\} = 2.94\, r_e$, where $r_e = \dfrac{25}{1 - 700/(51\, r_e)}$, $25 = r_e - 700/51 \rightarrow r_e = 25 + 700/51 = 38.7\Omega$, $\therefore\ r_{eq} = 38.7\,(2.94) = \mathbf{114\Omega}$.

SECTION 9.6: Power BJTs

9.14 At 30°C, the junction drop at current I is 630mV. At 10 times that current the drop would be 630 + 25 ln $(10I/I)$ = 687.6mV. Now at T°C, the junction drop is 500mV. The new temperature, T = 30°C + $\dfrac{(687.6 - 500)mV}{2mV/C°}$ = **123.8°C**. For a total dissipation of 45W, the thermal resistance, junction-to-ambient is (123.8 − 30)/45 = **2.08°C/W**. For a junction temperature of 180C°, total dissipation could be (180 − 30)/2.08 = **72.1W**, and the new current would be 72.1/45 × 10I = **16I**, that is 16 times the original test current. At 30°C, at this current, V_{BE} = 630 + 25 ln 16 = 699.3mV, or 700mV. At 180°C, V_{BE} = 699.3 − 2 (180 − 30) = **399.3mV**, or **400mV**.

9.15 $P_{max} \leq$ (150 − 55)/1.1°C/W = **86.4W**. For 86.4/2 = 43.2W, the junction-to-case rise is 1.1 (43.2) = 47.5°C, and for T_j = 150, T_C = 150 − 47.5 = **102.5°C**. For T_A = 30°, the thermal resistance of the heat sink required is (55 − 30)/86.4 = 0.289°C/W in the first case, and (102.5 − 30)/43.2 = 1.68°C/W in the second. For a heat-sink length L, the rating is 3/L°C/W. In the first case 3/L = 0.289 and L = 3/.289 = **10.4cm**. In the second case 3/L = 1.68, and L = 3/1.68 = **1.79cm**. Now for a potential error of 20% in all thermal measurements, but with 86.4W applied, T_C should be 150 − 86.4W (1.1°C/W) (1.2) = 35.95°, for which the thermal resistance of the sink must be (35.95 − 30)/86.4 = 0.0689°C/W, for which L = 3 (1+0.2)/0.0689 = **52.3cm**. Note the dramatic impact of measurement error on the adequacy of a design!

9.16 For a device dissipating W watts: 40° + 1/10 × W + 0.5W + 2W = 150°, W (0.1 + 0.5 + 2) = 110°, W = 110/2.6 = **42.3Watts**. For a heat sink twice as long, $W = \dfrac{110}{2.5 + 1/20}$ = **43.1W**. We now conclude that the heat sink is already quite large, the major problem lying in the transistor itself, with its dominating thermal resistance. For an infinite heat sink, the maximum rating would only be 44W!

9.17 For I_E = 5A, I_B = 0.2A, β = (5 − 0.2)/0.2 = 24, and r_e = 25mV/5A = 5mΩ. Thus r_π = (24 + 1) 5 × 10^{-3} = 125 × 10^{-3} = 0.125 Ω. For R_{ib} = .72Ω, r_x = .72 − .125 = **0.595Ω**. At I_E = 3A, r_e = 25/3 = 8.3 mΩ, r_π = 25 × 8.3 = 208.3 mΩ = 0.208Ω. Thus, R_{ib} = .595 + .208 = **0.80Ω**.

SECTION 9.7: Variations on the Class-AB Configuration

9.18 For $I_{L\,max}$ = 100 mA and a standing current i, the maximum base current occurring in the pnp transistor, is $(100 + i)/81$. Thus $I_{E2} \geq (100 + i)/81$, or $\geq$ 1.5mA. Thus, use $I = 2\,(100 + i)/81 = 2.47 + .025i$ – – – (1), or $I = (100 + i)/81 + 1.5 = 2.73 + .0123i$ – – – (2). Now for the quiescent state, $i = I_{E3} = I_{E4} = I_{E1} = I_{E2} = I$. Now, for (1) above $I = 2.47 + .025I$, or $I = 2.47/(1 - .025) = 2.53$mA, for which $I_{E2} = (100 + 2.53)/81 < 1.5$, and thus use (2), for which $I = 2.73 + .0123\,(I)$, for which $I = 2.73/(1 - .0123)$ = **2.76mA**. Now, since the output transistors are 5 times larger than the bias transistors, $V_{R3} = V_{R4} = 25 \ln 5 = 40.2$mV at a current I = 2.76mA. That is, $R_3 = R_4 = 40.2\text{mV}/2.76\text{mA} = 14.6\Omega$; Use **15Ω**. Now for outputs near zero volts, $I_{E3} = I_{E4} = 2.8$mA, and $r_e = 25\text{mV}/2.8\text{mA} \approx 9\Omega$. Thus, $R_{out} = (9 + 15)/2 = 12\Omega$. For a gain of 0.90 (dominated by the output coupling) $\dfrac{R_L}{R_L + 12} = 0.90$, where $R_L = 12/(1 - .90) = 120\Omega$. Near ±10V, where the situations are essentially the same for a particular R_L, one transistor is likely to be cut off and $I_L = 10/R_L$. Thus $r_e = \dfrac{25mV}{10/R_L} = 2.5R_L\ \Omega$, with R_L in kΩ, and $R_{out} = (2.5R_L + 15)\ \Omega$. Thus, $\dfrac{R_L}{R_L + (2.5R_L + 15) \times 10^{-3}} = 0.90$, or $R_L = 0.90R_L + 0.00225\,R_L + 0.0135$. Thus $R_L = 0.0135/(1 - 0.90225) = 138\Omega$. Thus for loads in excess of **140Ω**, the gain of the output stage can exceed 0.90, for outputs of ±10V.

9.19 For a standing current of 10.0mA in the output:

(a) $I_{E2} \approx I_{E4}$ = 10mA, and $I_{E1} = 10/100 = 0.1$mA, and $I_{E3} = 10/100 = 0.1$mA. That is $V_{BB} = 700 + 25 \ln (10/100) + 700 + 25 \ln (0.1/1) + 700 + 25 \ln (0.1/1) = 2100 - 57.6 - 57.6 - 57.6 = (2100 - 172.8)$mV $\equiv$ **1.93V**. Now, for all β increased by 10 and V_{BB} = 1.93V, all currents will increase to maintain the voltage. The current, say 10k mA, will be such that the currents in Q_2, Q_1, Q_3 will change to 10k, 10k/(1000) = .01k, and 10k/(100 × 10) = 0.01k respectively. Now – 172.8/25 = (ln 10k/100 + ln 0.01k/1 + ln 0.01k) = –6.91. Try k = 10: ln 1 + ln .1 + ln .1 = –2.3 – 2.3 = –4.60. Try k = 3: ln 0.3 + ln .03 + ln .03 = –1.20 – 3.51 – 3.51 = –8.22. Try k = 5: ln 0.5 + ln 0.05 + ln 0.05 = –0.693 – 2.995 – 2.995 = –6.68. Thus, the standing current increases by more than 5 times!.

(b) $I_{E4} \approx I_{E2} \approx 10 - 1 = 9$mA, $I_{E1} \approx I_{E3}$ = 1mA. Thus, $V_{BB} = 700 + 25 \ln (9/100) + 2\,(700)$ = **2.04V**. Now for all β increased by 10, the output current increases slightly, but the base-shunt current, established by resistors and a V_{BE} which changes only slightly, stay essentially the same. Thus for a factor-of-10 change in I_{B2}, from 0.1mA to .01mA, I_{E1} changes from 1.00 to .91 mA for the same output current. Thus it is likely that the standing current changes by a few tens of %. A great improvement!

9.20 For $I_O = 25\text{mA} = I_{E1}$, $I_{B1} = 25/100 = .25$mA. Thus $I_{C5} = 1 - .25 = .75$mA, ∴ $v_{BE5} \approx 700 + 25 \ln (0.75/1)$ = 693mV. Thus $R_{E1} \approx 693\text{mV}/25\text{mA} = 27.7\Omega$. Use **27Ω**. Without Q_5, the peak load current could be 1mA × 100 = **100mA**.

9.21 For both devices having β = 50, I_{E2} = 10mA, $I_{B2} = I_{E1} = 10/50 = .2$mA, $r_{e2} = \dfrac{25}{(51/50)(10)} = 2.45\Omega$, $r_{\pi2}$ = 51 (2.45Ω) = 125 Ω, $r_{e1} = 25/.2 = 125\Omega$, $r_{\pi1}$ = 51 (125) = 6.375kΩ. For v_2 at the base of Q_2, $i_{b2} = v_2/127.5 = i_{e1}$ and $i_{c2} = \beta\, v_2/r_{\pi2} = 50\, v_2/127.5$. Now voltage v_1 at the base of $Q_1 = v_2 + i_{e1}\, r_{e1}$, or $v_1 = v_2 + v_2/125 \times 125 = 2\,v_2$, and $i_{c1} = 50/51\ i_{e1} = 50/51 \times v_2/125$. Thus, $g_{m\,eq} = \dfrac{i_{c1} + i_{c2}}{v_1}$ =

$$\frac{\dfrac{50}{51}\dfrac{v_2}{125} + \dfrac{50\,v_2}{125}}{2\,v_2} = \frac{50}{2\,(125)}(1/51 + 1) = \mathbf{203.9mA/V}.$$

For Q_2, $r_{02} = \dfrac{100}{(.98)\,(10)} = 10.2\text{k}\Omega$, $r_{\mu2}$ = 10 (50) 10.2 = 5.1MΩ, $r_{01} \approx 100/.2 = 500$kΩ, $r_{\mu1} \approx 250$MΩ. Now, for a rise in output of v, with the input *short-circuited*, the total current is approximately $i = (v/1\text{M}\Omega) + (v/250\text{M}\Omega) + (v/(0.5\text{M}\Omega \| 5.1\text{M}\Omega)\ 50 + (v/10.2\text{k}\Omega)$. Thus $1/R_{out} = 1 + .004 + 109.8 + 98.0 = 208.8\mu$A/V, and R_{out} = **4.79kΩ**. Thus the gain $v_o/v_i = -g_{m\,eq}\, R_{out} = -202 \times 10^{-3} \times 4.79 \times 10^3$ = **–977V/V**, and $R_{in} \approx (1\text{M}\Omega/(1 + 977)) \| (51\,(2\,(125))) = 1.022\text{k}\Omega \|$

12.75kΩ = **0.946kΩ**.

For **both** devices having $\beta = 150$, $I_{B2} = I_{E1} = 10/150 = 0.066$mA. Thus $r_{\pi 2} = r_{e1} = 25/.066 = 375\Omega$, $r_{\pi 1} = 151\ (375) = 56.6k\Omega$. Thus $g_{m\,eq} = \dfrac{\frac{150}{151}\frac{\upsilon_2}{375} + \frac{150\,\upsilon_2}{375}}{\upsilon_2 + \upsilon_2} = \dfrac{150}{2(375)}\left(1 + \dfrac{1}{151}\right) =$ **206.6mA/V**, that is, almost the same as with $\beta = 50$.

For Q_2, $r_{02} = \dfrac{100}{\frac{150}{151}(10)} = 10.07k\Omega$, $r_{\mu 2} = 10\ (150)\ (10.07) = 15.1M\Omega$, $r_{01} \approx 100/.066 = 1.501M\Omega$.

$R_{out} = 1M\Omega \,||\, \infty \,||\, \dfrac{(1.5M\Omega \,||\, 15M\Omega)}{150} \,||\, 10.07k\Omega = 1M\Omega \,||\, 10k\Omega \,||\, 100k\Omega \,||\, 10.07k\Omega =$ **4.75kΩ**.

Now the gain = $-206.6 \times 10^{-3} \times 4.75k\Omega =$ **−981V/V**, and $R_{in} \approx 1M\Omega/[(1 + 981)\ ||\ (151\ (2\ (375)))] = 1.018k\Omega \,||\, 113.3k\Omega =$ **1.01kΩ**. Overall, for $\beta = 50$ to 150, g_m ranges from 204mA/V to 207mA/V, the gain ranges from −977V/V to 981V/V, R_{in} ranges from 946Ω to 1010Ω, R_{out} ranges from 4.75kΩ to 4.79kΩ. That is, there is *very little effect*.

9.22 At 125°C, and 100μA, $V_{BE} = 700 - 2\ (125 - 25) = 500$mV . Thus $R_1 = \dfrac{10V - 0.5}{100\mu A} =$ **95kΩ**, and $i_B = 100\mu A/100 = 1\mu A$. Thus $i_{R2} = 100 - 1 = 99\mu A$, and $R_2 = .500/99\mu A =$ **5.05kΩ**. Now, at 25°C (with i_{C2} low) and $i_{B2} \approx 0$, $\upsilon_B = \dfrac{5.05}{5.05 + 95.0} \times 10 = 504.7$mV. Now $i_2 = 100 \times 10^{-6}\, e^{\frac{-700 + 504.7}{25}} =$ **0.0405μA**. For doubling, that is $i_2 = 0.0810\mu A$, $\upsilon_B = 504.7 + 25 \ln 2 = 522$mV. Thus the supply voltage = (0.522/5.05) (5.05 + 95.0) = **10.34V**. At $i_{C2} = 50\mu A$, $V_{BE} = 700 + 25 \ln \dfrac{50 \times 10^{-6}}{100 \times 10^{-6}} = 682.7$mV. Now at 100°, $V_{BE} = 682.7 - (100 - 25)2 = 532.7$mV, for which the supply voltage ≈ (0.533/5.05)(100.05) = **10.56V**.

SECTION 9.8: IC Power Amplifiers

9.23 For the circuit shown in Fig. 9.30, $I_{R1} \approx \dfrac{25 - 0.7 - 2(0.7)}{25k\Omega + 25k\Omega} = 0.458$mA. Thus $I_{bias} = \dfrac{0.458}{20 \times 20} =$ **1.15μA**. To reduce this to 0.5μA, raise R_1 to (1.15/0.5) × 50 = **115kΩ** with **57.5kΩ** in each half. Now for the same gain and to maintain the same assumptions for the gain calculation, raise R_2 and R_3 by the same factor [(= 1.15/0.5 = 2.3)] to **2.3kΩ** and **57.5kΩ** respectively. Because of the change, the current in Q_{10}, Q_{11} and Q_{12} all reduce by a factor of 2.3, that in Q_9 reduces, but not by as large a factor due to R_6, R_7.

9.24 For the calculation of A, include Q_{12}, Q_{11}, Q_7, Q_8, Q_9, as driven by the output resistance of Q_6 and Q_4. With a 27V supply, bias current = (27 − 3(0.7))/50 = 0.5mA. Thus $I_{C4} = I_{C6} = I_{C10} = I_{C11} = I_{C12} = 0.5$mA, $I_{C9} = 10\ I_{C11} = 5$mA. Now $r_{e12} = 25mV/0.5mA = 50\Omega$, $r_{o12} = 100V/0.5mA = 200k\Omega = r_{o11}$. Now Q_8 operates as a follower with $\beta = 100 \times 20 = 2000$, while Q_7 operates as a follower with $\beta = 100$, where at the output the only load is $r_{o7} \,||\, r_{o9} \approx r_{o7}/2$ where $r_{o7} = 100V/5mA = 20k\Omega$. Thus the net load on the collector of Q_{12} is $r_{o12} \,||\, r_{o11} \,||\, ((101)\ (20k\Omega/2)) = 200k\Omega \,||\, 200k\Omega \,||\, 1M\Omega = 90.9k\Omega$. The gain from the base of Q_{12} to the collector of Q_{12} is about −90.9kΩ/50Ω = −1818V/V. Follower gain for no load is nearly 1V/V. Thus the overall gain A is − 1818V/V. Equivalent input capacitance is $C_T = 10 \times 10^{-12}\ (1 + 1818) = 1.82 \times 10^{-8} F$. Corresponding input resistance is $R_T = r_{o6} \,||\, r_{o4}$, where at 0.5mA, $r_o = 100V/0.5 = 200k\Omega$. Thus, the cutoff frequency = $\dfrac{1}{2\pi\ (200 \times 10^3/2)\ (1.82 \times 10^{-8})} =$ **87.5Hz**.

9.25 For equal sharing, each conducts 50/2 = 25mA. For $V_{EB5} = 0.70V$, $R_3 = 700mV/25mA =$ **28Ω**. Note that a specification of 1.0V at 1A is given for Q_3. However a lot of this V_{EB} is likely due to resistive effects in the base. Thus we use 0.7V as above. (Note, that we get a higher result for R_3 if we use the 0.1V/decade idea, in which case $V_{EB5} = 1.00 - 0.1 \log 25/1000 = 0.84V$). Now at $I_{out} = 1A$, $V_{BE} = 1.00V$, and $I_{R3} = 1.00/28 =$

35.7mA. As well, I_{B5} = 1A/30 = 33.3mA. Thus I_{C3} = 35.7 + 33.3 = 69mA. For a load change from 50mA to 1A, a factor of 20, the current in Q_3 varies from 25mA to 69mA, a factor of **2.76**. For Q_1, Q_2 operating at 1mA, $|V_{BE}|$ = 0.700V. For Q_3, Q_4 operating at 2mA, $|V_{BE}|$ = 700 + 0.1 log (2/10) = 0.630V. Thus R_5 = R_6 = (0.700 – 0.630)/2mA = **35Ω**.

9.26 For ±12V supplies and 2V saturation, outputs of ±10V are available. Thus a 20V peak signal is possible. Input provided is 0.1V peak. Required input resistance = 10kΩ. Thus, R_3 = 10kΩ, and R_4 = 10V/0.1V × 10kΩ = 1MΩ. For the highest possible input resistance, use R_4 = **10MΩ** and R_3 = **100kΩ** for a 100kΩ input resistance. For the positive side, 10/0.1 = 1 + R_2/R_1, or $R_2 = 99R_1$. Use $R_2 = R_4$ = 1MΩ and R_1 = 1MΩ/99 = **101kΩ**, a 100kΩ and 1kΩ in series, as a quick solution.

9.27 There are several choices:

(a) One is to drive A_1 as shown, but with R_3 connected to the output of A_1, with $R_4 = R_3$ and $1 + R_2/R_1$ = 20/2 = 10, $R_2 = 9R_1$, or $R_1 = R_2/9$. Use R_1 = 10kΩ, R_2 = 90kΩ, $R_4 = R_3$ = 100kΩ.

(b) Modify (a) above to merge R_3 and R_1 into R_{13} = 10kΩ, with R_2 = 90kΩ, and R_4 = 100kΩ using only 3 resistors in all.

SECTION 9.9: MOS Power Transistors

9.28 $K = 1/2\mu_n\ C_{ox}\ W/L = 1/2 \times 30 \times 10^{-6} \times 10^5/5 = 0.3\text{A/V}^2$. At low υ_{GS}, $i_D = K(\upsilon_{GS} - V_t)^2 = 0.3(\upsilon_{GS} - V_t)^2$. At high υ_{GS}, $i_D = 1/2\ C_{ox}\ W\ U_{sat}\ (\upsilon_{GS} - V_t) = 1/2 \times \frac{30 \times 10^{-6}}{5 \times 10^{-2}} \times 10^5 \times 10^{-6} \times 5 \times 10^4\ (\upsilon_{GS} - V_t) = 1.5 \times (\upsilon_{GS} - V_t)$. These currents are equal when $0.3(\upsilon_{GS} - V_t)^2 = 1.5 \times (\upsilon_{GS} - V_t)$, or $\upsilon_{GS} - V_t = 5$, or $\upsilon_{GS} = 5 + 2 =$ **7V**. For υ_{GS} = 7V, $i_D = K(\upsilon_{GS} - V_t)^2 = 0.3(7-2)^2$ = **7.5A**, for which g_m = 2 (0.3) (7 – 2) = **3A/V**. For υ_{GS} = 3.5V, $i_D = 0.3(3.5-2)^2$ = **675mA** or **.675A**, for which g_m = 2 (0.3) (3.5 – 2) = **0.9A/V**. For υ_{GS} = 14V, i_D = 1.5 (14 – 2) = **18A**, for which g_m = $1.5(\upsilon_{GS} - V_t)$ = 1.5(14 – 2) = 18 A/V. Note that from the velocity-saturated relationship at 7V, the g_m would be 1.5A/V (rather than 3A/V). In practice, the transition between modes begins to occur at lower values of υ_{GS} and g_m.

9.29 For 5mA bias: $i_D = 5 = K(\upsilon_{GS} - V_t)^2 = 200(\upsilon_{GS} - 2)^2$, $\upsilon_{GS} = (5/200)^{1/2} + 2$ = 2.158V. Thus V_{14} = 4 (0.7) + 2 (2.158) = **7.12V**, R = 2(2.158)/5mA = **863Ω**. Total FET TC = 2 (–3) = –6mV/°C. Total BJT TC = 4 (–2) = –8mV°C. Total TC = –14mV/°C. Thus 6/14 = **0.429** of V_{14} must appear across Q_6.

For outputs around zero: $g_m = 2K(\upsilon_{GS} - V_t)$ = 2 (200 (2.158 – 2)) = 63.2mA/V. Thus, R_{out} of follower = $(1/g_m) \| (/g_m)$ = (1/63.2)/2 = **7.9Ω**. The gain for 100Ω load ≈ 100/(7.9 + 100) = **0.927V/V**.

For outputs around +20V: I_L ≈ 20/.100 = 200mA, whence 200 = $200(\upsilon_{GS} - 2)^2$, $\upsilon_{GS} = (200/200)^{1/2} + 2$ = 3V, and g_m = 2 (200) (3 – 2) = 400mA/V. Thus R_{out} = (1/400)/2 = 1.25. The incremented gain = 100/(1.25 + 100) = **0.988V/V**. Alternatively, note that the input must be 20 + 3 – 2.158 = 20.84V at the peak. The corresponding overall gain = 20/20.84 = **0.960V/V**.

9.30 See that Q_3, Q_4 are 100 times larger than Q_1, Q_2. Thus for 10mA in Q_3, Q_4 require I = 10/100 = **0.1mA** in Q_1, Q_2. For I_Q = 5mA, 5 = $100(\upsilon_{GS} - 1)^2$. Thus $\upsilon_{GS} = (5/100)^{1/2} + 1$ = 1.224V, g_m = 2 (100) (1.224 – 1) = **44.8mA/V**, $R_{out} = 1/g_m \| 1/g_m = (\frac{1}{44.8})/2$ = 11.2Ω. *For* V_O = 0V, the gain with 100Ω load is 100/(100 + 11.2) = **0.899V/V**, with V_D = **0V**, V_B = **1.22V**, V_C = **–1.22V**, V_A = **0V**. See that $\upsilon_{GD1} = \upsilon_{GD2} = 0$ and all transistors operate in saturation mode. *For* V_O = +10V, with 100Ω load, i_{D3} = 10/0.1 = 100mA, that is 100 = $100(\upsilon_{GS} - 1)^2$, υ_{GS} = 1 + 1 = 2V. Thus V_D = **+10V**, and V_B = **+12V**. Now Q_1 continues to conduct 0.1mA with V_{GS} = 1.224V, that is V_A = 12 – 1.224 = **10.78V**. For Q_2, V_D = 10V, V_G = 10.78V, and operation is in triode mode with υ_{GD} = 0.78V, where, 0.1 = 1 $(2(\upsilon_{GS} - 1)(\upsilon_{DS}) - \upsilon_{DS}^2)$. Now, since $\upsilon_{GD} = \upsilon_{GS} - \upsilon_{DS}$ = 0.78, and $\upsilon_{GS} = 0.78 + \upsilon_{DS}$, 1 = 20 $((.78 + \upsilon_{DS} - 1)\upsilon_{DS} - 10\upsilon_{DS}^2) = -4.4\upsilon_{DS} + 20\upsilon_{DS}^2 - 10\upsilon_{DS}^2$, or 10

$\upsilon_{DS}^2 - 4.4\ \upsilon_{DS} - 1 = 0$. Thus $\upsilon_{DS} = \dfrac{4.4 \pm \sqrt{4.4^2 - 4(-1)(10)}}{2(10)} = 0.605\text{V}$, and $\upsilon_{GS} = .78 + .605 = 1.38\text{V}$. Correspondingly, $V_C = +10 - 0.605 \simeq$ **9.40V**. Thus Q_1, Q_3 operate in saturation, Q_4 cuts off, and Q_1 is in triode mode. Overall, the gain is $V_D/V_A = 10/10.78 =$ **0.928V/V**. Incrementally, for Q_3, $g_m = 2(100)(2 - 1) = 200\text{mA/V}$, and $R_{out} = 1/g_m = 1/200 = 5\Omega$. Thus the gain $\upsilon_d/\upsilon_a = 100/(100 + 5) =$ **0.952V/V**.

9.6 (Continued)

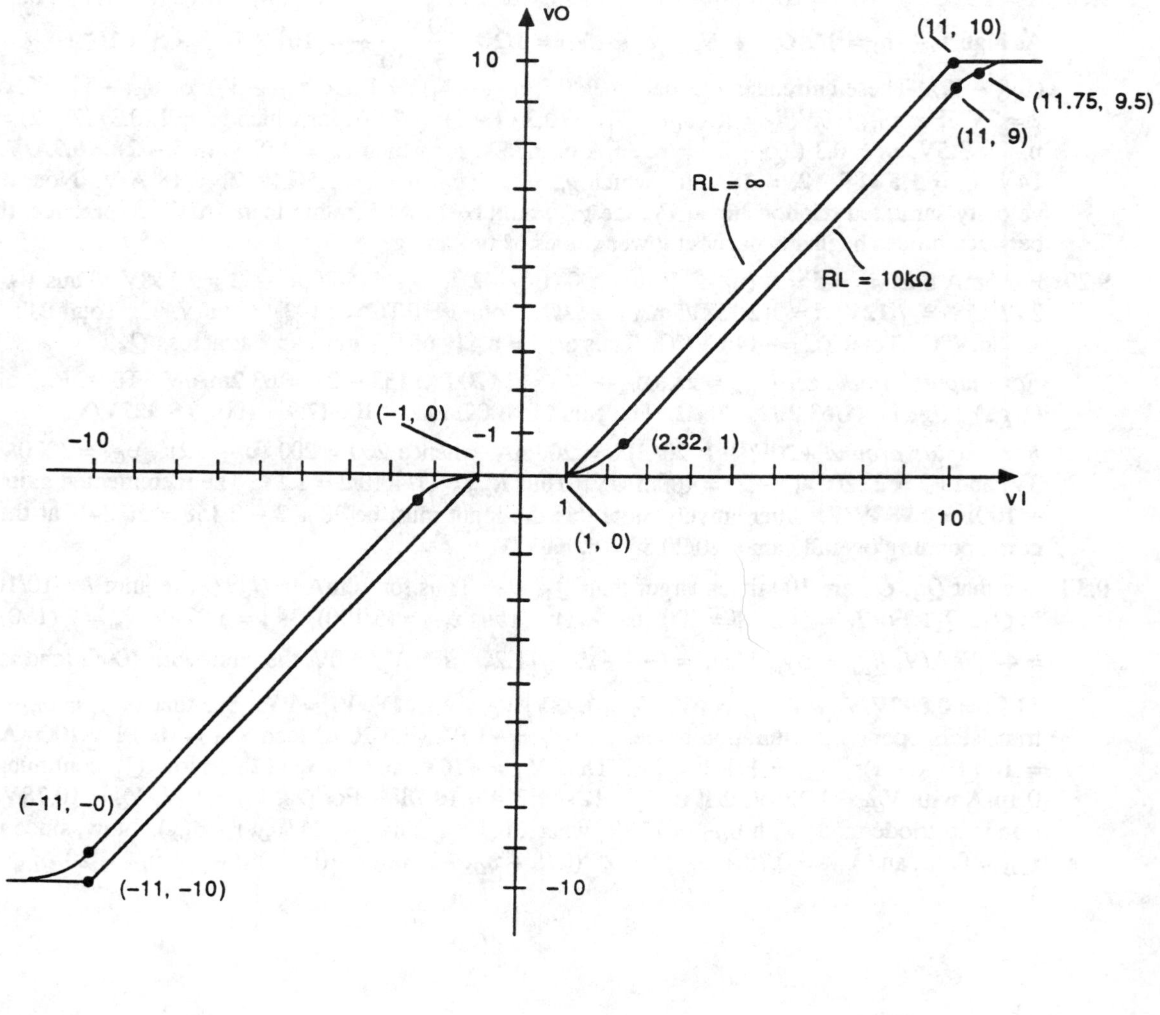

Chapter 10

ANALOG INTEGRATED CIRCUITS

SECTION 10.1: The 741 Op-Amp Circuit

10.1 For ±15V supplies, I_{REF} = [+15 − −15 − 0.7 − 0.7]/39kΩ = **0.733mA**, or **733μA**. For ±5V supplies, I_{REF} = [+5 − −5 − 1.4]/39kΩ = **0.2205mA**. For ±5V supplies and I_{REF} = .73mA, R_5 = (10 − 1.4)/0.733 = **11.73kΩ**. One could use **12kΩ** as a standard value.

10.2 For ±15V supplies, I_{REF} = 733μA. Thus $\upsilon_{BE11} = n\ V_T \ln (I/I_S) = 25 \ln (.733 \times 10^{-3}/10^{-4})$ = 625.5mV. Now $\upsilon_{BE10} = \upsilon = 625.5 + 25 \ln (i/.733)$, with i in mA, where $iR_4 = -25 \ln i/.73 = 5000i$, or $i = -.005 \ln i/0.73$.

Iterate: For i = 0.1 (mA), i = −.005 ln 0.1/.73 = .01mA. For i = .01 (mA), i = −.005 ln .01/.73 = .0215mA. For i = .015 (mA), i = .005 ln .015/.73 = .0195mA. For i = .018 (mA), i = .005 ln .018/.73 = .0185mA. Thus $i \approx$ **18.3μA**.

For ±5V supplies, I_{REF} = .2205mA, $\upsilon_{BE11} = 25 \ln ((.2205/10^{-14}) \times 10^{-3})$ = 595.4mV, and $\upsilon_{BE10} = 595.4 \times 25 \ln i/.2205$, where $iR_1 = -25 \ln i/.2205 = 5000i$, or $i = -.005 \ln i/.2205$.

Iterate: For i = .01mA, i = .005 ln .01/.2205 = .0155mA. For i = .015mA, i = .0133mA. For i = .014mA, i = .0138mA. Thus $i \approx$ **13.9μA** (reduced from 18.3μA).

For i = 18.3μA as before, υ_{BE10} = 625.5 + 25 ln (.0183/.733) = 533.2mV, (or, υ_{BE10} = 595.4 + 25 ln (.0183/.2205) = 533.2mV) and, for ±5V supplies, υ_{BE11} = 595.4mV. Thus $R_4 = \dfrac{(595.4 - 533.2)mV}{18.3\mu A}$ = **3.40kΩ**.

10.3 Replace R_a by a transistor Q_{25} whose collector is connected to the emitter of Q_{16}, emitter to a resistor R_{12} connected to $-V_{EE}$, and base to either the base of Q_{11} or the emitter of Q_7. From page 707 of the Text, note that the voltage $V_{B17} \approx$ 618mV + 550μA (100Ω) = 618 + 55 = 673mV, and I_{R9} = 673mV/50kΩ = 13.5μA. Now at 13.5μA, $V_{BE25} = 25 \ln (13.5\mu A/10^{-14} A)$ = 525.6mV. *For the connection to the base of* Q_{11}, $V_{BE11} = 25 \ln \dfrac{730 \times 10^{-6}}{10^{-14}}$ = 625.3mV. Thus $R_{12} = \dfrac{(625.3 - 525.6)mV}{13.5\mu A}$ = **7.4kΩ**. *For the connection to the emitter of* Q_7, $V_{E7} = V_{BE6} + R_2\ (I_{E6})$ and, (from page 706), $V_{E7} = 517 + 9.5 \times 10^{-6} \times 1 \times 10^3 = 517 + 9.5$ = 526.5mV, for which $R_{12}' = \dfrac{(525.6 - 526.5)mV}{13.5\mu A} \approx 0$. To avoid an unusual load on Q_7, it would be best to include a resistor R_{12} = **1kΩ** say (= R_1, R_2), in which case the current extracted from Q_7 is about 9.5μA. The latter has the advantage of using a smaller resistor and, as well provides a (small) signal component of a reinforcing polarity from the collector of Q_5 via Q_7.

10.4 *For inputs limited to the supply range*, the worst case is for one of I_n+ and I_n− connected to +15V (say I_n+) and the other (say I_n−) to −15V, in which situation the collector junction of Q_2 has 15 − 0.7 − −15 = 29.3V reversed across it. OK. Also the EBJ of Q_2 and Q_4 are reversed in series, with a combined rating of 7 + 50 = 57 > 30. OK.

Now for inputs outside the supply range: For I_n+ or I_n− positive, the base-collector diode of Q_2 (say) conducts, reversing the EBJ of Q_8, Q_9. The greatest allowed voltage is 7 + 2 (0.7) = **8.4V** above the positive supply.

For I_n+ or I_n− negative, the most-negative input stresses the CB junction of Q_2 (say) to 50V when V_{in} = 30 − 0.7 − 50 = **−20.7V** below the negative supply.

SECTION 10.2: DC Analysis of the 741

10.5 From the preamble to Eq. 10.1 on page 704 of the Text, $V_{BE11} - V_{BE10} = I_{C10} R_4$, or $V_T \ln \frac{R_{REF}}{I_5} - V_T \ln \frac{I_{C10}}{k I_S} = I_{C10} R_4$, or $V_T \ln \frac{k I_{REF}}{I_{C10}} = \mathbf{I_{C10} R_4}$ – – – (1). For $I_{REF} = 730\mu A$, $R_4 = 5000\Omega$, and $k = 0.5$, $25 \times 10^{-3} \ln \frac{0.5(730 \times 10^{-6})}{I_{C10}} = I_{C10} \times 5000$. Now, for $I_{C10} = i$ in μA, ln 365/*i* = .2*i*, or *i* = 5 ln 365/*i*. Use a process of *trial and success*: *i* = 10μA → *i* = 5 ln 365/10 = 17.98, *i* = 15μA → *i* = 5 ln 365/15 = 15.95, *i* = 15.5μA → *i* = 5 ln 365/15.5 = 15.8, *i* = 15.7μA → *i* = 5 ln 365/15.7 = 15.7. Thus, I_{C10} = **15.7μA**.

Normally for *k* = 1, 25 ln 730/*i* = 5*i*, or *i* = 5 ln 730/*i*, *i* = 19 → *i* = 5 ln 730/19 = 18.24, *i* = 18.5 → *i* = 5 ln 730/18.5 = 18.4. Thus for I_{REF} = 730.00μA, I_{C10} = 18.45μA (or so). Thus for *k* = 0.5, from (1). $R_4 = \frac{25 \times 10^{-3}}{18.45 \times 10^{-6}} \ln \frac{0.5(730)}{18.45} = 4.04k\Omega$. Use **4kΩ**.

10.6 For Q_1 through Q_4, $I_S = 10^{-14}$A, *n* = 1, and $I_C = 19/2 = 9.5\mu A$. Thus $V_{BE} = V_T \ln I/I_S = 25 \ln \frac{9.5 \times 10^{-6}}{10^{-14}}$ = 516.8V. Accordingly, the voltage on the bases of Q_3, Q_4 is –2 (516.8) = **–1.034V**.

10.7 Use the fact that V_{BE6} = 517mV and $I_{E6} \approx I_{E5}$ = 9.5μA. Correspondingly, $V_{E7} = 0.517 + 9.5 \times 10^{-6} \times 1 \times 10^3$ = **0.5265V**, $I_{E7} = (I_{C6} + I_{C5})/\beta + .5265/50k\Omega = I_{C6}/2\beta + 11.3\mu A$, and $I_{B7} = I_{C6}/(2\beta^2) + 11.3/\beta$, $I_{C3} = I_{C5} + I_{B7} = I_{C6} + I_{B7} = I_{C6} + I_{C6}/(2\beta^2) + 11.3/\beta$, whence $I_{C3}/I_{C6} = 1 + 1/(2\beta^2) + \frac{11.3}{\beta I_{C6}} \approx 1 + 1/(2\beta^2) + \frac{1}{\beta}(11.3/9.5) = 1 + 1/2\beta^2 + 1.19/\beta$. Thus $\mathbf{I_{C6}/I_{C3} = 1/(1 + 1/(2\beta^2) + 1.19/\beta)}$. For β = 200,

$$I_{C6}/I_{C3} = \left(1 + \frac{1}{2\,(200)^2} + \frac{1.19}{200^{-1}}\right) = 1/(1 + (.0025 + 1.19)/200) = \mathbf{0.994}.$$

10.8 For high β, ignore the base current of Q_7. For $I_{C5} = I$, $I_{C6} = 19 - I$, for *I* in μA. Now, $V_{BE5} = 25 \ln I/I_S = 25 \ln 10^{14} I$, and $V_{BE6} = 25 \ln (19 - I)\, 10^{14}$.

For R_1 shorted: $25 \times 10^{-3} \ln (10^{14} I) = 25 \times 10^{-3} \ln ((19 - I)10^{14}) + 1 \times 10^3 (19 - I)$, whence *I* = 19 –25 ln (*I*/(19 – *I*)) with *I* in μA. Try *I* = 10 → *I* = 19 – 25 ln (10/(19 – 10)) = 16.37. Try *I* = 12 → *I* = 19 – 25 ln 12/7 = 5.52, try *I* = 11 → *I* = 19 – 25 ln 11/8 = 11.04. Thus I_{C5} = 11.0μA, for which I_{C6} = 19 – 11.0 = 8.0μA, and I_{C6}/I_{C5} = 8/11 = **0.727**.

For R_2 shorted: $V_{BE5} + R_1 I = V_{BE6}$, $25 \ln (I/I_S) + R_1 I = 25 \ln ((19 - I)/I_S)$, $R_1 I = 25 \ln ((19 - I)/I)$, *I* = 25 ln ((19 – *I*)/*I*), for *I* in μA. Try *I* = 8μA → *I* = 25 ln (19 – 8)/8 = 7.96. Thus I_{C5} = 8.0μA, for which I_{C6} = 19 – 8 = 11.0μA, and I_{C6}/I_{C5} = 11/8 = **1.375**.

10.9 For 8μA in Q_1, and 11μA in Q_2, $V_{BE1} = 25 \ln \frac{8 \times 10^{-6}}{1 \times 10^{-14}}$ = 512.5mV, $V_{BE2} = 25 \ln \frac{11 \times 10^{-6}}{1 \times 10^{-14}}$ = 520.5mV. Now the offset (due npn + pnp devices) = 2 (520.5 – 512.5) = **16mV**, the negative input being higher with R_2 shorted. The offset is **–16mV** for R_1 shorted.

10.10 At present, V_{BE17} = 618mV, and $I_{E16} \approx$ 16.2μA. Thus, the revised $R_9{}' = \frac{618 \times 10^{-3}}{16.2 \times 10^{-6}}$ = **38.1kΩ**. Now, for I_{C17} = 550μA, I_{B17} = 550/200 = 2.75μA, for I_{R9} = 4 (2.75) = 11μA, $R_9{}'' = \frac{618 \times 10^{-3}}{11 \times 10^{-6}}$ = **56kΩ**.

10.11 For the current in Q_{14}, Q_{20}, increased to 1.5 (154) = 231μA, $V_{BE} = 25 \ln \frac{231 \times 10^{-6}}{3 \times 10^{-14}}$ = 569.1mV, and $V_{R6} = 27 \times 231 \times 10^{-6}$ = 6.23mV. For $R_6 = R_7 = 0\Omega$, V_{BB} = 2 (569.1) = 1138.2mV. For $R_6 = R_7 = 27$, V_{BB} = 2 (6.23 + 569.1) = 1150.7mV. Now $V_{BB} = V_{BE18} + V_{BE19}$, and with high β, $V_{BB} = 25 \ln \frac{(180 \times 10^{-6} - V_{BE18}/R_{10})}{10^{-14}} + 25 \ln \frac{(V_{BE18}/R_{10})}{10^{-14}}$, or for V_{BE18}/R_{10} in μA, $V_{BB} = 25 \ln \frac{(180 - V_{BE18}/R_{10})}{10^{-8}}$ +

$25 \ln \frac{(V_{BE18}/R_{10})}{10^{-8}}$ - - - (1).

Now *for* $R_6 = R_7 = 0\Omega$, $V_{BB} = 1138$mV, $V_{BE18} = \upsilon$, $R_{10} = R$. From (1), $10^{-16} \ln^{-1} 1138/25 = (180 - \upsilon/R) \times (\upsilon/R) = 5876$, or $(\upsilon/R)^2 - 180(\upsilon/R) + 5876 = 0$, $\frac{\upsilon}{R} = \frac{+180 \pm \sqrt{180^2 - 4(5876)}}{2} = \frac{180 - 94.3}{2} = 42.8\mu A$.

Thus $I_{E18} = 180 - 42.8 = 138.2\mu A$, $V_{BE18} = 25 \ln \frac{138.2 \times 10^6}{10^{-14}} = 583.7$mV. Thus $R_{10} = 583.7/42.8 =$ **13.6kΩ**. *Check*: $V_{BE19} = 25 \ln 42.8 \times 10^8 = 554.4$mV, and $V_{BB} = 583.7 + 554.4 = 1138.1$mV, as required.

Now for $R_6 = R_7 = 27\Omega$, $V_{BB} = 1151$mV, $V_{BE18} = \upsilon$, $R_{10} = R$. From (1) $10^{-16} \ln^{-1} 1151/25 = (180 - \upsilon/R) \times (\upsilon/R) = 9766$. Thus $(\upsilon/R)^2 - 180(\upsilon/R) + 9766 = 0$, and $\frac{\upsilon}{R} = \frac{180 \pm \sqrt{180^2 - 4(9786)}}{2} = \frac{180 \pm j81}{2}$. See the result is **imaginary**, that is it is not possible to provide the desired result by varying R_{10}. To check this fact, note that the largest real value of $\upsilon/R = 180/2 = 90\mu A$, at which $V_{BE} = 25 \ln(90 \times 10^{+8}) = 573$mV, for which $V_{BB\,max} = 2(573) = 1146$mV < 1151mV required.

SECTION 10.3: Small-Signal Analysis of the 741 Input Stage

10.12 From Eq. 10.4 of the Text, $R_{id} = 4(\beta_N + 1) r_e$, whence $r_e = \frac{3.6 \times 10^6}{4(180+1)} = 4.97$kΩ, for which $I_E = \frac{25 \times 10^{-3}}{4.97 \times 10^3} =$ **5.03μA**, reduced from the present value of 9.5μA. For this change, G_{m1}, $= \frac{\alpha}{2r_e} = \frac{180/181}{2(4.97)}$ **= 0.1mA/V**.

10.13 Generally, $R_0 = r_o[1 + g_m(R_E \| r\pi)]$. Here, $r_e = 2.63$kΩ, and $r_\pi = (\beta + 1)r_e = 201(2.63) = 528.6$kΩ, with $r_o = 5.26$MΩ, where $g_m = \frac{\alpha}{r_e} = \frac{200}{201(2.63) \times 10^3} = 0.378$mA/V. Thus, $R_{06} = 10.5 \times 10^6 = 5.26 \times 10^6 [1 + .378 \times 10^{-3}(R_2 \| 528.6k\Omega)]$. $\therefore R_2 \| 529k\Omega = (\frac{10.5}{5.26} - 1)\frac{1}{.378 \times 10^{-3}} = 0.996/(0.378 \times 10^{-3}) = 2.635$kΩ, and $R_2 = 2.635 \| (-529)$kΩ **= 2.65kΩ**. Thus $R_{01} = R_{06} \| R_{04} = 10.5/2 =$ **5.25MΩ**. For $G_{m1} = 1/5.26$mA/V, and for the new situation, $A'_{\upsilon 1} = \frac{1}{5.26} \times 10^{-3} \times 5.25 \times 10^6 \approx$ **1000V/V**. For the old, $R_{06} = 5.26 \times 10^6 [1 + .378 \times 10^{-3}(1k\Omega \| 528k\Omega)] = 7.25$MΩ, and $R_{01} = R_{06} \| R_{04} = 10.5 \| 7.25 = 4.29$MΩ, for which $A_{\upsilon 1} = \frac{1}{5.26} \times 10^3 \times 4.29 \times 10^6 =$ **815V/V**.

10.14 Use $R_1 = R_2 = 2.65$kΩ $= R$. From Example 10.1, $V_{OS} = \frac{\Delta R(I/G_{m1})}{R + \Delta R + r_e}$, with $\Delta R = .02(2.65) = 53\Omega$, and $r_e = \frac{25 \times 10^{-3}}{9.5 \times 10^{-6}} = 2.63$kΩ, where $G_{m1} = 1/5.26$mA/V. Thus $V_{OS} = \frac{53 \times 5.26 \times 10^3 \times 9.5 \times 10^{-6}}{(2.65 + .053 + 2.63) \times 10^{13}} =$ **0.497mV**. This is $.497/.3 \approx 1.67$ of the original, or **67% larger**, although the resistors are only (1 − 2.65/2), or 33% larger.

10.15 From Ex. 10.9 of the Text, $G_{mcm} = \frac{\beta_p}{2R_0} \times \frac{\Delta R}{R + r_{e5}}$, where $\beta_p = 50$, $r_{e5} = 25$mV/9.5μA = 2.63kΩ, and $R_0 = R_{09} \| R_{010} = 2.43$MΩ (From Ex. 10.10), and $R_1 = R = 1$kΩ. For CMRR = 80dB, $G_{m1}/G_{mcm} = 10^4$, or $G_{mcm} = 10^{-4} G_{m1}$, where $G_{m1} = (1/5.26)$mA/V, or $G_{mcm} = \frac{10^{-4}}{5.26} = \frac{1}{5.26 \times 10^4}$ mA/V, $\therefore \frac{1}{5.26 \times 10^4} = \frac{50}{2(2.43 \times 10^6)} \times \frac{\Delta R}{(1 + 2.63) \times 10^3}$, $\therefore \Delta R = \frac{3.63 \times 10^3 \times 2(2.43 \times 10^6)}{5.26 \times 10^6 \times 50} = 67.1\Omega$. The corresponding tolerance is $67/1000 \times 100 =$ **6.7%** or **±3.4%**.

10.16 Add resistors of value R_E in series with the emitters of both Q_8 and Q_9. Now, for the p-channel devices, $V_A = 50$V, $\beta = 50$, (rather than 125Vand 200 for Q_{10}). Now, since $I_{C10} = I_{C9} = I_{C8} = 19\mu A$, $r_e = \frac{V_T}{I_E}$

$\approx \dfrac{25mV}{19\mu A\ (51/50)} = 1.29k\Omega$, $r_\pi = (\beta + 1)\, r_e = 51\ (1.29) = 65.8k\Omega$, $g_m = \alpha / r_e = (50/51)/1.29 = 0.760mA/V$, $r_o = V_A/I_C \approx (50/(19) \times 10^{-6}) = 2.63M\Omega$. From Equation 10.7, want $R_O = r_o\ (1 + g_m\ (R_E \| r_\pi))$ to be $31.1M\Omega$. Thus $31.1 = 2.63\ (1 + .760 \times 10^{-3}\ (R_E \| 65.8 \times 10^3))$, $\dfrac{R_E\ (65.8)\ (.760)}{R_E + 65.8 \times 10^3} = 31.1/2.63 - 1 = 10.83$. Thus $50\ R_E = 10.83\ R_E + 712.6$, and $R_E = 712.6/(50 - 10.83) =$ **18.2kΩ**. Thus use resistors $R_E =$ **18.2kΩ**. In this case, $R_{09} = R_{010} = 31.1M\Omega$, and $R_0 = 31.1/2 =$ **15.6MΩ**. Now $G_{mcm} = \dfrac{\beta_p}{2R_0} \dfrac{\Delta R}{R + r_{e5}}$, where $\beta_p = 50$, $\dfrac{\Delta R}{R} = .02$, whence $\Delta R = .02\ (1k\Omega) = 0.02k\Omega$. Now $I_{C5} = 9.5\mu A$, and $r_e \approx 25mV/9.5\mu A = 2.63k\Omega$. Thus $G_{mcm} = \dfrac{50}{2\ (15.6 \times 10^6)} \times \dfrac{.02 \times 10^{-3}}{(1 + 2.63) \times 10^{-3}} =$ **0.0088μA/V**. Now $G_{m1} = 1/5.26mA/V$ as noted below Eq. 10.6 in the Text. Thus CMRR $= \dfrac{G_{m1}}{G_{mcm}} = \dfrac{1/(5.26 \times 10^{-3})}{0.0088 \times 10^{-6}} = 21.6 \times 10^3 \equiv$ **86.7dB**.

SECTION 10.4: Small-Signal Analyses of the 741 Second Stage

10.17 The situation is one in which the base of Q_{25} is joined to the emitter of Q_7, with $1k\Omega$ connecting the emitter of Q_{25} to $-V_{EE}$, and the collector of Q_{25} connected to the emitter of Q_{16} joined to the base of Q_{17}. Here, the collector current of Q_{25} is the same as that in Q_6, namely $9.5\mu A$, for which $r_{o25} = V_A/I_C = 125/(9.5 \times 10^{-6}) = 13.2M\Omega$. Now $I_{C11} = 730\mu A$, $I_{B11} = 730/200 = 3.65\mu A$. Thus the emitter current of $Q_{16} = 9.5 + 3.65 = 13.15\mu A$, for which $r_{e16} = 25mV/13.15\mu A = 1.90k\Omega$. Also $r_{e7} = 25mV/730\mu A = 34.2\Omega$. From Eq. 10.12, $R_{i2} = (\beta + 1)\ [r_{e16} + r_{o25} \| [(\beta + 1)\ (r_{e17} + R_8)]] = 201\ [(1.90 \times 10^3 + (13.2 \times 10^6) \|\ (201\ (34.2 + 100))]$, or $R_{i2} = 201\ (1.9 \times 10^3 + 27.0 \times 10^3) =$ **5.81MΩ**, (rather than the 4.0MΩ found previously).

10.18 For $R_{i2} = (\beta + 1)\ [r_{e16} + r_{o25} \| ((\beta + 1)\ (r_{e17} + R_8))]$, or $R_{i2} = 4 \times 10^6 = 201\ [1.90 \times 10^3 + 13.2 \times 10^6 \| ((201\ (34.2) + R_8))]$, $19.9 \times 10^3 = 1.90 \times 10^3 + 6.87 \times 10^3 + 201\ R_8$, $R_8 = \dfrac{(19.9 - 1.9 - 6.87) \times 10^3}{201} =$ **55.4Ω**.

Now $G_{m2} = \dfrac{i_{C17}}{v_{i2}} = \dfrac{\alpha \dfrac{R_9 \| R_{i17}}{R_9 \| R_{17} + r_{e16}}}{r_{e17} + R_8}$, where $R_{i17} = (\beta_{17} + 1)\ (r_{e17} + R_8)$. Here, $r_{e16} = 1.90k\Omega$ (from the solution to P.10.17 above). Thus, $R_{i17} = 201\ (34.2 + 55.4) = 18.01k\Omega$, and $G_{m2} = \dfrac{\dfrac{200}{201} \times \dfrac{18.01}{18.01 + 1.90}}{34.2 + 55.4} = 0.01A/V =$ **10.0mA/V**. This is to be compared to 6.5mA/V found previously. Now $R_{02} = R_{013B} \| R_{017}$, where $R_{013B} = r_{o13B} = 90.9k\Omega$, and $R_{017} = r_{o17}\ (1 + g_{m17}\ (R_8 \| r_{\pi 17}))$, where $r_{017} = \dfrac{125V}{550 \times 10^{-6}} = 0.227M\Omega$, such that $R_{017} = .227 \times 10^6\ (1 + 200/201 \times (1/34.2)\ (55.4 \| (201\ (34.2))) =$ **00.59MΩ**. Thus $R_{02} = 90.9k\Omega \| 590k\Omega =$ **78.8kΩ**, and the open-circuit voltage gain is $-G_{m2}\ R_{02} = -10.0 \times 10^{-3} \times 78.8 \times 10^3 =$ **−788V/V**, compared to −526.5V/V found previously). Thus the change of bias network produces a gain increase of $(788/526.5 - 1) = .50$, or **50%**!

SECTION 10.5: Analysis of the 741 Output Stage

10.19 For the basic design, $R_{02} = 81k\Omega$. Now, for $R_L = 2k\Omega$, $R_{i3} = \beta_{23}\ (74k\Omega) = 4(81k\Omega)$ for which, $\beta_{23} = (81/74)\ (4) =$ **4.38**. Now the second-stage gain is $A_2 = -G_{m2}\ R_{02} \dfrac{R_{i3}}{R_{i3} + R_{02}}$, where (from the bottom of page 716 of the Text) $G_{m2} = 6.5mA/V$, and $R_{02} = 81k\Omega$. Thus $A_2 = -6.5 \times 10^{-3} \times 81 \times 10^3 \times \dfrac{4\ (81 \times 10^3)}{4\ (81 \times 10^3) + 81 \times 10^3} = -6.5 \times 81 \times 4/5 =$ **−421V/V** (as contrasted with −515V/V available with high β_{23}). Now $R_O = \dfrac{\dfrac{R_{02}}{\beta_{23} + 1} + r_{e23}}{\beta_{20} + 1} + r_{e20} + R_7$, where (using the data at the bottom of page 720 and top of page 721), R_{02} = 81kΩ, β_{23} = 4.38, r_{e23} = 139Ω, β_{20} = 50, r_{e20} = 5Ω, R_7 = 27Ω. ∴ R_0 =

$$\frac{\frac{81\times10^3}{4.38+1}+139}{51}+5+27=330\Omega.$$

10.20 Assume the base current of Q_4 to be i. Thus the collector current in Q_{15} is $180 - i$ and the base current of Q_{15} is $(180 - i)/\beta$. Thus the load current is $(\beta + 1)\,i + 180 - i = \beta i + 180$, while the current in R_6 is $(\beta+1)\,i - (\frac{180-i}{\beta}) \approx (\beta+1)\,i$, for which $V_{BE\,15} = R_6\,(\beta+1)\,i$, with $I_{C15} = 180 - i$. Thus $27\,(\beta+1)\,i \times 10^{-6} = 25\times10^{-3}\ln\left[\frac{(180-i)(10^{-6})}{10^{-14}}\right]$, with i in μA.

For $\beta = 400$, i in μA: $10^{-6}\times 27\,(401)\,i = 25\times10^{-3}\,(\ln(180-i)+\ln 10^8)$, or $10.83\times10^{-3}\,i = 25\times10^{-3}\,(\ln(180-i)+18.42)$, $i = 2.31\ln(180-i)+42.6$.

Try $i = 100$, $i = 2.31\ln(100)+42.6 = 52.2\mu A$;

Try $i = 52$, $i = 2.31\ln(127.8)+42.6 = 53.8\mu A$;

Try $i = 53.8$, $i = 2.31\ln(126.2)+42.6 = 53.8\mu A$, for which $I_L = 400\,(53.8)+180 =$ **21.7mA**.

Note that for low β, below about 100, the output will be β-limited:

For $\beta = 200$: $10^{-6}\times27\,(201)\,i = 25\times10^{-3}\,(\ln(180-i)+\ln10^8)$, $i = 4.60\ln(180-i)+87.74$.

Try $i = 50$, $i = 4.6\ln(130)+87.7 = 110$;

Try $i = 110$, $i = 4.6\ln(70)+87.7 = 107.2$;

Try $i = 107.2$, $i = 4.6\ln(72.8)+87.7 = 107.4$, for which $I_L = 200\,(107.4)+180 =$ **21.6mA**.

For $\beta = 100$: $10^{-6}\,(27)\,(101)\,i = 25\times10^{-3}\,(\ln(180-i)+\ln10^8)$, $i = 9.17\ln(180-i)+168.9$

Try $i = 150$, $i = 9.17\ln(30)+168.9 = 200$;

Try $i = 175$, $i = 9.17\ln(5)+168.9 = 183.6$;

Try $i = 179$, $i = 9.17\ln(1)+168.9 = 168.9$;

Try $i = 178$, $i = 9.17\ln(2)+168.9 = 175.2$;

Try $i = 177$, $i = 9.17\ln(3)+168.9 = 178.9$;

Try $i = 177.5$, $i = 9.17\ln(2.5)+168.9 = 177.3$, for which $I_L = 100\,(177.3)+180 =$ **17.9mA**.

SECTION 10.6: Gain and Frequency Response of the 741

10.21

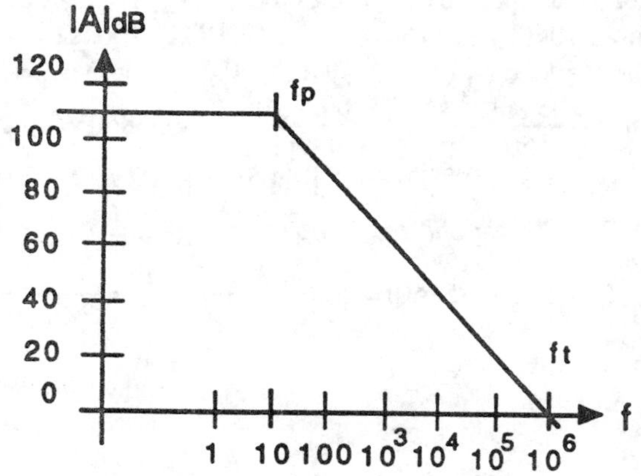

Using the results of the solution of P10.18 above, the overall gain is $A_v = -G_{m1}\,(R_{01}\,||R_{i2}) \times (-G_{m2}\,R_{02})\,\mu\,\frac{R_L}{R_L+R_0}$, assuming that R_{i3} is very large, where G_{m1} = 1/5.26mA/V, R_{01} = 6.7MΩ, R_{i2} = 4.0MΩ, G_{m2} = 10.0mA/V, R_{02} = 78.8kΩ, μ = 1.0, R_L = 2kΩ, R_0 = 66Ω (from page 721 of the Text). Thus $A_v = +\frac{1}{5.26}\times10^{-3}\,(6.7\times10^6\,||4\times10^6)\,(10.0\times10^{-3}\times78.8\times10^3)\times1\times\frac{2\times10^3}{2\times10^3\times66} = 362.6\times10^3 \equiv$ **111.2dB**. For the pole associated with C_C, at the base of Q_{16}, $C_i = C_C\,(1 - \text{gain}) = 30 \times 10^{-12}\,(1 - -G_{m2}\,R_{02}) = 30 \times 10^{-12}\,(1 + 10 \times 10^{-3} \times 78.8 \times 10^3) = 23 \times 10^{-9}$F, $R_i = R_{01}\,||R_{02} = 6.7\text{M}\Omega\,||\,4.0\text{M}\Omega = 2.5 \times 10^6\Omega$. Thus $f_p = \dfrac{1}{2\pi \times 23 \times 10^{-9} \times 2.5 \times 10^6}$ = **2.77Hz**, for which $f_t = 363 \times 10^3 \times 2.77$ = **1.00MHz**. Note that f_t is the same, since the increased gain is in the Miller compensation stage!

10.22 Using the results on page 723 of the Text, see that for customized compensation with capacitor C_C, $C_i = C_C\,(1 - -\text{gain}) = C_C\,(1 + 515) = 516C_C$, where $f_p = \dfrac{1}{2\pi\,C_i\,R_i}$, or $516\,C_C = \dfrac{1}{2\pi\,f_p\,R_i} = \dfrac{1}{2\pi\,f_p\,(2.5 \times 10^6)}$, whence $C_C = 1.23 \times 10^{-10}/f_p$.

(a) 45° *phase margin* occurs when the first pole contributes 90° and the second 45°, where $f = f_2 \tan 45° = f_2$, at the frequency of the second pole, *say at* $f_t = 1MHz$. *Now for* $A_f = 10^3$, for which $\beta \simeq 10^{-3}$, the first pole should be at $f_p = \dfrac{10^6}{10^6/10^3}$ = 1kHz, for which $C_C = 1.23 \times 10^{-10}/10^3$ = **0.123pF**. The corresponding 3dB frequency is at $10^3\,(1 + A\beta) = 10^3\,(1 + 243 \times 10^3 \times 10^{-3})$ = **244kHz**. *Now for* $A_f = 10^4$, for which $\beta \simeq 10^{-4}$, the first pole should be at $f_p = \dfrac{10^6}{10^6/10^4} = 10^4$Hz, for which C_C = **0.0123pF**. The corresponding 3dB frequency is at $10^4\,(1 + 243 \times 10^3 \times 10^{-4})$ = **253kHz**.

(b) 60° *phase margin* occurs when the first pole contributes 90° and the second 30°, where $f = f_2 \tan 30° = 0.58\,f_2$, or $f = 0.58\,f_t$ = 580kHz. *Now for* $A_f = 10^3$, for which $\beta \simeq 10^{-3}$, the first pole should be at $f_p = \dfrac{.58 \times 10^6}{10^6/10^3}$ = 0.58kHz, for which $C_C = 1.23 \times 10^{-10}/.58 \times 10^3$ = **0.212pF**, and $f_{3dB} = 0.58 \times 10^3\,(244)$ = **142kHz**. *Now for* $A_f = 10^4$, for which $\beta \simeq 10^{-4}$, the first pole should be at $f_p = \dfrac{.58 \times 10^6}{10^6/10^4}$ = 5.8kHz, for which C_C = **0.0212pF**, and f_{3dB} = 5.8kHz (25.3) = **146kHz**.

10.23 $SR = 2I/C_C$, where $I = 9.5\mu$A or C_C = 0.123pF and .0123pF, for which $SR = \dfrac{2\,(9.5) \times 10^{-6}}{.123 \times 10^{-12}}$ = **154V/μsec.** and **1540V/μsec.** respectively. Now, from Eq. 2.34 on page 89 of the Text, the full-power bandwidth is $f_M = \dfrac{SR}{2\pi\,V_{O\,max}} = \dfrac{154 \times 10^6}{2\pi \times 10}$ = **2.45MHz**, and **24.5MHz** respectively.

10.24 The output stage is a **Class AB** type. Its standing current is defined by the current in Q_5, being I_{C5} = 500μA, and the fact that Q_6 is four times larger than Q_5, being $I_{C6} = 5 \times 500\mu\text{A}$ = **2.5mA**.

For the gain: $I_1 = I_2 = I_3 = I_4 = 50\mu\text{A}/2 = 25\mu\text{A}$, $r_{e1} = r_{e2} = 25\text{mV}/25\mu\text{A} = 1\text{k}\Omega$, $r_{02} = r_{04} = 200\text{V}/25 \times 10^{-6} = 8\text{M}\Omega$. $I_5 = 500\mu\text{A}$, $r_{e5} = 25 \times 10^{-3}/500 \times 10^{-6} = 50\Omega$, $r_{\pi 5} = 121\,(50) = 6.05\text{k}\Omega$, $r_{05} = 200\text{V}/500\mu\text{A} = 400\text{k}\Omega$. $I_6 = 1.0\text{mA}$, $r_{e6} = 25\text{mV}/1.0\text{mA} = 25\Omega$, $r_{\pi 6} = 121\,(25) = 3.025\text{k}\Omega$, $r_{06} = 200\text{V}/1.0\text{mA} = 200\text{k}\Omega$. For output-stage operation, assume the gain to be controlled primarily through the Q_5, Q_7 connection. Now with $I_7 = 1.0\text{mA}$, $r_{e7} = 25\Omega$, $r_{06} = 200\text{k}\Omega$, and with a 10kΩ load, $R_L{}' = 10\text{k}\Omega\,||\,200k\Omega\,||\,200k\Omega = 9.90\text{k}\Omega$. At the base of Q_7, $R_{in} = 121\,(9.09\text{k}\Omega) = 1.1\text{M}\Omega$, and the load on $Q_5 = r_{05}\,||R_{in} = 400\text{k}\Omega\,||\,1.1\text{M}\Omega = 283\text{k}\Omega$. Thus, the gain from the base of $Q_5\,Q_6$ to the output $\simeq \dfrac{293 \times 10^3}{50} \times \dfrac{9.09 \times 10^3}{10 + 9.09 \times 10^3} = -5860$V/V. Now at the base of Q_5, the resistance is $R_T = r_{04}\,||\,r_{02}\,||\,r_{\pi 5}\,||\,r_{\pi 6} = 8 \times 10^6\,||\,8 \times 10^6\,||\,6.05\text{k}\Omega\,||\,3.025\text{k}\Omega = 2.017\text{k}\Omega$. Thus the gain of stage 1 = $\dfrac{2(2.017 \times 10^3)}{1k\Omega + 1k\Omega}$ = 2.02V/V. Correspondingly, the overall gain = 2.02 × 5860 = **11.8 × 10³V/V**. At the base of Q_5, $R_T = 2.017\text{k}\Omega$, $C_T = C(1 - -5860) = 5861C$. Since $f_p = \dfrac{1}{2\pi\,R_T\,C_T}$, $C_T = 5861C = \dfrac{1}{2\pi \times 2.017 \times 10^3 \times 1 \times 10^3}$, whence C = **13.5pF**.

SECTION 10.7: CMOS Op Amps

10.25 For Q_8, Q_5, Q_7, I_D = **25μA**, K = 1/2 (10 × 10^{-6}) (200/10), = 100μA/V^2, $r_0 = V_A/I_D$ = 25V/25μA = **1MΩ**. 25 = 100 (υ_{GS} – 1)^2, or υ_{GS} = (25/100)^{½} + 1 = **1.504V**, g_m = 2K ($\upsilon_{GS} - V_t$) = 2 (100) (0.50) = **100μA/V**.

For Q_1, Q_2, I_D = 25/2μA = **12.5μA**. K = 100μA/V^2, r_0 = 25/12.5μA = **2MΩ**. 12.5 = 100 (υ_{GS} – 1)^2, or υ_{GS} = (12.5/100)^{½} + 1 = **1.354V/V**, g_m = 2 (100) (.354) = **70.8μA/V**.

For Q_3, Q_4, I_D = **12.5μA**, K = 1/2 (20 × 100/10) = 100μA/V^2, r_0 = 25/12.5μA = **2MΩ**, 12.5 = 100 (υ_{GS} – 1)^2, or υ_{GS} = **1.354V**, and g_m = **70.8μA/V**.

For Q_6, I_D = **25μA**, K = 1/2 (20 × 10^{-6}) × 200/10 = **200mA/V^2**, r_0 = 25V/25μA = **1MΩ**, 25 = 200 (υ_{GS} – 1)^2, or υ_{GS} = (25/200)^{½} + 1 = **1.354V**, g_m = 2 (200) (.354) = **141.6μA/V**.

	Q_1	Q_2	Q_3	Q_4	Q_5	Q_6	Q_7	Q_8
I_D (μA)	12.5	12.5	12.5	12.5	25	25	25	25
$\lvert V_{GS}\rvert$(V)	1.35	1.35	1.35	1.35	1.50	1.35	1.50	1.50
g_m (μA/V)	70.8	70.8	70.8	70.8	100	141.6	100	100
r_o (MΩ)	2	2	2	2	1	1	1	1

For Gains: $A_1 = \dfrac{r_{04} \| r_{02}}{1/g_{m1} + 1/g_{m2}} = \dfrac{2 \times 10^6/2}{2\,(1/70.8 \times 10^{-6})} = \dfrac{10^6\,(10^{-6})\,70.8}{2}$ = **35.4V/V**. *For no load*, $A_2 = g_{m6}$ ($r_{06} \| r_{07}$) = 141.6 × 10^{-6} × 1 × 10^6/2 = **70.8V/V**. Overall, the open-loop gain = $A_1 A_2$ = 35.4 (70.8) = **2506V/V**.

For the Input Common-Mode Range: Input High: V_{G5} = 5 – 1.5 + 1 = 4.5V, V_{GS1} = 1.35. Thus, V_I ≤ 4.5 – 1.35 = **+3.15V**. Input Low: V_{D3} = –5 + 1.35 = –3.65V, V_{GD1} = –1V, V_I ≥ –3.65 – 1 = **–4.65V**

For the Output Common-Mode Range: For triode operation of the output devices, the output range is ±5V. For saturated-mode operation, V_o ≤ 5 –1.5 + 1 = **4.5V**, V_o ≥ –5 + 1.35 –1 = **–4.65V**.

10.26 For I_{REF} = 12μA: Reduce all currents in the Table on page 729 of the Text, by the factor 12/25 = 0.48.

	Q_1	Q_2	Q_3	Q_4	Q_5	Q_6	Q_7	Q_8
I_D (μA)	6	6	6	6	12	12	12	12
$\lvert V_{GS}\rvert$(V)	1.28	1.28	1.35	1.35	1.4	1.35	1.4	1.4
g_m (μA/V)	42.5	42.5	34.6	34.6	60	69.2	60	60
r_o (MΩ)	4.2	4.2	4.2	4.2	2.1	2.1	2.1	2.1

For Q_8, Q_5, Q_7, K = 1/2 (10 × 10^{-6}) (150/10) = 75μA/V^2, I_D = 12μA = 75μA/V^2 (υ_{GS} – 1)^2, υ_{GS} = (12/75)^{½} + 1 = **1.4V**, g_m = 2 (75) (1.4 – 1) = **60μA/V**, r_o = 25/12 = **2.08MΩ**.

For Q_1, Q_2, $K = 1/2\ (10 \times 10^{-6})\ (120/8) = 75\mu A/V^2$, $I_D = 12/2 = 6\mu A = 75\ (\upsilon_{GS} - 1)^2$, $\upsilon_{GS} = (6/75)^{1/2} + 1 =$ **1.283V**, $g_m = 2\ (75)\ (.283) =$ **42.5μA/V**, $r_o = 25/6 =$ **4.17MΩ**.

For Q_3, Q_4, $K = 1/2\ (20 \times 10^{-6})\ (50/10) = 50\mu A/V^2$, $I_D = 6\mu A = 50\ (\upsilon_{GS} - 1)^2$, $\upsilon_{GS} = (6/50)^{1/2} + 1 =$ **1.346V**, $g_m = 2\ (50)\ (\upsilon_{GS} - 1) = 2\ (50)\ (.346) =$ **34.6μA/V**. $r_o = 25/6 =$ **4.17MΩ**.

For Q_6, $K = 1/2\ (20 \times 10^{-6})\ (100/10) = 100\mu A/V^2$, $I_D = 12\mu A = 100\ (\upsilon_{GS} - 1)^2$, $\upsilon_{GS} = (12/100)^{1/2} + 1 =$ **1.346V**, $g_m = 2\ (100)\ (.346) =$ **69.2μA/V**, $r_o = 25/12 =$ **2.08MΩ**.

Now $A_1 = -g_{m1}\ (r_{02} \| r_{04}) = -42.5 \times 10^{-6}\ (4.2 \| 4.2) \times 10^6 =$ **−89.25V/V**, and $A_2 = -g_{m6}\ (r_{06} \| r_{07} = -69.2 \times 10^{-6}\ (2.1/2) \times 10^6 =$ **−72.7V/V**. Thus $A_0 = A_1 A_2 = (89.25)\ (72.7) =$ **6488V/V**. Also $\upsilon_{I\,CM\,max} = V_{DD} - |V_{GS5}| + |V_t| - |V_{GS1}| = 5 - 1.4 + 1 - 1.28 =$ **3.32V**, $\upsilon_{I\,CM\,min} = -V_{SS} + |V_{GS3}| - |V_t| = -5 + 1.35 - 1 =$ **−4.65V**, $\upsilon_{O\,max} = V_{DD} - |\upsilon_{GS7}| + |V_t| = 5 - 1.4 + 1 =$ **4.60V**, $\upsilon_{O\,min} = -V_{SS} + |\upsilon_{GS6}| - |V_t| = -5 + 1.35 - 1 =$ **−4.65V**.

10.27 For Q_6, $(W/L)_6 = 50/10$, $K = 1/2\ (20)\ 50/10 = 50\mu A/V^2$, $I_D = 25 = 50\ (\upsilon_{GS} - 1)^2$, $\upsilon_{GS} = (1/2)^2 + 1 =$ **1.707V**, $g_m = 2\ (50)\ (.707) = 70.7\mu A/V$, $r_o = 25/25 = 1M\Omega$. $A_2 = -70.7\ (1/2) =$ **−35.4V/V**, $A_1 = -62.5V/V$ (from Example 10.2), $A_0 = A_1 A_2 = -62.5\ (-35.4) =$ **2212V/V**. Now for $I_6 = I_7 = 25\mu A$, $\upsilon_{GS6} = 1.707V$. But $\upsilon_{GS4} = 1.50V$. Thus, the input offset $= \dfrac{\upsilon_{GS6} - \upsilon_{GS4}}{A_1} = \dfrac{1.707 - 1.5}{62.5} =$ **3.3mV**.

10.28 From the solution of P10.25 above, and the development following Eq. 10.47 of the Text, $R_1 = r_{02} \| r_{04} = 4.2/2 = 2.1M\Omega$, $R_2 = r_{07} \| r_{06} = 2.1/2 = 1.05M\Omega$, $G_{m1} = g_{m1} = 42.5\mu A/V$, $G_{m2} = g_{m2} = 69.2\mu A/V$, $f_t = \dfrac{1}{2\pi} \times \dfrac{G_{m1}}{C_C}$, $C_C = \dfrac{42.5 \times 10^{-6}}{2\pi \times 10^6} =$ **6.76pF**. For a zero at ∞, $R = 1/G_{m2} = 1/g_{m2} = 1/69.2 \times 10^{-6} =$ **14.5kΩ**. For C_2, the 10pF output capacitance, and $C_2 >> C_1$, the second pole is at $f_2 \approx \dfrac{G_{m2}}{2\pi\ C_2} = \dfrac{69.2 \times 10^{-6}}{2\pi \times 10 \times 10^{-12}} =$ **1.10MHz**. Excess phase at 1MHz is $\tan^{-1}$ (1/1.1) = **42.3°**. For 6° excess phase at f_t', $\tan^{-1} \dfrac{f_t'}{1.1} = 6°$, or $f_t' = 1.1 \times .1051 =$ **0.116MHz**, for which $C_C = \dfrac{G_{m1}}{2\pi\ f_t'} = \dfrac{42.5 \times 10^{-6}}{2\pi \times .116 \times 10^6} =$ **58.3pF** is required. *Slew rate, SR* $= 2I/C_C$ where $2I = 12\mu A$. For the case, of 42.3° excess phase, $SR = \dfrac{12 \times 10^{-6}}{6.76 \times 10^{-12}} =$ **1.78V/μs**. For 6° excess phase, $SR = \dfrac{12 \times 10^{-6}}{58.3 \times 10^{-12}} =$ **0.206V/μs**.

SECTION 10.8: Alternative Configurations for CMOS and BiCMOS Op Amps

10.29 For the Wilson mirror (Fig. 10.25), $V_{BIAS2} - |V_t|_{1C} + V_{GS\,3C} + V_{GS4} = -V_{SS}$, or $V_{BIAS2} = -V_{SS} + V_t - 2\ V_{GS}$. Thus the minimum voltage between V_{BIAS2} and V_{SS} is $\mathbf{2V_{GS} - V_t}$. Now from Eq. 10.57, $R_0 = g_{m4C}\ r_{04C}\ r_{03} = \mathbf{g_m\ r_0^2} = 2K\ (V_{GS} - V_t)\ r_0^2$. Now for the cascode mirror (Fig. 6.32b)), the bias situation is essentially the same, with $-V_t + V_{GS\,3C} + V_{GS\,3}$ between V_{BIAS2} and $-V_{SS}$, ie $\mathbf{2V_{GS} - V_t}$. As well, R_0 is the same: $R_0 = g_{m4\,C}\ r_{04\,C}\ r_{04} = \mathbf{g_m\ r_0^2}$, as before. Thus as measured by the output resistance and output-voltage overhead, the cascode and (modified) Wilson are the same. However, if Q_{3C} were eliminated, V_{BIAS2} could be reduced to $2V_{GS} - 2V_t$, but the current transfer ratio would be reduced.

10.30 For $2I = 10\mu A$, $I_D = 5\mu A$, $K_n = 1/2 \times 20 \times 60/8 = 75 = K$, $K_p = 1/2 \times 10 \times 120/8 = 75 = K$, $r_0 = 25/5\mu A = 5.0M\Omega$, $5 \times 10^{-6} = 75 \times 10^{-6}\ (\upsilon_{GS} - 1)^2$, $\upsilon_{GS} = 1 + (5/75)^{1/2} = 1.258V$, $g_m = 2K\ (\upsilon_{GS} - V_t) = 2\ (75)\ (.258) = 38.7\mu A/V$. Now $R_0 = 1/2\ (g_m\ r_0^2) = 1/2\ (38.7 \times 10^{-6} \times 5 \times 10^6 \times 5 \times 10^6) =$ **484MΩ**, and $A_1 = g_m R_0 = 38.7 \times 10^{-6} \times 484 \times 10^6 =$ **18.7 × 10³V/V**.

10.31 For the double cascode, $R_{04\,CC} \approx (g_{m4\,CC}\ r_{04\,CC})\ R_{04C} \approx \mathbf{g_{m4\,CC}\ r_{04\,CC}\ g_{m4C}\ r_{04C}\ r_{03}}$, $R_{02C} \approx \mathbf{(g_{m2C}\ r_{02C})\ r_{02}}$. Now $R_0 = R_{02C} \| R_{04\,CC}$. For conditions as in Ex. 10.28, $2I = 25\mu A$, $I = 12.5\mu A$, $K_n = K_p = 1/2\ (20)\ (60/8) = 75\mu A/V^2$, $12.5 = 75\ (\upsilon_{GS} - 1)^2$, $\upsilon_{GS} = (12.5/75)^{1/2} + 1 = 1.408V$, $g_m = 2\ (75 \times 10^{-6})\ .408 = 61.2\mu A/V$, $r_0 =$

$25/12.5 \times 10^{-6} = 2M\Omega$, $R_{04\,CC} = g_m^2\, r_0^3 = (61.2 \times 10^{-6})^2 \times (2 \times 10^6)^3 = 29.96G\Omega$, $R_{02C} = g_m\, r_0^2 = 61.2 \times 10^6 \times (2 \times 10^6)^2 =$ **245MΩ**, $R_0 = (.245 \,||\, 30)G\Omega =$ **243MΩ**, $G_{m1} = g_m = 61.2 \times 10^{-6}$A/V. Thus $A_1 = 61.2 \times 10^{-6} \times 243 \times 10^6 =$ **14.9 × 10³V/V**. Total voltage from input to supply is $-1 + 1.41 + 1.41 + 1.41 =$ **3.23V**.

10.32 Here, $2I = I_B = 10\mu A$, $|V_t| = 1V$.

For Q_6, Q_7: $I_6 = 10\mu A$, $K_6 = 1/2 \times 20 \times 8/8 = 10\mu A/V^2$, $10 = 10\,(\upsilon_{GS} - 1)^2$, or $\upsilon_{GS} = 2V$. Thus $V_{BIAS3} = 2 + -5 =$ **−3V**.

For Q_{1C}, Q_{2C}: $I_{1C} = 10 - 10/2 = 5\mu A$, $K_{1C} = 1/2 \times 20 \times 60/8 = 75\mu A/V$, $5 = 75\,(\upsilon_{GS} - 1)^2$, $\upsilon_{GS} = (5/75)^{1/2} + 1 = 1.258V$, also $g_m = 2\,(75)\,(.258) =$ **38.7μA/V**. Thus, $V_{BIAS2} = -3 - 1 + 1.258V =$ **−2.74V**. Use **−2.75V**.

For Q_5: $I_5 = 10\mu A$, $K_5 = 1/2 \times 20 \times 150/10 = 150\mu A/V^2$, $10 = 150\,(\upsilon_{GS} - 1)^2$, $\upsilon_{GS} = (1/15)^{1/2} + 1 = 1.258V$, $V_{BIAS1} = +5 - 1.26 =$ **3.74V**. Use **3.75V**.

For Q_1, Q_2: $I_1 = 5\mu A$, $K_1 = 1/2 \times 10 \times 120/8 = 75\mu A/V^2$, $5 = 75\,(\upsilon_{GS} - 1)^2$, $\upsilon_{GS} = (5/75)^{1/2} + 1 = 1.258V$, $g_{m1} = 2\,(75)\,(\upsilon_{GS} - 1) = 2\,(75)\,(.258) =$ **38.7μA/V**.

For Q_{4C}: $I_{4C} = 5\mu A$, $K_{4C} = 1/2 \times 10 \times 120/8 = 75\mu A/V^2$, $5 = 75\,(\upsilon_{GS} - 1)^2$, $\upsilon_{GS} = 1.258$, $g_m = 2\,(75)\,.258 =$ **38.7μA/V**.

Output resistance: $r_{01} = r_{02} = 25/5\mu A = 5M\Omega$, $r_{06} = r_{07} = 25/10\mu A = 2.5M\Omega$, $r_{01C} = r_{02C} = r_{04C} = r_{03} = 25/5 = 5M\Omega$. Thus, $R_{04C} = g_{m4C}\, r_{04C}\, r_{03} = 38.7 \times 10^{-6} \times 5 \times 10^6 \times 5 \times 10^6 =$ **967MΩ**, and $R_{02C} = g_{m2C}\, r_{02C}\, r_{07} \,||\, r_{02} = 38.7 \times 10^{-6} \times 5 \times 10^6 \times (2.5 \times 10^6 \,||\, 5 \times 10^6) =$ **322MΩ**. Correspondingly, $R_0 = 967 \,||\, 322 =$ **242MΩ**. Gain $A_0 = g_{m1}\, R_0 = 38.7 \times 10^{-6} \times 242 \times 10^6 =$ **9.36 × 10³V/V**.

10.33 For all ($Q_1, Q_2, Q_{1C}, Q_{2C}, Q_{3C}, Q_{4C}$), I_D, K, V_A, and $r_0 = 25/10 = 2.5M\Omega$ are the same. From Eq. 10.64, $f_t = \dfrac{g_{m1}}{2\pi\, C_L}$, or $g_{m1} = 2\pi\, C_L\, f_t = 2\pi \times 10 \times 10^{-12} \times 10^6 =$ **62.8μA/V**. Now $I_{D7} = 10 + 10 = 20\mu A$, and $r_{07} = 25/20 = 1.25M\Omega$. Thus $R_{04C} = g_m\, r_0^2 = 62.8 \times 10^{-6} \times (2.5 \times 10^6)^2 = 393M\Omega$, $R_{03C} = 62.8 \times 10^{-6} \times 2.5 \times 10^6 \times (1.25 \times 10^6 \,||\, 2.5 \times 10^6) = 131M\Omega$. Correspondingly, $R_0 = 131M\Omega \,||\, 393M\Omega = 393/4 =$ **98.25MΩ**. Also, $A_0 = g_{m1}\, R_0 = 62.8 \times 10^{-6} \times 98.25 \times 10^6 =$ **6170V/V**. The dominant-pole frequency, $f_D = \dfrac{1}{2\pi\, C_L\, R_0} = \dfrac{1}{2\pi \times 10 \times 10^{-12} \times 98.25 \times 10^6} =$ **162Hz** Now $SR = \dfrac{2I}{C_L} = \dfrac{2 \times 10 \times 10^{-6}}{10 \times 10^{-12}} =$ **2V/μs**.

10.34 Here, $I_B = 2I = 800\mu A$, $I_{D1} = I_{D2} = 400\mu A$. Assume $\mu_n\, C_{ox} = 2\mu_p\, C_{ox} = 20\mu A/V^2$, $|V_t| = 1V$, $V_A = 25V$. Thus $K_1 = K_2 = 1/2 \times 10 \times 600/10 = 300\mu A/V^2$, $400 = 300\,(\upsilon_{GS} - 1)^2$, $\upsilon_{GS} = (4/3)^{1/2} + 1 = 2.155V$, $g_{m1} = 2\,(300)\,(2.155 - 1) = 693\mu A/V$, and the output pole is at $f_t = \dfrac{g_{m1}}{2\pi\, C_L} = \dfrac{693 \times 10^{-6}}{2\pi \times 2 \times 10^{-12}} =$ **55.15MHz**. For the parasitic pole located at the folding node, $I_{E1C} \approx 400\mu A$ and $g_{m1C} = \dfrac{400 \times 10^{-6}}{25 \times 10^{-3}} = 16mA/V$. Now at the emitter of Q_{1C}, the total capacitance is $C_{\pi 1C} + C_{\mu 6} = C_\pi + C_\mu$, with the corresponding pole at $f_{t1C} = \dfrac{g_{m1C}}{2\pi\,(C_\pi + C_\mu)}$, and $C_\pi + C_\mu = \dfrac{16 \times 10^3}{2\pi\, f_{t1C}}$. For this parasitic pole to be $10 \times$ higher than the output pole $f_{t1C} = 10 \times 55.2 =$ **550MHz**, a BJT unity-gain frequency which is relatively easy to achieve.

SECTION 10.9: Data Converters - An Introduction

10.35 For a 100kHz sampling frequency, the highest frequency signal component that can be sampled "adequately", as noted by Shannon, is at $f = 100kHz/2 =$ **50kHz**. This means that for a square wave at 50kHz, the fundamental would be adequately represented, but the waveshape-specific harmonics would not. Note that for sampling at frequency $f_s = f$, that for an input signal at f, output is at f, for input at $0.5f$, output is at $f/2$, at $2f/2$, output is dc, and at $1.1f/2$, output is at $1.1f/2$ for 9 or so cycles with a break and corresponding phase reversal occurring at a rate of $(1.1 - 1.0)/2 = 0.1f/2$. The figure illustrates input and output waves at various frequencies f_i with sampling at f_s = 100kHz. For sampling in a 10ns interval, with source resistance R and capacitor C, $1 - e^{-t/RC} = 0.99$, or $e^{-t/RC} = 0.01$, $-t/RC = -4.6$, $RC = \frac{10 \times 10^{-9}}{4.6}$, $R = \frac{10 \times 10^{-9}}{4.6 \times 100 \times 10^{-12}}$ = **21.7Ω**. Thus the switch resistance should not exceed 21.7Ω.

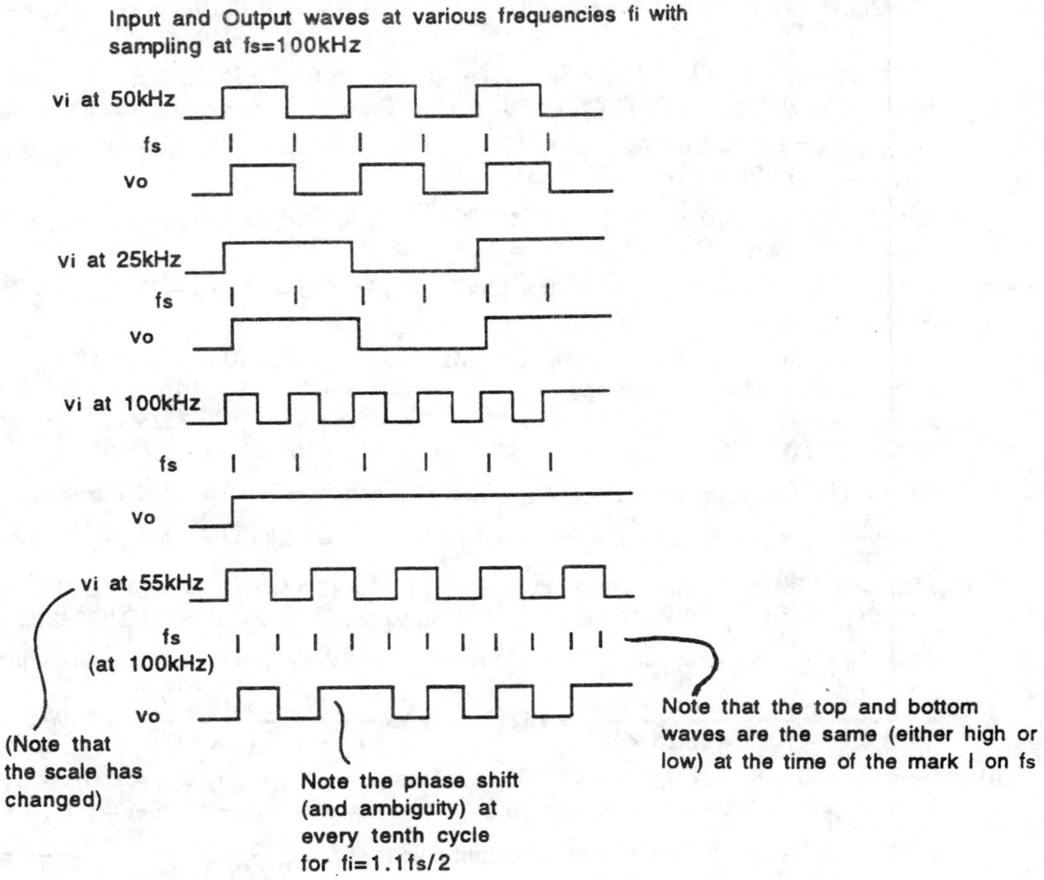
Input and Output waves at various frequencies fi with sampling at fs=100kHz
vi at 50kHz
fs
Vo
vi at 25kHz
fs
Vo
vi at 100kHz
fs
Vo
vi at 55kHz
fs
(at 100kHz)
Vo
(Note that the scale has changed)
Note the phase shift (and ambiguity) at every tenth cycle for fi=1.1fs/2
Note that the top and bottom waves are the same (either high or low) at the time of the mark I on fs

10.36 See that the required resolution is 0.1V in 2(5) = 10V or 1 in 100. Thus $2^n > 100$. Now, $2^7 = 128 > 100$. Thus need **7 bits**. For a 10-bit converter, $2^{10} = 1024$, and the resolution would be 10V/1024 = **9.77mV**.

SECTION 10.10: D/A Converter Circuits

10.37 For R = 1kΩ and n = 8, $2^{n-1}R = 2^{8-1}R = 2^7 R$ = **128kΩ**. *But* for $R/2$ = 1kΩ, the largest resistor required is 2 (128) = **256kΩ**. The LSB current is $V/(2^7 R)$ = 7.81VμA. Correspondingly, the allowed error is 1/2 (7.81V) = 3.91V μA (or 1.95V μA for the second view of R). Here, the MSB current is V/R nominally, or $\frac{V}{R + \Delta R}$, for a switch resistance ΔR. Now $\frac{V}{R} - \frac{V}{R + \Delta R} < \frac{V}{2\,(2^7)\,R}$, or $\frac{1}{R} - \frac{1}{R + \Delta R} \le \frac{1}{2^8\,R}$, $1 - \frac{R}{R + \Delta R} = \frac{1}{2^8}$, $\frac{R}{R + \Delta R} = 1 - 2^{-8}$, $R = R + \Delta R - 2^{-8} R - 2^{-8} \Delta R\, \Delta R = 2^{-8} R + 2^{-8} \Delta R$. Thus, $\frac{\Delta R}{R} = \frac{2^{-8}}{1 - 2^{-8}} \approx 2^{-8}$.
Now, for an MSB resistor of 1kΩ, $\Delta R < 2^{-8}\, 10^3$ = **3.91Ω**, or for an MSB resistor of 2kΩ, ΔR < **7.81Ω**. Now for both resistor error and switch resistance each contributing a half, switch resistances less than 3.91/2

= **1.95** Ω, (or 7.81/2 = **3.91** Ω) are acceptable. Also, for a perfect R_f (say it is trimmed to the correct value), resistor tolerance allowed is $(\frac{100}{2^8})/2 = 0.39\%$. For R_f also variable, allowed resistor-tolerance = 0.39/2 ≃ **0.2%**.

10.38 The resistance of an $R - 2R$ ladder as seen from the supply is $2R \| 2R = R$. For a 10V reference, and 1mA, R = 10V/1mA = **10kΩ**. See that the current in the LSB switch is $\frac{1}{2^{n-1}} = \frac{1}{128}$ that in the MSB switch. Thus $\frac{1}{2R} - \frac{1}{2R + \Delta R} = \frac{1}{2R}(\frac{1}{128})/2$, $1 - \frac{1}{1 + \Delta R/2R} = \frac{1}{256}$, $\frac{1}{1 + \Delta R/2R} = 1 - \frac{1}{256} = \frac{255}{256}$, or $1 + \Delta R/2R = \frac{256}{255} = 1 + \frac{1}{255}$, that is $\Delta R/2R = \frac{1}{255}$, $\Delta R = \frac{2R}{255} = \frac{2\,(10k\Omega)}{255}$ = **78.4Ω**. Now if $2R$ is reduced by 78.4Ω to compensate, doubling of the nominal switch resistance of **78Ω** to 156Ω would again produce an 1/2 LSB error.

10.39 For device junction area 1% in error, in both Q_{ref} and Q_1, the output current error may be as much as 2% I_{ref}. For n bits, the LSB current = $\frac{I_{ref}}{2^{n-1}}$. Now, $\frac{2}{100} = \frac{1}{2^{n-1}} \times \frac{1}{2}$, when $2^n = \frac{100}{2} = 50$, for which **n < 5 bits**. If the absolute value of the output current is not critical, 6 bits is available.

SECTION 10.11: A/D Converter Circuits

10.40 The requirement is for ±1V signals ≡ 2V range, with 2 bits. Now, the range is divided into 2^2 = 4 parts, each of 1/2 volt. Thus, use **3 comparators**, with references at **–1/2V, 0V, +1/2V**, as shown, first in a parallel connection, or alternatively, as 2 in cascade. Output codes for the two circuits are shown in the Table.

V_I	B_1	B_0	H_2	H_1	H_0
+0.75	1	1	1	1	1
+0.25	1	0	0	1	1
–0.25	0	1	0	0	1
–0.75	0	0	0	0	0

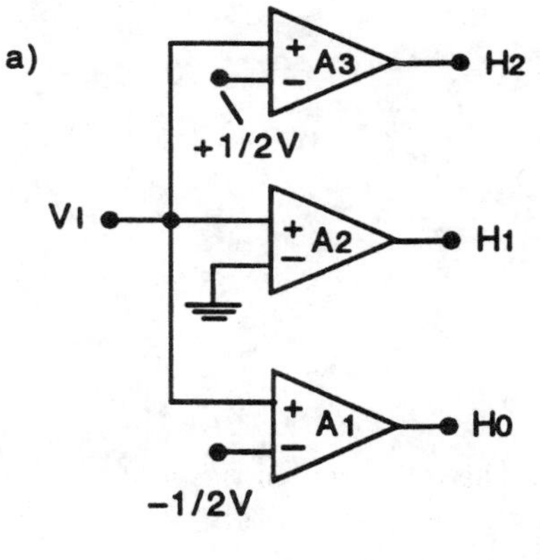

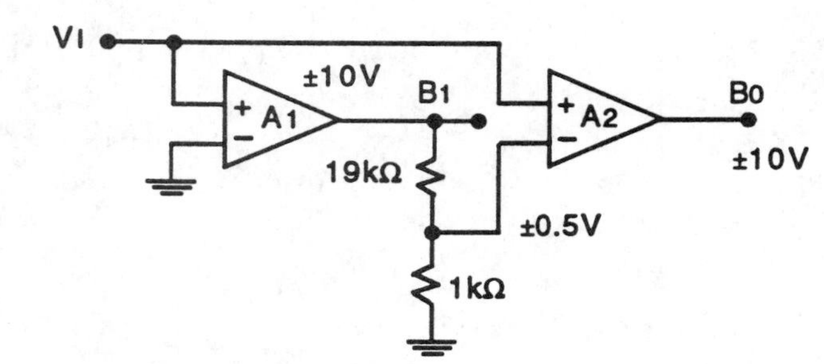

10.41 (a) During Φ_A, $V_X = V_A$, $V_O = V_Y = 0V$.

(b) Following Φ_A, before Φ_B, $V_X = V_A$, $V_Y = 0$, $V_O = 0V$, held by V_Y (and by stray feedback capacitance across the switch connecting output to input).

(c) During Φ_B, $V_X = V_{REF}$, $V_Y = 0 + \frac{V_A - V_{REF}}{2}$, V_O saturates at ±10V with the sign reversed from that at node Y.

(d) Following Φ_B, $V_X = V_{REF}$, $V_Y = \frac{V_A - V_{REF}}{2}$, V_O stays saturated.

Specifically, for

i) **$V_A > V_{REF}$, (a) $V_O = 0$, (b) $V_O = 0$, (c) $V_O = -10V$, (d) $V_O = -10V$.**

ii) **$V_A < V_{REF}$, (a) $V_O = 0$, (b) $V_O = 0$, (c) $V_O = +10V$, (d) $V_O = +10V$.**

Thus the circuit operates as a comparator of V_A against V_{REF}.

Chapter 11

FILTERS AND TUNED AMPLIFIERS

SECTION 11.1: Filter Transmission, Types, and Specification

11.1 $T(s) = \frac{s}{s + \omega_o}$, $T(j\omega) = \frac{j\omega}{j\omega + \omega_o}$, $|T| = |T(j\omega)| = \frac{\omega}{(\omega^2 + \omega_o^2)^{1/2}}$, $\Phi(\omega) = 90 - \tan^{-1} \omega/\omega_o$, $G = 20 \log |T|$ dB, $A = -20 \log |T|$ dB.

For $\omega = \infty$, $|T| = 1$, $\Phi = 90 - 90 = 0°$, $G = 20 \log(1) = 0$**dB**, $A = 0$**dB**.

For $\omega = 2\omega_o$, $|T| = \frac{2\omega_o}{((2\omega_o)^2 + \omega_o^2)^{1/2}} = \frac{2}{(5)^{1/2}} = 0.894$, $\Phi = 90 - \tan^{-1} \frac{2\omega_o}{\omega_o} = 26.6°$, $G = 20 \log(.894) = -.969$**dB**, $A = -G = .969$**dB**.

For $\omega = \omega_o$, $|T| = \frac{\omega_o}{(\omega_o^2 + \omega_o^2)^{1/2}} = \frac{1}{\sqrt{2}} = 0.707$, $\Phi = 90 - \tan^{-1} \frac{\omega_o}{\omega_o} = 45°$, $G = -3$**dB**, $A = 3$**dB**.

ω rad/s	$\lvert T\rvert$ V/V	Φ °	G dB	A dB
∞	1	0	0	0
$2\omega_o$	.894	26.6	−0.969	+0.969
ω_o	.707	45	−3	+3
$\omega_o/2$	.447	63.4	−6.99	6.99
$\omega_o/5$	.196	78.7	−14.2	14.2
$\omega_o/10$	.0995	84.3	−20	20
$\omega_o/100$	.010	89.4	−40	40
$\omega_o/1000$	.001	89.94	−60	60

11.2

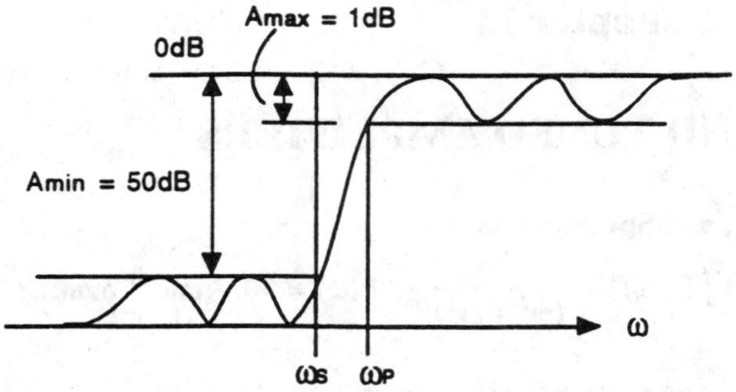

For $\omega = \infty, |T| = $ **0 dB,**
$\omega = \omega_p, |T| =$ **−1 dB,**
$\omega = \omega_s, |T| =$ **−50 dB.**

11.3 Here, ±5% transmission variation ≡ 0.95 ± 0.05. Thus $A_{max} = |20 \log (0.9)| =$ **0.915 dB**. Now, $A_{min} = |20 \log 0.05| =$ **26 dB**. The selectivity factor (high pass) = $\dfrac{\omega_p}{\omega_s} = \dfrac{f_p}{f_s} = 4/3.2 =$ **1.25**.

11.4 See $T(s) = \dfrac{s}{s + 1/\tau} = \dfrac{s}{s + 10^3}$, for a high-pass filter with $\omega_o = 10^3$rad/s. Now, $|T(\omega)| = \dfrac{\omega}{(\omega^2 + (10^3)^2)^{1/2}}$. For $A_{max} = 0.5$ dB, $|T| = 10^{-0.5/20} = 0.944$. Thus, $\dfrac{\omega}{(\omega^2 + (10^3)^2)^{1/2}} = 0.944$, $\omega^2 = 0.891\ (\omega^2 + 10^6)$, $\omega^2 = 8.195 \times 10^6$, or $\omega_p =$ **2.86×10^3 rad/s**. For $A_{min} = 20$ dB, $|T| = 0.1$. Thus $\dfrac{\omega}{(\omega^2 + (10^3)^2)^{1/2}} = 0.1$, $100\ \omega^2 = \omega^2 + 10^6$, $99\omega^2 = 10^6$, $\omega^2 = 10.1 \times 10^3$, $\omega_s =$ **100.5 rad/s**. The selectivity factor (high-pass) = $\dfrac{\omega_p}{\omega_s} = \dfrac{2.86 \times 10^3}{100.5} =$ **28.5**.

Now, for a modified filter for which $f_p = 10^3$ Hz, $\omega_p = 2\pi f_p = 6.283 \times 10^3$ rad/s. Formerly, for the same shape and $\omega_o = 10^3$ rad/s, $\omega_p = 2.86 \times 10^3$ rad/s. Thus the revised ω_o is $\omega_o = \dfrac{6.283 \times 10^3}{2.86 \times 10^3} \times 10^3 = 2.197 \times 10^3$ rad/s, for which $\tau = 1/(2.197 \times 10^3) =$ **0.455×10^{-3}s**, and $f_{3dB} = f_o = \dfrac{\omega_o}{2\pi} = \dfrac{2.197 \times 10^3}{2\pi} =$ **0.35 kHz.**
At 100Hz, $|T(f)| = \dfrac{f}{(f_o^2 + f^2)^{1/2}} = \dfrac{100}{(350^2 + 100^2)^{1/2}} = .275 \equiv 20_{10}\ .275 = -\ 11.2$ dB. Thus, $A_{100} =$ **11.2 dB.**

SECTION 11.2: The Filter Transfer Function

11.5 $T(s) = \dfrac{s^2\ (s - -0.1)}{(s - -1)\ (s - (-0.5 - j0.8))\ (s - (-0.5 + j0.8))} = \dfrac{s^2\ (s + 0.1)}{(s + 1)\ (s + 0.5 + j0.8)\ (s + 0.5 - j0.8)} =$
$\dfrac{s^2\ (s + 0.1)}{(s + 1)\ (s^2 + 0.5s - j0.8s + 0.5s + 0.25 - j0.4 + j0.8s + 0.4j + 0.64)} = \dfrac{s^2\ (s + 0.1)}{(s + 1)\ (s^2 + s + 0.89)}$,
or $\mathbf{T(s)} = \dfrac{\mathbf{s^3 + 0.1s^2}}{\mathbf{s^3 + 2s^2 + 1.89s + 0.89}}$. Note that $|T| = 1$ as $s \to \infty$.

11.6 Following the preamble to equation 11.9:

$$T(s) = \frac{a_9\ s\ (s^2 + 1 \times 10^6)\ (s^2 + 4 \times 10^6)\ (s^2 + 36 \times 10^6)\ (s^2 + 144 \times 10^6)}{s^{10} + b_9\ s^9 + b_8\ s^8 + - - - + b_0},$$

where the filter order is $N = 10$. From Fig. 11.4:

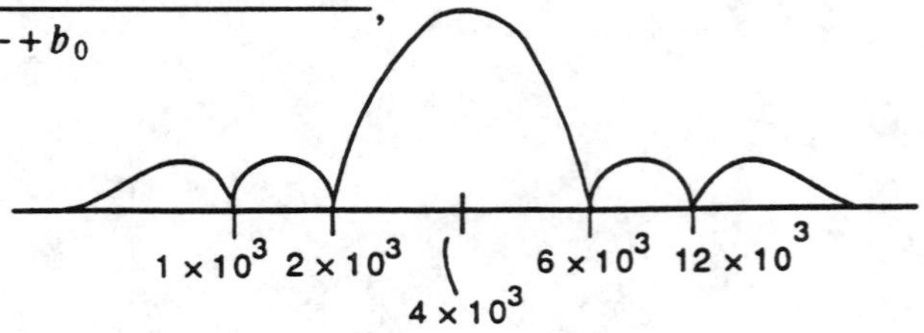

SECTION 11.3: Butterworth and Chebyshev Filters

11.7 For a Butterworth filter of order N, $\varepsilon = (10^{A_{max}/10} - 1)^{1/2} = (10^{0.5/10} - 1)^{1/2} =$ **0.349**. Now, at the edge of the stopband $A(\omega_s) = -20 \log (1 + \varepsilon^2 (\omega_s/\omega_p)^{2N})^{-1/2} = 10 \log (1 + \varepsilon^2 (\omega_s/\omega_p)^{2N})$, or $40 = 10 \log (1 + .349^2 (1.6)^{2N})$, $1 + .1218 (1.6)^{2N} = 10^4$, $(1.6)^{2N} = 8.21 \times 10^4$. Try $N = 10$, or $2N = 20$, whence $(1.6)^{20} = 1.209 \times 10^4$ rather than 8.21×10^4. Try $N = 12$, $2N = 24$, $(1.6)^{24} = 7.92 \times 10^4$. Try $N = 13$, $2N = 26$, $(1.6)^{26} = 20.28 \times 10^4$. That is, 13th order will clearly do the job. For **N = 13**, $A(\omega_s) = -20 \log (1 + \varepsilon^2 (\omega_s/\omega_p)^{2N})^{-1/2} = -20 \log (1 + .349^2 (1.6)^{26})^{-1/2} = -20 \log (1 + .1218 \times 20.28 \times 10^4)^{-1/2} =$ **43.93 dB**.

Now, for $A_{min} = 40$ dB exactly, with $N = 13$, $A_{min} = 10 \log (1 + \varepsilon^2 (\omega_s/\omega_p)^{2N})$, or $40 = 10 \log (1 + \varepsilon^2 (1.6)^{26})$, $1 + \varepsilon^2 (1.6)^{26} = 10^4$, $\varepsilon = (10^4/(1.6)^{26})^{1/2} = 10^2/(1.6)^{13} =$ **.222**. Now $.222 = (10^{A_{max}/10} - 1)^{1/2}$, $10^{A_{max}/10} = .0493 + 1 = 1.0493$, $A_{max} = 10 \log 1.0493 =$ **.209 dB**. *Alternatively*, if A_{max} is raised from 0.5 dB to 0.6 dB, $\varepsilon = (10^{0.6/10} - 1)^{1/2} = .385$, and $A_{min} = 10 \log (1 + \varepsilon^2 (\omega_s/\omega_p)^{2N}) = 10 \log (1 + .385^2) (1.6)^{26}) =$ **44.8 dB**. Now, we can check whether the filter order could be reduced for $A_{min} = 40$ dB. See that $40 = 10 \log (1 + .385^2 (1.6)^{2N})$, or $1.6^{2N} = 6.75 \times 10^4$. Taking logs, $2N = \log (6.75 \times 10^4)/\log 1.6 = 4.829/.204 = 23.6$. Thus, *12th order would suffice*!

11.8 Now, the (low-pass) selectivity ratio is $f_s/f_p = 30/20 = 1.5$, $A_{max} = 1$ dB, $\varepsilon = (10^{1/10} - 1)^{1/2} = 0.5088$, and $A(f_s) = 10 \log (1 + \varepsilon^2 (f_s/f_p)^{2N})$ or $20 = 10 \log (1 + (.509)^2 (1.5)^{2N})$, $1 + .259 (1.5)^{2N} = 100$, $(1.5)^{2N} = 99/.259 = 382.3$. Try $N = 6$, $2N = 12$, $(1.5)^{12} = 129.7$. Try $N = 7$, $2N = 14$, $(1.5)^{14} = 292$. Try $N = 8$, $2N = 16$, $(1.5)^{16} = 657$. Thus use 8th order, **N = 8**. The poles all have the same frequency $\omega_o = 2\pi f_p (1/\varepsilon)^{1/n}$, or $\omega_o = (2\pi)\ 20 \times 10^3 (\frac{1}{.5088})^{1/8} =$ **136.7 krad/s**. The first pole (or natural mode) p_1 is given by $p_1 = \omega_o (-\cos (90 - 11.25) + j\sin (90 - 11.25)) = \omega_o (-.1951 + j0.9808)$. Combining p_1 with its conjugate p_8 yields the factor $(s^2 + s0.3902\omega_o + \omega_o^2)$. Likewise $p_2 = \omega_o (-\cos (90 - \frac{3\pi}{2(8)}) + j \sin (90 - 33.75)) = \omega_o (-0.5556 + j\,0.8315)$ with factor $(s^2 + s1.1111\omega_o + \omega_o^2)$ and $p_3 = \omega_o (-\cos (33.75) + j \sin (33.75)) = \omega_o (-.8315 + j\,0.5556)$ with factor $(s^2 + s1.663\ \omega_o + \omega_o^2)$, and $p_4 = \omega_o (-\cos (11.25) + j \sin (11.25)) = \omega_o (-0.9808 + j\,0.1951)$ with factor $(s^2 + s1.9616\ \omega_o + \omega_o^2)$. Thus

$$T(s) = \frac{\omega_o^8}{(s^2 + 0.3902\ \omega_o s + \omega_o^2)(s^2 + 1.111\ \omega_o\ s + \omega_o^2)(s^2 + 1.663\ \omega_o\ s + \omega_o^2)(s^2 + 1.962\ \omega_o\ s + \omega_o^2)}$$

Generally, $|T| = (1 + \varepsilon^2 (\omega/\omega_p)^{2N})^{-1/2}$. Now with $f_p = 20$kHz, $N = 8$, $\varepsilon = .5088$. At 25kHz, $|T| = (1 + (.5088)^2 (25/20)^{16})^{-1/2} = (1 + .2589 \times 35.53)^{-1/2} = 0.313 \equiv 20 \log .313 = -10.1$ dB, or $A =$ **10.1dB** at 25kHz. At 40kHz, $|T| = (1 + .2589 \times (40/20)^{16})^{-1/2} = .00768 \equiv 20 \log .00268 = -42.3$ dB. Thus $A =$ **42.3dB** at 40kHz.

11.9

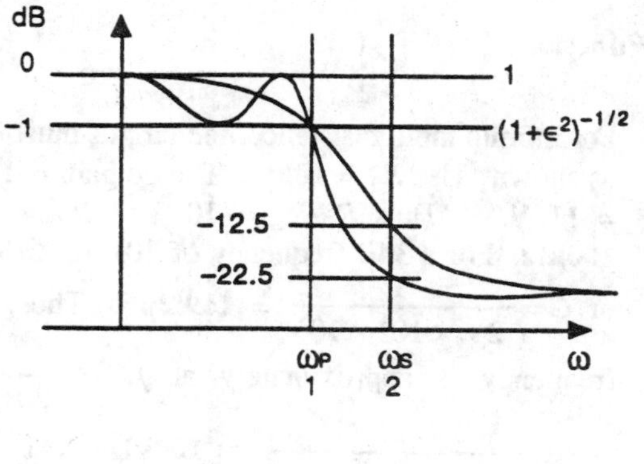

Here, at $\omega_s = 2\omega_p$, $\omega_s/\omega_p = 2$, $N = 3$. For Butterworth and Chebyshev, $\varepsilon = (10^{A_{max}/10} - 1)^{1/2} = (10^{1/10} - 1)^{1/2} = .509$. For Butterworth, $A(\omega_s) = 10 \log [1 + \varepsilon^2 (\omega_s/\omega_p)^{2N}] = 10 \log [1 + .509^2 (2)^6] =$ **12.45 dB**. For Chebyshev, $A(\omega_s) = 10 \log [1 + \varepsilon^2 \cosh^2 (N \cosh^{-1} \omega_s/\omega_p)] = 10 \log [1 + .509^2 \cosh^2 (3 \cosh^{-1} 2)] = 10 \log [1 + .259 \cosh^2 (3 (1.317))] = 10 \log [1 + .259 \cosh^2 3.951] = 10 \log [1 + .259 (26.0)^2] =$ **22.5 dB**.

11.10 Required that A_{max} = 0.5dB, $A_{min} \geq$ 40 dB, ω_s/ω_p = 1.6 for a Chebyshev filter.
From Equation 11.21: $\varepsilon = \sqrt{10^{A max/10} - 1} = \sqrt{10^{0.5/10} - 1} = \sqrt{1.122 - 1} = \sqrt{.122}$ = 0.3493. From Equation 11.22, at the stopband edge, where $\omega = \omega_s$, $A(\omega_s)$ = 10log [1 + $\varepsilon^2 \cosh^2$ ($N \cosh^{-1} \omega_s/\omega_p$)] = 10log [1 + $.3493^2 \cosh^2$ ($N \cosh^{-1}$ 1.6)] = 10log [1 + 0.122 $\cosh^2$ (1.04697N). Now for N = 10, A = 10 log [1 + 0.122 $\cosh^2$ (1.047 × 10)] = 75.8dB, much greater than required. For N = 6, A = 10 log [1 + .122 $\cosh^2$ (1.047 × 6)] = 39.4dB < A_{min}. For N = 7, A = 10 log [1 + .122 $\cosh^2$ (1.047 × 7)] = 48.5dB > A_{min}. Thus, use **N = 7**, for which A_{min} = **48.5dB**. Now for N = 7 and A_{min} = 40dB, 40 = 10 log [1 + $\varepsilon^2 \cosh^2$ (7 $\cosh^{-1}$ 1.6)]. Thus 1 + $\varepsilon^2 \cosh^2$ (7.329) = 10^4, or $\varepsilon^2 = (10^4 - 1)/0.5802 \times 10^6$ = 0.01723, or ε = 0.1312, for which A_{max} = 10 log (1 + ε^2) = **0.074dB** is possible. *Check*: A_{min} = 10 log [1 + $\varepsilon^2 \cosh^2$ ($N \cosh^{-1} \omega_s/\omega_p$)] = 10 log [1 + .01723 $\cosh^2$ (7 $\cosh^{-1}$ 1.6)] = 40dB; OK. Now for A_{max} raised to 0.074 + 0.1 = 0.174dB, $\varepsilon = (10^{A max/10} - 1)^{1/2}$ = 0.202, and A_{min} = 10 log [1 + $0.202^2 \cosh^2$ (7.329)] = **43.8dB**, an increase of nearly 4dB of stopband attenuation in return for a 0.1dB increase in passband ripple!!

11.11 Consider the question to refer to the initial specification in P11.7 and P11.10, for which A_{max} = 0.5dB and $A_{min} \geq$ 40dB, with ω_s/ω_p = 1.6. Here $\omega_p = 10^3$rad/s, for which $\omega_s = 1.6 \times 10^3$rad/s, and, the dc gain is 1.
Now for the Butterworth filter, N = 13, with ε = 0.349, the poles are on a circle with radius $\omega_o = \omega_p (1/\varepsilon)^{1/N}$ = $10^3 (1/0.349)^{1/13} = 1.084 \times 10^3$rad/s, at an angular separation of π/N = 13.85°, at angles (from the negative real axis) of 0°, ±13.85°, ±27.69°, ±41.54°, ±55.38°, ±69.23°, ±83.08°. Now $p_1 = \omega_o$ (– cos 83.08° + j sin 83.08°) = 1.084 × 10^3 (– 0.1205 + j 0.9927) = 10^3 (– 0.131 + j 1.076). Correspondingly, p_1, p_{13} = **10^3 (– 0.131 ± j 1.076) rad/s**; $p_2, p_{12} = \omega_o$ (– cos 69.23° ± j sin 69.23°) = 1.084 × 10^3 (– .3546 ± j 0.9350) = **10^3 (– 0.384 ± j 1.013) rad/s**; $p_3, p_{11} = \omega_o$ (– cos 55.38° ± j sin 55.38°) = **10^3 (– 0.616 ± j 0.892) rad/s**; $p_4, p_{10} = \omega_o$ (– cos 41.54° ± j sin 41.54°) = **10^3 (– 0.811 ± j 0.719) rad/s**; $p_5, p_9 = \omega_o$ (– cos 27.69° ± j sin 27.69°) = **10^3 (– 0.960 ± j 0.504) rad/s**; $p_6, p_8 = \omega_o$ (– cos 13.85° ± j sin 13.85°) = **10^3 (– 1.052 ± j 0.259) rad/s**; **p_7 = 1.08 × 10^3 rad/s.**
Now for the Chebyshev filter, N = 7, with ε = 0.349, and $\omega_p = 10^3$ rad/s, the poles are (for k = 1 to 7):

$$p_k = \omega_p \left[-\sin\left(\frac{2k-1}{N}\,\frac{\pi}{2}\right) \sinh\left(\frac{1}{N}\sinh^{-1} 1/\varepsilon\right) + j\cos\left(\frac{2k-1}{N}\,\frac{\pi}{2}\right)\cosh\left(\frac{1}{N}\sinh^{-1} 1/\varepsilon\right)\right]$$

$$= 10^3 \left[-\sin\left(\frac{2k-1}{7}(90°)\right) \sinh\left[\frac{1}{7}\sinh^{-1}\left(1/.349\right)\right] + j\cos\left(\frac{2k-1}{7}(90°)\right)\cosh\left[\frac{1}{7}\sinh^{-1}\left(1/.349\right)\right]\right]$$

$$= 10^3 \left[-.2563 \sin\left((2k-1)(12.86°)\right) + 1.032 j \cos\left((2k-1)(12.86°)\right)\right].$$

Now, p_1, p_7 = **10^3 [– 0.057 ± j 1.006] rad/s**; p_2, p_6 = **10^3 [– 0.160 ± j 0.807] rad/s**; p_3, p_5 = **10^3 [– 0.231 ± j 0.448] rad/s**; p_4 = **10^3 [– 0.256] rad/s.**

SECTION 11.4: First-Order and Second-Order Filter Functions

11.12

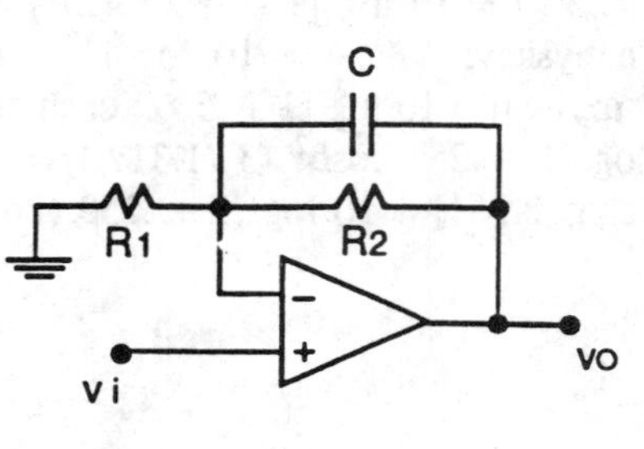

For infinite input resistance, the circuit must be driven as shown. Use R_1 = **10kΩ**. The dc gain is 1 + R_2/R_1 = 11 V/V. Thus R_2/R_1 = 10, and $R_2 = 10\,R_1$ = **100kΩ**. For a 3dB frequency of 10kHz, $CR_2 = 1/\omega_o$, or $C = \dfrac{1}{2\pi \times 10^4 \times 10^5}$ = **159.2pF**. Thus, the zero frequency is approximately at $f_z = \dfrac{1}{2\pi C R_1} = \dfrac{1}{2\pi \times 159.2 \times 10^{-12} \times 10^4} \approx$ **100kHz**. At very high frequencies, C is a short circuit, and υ_o follows υ_i, such that υ_o/υ_i = **+1V/V**.

From First Principles: $V_o/V_i = 1 + \dfrac{R_2 \| \frac{1}{Cs}}{R_1} = \dfrac{1}{R_1}\left[R_1 + \dfrac{\frac{R_2}{Cs}}{R_2 + 1/Cs}\right] = \dfrac{1}{R_1}\left[R_1 + \dfrac{R_2}{R_2\,Cs + 1}\right] =$

$\dfrac{R_2 + R_1 + R_1 R_2 Cs}{R_1 + R_1 R_2 Cs} = \dfrac{s + \left(\frac{R_1 + R_2}{R_1 R_2}\right)\frac{1}{C}}{s + 1/R_2 C}$. Thus the zero is more precisely at $f_z = \dfrac{1}{2\pi C\,(R_1 \| R_2)} =$

$\dfrac{1}{2\pi \times 159.2 \times 10^{-12}\,(10^4 \| 10^5)}$ = **110kHz**. *Note* that $A_L f_p = A_H f_z$, ie 11 (10kHz) = 1 (110kHz). Now for $A_H = 1$V/V, $f_z/f_p = 100$, $A_L = A_H f_z/f_p$ = **100V/V**, for which R_1 = **10kΩ**, and $R_2 = R_1\,(100–1)$ = **990kΩ**.

11.13 The Bandpass requirement suggests the combination of Fig. 11.13a and 11.13b as shown, with

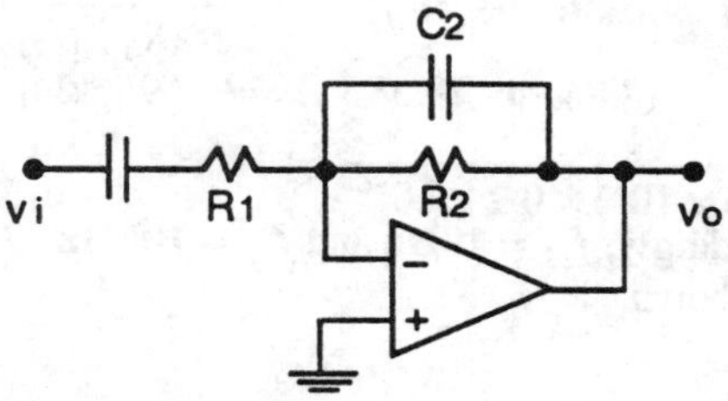

$C_1 R_1 = \dfrac{1}{2\pi\,(100)}$, and $C_2 R_2 = \dfrac{1}{2\pi\,(1000)}$. The midband gain is $-R_2/R_1$ = 1V/V, with $R_1 \approx R_{in} \approx$ 10kΩ. Now, if use $R_1 = R_2$ = 10kΩ, $C_1 = \dfrac{1}{2\pi\,(100)\,(10^4)}$ = 0.159μF. Chose C_1 = **0.1μF**, and C_2= **0.01μF**. Now $R_1 = \dfrac{1}{2\pi \times 0.1 \times 10^{-6} \times 100}$ = **15.9kΩ**. For gain = –1V/V, R_2 = 15.9kΩ, as well, with an input resistance at midband of about **15.9kΩ**.

11.14 $T(s) = -\dfrac{R_{2a} + R_{2b} \| \frac{1}{C_2 s}}{R_{1a} + R_{1b} \| \frac{1}{C_1 s}} = -\dfrac{R_{2a} + \frac{R_{2b}}{C_2 s\,(R_{2b} + 1/C_2 s)}}{R_{1a} + \frac{R_{1b}}{C_1 s\,(R_{1b} + 1/C_1 s)}} = -\dfrac{R_{2a} + \frac{R_{2b}}{1 + R_{2b} C_2 s}}{R_{1a} + \frac{R_{1b}}{1 + R_{1b} C_1 s}}$, or

$$\mathbf{T(s)} = \frac{1 + R_{1b} C_1 s}{1 + R_{2b} C_2 s} \times \frac{R_{2a} + R_{2b} + R_{2a} R_{2b} C_2 s}{R_{1a} + R_{1b} + R_{1a} R_{1b} C_1 s} = \frac{R_{1b} C_1}{R_{2b} C_2} \times \frac{R_{2a} R_{2b} C_2}{R_{1a} R_{1b} C_1} \times \left[\frac{s + \frac{1}{R_{1b} C_1}}{s + \frac{1}{R_{2b} C_2}}\right] \times$$

$$\left[\frac{s + \frac{1}{(R_{2a} \| R_{2b})\,C_2}}{s + \frac{1}{(R_{1a} \| R_{1b})\,C_1}}\right] = \frac{\mathbf{R_{2a}}}{\mathbf{R_{1a}}} \times \left[\frac{\mathbf{s} + \frac{1}{\mathbf{R_{1b}\,C_1}}}{\mathbf{s} + \frac{1}{\mathbf{R_{2b}\,C_2}}}\right] \times \left[\frac{\mathbf{s} + \frac{1}{(\mathbf{R_{2a} \| R_{2b}})\,\mathbf{C_2}}}{\mathbf{s} + \frac{1}{(\mathbf{R_{1a} \| R_{1b}})\,\mathbf{C_1}}}\right]$$

, with zeros at $\dfrac{1}{R_{1b} C_1}$ and $\dfrac{1}{(R_{2a} \| R_{2b})\,C_2}$, with poles at $\dfrac{1}{R_{2b} C_2}$ and $\dfrac{1}{(R_{1a} \| R_{1b})\,C_1}$, and with a high-frequency gain of R_{2a}/R_{1a}, a low-frequency gain of $(R_{2a} + R_{2b})/(R_{1a} + R_{1b})$, a mid-band gain of $(R_{2a} + R_{2b})/R_{1a}$ or $R_{2a}/(R_{1a} + R_{1b})$, depending on the relative locations of the poles and zeros. For a midband gain of –10V/V, gains at low and high frequencies of –1V/V, and 3dB points at 100Hz and 1000 Hz, the corresponding Bode plot is as shown:

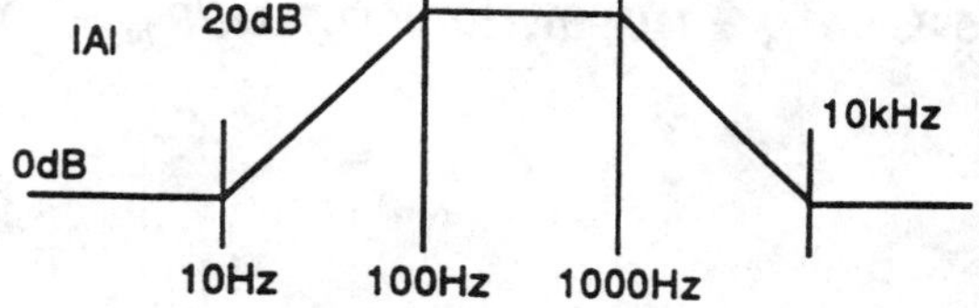

Now, at low frequencies, $\dfrac{R_{2a} + R_{2b}}{R_{1a} + R_{1b}} = 1$; at high frequencies, $\dfrac{R_{2a}}{R_{1a}} = 1$; at midband, $\dfrac{(R_{2a} + R_{2b})}{R_{1a}} = 10$. Thus $R_{2a} + R_{2b} = R_{1a} + R_{1b} = 10\,R_{1a} = 10\,R_{2a}$, and $R_{1b} = 9R_{1a}$, $R_{2b} = 9R_{2a}$, $R_{1a} = R_{2a}$, and $R_{1b} = R_{2b}$. For $R_{in} \approx$ 10kΩ at midband, $R_{1a} \approx$ 10kΩ and $R_{1b} \approx$

90kΩ.

For the zero at 10Hz, $\frac{1}{2\pi R_{1b} C_1} = 10$, $C_1 = \frac{1}{2\pi R_{1b} \times 10} = \frac{1}{2\pi \times 10 \times 90 \times 10^3} = 0.177\mu F$.

Use $C_1 =$ **0.2μF**, for which $R_{1b} = \frac{1}{2\pi C_1 \times 10} = \frac{1}{2\pi \times 0.2 \times 10^{-6} \times 10} =$ **79.58kΩ** $= R_{2b}$, and $R_{1a} = R_{2a} =$ 79.58/9 = **8.84kΩ**. Now for pole the at 1kHz: $\frac{1}{2\pi C_2 R_{2b}} = 10^3$, $C_2 = \frac{1}{2\pi \times 10^3 \times 79.58 \times 10^3} =$ **0.002μF**. Now, check the second pole, to be at $f_{p2} = \frac{1}{2\pi (R_{1a} \| R_{1b}) C_1}$. Here $R_{1a} \| R_{1b}$ = (8.84 ‖ 79.58)kΩ = 7.956kΩ, and $f_{p2} = \frac{1}{2\pi (7.956 \times 10^3) \times 0.2 \times 10^{-6}}$ = **100Hz**, with f_{p1} at **1kHz**. Correspondingly, f_{z1} = **10Hz** and f_{z2} = **10kHz**. The pole-zero plot is as shown:

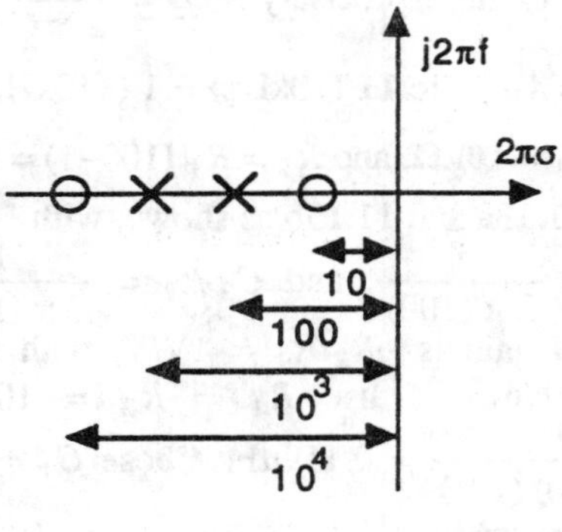

11.15 The required response is as shown in the Bode plot below:

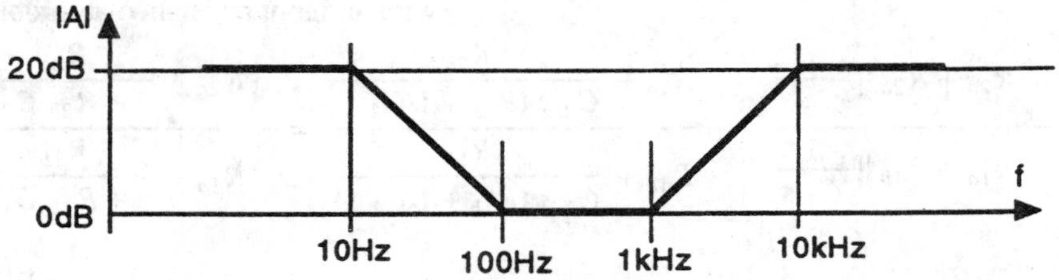

For the circuit in P11.14 to have a gain of −10V/V at dc, $\frac{R_{2a} + R_{2b}}{R_{1a} + R_{1b}} = 10$ – – – (1). For a gain of −10V/V at high frequencies, $\frac{R_{2a}}{R_{1a}} = 10$ – – – (2). Now for a *lower* gain at midband, C_2 must provide a zero at 100Hz, and C_1 a zero at 1kHz, with R_{2b} shorted there (while C_1 is still (relatively) open). Thus in the midband, the gain is $\frac{R_{2a}}{R_{1a} + R_{1b}} = 1$ – – – (3).

The pole associated with C_2 must be at 100Hz/10 = 10Hz. The pole associated with C_1 must be at (1kHz) × 10 = 10kHz. Now, $R_{in} \geq 10k\Omega$, $R_{1a} \geq 10k\Omega$. For $R_{1a} = 10k\Omega$, R_{2a} = 10 (10) = 100kΩ. From (3), $R_{2a} = R_{1a} + R_{1b}$, or $R_{1b} = R_{2a} - R_{1a}$ = 100kΩ − 10kΩ = 90kΩ. From (1), $R_{2a} + R_{2b} = 10\,R_{1a} + 10\,R_{1b}$, or R_{2b} = 10 (10) + 10 (90) − 100 = 100 + 900 − 100 = 900kΩ. That is (tentatively), R_{1a} = 10kΩ, R_{1b} = 90kΩ, R_{2a} = 100kΩ, R_{2b} = 900kΩ. For a zero at 1kHz, $\frac{1}{2\pi R_{1b} C_1}$ = 1kHz, or $C_1 = \frac{1}{2\pi \times 10^3 \times 90 \times 10^3}$ = .00177μF. Now, for a one-significant-digit capacitor, use C_1 = **.001μF**, or 1nF, to maintain $R_{in} > 10k\Omega$. *Conclude*: $R_{1b} = \frac{1}{2\pi \times 10^3 \times .001 \times 10^{-6}}$ = **159kΩ**, R_{1a} = (10/90) 159 = **17.7kΩ**, R_{2a} = 10 (17.7) = **177kΩ**, R_{2b} = 10 (159) = **1.591MΩ**.

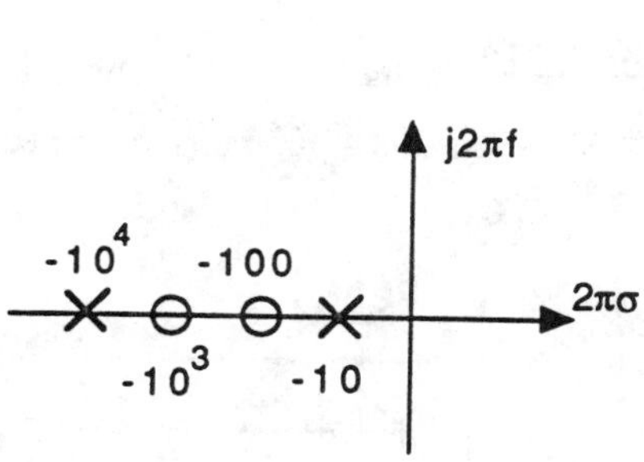

Now for a zero at 100Hz associated with C_2, use $C_2 = \dfrac{1}{2\pi \times 10^2\,(R_{2a} \| R_{2b})} = \dfrac{1}{2\pi \times 10^2\,(.177 \| 1.591)\,10^6} = .999 \times 10^{-8}$ = **10nF**. *Check*: The pole from C_2 is $\dfrac{1}{2\pi C_2 R_{2b}} = \dfrac{1}{2\pi \times 10 \times 10^{-9} \times 1.591 \times 10^6}$ = **10Hz**, and the pole from C_1 is $\dfrac{1}{2\pi C_1\,(R_{1a} \| R_{1b})} = \dfrac{1}{2\pi \times .001 \times 10^{-6} \times (17.7 \| 159) \times 10^3} = .0099 \times 10^6$ = **10kHz**. Overall, there are zeroes at **100Hz, 1kHz,** and poles at **10Hz, 10kHz,** for which the pole-zero plot is shown.

11.16 From Fig. 11.14 modified:

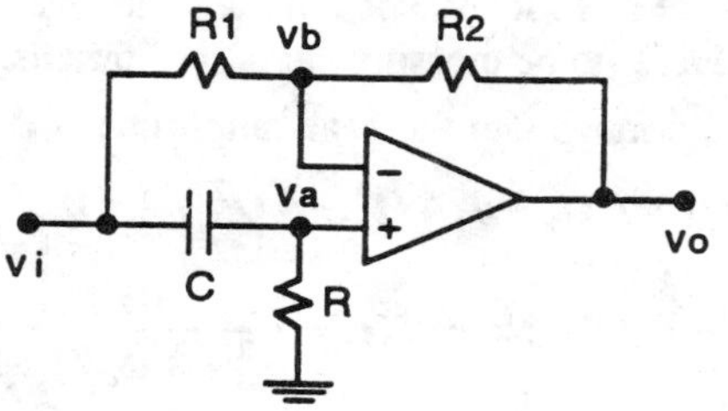

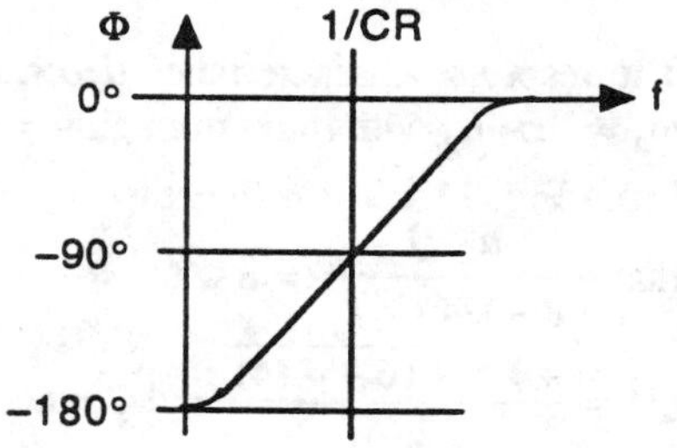

$v_a = \dfrac{R}{R + 1/Cs}\, v_i = v_b$, $v_o = v_b + v_b - v_i = 2v_b - v_i$, and $T(s) = \dfrac{v_o}{v_i} = \dfrac{2R}{R + 1/Cs} - 1 = \dfrac{2R - R - 1/Cs}{R + 1/Cs} = \dfrac{RCs - 1}{RCs + 1}$, or $\mathbf{T(s)} = \dfrac{\mathbf{s - 1/RC}}{\mathbf{s + 1/RC}}$. See a pole at $-\mathbf{1/RC}$ and a zero at $\mathbf{1/RC}$. Now at $\omega_o = 1/RC$, phase shift is $\mathbf{+\tan^{-1}(-1) - \tan^{-1}(1) = -90°}$.

Now, for $f_o = 10^4$, $C = 1.59 \times 10^{-9}$F, and $|\Phi| = 90°$, $R = \dfrac{1}{2\pi \times 1.59 \times 10^{-9} \times 10^4} = 10^4\Omega$ = **10kΩ**. For Φ in general, $\Phi = \tan^{-1}\left[-\dfrac{\omega_o}{1/RC}\right] - \tan^{-1}\left[\dfrac{\omega_o}{1/RC}\right] = -2\tan^{-1}(\omega_o RC)$, or $R = \dfrac{1}{2\pi f_o C}\tan(-\Phi/2)$. Now, at $f_o = 10^4$Hz, for $\Phi = -90°$, $R = \dfrac{1}{2\pi \times 10^4 \times 1.59 \times 10^{-9}}\tan(--90/2) = 10^4(1) = 10^4\Omega$.

For $|\Phi| = 6°$, $R = 10^4\tan(6/2)$ = **524Ω**; for 12°, $R = 10^4\tan(12/2)$ = **1051Ω**; for 30°, $R = 10^4\tan(30/2)$ = **2.68kΩ**; for 60°, $R = 10^4\tan(60/2)$ = **5.77kΩ**; for 90°, $R = 10^4\tan(90/2)$ = **10kΩ**; for 120°, $R = 10^4\tan(120/2)$ = **17.3kΩ**; for 150°, $R = 10^4\tan(150/2)$ = **37.3kΩ**; for 168°, $R = 10^4\tan(168/2)$ = **95.1kΩ**; for 174°, $R = 10^4\tan(174/2)$ = **190.8kΩ**.

11.17 For $\omega_o = 10^3$rad/s, and 3dB bandwidth is of $\omega_o/Q = 200$ rad/s, see $Q = 10^3/200 = 5$. Now peak gain is $\dfrac{a_1 Q}{\omega_o} = 1$. Thus, $a_1 = \omega_o/Q = 10^3/5 = 200$, and $\mathrm{T(s)} = \dfrac{a_1 s}{s^2 + \dfrac{s\,\omega_o}{Q} + \omega_o^2} = \dfrac{\mathbf{200s}}{\mathbf{s^2 + 200s + 10^6}}$. At ω, $T(j\omega) = \dfrac{200\,j\omega}{200\,j\omega + 10^6 - \omega^2}$. At $A_{\min} = -20$dB, $\dfrac{200\,\omega}{(200\,\omega^2 + (10^6 - \omega^2)^2)^{1/2}} = \dfrac{1}{10}$, $2000^2\,\omega^2 = 200\,\omega^2 + 10^{12} - 2 \times 10^6\,\omega^2 + \omega^4$, or $\omega^4 + \omega^2(-4 \times 10^6 - 2 \times 10^6 + 200) + 10^{12} = 0$, or $\omega^4 - 6 \times 10^6\,\omega^2 + 10^{12} = 0$, or $\omega^2 = \dfrac{+6\times10^6 \pm \sqrt{(6\times10^6)^2 - 4\times10^{12}}}{2} = \dfrac{6\times10^6 \pm 5.657\times10^6}{2} = .1716\times10^6$, and 5.829×10^6, where ω = **414 rad/s and 2.414 k rad/s.** *Check*: $0.414\times2.414 = .999$ krad/s $\approx 10^3$ rad/s, as expected.

11.18

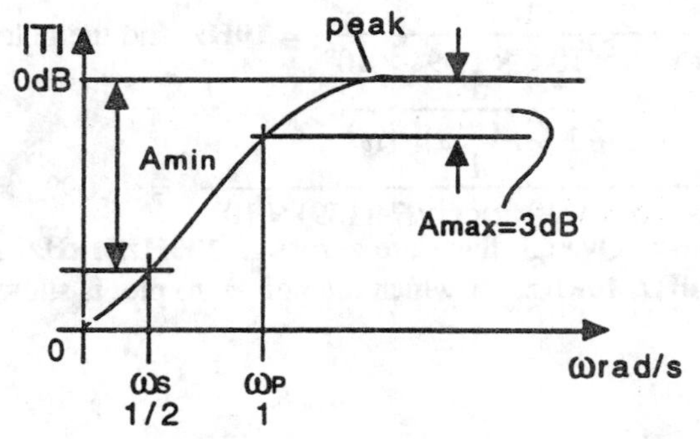

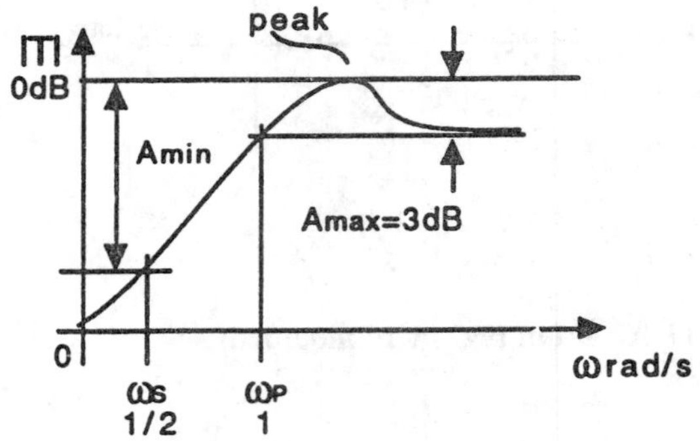

The response characteristics described above are as follows: For both designs, we require A_{max} = 3dB at ω_p = 1 rad/s, with maximum gain = 1V/V, and A_{min} to be measured at ω_s = 0.5 rad/s.

a) From Fig. 11.16b for the response shown first, arrange that the peak amplitude is such that $a_2 = 1$, that is that $\dfrac{a_2 Q}{(1 - 1/4\,Q^2)^{1/2}} = a_2$, $Q^2 = 1 - \dfrac{1}{4Q^2}$, $4Q^4 = 4Q^2 - 1$, $4Q^4 - 4Q^2 + 1 = 0$, $Q^2 = \dfrac{+4 \pm \sqrt{16 - 4(4)}}{2(4)} = 1/2$, and $Q = \dfrac{1}{\sqrt{2}} = 0.707$. Now, $T(s) = \dfrac{a_2 s^2}{s^2 + s\,\omega_o/Q + \omega_o^2}$ in general, For a_2 = 1, Q = 0.707, $\omega = \omega_p$ = 1 for $|T|$ = –3dB ≡ 0.707. Thus $T(j1) = \dfrac{1\,(j1)^2}{(j1)^2 + j1\,\omega_o/.707 + \omega_o^2}$ and $\dfrac{1}{((\omega_o^2 - 1)^2 + 2\omega_o^2)^{1/2}} = .707$, or $\omega_o^4 + 2\omega_o^2 + 1 - 2\,\omega_o^2 = 2$, or $\omega_o^4 = 1$, and $\omega_o = 1$. Thus $\mathbf{T(s)} = \dfrac{\mathbf{s^2}}{\mathbf{s^2 + 1.414\,s + 1}}$, with Q = **0.707**, and $\omega_o = \omega_{3dB}$ = **1 rad/s.** Now, for $A_{min} = -|T(j\,1/2)|$, $T(j/2) = \dfrac{(j/2)^2}{(j/2)^2 + 1.414\,(j/2) + 1} = \dfrac{-1}{4\,[(-1/4) + 1 + .707j]} = \dfrac{-1}{2.828j + 3}$. $A_{min} = -20\log\left(\dfrac{1}{((2.828)^2 + 3^2)^{1/2}}\right) = -20\log .2425 =$ **12.3dB.**

(b) From the second response, see a_2 = 0.707, and at the peak, $\dfrac{a_2 Q}{\sqrt{1 - 1/(4Q^2)}} = 1$, or $(0.707\,Q)^2 = 1 - \dfrac{1}{4Q^2}$, $\dfrac{Q^2}{2} = 1 - \dfrac{1}{4Q^2}$, $Q^2 = 2 - \dfrac{1}{2Q^2}$, $2Q^4 = 4Q^2 - 1$, $Q^4 - 2Q^2 + 0.5 = 0$, $Q = \dfrac{--2 \pm \sqrt{2^2 - 4(0.5)}}{2} = \dfrac{2 \pm \sqrt{2}}{2}$ = **1.707** for $Q > 1$. Now, in general, $T(s) = \dfrac{a_2 s^2}{s^2 + s\,\omega_o/Q + \omega_o^2}$, and for Q = 1.707, a_2 = 0.707, $\omega = \omega_p$ = 1 for $|T|$ = .707. $T(j1) = \dfrac{.707\,(j1)^2}{(j1)^2 + j1\,\omega_o/1.707 + \omega_o^2} = \dfrac{.707}{(\omega_o^2 - 1) + .5858j\,\omega_o}$. Now, $0.707 = \dfrac{.707}{[((\omega_o^2 - 1)^2 + .5858^2\,\omega_o^2)]^{1/2}}$, or $\omega_o^4 - 2\,\omega_o^2 + 1 + .343\,\omega_o^2 = 1$, $\omega_o^4 = 1.657\,\omega_o^2$, $\omega_o^2 = 1.657$, ω_o = **1.287 rad/s.** Thus $T(s) = \dfrac{.707\,s^2}{s^2 + s\,1.287/1.707 + (1.287)^2}$, or $\mathbf{T(s)} = \dfrac{\mathbf{.707\,s^2}}{\mathbf{s^2 + 0.754s + 1.657}}$, with ω_o = 1.287 rad/s, and ω_{3dB} = **1 rad/s.** Now $T(j/2) = \dfrac{-.707\,(1/4)}{-1/4 + .754\,j/2 + 1.657} = \dfrac{-.707}{5.628 + 1.508j} = \dfrac{-1}{7.960 + 2.133j}$, and $A_{min} = -20\log\dfrac{1}{(7.96^2 + 2.133^2)^{1/2}} = -20\log(.1213)$ = **18.3dB**, an improvement of 6dB over the arrangement in (a).

11.19 Use the description in Fig. 11.16d for the notch. Here f_o = 60Hz, and $\omega_o = 2\pi(60) =$ **377.0 rad/s.** Now $T(s) = \dfrac{s^2 + 377^2}{s^2 + s\dfrac{377}{Q} + 377^2}$. Now, assuming the required 1Hz band to be centered, arrange that the attenuation is 20dB at 60 + 1/2 = 60.5Hz, or 380.1 rad/s. Thus,

$$\left|\frac{(j380.1)^2 + 377^2}{(j380.1)^2 + j380.1\,(377/Q) + 377^2}\right| = \frac{1}{10}, \text{ or } \left|\frac{-2347}{-2347 + 143298j/Q}\right| = \left|\frac{1}{1 - \dfrac{61.06j}{Q}}\right| = \frac{1}{10}, \text{ or}$$

$\left[1 + (61.06/Q)^2\right] = 10^2$, $(\frac{61.06}{Q})^2 = 99$, $\frac{61.06}{Q} = 9.95$, $Q = \frac{61.06}{9.95} =$ **6.136**. Now the 3dB bandwidth is $\frac{f_o}{Q} = \frac{60}{6.14} =$ **9.78Hz.**

For the 3dB frequencies: $\left|\dfrac{(j\omega)^2 + 377^2}{(j\omega)^2 + j\omega\dfrac{377}{6.14} + 377^2}\right| = 0.707$, $\left|\dfrac{377^2 - \omega^2}{377^2 - \omega^2 + j\omega\, 61.4}\right| = 0.707$,

$\left|\dfrac{1}{1 + \dfrac{j\omega\, 61.4}{377^2 - \omega^2}}\right| = 0.707$, $1^2 + \left[\dfrac{61.4\omega}{377^2 - \omega^2}\right]^2 = \left[\dfrac{1}{.707}\right]^2 = 2$, $\dfrac{61.1\,\omega}{377^2 - \omega^2} = (2-1)^{½} = \pm 1$ - - - (1).

Now, $377^2 - \omega^2 = \pm 61.1\omega$, $\omega^2 \pm 61.1\omega - 377^2 = 0$, $\omega = \dfrac{\pm 61.1 \pm \sqrt{61.1^2 - -4\,(377)^2}}{2} = \dfrac{\pm 61.1 \pm 756.5}{2}$, of which the relevant solutions are $\dfrac{756.5 - 61.1}{2} = 347.7$ rad/s, and $\dfrac{756.5 + 61.1}{2} = 408.8$ rad/s, or **55.3Hz** and **65.06Hz**, for a 3dB bandwidth of 65.1 – 55.3 = **9.8Hz.** (As noted.)

For the 1dB frequencies: 20 log $x = -1 \rightarrow x = .891$V/V, and $\left|\dfrac{(j\omega)^2 + 377^2}{(j\omega)^2 + j\omega\dfrac{377}{6.14} + 377^2}\right| = 10^{-1/20} = 0.891$,

$\left|\dfrac{377^2 - \omega^2}{377^2 - \omega^2 + 61.4\,j\omega}\right| = .891$, $\left|\dfrac{1}{1 + \dfrac{61.4\omega}{377^2 - \omega^2}j}\right| = .891$, $1^2 + \left[\dfrac{61.4\omega}{377^2 - \omega^2}\right]^2 = \left[\dfrac{1}{.891}\right]^2 = 1.2596$,

$61.4\omega = \pm(377^2 - \omega^2)\,(.2596)^{½} = \pm(377^2 - \omega^2)\,.5096$, $\omega^2 \pm 120.5\omega - 377^2 = 0$,

$\omega = \dfrac{\pm 120.5 \pm \sqrt{120.5^2 + 4\,(377)^2}}{2} = \dfrac{\pm 120.5 \pm 763.6}{2}$, for which relevant solutions are $\dfrac{120.5 + 763.6}{2} =$ 442rad/s or **70.35Hz**, and $\dfrac{736.6 - 120.5}{2} = 308.1$ rad/s or **49.0Hz**.

For the 1% frequencies (from (1) above) $\dfrac{61.1\omega}{377^2 - \omega^2} = \left[\left(\dfrac{1}{.99}\right)^2 - 1\right]^{½} = \pm .142$, or $\omega^2 \pm 428.8\omega - 377^2 = 0$, $\omega = \dfrac{\pm 428.8 \pm \sqrt{428.8^2 + 4\,(377^2)}}{2} = \dfrac{\pm 428.8 \pm 754.3}{2}$, of which appropriate solutions are 591.6 rad/s or **94.1Hz**, and 162.8 rad/s or **25.9Hz**.

SECTION 11.5: The Second-Order LCR Resonator

11.20 Equation 11.29 indicates that: $\omega_1, \omega_2 = \omega_o \left[\left[1+\frac{1}{4Q^2}\right]^{½} \pm \frac{1}{2Q}\right]$.

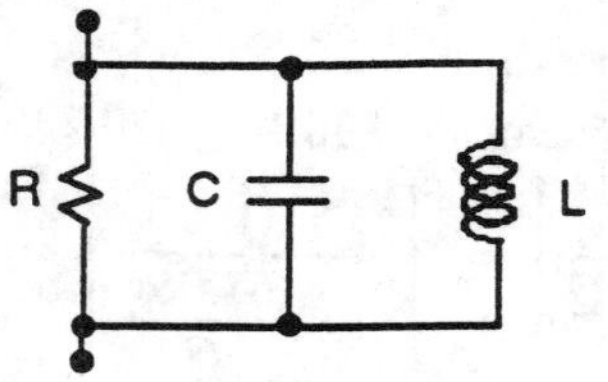

Now $BW = \frac{\omega_o}{Q} = 2\,\pi\,(20 \times 10^3)$, whence $Q = \frac{10^6}{20 \times 10^3} = 50$. Now $Q = \omega_o\,CR$, and $C = \frac{Q}{\omega_o\,R} = \frac{50}{2\,\pi \times 10^6 \times 10^4} =$ **796pF**. Accordingly, $L = \frac{1}{\omega_o^2\,C} = \frac{1}{(2\,\pi \times 10^6)^2 \times 796 \times 10^{-12}} =$ **264μH**. For a 1mA rms input at the 1MHz resonant frequency, $\upsilon_o = 10^4 \times 10^{-3} =$ **10V rms**.

11.21 Here, $\omega_o^2 = \frac{1}{LC} = (2\,\pi \times 99.9 \times 10^6)^2 = (6.2769 \times 10^8)^2 = 3.940 \times 10^{17}$, $L = \frac{1}{C \times 3.94 \times 10^{17}}$. Now the response is 3dB down at 99.9 −2 = 97.9MHz. Thus $\frac{f_o}{Q} = 2\,(2 \times 10^6)$, $Q = \frac{99.9 \times 10^6}{4 \times 10^6} = 24.98$. Now $Q = \omega_o\,CR$, $C = \frac{Q}{\omega_o\,R} = \frac{24.98}{2\,\pi \times 99.9 \times 10^6 \times 75} =$ **530.6pF**, and $L = \frac{1}{530.6 \times 10^{-12} \times 3.94 \times 10^{17}} =$ **0.0048μH**. For off-tuning by 100kHz, $\omega_1 = 2\,\pi\,(99.9 \times 10^6 + 0.1 \times 10^6) = 200\,\pi \times 10^6 = 6.2822 \times 10^8$rad/s. From Eq. 11.40 for the notch (with $a_2 = 1$), $T(s) = \frac{s^2+\omega_o^2}{s^2+s(\omega_o/Q)+\omega_o^2} = \frac{1}{1+\frac{s\,\omega_o/Q}{s^2+\omega_o^2}}$, and $T(\omega_1) =$

$$\frac{1}{1+\frac{j6.2832 \times 10^8\,(6.2769 \times 10^8)/24.98}{-(6.2832 \times 10^8)^2+(6.2769 \times 10^8)^2}} = \frac{1}{1-\frac{j1.579 \times 10^{16}}{.07913 \times 10^{16}}} = \frac{1}{1-j19.95}.$$

Now, $|T(\omega_1)| = \frac{1}{(1^2+19.95^2)^{½}} = .050$. Thus the attenuation expected would be −20 log (.05) = **26dB**.

11.22 For a maximally flat response, $Q = 1/\sqrt{2} = 0.707$. Now, for the high-pass filter, $T(s) = \frac{a_2\,s^2}{s^2+s\,(\omega_o/Q)+\omega_o^2}$ where $a_2 = 1$, and $T(j\omega) = \frac{-\omega^2}{-\omega^2+1.414j\omega\,\omega_o+\omega_o^2}$. For a 3dB frequency of 100kHz:

$$\left|\frac{-(10^5)^2}{-(10^5)^2+1.414j\,10^5\,f_o+f_o^2}\right| = \frac{1}{\sqrt{2}},\ \sqrt{2} \times 10^{10} = ((f_o^2-10^{10})^2+(1.414 \times 10^5)^2\,f_o^2)^{½},\text{ or } 2 \times 10^{20} =$$

$f_o^4 - 2 \times 10^{10}\,f_o^2 + 10^{20} + 2 \times 10^{10}\,f_o^2$, or $f_o^4 = (2-1)\,10^{20} = 10^{20}$, or $f_o = 10^5$Hz (as could be seen directly). Now, $LC = 1/(2\,\pi \times 10^5)^2 = 2.533 \times 10^{-12}$. For an ideal coil, $Q = \omega_o\,CR$, and $C = \frac{0.707}{10^4 \times 2\,\pi \times 10^5} =$ **112.5pF**, with $L = \frac{1}{\omega_o^2 \times C} = \frac{1}{(2\,\pi \times 10^5)^2 \times 112.5 \times 10^{-12}} =$ **22.5mH**. ow for the coil available, $Q = \frac{\omega_o\,L}{r}$, or $r = \frac{2\,\pi \times 10^5 \times 22.5 \times 10^{-3}}{50} =$ **283Ω** (in series with L). The equivalent parallel resistance is $R_p = \frac{Q}{\omega_o\,C} = \frac{50}{2\,\pi \times 10^5 \times 112.5 \times 10^{-12}} =$ **707kΩ**. Since $R_p >> R$, one can ignore it.

SECTION 11.6: Second-Order Active Filters Based on Inductor Replacement

11.23 For the Inductance-Simulator of Fig. 11.20, $L = C_4\,R_1\,R_3\,R_5/R_2$. Use $R_1 = R_3 = R_2 = 10$kΩ, and $R_5 \simeq 10$kΩ to accommodate the lack of capacitor choice. Now for $L = 10$H and $R_5 = 10$kΩ, $C_4 = \frac{L \times 10^4}{10^4 \times 10^4 \times R_5} = \frac{L}{10^4\,R_5} = \frac{L}{10^4 \times 10^4} = 10 \times 10^{-8} =$ **0.1μF**, and for $L = 0.1$H, $C_4 = \frac{0.1}{10} \times 0.1 =$ **1nF**.

Alternatively, for a fixed C, where $\frac{1}{2\pi fC} = 10\text{k}\Omega$, $C_4 = \frac{1}{2\pi \times 10^4 \times 10^3} = .0159\mu\text{F}$. Use **0.01μF = 10nF**. Now select $R_1 = R_3 = R_2 = 10\text{k}\Omega$ in which case $L = C_4 R_1 R_3 R_5/R_2 = 10^4 C_4 R_5$. *For* $L = 10\text{H}$, $R_5 = \frac{L}{10^4 C_4} = \frac{10H}{10^4 \times .01 \times 10^{-6}} =$ **100kΩ**. *For* $L = 0.1H$, $R_5 = \frac{0.1}{10^4 \times .01 \times 10^{-6}} =$ **1kΩ**.

11.24 From the solution to P11.23 above, for $L = 10\text{H}$, $R_1 = R_2 = R_3 = 10^4\Omega$, $R_5 = 10^5\Omega$, $C_4 = 10^{-8}\text{F}$, and for $L = 0.1\text{H}$, $R_1 = R_2 = R_3 = 10^4\Omega$, $R_5 = 10^3\Omega$, $C_4 = 10^{-8}\text{F}$. For $f_o = 1\text{kHz}$, $\omega_o = 2\pi \times 10^3 = 6.28 \times 10^3$ rad/s. Using $R_S = 20\text{k}\Omega$, with $\omega_o = (1/LC)^{1/2}$, or $C = \frac{1}{L\omega_o^2}$, and $Q = \omega_o CR$. *For* $L = 10\text{H}$, $C = \frac{1}{10 \times (6.28 \times 10^3)^2} = .00254\mu\text{F}$, and $Q = \omega_o CR = 6.28 \times 10^3 \times .00254 \times 10^{-6} \times 20 \times 10^3 = 0.319$. *For* $L = 0.1\text{H}$, $C = \frac{1}{0.1 \times (6.28 \times 10^3)^2} = .254\mu\text{F}$, and $Q = \omega_o CR = 6.28 \times 10^3 \times 0.254 \times 10^{-6} \times 20 \times 10^3 = 31.9$.

(a) Use $R_1 = R_2 = R_3 = \mathbf{10^4\Omega}$, $R_5 = \mathbf{10^3\Omega}$, $C_4 =$ **10nF**, and $C =$ **254nF**, to obtain a Q of **31.9**.

(b) Use $R_1 = R_2 = R_3 = \mathbf{10^4\Omega}$, $R_5 = \mathbf{10^5\Omega}$, $C_4 =$ **10nF**, and $C =$ **2.54nF**, to obtain a Q of **0.32**.

(c) For a design with equal-valued capacitors: For the simulated inductor $L = C_4 R_1 R_3 R_5/R_2$, and for $R_1 = R_3 = R_2 = \mathbf{10k\Omega}$, $L = 10^4 C_4 R_5$. Now with $C = C_4$, the resonant frequency is $\omega_o = (\frac{1}{L C_4})^{1/2} = 2\pi \times 10^3$ rad/s.

Thus $(C_4^2 R_5 \times 10^4)^{1/2} = \frac{1}{2\pi \times 10^3}$, and $R_5 = (\frac{1}{2\pi \times 10^3})^2 \times \frac{10^{-4}}{C_4^2} = \frac{2.533 \times 10^{-12}}{C_4^2}$, or $C_4 = (\frac{2.533 \times 10^{-12}}{R_5})^{1/2}$. Now, as $Q = \omega_o CR$, to raise Q, keep C_4 relatively large. Now, if R_5 is limited to $10^3\Omega$, $C_4 = (\frac{2.533 \times 10^{-12}}{10^3})^{1/2} = .0503\mu\text{F}$. (Use .05μF). Now for $C_4 =$ **.05μF**, $R_5 = \frac{2.533 \times 10^{-12}}{(.05 \times 10^{-6})^2} =$ **1.013kΩ**. In which case, $Q = \omega_o C_4 R = 2\pi \times 10^3 \times .05 \times 10^{-6} \times 20 \times 10^3 =$ **6.28**. It is apparent that if a smaller value of R_5 were allowed, $C_4 = C$ could be higher, and Q would be raised. For example, with $C_4 = C = 0.1\mu\text{F}$, $R_5 = \frac{2.533 \times 10^{-12}}{(0.1 \times 10^{-6})^6} =$ **253Ω**, and $Q = 2\pi \times 10^3 \times .1 \times 10^{-6} \times 20 \times 10^3 =$ **12.6**.

11.25 For a 5th-order Butterworth with 3dB bandwidth of 10^4Hz, $\varepsilon = 1$, and $\omega_p = 2\pi \times 10^4$rad/s, with a pole radius $\omega_o = \omega_p (1/\varepsilon)^{1/N} = 2\pi \times 10^4$rad/s. Now poles are at $90° - \frac{\pi}{2(5)} = 82°$, and $82° - \frac{\pi}{5} = 46°$, and 0°. Thus the first complex pole pair has a Q such that $\cos 82° = \frac{\omega_o/2Q}{\omega_o} = \frac{1}{2Q}$, or $Q = \frac{1}{2\cos 82°} = 3.593$. For the second complex pole pair, $Q = \frac{1}{2\cos 46°} = 0.7198$. For the 5th pole, $Q = 0.5$. Use a cascode of two circuits of the form shown in Fig. 11.22a with one of the form shown in Fig. 11.13a on the right. For a straightforward design, seven op amps would be needed. To achieve a low-frequency gain of 10, arrange a) a gain of 1 in one second-order section using a wire. b) a gain of 5 in the other second-order section using 2 series 33kΩ in the feedback path and 2 parallel 33kΩ to ground. c) a gain of 2 in the first-order section with one 33kΩ in the feedback and 2 parallel 33kΩ at the input.

Now for the first-order section (Fig. 11.13a), for which $\omega_o = \frac{1}{2\pi 10^4}$ and $Q = 0.5$, $R_1 =$ **33kΩ||33kΩ**, $R_2 =$ **33kΩ**, $C = \frac{1}{2\pi \times 10^4 \times 33 \times 10^3} = 482.2\text{pF}$. Use $C =$ **400pF||80pF**.

Now for one first second-order section (Fig. 11.22a), for which $\omega_o = \frac{1}{2\pi 10^4}$ and $Q = 3.593$, use $C_4 = C_6 = C =$ **400pF||80pF**, and $R_1 = R_2 = R_3 = R_5 =$ **33kΩ**, with $R_6 = QR = 3.593\ (33) = 118.6\text{k}\Omega$. Use **120kΩ**.

Now for the other second-order section (Fig. 11.22a), for which $\omega_o = \dfrac{1}{2\pi\, 10^4}$ and $Q = .7198$, use $C_4 = C_6 = C =$ **400pF || 80pF**, and $R_1 = R_2 = R_3 = R_5 =$ **33kΩ**, with $R_6 = QR = .7198\,(33) = 23.75\text{k}\Omega$. Use **24kΩ**.

11.26 From Table 11.1 for the BP filter, $T(s) = \dfrac{Ks/C_6 R_6}{s^2 + \dfrac{s}{C_6 R_6} + \dfrac{R_2}{C_4 C_6 R_1 R_3 R_5}}$. Now, $\dfrac{\omega_o}{Q} = \dfrac{1}{C_6 R_6}$, $\omega_o^2 = \dfrac{R_2}{C_4 C_6 R_1 R_3 R_5}$. Use $C_4 = C_6 = C$, with **R_6 to control Q** (and ω_o), and $R_1 = R_2 = R_3 = R$, with **R_5 to control ω_o**. Thus $\omega_o^2 = \dfrac{1}{C^2 R R_5}$, $\omega_o = \dfrac{1}{C\sqrt{R R_5}}$ – – – (1), and $Q = \omega_o C R_6 = \dfrac{R_6}{\sqrt{R R_5}}$ – – – (2). For capacitors using single digits 1, 2, 3, 5, the largest range of C to be accommodated by R_5 in establishing ω_o is 2/1 = 2. Since from (1), $R_5 = \dfrac{1}{\omega_o^2 C^2 R}$, the compensating range of R_5 will be $2^2 =$ **4**. Now from (2), $R_6 = \sqrt{R R_5}\, Q$, and for Q varying from 0.5 to 50 (that is by a factor of 100), and R_5 by a factor of 4, R_6 must vary by a factor of $\sqrt{4}\,(100) =$ **200**.

SECTION 11.7: Second-Order Active Filters Based on the Two-Integrator-Loop Topology

11.27 From Fig. 11.16c) for a BP filter, see that the 3dB bandwidth is $\dfrac{\omega_o}{Q}$. Here, $\omega_o = 2\pi \times 10^4$, and $\dfrac{\omega_o}{Q} = 500 \times 2\pi$, or $Q = \dfrac{10^4}{500} = \dfrac{100}{5} = 20$. Now for Fig. 11.24a, $CR = \dfrac{1}{\omega_o}$, $\dfrac{R_3}{R_2} = 2Q - 1$, and $\dfrac{R_f}{R_1} = 1$. For C = **1nF**, $R = \dfrac{1}{2\pi \times 10^4 \times 1 \times 10^{-9}} =$ **15.92kΩ**. From Eq.11.59, $T_{bp}(s) = \dfrac{-K\,\omega_o\, s}{s^2 + s\,(\omega_o/Q) + \omega_o^2}$. Thus at ω_o, $\dfrac{V_{bp}}{V_i} = \dfrac{-K\,\omega_o\, j\omega_o}{-\omega_o^2 + j\omega_o\,(\omega_o/Q) + \omega_o^2} = -KQ$. Now, from Eq. 11.64, $K = 2 - (1/Q)$. Thus $\dfrac{V_{bp}}{V_i} = -Q\,(2 - 1/Q) = -2Q + 1 = -2\,(20) + 1 = -39$V/V. Now using Miller's theorem at resonance (ω_o), $R_{in} = \dfrac{(R_2 + R_3)}{1 - gain} = \dfrac{R_2 + R_3}{1 - -39} = \dfrac{R_2 + R_3}{40}$. Now, $\dfrac{R_2 + R_3}{40} = 100\text{k}\Omega$ by specification. Thus $R_2 + R_3 = 4\text{M}\Omega$. Also $\dfrac{R_3}{R_2} = 2Q \times 1 = 39$ as well, and $R_3 = 39\,R_2$. Thus $R_2 + 39\,R_2 = 4\text{M}\Omega$, whence $R_2 = 100\text{k}\Omega$ and $R_3 = 3.9\text{M}\Omega$, unfortunately too large. To reduce R_3, replace the original $R_2\, R_3$ circuit by a network.

In the original situation, $V_x = V_i - \dfrac{R_2}{R_2 + R_3}(V_i - V_{bp})$

$= V_i \left[1 - \dfrac{R_2}{R_2 + R_3}(1 - -39)\right] = V_i \left[1 - \dfrac{100k\Omega}{4M\Omega}(40)\right] = 0$. Thus the positive input node of the leftmost amplifier is a virtual ground (as *might* have been obvious already). Thus use the network as shown:

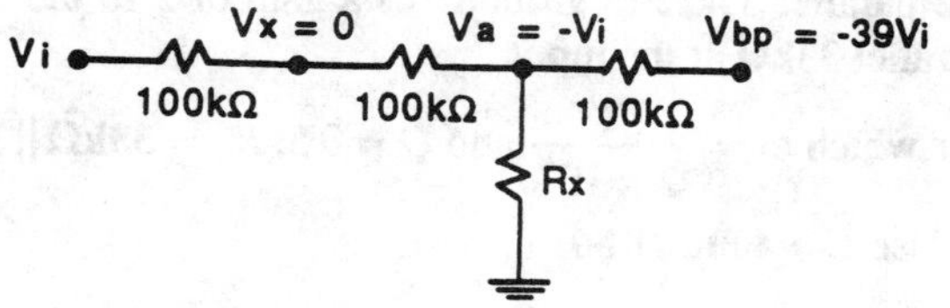

Note that since $V_x = 0$, the voltage $V_a = -V_i$, $\dfrac{R_x \| 100}{100 + R_x \| 100} = \dfrac{1}{39}$, $100 + R_x \| 100 = 39\,(R_x \| 100)$, $\dfrac{38\,R_x\,(100)}{100 + R_x} = 100$, $38\,R_x = 100 + R_x$, $37\,R_x = 100$, $R_x = 100/37 =$ **2.7kΩ**. Use $R_2 = R_{3a} = R_{3b} = R_1 = R_f =$ **100kΩ**. The center-frequency gain is **–39V/V**.

11.27 (Contined)

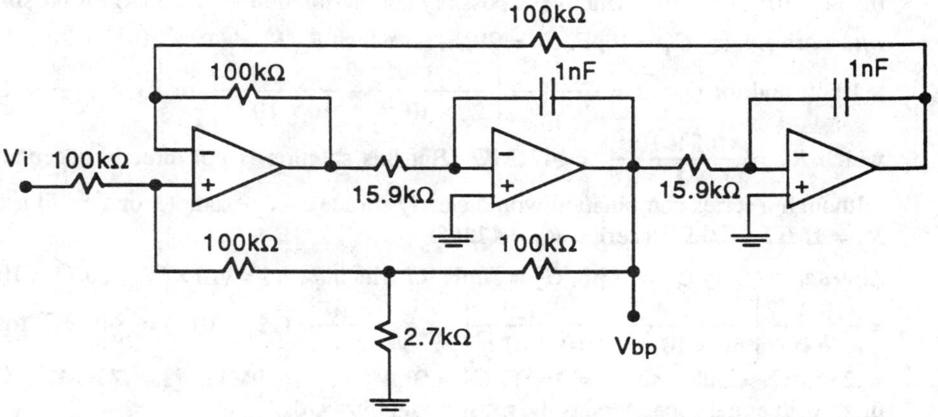

11.28 From Ex. 11.22, for f_o = 5kHz, f_n = 8kHz, Q = 5, dc gain = 3V/V, and C = 1nF, one used R = 31.83kΩ, R_1 = R_f = R_2 = 10kΩ, R_3 = 90kΩ, R_H = 25.6kΩ, R_F = 42.7kΩ, R_B = ∞, R_L = 10kΩ. Here, f_o = 5kHz, f_n = 7.5kHz, Q = 10, and dc gain = 3V/V. From Eq. 11.67, $\frac{R_H}{R_L} = (\frac{\omega_n}{\omega_o})^2 = (\frac{7.5}{5})^2 = 2.25$. For R_L = **10kΩ**, R_H = **22.5kΩ**, R_B = ∞. From Eq. 11.6, dc gain = $\frac{-K\,R_F/R_L\,\omega_o^2}{\omega_o^2} = \frac{-K\,R_F}{R_L}$. Here $K = 2 - 1/Q$, with Q = 10. Thus $K = 2 - 1/10 = 1.9$, and $R_F = \frac{3R_L}{K} = \frac{3 \times 10k\Omega}{1.9}$ = **15.79kΩ**. Use **C = 1nF**, $R = \frac{1}{2\pi \times 5 \times 10^3 \times 10^{-9}}$ = **31.83kΩ**. Use $R_1 = R_f = R_2$ = **10kΩ**, and $R_3 = R_2\,(2Q - 1) = 10\,(2\,(10) - 1)$ = **190kΩ**.

11.29 Required to design a bandpass filter with f_o = 10kHz, f_{3dB} = 0.5kHz, R_{in} = 100kΩ using C = 1nF, with a center-frequency gain of −39V/V. Thus, for the Tow-Thomas circuit of Fig. 11.26, chose the negative bandpass version, for which $C_1 = 0$, $R_2 = R_3 = \infty$, and $R_1 = \frac{QR}{\text{center-frequency gain}}$. Here, from Fig. 11.16c, $f_{3dB} = \frac{f_o}{Q}$, whence $Q = \frac{f_o}{f_{3dB}} = \frac{10}{0.5} = 20$. Thus $R_1 = \frac{20R}{39}$, but the input resistance = $R_1 \geq$ 100kΩ. Thus $R = \frac{39}{20} R_1$ = 195kΩ. Use R = **200kΩ** for a slightly higher value of $R_{in} = R_1 = \frac{20}{39}(200)$ = **102.6kΩ**. Now $\omega_o = \frac{1}{RC}$, and $C = \frac{1}{2\pi \times 10^4 \times 20 \times 10^3}$ = **769pF**. As well, use r = **100kΩ**, and QR = 4MΩ.

SECTION 11.8: Single-Amplifier Biquadratic Active Filters

11.30 Eq. 11.73 and 11.74 indicate that $\omega_o = (C_1\,C_2\,R_3\,R_4)^{-1/2}$ and $Q = \left[\frac{(C_1\,C_2\,R_3\,R_4)^{1/2}}{R_3}\,(1/C_1 + 1/C_2)\right]^{-1}$

Try $R_3 = R_4$ = 1MΩ and $C_1 = C_2 = C$. Now for $\omega_o = 10^5$ rad/s, $C = \frac{1}{10^5 \times 10^6}$ = 10pF, $Q = 10^5 \times 10^6$ $[1/10^{-11} + 1/10^{-11}]^{-1} = 10^{11} \times [2 \times 10^{11}]^{-1} = 1/2$ (as expected). Now since the required Q is 0.707, C must be raised to allow resistor values which are different, yet no larger than 1MΩ. Thus we could try C = 20pF and proceed to find $R_3\,R_4$, then each separately, as follows: See $R_3\,R_4 = (C_1\,C_2\,\omega_o^2)^{-1} = (20 \times 10^{-12} \times 10^5)^{-2} = (20 \times 10^{-7})^{-2} = \frac{10^{14}}{20 \times 20} = .25 \times 10^{12}$. Now for $Q = 1/\sqrt{2}$, $\sqrt{2} = \frac{(10^5)^{-1}}{R_3}(\frac{2}{20 \times 10^{-12}})$, or $R_3 = \frac{10^{-5}}{\sqrt{2}} \times 10^{11}$ = .707MΩ, for which $R_4 = \frac{.25 \times 10^{12}}{.707 \times 10^6}$ = .354MΩ. Thus, $C_1 = C_2$ = **20pF**, R_3 = **.707MΩ** and $R_4 = R_3/2$ = **.354MΩ** is a possible solution. *Check*: $\omega_o = (20 \times 10^{-12} \times 20 \times 10^{-12} \times .707 \times 10^6 \times$

$0.354 \times 10^6)^{-1/2} = 10^5$. But this is possibly not the solution with the largest possible resistors! *Alternatively*, try C_1 = 10pF, C_2 = 20pF, for which $R_3 R_4 = (10 \times 10^{-12} \times 20 \times 10^{-12} \times 10^5 \times 10^5)^{-1} = 0.5 \times 10^{-12}$, and for Q, $\sqrt{2} = \frac{(10^5)^{-1}}{R_3}\left[\frac{1}{20\times 10^{-12}} + \frac{1}{10\times 10^{-12}}\right]$, or $R_3 = \frac{10^{-5}}{\sqrt{2}} \times 10^{11}\,(1.5) = 1.061\text{M}\Omega$, for which $R_4 = \frac{0.5\times 10^{12}}{1.061\times 10^6} = .471\text{M}\Omega$. But this solution is not directly acceptable since R_3 is too large (although a series combination would clearly suffice). For example, one could use C_1 = **10pF**, C_2 = **20pF**, R_3 = **1MΩ + 62kΩ** in series, R_4 = **470kΩ**.

Alternatively, try C_1 = 10pF, C_2 = 50pF, for which $R_3 R_4 = (10 \times 50 \times 10^{-24} \times 10^{10}) = 0.2 \times 10^{12}$, and $\sqrt{2} = \frac{(10^5)^{-1}}{R_3}\left[\frac{1}{50\times 10^{-12}} + \frac{1}{10\times 10^{-12}}\right]$, or $R_3 = \frac{10^{-5}}{\sqrt{2}}(1.2 \times 10^{11}) = .849\text{M}\Omega$, for which $R_4 = \frac{0.2\times 10^{12}}{.849\times 10^6} = .236\text{M}\Omega$. Could use C_1 = **10pF**, C_2 = **50pF**, R_3 = **.849MΩ**, R_4 = **.236MΩ**. Of the three solutions, the first, with equal capacitors, is the most straightforward.

11.31 Now, for $\omega_o = 2\pi \times 10^4$rad/s, $Q = \frac{10^4}{5\times 10^2} = 20$. Use $C_1 = C_2$ = 1nF = 10^{-9}F. From $\omega_o = (C_1 C_2 R_3 R_4)^{-1/2}$, $R_3 R_4 = \frac{1}{(2\pi\times 10^4 \times 10^{-9})^2} = 2.533 \times 10^8$. From $Q = \omega_o R_3 [1/C_1 + 1/C_2]^{-1}$, $R_3 = \frac{20}{10^4 \times 2\pi}(\frac{2}{10^{-9}}) = 6.366 \times 10^5$ = **637Ω**, and $R_4 = \frac{2.533\times 10^8}{6.366\times 10^5}$ = **398Ω**. *Check*:

$$Q = \left[\frac{(C_1 C_2 R_3 R_4)^{1/2}}{R_3}(1/C_1 + 1/C_2)\right]^{-1} = \left[\frac{(10^{-9}\times 10^{-9}\times 6.37\times 10^5\times 398)^{1/2}}{6.36\times 10^5}(1/10^{-9} + 1/10^{-9})\right]^{-1} =$$

$\left[\frac{159.2\times 10^{-7}}{6.36\times 10^5}\times\frac{2}{10^{-9}}\right]^{-1}$ = 19.97: **OK**. Now, *for the center frequency gain*: For V_A (at the common node A of C_1, C_2, R_4/α and $R_4/(1-\alpha)$), with the voltage at the amplifier's negative input being zero volts, see $V_A = 0 - \frac{1}{sC_2}\frac{V_o}{R_3} = -\frac{V_o}{s C_2 R_3}$. Writing a node equation at node A yields: $\frac{V_o}{R_3} + s C_1\left[V_o + \frac{V_o}{s C_2 R_3}\right] - \frac{V_A}{R_4/(1-\alpha)} + \frac{V_i - V_A}{R_4/\alpha} = 0$, or $\frac{V_o}{R_3} + s C_1 V_o + V_o\frac{C_1}{C_2 R_3} + \frac{(1-\alpha)V_o}{s C_2 R_3 R_4} + \frac{\alpha V_o}{s C_2 R_3 R_4} = \frac{-V_i}{R_4/\alpha}$, or $\frac{V_o}{V_i} = \frac{-\alpha/R_4}{s C_1 + \frac{1}{R_3} + \frac{C_1}{R_3 C_3} + \frac{1}{s C_2 R_3 R_4}}$

$$= \frac{-s\frac{\alpha}{C_1 R_4}}{s^2 + s(\frac{1}{C_1 R_3} + \frac{1}{C_2 R_3}) + \frac{1}{C_1 C_2 R_3 R_4}}$$, for which the center-frequency gain is

$-\left[\frac{\alpha}{C_1 R_4}\right]\left[\frac{1}{C_1 R_3} + \frac{1}{C_2 R_3}\right] = \frac{-\alpha R_3 C_2}{R_4 (C_1 + C_2)}$. Here, $1 = \frac{\alpha R_3 C_2}{R_4(C_1 + C_2)}$, whence $\alpha = \frac{R_4(C_1 + C_2)}{R_3 C_2} = \frac{398\,(10^{-9} + 10^{-9})}{637\times 10^3\,(10^{-9})} = 1.25 \times 10^{-3}$. Now, $\frac{R_4}{\alpha} = \frac{398}{1.25\times 10^{-3}}$ = **318.4kΩ**, and $\frac{R_4}{1-\alpha} = \frac{398}{1 - 1.25\times 10^{-3}}$ = **398.5Ω**. As a result, at very high frequencies (where the capacitors are short circuits), $R_{in} \approx$ **318kΩ**, and at very low frequencies (where the capacitors are open), R_{in} = 318kΩ + 398Ω ≈ **318kΩ**, as well.

11.32 From the derivation in P11.31 above: $$\frac{V_o}{V_i} = T(s) = \frac{-s\frac{\alpha}{C_1 R_4}}{s^2 + s\left[\frac{1}{C_1 R_3} + \frac{1}{C_2 R_3}\right] + \frac{1}{C_1 C_2 R_3 R_4}} \quad \text{---(1)},$$

where the voltage at the join-point A of C_1, C_2, $\frac{R_4}{\alpha}$, and $\frac{R_4}{(1-\alpha)}$, is V_A. Note that the current in R_3 establishes the voltage across C_2, since their join-point is virtual-ground point with no current loss. Thus $V_A = \frac{-V_o}{s C_2 R_3}$. Now the input current is $I_i = \frac{V_i - V_A}{R_4/\alpha} = \frac{V_i + \frac{V_o}{s C_2 R_3}}{R_4/\alpha}$, and the input impedance is $Z_i(s) = \frac{V_i}{I_i} = \frac{R_4/\alpha}{1 + \frac{T(s)}{s C_2 R_3}}$. Thus $Y_i(s) = 1/Z_i(s)$

$$= \frac{\alpha}{R_4}\left[1 - \frac{\alpha/(C_1 C_2 R_3 R_4)}{s^2 + \frac{s}{R_3}\left[\frac{1}{C_1} + \frac{1}{C_2}\right] + \frac{1}{C_1 C_2 R_3 R_4}}\right] = \frac{\alpha}{R_4}\,\frac{s^2 + s\left[\frac{1}{R_3 C_1} + \frac{1}{R_3 C_2}\right] + \frac{1-\alpha}{C_1 C_2 R_3 R_4}}{s^2 + \frac{s}{R_3}\left[\frac{1}{C_1} + \frac{1}{C_2}\right] + \frac{1}{C_1 C_2 R_3 R_4}}$$

$$\text{Thus } \mathbf{Z_1(s)} = \frac{\mathbf{R_4}}{\alpha} \times \frac{\mathbf{s^2} + \frac{\mathbf{s}}{\mathbf{R_3}}\left[\frac{1}{\mathbf{C_1}} + \frac{1}{\mathbf{C_2}}\right] + \frac{1}{\mathbf{C_1 C_2 R_3 R_4}}}{\mathbf{s^2} + \frac{\mathbf{s}}{\mathbf{R_3}}\left[\frac{1}{\mathbf{C_1}} + \frac{1}{\mathbf{C_2}}\right] + \frac{1-\alpha}{\mathbf{C_1 C_2 R_3 R_4}}}$$

Now (from Equations 11.73, 11.74), $\omega_o = 1/(C_1 C_2 R_3 R_4)^{1/2}$, $Q = \left[\frac{(C_1 C_2 R_3 R_4)^{1/2}}{R_3}\left[\frac{1}{C_1} + \frac{1}{C_2}\right]\right]^{-1}$, or

$1/Q = \frac{1}{\omega_o R_3}\left[\frac{1}{C_1} + \frac{1}{C_2}\right]$. Thus $\mathbf{Z_1(s)} = \frac{\mathbf{R_4}}{\alpha} \times \frac{\mathbf{s^2} + \frac{\omega_o}{\mathbf{Q}}\mathbf{s} + \omega_o^2}{\mathbf{s^2} + \mathbf{s}\left(\frac{\omega_o}{\mathbf{Q}}\right) + (1-\alpha)\omega_o^2}$. Now at very low frequencies,

as $s \to 0$, $Z_i = \frac{R_4}{\alpha} \times \frac{\omega_o^2}{(1-\alpha)\omega_o^2} = \frac{\mathbf{R_4}}{\alpha\,(1-\alpha)}$, as can be seen quite directly with capacitors open, noting that $\frac{R_4}{\alpha} + \frac{R_4}{1-\alpha} = \frac{R_4}{\alpha\,(1-\alpha)}$. Now, at very high frequencies, as $s \to \infty$, $Z_i = \mathbf{R_4}/\alpha$, with capacitors shorted.

Now at the center frequency, $s = -j\omega_o$, $Z_i = \frac{R_4}{\alpha} \times \frac{-\omega_o^2 + j\frac{\omega_o^2}{Q} + \omega_o^2}{-\omega_o^2 + j\frac{\omega_o^2}{Q} + (1-\alpha)\,\omega_o^2} = \frac{R_4}{\alpha}\,\frac{j\,\omega_o^2/Q}{j\,\omega_o^2/Q - \alpha\,\omega_o^2} =$ $\frac{\mathbf{R_4}}{\alpha} \times \frac{1}{1 - \mathbf{j}\alpha\mathbf{Q}} = \frac{R_4\,(1 + j\,\alpha\,Q)}{\alpha\,(1 + \alpha^2 Q^2)}$, which, for $\alpha\,Q$ large, is $\frac{R_4}{\alpha}\left[\frac{1}{\alpha^2 Q^2} + \frac{j}{\alpha\,Q}\right]$, of relatively low magnitude, and with a 90° (inductor-like) phase.

11.33 For this Butterworth, $N = 7$, $A_{max} = 3\text{dB}$, $A_{dc} = 0\text{dB}$. Here $f_p = 5\text{kHz}$, and $\omega_p = 2\pi(5) = 10^5\pi$ rad/s, and $\varepsilon = (10^{A_{max}/10} - 1)^{1/2} = (10^{3/10} - 1)^{1/2} = 0.998$. (It should actually be 1.000. Why? Why is it not?). Thus for each filter stage, $\omega_o = \omega_p \left[\frac{1}{\varepsilon}\right]^{1/7}$, or $\omega_o = 10^5\pi \left[\frac{1}{.998}\right]^{1/7} = 3.14 \times 10^5$ rad/s. The 7 poles are located at 0°, ± 180/7 = ±25.7°, ±51.4°, and ±77.14°. For each pole pair, $\cos\Theta = \frac{\omega_o/2Q}{\omega_o} = 1/2Q$, whence $Q = 0.5/\cos\Theta =$ 0.5/cos 0 = 0.5, 0.5/cos 25.7 = .555, 0.5/cos 51.4 = 0.801, and 0.5/cos 77.14 = 2.25 respectively. From Ex. 11.28, see that for the Sallen and Key circuit of Fig. 11.34(c), that the dc gain is 1, and from Eq.11.77 and

Eq. 11.78 that $\omega_o = (R_1R_2C_3C_4)^{-1/2}$, and $1/Q = \frac{1}{\omega_o C_4}\left(\frac{1}{R_1}+\frac{1}{R_2}\right)$. Use C = 3.3nF for C_1, C_2 for all sections.

Now for the first-order section, $\omega_o = 1/RC$, and $R = \frac{1}{\omega_o C} = \frac{1}{3.14\times10^5\times3.3\times10^{-9}} = 965\Omega$ or **0.965kΩ.**

For the lowest-Q Sallen and Key Section, $1/Q = 1/.555 = \frac{1}{3.14\times10^5\times3.3\times10^{-9}}\left(\frac{1}{R_1}+\frac{1}{R_2}\right)$, or $1/R_1 + 1/R_2 = \frac{1}{.555}\times\frac{1}{965} = .001868$ – – – (1). Also $\omega_o = (R_1R_2C^2)^{-1/2}$ implies that $R_1 R_2 = \frac{1}{\omega_o^2 C^2} = R^2 = 965^2$, or $R_1 = \frac{965^2}{R_2}$. Thus in (1) $\frac{R_2}{965^2}+\frac{1}{R_2} = \frac{1}{.555}\times\frac{1}{965}$. Therefore $R_2^2 + 965^2\,R_2 - \frac{965}{.555} = 0$ - - - (2), or for R_2 in kilohm, $R_2^2 + .9312\,R_2 - 1.739 = 0$, $R_2 = \frac{-.9312\pm\sqrt{.9312^2+4(1.739)}}{2} = \frac{.9312\pm2.797}{2}$ = **.933kΩ**, for which $R_1 = \frac{965^2}{.933}$ = **.998kΩ.**

Now for the second Sallen and Key (by analogy from (2)), $R_2^2 + .965^2\,R_2 - \frac{.965}{.801} = 0$, $R_2^2 + .9312\,R_2 - 1.205 = 0$, $R_2 = \frac{-.9312\pm\sqrt{.9312^2+4(1.205)}}{2}$ = **.727kΩ**, for which $R_1 = \frac{.965^2}{.727}$ = **1.281kΩ.**

Now for the third Sallen and Key, $R_2^2 + .965^2\,R_2 - \frac{.965}{2.25} = 0$, $R_2^2 + .9312\,R_2 - .429 = 0$, $R_2 = \frac{-.931\pm\sqrt{.931^2+4(.429)}}{2}$ = **0.348kΩ**, for which $R_1 = \frac{.965^2}{.348}$ = **2.68kΩ**, with all capacitors of 3.3nF value.

SECTION 11.9: Sensitivity

11.34 From Eq. 11.77 and 11.78, $\omega_o = (C_3\,C_4\,R_1\,R_2)^{-1/2}$, and $Q = \left[\frac{(C_3\,C_4\,R_1\,R_2)^{1/2}}{C_4}\left(\frac{1}{R_1}+\frac{1}{R_2}\right)\right]^{-1}$. **Now for** ω_o, see $\frac{\partial\,\omega_o}{\partial\,C_3} = -1/2\,(C_4\,R_1\,R_2)^{-1/2}\,C_3^{-3/2} = -\frac{\omega_o}{2\,C_3}$, and $S_{C_3}^{\omega_o} = \frac{\partial\,\omega_o}{\partial\,C_3}\times\frac{C_3}{\omega_o} = -\frac{\omega_o}{2\,C_3}\times\frac{C_3}{\omega_o} = -1/2$. Now $C_3 = C_{3a} + C_{3b}$ and $C_{3b} = k_3\,C_{3a}$, nominally. However, note that C_{3a} and C_{3b} are independent from a sensitivity point of view. Thus $\frac{\partial\,C_3}{\partial\,C_{3a}} = 1$. Now $S_{C_{3a}}^{\omega_o} = \frac{\partial\,\omega_o}{\partial\,C_{3a}}\times\frac{C_{3a}}{\omega_o} = \frac{\partial\,\omega_o}{\partial\,C_3}\times\frac{\partial\,C_3}{\partial\,C_{3a}}\times\frac{C_{3a}}{\omega_o} = -\frac{\omega_o}{2\,C_3}\times1\times\frac{C_{3a}}{\omega_o} = -\frac{1}{2}\,\frac{C_{3a}}{C_3} = -\frac{1}{2}\,\frac{C_{3a}}{(1+k_3)\,C_{3a}} = \frac{-1}{2\,(1+k_3)}$, and $S_{C_{3b}}^{\omega_o} = -\frac{1}{2}\,\frac{C_{3b}}{C_3} = -\frac{1}{2}\,\frac{k_3\,C_{3a}}{(1+k_3)\,C_{3a}} = \frac{-k_3}{2\,(1+k_3)}$. Correspondingly, $S_{C_{4a}}^{\omega_o} = -\frac{1}{2\,(1+k_4)}$, and $S_{C_{4b}}^{\omega_o} = -\frac{k_4}{2\,(1+k_4)}$. When $k_3 = k_4 = 1$, all these sensitivities become **–1/4.**

Now for Q, see $\frac{\partial\,Q}{\partial\,C_3} = \left[\frac{(C_3\,C_4\,R_1\,R_2)^{1/2}}{C_4}\left(\frac{1}{R_1}+\frac{1}{R_2}\right)\right]^{-1}\left[\frac{-1/2}{C_3}\right] = -1/2\,\frac{Q}{C_3}$. Thus $S_{C_3}^{Q} = -1/2$, where $S_{C_{3a}}^{Q} = -\frac{1}{2\,(1+k_3)}$, and $S_{C_{3b}}^{Q} = -\frac{k_3}{2\,(1+k_3)}$, both being **–1/4** for $k_3 = 1$. See also $\frac{\partial\,Q}{\partial\,C_4} = -\frac{-1/2\,Q}{C_4}$. Thus $S_{C_4}^{Q} = 1/2$, and $S_{C_{4a}}^{Q} = \frac{1}{2\,(1+k_4)}$, $S_{C_{4b}}^{Q} = \frac{k_4}{2\,(1+k_4)}$, both being **1/4** for $k_4 = 1$.

11.35 From Equations 11.77 and 11.78, $\omega_o = (C_3\,C_4\,R_1\,R_2)^{-1/2}$, and $Q = \left[\frac{(C_3\,C_4\,R_1\,R_2)^{1/2}}{C_4}\left(\frac{1}{R_1}+\frac{1}{R_2}\right)\right]^{-1}$.

Now for ω_o, see $\dfrac{\partial \omega_o}{\partial R_1} = -1/2\,(C_3\,C_4\,R_2)^{-1/2} \times R_1^{-3/2} = -\dfrac{\omega_o}{2R_1}$, and $S^{\omega_o}_{R_1} = \dfrac{\partial \omega_o}{\partial R_1} \times \dfrac{R_1}{\omega_o} = -\dfrac{\omega_o}{2R_1} \times \dfrac{R_1}{\omega_o}$ $= -\,1/2$. Likewise $s^{\omega_o}_{R_2} = -\,1/2$ as well.

Now for a fixed temperature, $R_1, R_2 = (1 \pm k)R$ with $\dfrac{\partial R_1, R_2}{\partial k} = \pm R$, and $S_k^{R_1, R_2} = \pm R \times \dfrac{k}{R} = \pm k$. Thus $S_k^{\omega_o} = S^{\omega_o}_{R_1}\,S_k^{R_1} + S^{\omega_o}_{R_2}\,S_k^{R_2} = -1/2\,(+k + (-k)) = \mathbf{0}$.

Now for small k, $R_1 \approx R_2 = R_0\,(1 + a\,(T_o - T))$, whence $\dfrac{\partial R_1}{\partial T} = -\,a\,R_o$ and $S_T^{R_1} = \dfrac{\partial R_1}{\partial T} \times \dfrac{T}{R_1} = -\,aR_o \times$ $\dfrac{T}{R_o\,(1 + a\,(T_o - T))} = -\dfrac{a\,T}{1 + a\,(T_o - T)} = S_T^{R_2}$. Thus $S_T^{\omega_o} = S^{\omega_o}_{R_1} \times S_T^{R_1} + S^{\omega_o}_{R_2} \times S_T^{R_2}$ $= 2\left[-\,1/2 \times \dfrac{-\,a\,T}{1 + a\,(T_o - T)}\right]$, which around $T = T_o$, is **a $\mathbf{T_o}$**.

Now for Q, see $Q = \left[\dfrac{C_4}{C_3\,R_1\,R_2}\right]^{1/2}\left[\dfrac{R_1\,R_2}{R_1 + R_2}\right] = \left[\dfrac{C_4}{C_3}\right]^{1/2}\dfrac{(R_1\,R_2)^{1/2}}{R_1 + R_2}$. Thus $\dfrac{\partial Q}{\partial R_1} = \left[\dfrac{C_4\,R_2}{C_3}\right]^{1/2} \times$ $\left[\dfrac{1/2}{R_1 + R_2} \times \dfrac{1}{R_1^{1/2}} + (-1) \times \dfrac{R_1^{1/2}}{(R_1 + R_2)^2}\right] = \left[\dfrac{C_4}{C_3}\right]^{1/2}\dfrac{(R_1\,R_2)^{1/2}}{R_1 + R_2}\left[\dfrac{1}{2R_1} - \dfrac{1}{R_1 + R_2}\right] = Q \times$ $\dfrac{(R_2 - R_1)}{2R_1\,(R_1 + R_2)}$. Now for $R_2 = R_1$, $\dfrac{\partial Q}{\partial R_1} = 0$, and $S^Q_{R_1} = S^Q_{R_2} = 0$. But for $R_1, R_2 = (1 \pm k)R$ with $k << 1$, $\dfrac{\partial Q}{\partial R_1} \approx Q \times \dfrac{-2\,kR}{2R\,(2R)} = -\dfrac{k\,Q}{2R}$, with $S^Q_{R_1} = \dfrac{-k\,Q}{2R} \times \dfrac{R}{Q} = -\,k/2$. Likewise $\dfrac{\partial Q}{\partial R_1} = \dfrac{k\,Q}{2R}$, with $S^Q_{R_2} = +k/2$.

Now, $S_k^Q = S^Q_{R_1}\,S_k^{R_1} + S^Q_{R_2}\,S_k^{R_2}$, where here, $R_1, R_2 = (1 \pm k)R$, and $\dfrac{\partial R_1, R_2}{\partial k} = \pm R$, with $S_k^{R_1, R_2} = \pm R \times \dfrac{k}{R}$ $= \pm k$. Thus $S_k^Q = -k/2(k) + k/2\,(-k) = -\,\mathbf{k^2}$. Also $S_T^Q = S^Q_{R_1}\,S_T^{R_1} + S^Q_{R_2}\,S_T^{R_2}$, where $R = R_o\,(1 + a\,(T_o - T)$, and $\dfrac{\partial R_1, R_2}{\partial T} = -\,(1 \pm k)\,R_o\,a$, with $S_T^{R_1, R_2} = -\,(1 \pm k)\,R_o\,a \times \dfrac{T}{(1 \pm k)\,R_o\,(1 + a\,(T_o - T)}$ $= -\dfrac{a\,T}{1 + a\,(T_o - T)}$. Thus $S_T^Q = -\,k/2\left[\dfrac{-a\,T}{1 + a\,(T_o - T)}\right] + k/2\left[\dfrac{-a\,T}{1 + a\,(T_o - T)}\right] = \mathbf{0}$.

SECTION 11.10: Switched - Capacitor Filters

11.36 For Φ_1 high, C_1 is discharged to zero (that is, virtual ground (VG) via the second switch. Then, for Φ_1 high, the first switch closes, charging C_1 to 1V, for a total charge transfer of $Q = CV = 0.1 \times 10^{-1/2} \times 1 =$ **0.1pC**. For 1MHz operation, the corresponding average current is $I = 0.1 \times 10^{-12} \times 10^6 =$ **0.1μA**. Equivalent input resistance is $\dfrac{V}{I} = \dfrac{1V}{0.1\mu A} =$ **10MΩ**. For a 2pF feedback capacitance, 0.1pC charge produces an output change of $V = \dfrac{Q}{C} = \dfrac{0.1 \times 10^{-12}}{2 \times 10^{-12}} =$ **0.05V**. Thus the output change per cycle would be **50mV**. For $\upsilon_i = +1$V, C_1 charges positive when Φ_1 occurs. When Φ_2 occurs, VG tends to go positive, forcing the output negative to compensate. Thus the output changes in the negative direction. For saturation at ±10V, the maximum output change is 20 volts, requiring $\dfrac{20V}{50mV} =$ **400 cycles**. The average slope is −20V in 400 μs, or $-\dfrac{20}{400} = -.05$V/μs, or −50V/ms, or **−50,000V/s**. For a −0.1V input, the output slope becomes +5V/ms or **5000V/s**.

11.37 Require $f_{3dB} = f_o = 10^5$Hz, $Q = 1/\sqrt{2} = 0.707$, and $A_o = 1$V/V for $f_c = 1/T_c = 10^6$Hz, $C_1 = C_2 = 2$pF.

Now $\omega_o = \frac{1}{T_c}\sqrt{\frac{C_3 C_4}{C_2 C_1}}$. Thus $\sqrt{C_3 C_4} = \omega_o T_c \sqrt{C_2 C_1} = 2\pi \times 10^5 \times 10^{-6} \times 2 \times 10^{-12} = 1.257 \times 10^{-12}$. Now for $\frac{T_c}{C_3} C_2 = \frac{T_c}{C_4} C_1$, with $C_1 = C_2$, then $C_3 = C_4 =$ **1.257pF**. Now $Q = \frac{C_4}{C_5}$. Thus $C_5 = \frac{C_4}{Q} = \frac{1.257 \times 10^{-12}}{.707} =$ **1.777pF**. *Check*: $C_5 = \omega_o T_c C/Q = 2\pi \times 10^5 \times 10^{-6} \times 2 \times 10^{-12} \times 1.414 = 1.7772 \times 10^{-12}$, as before. Now A_o of the low-pass function is $R_4/R_6 = C_6/C_5$. Thus $C_6 = 1 \times C_5 =$ **1.777pF**. In summary, $C_1 = C_2 =$ **2.000pF**, $C_3 = C_4 =$ **1.257pF**, $C_5 = C_6 =$ **1.777pF**. The output of the first integrator is the **bandpass output**. Its center frequency is $f_o = \mathbf{10^5 Hz}$. Its 3dB bandwidth is $\frac{f_o}{Q} = \frac{10^5}{.707} = \mathbf{1.41 \times 10^5 Hz}$. Its maximum gain (from Eq. 11.100) is $\frac{C_6}{C_5} = \frac{1.777}{1.777} =$ **1V/V**.

SECTION 11.11: Tuned Amplifiers

11.38 For $I_E = 1$mA, $r_e = \frac{25mV}{1mA} = 25\Omega$. *For* $R_E = r_e = 25\Omega$: $R_{in} = (\beta + 1)(r_e + R_E) = 201(25 + 25) = 10.05k\Omega$. Gain from base to collector, $A_{bc} = \frac{-R_L}{r_e + R_E} = \frac{-5 \times 10^3}{25 + 25} = -100$V/V. Gain from base to emitter, $A_{be} = \frac{R_E}{r_e + R_E} = \frac{25}{25 + 25} = 0.5$V/V. Equivalent input capacitance $C_{eg} = 10(1 - 0.5) + 1(1 - (-100)) = 5 + 101 = 106$pF. The total tuning capacitance is $C = 200 + 106 = 306$pF. Now $\omega_o = 1/\sqrt{LC} = (1 \times 10^{-6} \times 306 \times 10^{-12})^{-1/2} = \mathbf{57.17 \times 10^6 rad/s.} \equiv$ **9.1MHz**. $B = \frac{1}{CR} = \frac{1}{306 \times 10^{-12}(10.05 \times 10^3 \,||\, 10 \times 10^3)} =$ **652krad/s.** $Q = \frac{\omega_o}{B} = \frac{57.17 \times 10^6}{652 \times 10^3} =$ **87.7**. Center-Frequency Gain $A = \frac{10.05}{10 + 10.05} \times (-100) =$ **−50.1V/V**.

For $R_E = 9\, r_e$: $R_{in} = 201(10)(25) = 50.25k\Omega$, $A_{bc} = -\frac{5 \times 10^3}{10(25)} = -20$V/V, $A_{be} = \frac{9(25)}{10(25)} = 0.9$V/V, $C = 200 + 10(1 - 0.9) + 1(1 - -20) = 200 + 1 + 21 = 222$pF, $R = 10$k$\Omega \,||\, 50.25$k$\Omega = 8.34$kΩ, $\omega_o = (1 \times 10^{-6} \times 222 \times 10^{-12})^{-1/2} = \mathbf{67.1 \times 10^6 rad/s}$, $B = \frac{1}{222 \times 10^{-12} \times 8.34 \times 10^3} = \mathbf{540 \times 10^3 rad/s}$, $Q = \frac{67.1}{0.54} =$ **124**, $A = \frac{50.25}{50.25 + 10} \times (-20) =$ **−16.7V/V**.

11.39 Here, $I_E = 1$mA. Thus $r_e = 25\Omega$ and $r_\pi = 201(25) = 5.025$kΩ. Gain from base to collector $= -5000/25 = -200$V/V, $C_{in} = 10$pF + 1pF (1 − −200) = 211pF,

For direct connection (as in P11.72 of the Text): $C = 200 + 211 = 411$pF, $R = 10$k$\Omega \,||\, 5.025$k$\Omega = 3.34$kΩ, $\omega_o = (LC)^{-1/2} = (1 \times 10^{-6} \times 411 \times 10^{-12})^{-1/2} = \mathbf{49.3 \times 10^6 rad/s}$, $B = (CR)^{-1} = (411 \times 10^{-12} \times 3.34 \times 10^3)^{-1} = \mathbf{0.728 \times 10^6 rad/s}$, $Q = \frac{\omega_o}{B} = \frac{49.3}{0.728} =$ **67.7**, $A = \frac{5.025}{5.025 + 10.0} \times (-200) =$ **−66.9V/V**.

For a tapped coil with k = 0.5: $R_{in} = 5.025$kΩ is transformed to $\frac{5.025 \times 10^3}{(0.5)^2} = 20.1k\Omega$, $C_{in} = 211$pF is transformed to $211 \times 0.5^2 = 52.8$pF, $C = 200 + 52.8 = 252.8$pF, $R = 10$k$\Omega \,||\, 20.1$k$\Omega = 6.68$kΩ, $\omega_o = (1 \times 10^{-6} \times 252.8 \times 10^{-12})^{-1/2} = \mathbf{62.9 \times 10^6 rad/s}$, $B = (252.8 \times 10^{-12} \times 6.68 \times 10^3)^{-1} = \mathbf{0.592 \times 10^6 rad/s}$, $Q = \frac{62.8}{0.592} =$ **106**, $A = \frac{20.1}{10 + 20.1} \times 0.5 \times (-200) =$ **−66.8V/V**.

For a tapped coil with k = 0.1: $R_T = \frac{5.025 \times 10^3}{(0.1)^2} = 502k\Omega$, $C_T = 211 \times 10^{-12} \times (0.1)^2 = 2.11$pF, $C = 200 + 2.1 = 202.1$pF, $R = 10$k$\Omega \,||\, 502$k$\Omega = 9.8$kΩ, $\omega_o = (1 \times 10^{-6} \times 202.1 \times 10^{-12})^{-1/2} = \mathbf{70.3 \times 10^6 rad/s}$, $B = (202.1 \times 10^{-12} \times 9.80 \times 10^3)^{-1} = \mathbf{0.505 \times 10^6 rad/s}$, $Q = \frac{70.3}{0.505} =$ **139.2**, $A = \frac{502}{10 + 502} \times 0.1 \times (-200) =$

–19.6V/V.

Comparison:

	ω_o Mrad/s.	Q	A V/V
Basic Circuit	49.3	68	–66.9
$R_E = r_e$	57.2	88	–50.1
$R_E = 9\, r_e$	67.1	124	–16.7
$k = 0.5$	62.9	106	–66.8
$k = 0.1$	70.3	139	–19.6

Thus the tapped coil gives the best results in general.

11.40 For the coil, $Q_o = \frac{\omega_o L}{r_s} \approx \frac{R_p}{\omega_o L}$. Here, at 10MHz, $\omega_o L = 2\pi \times 10^7 \times 2 \times 10^{-6} = 125.7\Omega$. For $Q_o = 200$, $r_s = \frac{125.7}{200} = \mathbf{0.63\Omega}$, and $R_p = 200\,(125.7) = \mathbf{25.1k\Omega}$. For resonance at 10MHz, $\omega_o = (LC)^{-1/2}$, or $C = \frac{1}{\omega_o^2 L} = \frac{1}{(2\pi \times 10^7)^2 \times 2 \times 10^{-6}} = \mathbf{126.7pF}$. Bandwidth $B = \frac{1}{2\pi CR}$, for which $R = \frac{1}{2\pi \times 200 \times 10^3 \times 126.7 \times 10^{-12}} = 6.281k\Omega$. The resistor to be added is $6.281 \parallel (-25.1)k\Omega = \frac{6.281 \times 25.1}{25.1 - 6.28} = \mathbf{8.38k\Omega}$.

11.41 For a single LC circuit, $Q = \frac{\omega_o}{B} = \frac{f_o}{f_b} = \frac{10^6}{10^5} = \mathbf{10}$. For synchronous tuning of N stages, from Eq. 11.110, $f_B = \frac{f_o}{Q}\sqrt{2^{1/N}-1}$. Here, $50 \times 10^3 = \frac{10^6}{10}\sqrt{2^{1/N}-1}$, or $1 = 2\sqrt{2^{1/N}-1}$, $2^{1/N}-1 = (\frac{1}{2})^2 = .25$, $2^{1/N} = 1.25$, $1/N \log 2 = \log 1.25$, $N = \frac{\log 2}{\log 1.25} = \frac{.301}{.0969} = 3.1$. Thus, use **3 stages**: *Check*: $f_B = \frac{10^6}{10}\sqrt{2^{1/3}-1} = 10^5\sqrt{1.2599-1} = 10^5\sqrt{.2599} = .51 \times 10^5$Hz. OK.

For the 30dB bandwidth, (from problem 11.77 on page 840 of the Text, part b)), for synchonous tuning:

$$|T(j\omega)| = \frac{|T(j\,\omega_o)|}{[1 + 4\,(2^{1/N}-1)\,(\delta f/f_b)^2]^{N/2}}$$

For 1 stage: $N = 1$, $[1 + 4\,(2-1)\,(\delta f/f_b)^2]^{1/2} = \frac{|T_o|}{|T|} = 10^{+30/20} = 31.6$, or $1 + 4\,(\delta f/f_b)^2 = 1000$, $\delta f/f_b = (\frac{1000-1}{4})^{1/2} = 15.8$. Thus the skirt selectivity is $\frac{S}{B} = \frac{2\,\delta f}{f_b} = 2\,(15.8) = \mathbf{31.6}$, and the 30dB bandwidth is

31.6 (100kHz) = **3.16MHz**. For the synchronously-tuned cascade, with $N = 3$, $[1 + 4\,(2^{1/3} - 1)\,(\delta f/f_b)^2]^{3/2}$ = 31.6, $1 + 4\,(.2600)\,(\delta f/f_b)^2 = 9.995$, that is $\delta f/f_b = \left[\dfrac{9.995 - 1}{4\,(.26)}\right]^{1/2} = 2.94$. Thus the skirt selectivity factor is 2 (2.94) = **5.88**, and the 30dB bandwidth is 5.88 (50kHz) = **0.294MHz**.

11.42 Using Equations 11.115 and 11.116 of the Text: $f_{01} = f_o + \dfrac{B}{2\sqrt{2}} = 10.7 + \dfrac{0.1}{2\sqrt{2}} = 10.735\text{MHz}$, $f_{02} = f_o - \dfrac{B}{2\sqrt{2}} = 10.7 - \dfrac{0.1}{2\sqrt{2}} = 10.665\text{MHz}$, $B_1 = B_2 = \dfrac{B}{\sqrt{2}} = \dfrac{.100}{\sqrt{2}} = .0707\text{MHz}$, $Q_1 = Q_2 = \dfrac{\sqrt{2} \times 10.7}{0.1} = 151.3$. Now $f_o = \dfrac{1}{2\pi\sqrt{LC}}$, and $C_1 = \dfrac{1}{(2\pi f_{01})^2 L} = \dfrac{1}{(2\pi \times 10.735 \times 10^6)^2 \times 3 \times 10^{-6}} = $ **73.27pF**, and $C_2 = \dfrac{1}{(2\pi \times 10.665 \times 10^6)^2 \times 3 \times 10^{-6}} = $ **74.23pF**. Now $R = 2\pi f_o L Q$, and $R_1 = 2\pi \times 10.735 \times 10^6 \times 3 \times 10^{-6} \times 151.3 = $ **30.616kΩ**, $R_2 = 2\pi \times 10.665 \times 10^6 \times 3 \times 10^{-6} \times 151.3 = $ **30.416kΩ**. Since the voltage gain at resonance is proportional to R (see Ex. 11.37), the relative peak gain of each of the two is $\dfrac{R_1}{R_2} = \dfrac{30.616}{30.416} = $ **1.007**.

Chapter 12

SIGNAL GENERATORS AND WAVEFORM - SHAPING CIRCUITS

SECTION 12.1: Basic Principles of Sinusoidal Oscillators

12.1 Look ahead to the diagram in the solution to P12.2 following only if you need to! Convert all resistive elements to equivalent values across the tank inductor L. Now, $Q = \frac{\omega_o L}{R_{ls}} \approx \frac{R_{lp}}{\omega_o L}$. Thus $R_{lp} = \frac{\omega_o^2 L^2}{R_{ls}}$ is the equivalent resistor across L due to inductor-wire resistance. Now the amplifier input resistance reflected through the turns ratio is $R_i = n^2 R_{in}$. Thus the total load on the tank is $R = R_o \| R_{lp} \| R_{cp} \| R_i = R_o \| \omega_o^2 L^2 / R_{ls} \| R_{cp} \| n^2 R_{in}$. Loop gain (from the active end of the coil and back) is $G_m R/n$. Oscillation will occur when this is unity, that is when $\frac{G_m R}{n} = 1$, or $\frac{G_m}{n} = \frac{1}{R} = \frac{1}{R_o} + \frac{R_{ls}}{\omega_o^2 L^2} + \frac{1}{R_{cp}} + \frac{1}{n^2 R_{in}}$, at a frequency $\omega_o = \frac{1}{\sqrt{LC}}$. Now $\omega_o^2 = \frac{1}{LC}$, and $\frac{1}{\omega_o^2 L^2} = \frac{LC}{L^2} = \frac{C}{L}$, with oscillation occuring for $\frac{G_m}{n} = \frac{1}{R_o} + \frac{R_{ls} C}{L} + \frac{1}{R_{cp}} + \frac{1}{n^2 R_{in}}$, or for $G_m = \frac{n}{R_o} + \frac{n R_{ls} C}{L} + \frac{n}{R_{cp}} + \frac{1}{n R_{in}}$.

12.2

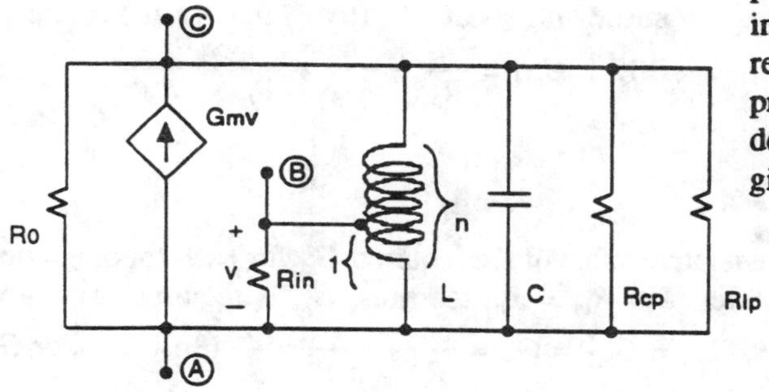

(a) Showing the inductor's parasitic resistance in its parallel form, the topology shown results for a non-inverting amplifier. Three classical circuit forms result depending on which of A, B, C is grounded in a practical implementation. Grounding A produces the design that is usually drawn from the description given.

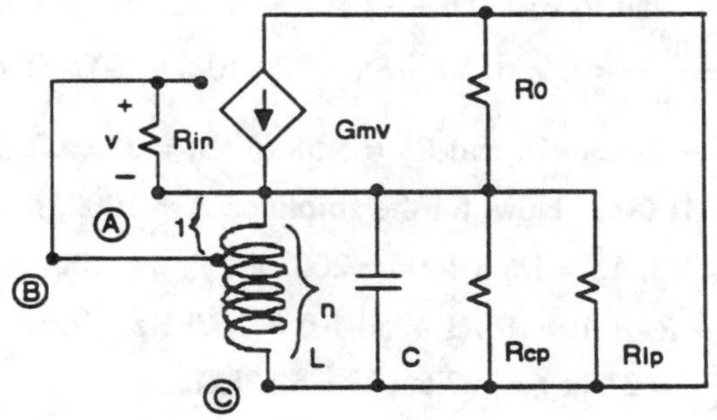

(b) For an inverting transconductance element, the topology shown results, again with three classical variations depending on where ground is connected. The one with C grounded, in which the device acts as a follower, is quite common in practice.

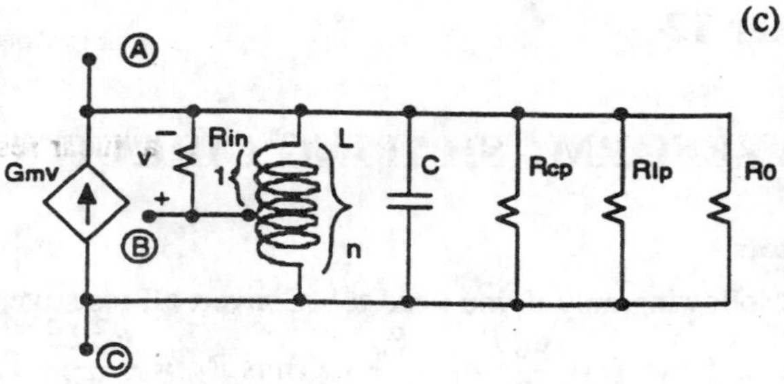

(c) This may be redrawn to resemble (a) more closely to illustrate that the change of amplifier sign is compensated simply by shifting the tap from one end of the autotransformer to the other. For the negative-gain version, the loop-gain magnitude (as measured across L) is exactly as before, with the same conditions for oscillation. Note that the amplifier inversion is accounted for by the coil-tap-connection reversal.

12.3

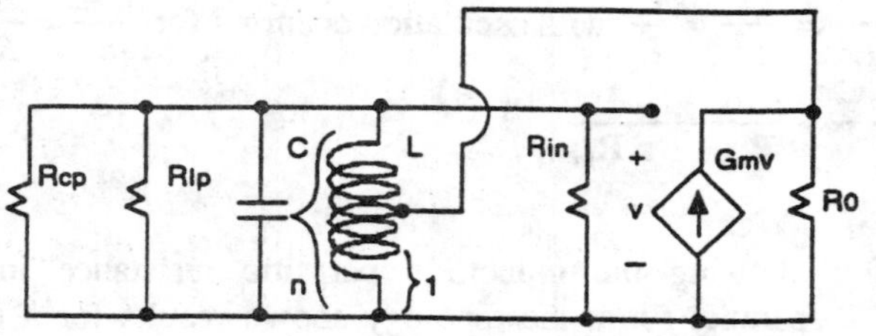

Here, at resonance, the load on the transconductor is $R = R_o \,||\, \frac{1}{n^2}(R_{cp} \,||\, R_{lp} \,||\, R_{in})$, and the voltage across the tap is $G_m R$, with the loop gain equal to $n\, G_m R$. For oscillation to occur, $n\, G_m R = 1$, or $G_m = \frac{1}{n}\frac{1}{R} = \frac{1}{n}\left[\frac{1}{R_o} + \frac{n^2}{R_{cp}} + \frac{n^2}{R_{lp}} + \frac{n^2}{R_{in}}\right]$, or $G_{m2} = G_m = \frac{1}{n R_o} + \frac{n}{R_{cp}} + \frac{n}{R_{lp}} + \frac{n}{R_{in}}$, with oscillation at a frequency $\omega_o = (LC)^{-½}$. Recall that for the connection in P.12.1, $G_{m1} = \frac{n}{R_o} + \frac{n}{R_{cp}} + \frac{n}{R_{lp}} + \frac{1}{n R_{in}}$.

Note that for a particular coil and capacitor, the relative value of the required G_m for each topology depends on the relative size of R_o and R_{in} of the amplifier. For $R_{in} = R_o$, the same G_m is required for a given n. Thus, for example, for $R_o = R_{in} = R_{cp} = R_{lp} = R$, $G_{m1} = G_{m2} = G_m = \frac{3n}{R} + \frac{1}{nR} = \frac{1}{R}(3n + 1/n)$, or $G_m R = 3n + 1/n$.

12.4 For an input resistance of 10kΩ, R_1 = **10kΩ**, and $R_f = 5\,(R_1)$ = **50kΩ**. Negative clamping occurs with D_1 conducting with 0.6V drop, the negative op-amp input at 0 volts, and V_A at –0.6V correspondingly. Thus at the edge of D_1 conduction, $V_{R2} = 10 - -0.6 = 10.6$V, and $V_{R3} = -0.6 - -2.5 = 1.9$V. Correspondingly, $R_2 = \frac{10.6}{1.9} R_3 = 5.58\, R_3$. For a limiting gain of 0.5, $\frac{R_f \,||\, R_3}{R_1} = 0.5$, or $50\text{k} \,||\, R_3 = 0.5\,(10\text{k}) = 5\text{k}\Omega$. Thus, $R_3 = 50\text{k} \,||\, R_3 \,||\, (-50\text{k}) = 5\text{k} \,||\, (-50\text{k}) = \frac{5\,(-50)}{-50+5} = \frac{5\,(50)}{45}$ = **5.56kΩ**, and $R_2 = 5.58\,(5.56)$ = **31.0kΩ**. Overall, use R_1 = 10kΩ, R_f = 50kΩ, R_3 = 5.6kΩ, and R_2 = 31.0kΩ. Now, for the amplifier, $A = 1000$, $f_t = 10^6$Hz, and thus $f_{3dB} = \frac{10^6}{1000} = 10^3$Hz. For a gain of 5, $\beta = 1/5$, $A\beta = 1/5 \times 1000 = 200$, and $f_{3dB} \approx 200$kHz. *Alternatively*, from page 80 of the Text, $f_{3dB} = 10^6/(1 + R_f/R_1) = 10^6/(1 + 50/10) = 167$kHz. For a 3dB frequency of 167kHz, 2° phase shift occurs at $\tan^{-1} \frac{f}{167} = 2°$, or $f = 167 \tan 2°$ = **5.83kHz**.

12.5 For low input voltages, the gain is $-\frac{R_2}{R_1} = -\frac{10}{7.5} = -1.33\text{V/V}$. For the zeners conducting, the gain is $-\frac{R_2 \| R_3}{R_1} = -\frac{10\|10}{7.5} = -\frac{5}{7.5} = -0.67\text{V/V}$. Assume that the zener can be characterized by a linear resistor of value $R_2 = \frac{V_Z}{I_{ZK}}$ for voltages below the knee.

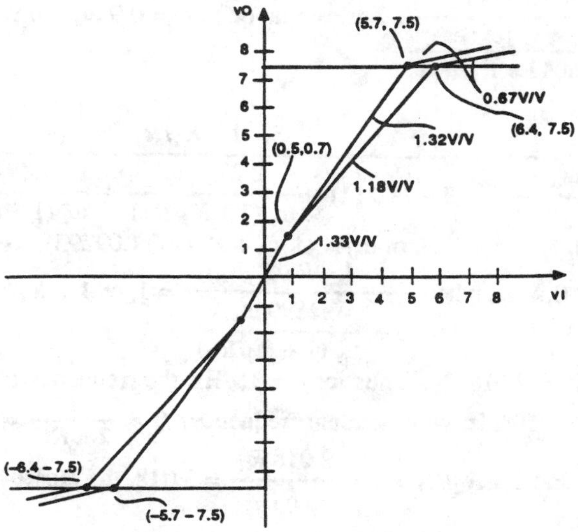

Here, $R_{Z1} = \frac{6.8}{100\mu A} = 68\text{k}\Omega$, and $R_{Z2} = \frac{6.8}{10\mu A} = 680\text{k}\Omega$. For the poorer zener, for output voltages beyond 0.7V (with corresponding inputs of $\frac{0.7}{1.33} = 0.53\text{V}$), $R_{Z1} = 68\text{k}\Omega$, and the gain is $-\frac{R_2 \| (R_3 + R_{Z1})}{R_1} = -\frac{10\|(10+68)}{7.5} = -1.18\text{V/V}$. For the better zener it is $-\frac{10\|(10+680)}{7.5} = -1.32\text{V/V}$. For outputs greater than 6.8 + 0.7 = 7.5V, the gain is –0.67V/V as calculated earlier. For the better zener, the corresponding input is about $\frac{7.5}{1.32} = 5.68\text{V}$, and for the poorer zener, it is about $\frac{7.5}{1.18} = 6.36\text{V}$.

SECTION 12.2: Op-Amp-RC Oscillators

12.6 Now, $\omega_o = \frac{1}{CR}$. For C = 10nF, $R = \frac{1}{2\pi f_o C} = \frac{1}{2\pi \times 10^4 \times 10 \times 10^{-9}} = 1.59\text{k}\Omega$. For oscillation, $R_2 \geq 2R_1$. For 2V peak-to-peak output, peak output is 1V. At the threshold of oscillation the voltage at the positive op amp input must be 2/3V. Thus the voltage across the regulating diode plus its series resistor (R_2) must be 2/3V, as must be the voltage across R_2, with 1/3V across R_1. Thus the current in $R_1 = I = \frac{1/3}{R_1} = \frac{2/3}{R_2} + I_D$. Now $I_D = I_s e^{V_D/2V_T}$ where $V_D + I_D R_2$ = 2/3V. Now, $V_D = 0.70 + 0.05 \ln I_D/1$ mV, and $0.700 + .05 \ln I_D + I_D R_2 = 0.667$, or $I_D = \frac{1}{R_2}(-.05 \ln I_D - .033)$ – – – (1). Also $I_D = \frac{0.333}{R_1} - \frac{0.667}{R_2}$ – – – (2), with $R_2 > 2R_1$ – – – (3). With (1), try I_D = 0.1mA, whence $R_2 = \frac{1}{0.1}(-.05 \ln 0.1 - .0333) = .818\text{k}\Omega$. Try $I_D = 10\mu$A: $R_2 = \frac{1}{.01}(-.05 \ln .01 - .0333) = 19.7\text{k}\Omega$. Try $I_D = 20\mu$A: $R_2 = \frac{1}{.02}(-.05 \ln .02 - .0333) = 8.11\text{k}\Omega$. Use $R_2 = 8.2\text{k}\Omega$. From (2), $.01 = \frac{.3333}{R_1} - \frac{.6666}{8.2}$, or $R_1 = \frac{.3333}{.010 + .0813} = 3.65\text{k}\Omega$. Use 3.6kΩ. *Check*: from (3), $R_2 = 8.2 > 2R_1 = 2(3.65) = 7.3$ OK. *Check*: From (2), $I_D = \frac{.3333}{3.6} - \frac{.6666}{8.2} = .0926 - .0813 = .0113\text{mA}$. Now the drop across the diode plus the extra R_2 is 0.0113 (8.2) + .05 ln .0113 + 0.70 = .0927 – .2241 + 0.70 = .569V. Current in the feedback $R_2 = \frac{.569}{8.2} = .0693\text{mA}$. Voltage across R_1 = (.0693 + .0113) (3.6) = 0.200V, and the output voltage is .290 + .569 = .859V. Too low! Now, reduce R_1 slightly to R_1 = 3.3kΩ. From (2), $I_D = \frac{.3333}{3.3} - \frac{.6666}{8.2} = .1010 - .0823 = .0187\text{mA}$. Now the drop across the diode plus the extra R_2 is .0187 (8.2) + .05 ln .0187 + 0.70 = .1534 − .199 + 0.70 = .654V. Current in the feedback $R_2 = \frac{.654}{8.2}$ = .0800mA. Voltage across R_1 = (.0800 + .0187) (3.6) = .355V. Thus the output voltage is .355 + .654 = **1.01V**. Use **C = 10nF, R = 1.59kΩ, R_1 = 3.3kΩ, R_2 = 8.2kΩ.**

12.7 From Eq. 12.11 of the Text, using an ideal amplifier, $L(j\omega) = \frac{1 + R_2/R_1}{3 + j(\omega CR - 1/(\omega CR))}$, where oscillation occurs when $\Phi_L = 0$, and $|L| = 1$ at $\omega_o = \frac{1}{CR}$. From Eq. 2.20 of the Text, for an amplifier with $\omega_t = 2\pi$rad/s and $A_o >> (1 + R_2/R_1)$, $\frac{V_o}{V_i}(j\omega) = \frac{1 + R_2/R_1}{1 + \frac{j\omega}{\omega_t/(1 + R_2/R_1)}}$. For oscillation at $\omega = 0.9\ \omega_o = 0.9/CR$ with this amplifier:

$$L\,(j\,0.9\,\omega_o) = \frac{1 + R_2/R_1}{(3 + j\,(.9 - \frac{1}{0.9}))\,(1 + \frac{j\,0.9\,\omega_o}{\omega_t\,(1 + R_2/R_1)})} = \frac{1 + R_2/R_1}{3 - j\,(.211) + \frac{2.27\,j\,\omega_o}{\omega_t\,(1 + R_2/R_1)} + \frac{.1899\,\omega_o}{\omega_t\,(1 + R_2/R_1)}}$$

Oscillation will occur when $0.211 = 2.27\omega_o/(\omega_t\,(1 + R_2/R_1))$, or $\omega_o = \omega_t\,(1 + R_2/R_1)\,(.09295)$. Now for a nominal frequency ω_o, oscillation will occur at 0.9 ω_o, when $\frac{1 + R_2/R_1}{3 + \frac{.1899\,\omega_o}{\omega_t\,(1 + R_2/R_1)}} = 1$, or $1 + R_2/R_1 = 3 +$ .1899 (.09295) = 3.018, or $R_2/R_1 = 2.018$, or $R_2 = 2.018\,R_1$. Thus for f_t = 1MHz, the frequency of oscillation will be 0.9 × 10^6 (3.018) (.09295) = **0.252 × 10^6Hz**, with nominal frequency, $f_o = \frac{1}{2\pi RC} = \frac{.252}{0.9}$ = **0.280MHz**. Thus, for operation at 0.252MHz, and $1 + R_2/R_1 = 1 + \frac{2.018\,R_1}{R_1}$ = 3.018, the closed-loop op-amp gain must be $\frac{3.018}{1 + \frac{j\,.25 \times 10^6}{1 \times 10^6/3.018}} = \frac{\mathbf{3.02}}{\mathbf{1 + 0.755\,j}}$, a value which seems quite reasonable.

12.8 From Eq. 2.20 of the Text, for an op amp with A_o large and a unity-gain frequency ω_t, the gain of the non-inverting topology is: $G(\omega) = \frac{1 + R_2/R_1}{1 + \frac{j\omega}{\omega_t\,(1 + R_2/R_1)}}$, and of the network is:

$$T(\omega) = \frac{R}{R + \frac{1}{j\omega C}} = \frac{1}{1 + \frac{1}{j\omega RC}}.$$ Overall,

$$L(\omega) = \frac{1 + R_2/R_1}{(1 + \frac{j\omega}{\omega_t\,(1 + R_2/R_1)})(1 + \frac{1}{j\omega RC})} = \frac{\mathbf{1 + R_2/R_1}}{1 + \frac{j\omega}{\omega_t\,(1 + R_2/R_1)} - \frac{j}{\omega RC} + \frac{1}{RC\,\omega_t\,(1 + R_2/R_1)}}$$

Oscillation will occur when the net phase shift is 0°, when $\frac{\omega}{\omega_t\,(1 + R_2/R_1)} = \frac{1}{\omega RC}$ or $\omega = \left[\frac{\omega_t\,(1 + R_2/R_1)}{RC}\right]^{1/2}$, provided the gain is at least one, that is $\frac{1 + R_2/R_1}{1 + \frac{1}{RC\,\omega_t\,(1 + R_2/R_1)}} \geq 1$, or

$\mathbf{R_2/R_1} = \frac{1}{RC\,\omega_t\,(1 + R_2/R_1)}$, or $(R_2/R_1)^2 + (R_2/R_1) - 1/(RC\omega_t) = 0$. Now for $\omega = 4/RC = \left[\frac{\omega_t\,(1 + R_2/R_1)}{RC}\right]^{1/2}$, $\omega_t\,(1 + R_2/R) = 16/(RC)$, and $R_2/R_1 = \frac{1}{RC\,(16/RC)} = \frac{1}{16}$. ∴ $\mathbf{R_2 = R_1/16 = .0625}$ R_1, for $\frac{1}{RC} = \omega_t$.

More generally, for $\frac{1}{RC} = \frac{\omega_t}{a}$, $\omega = \frac{\omega_t}{\sqrt{a}}(1 + R_2/R_1)^{1/2}$, with $(1 + R_2/R_1) = 1 + 1/(a\,(1 + R_2/R_1))$, or $x = 1 + 1/(ax)$, $ax^2 - ax - 1 = 0$, $x^2 - x - 1/a = 0$, for which $x = \frac{-1 \pm \sqrt{1 - 4(-1/a)}}{2}$, or $1 + R_2/R_1 = \frac{1 \pm \sqrt{4/a + 1}}{2}$. Now, for example: For $a = 1$, $1 + R_2/R_1 = \frac{1 + \sqrt{5}}{2} = 1.618$, $R_2 = .618\,R_1$ and $\omega = \omega_t\,(1.618)^{1/2} = 1.27\omega_t$. For $a = 2$, $1 + R_2/R_1 = \frac{1 \pm \sqrt{2+1}}{2} = 1.366$, $R_2 = .366\,R_1$ and $\omega = \omega_t\,(\frac{1.366}{2})^{1/2} = .826\omega_t$. For $a = 4$, $1 + R_2/R_1 = \frac{1 \pm \sqrt{2}}{2} = 1.2707$, $R_2 = .2707\,R_1$ and $\omega = \omega_t\,(\frac{1.2707}{4})^{1/2} = 0.564\omega_t$. For $a = 16$, $1 + R_2/R_1 = \frac{1 \pm \sqrt{1.25}}{2} = 1.059$, $R_2 = .059\,R_1$, and $\omega = \omega_t\,(\frac{1.059}{16})^{1/2} = 0.257\omega_t$.

Generalizing, we see that for frequencies significantly lower than ω_t for an amplifier with idealized single-pole rolloff, oscillation is possible for R_2 nearly zero, at a frequency which is the geometric mean of ω_t and $1/RC$. For excess phase shift due to additional poles near and above ω_t, more compensating phase shift will be required from the RC network, implying operation at a frequency lower than in the simple case. Note that this type of oscillator, operating at ω_o with a network for which $\omega = 1/RC$, allows an estimate of ω_t, and associated excess phase, if R_2 is adjusted to the maximum value for which oscillation is sustained. For $1/RC << \omega_o$, the oscillation frequency ω_o can become a sensitive function of various things, such as construction, but not of excess phase. For $1/RC$ very near ω_o, the frequency of oscillation is quite sensitive to excess phase. By varying R, find ω_o in the range two to four times $1/RC$ in order to evaluate ω_t. $\omega_o < 1/RC$ is an indication of excess phase, where operation also requires a higher value of R_2.

12.9 From the right, label the components $C_1, R_1, C_2, R_2, C_3, R_3$, with joining nodes N_1, N_2, N_3 respectively. Assume a virtual ground at the op-amp input into which a current flows from C_1. At N_1, $i_{C1} = i$, $v_1 = i/C_1 s$. At N_2, $v_2 = i_{C1} R_1 + v_1 = i\,(R_1 + 1/C_1 s)$, $i_{C2} = v_2/(1/C_2 s) = v_2\, C_2 s = i\,(R_1 C_2 s + C_2/C_1)$. At N_3, $v_3 = (i_{C1} + i_{C2})\,R_2 + v_2 = i\,(1 + R_1 C_2 s + C_2/C_1)\,R_2 + i\,(R_1 + 1/C_1 s) = i\,(R_1 R_2 C_2 s + R_2 C_2/C_1 + R_2 + R_1 + 1/C_1 s)$, $i_{C3} = v_3 C_3 s = i\,(R_1 R_2 C_2 C_3 s^2 + R_2 C_2 C_3 s/C_1 + R_2 C_3 s + R_1 C_3 s + C_3/C_1$. At x, $v_x = (i_{C1} + i_{C2} + i_{C3})\,R_3 + v_3 = i\,(1 + R_1 C_2 s + C_2/C_1 + R_1 R_2 C_2 C_3 s^2 + R_2 C_2 C_3 s/C_1 + R_2 C_3 s + R_1 C_3 s + C_3/C_1)\,R_3 + i\,(R_1 R_2 C_2 s + R_2 C_2/C_1 + R_2 + R_1 + 1/C_1 s) = i\,(R_1 + R_2 + R_3 + R_3 C_2/C_1 + R_3 C_3/C_1 + R_2 C_2/C_1 + s\,(R_1 R_3 C_2 + R_1 R_3 C_3 + R_2 R_3 C_3 + R_1 R_2 C_2 + R_2 R_3 C_2 C_3/C_1) + R_1 R_2 R_3 C_2 C_3\, s^2 + 1/C_1\, s$. Now the loop gain $L = -i\, R_f/v_x$. Thus, for $C_1 = C_2 = C_3 = C$, $L(s) = -R_f/[R_1 + 2\,R_2 + 3\,R_3 + sC\,(R_1 R_2 + 2\,R_2 R_3 + 2\,R_1 R_3) + R_1 R_2\, R_3 C^2 s^2 + 1/(Cs)]$, and $L(s) = -R_f Cs/[1 + (R_1 + 2\,R_2 + 3\,R_3)\,Cs + (R_1 R_2 + 2\,R_2 R_3 + 2\,R_1 R_3)\,C^2 s^2 + R_1 R_2\, R_3 C^3 s^3]$. Now substituting $s = j\omega$ and multiplying top and bottom by j, $L(j\omega) = +R_f C\omega/[j - (R_1 + 2R_2 + 3R_3)\,C\omega - (R_1 R_2 + 2R_2 R_3 + 2R_1 R_3)\,C^2\omega^2\, j + R_1 R_2\, R_3 C^3 \omega^3]$. For oscillation, the phase angle = zero, that is $1 = (R_1 R_2 + 2\,R_2 R_3 + 2\,R_1 R_3)\,C^2\,\omega^2$, where the frequency of oscillation is $\omega_o = \frac{1}{C\sqrt{R_1 R_2 + 2\,R_2 R_3 + 2\,R_1 R_3}}$, which for $R_1 = R_2 = R_3 = R$ is $\omega_o = \frac{1}{CR\sqrt{5}}$, with $\frac{R_f\, C\,\omega_o}{(R_1 + 2R_2 + 3R_3)\,C\,\omega_o + R_1\, R_2\, R_3\, C^3\,\omega_o^3} = 1$, or $R_f = (R_1 + 2R_2 + 3R_3) + R_1 R_2 R_3 C^2 \omega_o^2 = (R_1 + 2R_2 + 3R_3) + (R_1 R_2 R_3)\,(R_1 R_2 + 2R_2 R_3 + 2R_1 R_3)^{-1}$, which for $R_1 = R_2 = R_3 = R$ is $R_f = 6\,R + R/5$ $= \mathbf{6.2R}$.

Now for sensitivities (with $R_1 \approx R_2 \approx R_3 \approx R$):

$\frac{\partial\,\omega_o}{\partial\,R_1} = -\frac{1}{2}\left[\frac{1}{C\,(R_1\,R + 2R^2 + 2R_1\,R)^{3/2}}\right](3R) = -\frac{3}{2}\,\omega_o\,\frac{R}{3R\,R_1 + 2R^2} = -\frac{3}{2}\,\omega_o\,\frac{1}{3R_1 + 2R}$, and $\frac{\partial\,\omega_o}{\partial\,R_2} = -\frac{3}{2}\,\omega_o\,\frac{1}{3R_2 + 2R}$, and $\frac{\partial\,\omega_o}{\partial\,R_3} = -\frac{4}{2}\,\omega_o\,\frac{1}{4R}$. Now $S^{\omega_o}_{R_1,\,R_2} = \frac{\partial\,\omega_o}{\omega_o} \times \frac{R_1}{\partial\,R_1} = -\frac{3}{2}\,\omega_o\,\frac{1}{3R_1 + 2R} \times \frac{R_1}{\omega_o}$, which, for $R_1 \approx R$, is $-\frac{3}{2(5)} = \mathbf{-0.3}$. Similarly $S^{\omega_o}_{R_3} = \mathbf{-0.5}$. Also $\frac{\partial\,R_f}{\partial\,R_1} = 1 + R_2 R_3\,(R_1 R_2 + 2R_2 R_3 +$

$2R_1R_3)^{-1} + (-1)\, R_1R_2R_3\, (R_1R_2 + 2R_2R_3 + 2R_1R_3)^{-2}\, (R_2 + 2R_3)$, which for $R_2 = R_3 = R$, is $\frac{\partial R_f}{\partial R_1} = 1 + R^2\, (R_1R + 2R^2 + 2R_1R)^{-1} - R_1R^2\, (R_1\, R + 2R^2 + 2R_1\, R)^{-2}\, (\,3\,R) = 1 + R\,(3R_1 + 2R)^{-1} - 3R_1\, R\,(3R_1 + 2R)^{-2} = \frac{9R_1^2 + 4R^2 + 12R_1\,R + 3R\,R_1 + 2R^2 - 3R_1\,R}{(3R_1 + 2R)^2} = \frac{9R_1^2 + 12R_1\,R + 6R^2}{(3R_1 + 2R)^2}$, and for $R_1 = R$, is $27/25 = 1.08$. Thus, $S_{R_1}^{R_f} = \frac{\partial R_f}{\partial R_1} \times \frac{R_1}{R_f}$, for $R_1 \simeq R$, is $1.08 \times \frac{R}{6.2R} =$ **0.174**. Also with $R_1 \simeq R_3 \simeq R$, $\frac{\partial R_f}{\partial R_2} = 2 + R^2\,(R\,R_2 + 2R\,R_2 + 2R^2)^{-1} - R^2\,R_2\,(R\,R_2 + 2R\,R_2 + 2R^2)^{-2}\,(3R) = 2 + R\,(3R_2 + 2R)^{-1} - 3R\,R_2\,(3R_2 + 2R)^{-2}$, which for $R_2 = R$, is $2 + R\,(5R)^{-1} - 3R^2\,(5R)^{-2} = 2 + 1/5 - 3/25 = 2.08$. Thus $S_{R_2}^{R_f}$ is $2.08 \times \frac{R}{6.2R} =$ **0.335**. Similarly, $S_{R_3}^{R_f}$ is $3.08/6.2 =$ **0.497**.

12.10

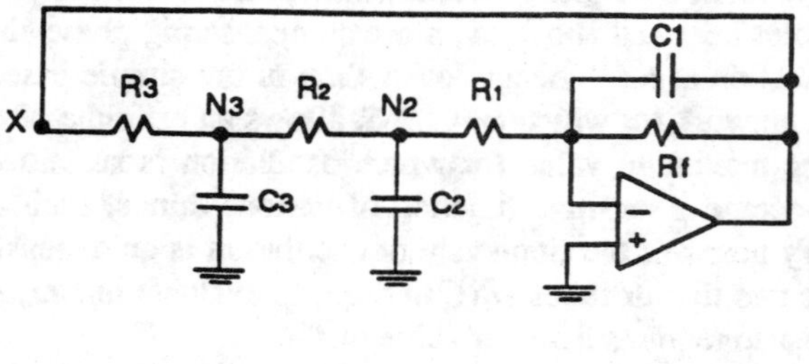

Here, $R_1 = R_2 = R_3 = R$, and $C_1 = C_2 = C_3 = C$. For a current i in R_1, at node N_2, $\upsilon_2 = i\,R_1 = iR$, $i_{C2} = \upsilon_2/(1/C_2 s) = iRCs$. At node N_3, $i_{R2} = i_{R1} + i_{C2} = i\,(1 + RCs)$, $\upsilon_3 = \upsilon_2 + R_2\,(i_{R2}) = iR + R\,(i)\,(1 + RCs) = i\,(2R + R^2\,Cs)$, $i_{C3} = \upsilon_3 C_2\, s = i\,(2R\,Cs + R^2C^2\,s^2)$. At node x, $i_{R3} = i_{R2} + i_{C3} = i\,(1 + RCs) + i\,(2RCs + R^2C^2s^2) = i\,(1 + 3RCs + R^2C^2s^2)$, $\upsilon_x = \upsilon_3 + R_3\, i_{R3} = i\,(2R + R^2Cs) + Ri\,(1 + 3RCs + R^2C^2s^2) = iR\,(3 + 4RCs + R^2C^2s^2)$. Now, i flows in $R_f \| C_1$. Thus the op-amp output voltage $= \upsilon_o = -i\,\frac{R_f\,(1/Cs)}{R_f + 1/Cs}$.

Loop gain $\mathbf{L(s)} = \frac{\upsilon_o}{\upsilon_x} = \frac{-\mathbf{R_f}}{\mathbf{R_f\,Cs + 1}} \times \frac{\mathbf{1}}{\mathbf{R\,(3 + 4RCs + R^2C^2s^2)}}$.

Now $\mathbf{L(s)} = -\,(R_f/R)\,[3 + 4R\,Cs + R^2C^2s^2 + 3\,R_fCs + 4R\,R_fC^2s^2 + R^2R_fC^3s^3]^{-1} = -\,\mathbf{(R_f/R)\,[3 + s\,(4\,RC + 3R_fC) + s^2\,(R^2\,C^2 + 4R\,R_fC^2) + R^2\,R_f\,C^3\,s^3]^{-1}}$. Substituting $s = j\omega_o$ and requiring the imaginary part to be zero, and $L(\omega_o) = 1$, see $(4RC + 3R_f\,C)\,\omega_o = R^2R_fC^3\omega_o^3$ – – – (1), and $1 = -(R_f/R)\,(3 - R^2C^2\omega_o^2\ 4R\,R_fC^2\,\omega_o^2)^{-1}$ – – – (2). From (1), $\omega_o = \frac{1}{RC}\left[\frac{4R + 3\,R_f}{R_f}\right]^{1/2}$. From (2), $R_f = R\,((R^2C^2 + 4R\,R_fC^2)\,\omega_o^2 - 3) = R\,(R^2C^2 + 4R\,R_fC^2)\,(\frac{4R + 3R_f}{R_fC^2R^2}) - 3R = (4R + 3R_f)\,(\frac{R}{R_f} + 4) - 3R$. Thus $R_f = \frac{4R^2}{R_f} + 3R + 16R + 12R_f - 3R$, $\frac{4R^2}{R_f} + 16R + 11R_f = 0$, $4R^2 + 16R\,R_f + 11R_f^2 = 0$, whence $R_f = \left[\frac{-16 \pm \sqrt{16^2 - 4\,(4)\,(11)}}{2\,(11)}\right]R = \left[\frac{-16 \pm 8.94}{22}\right]R$. Note that R_f/R is always negative, implying that oscillation is not possible. Why?

12.11 For each section, a maximum of 90° phaseshift is possible. Now with a positive-gain amplifier, the network must shift by 360°. Thus, realistically, for operation on a phase slope of more than 360/90 = 4%, **five sections are needed**. Label the sections 1 to 5 from right to left with the RC join nodes called N_1 through N_5 respectively. Now for $\upsilon_1 = \upsilon = \upsilon(1)$, $i_{C1} = \upsilon/R = \upsilon/R(1)$, $\upsilon_2 = i_{C1}/Cs + \upsilon_1 = \upsilon/RCs + \upsilon = \upsilon\,(1/RCs + 1)$, $i_{C2} = \upsilon_2/R + i_{C1} = (\upsilon/RCs + \upsilon)/R + \upsilon/R = \upsilon/R\,[1/RCs + 2]$, $\upsilon_3 = i_{C2}/Cs + \upsilon_2 = \upsilon\,[(\upsilon/RCs + 2)/RCs + 1/RCs + 1] = \upsilon\,(1/R^2C^2s^2 + 3/RCs + 1)$,

$i_{C3} = \upsilon_3/R + i_{C2} = \upsilon\,[(1/R^2C^2s^2 + 3/RCs + 1)/R + (1/RCs + 2)/R] = \upsilon/R\,[1/R^2C^2s^2 + 4/RCs + 3\,]$,

$\upsilon_4 = i_{C3}/Cs + \upsilon_3 = \upsilon\,[(1/R^2C^2s^2 + 4/RCs + 3)/RCs + 1/R^2C^2s^2 + 3/R\,Cs + 1] = \upsilon\,(1/R^3C^3s^3 + 5/R^2C^2s^2 + 6/R\,Cs + 1)$,

$i_{C4} = \upsilon_4/R + i_{C3} = \upsilon/R\,(1/R^3C^3s^3 + 5/R^2C^2s^2 + 6/RCs + 1 + 1/R^2C^2s^2 + 4/RCs + 3) = \upsilon/R\,(1/R^3C^3s^3 + 6/R^2C^2s^2 + 10/R\,Cs + 4)$,
$\upsilon_5 = i_{C4}/Cs + \upsilon_4 = \upsilon\,[1/R^4C^4s^4 + 6/R^3C^3s^3 + 10/R^2C^2s^2 + 4/RCs + 1/R^3C^3s^3 + 5/R^2C^2s^2 + 6/RCs + 1] = \upsilon\,(1/R^4C^4s^4 + 7/R^3C^3s^3 + 15/R^2C^2s^2 + 10/RCs + 1)$,
$i_{C5} = \upsilon_5/R + i_{C4} = \upsilon/R\,(1/R^4C^4s^4 + 7/R^3C^3s^3 + 15/R^2C^2s^2 + 10/RCs + 1 + 1/R^3C^3s^3 + 6/R^2C^2s^2 + 10/RCs + 4) = \upsilon/R\,(1/R^4C^4s^4 + 8/R^3C^3s^3 + 21/R^2C^2s^2 + 20/RCs + 5)$,
$\upsilon_6 = i_{C5}/Cs + \upsilon_5 = \upsilon\,(1/R^5C^5s^5 + 8/R^4C^4s^4 + 21/R^3C^3s^3 + 20/R^2C^2s^2 + 5/RCs + 1/R^4C^4s^4 + 7/R^3C^3s^3 + 15/R^2C^2s^2 + 10/RCs + 1) = \upsilon\,(1/R^5C^5s^5 + 9/R^4C^4s^4 + 28/R^3C^3s^3 + 35/R^2C^2s^2 + 15/RCs + 1)$.
Now $L(s) = \dfrac{K\upsilon}{\upsilon_6} = K\,(1/R^5C^5s^5 + 9/R^4C^4s^4 + 28/R^3C^3s^3 + 35/R^2C^2s^2 + 15/RCs + 1)^{-1}$. Now substituting $s = j\omega$, the condition for zero phase angle is $1/R^5\,C^5\,\omega^5 - 28/R^3\,C^3\,\omega^3 + 15/R\,C\omega = 0$, or $1/R^4\,C^4\,\omega^4 - 28/R^2\,C^2\,\omega^2 + 15 = 0$. Now $1/R^2\,C^2\,\omega^2 = -\dfrac{-28 \pm \sqrt{28^2 - 4\,(15)}}{2} = \dfrac{28 \pm 26.91}{2} = 27.45$ or $.545$. Now at $\omega = 1/RC$, the phase contribution of each section is 45°, and the total phase is about 225°. Thus operation must be at $\omega < 1/RC$. Thus choose the 27.45 solution, where $1/R^2\,C^2\,\omega^2 = 27.45$, or $\omega_o = \dfrac{1}{RC} \times \dfrac{1}{\sqrt{27.45}} = \dfrac{\mathbf{0.191}}{\mathbf{RC}}$. For this condition, $|L| = K\,(9/R^4\,C^4\,\omega^4 - 35/R^2\,C^2\,\omega^2 + 1)^{-1}$. Now for $|L| = 1$, $K = 9/(.191)^4 - 35/(.191)^2 + 1 = 6762.5 - 959.4 + 1 =$ **5804V/V**.

12.12 The modified amplifier has 4 sections, with all 4 capacitors of value C, and all 3 resistors of value R. The feedback resistor is R_f. Perhaps from the results of Ex. 12.5, we could reason that $L = \dfrac{\omega^2\,C^2\,R\,R_f}{5 + j\,(4\omega\,CR - 1/\omega\,CR)}$, (incorrectly!), but we must check: Label the network nodes, from the op-amp negative input toward the left, N_0, N_1, N_2, N_3, N_4. Note that $N_4 = x$. Now for a current i flowing into node N_0 from C_1, $\upsilon_o = 0\text{V}$, $i_{C1} = i$, $\upsilon_1 = i_{C1}/Cs = i/Cs$, $i_{C2} = \upsilon_1/R + i_{C1} = i(1/RCs + 1)$ $\upsilon_2 = i_{C2}/Cs + \upsilon_1 = i\,(1/R\,Cs + 1)/Cs + i/Cs$, $i_{C3} = \upsilon_2/R + i_{C2} = i\,[1/R^2C^2s^2 + 2/RCs + 1/RCs + 1] = i\,[1/R^2C^2s^2 + 3/RCs + 1]$, $\upsilon_3 = i_{C3}/Cs + \upsilon_2 = i\,(1/R^2C^2s^2 + 3/RCs + 1)/Cs + i\,(1/RCs + 2)/Cs = i\,(1/R^2C^2\,s^2 + 4/RCs + 3)/Cs$. Note that this corresponds to the result of Ex. 12.5. Now, $i_{C4} = \upsilon_3/R + i_{C3} = i\,[1/R^3C^3s^3 + 4/R^2C^2s^2 + 3/RCs + 1/R^2C^2s^2 + 3/RCs + 1] = i\,[1/R^3C^3s^3 + 5/R^2C^2s^2 + 6/RCs + 1]$, $\upsilon_4 = i_{C4}/Cs + \upsilon_3 = (i/Cs)\,[1/R^3C^3s^3 + 5/R^2C^2s^2 + 6/RCs + 1 + 1/R^2C^2s^2 + 4/RCs + 3] = (i/Cs)\,[1/R^3C^3s^3 + 6/R^2C^2s^2 + 10/RCs + 4]$. Now, $\upsilon_x = \upsilon_4$, and $\upsilon_o = -R_f\,i$. Thus $L(s) = \dfrac{\upsilon_o}{\upsilon_x} = -R_fCs\,[1/R^3C^3s^3 + 6/R^2C^2s^2 + 10/RCs + 4]$

$= \dfrac{\mathbf{-R_f\,C}}{\mathbf{1/R^3C^3s^4 + 6/R^2C^2s^3 + 10/RCs^2 + 4/s}}$. Now for $s = j\omega$, the phase of $L(j\omega) = 0$ when $6/R^2C^2\omega^3 = 4/\omega$, or $\omega = \sqrt{\dfrac{3}{2}}\,\dfrac{1}{RC} = \dfrac{\mathbf{1.224}}{\mathbf{RC}}$, where $|L|$ must be 1 (or more). For $1 = \dfrac{-R_f\,C}{1/R^3C^3\omega^4 - 10/RC\,\omega^2}$, $R_f = -1/R^3\,C^4\,\omega^4 + 10/R\,C^2\,\omega^2 = 10\,(2/3)\,R - 1\,(2/3)^2\,R = (6.666 - .444)\,R =$ **6.22R**.

Note that the result for υ_3 above allows the solution of the 3-capacitor circuit in Ex. 12.5 of the Text, namely $L(s) = \dfrac{\upsilon_o}{\upsilon_3} = \dfrac{-R_fi}{(i/Cs)\,[1/R^2C^2s^2 + 4/RCs + 3]} = \dfrac{-R_fs^2RC^2}{1/RCs + 4 + 3RCs}$, and $L(j\omega) = \dfrac{\omega^2RC^2R_f}{4 + j\,(3\omega RC - 1/\omega\,RC)}$, as on page 852 of the Text.

12.13 For the filter with $C_4 = C_6 = C$ and $R_1 = R_2 = R_3 = R_5 = R$, $f_o = \dfrac{1}{2\pi\,CR}$, with $Q = \dfrac{R_6}{R}$, from Eq. 11.53 and 11.54 on page 794 of the Text. Now for $f_o = 10\text{kHz}$ with $C = 10\text{nF}$, $R = \dfrac{1}{2\pi \times 10 \times 10^{-9} \times 10^4} =$ **1.59kΩ**. While the solution to satisfy the distortion specification is quite complex, we can simplify the process by assuming that the signal at υ_2 is a square wave of 0.7V peak amplitude, for which the fundamental is $\dfrac{4\,(0.7)}{\pi} = 0.89\text{V}$ and the 3rd harmonic is $1/3\,(0.89) = 0.297\text{V}$, (from page 4 of the Text). For this situation, since the gain from υ_2 to υ_1 is 2 times, the peak output at υ_1 will be $2\,(0.89\text{V})$ peak = **1.78V peak**. Now the 3rd harmonic component at υ_2 will be less than 1/3 of the fundamental there (since the wave at υ_2 is not very square. Thus a rejection of $\dfrac{1/3}{1/100} = 33.3 \equiv 20\log 33.30 = 30\text{dB}$ will be enough. Now for the second-order bandpass filter, $T(s) = \dfrac{(\omega_o/Q)s}{s^2 + (\omega_o/Q)\,s + \omega_o^2}$ with gain of 1 at ω_o. Now for gain of 1/33.3 at 3

ω_o, $T(j\omega) = \dfrac{(\omega_o/Q)\,j\omega}{-\omega^2 + j\omega\,(\omega_o/Q) + \omega_o^2} = \dfrac{(\omega_o/Q)\,\omega}{\omega\,(\omega_o/Q) + j\,(\omega^2 - \omega_o^2)}$, that is,

$T(3\omega_o) = \dfrac{(\omega_o/Q)\,3\omega_o}{(3\omega_o\,\omega_o/Q)^2 + (9\omega_o^2 - \omega_o^2)^2)^{1/2}} = \dfrac{3}{100}$, whence $100/Q = ((3/Q)^2 + (9 - 1)^2)^{1/2}$, or $\dfrac{10^4}{Q^2} = 9/Q^2 +$

64, $\dfrac{10^4 - 9}{Q^2} = 64$, $Q^2 = \dfrac{10^4 - 9}{64} = 156$. Thus, $Q = 12.5$, for which $R_6 = 12.5\,R = 12.5\,(1.59) = $ **19.9kΩ**. Use $R_1 = \dfrac{1.78 - 0.7}{1mA} = $ **1.1kΩ**.

SECTION 12.3: LC and Crystal Oscillators

12.14 For the FETs, operation is at $I_{DSS} = 4\text{mA}$, with $g_m = 2\,I_{DSS}/|V_p| = 2\,(4)/2 = 4$ mA/V. For each FET, $r_o = \dfrac{V_A}{I} = \dfrac{100V}{4mA} = 25\text{kΩ}$. For $L = 10\mu\text{H}$ and $Q = 100$ at 1MHz, the equivalent parallel resistance is $R_p = (2\,\pi \times 10^6 \times 10 \times 10^{-6}) \times 100 = 6.28\text{kΩ}$. Now at resonance, R_p loads the tank consisting of L and C_1 with C_2 in series. Now the voltage across C_1, ie from the follower output to ground, is $\dfrac{C_2}{C_1 + C_2}$ of that across R_p. Since the power supplied to R_p comes via the capacitors, the current must be correspondingly increased. Thus the corresponding load on R_p is $\left[\dfrac{C_2}{C_1 + C_2}\right]^2 \times R_p = \left[\dfrac{C_2}{C_1 + C_2}\right]^2 \times 6.28$. Thus the total load on Q_1

is $R_L = 25\text{kΩ} \,\|\, 25\text{kΩ} \,\|\, \left[6.28\left[\dfrac{C_2}{C_1 + C_2}\right]^2\right] \,\|\, 10\text{kΩ}$, or $R_L = 5.56 \,\|\, \left[6.28\left[\dfrac{C_2}{C_1 + C_2}\right]^2\right]$, and the voltage gain of the amplifier is $\dfrac{R_L}{R_L + 1/g_m} = \dfrac{R_L}{R_L + 1/4}$ V/V. Now, looking back from the $C_1\,C_2$ node to the gate of Q_1, the Thevenin gain equivalent is a gain of $\dfrac{R_L}{R_L + 1/g_m} = \dfrac{5.56}{5.56 + .25} = 0.957\text{V/V}$, with a resistance of $\dfrac{1}{g_m} \,\|\,$

$R_L = \dfrac{.25\,(5.56)}{.25 + 5.56} = .239\text{kΩ}$. Now when loaded by the $R_p\,L\,C_1\,C_2$ network, the gain to the gate end of C_2

becomes $\dfrac{(0.959) \times \left[\dfrac{C_2}{C_1 + C_2}\right]^2 6.28}{6.28\left[\dfrac{C_2}{C_1 + C_2}\right]^2 + 0.239} \times \dfrac{C_1 + C_2}{C_2}$. Now for oscillation, this must be unity, that is

$0.957\,(6.28)\,\dfrac{1}{1 + C_1/C_2} = 6.28\left[\dfrac{1}{1 + C_1/C_2}\right]^2 + .239$.

Now let $x = \dfrac{1}{1 + C_1/C_2}$. Thus $6.28\,x^2 - 6.01\,x + .239 = 0$, $x = \dfrac{6.01 \pm \sqrt{6.01^2 - 4\,(6.28)\,(.239)}}{2\,(6.28)} = \dfrac{6.01 \pm 5.49}{2\,(6.28)} = .915$ or $.0414$ (unlikely). Therefore $\dfrac{1}{1 + C_1/C_2} = .915$, $1 + C_1/C_2 = 1.0929$, $C_1/C_2 = .0929$, or $C_2/C_1 = $ **10.76**.

Alternatively, ignoring the effect of R_p and using Eq. 12.21, $C_2/C_1 = g_m R = 4 \times (25 \,||\, 25 \,||\, 10) = 22.22$. Now R_p reflected through the capacitor network becomes $\left[\frac{C_2}{C_1 + C_2}\right]^2 \times R_p = \left[\frac{22.22}{1 + 22.22}\right]^2 6.28 = 5.75\text{k}\Omega$, and $R = 25 \,||\, 25 \,||\, 10 \,||\, 5.75 = 2.82\text{k}\Omega$, and $C_2/C_1 = g_m R = 4\,(2.82) = 11.28$. Now R_p reflected becomes $\left[\frac{C_2/C_1}{1 + C_2/C_1}\right]^2 R_p = \left[\frac{11.28}{1 + 11.28}\right]^2 6.28 = 5.30\text{k}\Omega$, for which $R = 5.55 \,||\, 5.30 = 2.71\text{k}\Omega$, and $C_2/C_1 = 4\,(2.71) = \mathbf{10.84}$.

Alternatively, from Eq. 12.21, $C_2/C_1 = g_m R$, and here $R = 5.56 \,||\, 6.28 \left[\frac{C_2}{C_1 + C_2}\right]^2$. Now let $C_2/C_1 = x$.

Thus $x = 4 \times 5.56 \,||\, \left[6.28\left[\frac{x}{1+x}\right]^2\right] = \frac{4 \times 5.56 \times 6.28 \times (\frac{x}{1+x})^2}{5.56 + 6.28\,(\frac{x}{1+x})^2}$. Thus $5.56\,x + 6.28\,x\,(\frac{x}{1+x})^2 = 139.7\,(\frac{x}{1+x})^2$, or $x = (\frac{x}{1+x})^2 \frac{(139.7 - 6.28x)}{5.56}$. As a means for solution, try $x = 10$: $x = (\frac{10}{11})^2 \frac{(139.7 - 62.8)}{5.56} = 11.34$. Try $x = 10.6$: $x = (\frac{10.6}{11.6})^2 \frac{(139.7 - 6.28\,(10.6))}{5.56} = 10.98$. Try $x = 10.8$: $x = (\frac{10.8}{11.8})^2 \frac{(139.7 - 6.28\,(10.8))}{5.56} = 10.83$. Thus $\mathbf{C_2 = 10.83\,C_1}$, and, from Eq. 12.20, $\omega_o = \left[L\,(\frac{C_1 C_2}{C_1 + C_2})\right]^{-1/2}$, $\frac{C_1 C_2}{C_1 + C_2} = \frac{1}{\omega_o^2 L} = \frac{C_1\,(10.83\,C_1)}{C_1 + 10.83\,C_1}$, or $C_1 = \frac{11.83}{10.83} \times \frac{1}{(2\pi \times 10^6)^2 \times 10 \times 10^{-6}} = \mathbf{2.767nF}$, and $C_2 = 10.83\,(2.767\text{nF}) = \mathbf{29.97nF}$.

Check: for $C \simeq 2.5\text{nF}$, $f = \frac{1}{2\pi}(LC)^{-1/2} = \frac{(2.5 \times 10^{-9} \times 10 \times 10^{-6})^{-1/2}}{2\pi} \simeq 1.006\text{MHz}$.

For loop-gain > 1: With C_2 set 5% lower than calculated, the loop gain will be $\frac{C_1 + C_2}{C_2} = \frac{C_1 + 10.83\,C_1\,(.95)}{10.83C\,(.95)} = 1.097 = 1.1$, or 10% larger.

For diode limiting: There are several views of the limiting mechanism: One is to find the gate conducting-diode resistance, Miller-multiplied by the follower action which reduces the existing load at the gate (R_p) by 10%. This value is about $10\,R_p$. Now r_g at I_G is about $\frac{25mV}{I_G}$. Since the loaded follower gain is originally $\frac{C_2}{C_1 + C_2} = \frac{1}{1 + C_1/C_2} = (1 + 1/10.83)^{-1} = 0.915$. Thus, $10\,(6.28) = \frac{25 \times 10^{-3}}{I_G}\,\frac{1}{(1 - .915)}$, and $I_G = \frac{25 \times 10^{-3}}{10\,(6.28)}(11.76) = 4.68\mu\text{A}$. Now for $I_G = 4.68\mu\text{A}$, $V_G = 0.7 + 25 \ln \frac{4.68 \times 10^{-3}}{1} = 0.57\text{V}$. Now for υ_{GS} raised by 0.57V, $i_D = 4\,(1 - \frac{0.57}{-2})^2 = 6.60\text{mA}$, of which 4mA is absorbed by the current source.

Now for a load resistance of $5.56 \,||\, (6.28 \,||\, 6.28)\,(\frac{C_2}{C_1 + C_2})^2 = 5.56 \,||\, \left[5.71\left[\frac{1}{1 + C_1/C_2}\right]^2\right]$ $= 5.56 \,||\, \left[5.71\left[\frac{1}{1 + \frac{1}{(10.84)\,(0.95)}}\right]^{-2}\right] = 5.56 \,||\, (5.71\,(.831)) = 5.56 \,||\, 4.745 = 2.56\text{k}\Omega$, the peak output swing is $2.56\text{k}\Omega$ $(6.60 - 4) = \mathbf{6.66V}$. A second, simpler, view of finding the peak output is simply to use $\upsilon_{GS} \le 0.7\text{V}$, say 0.6V, and proceed, as above, to find $i_D = 4\,(1 - \frac{0.6}{-2})^2 = 6.76\text{mA}$, and a peak output of $(2.56)\,(6.76 - 4) =$

7.1V, or, using a much simpler view of the load as 5.56||6.28 = 2.95kΩ, find a peak = 2.95 (6.76 − 4) = **8.1V**.

For triode-region limiting: For $V_{GG} = 0$, $V_t = +3V$, and $v_{GS} \approx 0$, triode-mode operation for Q_1 begins for $v_{gd} = |V_p| = 2V$, that is for $v_{G1} = 3 - 2 = 1V$. Now for $v_{G1} = 1V$ for a 2.56KΩ load, $i_d = 1/2.56 = 0.39mA$, $i_D = 4 + .39 = 4.39mA$, and $4.39 = 4\left(1 - \frac{v_{GS}}{-2}\right)^2$, $v_{GS1} = 2\left(\left(\frac{4.39}{4}\right)^{1/2} - 1\right) = 0.2V$, for which $v_{DS1} = 3 - 1 = 2V$, and $v_{DG1} = 3 - 1 - 0.2 = 1.8V$. Triode-mode operation begins for Q_2, for $v_{DG2} = v_{DS2} = 2V$, that is for $V_- = -3V$, when $v_{S1} = v_{D2} = v_o = -1V$. Now it is apparent that the worst case occurs for Q_1, although that for Q_2 is easier to calculate. For simplicity, use the $v_{GS} = 0$ characteristic. Now to reduce the loop gain from 1.1 to 1, the triode-region resistance must reduce the equivalent load resistance from $5.56 \| \left[(6.28)\left(\frac{C_2}{C_1 + C_2}\right)^2\right] = 5.56 \| (6.28\,(.831)) = 5.56 \| 5.22 = 2.69k\Omega$ to $2.69/1.1 = 2.45k\Omega$ to reduce the loop gain to 1. Now $R_T \| 2.69 = 2.45$, or $R_T = 2.45 \| -2.69 = 27.5k\Omega$. Thus, the triode slope resistance must be about 27.5kΩ. Now, $i_D = K\,(2\,(v_{GS} - V_T)\,v_{DS} - v_{DS}^2)$, which for $I_{DSS} = K\,V_p^2$, and $V_T = -V_p$ is $i_D = I_{DSS}\left(2\left(1 - \frac{v_{GS}}{V_p}\right)\frac{v_{DS}}{-V_p} - \left(\frac{v_{DS}}{V_p}\right)^2\right)$. Now for $v_{GS} \approx 0$, $i_D = 4\left(2\,\frac{v_{DS}}{2} - \left(\frac{v_{DS}}{2}\right)^2\right)$. Now, $\frac{\partial i_D}{\partial v_{DS}} = 4\left(1 - \frac{2}{4}v_{DS}\right)$, and $R_T = \frac{\partial v_{DS}}{\partial i_D} = (4 - 2\,v_{DS})^{-1}k\Omega$. [*Aside*: Note that $R_T = 0$ at $v_{DS} = 4/2 = 2 = |V_p|$, since V_A is assumed infinite.] Now $1/(4 - 2\,v_{DS}) = 27.5k\Omega$, for $1 = 110 - 55\,v_{DS}$, or $v_{DS} = (110 - 1)/(55) = 1.98V$. Accordingly, for $v_{DS} = 1.98V$ with 3V supplies, a peak signal output of about 3 − 2 = **1V** would result.

12.15 $\omega_s = (L\,C_s)^{-1/2} = 2\,\pi\,(2.015) \times 10^6 = 12.661 \times 10^6$ rad/s − − − (1). $\omega_p = \left[\frac{LC_s\,C_p}{C_s + C_p}\right]^{-1/2} = 2\,\pi\,(2.018) \times 10^6 = 12.679 \times 10^6$ rad/s − − − (2). $Q = \omega_o\,L/r = 50 \times 10^3$. Now $\omega_o \approx \omega_s$. Thus $r = \frac{12.66 \times 10^6}{50 \times 10^3} \times L = 253.2\,L$ − − − (3). From (2), $L = \frac{1}{\omega_p^2} \times \frac{C_s + C_p}{C_s\,C_p} = \frac{1}{\omega_p^2}\left(\frac{1}{C_s} + \frac{1}{C_p}\right)$, where $C_p = 4 \times 10^{-12}$F. Thus $L = (12.679 \times 10^6)^{-2}\left[\frac{1}{4 \times 10^{-12}} + \frac{1}{C_s}\right]$ − − − (4). Now, from (1), $L = \frac{1}{\omega_s^2}\left(\frac{1}{C_s}\right)$ or $1/C_s = \omega_s^2\,L$. From (4), $L = 1.555 \times 10^{-3} + \left(\frac{2.015}{2.018}\right)^2 L$, $L = \frac{1.555 \times 10^{-3}}{.00297}$ = **0.523H**, and r = 253.2 (.523) = **132.5Ω**, with $C_s = \frac{1}{\omega_s^2\,L} = \frac{1}{(12.661 \times 10^6)^2\,(.523)}$ = **0.0119pF**. Note, as suggested, that the topology of Fig. 12.16 is the same as that in Fig. 12.13, in which oscillation is possible with the crystal acting as an inductor. In both cases the amplifier input voltage across C_2 is C_1/C_2 times that across C_1. Now, for an amplifier with input resistance R_{in} and input voltage V_2 (across R_{in} and C_2), the voltage across C_1 is $V_1 = C_1/C_2\,V_2$, where, for no power loss, V^2/R must be the same in each case. Thus $\frac{V_2^2}{R_{in}} = \frac{V_1^2}{R_{pg}} = \frac{(C_1/C_2)^2\,V_2^2}{R_{eq}}$, whence $R_{eq} = (C_1/C_2)^2\,R_{in} \ge 1/100\,R_{in}$, for $C_2 = 10pF$ and $C_1 \ge 1pF$. Now, the equivalent load on the amplifier is $R_L = (R_1 + R_{eq})\,\|\,R_f$ and its gain is $-(g_{mp} + g_{mn})\,R_L = -2\,R_L$. For this gain, $R_{in} = \frac{R_f}{1 - -2R_L} = \frac{1000k\Omega}{1 + 2R_L}$, and $R_L = \left[R_1 + \frac{1000}{1 + 2R_L} \times (C_1/C_2)^2\right] \| 1000$ − − − (4). Now, the loop gain, $L = 2\,R_L\,C_1/C_2$, must exceed 1 for oscillation. Thus for $2\,R_L\,C_1/C_2 = 1$, $2\left[\left(R_1 + \frac{1000}{1 + 2R_L} \times \left(\frac{C_1}{C_2}\right)^2\right) \| 1000\right]\frac{C_1}{C_2} = 1$ − − − (5). Now, to solve (4) and (5), try appropriate values of C_2/C_1 as a measure of the required gain. *First*, for $C_2 = 10pF$ and $C_1 = 1pF$, $C_2/C_1 = 10/1 = 10$. From (4) and (5), or directly, $2\left[\left(R_1 + \frac{1000}{1 + 2R_L} \times \frac{1}{10^2}\right) \| 1000\right] \times \frac{1}{10}$

$= 1$, or $\frac{2R_L}{10} = 1$, or $R_L = 5\text{k}\Omega$, whence from (4), $5 = \left[R_1 + \frac{1000}{100\,(1+2\,(5))}\right] \| 1000$, or $R_1 + \frac{10}{11} = 5$, $R_1 = 5 - 0.9 = 4.1\text{k}\Omega$. Use **3.9kΩ**. *Second,* for larger C_1, say 2pF, $C_1/C_2 = 1/5$ and $2R_L \times 1/5 = 1$, or $R_L = 2.5\text{k}\Omega$, whence $2.5 \simeq R_1 + \frac{1000}{25\,(1+2\,(2.5))}$, or $R_1 = 2.5 - 40/6 = -4.17\text{k}\Omega$. This implies that for $R_1 = 0$, a critical value of C_1/C_2 exists: Ignoring R_f, from (4), $R_L = \frac{1000}{1+2R_L} \times (\frac{C_1}{C_2})^2$, and from (5) $2R_L \frac{C_1}{C_2} = 1$, or $R_L = \frac{1}{2\,C_1/C_2} = \frac{C_2}{2C_1}$. Substituting, $\frac{C_2}{2C_1} = \frac{1000}{1+2\frac{C_2}{2C_1}} \times (\frac{C_1}{C_2})^2$, $\frac{C_2}{C_1} = \frac{2000}{1+C_2/C_1} \times (\frac{C_1}{C_2})^2$, $\frac{C_2}{C_1} + (\frac{C_2}{C_1})^2 = \frac{2000}{(C_2/C_1)^2}$, or $(\frac{C_2}{C_1})^4 + (\frac{C_2}{C_1})^3 = 2000$. Thus, $C_2/C_1 < (2000)^{1/4} = 6.69$. Try $C_2/C_1 = (2000 - (C_2/C_1)^3)^{1/4}$. For $C_2/C_1 = 6$, $C_2/C_1 = (2000 - 6^3)^{1/4} = 6.5$. For $C_2/C_1 = 6.5$, $C_2/C_1 = (2000 - 6.5^3)^{1/4} = 6.44$. For $C_2/C_1 = 6.3$, $C_2/C_1 = (2000 - 6.3^3)^{1/4} = 6.46$. Thus $C_2/C_1 =$ **6.4** for which $C_1 = \frac{10}{6.4} = 1.56\text{pF}$ is an estimate of the critical value for which $R_1 =$ **0kΩ**. Otherwise the largest value of R_1 occurs for the smallest value of C_1, that is for $C_1 =$ **1pF**, and is $R_1 \simeq$ **4.1kΩ**.

SECTION 12.4: Bistable Multivibrators

12.16 As specified, there is at most a 0.1V drop in the output device(s) with a voltage of $4.9 - 0 = 5.0 - 0.1 = 4.9\text{V}$ across the series connection of R_1 and R_2. For corresponding operation in triode mode, $i_D = K\,[2\,(v_{GS} - V_t)\,v_{DS} - v_{DS}^2]$, or $\frac{4.9}{R_1 + R_2} = 1\,[2\,(5-1)\,0.1 - 0.1^2] = 0.8 - .01 = 0.79$, $R_1 + R_2 = \frac{4.9}{0.79} = 6.20\text{k}\Omega$ (or more). Now assuming that the threshold of the symmetric input inverter is at 2.5V, V_{TL} and V_{TH} each lie 0.5V beyond the threshold. To establish this 0.5V offset, $\frac{4.9-2.5}{R_2} = \frac{0.5}{R_1}$, for example. Thus, $\frac{R_2}{R_1} = \frac{2.4}{0.5} = 4.8$, or $R_2 = 4.8\,R_1$. Now $R_1 + R_2 = (1 + 4.8)\,R_1 = 6.20$, for which $R_1 = \frac{6.20}{5.8} =$ **1.07kΩ**, and $R_2 = 4.8\,(1.07) =$ **5.13kΩ**. In practice, 5% resistors of higher values such as 2.7kΩ and 13kΩ would do the job, although 1% values of 5.11kΩ and 24.9kΩ would be better. In general, higher-valued resistors would save power, ensure a full output swing, and improve output transition times, while lower-valued ones would tend to reduce regeneration time.

Now for device variation, the highest value of V_{th} of the input device occurs for K_p = 20% high, K_n 20% low, V_{tn} 20% high, and V_{tp} 20% low. Now at V_{th}, $K_n\,(V_{th} - V_{tn})^2 = K_p\,(5 - V_{th} - V_{tp})^2$, or $0.8\,(V_{th} - 1.2)^2 = 1.2\,(5 - V_{th} - 0.8)^2$, or $V_{th} - 1.2 = (\frac{1.2}{0.8})^2\,(4.2 - V_{th}) = 9.45 - 2.25\,V_{th}$. Thus, $3.25\,V_{th} = 10.65$, or $V_{th} =$ **3.28V**. With $V_{th} = 3.28\text{V}$ and input at V_{IH}, $\frac{V_{IH} - 3.28}{R_1} = \frac{3.28 - 0.1}{R_2}$, or $V_{IH} = \frac{1.07}{5.13}\,(3.27) + 3.28 =$ **3.96V**. Likewise $V_{IL} = \frac{-1.07}{5.13}\,(4.9 - 3.28) + 3.28 =$ **2.94V**. Now for the reversed set of extremes, by symmetry, $V_{th} = 2.5 - (3.28 - 2.5) =$ **1.72V**, for which $V_{IH} = 5 - 2.94 =$ **2.06V**, and $V_{IL} = 5 - 3.96 =$ **1.04V**. Notice that for the two extreme versions that the two threshold sets do not overlap, although both lie within the power-supply range (0 to 5V).

Concerning the rise and fall times, a complete calculation is somewhat complex, involving estimates of the transition times of the internal node, the V_{IL} to V_{IH} range of the second inverter and its gain, as well as switching of the output inverter. Alternatively, a quick estimate of output rise and fall times is defined by the current available from each output transistor as $K\,(5-1)^2 = 16\text{K} = 16\text{mA}$. Thus t_r, between 0.5 and 4.5V, for a 1pF load, exceeds $\frac{(4.5-0.5) \times 1 \times 10^{-12}}{16 \times 10^{-3}} =$ **0.25ns**. A closer estimate could be found by including the effect of current reduction due to triode operation over a large part of the switching range. As an approximation, the current for $\upsilon_{DS} = 0.5V$ is $i_D = 1\,(2\,(5-1)\,0.5 - 0.5^2) = 3.75mA$. Thus a better estimate of the average current would be $\frac{16+3.75}{2} = 9.875mA$, and of the transition time as $\frac{4.5-0.5 \times 1 \times 10^{-12}}{9.9 \times 10^{-3}}$ **= 0.40ns.**

The delay requested includes a lot of factors, the most important of which are first the time for the first inverter to reach its threshold, and then for the second to do so. The former is dominated by the input itself which to rise (for example) from 0V to $V_{IH} = 3V$, by 3V, takes (3V/1V/μs) = 3μs!! Now the output of the first inverter changes at a rate defined by the input-signal rate and by its load capacitance. Likely the former dominates. Now, due to R_1, R_2 the signal at the gate of the first inverter pair is $\frac{R_2}{R_1+R_2} = \frac{5.13}{1.07+5.13} =$ 0.83 of the input. Now the gain of the first inverter at the middle of the switching region is $g_m\, r_o$ where, at the middle $i_D = K\,(\upsilon_{GS} - V_t)^2 = 1\,(2.5-1)^2 = 2.25mA$, $g_m = 2\,(2.5-1) = 3mA/V$, and $r_o = \frac{V_A}{i_D} = \frac{30}{2.25} =$ 13.3kΩ. Thus, the gain is $g_m\, r_o = 3 \times 13.3 = 40V/V$. Thus the rate of change of the output of the first inverter is limited to 1V/μs × .83 × 40 = 33.2V/μs, before regeneration. Thus the output of the first inverter will move from 0 or 5V to 2.5V in somewhat more than $\frac{2.5V}{33.2V/\mu s}$ **= 75.3ns.** After regeneration, the remainder of the transition will be faster, with the output current driving the 11pF load. Shortly after the transition begins as the circuit regenerates, the available current will be at least the 2.25mA middle bias current, for which the remaining transition time will be about $T = \frac{11 \times 10^{-12}\,(2.5-0.5)}{2.25 \times 10^{-3}} \simeq$ **10ns.** Now the final (isolated) conclusion about regeneration that can be easily made, is to estimate the time it takes for the input of the first inverter to cross the remaining half of its active region as driven through R_2 by the output changing by all (or part of) the output range. Now as implied earlier, the input active region is approximated by 5V/40 = .125V, and half is about 62mV. Now for a 5V change in output, the current change in R_2 is about 5V/5.13kΩ = 0.97mA, most of which supplies the 10pF input capacitance. For the input to move 62mV, takes about $\frac{10 \times 10^{-12} \times 62 \times 10^{-3}}{.97 \times 10^{-3}}$ **= 0.64ns.** Thus, regeneration, once it begins as the output of the second gate begins to move, can be very fast.

12.17 Now for υ_O (at node *B*) high, $\upsilon_A = 1.4V$. For $I_{D1} = 1mA$, $\frac{13.0-1.4}{R_2} - \frac{1.4}{R_1} = 1$, whence $\frac{11.6}{R_2} - \frac{1.4}{R_1} = 1$ – – – (1). Now for υ_o low, $V_A = V_{D4} + V_{D3} - V_{D2} =$ **0.7V**, and for $\Phi_{D2} = 1mA$, $\frac{0.7--13}{R_2} + \frac{0.7}{R_1} = 1$, or $\frac{13.7}{R_2} + \frac{0.7}{R_1} = 1$ – – – (2). Now adding (2) + (2) + (1) → $\frac{11.6}{R_2} + \frac{27.4}{R_2} + 0 = 3$, $R_2 = \frac{11.6+27.4}{3} = \frac{39}{3} =$ 13kΩ, and $\frac{11.6}{13} - \frac{1.4}{R_1} = 1$, and $R_1 = \frac{-1.4}{1-.892} = -13kΩ$. Now, the negative value of R_1, not available without active components, implies that the specifications are too tight. For example, allow the current in I_{D2} to increase, say to 2mA, for which change (2) becomes $\frac{13.7}{R_2} + \frac{0.7}{R_1} = 2$. Now combining the old (1) with the new (2) → $\frac{11.6}{R_2} - \frac{27.4}{R_2} = 5$, $R_2 = \frac{39}{5} =$ **7.8kΩ**, and $R_1 = \frac{-1.4}{1-11.6/7.8} = \frac{-1.4}{1-1.487} =$ **2.87kΩ.** Note that while this is a solution, that the latter formulation indicates the possibility of a solution for which $R_1 = \infty$, where the term in the denominator reaches zero. Examining the circuit with R_1 removed, one sees this possibility directly. Thus, for $R_1 = \infty$, either $\frac{13.0-1.4}{R_2} = 1$ for which $R_2 = 11.6kΩ$, or $\frac{0.7--13}{R_2} = 1$, for which $R_2 = 13.7kΩ$. Now, to ensure that the associated current exceeds 1mA, the smaller value (or one even smaller) should be chosen. For convenience, make $R_1 = \infty$ and $R_2 =$ **10kΩ**, for which $I_{D1} =$

$\frac{13.0-1.4}{10} = 1.16\text{mA}$, and $I_{D2} = \frac{13+0.7}{10} = 1.37\text{mA}$. Now for D_2 conducting, the current in D_3 and D_4 must exceed 1mA. Thus $R_3 = \frac{15-0.7-0.7}{1+1.37} = 5.74\text{k}\Omega$. Use $R_3 = \mathbf{5.6k\Omega}$. Now, the maximum current in D_4 is $\frac{15-1.4}{5.6} + 1.16 = 3.6\text{mA}$ (< 4mA, as required). Finally, note that the input thresholds are defined by the possible voltages at node A, namely $V_{TH} = \mathbf{+1.4V}$ and $V_{TL} = \mathbf{+0.7V}$. For inputs lower than 0.7V, the output is high, and $\upsilon_A = 1.4\text{V}$. Now, as υ_I rises and just exceeds 1.4V, υ_O falls, and υ_A falls to 0.7V. For υ_I > 1.4V, $\upsilon_O = -13\text{V}$. Now as υ_I falls, at $\upsilon_I = 0.7\text{V}$, the output reverses again, and υ_O goes to +13V.

12.18

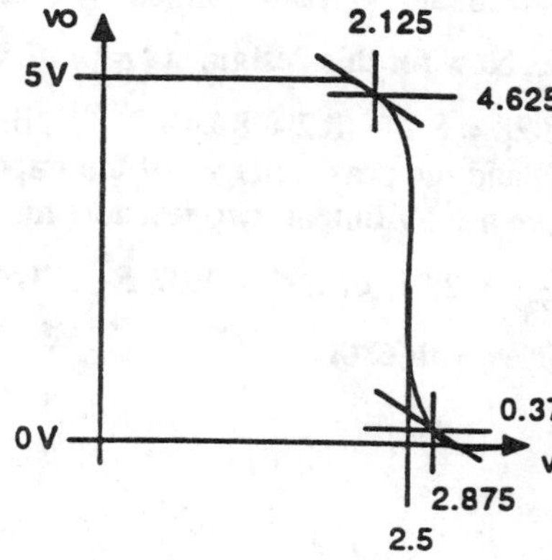

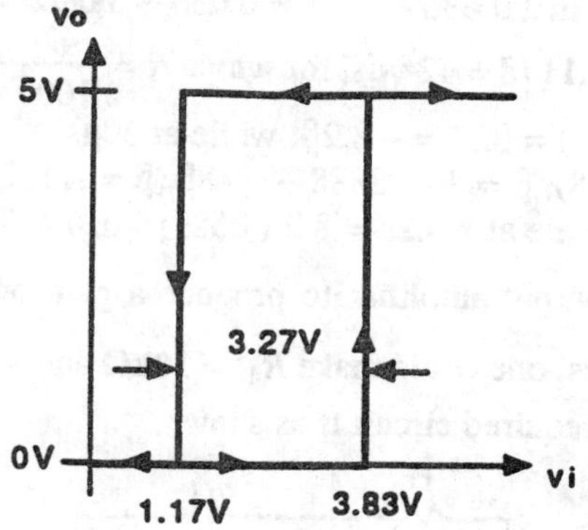

For the Q_3, Q_4 inverter, since $K_3 = K_4$ and $|V_t|$ are equal, $V_{th} = V_{DD}/2 = 5\text{V}$. As noted on page 934 of the Text, the transfer characteristic slope exceeds 1 in magnitude for υ_I between 5/8 $V_{DD} - V_t/4$, and 3/8 $V_{DD} + V_t/4$, or $V_{DD}/2 \pm (V_{DD}/8 - V_t/4) = 2.5 \pm (.625 - .25) = 2.5 \pm .375 = 2.875\text{V}$ and 2.125V, for which υ_O is $(V_{DD}/8 - V_t/4)$, and $V_{DD} - (V_{DD}/8 - V_t/4)$, or .375V and 4.625V respectively. While a relatively complete transfer characteristic is provided for interest, the value $V_{th} = 2.5\text{V}$ is the most essential feature. We will use this to estimate the overall characteristic υ_O/υ_I. Essentially, for the voltage at the internal node (A), υ_A < 2.5V, υ_O is high, which for $\upsilon_A > 2.5$, υ_O is low, in both of which cases, positive feedback via Q_5, Q_6 forces node A away from 2.5V in the direction to which it tends. Correspondingly, the thresholds at the input are the voltages υ_I for which the voltage υ_A is 2.5V with $\upsilon_I = V_{IL}$ when Q_6 conducts (for υ_O high (5V)), and $\upsilon_I = V_{IH}$ when Q_5 conducts (for υ_O low (0V)). For $V_{IL} = \upsilon$, $i_1 = i_2 + i_6$ for all devices in saturation, $K_1 (5-\upsilon-1)^2 = K_2 (\upsilon-1)^2 + K_6 (5-1)^2$. Now $K_1 = K_2 = 2K_6$. Thus $2(4-\upsilon)^2 = 2(\upsilon-1)^2 + 4^2$, $32 - 16\upsilon + 2\upsilon^2 = 2\upsilon^2 - 4\upsilon + 2 + 16$ $\upsilon(-16+4) = 16 + 2 - 32 = -14$, and $\upsilon = 14/12 = 1.167\text{V}$. Thus $V_{IL} = \mathbf{1.167V}$ and, symmetrically, $V_{IH} = 5 - 1.167 = \mathbf{3.83V}$. *Check*: $2(4-1.167)^2 = 2(1.167-1)^2 + 1(5-1)^2$, $16.052 = 0.056 + 16 \rightarrow$ **OK**.

SECTION 12.5: Generation of Square and Triangular Waveforms Using A Stable Multivibrator

12.19 Here, for a 6.8V zener, the voltage values at υ_{02} are $\pm(6.8 + 2(0.7)) = \pm 8.2\text{V}$. Now the υ_{03} output amplifier is a unity-gain follower. Thus the voltage at the RC common node is a triangle (approximately) of $\pm$ 1V peak amplitude. Now for the notation on pages 869, 870 of the Text, $L^+ = 8.2\text{V}$, $\beta = \frac{R_1}{R_1+R_2}$ and $\beta(8.2\text{V}) = 1\text{V}$, whence $\beta = .12195$, and $R_1 = .12195(R_1+R_2)$ and $R_1 = .1389 R_2$. Now, from Eq. 12.31, with $L^+ = -L^-$, $T_1 = \frac{1/10kHz}{2} = RC \ln\frac{(1-\beta(-1))}{(1-\beta)} = RC \ln\frac{1+\beta}{1-\beta} = RC \ln\frac{1+.12195}{1-.12195} = .245RC$. Thus $R = \frac{50\times10^{-6}}{.245\times1000\times10^{-12}} = 0.204\text{M}\Omega$. Use $R = \mathbf{200k\Omega}$, for which $RC = 200\mu\text{s}$, $R_2 = 200\text{k}\Omega$, $R_1 = .1389(200) = 27.8\text{k}\Omega$, or, better, $R_2 = \mathbf{240k\Omega}$ and $R_1 = 33.3\text{k}\Omega$, for which use **33kΩ**. For R_3, the current exceeds

$(1mA + 2\,(\frac{8.2}{200k\Omega})) \simeq 1.08mA$. Thus $R_3 = \frac{13-8.2}{1.08} = 4.4k\Omega$. Use **3.9kΩ**. Now, using the notation on page 870 of the Text, $\upsilon^- = L^+ - (L^+ - \beta L-)e^{-t/\tau}$, or $\upsilon^- = 8.2 - (8.2-1)e^{-t/RC}$, in general. Now $\frac{\partial\upsilon}{\partial t} = -7.2\,(\frac{1}{RC})e^{-t/RC} = \frac{7.2}{RC}e^{-t/RC}$. Thus at $t = 0$, slope is $\frac{7.2V}{200\times10^{-6}s}$ = **0.036V/μs**, and at $t = 50$ μs, slope is $.036\,e^{-\frac{50}{200}} = .036\,(.779)$ = **0.028V/μs**. Thus the average slope is $\frac{0.036+.028}{2}$ = **0.032V/μs**, and the slope change is .036 – .028 = 0.008V/μs. For the slope change reduced by half, reduce the voltage swing across C, so that $0.036\,e^{-50/RC} = 0.036 - .008/2 = 0.032$, or $e^{-50/RC} = .8888$, $50/RC = -\ln .889 = 0.1178$. Thus $RC = 50/.1178 = 424/\mu s$, for which $R = \frac{424\times10^{-6}}{1000\times10^{-12}}$ = **424kΩ**. Now for this design, at $t = 0$, $\upsilon^- = \upsilon = L^+ - L^+ - \beta L^-) = \beta L^- = -8.2\beta$, while at 50μs, $\upsilon^- = +8.2\beta$. Thus $8.2\beta = 8.2 - (8.2 + 8.2\beta)e^{-50/424}$, $\beta = 1 - (1+\beta)(.8888)$, $\beta = 1 - .8888 - .888\beta$, $\beta = .1112/1.8888 = 0.0589$, and the peak voltages of the improved triangle wave are at $\pm\,8.2\beta = 8.2\,(.0589) = \pm\,0.483V$. Now to achieve a ± 1V output, two resistors must be added to the output amplifier to produce a gain of $1 + \frac{R_2'}{R_1'} = \frac{1}{.483} = 2.07$, or $R_2' = 1.07\,R_1'$. To arrange such values, one could make R_1' = **10kΩ** and R_2', as **10kΩ in series with 680Ω**.

12.20 The required circuit is as shown:

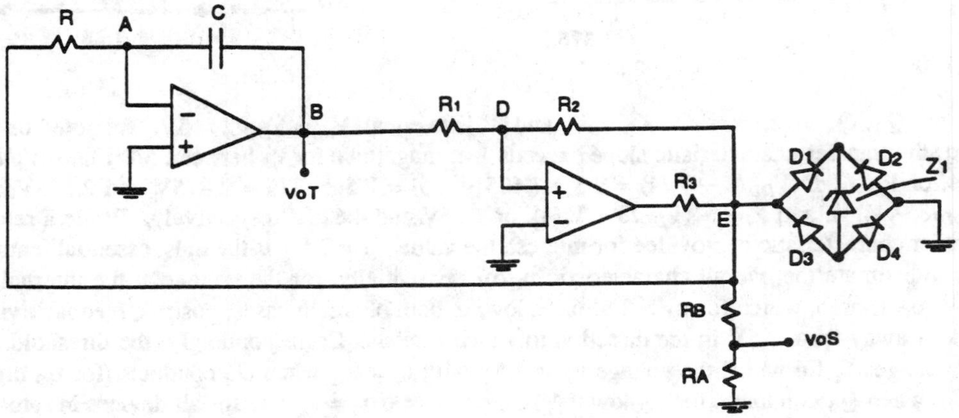

For a 6.8V zener and 0.7V diodes, the voltage at E is ± (6.8 + 2 (0.7)) = ± 8.2V. For 1mA drain, $R_A + R_B = \frac{8.2}{1mA} = 8.2k\Omega$. For ±1V square waves, $\frac{R_A}{R_A+R_B}\times 8.2 = 1$, that is R_A = **1kΩ** and R_B = **7.2kΩ**. For ±1V triangle waves, the thresholds at node B must be ±1V. Now, for R_1 = **10kΩ**, $R_2 = \frac{8.2}{1}R_1$ = **82kΩ**. Now for C = **1000pF**, to charge to 1V in $\frac{1/10kHz}{2} = 50\mu s$, the current in R must be $I = \frac{1000\times10^{-12}\times1}{50\times10^{-6}} = 20\mu A$. Thus $R = \frac{8.2V - 0V}{20\mu A}$ = **410kΩ**. Now the current in R_3 is 1mA for R_A and R_B, 1mA for the zener, 0.1mA for R_C, and 20μA for R. Thus $R_3 = \frac{13-8.2}{1+1+1+.02} = 1.589k\Omega$. Use R_3 = **1.5kΩ**. Note how easy this circuit is to design! This is because it has a desirable direct relationship between particular components and specific functions!

SECTION 12.6: Generation of a Standardized Pulse – The Monostable Multivibrator

12.21 See that for B more positive than A, $\upsilon_D = +10\text{V}$ and υ_A rises very slowly until $\upsilon_A > \upsilon_B$, at which point υ_D goes to -10V, υ_B falls to $-10 \times \dfrac{R_1 + R_5}{R_1 + R_5 + R_2}$, and A goes slowly to -0.7V.

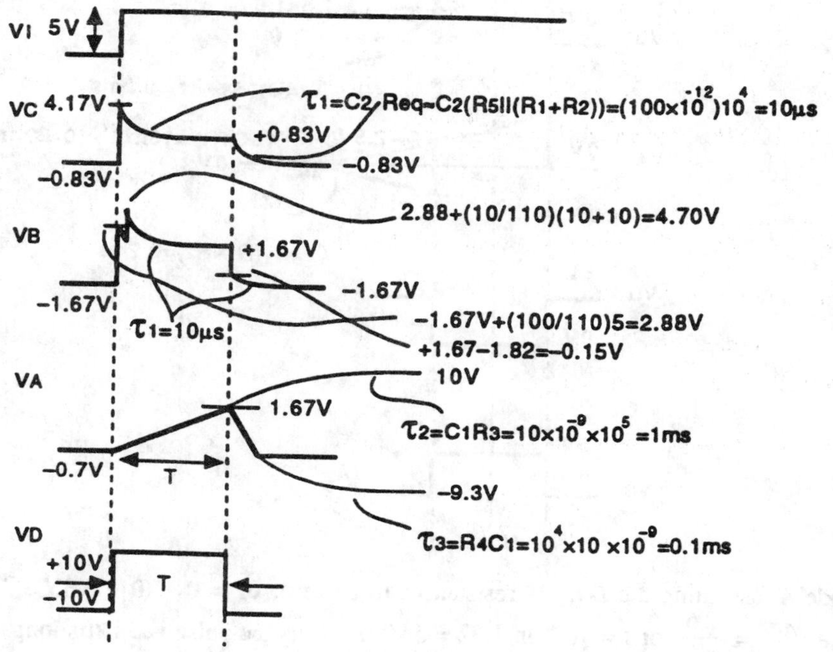

Since υ_A is limited to -0.7V, and υ_B can be made lower, the output remains at -10V in a stable state. In this state, for this design: $V_C = \dfrac{10R_5}{R_5 + R_1 + R_2} = \dfrac{-10\,(10)}{10 + 10 + 100} =$ **-0.833V**, $V_B = \dfrac{-10\,(R_5 + R_1)}{R_5 + R_1 + R_2} = \dfrac{-10\,(10 + 10)}{120} =$ **-1.67V**, $V_A =$ **-0.7V**, $V_D =$ **-10V**. To trigger the circuit, υ_A must rise to -0.7 or by $(1.67 - 0.7) = 0.97\text{V}$, and υ_C to υ where $\upsilon \dfrac{100}{10 + 100} = 0.97$, or $\upsilon = +0.97\ (1.1) = 1.067\text{V}$, that is from -0.833 to $+0.233\text{V}$. Thus the positive step at υ_I required is $\upsilon_I =$ **1.07V**. From waveform υ_A, see that the pulse ends when $10 - (10 - -0.7)\, e^{-t/1ms} = 1.67\text{V}$, or $10.7\, e^{t/10^{-3}} = 8.33$, $e^{-t/10^{-3}} = 0.778$, or $t = -10^{-3} \ln .778 = 0.25\text{ms}$. Thus the pulse length produced is **0.25ms**. *For a rate-limited input*: $i = \dfrac{C\, d\upsilon}{dt}$ and $i\ (R_5 \| (R_1 + R_2)) = 1.07\text{V}$. Thus $1000 \times 10^{-9} \dfrac{d\upsilon}{dt} \times 10^4 \,||\, (11 \times 10^4) = 1.07$, and $\dfrac{d\upsilon}{dt} = \dfrac{1.07}{0.91 \times 10^4 \times 1000 \times 10^{-9}} = 117.6\text{V/s}$ = **118V/s**. *For recovery*: See that for a positive input pulse shorter than the output pulse, that the longest recovery time constant, controlled by R_4 is $\tau_3 = R_4\, C_1 = 0.1\text{ms}$, where $\upsilon_A = -9.3 + (9.3 + 1.67)e^{-t/10^{-4}}$, and recovery is complete when $\upsilon_A = 0.7 = -9.3 + \ (10.97)e^{-t/10^{-4}}$, or $e^{-t/10^{-4}} = \dfrac{10}{10.97}$, or $t = 10^{-4}\ (.093) = 9.3\mu\text{s}$.

Thus it appears that retriggering could occur 10µs or so after the output falls (assuming zero recovery time for the amplifier). In practice, recovery of the amplifier itself (from limiting) might require somewhat more time.

12.22

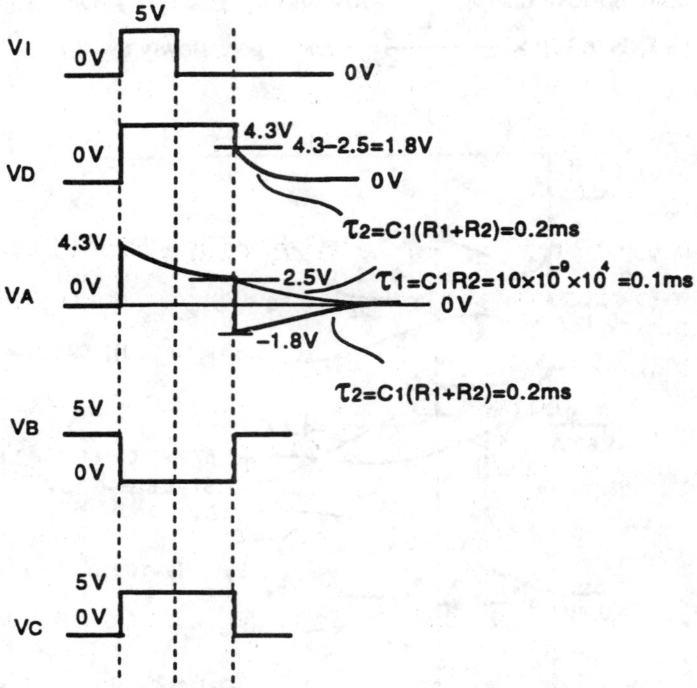

For node A, assuming the D_1, D_2 resistances to be zero, $\upsilon_A = 0 - (0 - 4.3\text{V})e^{-t/R_2C_1}$. Now, $\upsilon_A = 2.5$V where $e^{-t/10^{-4}} = \frac{2.5}{4.3}$, or $t = 10^{-4} \ln 1.72 = 54.2\mu$s. Thus the pulse is **54.2μs** long. Normally, υ_I would be very much shorter than that, a few tens or hundreds of ns, at most, the delay through the two inverters. For very long pulses at the input, the loop is held open, and the output fall time is an amplified version of that at node A. For gain of $40 \times 40 = 1600$V/V, and 5V output swing, the input active region is about $5/1600 = 3 \times 1 \times 10^{-3}$V. Now, $\upsilon_A = 4.3e^{-t/R_2C_1} = 4.3e^{-t/10^{-4}}$, and $\frac{\partial \upsilon_A}{\partial t} = \frac{-4.3}{10^{-4}} e^{-t/10^{-4}}$. At $t = 54.2\mu$s, $e^{-t/10^{-4}} = \frac{2.5}{4.3}$, $\frac{\partial \upsilon_A}{\partial t} = \frac{-4.3}{10^{-4}} \times \frac{2.5}{4.3} = \frac{2.5}{10^{-4}} = 2.5 \times 10^4$V/s. Thus the output fall time $\approx \frac{3.1 \times 10^{-3}V}{2.5 \times 10^4}$ = **124ns.**

SECTION 12.7: Integrated – Circuit Timers

12.23 From Eq. 12.39, $T = CR \ln 3 = 1.1 \times 10 \times 10^{-9} \times 10^4$ = **0.11ms**. The input pulse must be shorter than **0.11ms** by an amount which guarantees the relationship for component variation. For longer inputs, both comparator outputs are high, and the flip-flop is set and reset at the same time.

12.24 Extending Equation 12.38 for the case in which the capacitor voltage is not quite zero, but rather υ, $\upsilon_c = V_{CC} - (V_{CC} - \upsilon)\, e^{-t/RC} = 5 - (5 - \upsilon)e^{-t/RC}$. Now for $\upsilon_c = 2/3\, V_{CC} = 10/3$V, $10/3 = 5 - (5 - \upsilon)e^{-t/RC}$, $e^{-t/RC} = \frac{-3.333 + 5}{5 - \upsilon} = \frac{1.666}{5 - \upsilon}$, $t = RC \ln \frac{5 - \upsilon}{1.666} = RC \ln (3 - 0.6\upsilon)$, and $\frac{\partial t}{\partial \upsilon} = -0.6\, RC \frac{1}{(3 - 0.6\upsilon)}$. For υ small, $\frac{\partial t}{\partial \upsilon} = -\frac{0.6RC}{3} = -0.2\, RC$. Now for $T = 1.1\, RC$, for a 0.1V change in υ, the change in T is $-0.2\, RC\, (0.1)$, or $-\frac{0.2RC\,(0.1)}{1.1RC} \times 100 =$ **−1.8%**.

12.25 From page 878 of the Text, $T_H = 0.69\ C\ (R_A + R_B) = 0.69 \times 10 \times 10^{-9}\ (2 \times 10^4) = 138\mu s$, and $T_L = 0.69\ C\ R_B = 0.69 \times 10 \times 10^{-9}\ (10^4) = \frac{138}{2} = 69\mu s$. Thus the period is $T = 138 + 69 =$ **207**μ**s**, and the frequency $1/207 \times 10^{-6} =$ **4.83kHz**, with duty cycle $\frac{138}{207} \times 100 =$ **66.7%**. For $R_A = 10k\Omega$ and $R_B = 1k\Omega$, $T_H = 0.69\ C\ (R_A + R_B) = 0.69 \times 10^{-8}\ (10 + 1) \times 10^3 = 75.9\mu s$, and $T_L = 0.69\ C\ (R_B) = 0.69 \times 10^{-8}\ (10^3) = 6.9\mu s$, for which the frequency is $\frac{1}{75.9 + 6.9} =$ **12.1kHz**, and duty cycle is $75.9/(75.9 + 6.9) \times 100 =$ **91.7%**. For the same frequency, $T = 207 \times 10^{-6} = 6.9 \times 10^{-6} + .69 \times 10^{-8}\ (R_A + 1)\ 10^3$, $R_A = \frac{200 \times 10^{-6}}{0.69 \times 10^{-5}} - 1 =$ **28k**Ω. For 10kHz, $T = 100\mu s$, $100 \times 10^{-6} = .69\ C\ (R_A + 2R_B)$, $R_A + 2\,R_B = \frac{100 \times 10^{-6}}{.69 \times 10^{-8}} = 1.45 \times 10^4 =$ **14.5k**Ω. There is no combination of resistors which will produce 10% duty cycle. Use 90% duty cycle and an inverter!

SECTION 12.8: Nonlinear Waveform - Shaping Circuits

12.26 For a sine-wave of peak output υ, ie $\upsilon \sin \omega t$, the zero-crossing $|$slope$|$ is $\upsilon\omega$ volts per second. Thus the triangle wave reaches $\upsilon\omega \times T/4$, or $\upsilon\ (2\pi f) \times (1/4f) = 2\pi\upsilon/4 = \pi\upsilon/2 = 1.57\ \upsilon$, at the peak. Though the choice is arbitrary, let us assume a sine wave peak of 0.7V, with I_D peak = 1mA, such that the triangle input peak = $.7 \times 1.57 =$ **1.10V**, with the drop across R being (.57) (.7) = .400V. Thus $R = 0.400/1mA =$ **400**Ω. Now, in general, $\upsilon_i = 1.10\ (\Theta/90)$, or $\Theta = 81.8\ \upsilon_i$, over the range 0° to 90°. Also $\upsilon = 700 + 50 \ln i/1$, and $i = e^{(\upsilon-700)/50}$. *For* $\upsilon_o = 0.7V$, $\Theta = 90°$, $\upsilon_i = 1.10V$, $I = \frac{1.10 - 0.7}{400} = 1mA$. *For* $\upsilon_o = 0.65V$, $i = 1\ e^{(650-700)/.50} = 0.368mA$, $\upsilon_i = .65 + .368\ (.4) = 0.79$, $\Theta = 81.8\ (.797) = 65.2°$, and $0.7 \sin 65.2° = 0.635V$. *For* $\upsilon_o = 0.60V$, $i = e^{(+600-700)/50} = .135mA$, $\upsilon_i = .60 + .135\ (.4) = 0.654$, $\Theta = 81.8\ (.654) = 53.5°$, $0.7 \sin 49.5° = .563V$. *For* $\upsilon_o = 0.55V$, $i = e^{(550-700)/50} = .050\ mA$, $\upsilon_i = 0.55 + .05\ (.4) = 0.57V$, $\Theta = 81.8\ (.57) = 46.6°$, $0.7 \sin 46.6° = 0.509V$. *For* $\upsilon_o = 0.50V$, $i = e^{(500-700)/50} = .018mA$, $\upsilon_i = 0.50 + .018\ (.4) = .507$, $\Theta = 81.8\ (.507) = 41.5°$, $0.7 \sin 41.5° = .464V$. *For* $\upsilon_o = 0.45V$, $i = e^{(450-700)/50} = .007$, $\upsilon_i = 0.45 + .007\ (.4) = .453$, $\Theta = 81.8\ (.453) = 37.1°$, $0.7 \sin 41.5° = .422V$. *For* $\upsilon_o = 0.40V$, $i = e^{(400-700)/50} = .0025$, $\upsilon_i = 0.40 + .0025\ (.4) = .401$, $\Theta = 81.8\ (.401) = 32.8°$, $0.7 \sin 32.8° = .379V$. *For* $\upsilon_o = 0.35V$, $i = e^{(350-700)/50} = .0009$, $\upsilon_i = .35 + .0009\ (.4) = .350$, $\Theta = 81.8\ (.350) = 28.6°$, $0.7 \sin 28.6° = .335V$. *For* $\upsilon_o = 0.3V = \upsilon_i$, $\Theta = 81.8\ (.3) = 24.5°$, $0.7 \sin 24.5 = .291V$. *For* $\upsilon_o = 0.2V = \upsilon_i$, $\Theta = 81.8\ (.2) = 16.3°$, $0.7 \sin 16.3° = .197V$. *For* $\upsilon_o = 0.1V = \upsilon_i$, $\Theta = 81.8\ (.1) = 8.18°$, $0.7 \sin 8.18° = .099V$. *For* $\upsilon_o = 0V = \upsilon_i$, $\Theta = 0$, 0V. In summary:

$\Theta°$	90	65.2	53.5	46.6	41.5	32.8	24.5	16.3	8.2	0
υ_o, V	0.7	0.65	0.60	0.55	0.50	0.40	0.30	0.2	0.1	0
0.7 sinΘV	0.7	0.635	0.563	0.509	0.464	0.379	0.291	0.197	0.099	0
ε, mV	0	15	37	41	36	21	9	3	1	0
ε%	0	2.4	6.6	8.1	7.8	6.4	3.1	1.5	1	0

Note that the output wave is generally "fatter" than the sine wave.

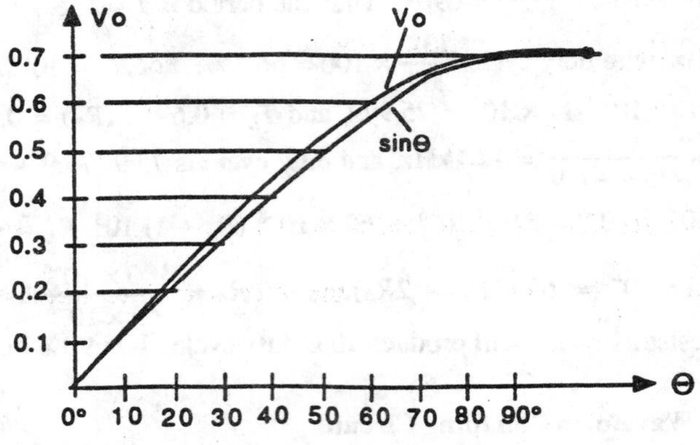

12.27 Here, $i = 0.1\upsilon^2$, with a match at υ = 2, 4, and 8V, using 0, 3, and 7V supplies. For υ = 2V, i = 0.4mA, R_1 = 2/0.4mA = **5kΩ** as before. Now chose $V_2 = 3 - 0.6 =$ **2.4V**, and $V_3 = 7 - 0.6 =$ **6.4V**. For υ = 4V, $i = 4/5 + \frac{4-0.6-2.4}{R_2+0.1} = 0.1(4)^2 = 1.6$mA. Thus, $\frac{1}{R_2+0.1} = 1.6 - .8 = .8$mA, $R_2 = \frac{1}{0.8} - 0.1 =$ **1.15kΩ**.

For υ = 8V, $i = \frac{8}{5} + \frac{8-3}{1.15+0.1} + \frac{8-0.6-6.4}{R_3+0.1} = 0.1(8)^2 = 6.4$mA. Thus $1/(R_3+0.1) = 6.4 - 1.6 - 4 =$ 0.8mA, $R_3 = 1/0.8 - 0.1 =$ **1.15kΩ**.

Now for the errors: At 3V, $i = \frac{3}{5} = 0.6$mA, rather than 0.1 (3^2) = 0.9mA, with an error = **– 0.3mA**. At 5V, $i = \frac{5}{5} + \frac{5-3}{1.15+.1} = 1 + 1.6 = 2.6$mA, rather than 0.1 $(5)^2$ = 2.5mA, for an error of **+0.1mA**. Thus at 7V, the error is **–0.3mA**, and at 10V, it is **0mA**, just as in Ex. 12.22.

Now for 1mA diodes with n = 2, with V_2 = **2.4V**, V_3 = **6.4V**. Now at υ = 2, i = 0.4mA and R_1 = **5kΩ** as before. Now at υ = 4V, i = 1.6mA, with $I_{D2} = 1.6 - 0.8 = 0.8$mA. For 0.8mA, $\upsilon_{D2} = 700 + 50 \ln \frac{0.8}{1} =$ 689mV. Thus $R_2 = \frac{4-2.4-.689}{0.8} =$ **1.139kΩ**. Now at υ = 8V, i = 6.4mA. Here $I_{R1} = \frac{8}{5} = 1.6$mA, and $I_{R2} \simeq \frac{8-2.4-0.7}{1.139} = 4.30$mA, $\upsilon_{D2} = 700 + 50 \ln 4.3/1 = 773$mV, $I_{D2} = \frac{8-2.4-.773}{1.139} = 4.24$mA (OK), Thus, $I_{D3} = 6.4 - 1.6 - 4.24 = 0.56$mA, $\upsilon_{D3} = 700 + 50 \ln .56 = 671$mV. Thus $R_3 = \frac{8-6.4-.671}{.56} =$ **1.659kΩ**.

Now for the errors: At 3V, $i = 3/5 + i_{D2}$, $\upsilon_D = 3 - 2.4 - I_{R2} = 0.6 - I_{R2}$. Try i_D = .05mA, $\upsilon_D = 700 + 50 \ln .05 = 550$mV, $\upsilon_{R2} = .05 \times 1.139 = .057$V, and $\upsilon_D + \upsilon_{R2} = .607$V (rather than .600V) (OK). Thus $i = .6 + .05 = .65$mA, rather than 0.9mA, for an error of **–0.25mA**. At 5V, the required i is 0.1 $(5)^2$ = 2.5mA, and the actual $i = \frac{5}{5} + \frac{5-2.4-V_{D2}}{1.139} \simeq 1 + \frac{5-2.4-0.7}{1.139} = 1 + \frac{1.9}{1.139} = 1 + 1.67 = 2.67$mA. Now for D_2, and I_{D2} = 1.67mA, $\upsilon_{D2} = 700 + 50 \ln 1.67 = 726$mV, $i_{D2} = \frac{5-2.4-.73}{1.139} = 1.64$mA. Overall, $i = 1 + 1.64 =$ 2.64mA with an error of **+0.14mA**. At 7V, required $i = (0.1)(7)^2 = 4.9$mA, and the actual $i = \frac{7}{5} + \frac{7-2.4-V_{D2}}{1.139} + \frac{7-6.4-V_{DS}}{1.659}$, which for $\upsilon_{D2} \simeq .75$V, $i_{D2} = \frac{7-2.4-0.75}{1.139} = 3.38$mA. Now $\upsilon_{D2} \simeq 700 + 50 \ln 3.38 = 761$mV, $i_{D2} = \frac{7-2.4-.76}{1.139} = 3.37$mA. Now $i_{D3} = \frac{7-6.4-.6}{1.659} \simeq 0$. Thus $i = 1.4 + 3.37 =$ 4.77mA, with an error of – **0.13mA**. At 10V, required $i = (0.1)\,10^2 = 10$mA. Actual $i = \frac{10}{5} +$

$\frac{10-2.4-V_{D2}}{1.139}+\frac{10-6.4-V_{D3}}{1.659}$. Now $i_{D2} \approx \frac{10-2.4-.78}{1.139} = 5.99\text{mA}$, for which $\upsilon_{D2} = 700 + 50 \ln 5.99 = .79\text{V}$, $i_{D2} = \frac{10-2.4-.79}{1.139} = 5.99\text{mA}$. Now $i_{D3} \approx \frac{10-6.4-0.72}{1.659} = 1.74\text{mA}$, for which $\upsilon_{D2} = 700 + 50 \ln 1.74 = .73\text{V}$, $i_{D2} = \frac{10-6.4-.73}{1.659} = 1.73\text{mA}$. Thus $i = 2.00 + 5.99 + 1.73 = 9.72$ with an error of **– 0.28mA**.

12.28 For each of the top two circuits to which υ_1 and υ_2 are applied as in Fig. 12.44, $i_D = \frac{\upsilon_I}{R} = I_S\, e^{-\upsilon_o/nV_T}$ where $\upsilon_O = -\upsilon_D$, $\upsilon_O = -nV_T \ln \frac{\upsilon_I}{R\,I_S}$, for $\upsilon_I > 0$. For the lower circuit to which we apply $-\upsilon_3$: $\upsilon_O = +nV_T \ln \frac{\upsilon_3}{R\,I_S}$. Thus $\upsilon_D = -\,nV_T\,(\ln \frac{\upsilon_3}{R\,I_S} - \ln \frac{\upsilon_1}{R\,I_S} - \ln \frac{\upsilon_2}{R\,I_S})\,(\frac{-R}{R}) = nV_T \ln \left[\frac{\frac{\upsilon_1 \times \upsilon_2}{(R\,I_S)^2}}{\frac{\upsilon_3}{R\,I_S}}\right] = nV_T \ln \left[\frac{\upsilon_1\,\upsilon_2}{\upsilon_3} \times \frac{1}{R\,I_S}\right]$. Now $i_{D4} = I_S\, e^{\frac{nV_T}{nV_T}\ln\left[\frac{\upsilon_1\,\upsilon_2}{R\,I_s\,\upsilon_3}\right]} = I_S \frac{\upsilon_1\,\upsilon_2}{R\,I_S\,\upsilon_3}$, and $\upsilon_O = -\frac{R\,I_S\,\upsilon_1\,\upsilon_2}{R\,I_S\,\upsilon_3} = -\frac{\upsilon_1\,\upsilon_2}{\upsilon_3}$. Thus to obtain $\upsilon_O = \frac{\upsilon_1\,\upsilon_2}{\upsilon_3}$, add one unity-gain inverter at the input of the lower logarithmic circuit (for υ_3), and a second unity-gain inverter at the output of the antilog circuit.

Now, as a check, for a 1mA diode with $n = 2$, 1kΩ inputs, and voltages of 0.5, 1, 2, 3V applied: at 1mA, υ = 0.700V; at 0.5mA, $\upsilon = 700 + 50 \ln 0.5 = 665.3\text{mV}$, at 2.0mA; $\upsilon = 700 + 50 \ln 2 = 734.6\text{mV}$; at 3.0mA, $\upsilon = 700 + 50 \ln 3 = 754.9\text{mV}$. Test $\frac{0.5 \times 0.5}{3}$: See $\upsilon_D = -1/1\,(-.6653 - .6653 + .7549) = .5757\text{V}$. Now $\upsilon_4 = 1\, e^{(-700+575.7)/50} = .0832\text{mA}$, $\upsilon_O = .0832 = 1/12$ as required. Test $\frac{3 \times 3}{.5}$: See $\upsilon_D = \frac{-1}{1}(-754.9 - 754{,}9 + 665.3) = 844.5$. Now $i_4 = e^{(-700+844.5)/50} \approx 17.993$, $\upsilon_O = 17.993\text{V} \approx 18$ as required, provided of course the supply voltages are high enough! Test $\frac{1 \times 1}{1}$: see directly OK, with $\upsilon_O = 1$.

SECTION 12.9: Precision Rectifier Circuits

12.29 This is called an absolute-value circuit.

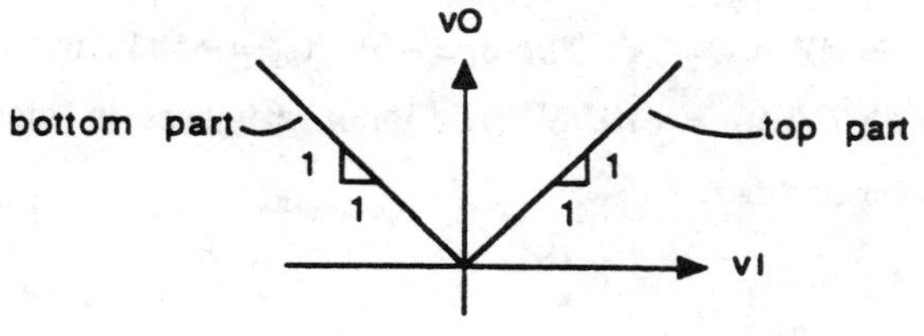

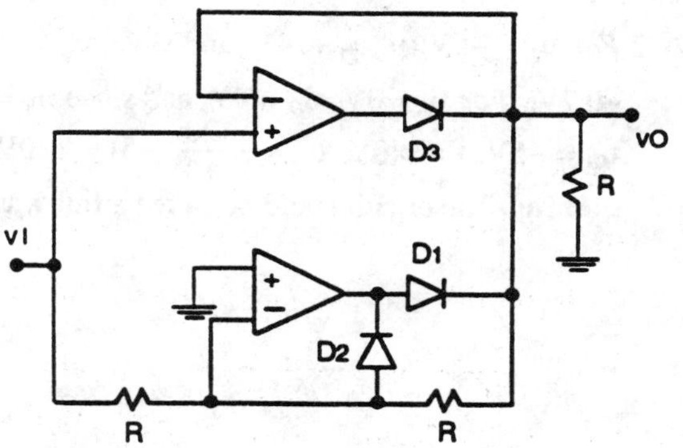

12.30

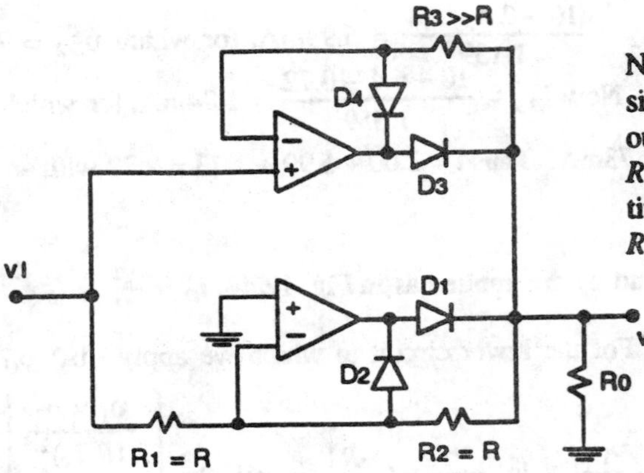

Note that while R_2 is a constant load on the output, since its leftmost end is always at ground, the non-output end of R_3 is connected to the ac input. Thus $R_3 \gg R_2$ for least effect, and usually $R_2 > R$ for relative efficiency. For equivalent offset current effects, $R_3 = R_1 \| R_2 = R/2$ with $R \gg R_0$.

12.31

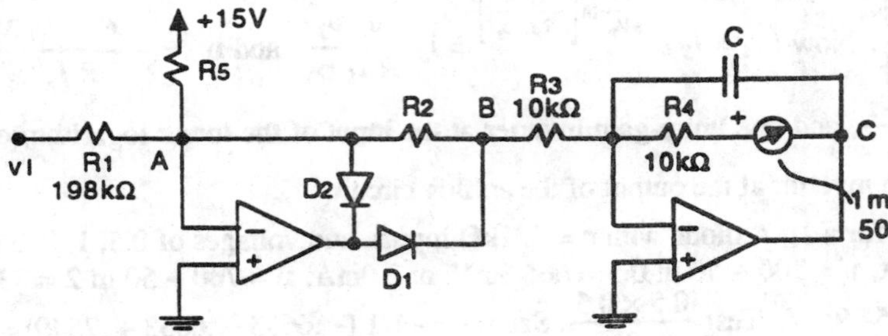

Here, 100Vrms = 141.4V peak, 140Vrms = 198.0V peak. Design for +10V at node B for 140Vrms input. Now R_5 supplies a current which cancels the effect of υ_I as υ_I goes negative, until it reaches 100V rms. Thus $\dfrac{R_5}{R_1} = \dfrac{15V}{141.4} \rightarrow R_5 = 0.1061\,R_1$. Chose R_1 so the change of input from 141.4V peak to 198V peak produces an extra 1mA. Thus $\dfrac{198 - 141.4}{R_1} = 1\text{mA}$, or $R_1 = 56.6\text{k}\Omega$ – use 56kΩ in series with 620Ω. For $\upsilon_B = 10$V full scale with 1mA, $R_2 = 10\text{k}\Omega$. Also $R_5 = 0.1061 \times 56.6\text{k}\Omega = 6.00\text{k}\Omega$ – use two 12kΩ resistors in parallel.

12.32 *For* $\upsilon_I = +5$V: $\upsilon_O = +10$V, and since $\upsilon_B = +5$V, $\upsilon_C = 0$V, $\upsilon_E = (1 + \dfrac{10}{10})\,5 = +10$V, $\upsilon_A = +10.7$V, $\upsilon_D = -0.7$V. *For* $\upsilon_I = 0$V: $\upsilon_O = 0$V, and since $\upsilon_B = 0$V, $\upsilon_C = 0$V, $\upsilon_E = 0$V. *For* $\upsilon_I = -5$V: $\upsilon_O = +10$V, and since $\upsilon_B = -5$V, $\upsilon_C = 0$V, $\upsilon_E = -\dfrac{20}{10}\,(-5) = +10$V, $\upsilon_A = -5.7$V, $\upsilon_D = +10.7$V. The input resistance is (ideally) **infinite**. The circuit could be called a full-wave doubler rectifier.

12.33

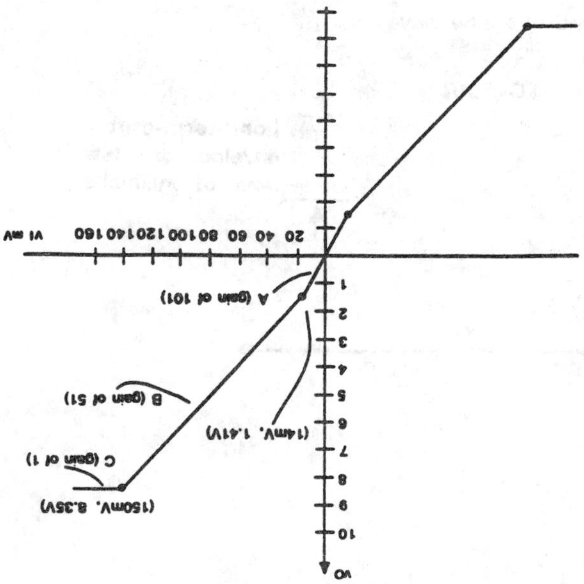

Note that the gain is generally $v_O/v_I = 1 + R_2/R_1$, for R_2/R_1 $v_I < (+0.7 + 0.7) = 1.4\text{V}$, $1 + (R_2 \| R_3)/R_1$ for $((R_2 \| R_3)/R_1)$ v_I between 1.4V and 1.4 + 6.8 = 8.2V when the zener conducts, and $1 + (R_2 \| R_3 \| 0)/R_1 =$ 1V/V beyond. In region *A*, for $0 < v_I < 1.4 \dfrac{R_1}{R_2} = 1.4 \dfrac{1}{100} = 14\text{mV}$, $\dfrac{v_O}{v_I} = 1 + \dfrac{R_2}{R_1} = 1 + \dfrac{100}{1} = \mathbf{101V/V}$, and v_O reaches $101 \times \dfrac{14}{1000} = \mathbf{1.414V}$. In region *B*, for $14\text{mV} < v_I < 14\text{mV} + 6.8\text{V} \times \dfrac{1}{100 \| 100} = \mathbf{150mV}$, $\dfrac{v_O}{v_I} = 1 + \dfrac{(R_2 \| R_3)}{R_1} = 1 + \dfrac{100 \| 100}{1} = \mathbf{51V/V}$, and v_O reaches $\dfrac{(150 - 14)}{1000} 51 + 1.414 = \mathbf{8.35V}$. In region *C*, for $150\text{mV} < v_I < \infty$, $\dfrac{v_O}{v_I} = \mathbf{1V/V}$. Due to the bridge connection, saturation is symmetric for positive and negative inputs.

12.34 The circuit is a dc restorer and rectifier, using the lower and upper amplifiers, respectively, to create ideal diodes. For a 100mV peak sine wave at the input, the voltage at the intermediate node (*B*) is a 100mV sine with lower peaks at 0V and upper peaks at 200mV. Correspondingly, the output v_o would be a dc level of 200mV, which would remain if the input is lowered or was removed. To return the output to zero from a peak-to-peak input v, one could use a resistor R_1 to ground at the output, or (better) to a negative supply connected to the output. For a ground connection, and a return to 0V in $10/f$, the time constant *RC* is such that $1/f < R_1 C < 10/f$. For 95% recovery, one must wait 3 time constants (since $e^{-3} = .05$). Thus, $3 R_1 C = 10/f$, and $R_1 = \dfrac{10}{3f C} = \dfrac{3.3}{fC}$. For input average drift at a low rate, the drift signal is coupled via the input *C* to node *B* where the resistance is infinite. Thus the average voltage at *B* would rise, and the output at *C* would rise to represent the maximum peak-to-peak value of the combined signal at *f* and *f*/100. To correct this, add a resistor R_2 from node *B* to ground (or to a negative supply). *Now*, from a filtering point of view, we want a high-pass filter with a pole at f_o, for which *f* is clearly in the passband and *f*/100 is rejected as much as possible. Now the transfer function from v_I to v_B is $T(s) = \dfrac{RCs}{1 + R\,Cs}$, and $T(j\omega) = \dfrac{jRC\omega}{1 + jw\,RC} = \dfrac{1}{1 + \dfrac{j}{RC\omega}}$, whence $|T| = \left[\dfrac{1}{1 + \dfrac{1}{(RC\omega)^2}}\right]^{½} = 0.95$. Now, at ω_o, $|T| = 1 + \dfrac{1}{(RC\omega_o)^2} = 1.108$, $\dfrac{1}{RC\omega_o} = (1.108 - 1)^{½} = 0.33$, or $\omega_o = \dfrac{3}{RC} = 2\pi f$, where $R_2 = \dfrac{3}{2\pi fC} = \dfrac{.48}{fC}$. Now at $\dfrac{f}{100}$, for $T(s) = \dfrac{1}{1 + \dfrac{1}{RCs}} = \dfrac{1}{1 + \dfrac{2\pi f}{3s}}$, $|T(\dfrac{f}{100})| = \left[\dfrac{1}{1 + \left[\dfrac{2\pi f}{3(2\pi f/100)}\right]^2}\right]^{½} = \left[\dfrac{1}{1 + (\dfrac{100}{3})^2}\right]^{½} = \dfrac{1}{33.3} \equiv 20 \log .03 = -30.5\text{dB}$. This implies that a "volt or so" of drift will be reduced to "1/30 volts or so", or to around 30mV, still significant, but about as good as can be done.

12.34

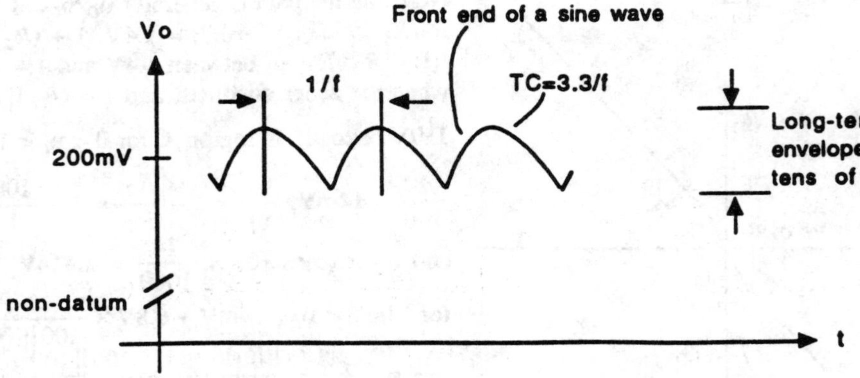
Vo
Front end of a sine wave
1/f
TC=3.3/f
200mV
Long-term-drift
envelope of a few
tens of millivolts
non-datum
t

Chapter 13

MOS DIGITAL CIRCUITS

SECTION 13.1: Logic Circuits – Some Basic Concepts

13.1 $V_{OH} = V^+ = \mathbf{3V}$, $V_{OL} = 50\text{mV} + (3.00 - .05)\ 100/(500 + 100) = \mathbf{0.542V}$

13.2 $NM_H = V_{OH} - V_{IH} = 3.5 - 2.2 = \mathbf{1.3V}$; $NM_L = V_{IL} - V_{OL} = 1.2 - 0.5 = \mathbf{0.7V}$. Transition region $= V_{IH} - V_{IL} = 2.2 - 1.2 = \mathbf{1.0V}$. Average gain $= -\left(\frac{V_{OH} - V_{OL}}{V_{IH} - V_{IL}}\right) = -\frac{3.5 - 0.5}{2.2 - 1.2} = -\frac{3.0}{1.0} = \mathbf{-3V/V}$.

13.3 *For* t_{PHL}: $C = 50\text{pF}$, $R = .50 \,||\, 0.1 = 0.083\text{k}\Omega$, $\tau = RC = 50 \times 10^{-12} \times 83.3 = 4.165\text{ns}$, $\upsilon_o = .542 + (3.00 - .542)\ e^{-t/4.165} = 1.5\text{V}$, whence $t_{PHL} = -4.165 \ln \frac{1.5 - .542}{3.0 - .542} = \mathbf{3.92ns}$.

For t_{PLH}: $C = 50\text{pF}$, $R = 0.5\text{k}\Omega$, $\tau = RC = 50 \times 10^{-12} \times 0.5 \times 10^3 = 25\text{ns}$, $\upsilon_O = 3.0 + (.542 - 3.0)\ e^{-t/25} = 1.5\text{V}$, whence $t_{PLH} = -25 \ln \frac{-1.5}{-2.458} = \mathbf{12.35ns}$.

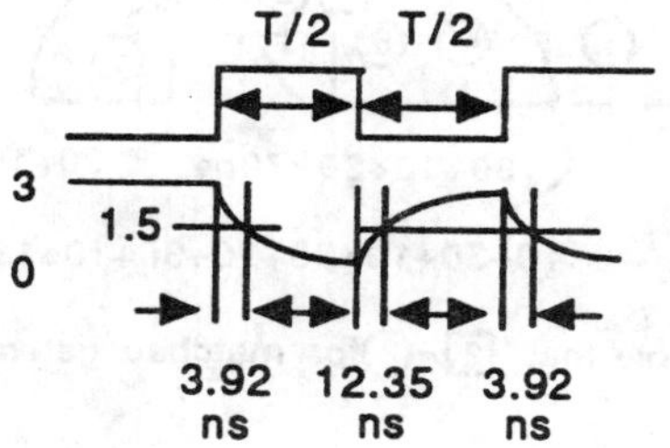

The output is above 1.5V for $T/2 - 12.35 + 3.92\text{ns} = T/2 - 8.43\text{ns}$. The requirement is that $T/4 = T/2 - 8.43$, $T/4 = 8.43$. Thus, $T = 33.72\text{ns}$, and $f = 1/T = \mathbf{29.7MHz}$. DC power $= 1/2 \left[0 + \frac{3 - 0.05}{500 + 100} \times 3\right] = 7.38\text{mW}$. AC power $= fCV^2 = 29.7 \times 10^6 \times 50 \times 10^{-12} \times 3^2 = 13.37\text{mW}$. Thus, the total power $= 13.37 + 7.38 = \mathbf{20.8mW}$, and the DP $= 20.8\text{mW} \times 12.35\text{ns} = \mathbf{256pJ}$.

13.4

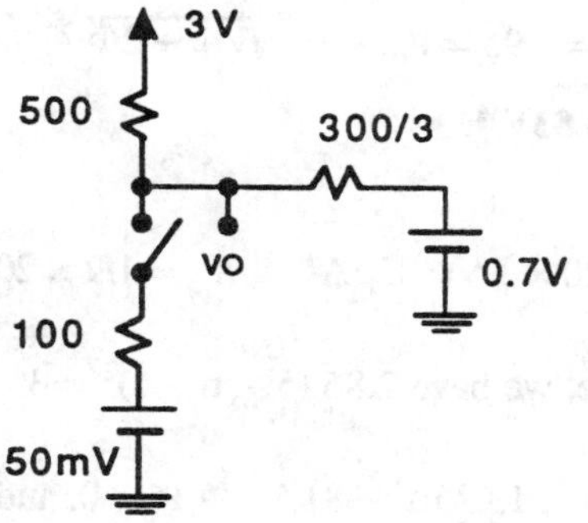

For the switch open, $V_{OH} = 0.7 + \frac{3 - 0.7}{100 + 500} \times 100 = \mathbf{1.083V}$. For the switch closed, $V_{OL} = \upsilon$, $\frac{3 - \upsilon}{0.5} + \frac{0.7 - \upsilon}{0.1} = \frac{\upsilon - .05}{0.1}$, and $6 - 2\upsilon + 7 - 10\upsilon = 10\upsilon - 0.5$, $22\upsilon = 13.5$, ie $V_{OL} = \upsilon = 13.5/22 = \mathbf{0.614V}$. For $V_{OL} = 0.614$, *I from each load* $= \frac{0.7 - 0.614}{0.3} = \mathbf{0.287mA}$. For $V_{OH} = 1.083$, *I to each load* $= \frac{1.083 - 0.7}{0.3} = \mathbf{1.28mA}$.

13.5 Dynamic Power $= f\,C\,(\Delta V)\,V_{DD} = f \times 10 \times 10^{-12} \times 4 \times 5 = 0.2 \times 10^{-9} f$. Static power $= 1/2\ (6 + 6/3)$ mW $= 4\text{mW}$, $DP = 50\text{pJ}$, $P = DP/t_p = 50\text{pJ}/5\text{ns} = \frac{50 \times 10^{-12}}{5 \times 10^{-9}} = 10\text{mW}$. Thus at f, $10 \times 10^{-3} = 4 \times 10^{-3} + 0.2 \times 10^{-9} f$, $0.2 \times 10^{-9} f = 6 \times 10^{-3}$, and $f = \frac{6 \times 10^{-3}}{0.2 \times 10^{-9}} = \mathbf{30 \times 10^6 Hz = 30MHz}$. Now, at 15MHz, $DP = 5 \times 10^{-9}\ (4 \times 10^{-3} + 0.2 \times 10^{-9} \times 15 \times 10^6) = 5 \times 10^{-9} \times 7 \times 10^{-3} = \mathbf{35pJ}$.

13.6 $t_p = (30 + 10)/2 = 20\text{ns}$. Total delay through 5 gates for 2 transitions each is $T = 5 \times 2 \times 20 = 200\text{ns}$. Frequency of oscillation = $1/T = 1/200\text{ns} = 5\text{MHz}$.

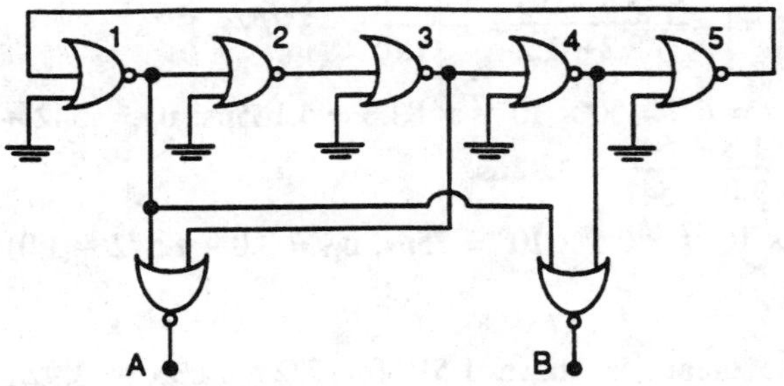

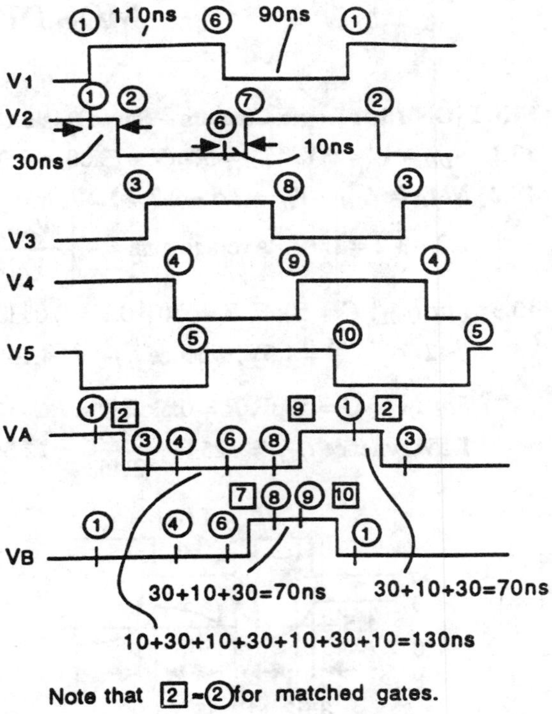

SECTION 13.2: NMOS Inverter with Enhancement Load

13.7 $V_{OH} = V_{DD} - V_t \rightarrow V_t = 5 - 3.05\text{V} = 1.95\text{V}$. Using Eq. 13.8: $V_t = 1.95 = V_{t0} + \gamma(\sqrt{V_{SB} + 2\Phi} - \sqrt{2\Phi})$ $= 1 + \gamma(\sqrt{3.05 + 0.6} - \sqrt{0.6})$. Thus $\gamma = \dfrac{1.95 - 1}{1.91 - .77} = \dfrac{0.95}{1.14} = \mathbf{0.83V^{1/2}}$.

13.8

(a) *No Body Effect*: $V_{OH} = 5 - 1 = \mathbf{4V}$; $V_{IL} = \mathbf{1V}$, $K_1 = 1/2 \times 20 \times 3.5 = 35\mu\text{A/V}^2$, $K_2 = 1/2 \times 20 \times 1/3.5 = 2.86\mu\text{A/V}^2$.

For $V_{OL} = \upsilon$: For Q_2 in saturation, and Q_1 in triode mode, we have $2.86\,(5 - \upsilon - 1)^2 = 35\,(2\,(4 - 1)\,\upsilon - \upsilon^2)$,

$(4 - \upsilon)^2 = 12.25\,(6\upsilon - \upsilon^2)$, $16 - 8\upsilon + \upsilon^2 = 73.5\text{V} = 12.25\upsilon^2$, $13.25\upsilon^2 - 81.5\,\upsilon + 16 = 0$, and

$\upsilon = \dfrac{+81.5 \pm \sqrt{81.5^2 - 4(16)\,(13.25)}}{2\,(13.25)}$. Thus, $V_{OL} = \dfrac{81.5 \pm 76.1}{2\,(13.25)} = \mathbf{0.204V}$. *Check*: (for υ very small) $2.86\,(4 - \upsilon)^2 = 35\,(6\upsilon) = \upsilon^2 - 8\upsilon + 16 = 73.4\upsilon$, $\upsilon^2 - 81.4\upsilon + 16 = 0$, $\upsilon = \dfrac{81.4 \pm \sqrt{81.4^2 - 4(16)}}{2} =$ 0.197V. **OK.**

For $V_{IH} = \upsilon$: For Q_2 in saturation, Q_1 in triode, υ_o small, we have $2.86\,(5 - \upsilon_O - 1)^2 = 35\,(2\,(\upsilon - 1)\,\upsilon_o - \upsilon_o^2)$ – – – (1). Now differentiating, $\dfrac{1}{2.86} \times \dfrac{\partial \upsilon_o}{\partial \upsilon}$, $\dfrac{1}{2.86} \times \dfrac{\partial \upsilon_O}{\partial \upsilon} \rightarrow 2\,(4 - \upsilon_O)\,(-\dfrac{\partial \upsilon_O}{\partial \upsilon}) =$ $12.25\,(2\upsilon_O + 2\,(\upsilon - 1)\,\dfrac{\partial \upsilon_O}{\partial \upsilon} - 2\upsilon_O\,\dfrac{\partial \upsilon_O}{\partial \upsilon})$. Now for $\dfrac{\partial \upsilon_O}{\partial \upsilon} = -1$, $2\,(4 - \upsilon_O) = 12.25\,(2\upsilon_O - 2\,(\upsilon - 1)$

$+ 2\upsilon_O$), or $8 - 2\upsilon_O = 49\upsilon_O - 24.5\upsilon + 24.5$, $51\upsilon_O = 24.5\upsilon - 16.5$, or $\upsilon = 2.08\upsilon_O + 0.673$ – – – (2). Substituting (2) in (1) $\rightarrow (4 - \upsilon_O)^2 = 12.25(2(2.08\upsilon_O + .673 - 1)\upsilon_O - \upsilon_O^2)$, $16 - 8\upsilon_O + \upsilon_O^2 = 50.96\upsilon_O^2 - 8.01\upsilon_O - 12.25\upsilon_O^2$, or

$37.71\upsilon_O^2 - .01\upsilon_O - 16 = 0$, $\upsilon_O = \dfrac{--.01 \pm \sqrt{.01^2 - 4(-16)(37.71)}}{2(37.71)} = \dfrac{+.01 \pm 49.13}{2(37.71)} = 0.652\text{V}$, for which $\upsilon = 2.08(.652) + 0.673 = 2.03\text{V}$. Thus, $V_{IH} = \mathbf{2.03V}$, for which $\upsilon_O = 0.652\text{V}$. Check in (1): $2.86(5 - .652 - 1)^2 = 35(2(1.03).652 - .652^2)$: 32.05 = 32.13; OK.

Now, as well, $NM_H = V_{OH} - V_{IH} = 4 - 2.03\text{V} = \mathbf{1.97V}$, and $NM_L = V_{IL} - V_{OL} = 1 - 0.204 = \mathbf{0.796V}$.

(b) Now, proceed using $V_{OH} = 4\text{V}$ and $V_{OL} = \mathbf{0.204V}$ in non-critical roles, but otherwise use the body effect.

For V_{OH}: See that $\upsilon_{OH} = V_{DD} = V_{to} - \gamma\sqrt{V_{OH} + 2\Phi_f} - \sqrt{2\Phi_f} = 5 - 1 - 0.5(\sqrt{4 + 0.6} - \sqrt{0.6})$. Thus, $V_{OH} = 4 - 0.5(2.14 - 0.77) = \mathbf{3.32V}$.

For V_{IH}: ($\approx$ 2.03, for which $\upsilon_o \approx 0.652\text{V}$). Now, for $\upsilon_o = .652$, $V_t = V_{t0} + \gamma(\sqrt{\upsilon_o + 2\Phi_f} - \sqrt{2\Phi_f}) = 1 + 0.5(\sqrt{.65 + 0.6} - \sqrt{0.6}) = 1 + 0.5(1.12 - .77) = 1.172$. Now, from (1): $2.86(5 - \upsilon_O - 1.17)^2 = 35(2(\upsilon - 1)\upsilon_O - \upsilon_O^2)$, where $\upsilon = V_{IH}$. Thus, $(3.83 - \upsilon_O)^2 = 12.24(2(\upsilon - 1)\upsilon_O - \upsilon_O^2)$ – – – (3).

Now, differentiating, $\dfrac{\partial}{\partial\upsilon} = 2(3.83 - \upsilon O)(-\dfrac{\partial\upsilon_O}{\partial\upsilon}) = 12.24(2\upsilon_O + 2(\upsilon - 1)\dfrac{\partial\upsilon_O}{\partial\upsilon} - \dfrac{2\upsilon_O}{\partial\upsilon}\partial\upsilon_O$.

Then, for $\dfrac{\partial\upsilon_O}{\partial\upsilon} = -1$, $2(3.83 - \upsilon_O) = 12.24(2\upsilon_O - 2\upsilon + 2 + 2\upsilon_O)$, $7.66 - 2\upsilon_O = 48.96\upsilon_O - 24.48\upsilon + 24.48$, $\upsilon = (50.96\upsilon_O + 16.82)/24.48 = 2.082\upsilon_O + .687$ – – – (4). Now, substituting (4) into (3): $(3.83 - \upsilon_O)^2 = 12.24(2(2.08\upsilon_O - .313)\upsilon_O - \upsilon_O^2)$, $14.67 - 7.66\upsilon_O + \upsilon_O^2 = 50.97\upsilon_O^2 - 7.66\upsilon_O - 12.24\upsilon_O^2$, $37.73\upsilon_O^2 - 14.67 = 0$, whence $\upsilon_O = (\dfrac{14.67}{37.73})^{1/2} = .623\text{V}$, for which $V_{IH} = 2.082(.623) + .687 = \mathbf{1.98V}$.

(c) *For V_{OL}*: ($\approx$ .204V), $V_t = V_{t0} + \gamma(\sqrt{0.204 + 0.6} - \sqrt{0.6}) = 1 + 0.5(.897 - .775) = 1.06\text{V}$. Thus, $2.86(5 - \upsilon - 1.06)^2 = 35(2(4 - 1)\upsilon - \upsilon^2)$, $(3.94 - \upsilon)^2 = 12.25(6\upsilon - \upsilon^2)$, $15.52 - 7.88\upsilon + \upsilon^2 = 73.5\upsilon - 12.25\upsilon^2$, $13.25\upsilon^2 - 81.38\upsilon + 15.52 = 0$. Thus, $\upsilon = \dfrac{81.38 \pm \sqrt{81.38^2 - 4(15.52)(13.25)}}{2(13.25)} = \dfrac{81.38 \pm 76.16}{2(13.25)}$. Correspondingly, $\mathbf{V_{OL} = 0.197V}$.

For V_{OH}: (Was 3.32V), $\upsilon_{OH} = 5 - 1 - 0.5(\sqrt{3.32 + 0.6} - \sqrt{0.6}) = 4 - 0.5(1.980 - 0.775)$, or $V_{OH} = \mathbf{3.40V}$.

For V_{IH}: (first, was 2.03V, with $\upsilon_O = .652\text{V}$, for $V_t = 1$, then, 1.98V, with $\upsilon_O = .623\text{V}$, for $V_t = 1.172\text{V}$). Now, for $\upsilon_O = .623$, $V_t = 1 + 0.5(\sqrt{.623 + .6} - \sqrt{.6}) = 1 + 0.5(1.106 - .775) = 1.166$. Thus, estimate $V_{IH} \approx \mathbf{1.99V}$, whence $NM_L = V_{OH} - V_{IH} = 3.40 - 1.99 = \mathbf{1.41V}$, and $NM_L = V_{IL} - V_{OL} = 1 - .197 = \mathbf{0.803V}$.

13.9 Here, $K_1 = 35\mu\text{A/V}^2$, and $K_2 = 2.86\mu\text{A/V}^2$.

(a) *Body effect neglected*: For $V_{th} = \upsilon$, $2.86(5 - \upsilon - 1)^2 = 35(\upsilon - 1)^2$, $4\upsilon = (\upsilon - 1)\sqrt{35/2.86} = 3.498$, $4 - \upsilon = 3.498\upsilon - 3.498$. Thus, $\upsilon = \dfrac{7.498}{4.498} = 1.67\text{V}$, and $V_{th} = \mathbf{1.67V}$.

(b) *For body effect included*, and $\upsilon_O < 1.67\text{V}$, say 1.6V, $V_t = V_{t0} + \gamma(\sqrt{\upsilon_O + 0.6} - \sqrt{0.6}) = 1 + 0.5(\sqrt{2.2} - \sqrt{0.6}) = 1 + 0.5(1.48 - .77) = 1.35\text{V}$. Thus, for $V_{th} = \upsilon$, $(5 - \upsilon - 1.35) = 3.498(\upsilon - 1)$, $3.65 - \upsilon = 3.498\upsilon - 3.498$, and $\upsilon = \dfrac{3.498 + 3.65}{4.498} = \mathbf{1.59V} = V_{th}$.

13.10 Here, $K_1 = 35\mu\text{A/V}^2$, $K_2 = 2.86\mu\text{A/V}^2$. For υ_O low, $i_{D2A} = 2.86(5 - 1)^2 = 45.8\mu\text{A}$. For υ_O in the middle, ie $\upsilon_O = 4 + 0.2/2 = 2.1$, $i_{D2M} \approx 2.86(5 - 2.1 - 1)^2 = 10.3\mu\text{A}$. Thus $I_{LH} = \dfrac{45.8 + 10.3}{2} = 28.1\mu\text{A}$, and t_{pLH}

$= 0.1 \times 10^{-12} \dfrac{(2.1-0.2)}{28.1 \times 10^{-6}}$ = **6.8ns**. For υ_O high, $i_{D2B} = 0$, $i_{D1D} = 35\,(4-1)^2 = 315\mu A$. For υ_O in the middle, $i_{D1N} = 35\,(2\,(4-1)\,2.1 - 2.1^2) = 287\mu A$, $i_{D2M} = 10.3\mu A$. Thus $I_{HL} = \dfrac{315+287-10.3}{2} = 296\mu A$, and $t_{pHL} = \dfrac{0.1 \times 10^{-12}\,(4-2.1)}{296 \times 10^{-6}}$ = **0.64ns**.

13.11 For the voltage levels essentially unaffected, reduce I_{HL} by a factor of 2, or raise I_{LH} by a factor of 2. For $I_{HL} = \dfrac{i_{D1D} + i_{D1N} - i_{D2M}}{2} = \dfrac{315+287-10.3}{2} = 296$. Thus, $\dfrac{K\,(315+287)-10.3}{2} = \dfrac{296}{2}$ or $K = \dfrac{296+10.3}{315+287} = 0.51$. Use $(W/L)_2 = 1/3.5 =$ **0.286**, and $(W/L)_1 = .51\,(3.5) =$ **1.79**. Alternatively, for $I_{LH} = \dfrac{(i_{D2A} + i_{D2M})}{2} = \dfrac{45.8+10.3}{2} = 28.0\mu A$, we can double $(W/L)_2$ to be $2/3.5 =$ **0.571**, while $(W/L)_1 =$ **3.5** as before.

For $V_{OL} = \upsilon$, $K_2\,(5-\upsilon-1)^2 = K_1\,(2\,(4-1)\,\upsilon - \upsilon^2)$ or $(4-\upsilon)^2 = K_1/K_2\,(6\,\upsilon - \upsilon^2)$. *For* $(W/L)_1$ *reduced*, $(4-\upsilon)^2 = \dfrac{1.79}{1/3.5}(6\,\upsilon-\upsilon^2) = 6.25\,(6\,\upsilon - \upsilon^2)$, $16 - 8\,\upsilon + \upsilon^2 = 37.5\,\upsilon - 6.25\,\upsilon^2$, $7.25\,\upsilon^2 - 45.5\,\upsilon + 16 = 0$, $\upsilon^2 - 6.28\,\upsilon + 2.21 = 0$, $\upsilon = \dfrac{--6.28 \pm \sqrt{6.28^2 - 4\,(2.21)}}{2} = \dfrac{6.28-5.53}{2} = 0.375V$, that is $V_{OL} =$ **.375V**.

For $(W/L)_2$ *increased*, $(4-\upsilon)^2 = 3.5/.571\,(6\,\upsilon - \upsilon^2) = 6.13\,(6\,\upsilon - \upsilon^2)$, $16 - 2\,\upsilon + \upsilon^2 = 36.78\,\upsilon - 6.13\,\upsilon^2$, $7.13\,\upsilon^2 - 38.78\,\upsilon + 16 = 0$, $\upsilon^2 - 5.44\,\upsilon + 2.24 = 0$, $\upsilon = -\dfrac{-5.44 \pm \sqrt{5.44^2 - 4\,(2.24)}}{2} = \dfrac{5.44-4.54}{2} =$ .449V, that is $V_{OL} =$ **.449V**. Note that for the original design, $V_{OL} =$ **0.204V**.

SECTION 13.3: NMOS Inverter with Depletion Load

13.12 Here, $K_1 = 1/2 \times 20 \times 7/2 = 35\mu A/V^2$, $K_2 = 1/2 \times 20 \times 2/7 = 2.86\mu A/V^2$. See $V_{OH} =$ **5V** and $V_{IL} \approx 1V$.

For $V_{OL} = \upsilon$, with Q_2 in saturation and Q_1 in triode operation, and no back-bias effect. $2.86\,(0-3)^2 = 35\,(2\,(5-1)\,\upsilon - \upsilon^2)$, $25.74 = 280\,\upsilon - 35\,\upsilon^2$, $\upsilon^2 - 8\,\upsilon + .735 = 0$, $\upsilon = \dfrac{+8 \pm \sqrt{8^2 - 4\,(.735)}}{2} = \dfrac{+8-7.81}{2} =$ .093. Thus $V_{OL} =$ **0.093V**. Check: since the operation is essentially linear, $9\,(2.86) = 280\,\upsilon$, $\upsilon = \dfrac{2.86(9)}{280}$ = .09V. OK.

For $V_{IH} = \upsilon$: Q_2 in saturation, Q_1 in triode operation, and υ_O small. From above, $25.74 = 35\,(2\,(\upsilon-1)\,\upsilon_o - \upsilon_O^2)$ --- (1). Differentiating, $0 = 35\,(2\,(\upsilon-1)\dfrac{\partial \upsilon_O}{\partial \upsilon} + 2\,\upsilon_O - 2\upsilon_O \dfrac{\partial \upsilon_O}{\partial \upsilon})$, For $\dfrac{\partial \upsilon_O}{\partial \upsilon} = -1 \rightarrow 0 = 35\,(2\,(1-\upsilon) + 2\,\upsilon_O + 2\,\upsilon_O)$. $0 = 70\,(1-\upsilon) + 140\,\upsilon_O$, $\upsilon - 1 = 2\,\upsilon_O$, $\upsilon = 1 + 2\,\upsilon_o$ --- (2). Now, substituting (2) in (1), $25.74 = 35\,(2\,(2\,\upsilon_O\,\upsilon_O) - \upsilon_O^2) = 35\,(3\,\upsilon_O^2)$. Thus, $\upsilon_O = .495V$, and $V_{IH} = 1 + 2\,(.495) =$ **1.99V**.

For $V_{IL} = \upsilon$: Q_2 in triode-mode operation, Q_1 in saturation, with υ_O large. Here $V_t = V_{t0} + \gamma\,(\sqrt{\upsilon_O + 0.6} - \sqrt{0.6}) = -3 + 0.5\,(\sqrt{5} - \sqrt{0.6} = -3 + 0.5\,(2.24 - .77) =$ **−2.26V**. Thus, $2.86\,(2\,(0 - -2.26)\,(5-\upsilon_O) - (5-\upsilon_O)^2) = 35\,(\upsilon-1)^2$. Thus $(5-\upsilon_O)^2 - (5-\upsilon_O)\,(4.52) + 12.24\,(\upsilon-1)^2 = 0$ --- (1). Differentiating, $2\,(5-\upsilon_O)\,(-\dfrac{\partial \upsilon_o}{\partial \upsilon}) + 4.52\dfrac{\partial \upsilon_O}{\partial \upsilon} + 12.24\,(2)\,(\upsilon-1) = 0$. Now, for $\dfrac{\partial \upsilon_o}{\partial \upsilon} = -1\ 2$,

$(5-\upsilon_O) - 4.52 + 24.48\,\upsilon - 24.48 = 0$, $10 - 2\upsilon_O - 4.52 + 24.48\,\upsilon - 24.48 = 0$, whence $\upsilon_O = (10 - 4.52 - 24.48 + 24.48\,\upsilon)\,/\,2$, and $\upsilon_O = 12.24\,\upsilon - 9.5$ --- (2). Now, substituting (2) in (1), $(14.5 - 12.24\,\upsilon)^2 - 4.52\,(14.5 - 12.24\,\upsilon) + 12.24\,(\upsilon-1)^2 = 0$, or $210.25 - 355\,\upsilon + 149.8\,\upsilon^2 - 65.54 + 55.32\,\upsilon + 12.24\,\upsilon^2 - 24.48\,\upsilon + 12.24 = 0$, or $162\,\upsilon^2 - 324.2\,\upsilon + 157 = 0$, $\upsilon^2 - 2.00\,\upsilon + 0.969 = 0$, whence

V_{IL} = **1.18V**. A second iteration could be used to correct for V_t. Assuming little change, $NM_H = V_{OH} - V_{IH} = 5 - 1.99 =$ **3.01V**, $NM_L = V_{IL} - V_{OL} = 1.18 - .093 =$ **1.09V**.

13.13 From P13.22 of the Text, $r_{DS1} = [(\mu_n\, C_{ox}\,(W/L)_1\,(V_{DD} - V_{tE})]^{-1} = [\,20 \times 10^{-6} \times 7/2\,(5-1)]^{-1} =$ **3.57kΩ**, and $r_{DS2} = (\mu_n\, C_{ox}\,(w/L)_2\,(V_{tD}))^{-1}$, where $V_{tD} = V_{t0} + \gamma(\sqrt{5+0.6} - \sqrt{0.6}) = -3 + 0.5\,(2.37 - .77) = -2.2\text{V}$. Thus, $r_{DS2} = (20 \times 10^{-6} \times 2/7\,(2.2))^{-1} =$ **79.5kΩ**.

13.14 Thus, from P13.23 of the Text, $V_{IH} = V_{tE} + 2|V_{tD}|/\sqrt{3K_R}$, and $\upsilon_O = |V_{tD}|/\sqrt{3K_R}$, where $K_R = \dfrac{K_1}{K_2} = \dfrac{7/2}{2/7} = \dfrac{49}{4} = 12.25$. Thus, $V_{IH} = 1 + \dfrac{2\,(3)}{\sqrt{3\,(12.25)}} =$ **1.99V**, for which $\upsilon_O = \dfrac{3}{\sqrt{3\,(12.25)}} =$ **0.495V**, for which $V_{tD} = -3 + 0.5\,(\sqrt{0.495+0.6} - \sqrt{0.6} = -3 + 0.5\,(1.046 - .774) =$ **2.86V**. For this value of V_{tD}, $V_{IH} = 1 + \dfrac{2\,(2.86)}{\sqrt{36.75}} =$ **1.94V**, and $\upsilon_O = \dfrac{2.86}{\sqrt{36.75}} =$ **0.472V**.

13.15 From P13.25 of the Text, $V_{OL} = V_{tDO}^2/(2K_R\,(V_{DD} - V_{tE})) = \dfrac{(-3)^2}{2\,\dfrac{7/2}{2/7}\,(5-1)} = \dfrac{9\,(4)}{2\,(49)\,(4)} =$ **0.092V**. For V_{OL} of half this value, K_R must be doubled. Thus W_1 must be doubled, other dimensions being unchanged. The factor is **2 times**.

13.16 Here, neglect the body effect. $K_1 = 35\mu\text{A/V}^2$, $K_2 = 2.86\mu\text{A/V}^2$, $V_{OH} = 5\text{V}$; $V_{OL} = 0.093\text{V}$;

For t_{pLH}: $I_{LH} = 2.86\,(0-3)^2 = 25.74\mu\text{A}$, $t_{pLH} = \dfrac{0.1 \times 10^{-12}\,(1/2\,(5 - .09) - .09)}{25.74 \times 10^{-6}} =$ **9.17ns**.

For t_{pHL}: At the top of the swing, $I_{D2} = 2.86\,(2\,(0-3)\,0 - 0^2) = 0\mu\text{A}$, $I_{D1} = 35(5-1)^2 = 560\mu\text{A}$. At the middle of the swing, $I_{D2} = 2.86\,(0-3)^2 = 25.74\mu\text{A}$, $I_{D1} = 35\,(2\,(5-1)\,2.5 - 2.5^2) = 481.25\mu\text{A}$. Thus, $I_{HL} = \dfrac{560 + 481 - 26}{2} = 507.5\mu\text{A}$, and $t_{pHL} = \dfrac{0.1 \times 10^{-12}\,(5 - 1/2\,(5 - .09))}{507.5} =$ **0.50ns**.

Thus $t_p = \dfrac{0.50 + 9.17}{2} =$ **4.84ns**.

From P13.29 of the Text: $t_{pHL} = \dfrac{4}{7}\,\dfrac{C}{K_1\,V_{DD}}$, and $t_{pLH} = \dfrac{C\,V_{DD}}{2\,\alpha\,K_2\,(V_{tDO})^2}$, where $\alpha = 1 + \dfrac{\gamma}{V_{tDO}}\sqrt{V_{DD}/2}$, and $DP = \dfrac{C\,V_{tDO}}{7\,K_R} + \dfrac{C\,V_{DD}^2}{8\,\alpha} \simeq \dfrac{1}{8\,\alpha}\,C\,V_{DD}^2$. See that $t_{pHL} = \dfrac{4}{7}\,\dfrac{0.1 \times 10^{-12}}{35 \times 10^{-6}\,(5)} =$ **0.33ns**, $\alpha = 1 + \dfrac{0.5}{-3}\sqrt{5/2} =$ **0.74**, $t_{pLH} = \dfrac{0.1 \times 10^{-12}\,(5)}{2\,(0.74)\,2.86 \times 10^{-6}\,(-3)^2} =$ **13.1ns**, and $DP = \dfrac{0.1 \times 10^{-12}\,(-3)^2}{7\,(35/2.86)} + \dfrac{0.1 \times 10^{-12}\,(5)^2}{8\,(.74)} = .0058 \times 10^{-12} + .422 \times 10^{-12} =$ **0.43pJ**.

SECTION 13.4: NMOS Logic Circuits

13.17 For the depletion-load inverter, $(W/L)_E = 4/2$ and $(W/L)_D = 2/4$,

(a) For a 4-input NOR, use 4 units 4/2, and 1 unit 2/4 for an input area of 4(4)2 = **32μm²**, and a total gate area of 5 (4) (2) = **40μm²**.

(b) For a 4-input NAND, 4 enhancement devices are connected in series to ground. Each must be 4 × 4 = 16μm wide. Thus use 4 units 16/2 and 1 unit 2/4 for an input area of 4 (16) 2 = **128μm²** and a total area of 128 + 2 (4) = **136μm²**.

(c) Alternatively: $Y = \overline{A\,B\,C\,D} = \bar{A} + \bar{B} + \bar{C} + \bar{D}$. Thus use a 4-input NOR with inverters at each input and at the output. For one NOR, use 4 units 4/2 and 1 unit 2/4 with a total gate area of 40μm². For 5 inverters, use 5 units 4/2 and 5 units 2/4 with a total gate area of (5 + 5) (2 × 4) = 80μm², for a total gate area of 40 + 80 = **120μm²**. Here the total input area = 4 (4) 2 = **32μm²**, just as with the

NOR.

13.18

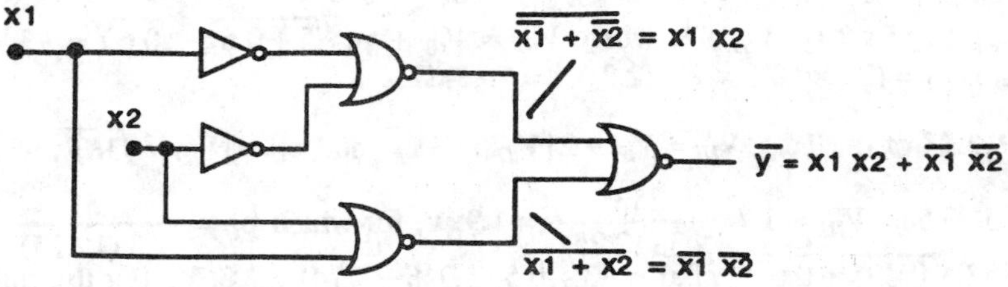

For the circuit in Fig. P13.32 of the Text, Q_1, Q_2, Q_4, are all 4/2, Q_5, Q_6 are both 8/2, Q_3, Q_7 are 2/4. Total device area = 3 (4) 2 + 2 (8) 2 + 2 (2) 4 = 24 + 32 + 16 = **72μm²**.

There are 5 load devices, and 8 switch devices. Total area = (8 + 5) (2) (4) = **104μm²**. Clearly the original design is quite area-efficient although it has 1.5 times as much input capacitance.

13.19

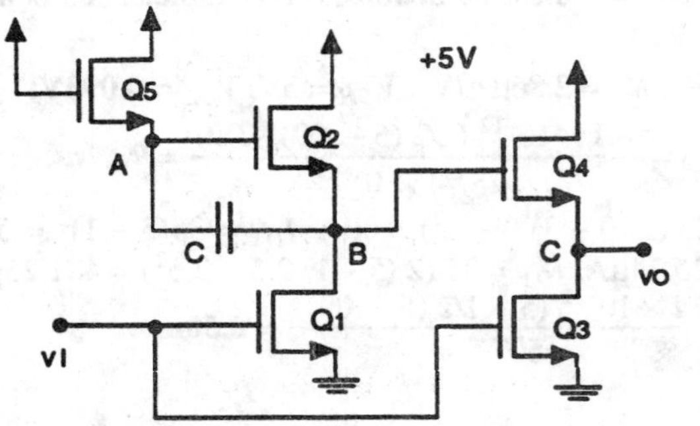

$K_1 = 1/2\ (20)\ 12/6 = 20\mu A/V^2$

$K_2 = 1/2\ (20)\ 6/12 = 5\mu A/V^2$

$K_3 = 1/2\ (20)\ 120/6 = 200\mu A/V^2$

$K_4 = 1/2\ (20)\ 240/6 = 400\mu A/V^2$

$K_5 = 1/2\ (20)\ 6/12 = 5\mu A/V^2$.

For υ_I at +4V: $V_{S5} = 5 - 1.0 = \mathbf{4V}$. For Q_2, Q_1, let $V_{S2} = \upsilon$, with Q_2 in saturation and Q_1 in triode mode.

Correspondingly, $K_1\ (2\ (\upsilon_{GS1} - V_t)\ \upsilon - \upsilon^2) = K_2\ (V_{S5} - \upsilon - 1)^2$, $20\ (2\ (4 - 1)\ \upsilon - \upsilon^2) = 5\ (4 - \upsilon - 1)^2$, $4\ (6\upsilon - \upsilon^2) = (3 - \upsilon)^2$, $24\upsilon - 4\upsilon^2 = 9 - 6\upsilon + \upsilon^2$, $5\upsilon^2 - 30\upsilon + 9 = 0$, $\upsilon = \dfrac{--30 \pm \sqrt{30^2 - 4\ (9)\ (5)}}{2\ (5)} = \dfrac{30 \pm 26.83}{10}$. Thus, $V_{S2} = \mathbf{0.317V}$, and $V_{S4} = \mathbf{0.00V}$. The supply current is defined by Q_2 to be $i_{D2} = 5 \times 10^{-6}\ (4 - .32 - 1)^2 = \mathbf{35.9\mu A}$. Capacitor C is charged to 4V – .32 = **3.68V**.

For υ_I *falling to* 0V: Node B rises, driving node A via capacitor C. Finally, node B reaches 5.0V and node A reaches 5.0 + 3.68 = 8.68V. Thus, $V_{S5} = \mathbf{8.68V}$, $V_{S2} = \mathbf{5V}$, $V_{S4} = 5 - 1 = \mathbf{4V}$. At V_{S2} = 5V, with V_{S4} = 0V initially, $I_{D4} = 400 \times 10^{-6}\ (5 - 0 - 1)^2 = \mathbf{6.4mA}$, is available to charge the load, at first.

For υ_I *rising to* 4V: V_{S2} falls, Q_4 cuts off, and $I_{D3} = 200 \times 10^{-6}\ (4 - 1)^2 = \mathbf{1.8mA}$, is available to discharge the load. As before, V_{S2} falls to **0.32V**, V_{S5} falls to **4.0V**, V_{S4} falls to **0.0V**.

SECTION 13.5: The CMOS Inverter

13.20 See $K_n = 1/2\ (20) \times 8/2 = 40\mu A/V^2$, $K_p = 1/2\ (10) \times 16/2 = 40\mu A/V^2$. From P13.35 of the Text, or directly from the triode relation, that is, $i_D = K\ (2\ (\upsilon_{GS} - V_{tt})\upsilon_{DS} - \upsilon_{DS}^2) \approx 2K\ (\upsilon_{GS} - V_t)\upsilon_{DS}$, see $r_{DS} = \dfrac{\upsilon_{DS}}{i_D} = \dfrac{1}{2K\ (\upsilon_{GS} - V_t)}$. *For input high,* $r_{DS} = \dfrac{1}{2\ (40) \times 10^{-6}\ (5 - 1)} = \mathbf{3.125k\Omega}$. *For input low,* $r_{DS} = \dfrac{1}{2\ (40) \times 10^{-6}\ (5 - 1)} = \mathbf{3.125k\ \Omega}$, the same.

13.21 Maximum currents are the same for both the p- and n-channel devices.

For the output connected to a contesting supply, $I_{MAX} = K\,(\upsilon_{GS} - V_t)^2 = 40 \times 10^{-6}\,(5-1)^2 = \mathbf{640\mu A}$.

For the output at $V_{DD}/2$, $I = K\,(2\,(\upsilon_{GS} - V_t)\upsilon_{DS} - \upsilon_{DS}^2) = 40 \times 10^{-6}\,(2\,(5-1)\,2.5 - 2.5^2) = 40 \times 10^{-6}\,(20 - 6.25) = \mathbf{550\mu A}$.

For the output $0.1V_{DD}$ *from the limit,* $I = 40 \times 10^{-6}\,(2\,(5-1)\,.5 - .5^2) = 40 \times 10^{-6}\,(4 - .25) = \mathbf{150\ \mu A}$.

13.22 For all inverters, $V_{OL} = 0\text{V}$, $V_{OH} = V_{DD}$, as V_{DD} varies from 3.75 to 6.25V, and $V_{tn} = V_{tp} = V_t$ varies from 0.75 to 1.25V, with $K_n = K_p$. From Eq. 13.37: $V_{IL} = 1/8\,(3\,V_{DD} + 2\,V_t)$. From Eq.13.36: $V_{IH} = 1/8\,(5\,V_{DD} - 2V_t)$, or $V_{IH} = V_{DD} - V_{IL}$ generally (for symmetry). Now for V_{IL} largest, $V_{DD} = 6.25$ and $V_t = 1.25$, whence $V_{IL} = 1/8\,(3\,(6.25) + 2\,(1.25)) = \mathbf{2.66V}$, for which $V_{IH} = 6.25 - 2.66 = \mathbf{3.59V}$. For V_{IL} smallest, $V_{DD} = 3.75$ and $V_t = .75$, $V_{IL} = 1/8\,(3\,(3.75) + 2\,(.75)) = \mathbf{1.59V}$, for which $V_{IH} = 3.75 - 1.59 = \mathbf{2.16V}$. Now V_{OL} is always **0V**, and V_{OH} ranges from **3.75V** to **6.25V**, V_{IL} from **1.59V** to **2.66V**, V_{IH} from **2.16V** to **3.59V**. See that noise margins between gates (with different supplies and different V_t) vary: Consider $NM_H = V_{OH} - V_{IH}$. Highest is $6.25 - 2.16 = \mathbf{4.09V}$. Lowest is $3.75 - 3.59 = \mathbf{0.16V}$ (see very bad). Consider $NM_L = V_{IL} - V_{OL}$. Highest is $2.66 - 0 = \mathbf{2.66V}$. Lowest is $1.59 - 0 = \mathbf{1.59V}$.

13.23 From P13.20 above, $K_n = K_p = 40\mu\text{A/V}$, $|V_t| = 1\text{V}$, $V_{DD} = 5\text{V}$. For $V_{in} = V_{DD}/2 = 2.5\text{V}$, $I_{peak} = i_D = 40\,(2.5 - 1)^2 = \mathbf{90\mu A}$. Assume (for the present purposes) that rise and fall times are measured from 0% to 100%. As the input goes from 1V to 4V, the current flow is triangular, with a peak value of 90µA. Over an interval of 3/5 × $T/4$, the average current is 90/2 = 45µA. The average current per cycle from self conductance is $\dfrac{2\,(3/5) \times T/4 \times 45}{T} =$ **13.5µA**. The average current due to capacitance load is $CV_{DD}f = 0.5 \times 10^{-12} \times 5 \times 20 \times 10^6 = \mathbf{50\mu A}$. Total Average Current = 50 + 13.5 = **63.5µA**. With load, $P_D = 63.5 \times 10^{-6} \times 5 = \mathbf{317.5\mu W}$. Without load, $P_D = 13.5 \times 5 = \mathbf{67.5\mu W}$.

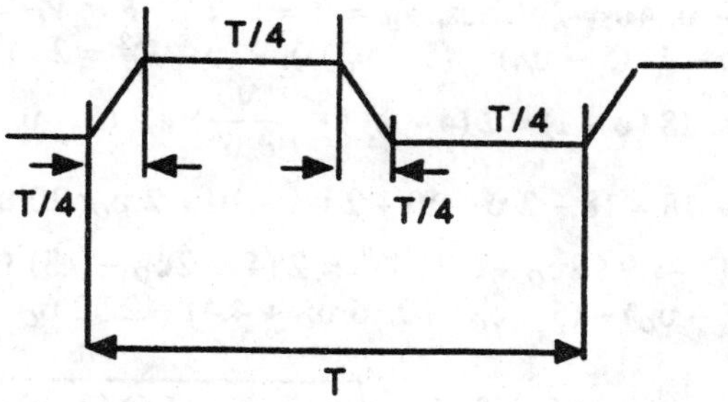

13.24 From Eq.13.44, $t_{PHL} = \dfrac{C}{K_n\,(V_{DD} - V_t)}\left[\dfrac{V_t}{V_{DD} - V_t} + 1/2 \ln\left[\dfrac{3V_{DD} - 4V_t}{V_{DD}}\right]\right]$. Here $K_n = K_p = 40\mu\text{A/V}$, $|V_t| = 1\text{V}$, $V_{DD} = 5\text{V}$, $t_{PHL} = t_{PLH} = t_p$. Thus $t_p = \dfrac{0.5 \times 10^{-12}}{40 \times 10^{-6}\,(5-1)}\left[\dfrac{1}{5-1} + 1/2 \ln \dfrac{(15-4)}{5}\right]$

= 3.125ns (1/4 + 1/2 (.788)) = **2.01ns**. For a 5-stage ring oscillator, $f = 1/10(2.01) = \mathbf{49.8MHz}$. Ignoring the transition-time peak-current flow per gate, $P_D = f\,CV_{DD}^2 = 49.8 \times 10^6 \times 0.5 \times 10^{-12}\,(5^2) = 623\mu\text{W}$. The Delay-Power product $DP = 2.01 \times 10^{-9} \times 623 \times 10^{-6} = 1.25 \times 10^{-12}J = \mathbf{1.25pJ}$.

13.25 For $V_{DD} = 1.3\text{V}$, $|V_t| = 0.8\text{V}$, (a) Q_n conducts for υ_I from 0.8 to 1.3V, and Q_p conducts for υ_I from 0 to 1.3 − 0.8 or 0 to 0.5V. For $\upsilon_I = 1.3/2 = 0.65\text{V}$, neither conducts and $i_D = \mathbf{0}$, (b) Output voltages range from 0 to 1.3V. $V_{OH} = \mathbf{1.3V}$; $V_{OL} = \mathbf{0.0V}$, (c) $V_{IL} = \mathbf{0.8V}$ and $V_{IH} = 1.3 - 0.8 = \mathbf{0.5V}$. Between V_{IL} and V_{IH}, *no current* flows.

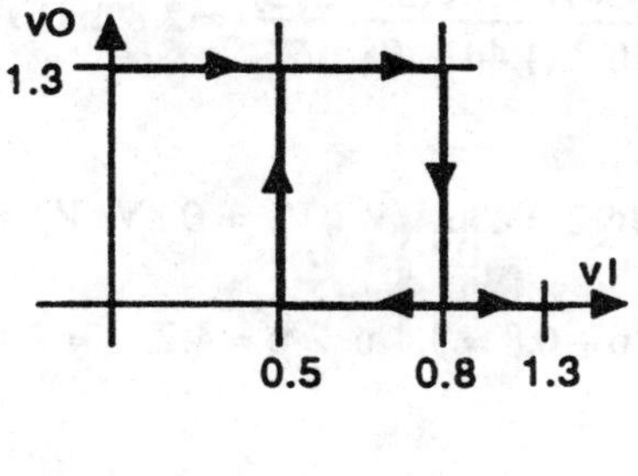

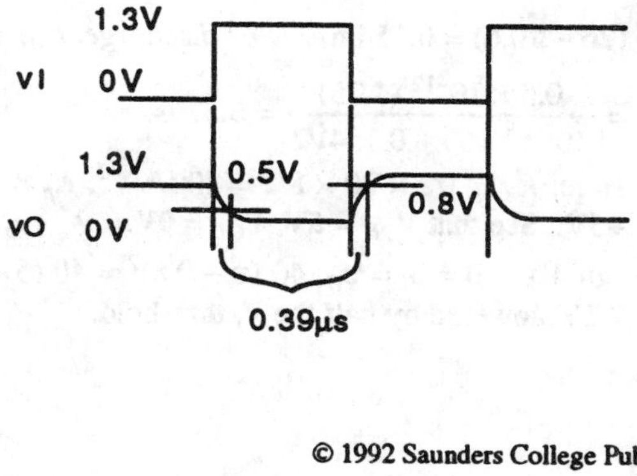

(d) (Note that the small capacitor at the output holds the output while neither transistor conducts.
(e) $i_D = K(1.3 - 0.8)^2 = 10 \times 10^{-6} (0.5)^2 = 2.5\mu A$ peak. For υ_O at 0.8V, $i_D = K(2(\upsilon_{GS} - V_t)\upsilon_{DS} - \upsilon_{DS}^2)$ $= 10 \times 10^{-6} (2(1.3 - 0.8)0.8 - 0.8^2) = 10 \times 10^{-6} (0.8 - .64) = 1.6\mu A$. Average current is $\frac{1.6 + 2.5}{2}$ = 2.05μA, and the time for 0.8V change is $\frac{CV}{I} = \frac{1 \times 10^{-12} \times 0.8}{2.05 \times 10^{-6}}$ = **0.39μs**. Conclude that propagation delay is **more than 0.39μs**. Now, for the output moving from 0.8V to 1.3V, average current available is (1.6+0)/2 = 0.8μA.

Thus, the time to reach 1.3V is about $1 \times 10^{-12} \times \frac{1.3 - 0.8}{0.8 \times 10^{-6}} = 0.625\mu s$. Considering the driven stage with input at 0.8V, the available output current is 0μA. At an input of 1.3V, the available output current is 2.5μA. Thus, the average current = $\frac{0 + 2.5}{2} = 1.25\mu A$, and the propagation delay $\approx \frac{1.0 \times 10^{-12} \times 0.8}{1.25 \times 10^{-6}}$ = **0.64μs**.

(f) Frequency oscillation of 5 gates = $\frac{1}{10(.64 \times 10^{-6})}$ = **156kHz**.

13.26 Here, $K_n = 1/2(20) \times 18/2 = 90\mu A/V^2$; $K_p = 1/2(20) \times 2/2 = 10\mu A/V^2$ $|V_t| = 1V$, $V_{DD} = 5V$. See V_{OH} = **5V** V_{OL} = **0V**. *For* $V_{th} = \upsilon$, $\upsilon_I = \upsilon_O$, both devices are in saturation and $90(\upsilon - 1)^2 = 10(5 - \upsilon - 1)^2$. Taking the square root, $3(\upsilon - 1) = 4 - \upsilon$, $3\upsilon - 3 = 4 - \upsilon$, $4\upsilon = 7$. Thus, $V_{th} = \upsilon$ = **1.75V**. *For* $V_{IL} = \upsilon$, Q_p in triode, Q_n in saturation, $90(\upsilon - 1)^2 = 10(2(5 - \upsilon - 1)(5 - \upsilon_O) - (5 - \upsilon_O)^2)$, $9(\upsilon - 1)^2 = 2(4 - \upsilon)(5 - \upsilon_O) - (5 - \upsilon_O)^2$ – – – (1). Now, taking derivatives, $18(\upsilon - 1) = 2(4 - \upsilon)(-\frac{\partial \upsilon_O}{\partial \upsilon}) + 2(5 - \upsilon_O)(-1) - 2(5 - \upsilon_O)\frac{-\partial \upsilon_O}{\partial \upsilon}$. Now with $\frac{\partial \upsilon_O}{\partial \upsilon} = -1 \rightarrow 18\upsilon - 18 = 18 - 2\upsilon - 10 + 2\upsilon_O - 10 + 2\upsilon_O$, $20\upsilon = 4\upsilon_O + 6$, $\upsilon = .2\upsilon_O + 0.3$ – – – (2). Substitute (2) in (1) $\rightarrow 9(.2\upsilon_O + 0.3 - 1)^2 = 2(4 - .2\upsilon_O - 0.3)(5 - \upsilon_O) - (5 - \upsilon_O)^2$, $9(.2\upsilon_O - 0.7)^2 = 2(3.7 - .2\upsilon_O - 0.3)(5 - \upsilon_O) - (5 - \upsilon_O)^2$, $0.36\upsilon_O^2 + 4.41 - 2.52\upsilon_O = 37 - 9.4\upsilon_O + .4\upsilon_O^2$

$-25 + 10\upsilon_O - \upsilon_O^2$, $0.96\upsilon_O^2 - 3.12\upsilon_O - 7.59 = 0$. Thus $\upsilon_O = \frac{3.12 \pm \sqrt{3.12^2 + 4(.96)(7.59)}}{2(.96)} = 4.87V$, and $V_{IL} = \upsilon = .2(4.87 + 0.3)$ = **1.275V**. *For* $V_{IH} = \upsilon$, Q_n in triode, Q_p in saturation, $10(5 - \upsilon - 1)^2 = 90(2(\upsilon - 1)(\upsilon_O) - \upsilon_O^2)$, $(4 - \upsilon)^2 = 18(\upsilon - 1)(\upsilon_O) - 9\upsilon_O^2$ – – – (1). $\frac{\partial}{\partial \upsilon} \rightarrow 2(4 - \upsilon)(-1) = 18(\upsilon - 1)\frac{\partial \upsilon_O}{\partial \upsilon} + 18\upsilon_O - 18\upsilon_O\frac{\partial \upsilon_O}{\partial \upsilon}$. For $\frac{\partial \upsilon_O}{\partial \upsilon} = -1 \rightarrow 2\upsilon - 8 = 18 - 18\upsilon + 18\upsilon_O + 18\upsilon_O$, $20\upsilon = 36\upsilon_O + 26$, $\upsilon = 1.8\upsilon_o + 1.3$ – – – (2). Now, substituting (2) in (1), $(4 - 1.8\upsilon_O - 1.3)^2 = 18(1.8\upsilon_O + 1.3 - 1)\upsilon_O - 9\upsilon_O^2$, $(2.7 - 1.8\upsilon_O)^2 = 18(1.8\upsilon_O + .3)\upsilon_O - 9\upsilon_O^2$, $7.29 - 4.86\upsilon_O + 3.24\upsilon_O^2 = 32.4\upsilon_O^2 + 5.4\upsilon_O - 9\upsilon_O^2$, $\upsilon_O^2(20.16) + \upsilon_O(10.26) - 7.29 = 0$, $\upsilon_O^2 + .509\upsilon_O - .362 = 0$, $\upsilon_O = \frac{-.509 \pm \sqrt{.509^2 + 4(.362)}}{2}$ = .400V, and $\upsilon = 1.8(.4) + 1.3 = 2.02V$. Thus V_{IH} = **2.02V**.

For t_p: For $\upsilon_O \approx 5V$, $I_n = 90(5 - 1)^2 = 1.44mA$. For $\upsilon_O \approx 0V$, $I_p = 10(5 - 1)^2 = .160mA$. For $\upsilon_O = 1.75V$, $I_n = 90(2(5 - 1)(1.75) - 1.75^2) = 0.984mA$, and $I_p = 10(2(5 - 1)(5 - 1.75) - (5 - 1.75)^2) = 10(26 - 10.6) = 0.154mA$. *For discharge*, $t_{pHL} = \frac{CV}{I} = \frac{0.5 \times 10^{-12} \times (5 - 1.75)}{10^{-3}(1.44 + .984)/2}$ = **2.89ns**. *For charge*, $t_{pLH} = \frac{0.5 \times 10^{-12}(1.75)}{10^3(0.160 + 0.154)/2}$ = **5.57ns**.

13.27 Here, $K_n = 1/2 \times 20 \times 8/2 = 40\mu A/V^2$, $K_p = 1/2 \times 10 \times 16/2 = 40\mu A/V^2$, $V_{tn} = 0.8V$, $V_{tp} = -1.6V$, V_{DD} = +5V. See that V_{OH} = **5V**, V_{OL} = **0V**.

For $V_{th} = \upsilon = \upsilon_I = \upsilon_O$, $40(\upsilon - 0.8)^2 = 40(5 - \upsilon - 1.6)^2$, $\upsilon - 0.8 = 3.4\upsilon$, $2\upsilon = 4.2$, $\upsilon = 2.1V$. Thus, V_{th} = **2.1V**, lowered by half the V_t threshold.

For $V_{IL} = \upsilon$, $(\upsilon - 0.8)^2 = (2\,(5 - \upsilon - 1.6)\,(5 - \upsilon_O) - (5 - \upsilon_O)^2)$, $(\upsilon - 0.8)^2 = 2\,(3.4 - \upsilon)\,(5 - \upsilon_O) - (5 - \upsilon_O)^2$ --- (1), $\frac{\partial}{\partial \upsilon} \rightarrow 2(\upsilon - 0.8) = 2\,(-1)\,(5 - \upsilon_O) + 2\,(3.4 - \upsilon)\,(-\frac{\partial \upsilon_O}{\partial \upsilon}) - 2\,(5 - \upsilon_O)\,(-\frac{\partial \upsilon_O}{\partial \upsilon})$. Now for $\frac{\partial \upsilon_O}{\partial \upsilon} = -1$, $2\upsilon - 1.6 = -10 + 2\upsilon_O + 6.8 - 2\upsilon - 10 + 2\,\upsilon_O$, $4\,\upsilon = 4\,\upsilon_O - 11.6$, $\upsilon = \upsilon_O - 2.9$ --- (2).

Substituting (2) in (1), $(\upsilon_O - 2.9 - 0.8)^2 = 2\,(3.4 - \upsilon_O + 2.9)\,(5 - \upsilon_O) - (5 - \upsilon_O)^2$, $(\upsilon_O - 3.7)^2 = (5 - \upsilon_O)(7.6 - \upsilon_O)$, $\upsilon_O^2 - 7.4\,\upsilon_O + 13.69 = 38.0 - 12.6\,\upsilon_O + \upsilon_O^2$, $5.2\,\upsilon_O = 24.31$, $\upsilon_O = 4.675\text{V}$,

and $V_{IL} = \upsilon = 4.675 - 2.9 =$ **1.78V**.

For $V_{IH} = \upsilon$, Q_n in triode, Q_p in saturation, $40\,(5 - \upsilon - 1.6)^2 = 40\,(2\,(\upsilon - 0.8)\,\upsilon_O - \upsilon_O^2)$, $(3.4 - \upsilon)^2 = 2\,(\upsilon - 0.8)\,\upsilon_O - \upsilon_O^2$ --- (1). $\frac{\partial}{\partial \upsilon} \rightarrow 2\,(3.4 - \upsilon)\,(-1) = 2\,(\frac{\partial \upsilon_O}{\partial \upsilon}) + 2\,(\upsilon_O) - 2\upsilon_o\,\frac{\partial \upsilon_O}{\partial \upsilon}$. Now for $\frac{\partial \upsilon_O}{\partial \upsilon} = -1$, $2\,\upsilon - 6.8 = -2 + 2\,\upsilon_O + 2\upsilon_O$, $2\,\upsilon = 4\,\upsilon_O + 4.8$, $\upsilon = 2\,\upsilon_O + 2.4$ --- (2).

Substituting (2) in (1), $(3.4 - 2\,\upsilon_O - 2.4)^2 = 2\,(2\,\upsilon_O + 2.4 - 0.8)\,\upsilon_O - \upsilon_O^2$, $(1.0 - 2\,\upsilon_O)^2 = 4\,\upsilon_O^2 + 3.2\,\upsilon_O - 3\upsilon_O^2$,

$1 - 4.0\,\upsilon_O + 4\,\upsilon_O^2 = 3\,\upsilon_O^2 + 3.2\,\upsilon_O$, $\upsilon_O^2 - 7.2\,\upsilon_O + 1.0 = 0$, $\upsilon_O = \frac{-7.2 \pm \sqrt{7.2^2 - 4(1)}}{2} = 0.14\text{V}$, for which $V_{IH} = \upsilon = 2\,(.14) + 2.4 =$ **2.68V**.

For t_p: For $\upsilon_O = 5\text{V}$, $I_n = 40\,(5 - 0.8)^2 = 0.7056\text{mA}$. For $\upsilon_O = 0\text{V}$, $I_p = 40\,(5 - 1.6)^2 = 0.4624\text{mA}$. For $\upsilon_O = 2.1$, $I_n = 40\,(2\,(5 - 0.8)\,(2.1) - 2.1^2) = 40\,(17.64 - 4.41) = 0.529\text{mA}$, $I_p = 40\,(2\,(5 - 1.6)\,(5 - 2.1) - (5.0 - 2.1)^2 = 40\,(19.72 - 8.41) = 0.452\text{mA}$. *For discharge*, $t_{pHL} = \frac{CV}{I} = \frac{0.5 \times 10^{-12} \times (5.0 - 2.1)}{10^{-3}\,(.706 + .529)/2} =$ **2.35ns**.

For charge, $t_{pLH} = \frac{0.5 \times 10^{-12}\,(2.1)}{10^{-3}\,(.4624 + .452)/2} =$ **2.30ns**.

SECTION 13.6: CMOS Gate Circuits

13.28

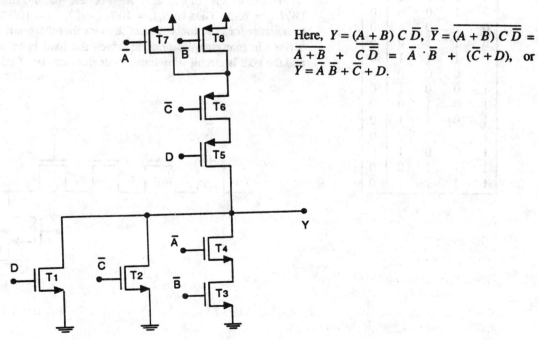

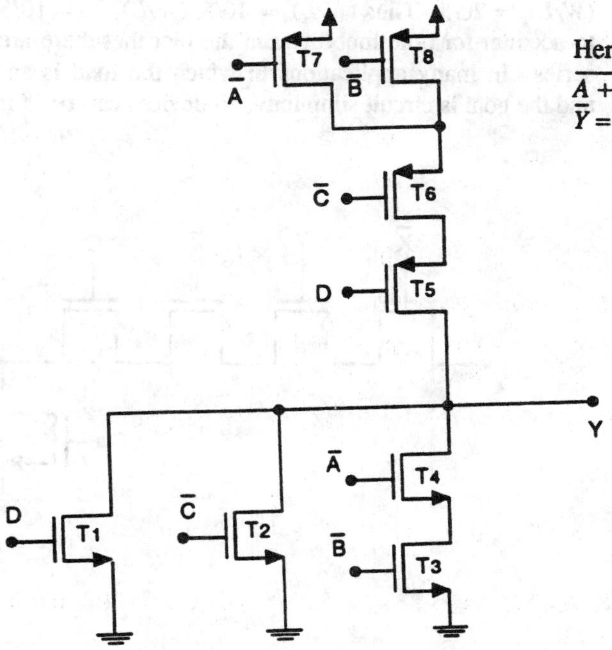

Here, $Y = (A + B)\,C\,\overline{D}$, $\overline{Y} = \overline{(A + B)\,C\,\overline{D}} = \overline{A + B} + \overline{C\,\overline{D}} = \overline{A} \cdot \overline{B} + (\overline{C} + D)$, or $\overline{Y} = \overline{A}\,\overline{B} + \overline{C} + D$.

13.29 For the *n*-channel devices, named by their input: *W/L* for *D* and $\overline{C}$ is 2, *W/L* for $\overline{A}$ and $\overline{B}$ is $2 \times 2 = 4$. For the *p*-channel devices, *W/L* for $\overline{A}, \overline{B}, \overline{C}, D$ is 3 (2) 2 = 12.

13.30 Use Fig. 13.22a) of the Text, and equations 13.49 and 13.50. *For one input active*: $I = K\,(2.5 - 1)^2 = K_p\,(V_1 - 2.5 - 1)^2$ and $I = K_p\,[2\,(5 - 1)\,(5 - V_1) - (5 - V_1)^2]$. Thus, $(V_1 - 3.5)^2 = 8\,(5 - V_1) - (5 - V_1)^2$, $V_1^2 - 7V_1 + 12.25 = 40 - 8\,V_1 - 25 + 10\,V_1 - V_1^2$, $2\,V_1^2 - 9\,V_1 - 2.75 = 0$,

$$V_1 = \frac{--9 \pm \sqrt{9^2 - 4\,(-2.75)\,(2)}}{2 \times 2} = \frac{9 \pm 10.15}{4} = 4.787\text{V},\ K_p = \frac{K\,(1.5^2)}{(4.787 - 3.5)^2} = \mathbf{1.36K}.$$

For two inputs active: $V_{th} = \upsilon$, $I = 2K\,(\upsilon - 1)^2 = K_p\,(V_1 - \upsilon - 1)^2$, and also $I = K_p\,[2\,(5 - \upsilon - 1)\,(5 - V_1) - (5 - V_1)^2]$. From the first pair, $\upsilon - 1 = \left[\frac{K_p}{2K}\right]^{1/2}(V_1 - \upsilon - 1) = \left[\frac{1.36}{2}\right]^{1/2}(V_1 - \upsilon - 1) = 0.82\,V_1 - .82\,\upsilon - .82$. Thus $1.82\,\upsilon = .82\,V_1 + .18$, $\upsilon = .4505\,V_1 + .2195$, or $V_1 = 2.22\,\upsilon - 0.220$.

From the second pair: $(V_1 - \upsilon - 1)^2 = (4 - \upsilon)\,(10 - 2\,V_1) - (5 - V_1)^2$, $(2.22\,\upsilon - .220 - \upsilon - 1)^2 = (4 - \upsilon)\,(10 - 4.44\,\upsilon + .44) - (5 - 2.22\,\upsilon + .22)^2$, or $(1.22\,\upsilon - 1.22)^2 = (4 - \upsilon)\,(10.44 - 4.44\,\upsilon) - (5.22 - 2.22\,\upsilon)^2$, or $1.488\,\upsilon^2 - 2.977\,\upsilon + 1.488 = 41.76 - 17.76\,\upsilon - 10.44\,\upsilon + 4.44\,\upsilon^2 - 27.25 + 23.18\,\upsilon - 4.93\,\upsilon^2$, $1.98\,\upsilon^2 + 2.043\,\upsilon - 13.02 = 0$, $\upsilon^2 + 1.032\,\upsilon - 6.575 = 0$, $\upsilon = \frac{-1.032 \pm \sqrt{1.032^2 - 4\,(-6.575)}}{2} = 2.10\text{V}$. Thus, the threshold is **2.10V** for which $V_1 = 2.22\,(2.10) - .220 = 4.44\text{V}$.

13.31

C	*X*	*Y*	*Z*	*F*
0	0	0	0	0
0	0	0	1	0
0	0	1	0	0
0	0	1	1	0
0	1	0	0	0
0	1	0	1	0
0	1	1	0	1
0	1	1	1	0
1	0	0	0	0
1	0	0	1	0
1	0	1	0	0
1	0	1	1	0
1	1	0	0	0
1	1	0	1	0
1	1	1	0	0
1	1	1	1	0

See that *F* goes high when *X* is high, $\overline{Y}$ is low, *Z* is low, as *C* goes low. That is, $F = \overline{C}\,X\,Y\,\overline{Z}$. If $\overline{X}$ is available, but *X* is not, and *Y* is available, replace *X* (as shown) by *Y*, and $\overline{Y}$ (as shown) by *X*. If $\overline{X}$ is available, but *X* is not and *Y* is not, use an additional *p*-channel MOSFET at the left connected to the supply and $\overline{X}$. For the basic symmetric inverter $(W/L)_p = 20/5$. Thus $(W/L)_4 = 10/5$, $(W/L)_{1,2,3} = 10/5 \times 2 \times 3 = 60/5$, to account for hole mobility and the fact that there are 3 transistors in series. In many applications in which the load is an inverting input, and the goal is circuit simplicity, all devices can be of minimum size.

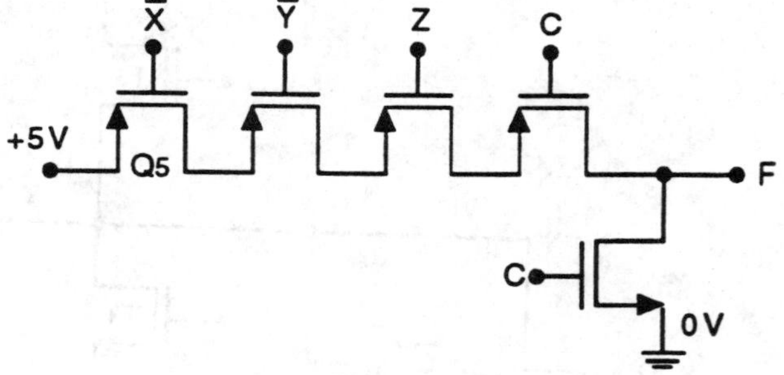

13.32 For V_{DD} = 10V, C_L = 50pF, I_{Dav} = 0.6μA/kHz. For the capacitor, the current at frequency 1kHz is $\frac{CV}{f}$ = $\frac{50 \times 10^{-12} \times 10}{10^3}$ = 0.5μA. Current with no load = 0.6 – 0.5 = 0.1μA/kHz. The origin of this current can be considered to be of two possible kinds: (a) It could be assumed to be due to the intrinsic self capacitance of the gate, which would apparently be as large as 10pF to provide the entire effect. (b) It could be assumed to be due entirely to current which flows between transistors in the switching transition: The peak transition current occurs with inputs of $V_{DD}/2$ and is proportional to $(V_{DD}/2 - V_t)^2$. That is, here, with a 10V supply, α (10/2 – 1)² = 0.1, whence α = 0.1/16. Correspondingly, with a 5V supply, the equivalent current would be α (5/2 – 1)² = 0.1/16 × (2.5 – 1)² = .014μA/kHz. Now, at 5V and 10MHz, with C_L = 30pF: For the equivalent-capacitance view: C_{Total} = 30 + 10pF = 40pF, $I_{Av} = CV/T = 40 \times 10^{-12} \times 5 \times 10 \times 10^6$ = **2mA**. For the transition-current view: $I_{Av} = 30 \times 10^{-12} \times 5 \times 10 \times 10^{-6} + .014 \times 10^{-6} \times \frac{10^7}{1000}$ = 1500 + 140 = **1.64mA**, with power dissipations of 5 × 2 = 10mW and 5 × 1.64 = 8.2 mW respectively.

13.33 *For a Buffered Inverter*: Total *n*-width is 1 + 3 + 9 = 13 units, while the total *p*-width is 2 + 2 (3) + 2 (9) = 26 units. Thus the total area is **39** units. *For a Buffered 4-Input NOR*: The input gate uses 4 *n*-channel devices of unit width, and 4 *p*-channel devices whose width is 2 units. Thus, the total area of the input stage is 4 (1) + 4 (2) = 12 units, and of the two buffer stages is 39 – 3 = 36. Thus, the total area of the buffered NOR is 12 + 36 = **48** units. *Input capacitance* is proportional to the area of the *p* and *n* device connected to each input. For the buffered inverter, it is 1 + 2 = **3 units**. For the buffered 4 input NOR, it is 1 + 4 (2) = **9 units**, that is, **3** times greater. *For input inverters and an intermediate NAND*, the total area of input inverters = 4 × (1 + 2) = 12. For the NAND, each *n*-channel device is 3 × 4 = 12 units wide, and each *p*-channel device is 3 × 2 = 6 units wide. Total area of the NAND is 4 (6 + 12) = 72. Total area of the output inverter is 9 (1 + 2) = 27 units. Overall area of the equivalent NOR is 12 + 72 + 27 = **111** units. For the equivalent NOR, the input capacitance is **3 units**, equal to that of an inverter and **1/3** that of the direct NOR.

SECTION 13.7: Latches and Flip-Flops

13.34

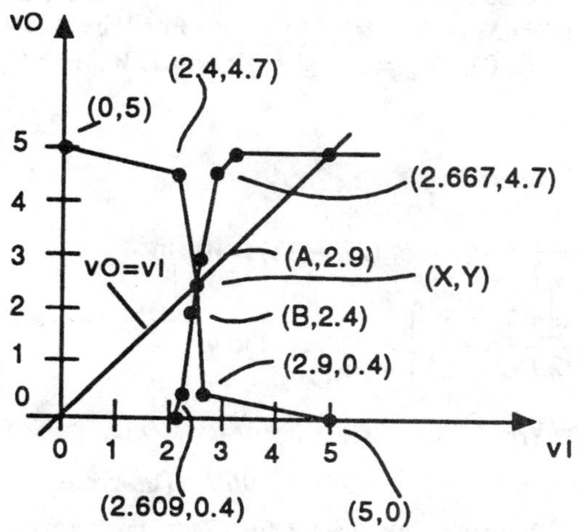

For the input coordinates corresponding to 2.9V and 2.4V output levels (ie A, B): $A = 2.4 + \frac{4.7 - 2.9}{4.7 - 0.4} \times (2.9 - 2.4) = 2.4 + \frac{1.8}{4.3}(0.5) = 2.609$V. $B = 2.4 + \frac{4.7 - 2.4}{4.7 - 0.4}(2.9 - 2.4) = 2.4 + \frac{2.3}{4.3}(0.5) = 2.667$V. There are 3 points for which input and output are at equal levels: Two of these are at **(0V, 0V)** and **(5V, 5V)** where the loop gain is 0, and the other (in the middle, more or less) is where, $v_O \approx \frac{2.609 + 2.667}{2} \approx$ **2.64V** $= v_I$, that is at **(2.64, 2.64)**. For each inverter, the gain there exceeds $\frac{4.7 - 0.4}{2.4 - 2.9} = -\frac{4.3}{0.5}$ = –8.6V/V. Thus the overall loop gain exceeds 8.6² or **74 V/V** at the (middle) unstable point, and is 0 V/V at the other two.

13.35

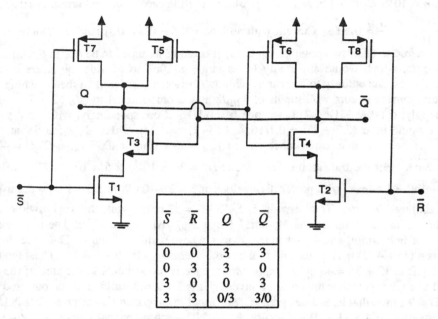

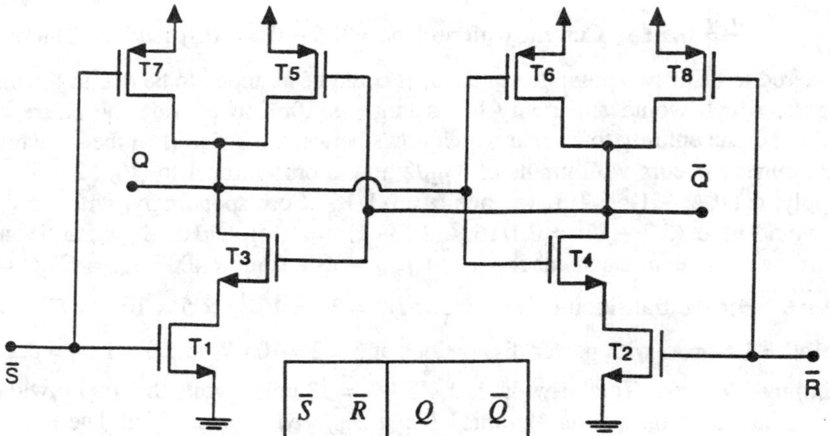

$\overline{S}$	$\overline{R}$	Q	$\overline{Q}$
0	0	3	3
0	3	3	0
3	0	0	3
3	3	0/3	3/0

Note that $|V_t| < 3/2$V is necessary in order to ensure that at least one transistor conducts for all input voltages in the range 0 to 3V.

13.36 Let the conductivity parameter of the enhancement devices be 1 and that of the depletion devices be K. Now, consider the gate of Q_3 rising toward the threshold V_{th} (with $\overline{Q}$ initially high). At the threshold, $\overline{Q}$ falls enough to release the drive on Q_2 enough to allow Q_6 to pull Q high enough to reach V_{th} of Q_1. For this situation, Q_2 is in triode mode with $\upsilon_{DS\,2} = V_{t1} = 1$V, and with Q_6 in pinchoff, conducting a current $i_{D6} = K\,(0 - -2)^2 = 4K$. For $\upsilon_{GS2} = \upsilon$, $4K = 1\,[(2\,(\upsilon - 1)\,1 - 1^2] = 2\upsilon - 2 - 1 = 2\,\upsilon - 1$, or $\upsilon = (4K+1)/2$. Now to reach this level, S must rise to V_{th}, with Q_3 providing a current equal to that in Q_5. Assume υ is low enough for Q_5 to operate in saturation, in which case $I_{D5} = \mathrm{K}\,(0 - -2)^2 = 4K$. Now for a particular input threshold V_{th}, and Q_3 in saturation, $1(V_{th} - 1)^2 = 4K$, or $K = 1/4\,(V_{th} - 1)^2$. For $V_{th} = 2.5$V, $K = 1/4\,(2.5 - 1)^2 = 0.5625$, for which $W_5 =$ **0.56**W. For $V_{th} = 2.0$V, $W_5 = 1/4\,(2 - 1)^2 =$ **0.25**W. For $V_{th} = 3.0$V, $W_5 = 1/4\,(3 - 1)^2 = W$. Use the same values for W_6.

SECTION 13.8: Multivibrator Circuits

13.37 Here, $T = C\,(R + R_{on}) \ln \left[\dfrac{R}{R + R_{on}}\,\dfrac{V_{DD}}{V_{DD} - V_{th}}\right]$, $T = 1500 \times 10^{-12}\,(22 + .18) \times 10^3 \ln \left[\dfrac{22}{22 + .18}\,\dfrac{5}{5 - 0.6\,(5)}\right] = 1.5 \times 10^{-6}\,(22.18) \ln \left[\dfrac{22}{22.18} \times \dfrac{1}{0.4}\right] =$ **30.2μs**. Now, $\Delta V_1 = V_{DD}\,\dfrac{R}{R + R_{on}} = 5\,\dfrac{22}{22 + .18} =$ **4.96V**, $\Delta V_2 = V_{DD} + V_{D1} - V_{th} = 5 + 0.7 - 0.6\,(5) =$ **2.7V**. During the interval T, υ_{i2} changes by $V_{th} - (V_{DD} - \Delta V_1) = 0.6\,(5) - (5 - 4.96) = 2.96$V. The current changes by $\dfrac{2.96V}{22k\Omega}$, and υ_{01} changes by $.18 \times \dfrac{2.96}{22} =$ **0.024V**. The peak sink current for G_1, occuring as υ_{01} falls at the beginning of the timing interval, is $i_1 = \dfrac{\Delta V_1}{R} = \dfrac{4.96}{22k\Omega} =$ **0.225mA**. The peak source current occurs at the end of the timing interval, where D_1 conducts, and υ_{01} (at the interior end of R_{0a}) changes from 0 to

+5V, while the gate of G_2 changes from V_{th} = 3V to $V_{DD} + V_{D1}$ = 5.7V, for which, $i_2 = \dfrac{5-(5.7-3)}{0.18k\Omega}$ = **12.8mA.**

13.38

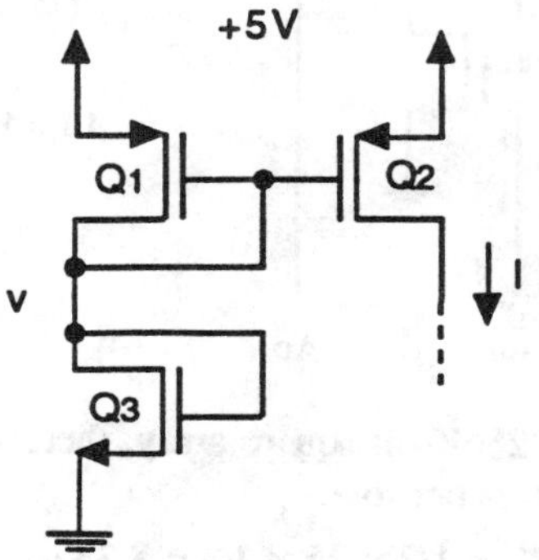

In the mirror, for Q_1, Q_2, $K = 1/2 \times 10 \times 2 = 10\mu A/V^2$, and for Q_3, $K = 1/2 \times 20 \times 2 \times 1/10 = 1\mu A/V^2$. For node voltage υ, $10\,(5-\upsilon-1)^2 = 1\,(\upsilon-1)^2$, $\upsilon - 1 = \sqrt{10}\,(4-\upsilon)$, $\upsilon - 1 = -3.16\,\upsilon + 12.65$, $7.16\,\upsilon = 13.65$, $\upsilon = 1.906$V, and $I = 1\,(1.906-1)^2$ = **0.822μA.** For G_1 loaded by I, at the beginning of the timing interval, the output voltage is υ, where: $I = .822 = 1/2 \times 20 \times 2\,[2\,(5-1)\,\upsilon - \upsilon^2] = 20\,[8\,\upsilon - \upsilon^2]$, $\upsilon^2 - 8\,\upsilon + .0411 = 0$, $\upsilon = \dfrac{--8 \pm \sqrt{8^2 - 4\,(.0411)}}{2}$, whence υ_{01} = 5mV. *For the Inverter Threshold*, at V_{th}, $\upsilon_O = \upsilon_I = \upsilon$, with $K_R = K_p/K_n = 1/2$, $I = K_p\,(5-\upsilon-1)^2 = K_n\,(\upsilon-1)^2$, $\sqrt{1/2}\,(4-\upsilon) = \upsilon - 1$, $2.83 - .707\,\upsilon = \upsilon - 1$, $1.707\,\upsilon = 3.83$, whence $V_{th} = \upsilon =$ **2.24V.**

Now, the time for υ_{I2} to rise to V_{th} from υ_{01} is $\dfrac{(2.24-.005)\times C}{.822\times 10^{-6}} = 10 \times 10^{-6}$ s, and $C = \dfrac{8.22\times 10^{-12}}{2.24}$ = **3.67pF.**

13.39 From P13.71, $T = CR \ln\left[\dfrac{2V_{DD}-V_{th}}{V_{DD}-V_{th}} \times \dfrac{V_{DD}+V_{th}}{V_{th}}\right]$, for added resistor 10 R. Thus, $10^{-6} = 100 \times 10^{-12}\,R \ln\left[\dfrac{2(5)-.44(5)}{5-.44(5)} \times \dfrac{5+.44(5)}{.44(5)}\right]$, $R = 10^4 \ln\left[\dfrac{10-2.2}{5-2.2} \times \dfrac{5+2.2}{2.2}\right] = 10^4 \ln\left[\dfrac{7.8}{2.8} \times \dfrac{7.2}{2.2}\right]$ = **22kΩ,** with the gate resistor used being **220kΩ.**

SECTION 13.9: Random-Access Memory (RAM)

13.40 Each decoder handles 1M/4 = 256K bits in an array of size $\sqrt{256K}$ or 512 by 512. Now $512 = 2^9$. Thus each decoder uses **9 bits,** such that the required address is 2 + 9 + 9 = 20 bits where $2^{20} = (1K)^2$ = 1M. For the address being two array-, nine row-, and nine column-bits, the address 102, 476 is in the zeroth quadrant. From the top (bit 19), the bits are found as follows by successive trial subtractions: Thus $102476 - 2^{16} = 102476 - 65536 = 36940$, $36940 - 2^{15} = 36940 - 32768 = 4172$, $4172 - 2^{12} = 4172 - 4096 = 76$, $76 - 2^6 = 76 - 64 = 12$, $12 - 2^3 = 12 - 8 = 4$, $4 - 2^2 = 0$.

Thus the total address is: 0 0 0 1 1 0 0 1 0 0 0 0 0 1 0 0 1 1 0 0, where 4 + 8 + 64 + 4096 + 32768 + 65536 = 102476, OK. Thus, the corresponding column address is the (9 bits on the right): **0 0 1 0 0 1 1 0 0** $\equiv 76_{10}$.

13.41

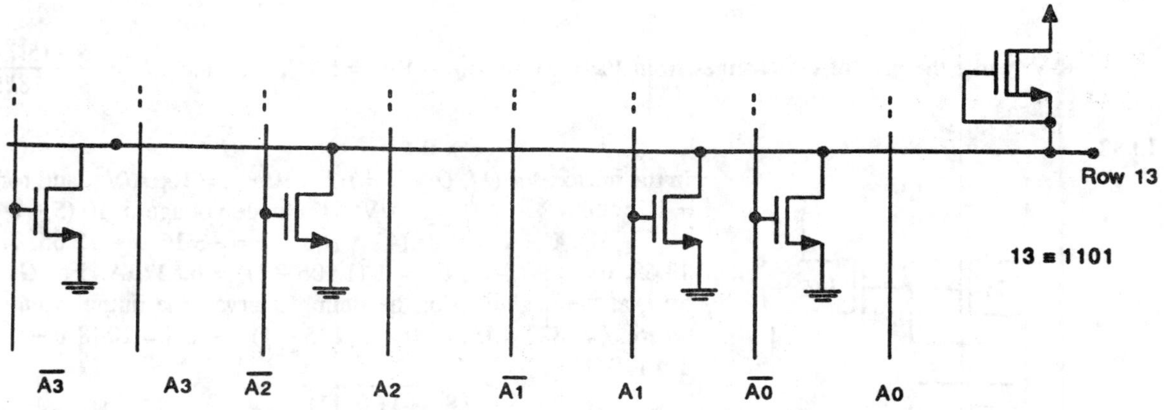

Here, $13_{10} \equiv (1101)_2$. Thus the decoding pattern is as shown. For a 256K-bit square array, there are two 512-line decoders, each with a 9-bit input requiring 9 + 1 = **10** transistors per row.

13.42 For each inverter, $V_{th} = \upsilon$ is the voltage at which $\upsilon_O = \upsilon_I = \upsilon$. Now $K_n = 1/2 \times 25 \times 2/3 = 8.3\dot{3}$ $\mu A/V^2$, $K_p = 1/2 \times 10 \times 2/3 = 3.3\dot{3} \mu A/V^2$. For both devices in pinchoff, $8.3\dot{3}\ (\upsilon - 1)^2 = 3.3\dot{3}\ (5 - \upsilon - 1)^2$, $\upsilon - 1 = (4 - \upsilon)\ (.632)$, $\upsilon - 1 = 2.53 - .632\ \upsilon$, $1.632\ \upsilon = 3.53$, and $\upsilon =$ **2.16V**. For input current i and $\upsilon_O = 2.16/2 = 1.08$V, with the n-channel device in triode mode with $\upsilon_{GS} = 5$V, $i = 8.3\dot{3}\ (2\ (5 - 1)\ (1.08) - 1.08^2) = 8.3\dot{3}\ (8\ (1.08 - 1.08^2))$. Thus the current to the cell = **62.3μA**. For the p-channel device, $i = 3.3\dot{3}\ [2(5-1)(\frac{5-2.16}{2}) - (\frac{5-2.16}{2})^2] = 3.3\dot{3}\ (8(1.42) - 1.42^2) =$ **31.1μA**. Now, during readout, the digit line is assumed held at $V_{DD}/2 = 2.5$V by its large capacitance, with a noise margin of half the threshold. Assume Q_5 and Q_6 are n-channel FETs and that the p-channel connection is the most sensitive. Thus, arrange the size of Q_5 (or Q_6) so that the current is 31.1μA (or less), with $\upsilon_{S5} = 2.5$V, $\upsilon_{D5} = 3.58$V, and $\upsilon_{G5} = 5$V. Thus, $31.1 = K\ (2\ (5 - 2.5 - 1)\ (3.58 - 2.5) - (3.58 - 2.5)^2) = K\ (2\ (1.5)\ (1.08) - 1.08^2) = 2.074K$, $K = 31.1/2.074 = 15.00 \mu A/V^2$. Since $K = 1/2\ (25) \times W/3 = 15$, $W = W_5 = \frac{3 \times 15 \times 2}{25} =$ **3.6μm**. For writing, the worst case is likely to be raising the output against the n-channel device. Here $\upsilon_{G5} = +5$V, $\upsilon_{D5} = +5$V and $\upsilon_{S5} \geq 2.16$V, while $\upsilon_{G1} = +5$V, $\upsilon_{D1} \geq 2.16$, $\upsilon_{S1} = 0$V. Correspondingly, $i_5 = (1/2 \times 25 \times \frac{3.6}{3})\ (5 - 2.16 - 1)^2 =$ **27.6μA**, $i_1 = (1/2 \times 25 \times \frac{2}{3})\ [2\ (5 - 1)\ (2.16) - 2.16^2] =$ **105.1μA**. Thus writing cannot occur by Q_5 overpowering Q_1. *However*, the other digit line at 0V may succeed. For this case, Q_4 and Q_6 compete, where $i_6 \approx (1/2 \times 25 \times 3.6/3)\ (5 - 1)^2 =$ **240μA**, while $i_4 \approx (3.3\dot{3})\ (5 - 1)^2 =$ **53.35μA**. Thus i_6 clearly exceeds i_4, and the design is viable. Therefore writing will occur, with a major role for one of inputs in pulling down and a secondary role for the other input in pulling up.

13.43 Here $Q = CV = IT$. Thus, $C = \frac{10 \times 10^{-15} \times 4 \times 10^{-3}}{1.5} = 26.7 \times 10^{-18} \equiv$ **0.027fF**.

SECTION 13.10: Read-Only Memory (ROM)

13.44 For the computation of X / Y as $F + Q + R$, where "1" in the 5 output columns represents the location of a ROM transistor.

x_1	x_0	y_1	y_0	f_1	q_1	q_0	r_1	r_0
0	0	0	0	1	0	0	0	0
0	0	0	1	0	0	0	0	0
0	0	1	0	0	0	0	0	0
0	0	1	1	0	0	0	0	1
0	1	0	0	1	0	0	0	0
0	1	0	1	0	0	1	0	0
0	1	1	0	0	0	0	0	1
0	1	1	1	0	0	0	1	1
1	0	0	0	1	0	0	1	0
1	0	0	1	0	1	0	0	0
1	0	1	0	0	0	1	0	0
1	0	1	1	0	0	0	1	0
1	1	0	0	1	0	0	1	1
1	1	0	1	0	1	1	0	0
1	1	1	0	0	0	1	0	1
1	1	1	1	0	0	1	0	0

Now, count the ones in the output columns of the table (the rightmost 5 columns). See that for the 5 column outputs, there are 20 transistors in the array itself, plus 5 loads, plus 5 inverters of 2 transistors each. Thus the design needs 20 + 5 = 25 transistors in the array, and 5(2) = 10 for the inverters for a total of 25 + 10 = 35. Without the inverters (and using transistors to represent logic zero), the design would use 5 (16) −20 +5 = 65 transistors total. For the 4-bit decoder, one needs 4 + 1 transistors per row (including the load), for 16 rows = 80 transistors in total.

SECTION 13.11: Gallium Arsenide Digital Circuits

13.45 For the load reduced with from 6µm to 5µm: $V_{DD} = 1.5V$, $V_{tD} = -1V$, $V_{tE} = 0.2V$, $\beta = 10^{-4} A/V^2$ per µm, $\lambda = 0.1V^{-1}$. For the input device, $W = 50µm$ and $\beta_I = 5mA/V^2$. For the load device, $W = 5µm$ and $\beta_L = 0.5mA/V^2$. $V_{OH} = 0.7V$. For $V_{OL} = \upsilon$: 5×10^{-3} [2 (0.7 − 0.2) $\upsilon - \upsilon^2$] (1 + 0.1 υ) = 0.5×10^{-3} [0 − 1 (−1)]^2 (1 + 0.1 (1.5 − υ), 10 (2 (0.5) $\upsilon - \upsilon^2$) (1 + 0.1 υ) = (1) (1 + 0.15 − 0.1 υ), or (10 $\upsilon - \upsilon^2$) (1 + 0.1 υ) = 1.15 −0.1 υ, 10 υ − 1υ^2 + 1 υ^2 − 0.1 υ^3 = 1.15 − 0.1 υ, 0.1 υ^3 − 10.1 υ + 1.15 = 0, and υ^3 − 101 υ + 11.5 = 0. For υ small, $\upsilon \simeq$ 11.5/101 = 0.114V. Thus $V_{OL} \simeq$ 0.114V.

For $V_{IL} = \upsilon$: **5×10^{-3} (υ − 0.2)^2 (1 + 0.1 (0.7)) = 0.5×10^{-3} [2 (1) (1.5 − 0.7) − (1.5 − 0.7)^2] (1 + 0.1 (1.5 − 0.7)), or 10 (υ − 0.2)^2 (1 + .07) = 1 [2 (0.8) − .8^2] [1.08], υ^2 − .4 υ + .04 = .0969, υ^2 − .4 υ − .0569 = 0,**

$$\upsilon = \frac{--.4 \pm \sqrt{.4^2 - 4(-.0569)}}{2} = \frac{.4 \pm .6226}{2} = 0.511, \text{ or } -.113V. \text{ Conclude } V_{IL} = \mathbf{0.51V}.$$

For $V_{IH} = \upsilon$: **Q_I in triode, Q_L in saturation, 5×10^{-3} [2 (υ − 0.2) $\upsilon_O - \upsilon_O^2$] [1 + 0.1 υ_O] = 0.5×10^{-3} (0 − − 1)^2 (1 + 0.1 (1.5 − υ_O)). Now, neglecting terms in 0.1 υ_O, 2 (υ − 0.2) $\upsilon_O - \upsilon_O^2$) = 0.1 (1 + .15), υ_O^2 − 2 υ υ_O + 0.4 υ_O + .115 = 0 − − − (1). Now taking $\frac{\partial}{\partial \upsilon} \rightarrow 2\upsilon_O \frac{\partial \upsilon_O}{\partial \upsilon} - 2\,\upsilon \frac{\partial \upsilon_O}{\partial \upsilon} - 2\upsilon_O + 0.4 \frac{\partial \upsilon_P}{\partial \upsilon}$ + .115 = 0, which for $\frac{\partial \upsilon_O}{\partial \upsilon}$ = − 1, becomes −2 υ_O + 2υ − 2υ_O −.4 + .115 = 0, 2υ = 4υ_O + .285, υ = 2υ_O +**

.1425 – – – (2).

Substituting (2) in (1) → $\upsilon_O^2 - 2\upsilon_O(2\upsilon_O + .1425) + .4\,\upsilon_O + .115 = 0$, $\upsilon_O^2 - 4\,\upsilon_O^2 - .285\,\upsilon_O + .4\,\upsilon_O + .115 = 0$, $-3\,\upsilon_O^2 - .115\,\upsilon_O + .115 = 0$, $\upsilon_O^2 + .0383\,\upsilon_O - .0383 = 0$, $\upsilon_O = \frac{-.0383 \pm \sqrt{.0383^2 - 4(-.0383)}}{2} = \frac{-.0383 + .3933}{2} = .177\text{V}$, whence $V_{IH} = \upsilon = 2(.177) + .1425 =$ **.497V**. Now $NM_H = V_{OH} - V_{IH} = 0.700 - .497 =$ **0.203V**, and $NM_L = V_{IL} - V_{OL} = 0.510 - .114 =$ **0.396V**.

13.46 From P13.45: $\beta_I = 5\text{mA/V}^2$; $\beta_L = 0.5\text{mA/V}^2$.

Output High: $I_{DD} = 0.5\text{mA/V}^2\,[2(0 - -1)(1.5 - 0.7) - (1.5 - 0.7)^2]\,[1 + 0.1(1.5 - 0.7)] = 0.5\,[2(0.8) - 0.8^2]\,[1 + .08] = 0.5184\text{mA}$.

Output Low: ($V_{OL} = 0.114\text{V}$) Q_L in saturation, $I_{DD} = 0.5 \times 10^{-3}\,(0 - -1)^2\,(1 + 0.1\,(1.5 - .114)) = 0.5693\text{mA}$. Thus average current $= \frac{.5184 + .5693}{2} =$ **0.544mA**. Average static power + 1.5 (.574) = **0.816mW**.

For t_{pHL}: Need to calculate the time to fall from V_{OH} to $\frac{V_{OH} + V_{OL}}{2}$, that is, from 0.7 to $\frac{0.7 + .11}{2} = 0.407\text{V}$. For i_I at $\upsilon_O = 0.7$, Q_I is in saturation (with $\upsilon_{GS} = 0.7\text{V}$), $i_I = 5 \times 10^{-3}\,(0.7 - 0.2)^2\,(1 + 0.1\,(0.7)) = 1.38\text{mA}$. For i_I at $\upsilon_O = 0.407$, Q_I is nearly in saturation (with $\upsilon_{GS} = 0.7\text{V}$), $i_I = 5 \times 10^{-3}\,(0.70 - 0.2)^2\,(1 + 0.1\,(0.41)) = 1.30\text{mA}$. Also i_L at $\upsilon_O = 0.7\text{V}$ is 0.518mA, and i_L at $\upsilon_O = .114\text{V}$ is 0.569mA, and (probably) $i_L = .518 + \frac{.569 - .518}{2} = 0.544\text{mA}$, at $\upsilon_O = 0.407$. The discharge current $= \frac{1.38 - .518 + 1.30 - .544}{2} = 0.809\text{mA}$, and $t_{pHL} = \frac{CV}{I}\ \frac{30 \times 10^{-15} \times (0.70 - .407)}{.809 \times 10^{-3}} =$ **10.9ps**.

For t_{pHL}: At $\upsilon_O = .114\text{V}$, $i_L = .569\text{mA}$. At $\upsilon_O = 0.407\text{V}$, $i_L \approx 0.544\text{mA}$. Thus the charging current $= \frac{.544 + .569}{2} = 0.556\text{mA}$, and $t_{pLH} = \frac{30 \times 10^{-15}\,(.407 - .114)}{.556 \times 10^{-3}} =$ **15.8ps**. Thus, overall, $t_p = \frac{15.8 + 10.9}{2} =$ **13.4ps**.

Dynamic Power at 2GHz: $P_D = 30 \times 10^{-15} \times (0.70 - .114)(2 \times 10^9) =$ **35.2μW**. Total power = 35.2μW + 816μW = **851μW**. Delay-power product $DP = 851 \times 10^{-6} \times 13.4 \times 10^{-12} =$ **0.011pJ**.

10.47 For the nominal design, $\beta_S = 10^{-4} \times 20 = 2\text{mA/V}^2$, $\beta_L = 2\text{mA/V}^2$, $\beta_{PD} = 10^{-4} \times 10 = 1\text{mA/V}^2$, $V_{tD} = -0.9\text{V}$, $\lambda = 0\text{V}^{-1}$, $V_{DD} = 3\text{V}$, $V_{SS} = 2\text{V}$, $V_D = 0.7\text{V}$. From Fig. 13.54 of the Text, for the nominal design, $V_{OH} = 0.7\text{V}$, $V_{OL} = -1.27\text{V}$, $V_{IL} = -0.26\text{V}$, $V_{IH} = -0.16\text{V}$, $NM_H = 0.86\text{V}$, $NM_L = 1\text{V}$.

(a) *For* $V_{tD} = -0.8\text{V}$, $V_{OH} =$ **0.7V**.

For V_{IL}, $i_S \simeq i_L - i_{PD}$, $i_{PD} = 1\,(0 - -0.8)^2 = .64\text{mA}$, $i_L = 2\,(0 - -0.8)^2 = 1.28\text{mA}$, $i_S = 1.28 - 0.64 = 0.64\text{ mA}$. Thus, $0.64 = 2\,(\upsilon - -0.8)^2$, $\upsilon + 0.8 = \sqrt{.64/2} = .566$, $\upsilon = .566 - 0.8 =$ **−0.234V**. Thus, $V_{IL} =$ **−0.234V**.

For V_{OL}, $i_{PD} = .64\text{mA}$, $i_L = 1.28\text{mA}$, $i_S = 1.28 - 0.64 = 0.64\text{mA}$. But $\upsilon_I = 0.7\text{V}$ with Q_S in triode mode, with $\upsilon_{DSS} = \upsilon_O$, $0.64 = 2\,[2\,(0.7 - -0.8)(\upsilon_O) - \upsilon_O^2]$, $0.32 = 3\upsilon_O - \upsilon_O^2$, $\upsilon_O^2 - 3\upsilon_O + 0.32 = 0$, $\upsilon_O = -\frac{-3 \pm \sqrt{3^2 - 4\,(.32)}}{2} = 0.111\text{V}$. Thus, $V_{OL} = 0.111 - 0.7 - 0.7 =$ **−1.29V**.

For $V_{IH} = \upsilon$: As above, but with input υ (rather than 0.7V), see $0.64 = 2\,[2\,(\upsilon - -0.8)\,\upsilon_O - \upsilon_O^2]$, $0.32 = 2\,(\upsilon + 0.8)\,\upsilon_O - \upsilon_O^2$, or $\upsilon_O^2 - 2\,(\upsilon + 0.8)\,\upsilon_O + 0.32 = 0$ – – – (1). $\frac{\partial}{\partial \upsilon} \rightarrow 2\,\upsilon_O \frac{\partial \upsilon_O}{\partial \upsilon} - 2\,(\upsilon + 0.8)\frac{\partial \upsilon_O}{\partial \upsilon} - 2\upsilon_O = 0$. Now, for $\frac{\partial \upsilon_O}{\partial \upsilon} = 1$, $-2\upsilon_O + 2\,(\upsilon + 0.8) - 2\upsilon_O = 0$, $2\upsilon = 4\upsilon_O - 1.6$, $\upsilon = 2\upsilon_O - 0.8$ – – – (2).

Substituting (2) in (1), $\upsilon_O^2 - 2\,(2\upsilon_O^2) + .32 = 0$, $3\upsilon_O^2 = .32$, $\upsilon_O = (.32/3)^{1/2} = 0.327\text{V}$, $\upsilon = 2\,(.327) - 0.8 = -0.146\text{V}$. Thus, $V_{IH} = \mathbf{-.146V}$. Now $NM_H = V_{OH} - V_{IH} = 0.7 - -.146 = \mathbf{0.85V}$, $NM_L = V_{IL} - V_{OL} = -.234 - -1.29\text{V} = \mathbf{1.06V}$.

(b) *For all* $V_{tD} = -1.0\text{V}$, $V_{OH} = \mathbf{0.7V}$.

For V_{IL}: $i_{PD} = 1\,(0 - -1.0)^2 = 1\text{mA}$, $i_L = 2\,(0 - -1.0)^2 = 2\text{mA}$, $i_S = 2 - 1 = 1\text{mA}$. Now $1 = 2\,(\upsilon - -1.0)^2$, $\upsilon = \sqrt{1/2} - 1 = -0.293\text{V}$, and $V_{IL} = \mathbf{-.293V}$.

For $V_{OL} = \upsilon_O - 1.4$: $1.0 = 2\,[2\,(0.7 - -1)\,(\upsilon_O) - \upsilon_O^2]$, $0.5 = 3.4\,\upsilon_O - \upsilon_O^2$, $\upsilon_O^2 - 3.4\,\upsilon_O + 0.5 = 0$,

$$\upsilon_O = -\frac{-3.4 \pm \sqrt{3.4^2 - 4\,(0.5)}}{2} = 0.308\text{V},\ V_{OL} = 0.308 - 1.4 = \mathbf{-1.09V}.$$

For $V_{IH} = \upsilon$, $1 = 2\,[2\,(\upsilon - -1)\,\upsilon_O - \upsilon_O^2]$, $0.5 = (2\upsilon + 2)\,\upsilon_O - \upsilon_O^2$, $\upsilon_O^2 - 2\,(\upsilon + 1)\,\upsilon_O + 0.5 = 0$ --- (1), $\frac{\partial}{\partial \upsilon} \to 2\,\upsilon_O \frac{\partial \upsilon_O}{\partial \upsilon} - 2\,(\upsilon + 1)\frac{\partial \upsilon_O}{\partial \upsilon} - 2\,(\upsilon_O) = 0$. Now, for $\frac{\partial \upsilon_O}{\partial \upsilon} = -1$, $-2\upsilon_O + 2\upsilon + 2 - 2\upsilon_O = 0$, $2\upsilon = 4\upsilon_O - 2$, $\upsilon = 2\upsilon_O - 1$ --- (2).

Substituting (2) in (1), $\upsilon_O^2 - 2\,(2\upsilon_O)\,\upsilon_O + 0.5 = 0$, $\upsilon_O = (0.5/3)^{1/2} = .408\text{V}$, $\upsilon = 2\,(.408) - 1 = -.184\text{V}$, $V_{IH} = \mathbf{-0.184V}$. Now, $NM_H = 0.7 - -.184 = \mathbf{0.88V}$, and $NM_L = -.293 - -1.09 = \mathbf{0.80V}$.

(c) *For* $W_L = 2W_{PD} = 5\mu\text{m}$, $\beta_S = 10^{-4} \times 20 = 2\text{mA/V}^2$, $\beta_L = 10^{-4} \times 5 = 0.5\text{mA/V}^2$, $\beta_{PD} = 10^{-4} \times 5/2 = .25\text{mA/V}^2$, $V_{tD} = -0.9\text{V}$, $\lambda = 0\text{V}^{-1}$, $V_{DD} = 3\text{V}$, $V_{SS} = 2\text{V}$. See $V_{OH} = \mathbf{0.7V}$.

For $V_{IL} = \upsilon$: $i_{PD} = 0.25\,(0 - -.9)^2 = .2025\text{mA}$, $i_L = 0.5\,(0 - -.9)^2 = .4056\text{mA}$, $i_S = .405 - .2025 = .2025\text{mA}$. Now $0.2025 = 2\,(\upsilon - -0.9)^2$, $\upsilon = (\frac{.2025}{2})^{1/2} - 0.9 = -.582\text{V}$, $V_{IL} = \mathbf{-0.582V}$.

For $V_{OL} = \upsilon_O$: $\upsilon_O^2 - 3.2\upsilon_O + .10125 = 0$, $\upsilon_O = \frac{3.2 \pm \sqrt{3.2^2 - 4\,(.10125)}}{2} = \mathbf{0.032V}$, $V_{OL} = .032 - 1.4 = \mathbf{-1.37V}$.

For $V_{IH} = \upsilon$, $0.2025 = 2\,[2\,(\upsilon - -0.9)\,\upsilon_O - \upsilon_O^2$, $\upsilon_O^2 - 2\,(\upsilon + 0.9)\,\upsilon_O + .10125 = 0$ --- (1). $\frac{\partial}{\partial \upsilon} \to 2\upsilon_O \frac{\partial \upsilon_O}{\partial \upsilon} - 2\,(\upsilon + 0.9)\frac{\partial \upsilon_O}{\partial \upsilon} - 2\upsilon_O = 0$. Now for $\frac{\partial \upsilon_O}{\partial \upsilon} = -1 \to -2\upsilon_O + 2\upsilon + 1.8 - 2\upsilon_O = 0$, $\upsilon = 2\upsilon_O - 0.9$ --- (2).

Substituting (2) in (1), $\upsilon_O^2 - 2\,(2\upsilon_O)\,\upsilon_O + .10125 = 0$, $\upsilon_O = (\frac{.10125}{2})^{1/2} = 0.184\text{V}$, $\upsilon = 2\,(.184 - 0.9) = -.532\text{V}$, $V_{IH} = \mathbf{-0.532V}$. Now, $NM_H = V_{OH} - V_{IH} = 0.7 - -.532 = \mathbf{1.23V}$, and $NM_L = V_{IL} - V_{OL} = -.582 - -1.37 = \mathbf{0.79V}$.

13.48 Use $V_{tD} = -0.9\text{V}$ for all devices, and use $W_S = W_L = 20\mu\text{m}$, and $W_{PD} = 10\mu\text{m}$ for easy equivalence to the *FL* design. However, actually, since Q_{PD} is the load presented by each gate, and since the loading of the Q_{PD} drain is very light (the gate of Q_S), then Q_{PD} can be made much smaller. For a fanout of 4, reduce the width of Q_{PD} to 2.5 µm or less. $\beta_S = \beta_L = 20 \times 10^{-4} = 2\text{mA/V}^2$, $\beta_{PD} = 10 \times 10^{-4} = 1\text{mA/V}^2$, $V_D = 0.7\text{V}$. For the *FL* gate, $V_{OH} = 0.7\text{V}$, $V_{IL} = -0.26\text{V}$, $V_{IH} = -0.16\text{V}$, $V_{OL} = -1.27\text{V}$. At the gate of Q_S of the connected circuit, $\upsilon_O = 0.7\text{V}$, $V_{OH} = 0.7 + 0.7 + 0.7 = \mathbf{2.1V}$, with a fanout of 1! $V_{OL} = -1.27 + 2\,(0.7) = \mathbf{0.13V}$, or lower with a fanout > 1. $V_{IL} = -0.26\text{V} + 1.4\text{V} = \mathbf{1.14V}$, $V_{IH} = -.16\text{V} + 1.4\text{V} = \mathbf{1.24V}$. Thus, $NM_H = 2.10 - 1.24 = \mathbf{0.86V}$, $NM_L = 1.14 - .13 = \mathbf{1.01V}$. For a design in which W_{PD} is reduced to 2.5µm, these results are an approximation for a fanout of 4. For a fanout of 1, V_{OL} increases, as do V_{IH} and V_{IL} slightly.

Chapter 14

BIPOLAR DIGITAL CIRCUITS

SECTION 14.1: The BJT as a Digital Circuit Element

14.1

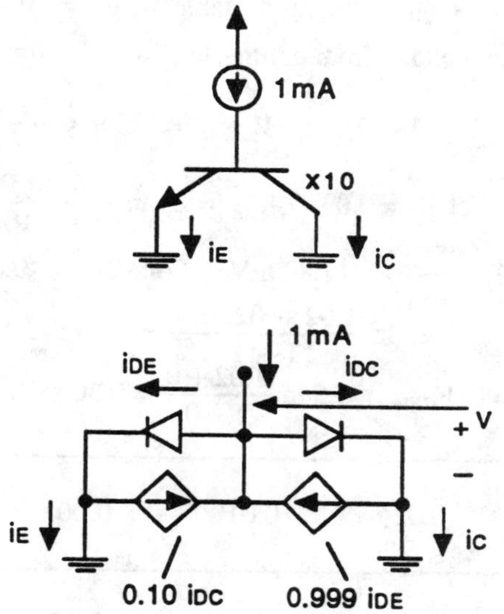

(a) For $\alpha_F = 0.999$, $\alpha_R = \dfrac{0.999}{10} = .0999 \approx 0.100$. From Eq. 14.8,
$i_E = \dfrac{I_S}{\alpha_F}\,(e^{v_{BE}/V_T}-1) - I_S\,(e^{v_{BC}/V_T}-1)$, or
$\dfrac{i_E}{I_S} \approx \dfrac{1}{0.999}\,(e^{v/V_T}) - (e^{v/V_T}) = .001\,e^{v/V_T}$ - - (1).
From Eq. 14.9, including a change of sign of i_C in the two situations, $-i_C \approx I_S\,(e^{v/V_T} - \dfrac{1}{0.1}\,e^{v/V_T})$ or $\dfrac{i_C}{I_S} \approx 9e^{v/V_T}$ - - (2). Thus, $\dfrac{i_C}{i_E} = \dfrac{9}{.001}\,e^{v/V_T} = 9000$, and i_C = **1mA, whence $i_E = 0.11\ \mu A$.**
Now if this transistor is a 1 mA device; that is $v_{BE} = 700$ mV at $i = 1$mA, $i_E \approx \dfrac{I_S}{1}\,(e^{700/25}) = 1$ mA, whence $I_S = 6.9\times10^{-16}\ A$. Now, for this (saturated) case: $i_C = 9\,I_S\,e^{v/V_T}$. Thus $v = 25\ln\dfrac{1\times10^{-3}}{9\times6.9\times10^{-16}}$ = **645mV**.

(b) **For α_F = 0.99,** $\alpha_R = \dfrac{0.99}{10} = 0.099 \approx 0.1$. Thus, as above, (as in (1)), $\dfrac{i_E}{I_S} \approx \left[\dfrac{1}{.99} - 1\right] e^{v/V_T} = 0.01e^{v/V_T}$, and (as in (2)), $\dfrac{i_C}{I_S} = \left[1 - \dfrac{1}{.01}\right] e^{v/V_T} = 9e^{v/V_T}$, whence $i_C = 900\ i_E$, $i_C + i_E$ = 1 mA = 901 i_E. Thus i_E = **1.1μA**, and i_C = **0.999 mA**, with $v = 25\ln\dfrac{1\times10^{-3}}{0.999\times10^{-16}}$ = **691mV**.

14.2 For the collector open, and I_B finite, $\beta_{forced} = 0$. From Eq. 14.16, $V_{CE\,sat} = V_T \ln \dfrac{1+(\beta_{forced}+1)/\beta_R}{1-\beta_{forced}/\beta_F}$. For $\beta_{forced} = 0$, $V_{CE\,sat} = V_T \ln(1+\dfrac{1}{\beta_R})$. Now, for β_F = 70 to 280, and, correspondingly, $\alpha_F = \dfrac{\beta_F}{\beta_F} = \dfrac{70}{71}$ to $\dfrac{280}{281}$ = 0.9859 to 0.9964, with $\alpha_R = \dfrac{\alpha_F}{10}$ = 0.0986 to 0.0996, and $\beta_R = \dfrac{\alpha_R}{1-\alpha_R} = \dfrac{.0986}{1-.0986}$ to $\dfrac{.0996}{1-.0996}$ = 0.109 to 0.111, Thus, $V_{CE\,sat} = 25\ln\left[1+\dfrac{1}{.109}\right]$ to $25\ln\left[1+\dfrac{1}{.111}\right]$ = **58.0mV to 57.7mV**.

14.3 For open-collector operation, β_{forced} = 0, and $V_{CE\,sat} = V_T \ln\left[1+\dfrac{1}{\beta_R}\right]$. For $V_{CE\,sat}$ = 100 mV, $\ln\left[1+\dfrac{1}{\beta_R}\right] = \dfrac{100}{25} = 4$, $1+\dfrac{1}{\beta_R} = 54.6$, and $\beta_R = 0.0187$. Now, for the circuit shown, assume V_{CE} to be

100 mV. Thus $i_C = -\frac{100\ mV}{1k + 1k} = -50\mu A$. Thus $\beta_{forced} = \beta_f = \frac{-50 \times 10^{-6}}{1 \times 10^{-3}} = -0.05$. Now, $V_{CE\,sat} = V_T \ln \frac{1 + (\beta_f + 1)/\beta_R}{1 - \beta_f/\beta_F} = 25 \ln \frac{1 + (1 - .05)/.0187}{1 - (-.05)/100} = 25 \ln \frac{51.8}{1.0005} = 98.67$ mV. Thus $\upsilon_O = \frac{98.7}{2} =$ **49.3mV**.

Now for $R_1 = R_2 = 500\Omega$, $i_C \approx -100\mu A$, and $\beta_f = -0.1$, and $V_{CE\,sat} = 25 \ln \frac{1 + (1 - 0.1)/.0187}{1 - (-0.1)/100} = 25 \ln \frac{49.13}{1.001} = 97.33$ mV, with $\upsilon_O = \frac{97.3}{2} =$ **48.7mV**.

See $R_{CE\,sat} \approx \frac{\Delta V}{\Delta I} = \frac{(49.3 - 48.7)\ (mV)}{(100 - 50)\ (\mu A)} = 12\Omega$

14.4 For Table 14.1, $\beta_F = 50$, and $\beta_R = 0.1$. For the required table, $\beta_F = 0.1$, $\beta_R = 50$, $\beta_{forced} = \frac{current\ in\ emitter\ lead}{current\ in\ base\ lead} = \beta_f < \beta_F =$ **0.1**. The voltage from emitter lead to collector lead (from Eq. 14.16), is $V_{EC\,sat} = V_T \ln \left\{ \frac{1 + (\beta_f + 1)/\beta_F}{1 - \beta_f/\beta_R} \right\}$ (like Eq. 14.17), or $V_{ECsat} = 25 \ln \left\{ \frac{1 + (\beta_f + 1)/50}{1 - \beta_f/0.1} \right\} = 25 \ln \frac{1.02 + .02\ \beta_f}{1 - 10\ \beta_f}$. Now, for $\beta_f = 0.1$, $V_{CE\,sat} = \infty$. For $\beta_f = 0.09$, $V_{CE\,sat} = 25 \ln \frac{1.02 + .02\ (.09)}{1 - 10(.09)} = 58.1$ mV. For $\beta_f = 0.05$, $V_{CE\,sat} = 25 \ln \frac{1.02 + .02(.05)}{1 - .5} = 17.8$ mV. For $\beta_f = 0.02$, $V_{CE\,sat} = 25 \ln \frac{1.02 + .02(.02)}{1 - .2} = 6.9$ mV. For $\beta_f = 0.01$, $V_{CE\,sat} = 25 \ln \frac{1.02 + .02(.01)}{1 - .1} = 3.4$ mV. For $\beta_f = 0.001$. $V_{CE\,sat} = 25 \ln \frac{1.02 + .02(.001)}{1 - .01} = 0.75$ mV. For $\beta_f = 0$, $V_{CE\,sat} = 25 \ln \frac{1.02 + 0}{1 - 0} = 0.50$ mV. In summary:

β_{forced}	0.1	0.09	0.05	0.02	0.01	0.001	0.000
$V_{EC\,sat}$(mV)	∞	58.1	17.8	6.1	3.1	0.75	0.50

14.5 Here, $\alpha_F = 0.995$ implies $\beta_F = \frac{.995}{1 - .995} = 199$, and $\alpha_R = \frac{0.995}{5} = 0.199 \approx 0.2 \rightarrow \beta_R = \frac{.199}{1 - .199} = 0.25$.

The limiting value of forced β is **199**. For $\beta_{forced} = \beta_f < 199$, $V_{CE\,sat} = 25 \ln \left[\frac{1 + (\beta_f + 1)/0.25}{1 - \beta_f/199} \right] = 25 \ln \left[\frac{5 + 4\beta_f}{1 - .005\beta_f} \right]$.

Now for $\beta_f = 199$, 180, 100, 40, 20, 2, 0: For $\beta_f = 199$: $V_{CE\,sat} = \infty$. For $\beta_f = 180$: $V_{CE\,sat} = 25 \ln \frac{5 + 4(180)}{1 - 180/199} = 223$ mV. For $\beta_f = 100$: $V_{CE\,sat} = 25 \ln \frac{5 + 4(100)}{1 - 100/199} = 168$ mV. For $\beta_f = 40$: $V_{CE\,sat} = 25 \ln \frac{5 + 4(40)}{1 - 40/199} = 133$ mV. For $\beta_f = 20$: $V_{CE\,sat} = 25 \ln \frac{5 + 4(20)}{1 - 20/199} = 114$ mV. For $\beta_f = 2$: $V_{CE\,sat} = 25 \ln \frac{5 + 4(2)}{1 - 2/199} = 64$ mV. For $\beta_f = 0$: $V_{CE\,sat} = 25 \ln = 40$ mV.

These results are summarized in the table on the next page:

β_{forced}	199	180	100	40	20	2	0
$V_{EC\,sat}$(mV)	∞	223	168	133	114	64	40

Now, for $I_B = 10$ mA, $I_C = 1$ mA, $\beta_f = \frac{1}{10} = 0.1$. **In normal mode**: $V_{CE\,sat} = 25 \ln \left(\frac{5 + 4(0.1)}{1 - \frac{0.1}{199}}\right) = \mathbf{42.2mV}$.

In inverted mode: $V_{EC\,sat} = 25 \ln \left(\frac{1 + (0.1+1)/199}{1 - 0.1/0.25}\right) = 25\ln \frac{1.0055}{0.6} = \mathbf{12.9mV}$.

14.6

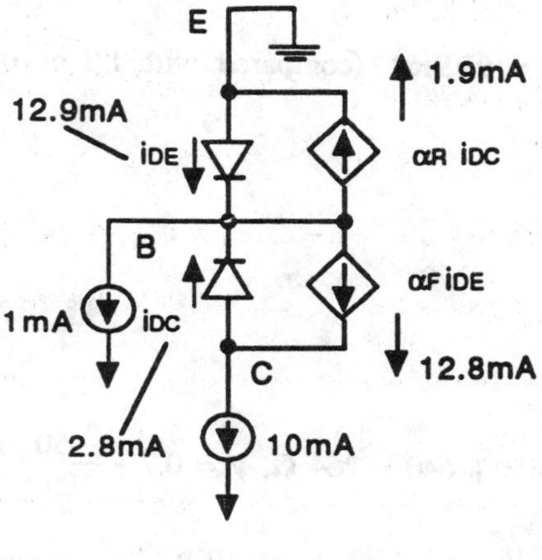

Here, $\alpha_F = \frac{\beta_F}{\beta_F + 1} = \frac{200}{201} = 0.995$, and $\alpha_R = \frac{2}{2+1} = 0.667$.

a) See $i_{DE} - 0.667\, i_{DC} = 10 + 1 = 11$ – – (1), and $0.995\, i_{DE} - i_{DC} = 10.0$ – – (2). From (1), $0.995\, i_{DE}$ - $0.663\, i_{DC} = 10.945$ – – (3). (2) – (3) → $-0.337\, i_{DC} = -0.945$, and i_{DC} = **2.804mA**. From (1), $i_{DE} = 0.667 (2.804) + 11$ = **12.87mA**. *Check* in (2): 0.995 (12.87) −2.804 = 10.00, with $\alpha_R\, i_{DC}$ = **1.87 mA**, and $\alpha_F\, i_{DE}$ = **12.8 mA**.

b) Now, $i = I_S\, e^{\upsilon/V_T} \rightarrow \upsilon = V_T \ln i/I_S$. Thus $\upsilon_{EB} = 25 \ln \frac{12.87 \times 10^{-3}}{10^{-14}}$ = **697mV**.

Now for the collector: $I_{SC} = 10^{-14} \times \frac{.995}{.667} = 1.49 \times 10^{-14}\, A$. Thus, $\upsilon_{CB} = 25 \ln \frac{2.804 \times 10^{-3}}{1.49 \times 10^{-14}}$ = **649mV** and $V_{EC\,sat} = 697 - 649 =$ **48mV**.

c) From Eq. 14.16: $V_{EC\,sat} = V_T \ln \frac{1 + (\beta_f+1)/\beta_R}{1 - \beta_f/\beta_F} = 25 \ln \frac{1+(10+1)/2}{1+10/200} = 25 \ln \frac{6.5}{1.05}$ = **45.6mV, nearly the same as in b).**

14.7 Here, $t_s = \tau_s \frac{I_{B2} - I_{C\,sat}/\beta}{I_{B1} + I_{C\,sat}/\beta}$. Now, for $I_{B1} = 0$, $t_s = 20 \times 10^{-9} \frac{1 - 10/200}{0 + 10/200} = 20 \left(\frac{1 - .05}{.05}\right) = 20(19) =$ **380ns**. For $I_{B1} = 1$ mA: $t_s = 20 \frac{(1-.05)}{1+.05}$ = **18.1ns**; and for $I_{B1} = 10$ mA: $t_s = 20 \frac{(1-.05)}{10+.05}$ = **1.89ns**.

14.8 Here, for $\upsilon_{BE} = 0.7$ V, $\upsilon_{CE\,sat} = 0.2$ V, $I_{B2} = \frac{5-0.7}{1k} = 4.3$ mA, $I_{B1} = \frac{0.7-0.2}{1k} = 0.5$ mA, and $I_{C\,sat} = \frac{5-0.2}{0.5k} = 9.6$ mA. Now, the original stored base charge = $\tau_s\, (4.3 - \frac{9.6}{\beta})$.

Without a Capacitor: charge removed = 80 ns $(0.5 + \frac{9.6}{\beta})$ mA = 80 (0.5 + 9.6/β) pC.

With a Capacitor: charge removed by capacitor = CV = 8 pF × (5 – 0.2) V = 38.4 pC. Subsequent charge removed = 30 (0.5 + 9.6/β). Now, 80(0.5 + 9.6/β) −38.4 = 30 (0.5 + 9.6/β). Thus 50(0.5 + 9.6/β) = 38.4,

$0.5 + 9.6/\beta = 38.4/50 = 0.768$, $9.6/\beta = 0.268$, or $\beta = 9.6/0.268 =$ **35.8**.

Also $\tau_s(4.3 - 9.6/\beta) = 80(0.5 + 9.6/\beta)$. Thus, $\tau_s = 80\,\dfrac{0.5+0.268}{4.3-0.268}$ = **15.2 ns**. Now, $t_s = 0$ when the charge provided via C just cancels the internal charge: Thus $CV = \tau_s\,(I_{B2} - \dfrac{I_{C\,sat}}{\beta})$, or $C = \dfrac{15.2}{5-0.2}(4.3 - \dfrac{9.6}{35.8}) = \dfrac{15.2}{4.8}(4.032) = 12.77\text{pF}$ or **12.8pF**.

Now for C = 12.8 pF and $\beta = 2\,(35.8) = 71.6$, $t_s = \dfrac{15.2\,(4.3 - \dfrac{9.6}{71.6}) - 12.8\,(5.0-0.2)}{0.5 + \dfrac{9.6}{71.6}}$

$= \dfrac{15.2\,(4.166) - 12.8\,(4.8)}{.634} = 2.97$, or **3ns**.

Now for $C = 0$, $t_s = \dfrac{15.2\,(4.3 - \dfrac{9.6}{71.6})}{0.5 + \dfrac{9.6}{71.6}} = \dfrac{15.2\,(4.166)}{.634}$ = **99.9ns!** (compared with 80 ns originally).

Check: $t_s = \dfrac{15.2\,(4.3 - \dfrac{9.6}{35.8})}{0.5 + \dfrac{9.6}{35.8}} = 79.8$ ns.

SECTION 14.2 Early Forms of BJT Circuits

14.9

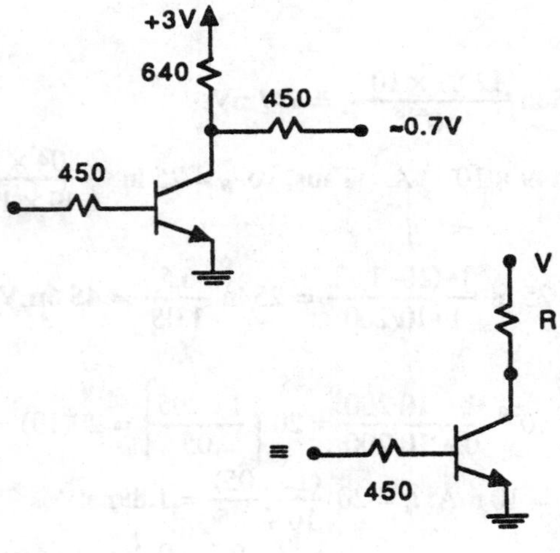

$R = 450 \parallel 640 = 264\ \Omega$, $V = 0.7 + \dfrac{450}{1090}(3 - 0.7) = 1.65$ V.

For gain = −1: Gain $= -\dfrac{\beta R_L}{R_S + r_\pi} = \dfrac{-50(264)}{450 + r_\pi} = -1$, $450 + r_\pi = 13200$, and $r_\pi = 12750$. Thus $r_e = \dfrac{12750}{50+1} = 250\ \Omega$, and $I_E = \dfrac{25mV}{250\Omega} = 0.1$ mA, for which $\upsilon_{BE} = 700 + V_t \ln \dfrac{0.1}{1} = 642$ mV. Thus $V_{IL} = 0.642 + \dfrac{0.1}{51} \times 0.45 =$ **0.643V** (rather than 0.60V, as assumed).

Now for two inputs, at gain of −1, equivalent r_π is the same; total current is the same; current in each transistor is half; and $\upsilon_{BE} = 700 + V_t \ln \dfrac{0.05}{1} = 625$ mV. Thus $V_{IL} =$ **0.625V**.

14.10

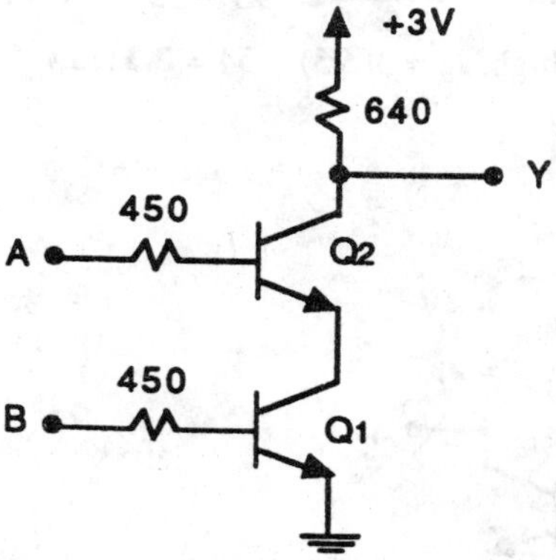

If for input A, $\beta_f = \dfrac{\beta_F}{2} = 25$, or $\dfrac{I_{C2}}{I_{B2}} = 25$, then $I_{C1} = 26\, I_{B2} = \dfrac{26}{25} I_{C2}$. For simplicity, let $I_{C1} = I_{C2}$, and $V_{CE\,sat} = 25 \ln \dfrac{1 + (\beta_f + 1)/\beta_R}{1 - \beta_f/\beta_F} = 25 \ln \left[\dfrac{1 + (25+1)/0.1}{1 - 25/50}\right] = 25 \ln (261/0.5) = 156\text{mV}$. Thus $V_Y = 0.156 + 0.156 = 0.313$ V, $I_{C2} = \dfrac{3 - 0.31}{640} = 4.20$ mA,

$I_{E2} = \dfrac{26}{25}(4.20) = 4.37$ mA, and $I_{E1} \approx (\dfrac{26}{25})^2 (4.20) = 4.54$ mA. Thus $\upsilon_{BE1} = 700 + 25 \ln 4.54/1 = 738$ mV, and $\upsilon_{BE2} = 700 + 25 \ln 4.37/1 = 737$ mV. Now, $V_{IH1} = 0.738 + 0.45 \times 4.54/25 =$ **0.820V**, $V_{IH2} = 0.737 + 0.156 + 0.45 \times 4.37/25 =$ **0.972V**.

14.11

(a) $I_C = \dfrac{4V - 2V}{4k\Omega} = 0.6$ mA; $I_B = \dfrac{0.5}{30} =$ **16.7μA**.

(b) $V_{BE} = 700 + 25 \ln 0.517/1 = 0.684\text{V}$, $I_{R2} = \dfrac{2 + .684}{5} = 0.537$ mA, $I_{D3,4} = 0.537 + .017 = 0.554$ mA, $V_{D3,4} = 700 + 25 \ln .554/1 = 0.685\text{V}$. Thus $V_X = 0.684 + 2\,(.685) = 2.054$ V, $I_{R1} = \dfrac{4 - 2.054}{2} = 0.973$ mA, $I_{D1} = 0.973 - 0.554 = 0.419$ mA, and $V_{D1} = 700 + 25 \ln 0.419/1 = 0.678\text{V}$. Thus $V_{th} = V_X - V_{D1} = 2.054 - 0.678 =$ **1.376V**.

(c) For diode drops as calculated, $V_X = 2.054\text{V}$, and $I_{R1} = 0.973$ mA. For A high, $I_{D3,4} = 0.973\text{mA}$, $V_{D3,4} \approx 700 + 25 \ln 0.973/1 = 699\text{mV}$, and $I_B \approx 0.973 - 0.537 = 0.436$ mA, for which Q is saturated with $I_C \approx \dfrac{4-0}{4} = 1$ mA, and $I_E \approx 1.44$ mA, with $V_B \approx 700 + 25 \ln 1.44/1 = 0.709\text{V}$. Thus the corrected value of $V_X = 0.709 + 2\,(.699) = 2.107$ V, for which $I_{R1} = \dfrac{4 - 2.107}{2} = 0.946$ mA, and $I_B \approx 0.946 - 0.537 =$ **0.409mA**.

(d) $I_{in} \approx \dfrac{4 - 0.7}{2} = 1.65$ mA, for which $\upsilon_D = 700 + 25 \ln 1.65/1 = 713\text{mV}$, and $I_{in} = \dfrac{4 - 0.713}{2} =$ **1.64mA**.

(e) Add another *diode in series with* D_3, D_4, which (for the same base drive) has a 0.678 V drop. Thus V_{th} is raised by 0.678 V, provided current levels remain the same. Now, for same maximum base drive, $V_X = 0.709 + 3\,(0.699) = 2.806\text{V}$, and $R_1 = \dfrac{4 - 2.806}{0.946} =$ **1.26kΩ**. Now, for $V_A = 0\text{V}$, $I_{in} \approx \dfrac{4 - 0.7}{1.26} = 2.62$ mA, for which $V_D = 700 + 25 \ln 2.62/1 = 724\text{mV}$, and $I_{in} = \dfrac{4 - .724}{1.26} =$ **2.60mA**.
For fanout, I_B max $= 0.436$ mA, and with $\beta_{forced} = \beta/2 = 30/2 = 15$, $I_{C\,sat} = 15\,(0.436) = 6.54$ mA. Thus the maximum load current $= 6.54 - (4-0)/4 = 5.54$ mA. Thus the maximum fanout $= 5.54/2.60 = 2.13$, or 2, conservatively. For a fanout $= 3$, $\beta_{forced} = \dfrac{3(2.60)+1}{0.436} = 20.2$. Thus $\beta_{forced} < \beta$, as required.

14.12

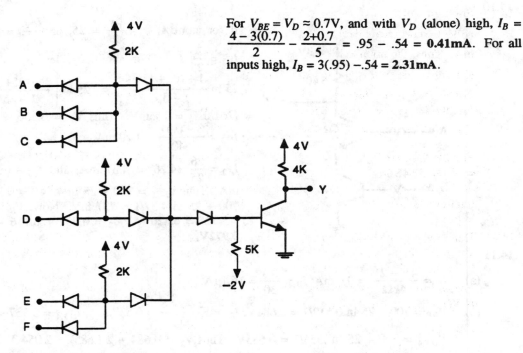

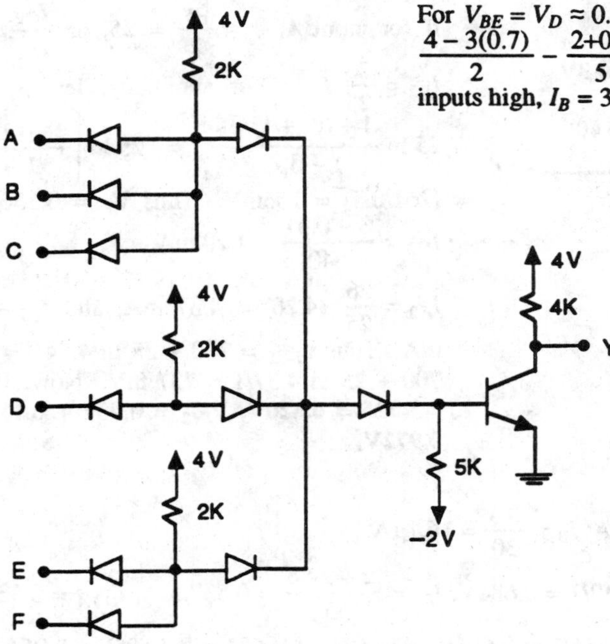

For $V_{BE} = V_D \approx 0.7$V, and with V_D (alone) high, $I_B = \frac{4-3(0.7)}{2} - \frac{2+0.7}{5} = .95 - .54 = \mathbf{0.41mA}$. For all inputs high, $I_B = 3(.95) - .54 = \mathbf{2.31mA}$.

SECTION 14.3: Transistor-Transistor Logic

14.13 Reduce R_B to 5/2 = 2.5 kΩ. From Exercise 14.8, the turn-on base current was 1.6 mA. Now, it is 1.6 – 0.7/5 = **1.46mA**. Now, for β = 50, $I_{C\,max}$ = 1.46(50) = 73 mA. The load current = 73 – 5/2 = 70.5 mA. Thus, for I_{in} = 1.1 mA, the maximum fanout is $\frac{70.5}{1.1}$ = 64.1, ie *N* = **64**. Now for *N* = 64/4 = 16, I_C = 16(1.1) + 2.5 = 20.1 mA, and β_{forced} = 20.1/1.46 = 13.8. Now, for β_F = 50, β_R = 50/100 = 0.5, $V_{EC\,sat}$ =

$$25 \ln \frac{1 + (\beta_f+1)/\beta_R}{1 - \beta_f/\beta_F} = 25 \ln \frac{1 + (13.8+1)/0.5}{1 - 13.8/50} = \mathbf{95.6mV} = V_{OL}.$$

14.14 *For N* = 0, I_{B2} = 1.46 mA, $I_{B1} = \frac{0.7V}{2.5k\Omega}$ = 0.28 mA, $I_C \approx 5/2$ = 2.5 mA. Now $t_s = \tau_s \frac{I_{B2} - I_C/\beta}{I_{B1} + I_C/\beta}$, which

for *N* = 0, I_C = 2.5 mA, becomes $t_s = 10 \frac{1.46 - \frac{2.5}{50}}{.28 + \frac{2.5}{50}} = \frac{10(1.41)}{.33} = \mathbf{42.7ns}$.

For N = 16, I_C = 5/2 + 16(1.1) = 20.1 mA, and $t_s = 10 \frac{1.46 - \frac{20.1}{50}}{0.28 + \frac{20.1}{50}} = \mathbf{15.5ns}$.

Now, for β = 100, *and N* = 0, $t_s = 10 \frac{1.46 - \frac{2.5}{100}}{.28 + \frac{2.5}{100}} = \mathbf{47.0ns}$.

For N = 16, $t_s = 10 \frac{1.46 - \frac{20.1}{100}}{0.28 + \frac{20.1}{100}} = \mathbf{26.2ns}$.

14.15 For a load of 2 kΩ, $I_C = \frac{5-0.2}{2k} = 2.4$ mA. For $\beta_{forced} = 10$, $I_B = 2.4/10 = 0.24$ mA. Thus, the maximum collector current available = 50 × 0.24 = 12 mA. Thus, the additional load current allowed = 12 – 2.5 = 9.5 mA. Thus, the minimum external load = $R_L = \frac{5-0.2}{9.5}$ = **505Ω**. Now consider a load of 2(505 Ω) = 1.01 kΩ, for which the total equivalent collector resistance = 2k || 1.01 k = 0.67 kΩ.

For the rise time: Here $\upsilon_O = 5.0 - (5-0.2)e^{-t/R_LC}$. Now the time for 10% = time for 0.2 V ≈ 0 ns. Now, for the time for 90%: 0.9(4.8) + 0.2 = 5.0 – 4.8 $e^{-t/R_L C}$, 4.32 + 0.2 = 5.0 – 4.8 $e^{-t/R_L C}$, $t = -R_L C \ln \frac{5-4.52}{4.8} = -0.67 \times 10^3 \times 10 \times 10^{-12} \times (-2.30) = 15.4$ns. Thus the rise Time ≈ **15.4ns**.

For the fall time: Here the output starts at 5.0 V and heads toward 5 – 12 mA(0.67k) = –3.04 V. Thus $\upsilon_O = -3.04 + (5+3.04)e^{-t/R_L C}$. Now, the fall time is complete when $-3.04 + 8.04e^{-t/R_L C} = 0.2 + 0.1(4.8)$, or t = –6.7 ns × ln 3.72/8.04 = **+5.16ns**. Because of the load, the rise and fall times have become more equal than they are in P14.25 of the Text. Note that if an external load of 505 Ω is used, t_r and t_f become equal. Why?

14.16 Here, (a) For I_C = 20 mA and $\beta_f = 40/2 = 20$, $I_B = 20/20 = 1$ mA, (b) $R_C = R$, (c) Supply = V is small, (d) $V_{OL} \approx 0.2$ V. Now, $V_{IL} = 0.6 - 0.2 = 0.4$V, and assuming Q_1 conducting in inverted saturation, $V_{IH} = 0.7 + 0.2 = 0.9$ V. Now, *for loading*:

For υ_O high: For reverse-current flow in the input transistor and fanout N, the total outward-directed load is $N(I_{B3})\beta_R$. For $N = 10$, $I_{LH} = 10(1)(0.1) = 1$ mA. Thus V_{OH} = V – R(1).

For υ_O low (=0.2V): The current required by each input is $\frac{V-0.7-0.2}{R}$. For N = 10, $I_{LL} = 10(\frac{V-0.9}{R}) \le \beta_f I_B = 20$. Thus, $V \le 2R + 0.9$ – – – (1). Now, $NM_L = V_{IL} - V_{OL} = 0.4 - 0.2 = 0.2$V, and $NM_H \ge 1.5\,NM_L$ = 1.5(0.2) = 0.3V. Now, since V_{IH} = 0.9V, $V_{OH} \ge$ 0.9 + 0.3 ≥ 1.2V. But, $V_{OH} = V - R(1) \ge 1.2$V. Thus $V \ge 1.2 + R$ – – – (2). Now from (a), for $I_B \ge 1$ mA, and for Q_3 saturated with Q_1 operating in reverse mode with β_R = 0.1, the current in the base of $Q_3 = \left[\frac{V-0.7-0.7}{R}\right](1+0.1) \ge 1$mA. Thus $V - 1.4 \ge R/1.1$, or $V \ge 1.4 + 0.91R$ – – – (3). Now, overall, there are three conditions: (1) $V \le 2R + 0.9$, (2) $V \ge 1.2 + R$, and (3) $V \ge 1.4 + 0.91\,R$.

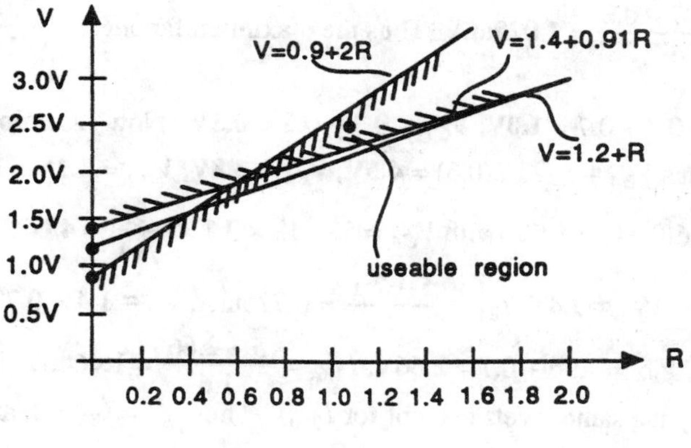

See that the minimum V occurs for the intersection of conditions (1) and (3), where 1.4 + 0.91 R = 2R + 0.9, or 1.09R = 0.5, whence R = 0.459kΩ, and V = 0.9 + 2(0.459) = 1.817V. Use (conservatively) V = **2.0V**, and (from (1)) $2R \ge 2 - 0.9$, $R > 0.55$ kΩ, and (from (3)) $0.91R \le 2.0 - 1.4 = 0.6$, or $R \le 0.66$K. Use R = **600Ω**. Here, it is apparent that the higher the voltage chosen (and the resistor chosen), the greater is the range of adequate operation.

Thus for V = 2.5V, $R \ge \frac{2.5-0.9}{2} = 0.8$ kΩ, and $R \le \frac{2.5-1.4}{0.91} = 1.21$ kΩ. Thus use **1.0kΩ**.

For V = 2.0, *R* = 0.6 kΩ: For υ_I high, $I_{C3} \le 40\left[\frac{2-0.7-0.7}{0.6}\right] = 40$ mA. For υ_I

low, $I_{in} = \frac{2-0.2-0.7}{0.6} = 1.83\text{mA}$. Thus, for the edge of saturation, $N \le 40/1.83 = 21.8$, say **21**. *For* $V = 2.5$, $R = 1.0\ \text{k}\Omega$: For v_I high, $I_{C3} \le 40\left[\frac{2.5-0.7-0.7}{1}\right] = 44$ mA. For v_I low, $I_{in} = \frac{2.5-0.2-0.7}{1} = 1.6$ mA. Thus $N \le 44/1.6 = 27.5$, say **27**.

14.17

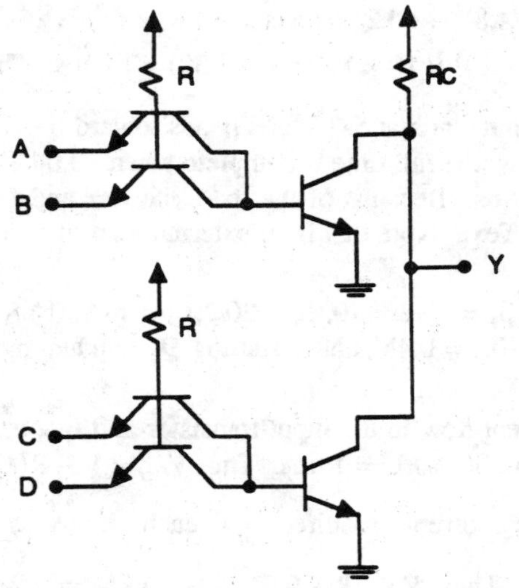

14.18 *For input high,* assuming $V_{BE} = 0.70$ V and $V_{CE\,sat} = 0.2$ V, see $V_{B3} = \mathbf{0.7V}$, $I_{1K} = 0.7$ mA, $V_{C2} = \mathbf{0.9V}$, $I_{C2} = \frac{5-0.9}{1.6} = \mathbf{2.56mA}$, $V_{B2} = \mathbf{1.4V}$, $V_{B1} \approx \mathbf{2.1V}$, $I_{B1} = \frac{5-2.1}{4} = \mathbf{0.725mA}$, $I_{E1} = 0.725(.05) = \mathbf{36\ \mu A}$. Thus $I_{E2} = 0.725 + 2.56 + .036 = \mathbf{3.33mA}$, and $I_{B3} = 3.33 - .70 = \mathbf{2.63mA}$.

Check Saturation: For Q_2, $\beta_f = \frac{2.56}{.725+.036} = 3.36$, $\beta_F = 9$, $\beta_R = .05$. Thus $V_{CE\,sat} = 25 \ln \frac{1+(\beta_f+1)/\beta_R}{1-\beta_f/\beta_F} = 25 \ln \frac{1+(3.36+1)/.05}{1-3.36/9} = \mathbf{0.124\ V}$.

For Q_3, $V_{C3} \approx 0.2\text{V}$, $I_{C3} = \frac{5-0.2}{1k\Omega} = 4.8$ mA, $\beta_f = 4.8/2.63 = 1.83$. Thus $V_{CE\,sat} = 25 \ln \frac{1+(1.83+1)/.05}{1-1.83/9} = \mathbf{107mV}$. Now for input low, $I_{in} \approx \frac{5-0.7-0.2}{4} = 1.025\text{mA}$. Thus the maximum fanout $N = \frac{2.63 \times 9}{1.025} = 23.1$, say **23**.

14.19 *For input low* at 0.3 V, $V_{E1} = \mathbf{0.3V}$, $V_{B1} = 0.3 + 0.7 = \mathbf{1.0V}$, $V_{C1} \approx 0.3 + 0.2 = \mathbf{0.5V}$. Now assuming $V_{out} \approx 3\text{V}$, $I_{out} \approx 3$ mA, $I_{B4} \approx \frac{3}{9+1} = \mathbf{0.3mA}$. Thus $V_{B4} \approx 5 - 1.6(0.3) = \mathbf{4.5V}$, $V_{E4} = \mathbf{3.8V}$, $V_{out} = \mathbf{3.1V}$. Thus $I_{out} = \mathbf{3.1mA}$, $I_{B4} = \mathbf{0.31mA}$, and $V_{B4} = 5 - 1.6(0.31) = \mathbf{4.5V}$, with $V_{C4} = 5 - .13 \times 3.1 \times \frac{9}{9+1} = \mathbf{4.64V}$.

14.20 Here, $V_{B3} = \mathbf{0.7V}$, $V_{B2} = \mathbf{1.4V}$, $V_{B1} \approx \mathbf{2.1V}$, $V_{E1} \approx \mathbf{1.4V}$, $I_{E1} \approx \frac{5-2.1}{4} = 0.725\text{mA}$, $V_O \approx 1.4 - 0.725(.200) = 1.26\text{V} = V_{C3}$, $V_{E4} = 1.26 + 0.7 = 1.96\text{V}$, $V_{B4} = 1.96 + 0.7 = 2.66\text{V}$, $I_{1.6k} = \frac{5-2.66}{1.6} = 1.46\text{mA}$. Now Q_3 and Q_4 will both conduct at approximately the same levels (except for I_{200}). Thus $I_{B4} = I_{B3} = I$, and $(1.46\ \text{mA} - I)\frac{10}{9} - \frac{0.7}{1k} = I$, $1.62 - 1.11I - 0.7 = I$, $2.11\ I = 0.92$. Thus $I = I_{B3} = I_{B4} = 0.44\text{mA}$, and $I_{C3} \approx$

9(0.44) = 3.92mA. Also I_{E2} = .44 + 0.7 = 1.14mA, $I_{B2} = \frac{1.14}{9+1}$ = 0.114mA. Thus I_{E1} = 0.725 – .114 = 0.61mA, with $V_O \approx$ 1.4 – 0.2(.61) = 1.29V = V_{C3}. Thus I_{E4} = 3.92 – .61 = 3.31mA, I_{B4} = 3.31/9+1 = 0.33mA, V_{B4} = 1.29 + 0.7 + 0.7 = 2.69V, $I_{1.6k} = \frac{5-2.69}{1.6}$ = 1.44mA, I_{C2} = 1.44 – .33 = 1.11mA, $I_{E2} = \frac{10}{9}$ (1.11) = 1.23mA, I_{B3} = 1.23 – 0.7 = 0.53mA, I_{C3} = 9(.53) = 4.77mA, I_{E4} = 4.77 – .61 = 4.16mA, $I_{B4} = \frac{4.16}{9+1}$ = 0.416mA, I_{C2} = 1.44 – .416 = 1.024mA, I_{E2} = (10/9) (1.024) = 1.14mA, I_{B3} = 1.14 – 0.7 = 0.44mA, I_{C3} = 9(.44) = 3.96mA, I_{B2} = 1.14/(9+1) = 0.114mA, I_{E1} = .725 – .114 = 0.61mA, I_{E4} = 3.96 – .61 = 3.35mA, I_{B4} = 3.35/(9+1) = 0.335mA, I_{C2} = 1.444 – .335 = 1.11mA, I_{E2} = 1.233mA, I_{B3} = 1.233 – 0.7 = 0.533mA, I_{C3} = 9(.533) = 4.80mA. Conclude $I_{C3} \approx \frac{4.80+3.96}{2}$ = **4.4mA**, I_{B3} = 4.4/9 = **0.49mA**, I_{E3} = 0.49 + 4.4 = **4.9mA**, I_{E2} = 0.49 + 0.7 = **1.19mA**, $I_{E1} \approx$ **0.61mA**, and V_O = **1.29V**.

For $V_{CE\,sat1}$: $I_{E1} \approx$ 0.61mA, $I_{B1} \approx$ 0.725mA, $I_{C1} \approx$ –0.114mA. Thus β_f = –0.114/0.725 = –0.157, whence $V_{CE\,sat} = 25\ln\frac{1+(\beta_f+1)/\beta_R}{1-\beta_f/\beta_F} = 25\ln\frac{1+.843/0.5}{1+.157/9}$ = **72mV**.

14.21 From the solution of P14.19, for no load, $V_O = V_{C3} \approx$ 1.29 V, $I_{E1} = I_{200}$ = 0.61 mA, V_{B4} = 2.69 V, $I_{1.6k}$ = 1.44 mA, $I_{C3} \approx$ 4.4 mA, with β_F = 9, V_{BE} = 0.7V, and β_F + 1 = 10. *Now iterate*:

(a) *With R_L = 200Ω to ground*, $I_L = \frac{1.29}{0.2}$ = 6.45 mA. Assume I_{C3} reduces, say to 2 mA. Thus I_{E4} = 6.45 + 2 = 8.5 mA, I_{B4} = 8.5/10 = 0.85mA. Thus $I_{C2} = I_{1.6k} - I_{B4} \approx$ 1.44 – 0.85 = 0.59 mA, I_{E2} = 10/9(0.59) = 0.66 mA, $I_{1k} \approx$ 0.7 mA. Thus $I_{B3} \approx$ 0mA. Thus $I_{C3} \approx$ 0. Thus $I_{E4} \approx I_L$ = 6.5 mA, I_{B4} = 6.5/10 = .65mA – – – (1), I_{C2} = 1.44 – .65 = 0.79 mA, I_{E2} = 10/9(0.79) = 0.88 mA, I_{B3} = 0.88 – 0.7 = 0.18 mA, I_{C3} = 9(0.18) = 1.62 mA, $I_{E4} = I_L + I_{C3} - I_{200}$ = 6.5 + 1.62 – 0.61 = 7.51 mA, I_{B4} = 7.51/10 = .75mA, I_{C2} = 1.44 –.75 = 0.69mA, I_{E2} = 10/9(.69) = 0.766mA, I_{B3} = 0.766 – 0.70 = .066mA, I_{C3} = 9(.066) = 0.6mA, I_{E4} = 6.5 + 0.6 – 0.61 = 6.5mA, I_{B4} = 6.5/10 = .65mA, I_{C2} = 1.44 – .65 = .79mA, I_{E2} = 10/9 (.79) = .88mA, and I_{B3} = .88 – .7 = .18, etc.

See that we begin a cycle initiated in (1) above, presumed to converge (with a more detailed analysis (eg exponential junctions and consideration of the effect of I_{B2} on I_{E1}, V_O, etc)) on the intermediate value. Thus $I_{C3} = \frac{0.60+1.62}{2}$ = **1.1mA**, I_{E4} = 6.5 + 1.1 –.6 = **7.0mA**, I_{B4} = 7.0/10 = **0.7mA**, I_{C2} = 1.44 –.7 = **0.74mA**, I_{E2} = (10/9)(.74) = **0.82mA**, I_{B3} = 0.82 –0.7 = **0.12mA**, I_{C3} = 0.12 × 9 = 1.08 = **1.1mA**, as conjectured, where $V_O = V_{C3}$ = **1.29V**, and I_L = **6.5mA**.

(b) *With R_L = 200Ω to +5V*: For $V_O \approx$ 1.29V, $I_{C3} = I_L + I_{E1} = \frac{5-1.29}{.200}$ + 0.61 = 19.2 mA. Likely Q_4 is nearly cutoff → $I_{B4} \approx$ 0mA. Thus $I_{C2} = I_{1.6k}$ = 1.44 mA, I_{E2} = (10/9) (1.44) = 1.6mA, I_{B3} = 1.6 – 0.7 = 0.9mA, I_{C3} = 9(0.9) = 8.mA (too small). Conclude that as V_O goes up, I_L reduces, I_{E1} reduces, I_{B2} increases, possibly Q_2 saturates.

For Q_2 saturated: $I_{C2} \approx \frac{5-0.7-0.2}{1.6k}$ = 2.56 mA, $I_{B1} \approx \frac{5-0.7-0.7-.7}{4}$ = 0.725mA, $I_{E2} \le$ 2.56 + .73 = 3.29mA, $I_{B3} \le$ 3.29 – 0.7 = 2.59mA, $I_{C3} \le$ 9(2.59) = 23.3 mA. Conclude that Q_2 is *not saturated*, but I_{E1} is small, $V_O \approx$ 1.4V, for which $I_{C3} \approx \frac{5-1.4}{200}$ = **18mA**, I_{B3} = 18/9 = **2mA**, I_{E2} = 2 + .7 = **2.7mA**, $I_{C2} \approx$ 9/10 × 2.7 = **2.43mA**, V_{C2} = 5 – 2.43(1.6) = **1.11V**, not quite saturated, as assumed. Thus I_{B2} = 2.7/10 = **0.27mA**, I_{E1} = .725 – .27 = **0.46mA**, and V_O = 0.7 + 0.7 + 0.7 – 0.7 – 0.46(0.2) = **1.3V**, with Q_4 cut off.

SECTION 14.4: Characteristics of Standard TTL

14.22 For $\upsilon_O(C) = 3.0$ V, $\upsilon_{C2}(C) = 3 + 0.65 + 0.65 = 4.3$ V. Now at C, $V_{BE3} \approx 0.6$V and $I_{E2} \approx I_{C2} = \frac{0.6}{R_2}$. Thus $\upsilon_{C2} = 5 - 1.6\left[\frac{0.6}{R_2}\right] = 4.3$V. Thus $R_2 = \frac{1.6(0.6)}{0.7} = \mathbf{1.37k\Omega}$. For this change, $\upsilon_I(C) = 1.2$V, $\upsilon_O(C) = 3.0$V, $\upsilon_I(B) = 0.5$V, $\upsilon_O(B) = 3.7$V. Thus gross slope $BC = -\frac{3.7-3.0}{1.2-0.5} = -\frac{0.7}{0.7} = \mathbf{-1\ V/V}$.

For Incremental gain: I_{E2} varies from 0 to 0.6/1.37 = 0.44mA, and $I_{E2\,av} = 0.22$ mA. Thus $r_{e2} \approx 25/.22 = 114\Omega$, and gain $= -0.98 \times \frac{1.6k}{114k + 1.37k} = -1.06$ V/V. For $R_2 = 1$kΩ and $\upsilon_O(C) = 3.0$V, $\upsilon_{C2} = 5 - R_1$ (0.6/1) = 4.3V, and $R_1 = \frac{0/7}{0.6} = \mathbf{1.17k\Omega}$.

For turnon of Q_3: Previously: $I_{B3} = I_{B2} + I_{C2} - I_{R2} = \frac{5-0.7-0.7-0.7}{4} + \frac{5-0.7-0.2}{1.6} - \frac{0.7}{1} = .725 + 2.563 - 0.7 = 2.59$ mA. Now: $I_{B3} = .725 + \frac{4.1}{1.17} - 0.7 = 3.53$ mA. Thus the increase = 3.53 − 2.59 = **0.94mA ≈ 36%**. Now the Storage Delay will increase even more, by a factor due to the $I_{B2} - \frac{I_C}{\beta_F}$ difference. For example, with $\beta_F = 50$, and $I_C = 10$, the increase is by $\frac{3.53-0.2}{2.59-0.2} = 1.39$, or 39%!

14.23 Use coordinates of points *C*, *D*:

Temp.	V_{OL}	V_{IL}	V_{OH}	V_{IH}
−55°C	0.1	1.52	2.16	1.72
125°C	0.1	0.80	3.46	1.0

For Source at −55°C, load at 125°C:

$NM_L = V_{IL}(125) - V_{OL}(-55) = 0.80 - 0.1 = \mathbf{0.7V}$,

$NM_H = V_{OH}(-55) - V_{IH}(125) = 2.16 - 1.0 = \mathbf{1.16V}$.

For Source at 125°C, load at −55°C:

$NM_L = V_{IL}(-55) - V_{OL}(125) = 1.52 - 0.1 = \mathbf{1.42V}$,

$NM_H = V_{OH}(125) - V_{IH}(-55) = 3.46 - 1.72 = \mathbf{1.74V}$.

For comparison, for the nominal circuit with source and load both at 25°C:

$NM_L = 1.2 - 0.1 = 1.1$V,

$NM_H = 2.7 - 1.4 = 1.3$V.

14.24 The maximum base current $I_{B3} = 4\left[\frac{5-2.1}{4k}\right] + \frac{5-0.7-0.2}{1.6} - \frac{0.7}{1} = 2.9 + 2.56 - 0.7 = \mathbf{4.76mA}$. For three inputs low, $I_{B3} = .725 + 2.56 - 0.7 = 2.585$ mA. Now $t_s = \tau_s \frac{I_{B2} - I_C/\beta}{I_{B1} + I_C/\beta} = \frac{I_{B2}}{I_{B1}}$, for $I_C = 0$mA. Thus the delay for four inputs dropping together is $t_s = 10 \times \frac{4.76}{2.585} = \mathbf{17.8ns}$.

14.25 For I_{C6} = 1 mA, the voltage drop in 4kΩ = 4V. Thus V_{C6} = 5 – 4 = 1 V. Now $I_{D2} \approx$ 1 mA, $V_{E6} \approx$ 0.7V. Thus Q_6 is barely saturated (if at all). Assume linear → $I_{B6} = \frac{1mA}{50}$ = 0.02 mA. Thus I_{E6} = 1.02 mA. Thus $\upsilon_{B6} \approx 2\left[(25)\ln\frac{1.02}{1} + 700\right]$ = 1.401V. *Check* $V_{CE\,sat3}$ for β_f = 49 (say): $V_{CE\,sat}$ = $25\ln\frac{(1+(49+1)/0.1)}{1-49/50}$ = 253 mV. Thus Q_6 operates linearly, and $I_{B5} \approx \frac{5-2.1}{4}$ = 0.725 mA, I_{C5} = –0.02 mA, I_{E2} = 0.705 mA, for which $\beta_f = -\frac{0.02}{.725}$ = –.028 → $V_{CE\,sat} = 25\ln\frac{(1+(.028+1)/0.1)}{1+.028/50}$ = 59 mV, and υ_I = 1.401 –.059 = **1.34V**.

14.26 $I_{B1} = \frac{5-3(0.7)}{4}$ = 0.725 mA. Via Q_7, V_{C2} = 0.7 + 0.2 = 0.9V. Now, from the symmetry of Q_7, D_2 and 4kΩ, with Q_2, V_{BE3} and 4kΩ, it is likely that $I_{C2} \approx \frac{I_{1.6k}}{2} = \frac{5-0.9}{2(1.6k)}$ = 1.28 mA. Thus $I_{B3} \approx$ 0.725 + 1.28 – 0.7 = **1.305mA**. That is, Q_3 is distinctly turned on. Thus we see the need for the connection from the tristate input to Q_1!

SECTION 14.5: TTL Families with Improved Performance

14.27

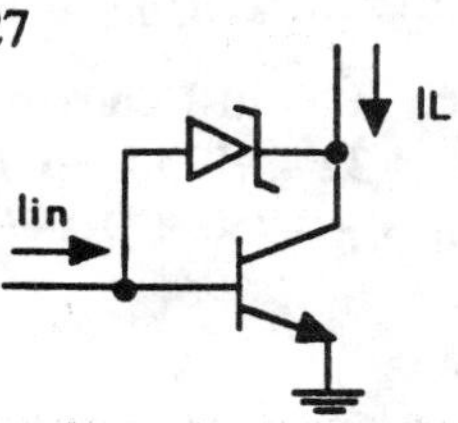

a) I_{in} = 1 mA, I_L = 0 mA. Thus I_E = **1mA**, $I_B = \frac{1}{\beta+1} = \frac{1}{51}$ = **19.6 μA**, I_D = 1 mA – 19.6μA = **0.98mA**. Thus V_B = 750 + 25 ln 1/1 = 750 mV = **0.750V**, and V_C = 750 – (500 + 25 ln .98/1) = **0.25V**.

(b) I_{in} = 1 mA, I_L = 1 mA. Thus I_E = **2mA**, I_B = **0.04mA**, I_D = **0.96mA**. Thus V_B = 750 + 25 ln 2 = 767 mV = **0.767V**, V_C = 767 – (500 + 25 ln $\frac{0.96}{1}$) = **0.266V**.

(c) I_{in} = 1 mA, I_L = 10 mA. Thus I_E = **11mA**, I_B = 11/51 = **0.216mA**, I_D = 1 – .216 = **0.784mA**. Thus V_B = 750 + 25 ln 11/1 = **0.810V**, V_C = 810 – (500 + 25 ln .784) = **0.316V**.

(d) I_{in} = 10 mA, I_L = 10 mA. Thus I_E = **20mA**, I_B = 20/51 = **0.392mA**, I_D = 10 – .392 = **9.61mA**. Thus V_B = 750 + 25 ln 20 = **0.825V**, V_C = 825 – (500 + 25 ln 9.61) = **0.268V**.

(e) I_{in} = 10 mA, I_L = 1 mA. Thus I_E = **11mA**, I_B = 11/51 = **0.216mA**, I_D = 10 – .46 = **9.78mA**. Thus V_B = 750 + 25 ln 11 = **0.810V**, V_C = 810 – (500 + 25 ln 9.78) = **0.256V**.

14.28 Use V_{BE} = 0.75 and V_D = 0.5V.

(a) Input is high: $I_{20k} = \frac{5-.75-.75}{20}$ = 0.175 mA, $I_{8k} = \frac{5-0.75-0.75+0.5}{8}$ = 0.5 mA. Thus I_{CC} = 0.175 + 0.5 = **0.675mA**, for output open or shorted to ground.

(b) Input is low; at υ_I = 0.75 – 0.5 = 0.25V. $I_{20k} = \frac{5-0.5-0.25}{20}$ = .2125 mA. Output shorted: Q_4 and Q_5 conducting with drop of 0.75 + 0.75 – 0.5 = 1V. Thus $I_S = \frac{5-1}{.120}$ = 33.3mA. Thus I_{CC} = **0.213mA** for output open, or I_{CC} = **33.5mA** for output grounded.

Now, Power Loss is 5 × .675 = **3.37 mW** for inputs high, and 5 × .213 = **1.07mW** for inputs low, output open, or 5 × 33.5 = **167.5mW** for inputs low, output grounded. The average dc power loss with no load = $\frac{3.37+1.07}{2}$ = 2.22 mW. Dynamic power with a 10 pF load at 30 MHz = fCV^2 = 30 × 10^6 × 10 × 10^{-12} × $(5 - 1.4 - 0.3)^2$ = 3.27mW. Thus total power = 3.27 + 2.22 = 5.49 mW. Now, DP product = 5.49 × 10^{-3} × 10 × 10^{-9} = **54.9 pJ**.

SECTION 14.6: Emitter-Coupled Logic (ECL)

14.29 (a) $V_{OH} = 0 - 0.75 = \mathbf{-0.75V}$. $V_{OL} = 0 - RI - 0.75 = \mathbf{-(RI + 0.75)V}$.

(b) $V_{th\,A} = -IR/2 - 0.75 = \mathbf{-(RI/2 + 0.75)V}$, for which $V_{BE} = 750 + 25 \ln i/I = 750 + 25 \ln \frac{I/2}{I} =$ **682.7mV**.

(c) For $i = 0.95I$: $V_{BE} = 750 + 25 \ln 0.95 = 750 - 1.3 = \mathbf{748.7mV}$.

(d) For $i = 0.05I$: $V_{BE} = 750 + 25 \ln .05 = 750 - 74.9 = \mathbf{675.1mV}$, and $\Delta V_{BE} = 748.7 - 675.1 = \mathbf{73.6mV}$.

(e) Thus $V_{IL} = -RI/2 - 0.750 - .0736 = \mathbf{-(RI/2 + 0.824)V}$, and $V_{IH} = -RI/2 - 0.750 + .0736 = \mathbf{-(RI/2 + 0.676)V}$.

(f) $NM_H = V_{OH} - V_{IH} = -0.750 + RI/2 + 0.676 = \mathbf{(RI/2 - 0.074)V}$, and $NM_L = V_{IL} - V_{OL} = -RI/2 - .824 + RI + .750 = \mathbf{(RI/2 - 0.074)V}$.

(g) Transition region = $2\,\Delta V_{BE} = 2(73.6) = 147.2$ mV. Thus $V_{IH} - V_{IL} = 147$ mV. Now $RI/2 - .074 = .147V \rightarrow RI = 2(.147 + .074) = \mathbf{0.442V} = IR$.

(h) Now for $IR = 0.442V$, $V_{OH} = \mathbf{-0.750V}$, $V_{OL} = \mathbf{-1.192V}$, $V_{IH} = \mathbf{-0.897V}$, $V_{IL} = \mathbf{-1.045V}$, $V_R = (-.897 - 1.045)/2 = \mathbf{-0.971V}$.

14.30 Here, $V_{BE} = 0.75$ @ 4 mA, and $V_{th} = -R/2 - 0.75 = -1.32V$. Thus $R = \frac{1.32 - .75}{2} = 0.285 = \mathbf{285\Omega}$. Also, $V_{OH} = 0 - 0.75 = \mathbf{-0.750V}$, and $V_{OL} = -0.750 - 0.285(4) = \mathbf{-1.89V}$. For a 1000-to-1 current split, $\Delta V_{BE} = 25 \ln 10^3 = 173$ mV. Thus $V_{IL} = -1.32 - .173 = \mathbf{-1.493V}$, and $V_{IH} = -1.32 + .173 = \mathbf{-1.147V}$. Thus $NM_L = V_{IL} - V_{OL} = -1.493 + 1.89$ V $= \mathbf{0.397V}$, and $NM_H = V_{OH} - V_{IH} = -0.750 + 1.147$ V $=$ **0.397V**.

14.31

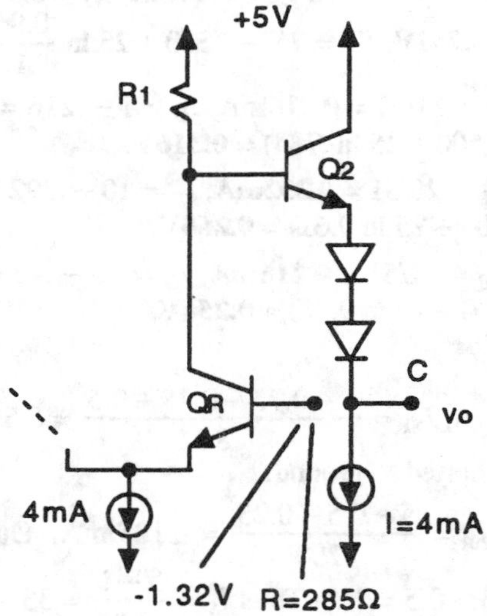

Here, $I = 4$ mA, $V_{th} = -(I_R/2 + 0.75) = -1.32$ V. Thus $R = \frac{1.32 - .75}{2} = \mathbf{285\Omega}$. Now $V_{OH} = 5 - 0.75 - N(0.75) \geq 2.7$ V – – – (1), and $V_{OL} = 5 - 0.75 - N(0.75) - R_1 I \leq 0.5$ – – – (2). *From* (1), $N \leq \frac{-2.7 + 5 - 0.75}{0.75} = 2.07$. Use **two diodes** with $V_{OH} = 5 - 3(0.75) = \mathbf{2.75V}$. *From* (2), $5 - 3(0.75) - R_1(4) \leq 0.5$, $R_1 > \frac{5 - 2.25 - .50}{4} = 0.5625$ kΩ. Use $R_1 = \mathbf{570\Omega}$, for which, $V_{OL} = 5 - 2.25 - .570(4) = \mathbf{0.47V}$.

14.32

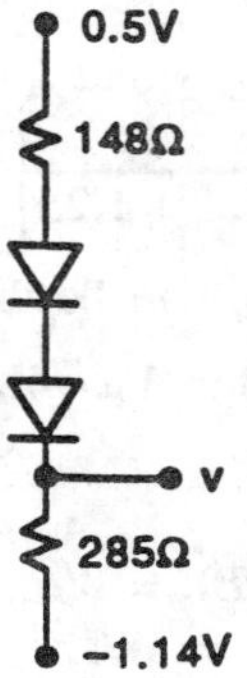

Use $I = 4$ mA, and $R = 285\Omega$. See with υ_I high, D_3 operates at 8 mA with $V_D = 750 + 25 \ln 8/4 = 0.767$V, while D_1 and D_2 operate at 4 mA. Thus $R_2 = \frac{2.70 - .767 - .75}{8} = 148\Omega$. For T^2L output low enough, I flows in R and diodes cut off: $V_{Base} = -4(.285) = -1.14$V. For V_{OL} (of T^2L) = 0.5V, voltage across two diodes (D_3, D_2) and R_2 is 1.14 + 0.5 = 1.64 V. Thus $I_{R2} \approx \frac{1.64 - 0.75 - 0.75}{.148} = \approx$ **0.95mA**, for which V_{Base} is higher. For 0.75 V diodes, $\upsilon = -1.14 + \frac{285}{148 + 285}$ (0.5 + 1.14 − 2(0.75)) = −1.04 V = V_{OL}. This is not ideal, but fairly good, and certainly OK. Now for $V_{O\,TTL} > V_{OH\,TTL}$, D_1 conducts excess current, causing the base of Q_2 to rise above ground, and V_{OH} of the converter to rise above −0.75 V. In summary, for the converter, V_{OH} = **−0.75V**, or slightly more positive, and V_{OL} = **−1.04V** to **−1.14V**.

14.33 For this smaller transistor, $V_{BE} = 0.75$ at $I_E = 0.5$ mA, with $\beta = 30$, and V_{OH} lowers. Iterate: Normally $V_{OH} = -0.88$V. Try $V_{OH} = -0.90$V. Thus $I_{50} = \frac{2-0.90}{50} = 22$ mA. Thus $V_{BE} = 750 + 25 \ln \frac{22}{0.5} = 0.845$V, $I_B = \frac{22}{31} = 0.710$ mA, and $V_B = 0.710 \times 245 = 0.174$V. Thus $V_{OH} = 0 - .174 - .845 = -1.02$V, for which $I_{50} = \frac{2-1}{50} = 20$ mA, and $V_{BE} = 750 + 25 \ln \frac{20}{0.5} = .842$V, and $V_B = \frac{20}{31} \times .245 = 0.158$V. Thus $V_{OH} = -0.158 - 0.842 = -1.000$ V, as conjectured. Now $NM_H = V_{OH} - V_{IH} = -1.000 - (-1.205) =$ **0.205V**.

14.34 Here, the signal velocity = 2/3 × 0.3 = 0.2 mm/ps. For a rise time of 1.2 ns, and rise/return ratio of 6, wave length = 1/2 × 1.2/6 = 0.1ns, or 100ps. A line whose end-to-end delay is 100 ps is 0.2 × 100 = 20 mm long. Thus the longest allowed interconnect is **2 cm!**

SECTION 14.7: BiCMOS Digital Circuits

14.35 For Q_1 in triode operation: $i_D = K_1(2(\upsilon_{GS} - V_t)\,\upsilon_{DS} - \upsilon_{DS}^2)$, and for small υ_{DS}, $i_D \approx 2K_1\,(\upsilon_{GS} - V_t)\,\upsilon_{DS}$, which for $\upsilon_{GS1} = V_{DD} = 5$ V, results in $r_{DS1} = \frac{\upsilon_{DS1}}{i_{DS1}} = \frac{1}{2\,\frac{K_2}{2}\,(5-1)} = \frac{1}{4K_2} = r$.

For no load: V_{OH} = **5.0V**. V_{OL} = **0.0V**. For V_{th}: with R_2 small, so that Q_4 does not conduct, V_{th} is reached with Q_1, Q_2 both in pinchoff and sharing the same current, ie $i_D = K_1\,(5 - V_{th} - 1)^2 = K_2(V_{th} - 1)^2$, or $\frac{K_2}{2}\,(4 - V_{th})^2 = K_2\,(V_{th} - 1)^2$, $4 - V_{th} = \sqrt{2}\,(V_{th} - 1)$, $2.414\,V_{th} = 5.414$, whence $V_{th} = \frac{5.414}{2.414}$ = **2.24V**, for which $i_D = K_2\,(2.24 - 1)^2 = 1.54\,K_2$. Now, the drop across $R_2 = r = \frac{1}{4K_2}$ is $\frac{1}{4K_2} \times 1.54K_2 = 0.25$V, and Q_4 does not turn on, as desired.

For a 5 kΩ load to 2.5 V: Assume, initially, that $V_{OH} = 5 - 0.7 = 4.3$V, and $V_{OL} = 0.7$V, for which $I_L = \frac{2.5 - 0.7}{5k\Omega} = 0.36$mA directed either in our out. *For V_{OH} in detail*: Q_1 conducts: $\upsilon_{DS1} = \upsilon$ is assumed small. Now, $i_{D1} = K_1\,(2(5 - 0 - 1)\,\upsilon - \upsilon^2) \approx \frac{K_2}{2}\,(2 \times 4\,\upsilon) = \frac{200}{2} \times 10^{-6} \times 8\upsilon = 800 \times 10^{-6}\,\upsilon$, and $r_{DS1} = \frac{\upsilon}{i_{D1}} = \frac{1}{800 \times 10^{-6}} = 1.25$ kΩ. Now $R_1 = r = \frac{1}{4K_2} = \frac{1}{4 \times 200 \times 10^{-6}} = 1.25$ kΩ, as well. Now for $\upsilon_O = +4.3$ V and $V_{B3} \approx 5$V, $I_L = 360$ mA, $I_{B3} \approx \frac{360}{100} = 3.6\mu$A, and $i_{D1} \approx \frac{V_{BE3}}{R_1} + I_{B3} \approx \frac{0.7}{1.25k} + 3.6\mu\text{A} = 0.564\mu$A. Now it is apparent that Q_3 will not conduct, but rather that Q_1 and R_1 in series will support the load, in which case $V_{OH} \approx 2.5 + \frac{5}{5+1.25+1.25}(5-2.5) = \mathbf{4.17V}$, with $V_{BE3} \approx \frac{5-4.17}{2} = 0.42\text{V}$.

For V_{OL} in detail: Assume Q_4 does not conduct, but that V_{OL} is sustained by R_1, r_{DS2} and R_2 in series. Now for $V_{S2} \approx 0$, $r_{DS2} \approx \frac{r}{2} = 0.625\ \text{k}\Omega$. Thus $V_{OL} = 2.5 - 2.5\left[\frac{5}{5+1.25+0.625+1.25}\right] = \mathbf{0.96V}$, with $V_{BE4} \approx \frac{1.25}{1.25+1.25+0.625} \times 0.96 = 0.384\text{V}$. Now with this centred load, V_{th} will be that voltage for which the output voltage is centred, that is where Q_1 and Q_2 conduct equally. Thus V_{th} will be essentially as before, $\approx$ **2.24V**.

14.36

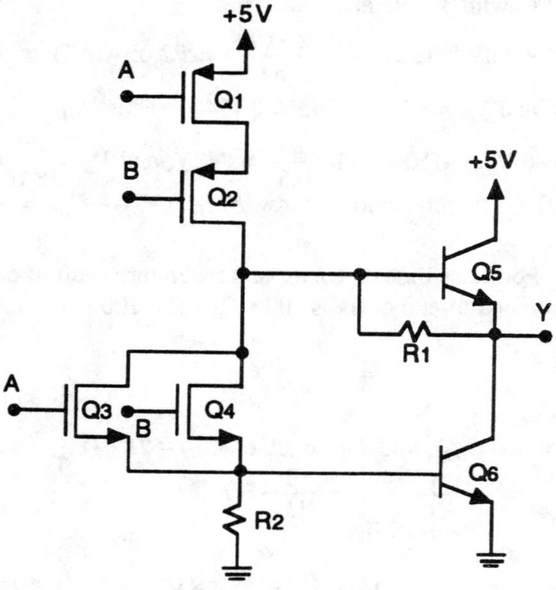

For the inverter $(W/L)_p = (W/L)_n = W/L$. For this NOR, $(W/L)_p = \mathbf{2W/L}$, $(W/L)_n = \mathbf{W/L}$. The threshold will obviously depend to a degree on R_2, but assuming the drop across R_2 to be small enough that Q_6 does not conduct, it can be ignored. Now for A low, Q_1 is in triode mode, Q_3 cut off, and Q_2 with Q_4 in pinchoff. Here $K_1 = K_2 = K_4/2 = K$, with $V_{th} = \upsilon$, and the voltage across $Q_1 = \upsilon_X$. Thus $i_4 = 2K(\upsilon - 1)^2$ – – – (1), $i_2 = K(5 - \upsilon_X - \upsilon - 1)^2$ – – – (2), $i_1 = K(2(5-\upsilon-1)\upsilon_X - \upsilon_X^2)$ – – – (3), where $i_4 = i_2 = i_1$. For (1) and (2): $\sqrt{2}\,(\upsilon - 1) = 4 - \upsilon_X - \upsilon$, $1.414\,\upsilon - 1.414 = 4 - \upsilon_X - \upsilon$, $2.414\,\upsilon + \upsilon_X = 5.414$, or $\upsilon = \frac{5.414 - \upsilon_X}{2.414} = 2.24 - 0.414\upsilon_X$.

For (2), (3): $(4 - \upsilon_X - \upsilon)^2 = (8 - 2\upsilon)\upsilon_X - \upsilon_X^2$, $(4 - \upsilon_X - 2.24 + .414\upsilon_X)^2 = (8 - 4.48 + .818\upsilon_X)\,\upsilon_X - \upsilon_X^2$. Thus $(1.76 - .586\upsilon_X)^2 = 3.52\ \upsilon_X - 0.182\ \upsilon_X^2$, $3.098 - 2.063\ \upsilon_X + 0.343\ \upsilon_X^2 = 3.52\ \upsilon_X - 0.182\ \upsilon_X^2$, $0.525\ \upsilon_X^2 - 5.583\ \upsilon_X + 3.098 = 0$, whence

$$\upsilon_X = \frac{5.583 \pm \sqrt{5.583^2 - 4(0.525)(3.098)}}{2(0.525)} =$$

0.529V, whence $\upsilon = 2.24 - .414(0.527) = \mathbf{2.02V} = V_{th}$.

PART III
ANSWERS

pages 353 to 376

ANSWERS

Chapter 1

Introduction to Electronics

1.1 See page 111 for sketches.

1.2 377 rad/s, 120 Hz, 400 Hz, 6.35×10^6 rad/s, 611×10^6 rad/s, 6.28 rad/s, 60 Hz, 0.159 Hz, 6.28×10^9 rad/s, 25.1×10^{11} rad/s.

1.3 1.67×10^{-2} s, 16.7 ms; 8.33×10^{-3} s, 8.33 ms; 2.50×10^{-3} s, 2.50 ms; 9.90×10^{-7} s, 990 ns; 1.03×10^{-8} s, 10.3 ns; 1.00 s, 1.00 s; 1.67×10^{-2} s, 16.7 ms; 6.29 s, 6.29 s; 1.00×10^{-9} s, 1.00 ns; 2.50×10^{-12} s, 2.50 ps.

1.4 7.01 ps.

1.5 a) 1 times, 1 times; b) 30 times, 50 times; c) 10 times, 1/500 times.

1.6 4.05%; 18.9%.

1.7 Between 5 kHz and 6 kHz.

1.8 a) 2.82 Vpp square wave; b) 2.82 Vpp square wave; c) 2 Vpp square wave; d) sequence of positive and negative pulses of amplitude 1.41 V, width 1/4 f, spaced 1/4 f apart; dc level of 1.41 V.

1.9 b) 1 mA, 20 mW, 2×10^3μA, 20 μW, 1 mA, 0.5 mW, 0.05 V/mV, 34 dB, 0.5×10^{-3} mA/μA, −6 dB, 0.025 mW/μW, 14 dB, 2.5%; c) 0.05 mA, 0.05 mA, 1 mW, 100 mV, 50 μW, 2 V, 0.2 kΩ, 0.02 V/mV, 26 dB, 0.01 mA/μA, 20 dB, 0.2 mW/μW, 23 dB; d) 10 mA, 10 mA, 14.1 mV, 1.41×10^3 μA, 2.82 V, 28.2 mA, 0.1 kΩ, 46 dB, 0.02 mA/μA, 26 dB, 4 mW/μW, 36 dB, 20%; e) 3.1×10^{-3} mA, 3.1×10^{-3} mA, 0.01 V/mV, 20 dB, 0.01 mA/μA, 20 dB, 20 dB.

1.10 140 mVp.

1.11 5.66 Vrms, −1 V, − 20 mV.

1.12 3.125%.

1.13 0.598 V, 5 V, 0.613 V, 1.89 Vp, −97.8 V/V.

1.14 430 Ω, 53.8 V/V.

1.15 2.0 kΩ.

1.16 $A_{v0} = 1 + R_o/R_i$.

1.17 $A_1 A_2$, 12.5 V/V, 0.05 V/V, 0.5 V/V, double the gain.

1.18 0.99×10^4 V/V, 0.5×10^4 V/V, 9.99 V/V, 1 A/V.

1.19 a) 0.066 V/V, 0.178 V/V, 0.016 V/V; b) 1, 2 or 3, 3 or 2; 3, 2, 1; 2, 1, 3; c) $A_1 A_2$, $A_1 A_3$; d) 30.3 V/V for $A_1 A_2$.

1.20 9 amplifier pairs: (2,1): 24752 V/V; (1,1): 4901 V/V; (1,2), (1,3): 4541 V/V; (2,2), (2,3), (3,2), (3,3): 4132 V/V; (3,1): 2500 V/V.

1.21 A_1, A_3, A_2; 4132 A/A.

1.22 10 A/V, 10^{-1} A/V, 10^{-2} A/V; $FM2 = g_m R_o R_i^2$; A_3, A_2, A_1.

1.23 200 mA/mA, 40 mA/V.

1.24 1 H, 20 Ω.

1.25 2000 V/V, 72 ° lagging.

1.26 100 kHz, 80 kHz.

1.27 high-pass output, low-pass output.

1.28 7.27 kHz, 80 MHz, 12.6 kHz.

1.29 1/(RC) rad/s, 0.0644/(RC) rad/s, k = 0.25.

1.30 60 dB, 0 dB, –20 dB, 40 dB; 60 dB, 0 dB, –3 dB, 57 dB; 60 dB, 0 dB, 0 dB, 60 dB; 60 dB, 0 dB, 0 dB, 60 dB; 60 dB, –3 dB, 0 dB, 57 dB; 60 dB, –20 dB, 0 dB, 40 dB; 10^5 Hz; from 100 Hz to 10^4 Hz.

1.31 10 Hz, 10^5 Hz, 10^3 V/V; The standard form is more straightforward.

1.32 $A_\upsilon = R_i/[R_s + R_i] \times G_m R_o$, $\omega_H = 1/(R_o C)$; 155 V/V; $GB = 100/(R_s C)$ for large I, independent of I; 15.9 V/V, 126 mA.

1.33 $[RCs-1]/[RCs+1]$.

Chapter 2
Operational Amplifiers

2.1 2 op amps, none unused; 4 op amps, none unused.

2.2 – 300 μV, 100.3 mV.

2.3 0.35 V.

2.4 – 10 V/V, – 0.1 V/V.

2.5 – 1 V, $10^{-7}A$, – 1.002 V.

2.6 2 solutions: a) 100 kΩ; b) 50 kΩ.

2.7 R_1 = 22 kΩ.

2.8 See page 135: Use three 100 kΩ resistors in various combinations, with one 950 kΩ resistor, or one 5.55 kΩ resistor.

2.9 a) – 10 V/V, – 0.1 V/V; b) – 9.009 V/V, – 0.0989 V/V.

2.10 909 V/V, 10^4 V/V.

2.11 R_1 = 100kΩ, R_2 = 1MΩ, R_3 = 50kΩ, R_4 = 1MΩ.

2.12 R_1 = 1MΩ in series with 1MΩ, R_2 = 1MΩ, R_3 = 2.5kΩ, R_4 = 1MΩ.

2.13 $R_1 = R_2 = R_4$ = 1MΩ, R_3 = 100kΩ || 100kΩ.

2.14 $\upsilon_O = -(R_4/R_3)\upsilon_2$, $\upsilon_O = -(R_2/R_2)(1 + R_4/R_2 + R_4/R_3)\upsilon_1 - (R_4/R_3)\upsilon_2$.

2.15 Negative-going ramp of slope 1000 V/s, falling from 10 V to 0 V in 10 ms.

2.16 An inverted sine-wave of 26.5 mV peak lagging by 90°.

2.17 Input falls at a rate of 200 V/s; for the rise, υ_O = – 5 V; for the fall, υ_O = +5 V.

2.18 See page 138: a) – 10 V, 0.1 ms, +1 V; b) 5 V, 0.1 ms, 0 V.

2.19 Use R_1 = 30kΩ, R_2 = 15kΩ, R_3 = 10 kΩ, R_f = 30kΩ.

2.20 See page 139: two op amps with R_1 = 20kΩ, R_2 = 10kΩ, R_3 = 15kΩ, R_4 = 30kΩ, R_5 = 10kΩ, R_8 = 10kΩ, with a total resistance of 95kΩ.

2.21 $v_O = V_O - 1000\int_0^t \left[v_1(t)+2v_2(t)\right]dt.$

2.22 $v_o(s)/v_i(s) = -(R_2/R_1)(1+R_1C_1s)/(1+R_2C_2s)$, independent of frequency if $R_1C_1 = R_2C_2$: a) – 10 V/V, independent of frequency; b) – 100 (1 + s / 100) / (1 + s / 10); c) – 1 (1 + s / 100) / (1 + s / 1000).

2.23 11 V/V, 1.10 V/V.

2.24 See page 141.

2.25 $R_1 = R_4 =$ 20kΩ, $R_2 = R_3$ 10kΩ, $R_f =$ 30kΩ.

2.26 $R_1 = R_2 =$ 10kΩ, $R_3 = R_f =$ 100 kΩ. This configuration, with $R_f/R_2 = R_3/R_1$, is called a difference amplifier.

2.27 $G =$ 0.909 V/V; $R_1 =$ 90kΩ, $R_2 =$ 10kΩ.

2.28 20kΩ to the – 10 V supply from the op amp negative input, with $R =$ 2.22kΩ.

2.29 See page 145: $R_1 =$ 10kΩ, $R_2 =$ 20kΩ; total power = 367.5 mW.

2.30 Gain = – 10 V/V.

2.31 Gain = – 5 V/V; Remove R_1 and R_2 and connect sources directly, or make $R_2 = R_4 =$ 200kΩ; Add an additional 2kΩ resistor in series with R_{S2} and R_3, or change R_4 to 180kΩ.

2.32 $v_O = -v_1(R_2/R_1)+v_2(1+R_2/R_1)/(1+R_3/R_4)+v_3(1+R_2/R_1)/(1+R_4/R_3)$; $v_O = (R_2/R_1)(v_2-v_1)+v_3$.

2.33 a) 5 V; b) 5 V; c) 0 V.

2.34 For $R_1 =$ 10kΩ, $R_2 =$ 45kΩ, and for $R_3 =$ 10kΩ, $R_4 =$ 100kΩ; $v_{O1} =$ 5.4 V, $v_{O2} =$ 4.5 V.

2.35 $v_I =$ 11 V, $v_O =$ 5.5 V.

2.36 The pulse width ranges from 10.0 to 11.1 ms.

2.37 10 Hz, 100 V/V.

2.38 0.909 MHz, 1.00 MHz, 90.9 kHz, 100 kHz.

2.39 10^3 V/V for a single amplifier; 4.14×10^5 V/V for two stages in cascade.

2.40 – 100 V/V, 30 kHz, 6 MHz.

2.41 4.55 MHz.

2.42 $R_{O1} = 1\times10^6$Ω, $C =$ 20 pF.

2.43 314 μA/V.

2.44 1.6 V.

2.45 0.83 MHz.

2.46 $I_{max} =$ 150 μA, $G_m =$ 1.88 mA/V.

2.47 $V_o = SR/(2\pi f_b)$, 0.64 Vpeak.

2.48 30 V/V, 0.075 V/V; CMRR = 400 V/V, or 52 dB.

2.49 10 Vpp.

2.50 10^{-5} V/V, 120 dB.

2.51 6.7 MΩ.

2.52 $R_1 =$ 1MΩ, $R_2 =$ 10kΩ, $C_2 =$ 15.9 nF; 10 Hz , 1 KHz.

2.53 $R = 10^5$Ω, $R_{out} =$ 110 Ω, $L =$ 8.75 mH, 20 KHz.

2.54 0.40 mV.

2.55 0.40 V; Use $R_3 =$ 100 kΩ for which $v_O =$ 0.13 V; For case a), bias current dominates; For case b), offset voltage dominates. For each effect halved, the output offset becomes 0.25 V and 0.08 V respectively.

2.56 a) 0.3 V; b) 0.03 V, with 10 MΩ to compensate.

2.57 $R =$ 15 kΩ, $R_{in} =$ 15 kΩ.

Chapter 3

Diodes

3.1 a) 0 V, 5 mA; b) – 5 V, 10 mA; c) – 10 V, 10 mA; d) 5 V, 5 mA; e) 5 V, 0 mA; f) – 5 V, 0 mA.

3.2 a) 5 V, logic OR (in positive logic), logic AND (in negative logic); b) 0 V, OR (in positive logic), AND (in negative logic); c) 5 V, AND (in positive logic), OR (in negative logic); d) 0 V, AND (in positive logic), OR (in negative logic); e) 0 V, AND (in positive logic), OR (in negative logic).

3.3 $Y = AE + BC + D$ in positive logic, 5 V.

3.4 82.8 mA, 13.5 mA, 49.5 mA, 6.2 mA.

3.5 5 mA, 3 mA, 1 mA, 1 mA, 0 V.

3.6 $n =$ 2.00, $I_S = 8.32\times10^{-17} A$.

3.7 0.758 V, 9.95 mA.

3.8 0.355 V, 9.2 mV.

3.9 0.445 V.

3.10 128 nA, 181 nA.

3.11 a) 0.75 V, 2.5 mA; b) 0.73 V, 1.7 mA; c) 0.74 V, 1.8 mA.

3.12 2.59 mA, 0.741 V.

3.13 2.92 mA, 0.708 V.

3.14 2.73 mA, 0.723 V, 2.5 mA, 0.75 V.

3.15 a) 0.7 V, 1 mA; b) 0.675 V, 1.25 mA; c) 0.701 V, 0.91 mA; d) 0.75 V, 0.50 mA.

3.16 600Ω

3.17 4.04 V.

3.18 0.752 V.

3.19 At 0.1 mA, $r =$ 500Ω; At 10 mA, $r =$ 5Ω; The geometric mean is likely to be best; $(r_1 + r_2)/2 =$ 252.5Ω, $(r_1 r_2)^{1/2} =$ 50Ω; At 1 mA, $r =$ 50Ω; At 5.05 mA, $r =$ 9.9Ω.

3.20 25Ω; 1 mA in each; This demonstrates that diode incremental resistance is independent of diode junction size.

3.21 0.005 V/V, 0.05 V/V, 0.33 V/V, 0.83 V/V.

3.22 Currents split equally; Thus all diode currents are equal; $R_T =$ 100/I; At 10 mA, 0.999 V/V; At 1 μA, 0.0909 V/V; Linearity is critical for small currents; v_S is limited to 22 mVpeak.

3.23 ± 40 mV, or ± 1%; – 50 mV, or – 1.25%; Combined as – 90 mV, or – 2.25%; 3.91 V.

3.24 6.70 V, 6.7 V, 11.3 mA, 7.9 V.

3.25 90.9 mV/V, – 18.2 mV/mA.

3.26 150Ω, 6.75 V, 7.19 V.

3.27 270Ω, 15.39 V, 296 mW.

3.28 10.6 V; Conduction for one-half cycle, or more precisely, 48% of a cycle; 3.60 V, 11.3 V; 9.97 V, 3.06 V.

3.29 See page 161: 2.41 V.

3.30 152.7 mA.

3.31 See page 162: 15.57 V, 0.876 ms, 9.40 V, 16.3 V.

3.32 16.3 V, 33.2 V.

3.33 41.7 μF, 167 μF, 28.8 mA, 81.8 mA.

3.34 20.9 μF, 83.4 μF, 14.4 mA, 40.9 mA, 33.2 V.

3.35 Use an 18.2 V rms centre-tapped secondary, 2500 μF, 25.1 V, 2.78 A.

3.36 1.05 A, 18.2 V; With a source resistance, the output drops by 0.3 V to 17.9 V on average.

3.37 +3.0 V for 6.0 V input; – 3.0 V for – 6.0 V input; $K =$ 0.5 V/V; 0.9 mA.

3.38 See page 165.

3.39 9.1 Vpp, 3.13Ω.

3.40 For light load, the output is a square wave of period T going from + 0.5 V to – 89.5 V; As R reduces, the negative side rises toward ground. For $RC =$ 2T, the waveform rises initially to 0.7 V, then drops to 0.55 V, then falls to – 89.5 V, then reduces to – 69.7 V, then rises to 0.7 V to begin a new cycle.

3.41 198.6 V, 193.6 V, 19.86 mA.

3.42 See page 167: 50 V, 75 V, 87.5 V, 96.9 V, 99.8 V.

3.43 88.4 V using 0.7 V diodes.

3.44 $p_{po} = 10^{-m}$, $n_{po} = 10^{-n}$.

3.45 Larger in the lighter-doped p region, by ten times.

3.46 5 nA.

3.47 0.1 pF.

3.48 $V_O =$ 0.71 V, $K =$ 8.7, $C_j =$ 15.3 pF.

3.49 0.42 mA, 4.2 mA.

3.50 2.08×10^{-8} pC, 20.8×10^{-8} pC, 0.025 pC.

3.51 $C_d = Ki/(nV_T)$.

3.52 $K =$ 1.344, $V_O =$ 0.7 V, $m =$ 1.10, $n =$ 2.0, $K_c =$ 8pF/mA, $C_T =$ 82 pF.

3.53 50Ω, 8 pF, 20 pF, 28 pF, 17.2 pF.

Chapter 4
Bipolar Junction Transistors (BJTs)

4.1 $\upsilon_{CB} \geq 0$ V, $\upsilon_{CE} \geq 700$ mV; $I_S = 6.91 \times 10^{-15}$ A, $\beta = 100$, $i_E = 10.1$ mA.

4.2 b) 20 μA, 1.02 mA, 0.980; c) 1.96 mA, 40 μA, 49; d) 1.99 mA, 2.00 mA, 199; e) 100 mA, 10 mA, 0.909; f) 1 mA, 1.001 mA, 0.999 mA.

4.3 See page 172: αi_E or βi_B.

4.4 0.47 V, 0.24 V.

4.5 10 nA, 1.28 μA.

4.6 0.01.

4.7 0.390 mA, 0.429 mA, 676 mV.

4.8 99, 1.46 A.

4.9 a) 1 mA, from – 0.6 V to – 0.8 V, from 0.909 mA to 0.997 mA; b) from 5.45 V to 5.02 V; V_E has no effect on I_C or V_C; c) 10 kΩ.

4.10 $i_C = \alpha(I + i_e) = 1.1$ mA for high beta, and 1 mA for $\beta = 10$; 9.90 kΩ, 1.82 Vpp; 1.65 Vpp.

4.11 9.3 kΩ, 1.021 mA, 9.7 kΩ, – 2.8 mV.

4.12 0.055 mA, 0.407 mA, 3.012 mA.

4.13 77.8 kΩ, 167 V; 1.7 MΩ; 17 KΩ.

4.14 122.5 μA.

4.15 a) – 3.3 V, – 5.39 V, 1 mA, 0.98 mA, 19.6 μA; b) – 5.3 V, – 5.5 V, 1.606 mA, 0.957 mA, 0.648 mA; c) – 1.3 V, – 3.39 V, 1 mA, 0.98 mA, 19.6 μA; d) 0 V, – 10 V, 0 mA; e) – 4.7 V, – 3.67 V, 1.128 mA, 1.105 mA, 22.1 μA; f) – 6.7 V, – 2.7 V, 0.702 mA, 0.688 mA, 13.8 μA.

4.16 a) 6.6 kΩ, 12 kΩ; b) 10.6 kΩ, 8 kΩ.

4.17 a) 0.930 mA, 1.023 mA, 93 μA, 8.14 V, 0 V, 0.7 V; b) 0.930 mA, 1.023 mA, 93 μA, 1.86 V, 10 V, 9.3 V; c) 0.230 mA, 0.253 mA, 23 μA, 9.54 V, 0 V, 0.7 V; d) 0 mA, 0 mA, 0 μA, – 10 V, 0 V, 0 V.

4.18 a) 1.55 V, – 0.423 V, – 1.123 V, 0.845 mA, 42.27 μA, 0.888 mA; b) 3.55 V, – 3.23 V, – 3.93 V, 0.645 mA, 32.26 μA, 0.677 mA; c) – 8.34 V, – 7.56 V, 8.26 V, 0.166 mA, 8.29 μA, 0.174 mA.

4.19 3 V, 2.3 V, 7.7 V; 2.465 V, 1.765 V, 8.23 V; 1.313 V, 0.613 V, 9.44 V.

4.20 120.

4.21 a) 2.3 mA, 2.1 V; b) 2.156 mA, 2.575 V; c) 1.432 mA, 4.964 V.

4.22 a) 0.582 mA, 3.86 V; b) 0.590 mA, 3.79 V; c) 0.571 mA, 3.96 V; d) 0.55 mA, 3.69 V.

4.23 5 V, 4.3 V, 10.7 V, 11.4 V, 3.6 V; 4.73 V, 4.03 V, 11.31 V, 12.07 V, 2.90 V.

4.24 40 μA/V, 4 mA/V, 40 mA/V, 4 A/V.

4.25 25 kΩ, 2.5 MΩ; 250 Ω, 25 kΩ; 25 Ω, 2.5 kΩ; 0.25 Ω, 250 Ω.

4.26 Gain = K/V_T, a constant.

4.27 37.8 kΩ, – 39.7 V/V.

4.28 – 4000 V/V, – 1000 V/V.

4.29 $g_m' = g_m r_e/(r_e + r_E)$, $r_\pi' = (\beta+1)(r_e + r_E)$, 9.9 mA/V, 10.1 kΩ.

4.30 a) 0.976 V/V; b) – 8 V/V; c) – 40 V/V; d) – 7.92 V/V; e) – 4.44 V/V.

4.31 – 297 V/V, 2.575 Vp, 8.67 mVp.

4.32 Increases by 49%, or reduces by 33%; 6.7 kΩ.

4.33 – 4000 V/V.

4.34 $r = \left[R_1R_2/(\beta+1)+r_e(r_1+R_2)\right]/(r_e+R_1)$: a) r_e; b) $2r_e$; c) $3r_\pi/(\beta+2) \approx 3r_e$.

4.35 $r = R_1+r_e \| R_2$; a) r_e; b) $2r_e$; c) $r_\pi/(\beta+2) \approx r_e$; d) $r_e+r_\pi/(\beta+2) \approx 2r_e$.

4.36 $Gain = -g_m(r_o \| R_f)$, $R_i = r_\pi \| (R_f/(1+g_m(r_o \| R_f)))$; $Gain = -g_m r_o/2$, $R_i = r_\pi \| (2/g_m) \approx 2r_e$.

4.37 See page 180: $r_\pi' = 2(\beta_1+1)r_{\pi_1}$, $g_m' = g_{m1}/2$.

4.38 See page 181: $i_B =$ 5 µA, $i_C =$ 1050 µA, $\upsilon_{EC} =$ 4.75 V; 2.9 V, 0.582 mA; 3.09 V, 0.618 mA; clipping for 50% of the cycle.

4.39 49.

4.40 30 Ω, 400 Ω, 200 Ω, 40 Ω; 101.2 mA, 4.995 V.

4.41 $R_E =$ 1.00 kΩ, $R_B =$ 68 kΩ.

4.42 For $R_B =$ 95.2 kΩ, I_E varies from 1.056 mA to 0.787 mA, and V_{CB} from 0.5 V to 1.47 V. For $R_B =$ 100 kΩ, I_E varies from 1.049 mA to 0.773 mA, and V_{CB} from 0.522 V to 1.516 V.

4.43 $R_\beta =$ 120 kΩ, $R_B =$ 91 kΩ; I_E varies from 1.047 mA to 0.700 mA and V_{BC} varies from 0.530 V to 1.78 V.

4.44 $R_\beta =$ 68 kΩ, $R_B =$ 47 kΩ; 0.484 V, 1.06 mA; 1.26 V, 0.844 mA.

4.45 – 0.84 V, – 1.54 V, 8.4 mA, +1.6 V, 336 mA/V, 2.95 Ω, 298 Ω, 11.9 kΩ; Note constant voltages, inversely-scaled currents, and directly-scaled parameters.

4.46 289 Ω, – 336 mA/V, 0.922 kΩ, – 310 V/V, – 97.1 A/A, 36.2 V/V, 46.7 A/A; Note that for resistance-scaled designs, parameters scale correspondingly and gains are constant.

4.47 Use $R_E =$ 10 kΩ: For $\beta = \infty$, – 0.7 V, +0.7 V, 0.83 mA, – 307 V/V, 3.26 mV; For $\beta =$ 90, – 1.52 V, 1.60 V, 0.74 mA, – 64.1 V/V, 3.66 mV, with $\upsilon_s =$ 15.6 mV.

4.48 – 37.5 V/V to – 79.0 V/V.

4.49 – 20.8 V/V to – 29.3 V/V.

4.50 – 199 V/V, 3.78 kΩ, – 109 V/V, 3.78 kΩ, 0.274 V/V, 5934 V/V.

4.51 21.4 Ω to 8.3 Ω; 6.14 V/V to 6.92 V/V.

4.52 71 Ω, – 4.55 V/V.

4.53 5.216 V, 4.516 V, 3.784 V, 0.153 mA; a) Source coupled to *B*, load to *E*, ground to *C*; b) Source to *B*, 10 kΩ from *E* to ground or 10/3 kΩ from *E*, with load to *C*; c) Source to *B*, ground to *E*, load to *C*; d) Source to *E*, ground to *B*, load to *C*.

4.54 a) 0.968 V/V; b) – 0.5 V/V or – 1.0 V/V for an added 10/3 kΩ resistor; c) – 20.4 V/V; d) 20.4 V/V.

4.55 0.685 V/V.

4.56 a) – 30.8 V/V, 193 kΩ; b) – 27.4 V/V, 221 kΩ; c) 27 kΩ.

4.57 2.7 kΩ, $R_B \le 0.448\beta$ kΩ.

4.58 a) Q_1 cut off and Q_2 saturated; b) Q_1 saturated and Q_2 cut off, with $\beta_{forced} =$ 1.12.

4.59 a) 2.05 V, 2.75 V, 2.25 V, 3.44, 0.561 mA; b) 4.8 V, 5.5 V, 5.0 V, 9, 10.1 mA.

4.60 2 MΩ, 1.2 GΩ.

4.61 a) 50 V; b) 30 V; c) 7 V.

4.62 50 Ω, 0.05 V.

4.63 109, 90, 60 V.

4.64 4.71 V, 4.85 V.

Chapter 5
Field-Effect Transistors (FETs)

5.1 $\upsilon_{GS} \geq 5$ V, $\upsilon_{DS} \geq 1.5$ V, $\upsilon_D \geq 1.5$ V, $\upsilon_D \leq 1.5$ V.

5.2 a) Saturated mode; b) Triode mode; c) Cut-off mode; d) Saturated mode; e) N-channel, $\upsilon_D \geq -2$ V; f) P-channel, saturated mode; g) P-channel, $\upsilon_G \geq 1$ V.

5.3 9 mA, 8 V, 8 V, 4 mA, 4 mA, 3 mA, 10 V, 4 V.

5.4 See page 190.

5.5 $V_D \leq 2$ V, 400 μA, 300 μA, 175 μA, 5 kΩ, 0.04 V, 0.36 V.

5.6 6 V to 1.005 V, 826 μA, 0.909 V, 0.43 nA, 0.45 mV.

5.7 $\upsilon_{GS} - V_t$, 0.859 ($\upsilon_{GS} - V_t$), 0.564 ($\upsilon_{GS} - V_t$), 0.134 ($\upsilon_{GS} - V_t$); 2 V, 1.72 V, 1.13 V, 0.26 V.

5.8 50 kΩ, 0.0093 V^{-1}, 107.5 V.

5.9 1.575 mA.

5.10 – 2 V, – 2.96 V.

5.11 a) 0.9 mA; b) 4 V; c) 0 mA; d) – 3 V.

5.12 a) – 4 V, 5 V, Cutoff, 0 mA; b) – 2 V, 3 V, Saturation, 4 mA; c) 0 V, 5 V, Saturation, 16 mA; d) 0 V, 2 V, Triode, 12 mA; e) – 1 V, 4 V, Edge of saturation, 25 mA; f) 2 V, 3 V, Triode, 35 mA; g) 2 V, 0 V, Triode, 0 mA; h) – 2 V, 2 V, Triode, 20 mA.

5.13 a) 0.4 mA; b) 0 V; c) 1 V; d) 0.9 mA, +1 V; e) 4.172 V.

5.14 0.296 mA, – 1.075 V.

5.15 10 mA, 10 mA, 7.5 mA, 0.586 V.

5.16 – 0.5 V, – 1.3 V.

5.17 100 Ω, 200 Ω, ∞Ω.

5.18 – 0.586 V, 15 kΩ, 0.004 V^{-1}, 250 V; or, alternatively 50.02 kΩ, 0.00403 V^{-1}, 248 V.

5.19 a) 4 mA; b) 1 V; c) 0.268 V; d) – 0.236 V.

5.20 0.4 mA, – 2 V, 7.5 kΩ.

5.21 0.4 mA/V^2, 2.43 V, 21.5%.

5.22 20 kΩ.

5.23 200 kΩ, 2.15 V.

5.24 2 V, V_t raised by 10.1%, K lowered by 18.4%.

5.25 4 mA, 11 V, 8 V, ≥ 14 V.

5.26 – 1.60 V.

5.27 1 mA, 0 V; 0.61 mA, 0 V.

5.28 5.0 V, 9 mA; 6 mA; – 2.8 V/V or (better) – 3 V/V; 1 Vpeak.

5.29 0.076 kΩ, 3.12 to 1.

5.30 9 mA, 7.5 V, 6 mA/V, 6 mA/V, – 3.0 V/V, ± 1.5 V, + 5.875 V and 8.875 V, versus 6.0 V and 9.0 V.

5.31 2.83 mA/V, 3.53 kΩ; Linear for υ_{gs}<< 5.66 V, for 1%, 0.06 V peak, or for 10%, 0.6 V peak.

5.32 0.240 mA/V^2, 127 V, 2.5 V.

5.33 $(I+1)^{1/2}$ V, I mA, $2\sqrt{I}$ mA/V, 50/I kΩ, – $2R\sqrt{I}$ V/V, $10^4/(1+2R\sqrt{I})$ kΩ; – 100 V/V, 99 kΩ; – 50 V/V, 196 kΩ; – 73.0 V/V, 135 kΩ.

5.34 10 mA, 5 V, 5 kΩ, – 50 V/V, 19.2 kΩ, – 25 V/V, 37 kΩ.

5.35 – 2.38 V/V, – 4.76 V/V, – 1 V/V.

5.36 0.189 mA, 5.22 V, 4.61 V, 1.89 V for cutoff, 0.276 mA, 3.48 V, 2.74 V, 2.76 V for cutoff.

5.37 10 MΩ, 8.6 MΩ, 1.75 kΩ, 5.33 kΩ.

5.38 0.311 mA, 2.79 V, 2 Vp; 0.358 mA, 1.85 V, 1 Vp.

5.39 0.189 mA, 5.22 V, 4.61 V, 0.276 mA, 3.48 V, 3.74 V.

5.40 See page 200: 3.0 kΩ.

5.41 10 MΩ, 2 MΩ, 2.1 kΩ.

5.42 0.25 kΩ, 0.375 kΩ, ∞ Ω, 10 MΩ.

5.43 10 mA, 2.5 V, 10 kΩ, 10 kΩ, 10 mA/V, – 50 V/V, 19.6 kΩ; – 50 V/V; – 25 V/V; – 33.1 V/V.

5.44 1 MΩ, – 8.33 V/V, 8.33 Ω, – 5.88 V/V.

5.45 1 kΩ; a) 8.33 V/V, 5.88 V/V; b) 4.17 V/V, 2.94 V/V.

5.46 1.002 mA/V, 20.62 kΩ, 20.6 V, ≥ 8.57 kΩ.

5.47 See page 202: 0.976 V/V, 0.61 kΩ, 0.277 kΩ.

5.48 See page 202: There are 2 enhancement and 4 depletion configurations, with 6 in total, for which current flows. Of these, four allow saturation operation.

5.49 a) 2.5 V, 0.25 mA; b) 2.414 V, 0.343 mA; c) 3.82 V, 3.31 mA; d) 2.5 V, 0.25 mA; e) 2.5 V, 16.25 mA.

5.50 – 3 V/V, 3.5 V, 1 V, for υ_I = 4 V, and 2.5 V for υ_I = 3.5 V.

5.51 3.48 V, for V_t = 1.25 V, and χ = 0.172 and g_m = 0 μA/V; 0 V at V_t = 0.9 V, and χ = 0.323 and g_m = 82 μA/V.

5.52 – 2.62 V/V.

5.53 – 8.34 V/V; The output range is from 3.0 V to about 1.2 V.

5.54 2.34 V, 1.62 μA; 1.45 V, 2 μA.

5.55 – 63.2 V/V, – 200 V/V, – 632 V/V.

5.56 225 μA, 2.06 V, – 47.1 V/V; 3.5 V, 1.06 V.

5.57 2.14 V, 707 Ω, 0.929 V/V, $R \geq$ 0.768 kΩ.

5.58 1963 μm.

5.59 2.4%, 3.2%, 2.4%.

5.60 38.4 mA/V to 25.6 mA/V, 347 Ω to 781 Ω, 13.3 V/V to 20.0 V/V.

5.61 1.21 V to 2.09 V, – 2.2 V/V.

5.62 15 mA, 0 V, 30 mA/V, – 15 V/V.

5.63 0.144 V, 29.7 mA/V, – 12.9 V/V.

5.64 0 V, $\upsilon_o = -2 \sin \omega t$ V.

Chapter 6

Differential and Multistage Amplifiers

6.1 a) 115 mV; b) 73.6 mV; c) 54.9 mV.

6.2 a) 0 V, 6 V; b) 1.3 V, 6 V; c) 2.0 V, 2.0 V, 10 V; d) 1 V, 2 V; e) 3.5 V, 10 V; f) – 4.7 V, 4 V; g) 4.0 V, 3.5 V, 10 V; h) 10 V, 3 V.

6.3 a) 0.0 V, 9.6 V; b) – 0.695 V, 9.52 V, 9.68 V; c) – 0.664 V, 9.90 V, 9.30 V; d) 0.037 V, 9.35 V, 9.85 V; e) – 1.05 V, – 1.714 V, 9.3 V; f) – 1.758 V, 6.00 V, 6.00 V; g) – 0.752 V, 5.21 V, 6.79 V; h) – 0.722 V, 9.05 V, 2.95 V; i) 0.951 V, 0.228 V, 9.00 V.

6.4 $i_{C1} = \alpha I/2 + \alpha I/(2V_T)(\upsilon_d/2)(1-(1/2)(\upsilon_d/(2V_T))^2-(1/8)(\upsilon_b/(2V_T^3)))$, 9% or 5%; 11 mV, 7.9 mV, 3.5 mV.

6.5 40 V/V, 20 V/V, 2.4 V.

6.6 400 V/V, 75.5 kΩ, 200 V/V.

6.7 50.25 kΩ; 56.94 V/V, 100 MΩ, 0 V/V, ∞ V/V, ∞ dB; 28.5 V/V, – 0.00995 V/V, 2864 V/V, 69.1 dB.

6.8 2×10^{-5} V/V, – 94 dB, 2.85×10^{6} V/V, 129 dB; 2×10^{-4} V/V, – 74 dB, 2.85×10^{5} V/V, 109 dB.

6.9 79.7 V/V, 12×10^{-4} V/V, 66.4×10^{3} V/V, 96.4 dB.

6.10 79.9 V/V, 1.5×10^{-4} V/V, 114.5 dB.

6.11 2.5 mV, 25 mV.

6.12 47.5 mV, 33.6 mV, 22.5 mV.

6.13 4 mV, 4.61 mV.

6.14 9.9 mV.

6.15 2.7 V.

6.16 250 Ω, 250 Ω, 500 Ω.

6.17 198, 1998.

6.18 60 Ω, 58.15 Ω; 0.990 A/A, 1.005 A/A; 0.982 A/A, 0.991 A/A, 0.996 A/A.

6.19 1.974 V, 0.474 V to 3.474 V.

6.20 20 kΩ, 2.7 V.

6.21 a) See page 214: 9 BJTs; b) See page 215: 10 BJTs; c) 9 BJTs.

6.22 $I_O/I_R = 1/(1+3/(\beta^2+\beta))$.

6.23 $I_O/I_R = 1/((\beta_1/\beta_2)(\beta_2+1)/(\beta_1+1)+((\beta+1)/(\beta_1+1)+1)/(\beta_2(\beta_3+1)))$; $\beta_1 = (1-k)\beta$, $\beta_2 = (1+k)\beta$, $\beta_3 = \beta$;
$I_O/I_R = (\beta^2+\beta+k\beta-k^2\beta^2)/(\beta^2+\beta-k\beta-k^2\beta^2+2 = 1$, for $k = 1/\beta$.

6.24 $I_O/I_R = 1/(1+2(\beta_2-\beta_3+1)/(\beta_2\beta_3+2\beta_3))$; Use $\beta_1 = \beta$, $\beta_2 = (1-k)\beta$, $\beta_3 = (1+k)\beta$;
$I_O/I_R = (\beta^2-k^2\beta^2+2k\beta+2\beta)/(\beta^2-k^2\beta^2-2k\beta+2\beta+2) = 1$ for $k = 1/(2\beta)$.

6.25 For 1 µA, $R_E =$ 115 kΩ; For 10 µA, $R_E =$ 5.75 kΩ.

6.26 2 mA/V, – 3000 V/V, 1.5 MΩ, 76 kΩ, – 144.7 V/V.

6.27 1 mA/V, 3 MΩ, – 3000 V/V.

6.28 2 mA/V, 51.8 MΩ, 103.5×10^3 V/V.

6.29 2 V, 0.25 mA/V, 2.83 V, 0.4 V.

6.30 See page 217: The term T is $(1-(\upsilon_{id}/V_p)^2(I_{DSS}/(2I)))^{1/2}$; $(\upsilon_{id}/V_p)(I_{DSS}/(2I))^{1/2}$ = 1/7 for 1% distortion and ½ for 10% distortion.

6.31 4.55 V/V, 25 V/V.

6.32 ± 0.2 V/V, ± 0.1 V/V, ± 0.02 V/V, for 10%, 5%, 1% drop, respectively.

6.33 1.35 V, 70.8 μA/V, 283.2 V/V, 141.6 V/V.

6.34 36 mV.

6.35 ± 1.25 mV, ± 1.25 mV, ± 0.6 mV; ± 3.1 mV, ± 1.87 mV.

6.36 1.95 V, 100 μA, 101 μA, 8 MΩ.

6.37 1.92 V, 91.9 μA, 92.9 μA, 8 MΩ.

6.38 $I_{O1} = I_{O2} =$ 50.8 μA, 5.6 kΩ, 800 kΩ; To improve, split both Q_3 and Q_2 into two parts, requiring a total device width of 4 W as in the original case, rather than 5 W as required by the initial modification.

6.39 – 20 V/V, 100 kΩ.

6.40 a) 400 μA/V, 250 kΩ, 10 MΩ, – 4000 V/V; b) 63.2 μA/V, ∞ Ω, 2 MΩ, – 126.4 V/V.

6.41 -55.6×10^3 V/V.

6.42 -17.6×10^3 V/V.

6.43 Normally, $R_{out} = 115\times10^9\Omega$; $0.889\times10^9\Omega$ with no Q_3, Q_6; $1.264\times10^9\Omega$ with no Q_5, Q_2.

6.44 a) 1.25 mA, 2.5 V; b) 2.95 mA, – 0.13 V; c) 3.0 mA, 1.5 mA, – 0.12 V, 2 V; d) 0.12 mA, 4.94 V; e) 1.5 mA, 0.37 V.

6.45 a) – 4.5 V; b) – 3 V; c) 0.55 mA; d) 880 kΩ; e) 4.5 μA.

6.46 a) – 4.3 V; b) – 3.3 V; c) 1.1 mA; d) 8.16 MΩ; e) 0.49 μA.

6.47 $W_1 = W_3 =$ 43.5 μm, $W_2 =$ 21.7 μm; Diode drop is 0.75 V at 2.5 mA; 0.926 V/V; 31.1 kΩ or 26.5 kΩ.

6.48 $W_1 =$ 4.67 μm, $W_2 =$ 45.2 μm; Use $i_D =$ 0.502 mA; 1.76 MΩ; 2 μA change is consistent.

6.49 0 V, 0.88 MΩ, – 822 V/V.

6.50 $W_1 = W_2 = W_3 =$ 16.3 μm; Use $I =$ 1.0 mA; Needs an 11 mV offset to keep υ_O at 2.25 V; 2450 V/V; Alternatively, with $W_1 = W_2 =$ 16.7 μm and $W_3 =$ 3.85 μm, and $I =$ 1.0 mA, the gain is – 1296 V/V.

6.51 a) ∞ Ω, 5 Ω, 80.9×10^3 V/V; b) 10.2 kΩ, 283 Ω, 4040 V/V.

6.52 Reduce R_4 to 1.15 kΩ, R_5 to 7.85 kΩ, R_6 to 0.75 kΩ, for which $A_3 =$ – 6.12 V/V, $A_4 =$ 0.998 V/V, $A =$ 8099 V/V, with $R_o =$ 71.4 Ω; With $R_L =$ 286 Ω, swing is from 9.8 V to – 4.1 V.

Chapter 7

Frequency Response

7.1 $T(s) = (1+R_1C_1s)/(1+R_1(C_1+C_2)s)$; a) Yes, low pass; b) See page 225: Pole at 182 rad/s, 0 at 2000 rad/s.

7.2 a) $T(s) = (10^2 s(1+s/10))/((1+s)(1+s/100)(1+s/10^5)(1+s/10^6))$; b) 0 V/V, 90°; 0 V/V, – 180°; c) Poles at $s =$ – 1, – 100, -10^5, -10^6 rad/s; Zeros at $s =$ 0, – 10, ∞, ∞ rad/s; d) Gain is 10^3 V/V and phase 0° from about 10^3 to 10^4 rad/s; e) 60 dB, 57 dB; Better to prepare plots earlier, certainly by the end of part a), easiest after c), most useful before d).

7.3 57 dB, 39.8°; 52.8 dB, – 74.7°.

7.4 $A_m = 10^3$, $F_L(s) = s(s+10)/((s+1)(s+100))$, $F_H(s) = 1/((1+s/10^5)(1+s/10^6))$, $A_L(s) = 10^3 s(s+10)/((s+1)(s+100))$, $A_H(s) = 10^3/((1+s/10^5)(1+s/10^6))$.

7.5 $F_L(s) = s/(s+100)$, $F_H(s) = 1/(1+s/10^5)$, $A(s) = 10^3 s/((s+100)(1+s/10^5))$.

7.6 a) 100 rad/s; b) 99 rad/s; c) 99.00 rad/s.

7.7 a) 10^5 rad/s; b) 0.995×10^5 rads/s; c) 0.990×10^5 rad/s.

7.8 $T(s) = 10^3 s(s+10)/((s+1)(s+100)(1+s/10^6)(1+s/2\times10^6))$; 0.894×10^6 rad/s, 0.856×10^6 rad/s.

7.9 a) 0.333×10^6 rad/s, 0.447×10^6 rad/s, 0.374×10^6 rads/s; b) 0.476×10^6 rad/s, 0.673×10^6 rad/s, 0.601×10^6 rad/s; c) 0.833×10^6 rad/s, 0.905×10^6 rad/s, 0.895×10^6 rad/s.

7.10 a) 300 rad/s, 224 rad/s; b) 210 rad/s, 147 rad/s; c) 1200 rad/s, 1054 rad/s.

7.11 – 13.14 V/V, 2.28 Hz, 53.1 Hz, 334.3 Hz; 15.9 Hz.

7.12 $C_S =$ 30 μF, $C_C =$ 0.02 μF, $C_{C2} =$ 5 μF; 11.1 Hz, 1.14 Hz, 1.08 Hz; 0.53 Hz.

7.13 $\omega_z = 1/(C_S(R_S+r_S))$, $\omega_p = 1/(C_S(R_S+r_S(1+g_mR_S))/(1+g_mR_S))$; The equivalent transconductance is $g_m(R_S+r_S)/(g_mR_Sr_S+R_S+r_S)$; $r_S =$ 526 Ω; $f_{ps} =$ 159 Hz, $f_{zs} =$ 15.1 Hz.

7.14 a) 144.7 MHz; b) 155.3 MHz; c) 9.65 GHz.

7.15 62.1 fF, 8.1 fF, 2.04 GHz.

7.16 240 fF, 2220 fF, 140 fF, 120.2 fF.

7.17 4.76 pF, 0.577 pF, 0.368 MHz, 42.3 MHz, 0.37 MHz, 42.3 MHz, 3.87 Ω.

7.18 0.365 MHz, 234 MHz, 318 MHz, $f_H =$ 0.365 MHz; 19.8 MHz, 392 MHz, $f_H =$ 19.8 MHz.

7.19 0.477 pF, 500 MHz.

7.20 760 kΩ.

7.21 A positive pulse of 2.5 V amplitude, of 50 μs duration, with transitions of 7 ns, and sag of 1.6%.

7.22 $r_o =$ 50 kΩ, $g_m =$ 80 mA/V, $r_\pi =$ 1.25 kΩ, $r_x =$ 50 Ω, $r_\mu =$ 50 MΩ, $\mu =$ 4000 V/V.

7.23 39.6 mA/V, 2.5 Ω, 250 Ω, 99, 40 Ω; To find *n*, add another measurement with another resistor in the base, say 100 Ω, to get a third equation.

7.24 400 mA/V, 0.5 kΩ, 50 Ω, 5 MΩ, 6.25 kΩ, 62.5 V.

7.25 2.54 GHz, 4.51 pF, 13.8 MHz, 22.6 pF; $I_C \geq 0.22$ mA.

7.26 a) 1.0 mA, 39.8 mA/V, 5.025 kΩ, 5.83 pF, 5 MHz; b) 0.1025 mA, 244 Ω, 40, 4 GHz, 0.059 pF; c) 2500 Ω, 0.397 mA/V, 378 kΩ, 0.03 GHz, 0.1 pF.

7.27 – 18 V/V, 3.0 MHz.

7.28 6.24 Hz, 8.4 Hz, 74.4 Hz, 1.94 Hz; 75.1 Hz.

7.29 C_E = 37.2 μF, C_{C1} = 12.0 μF, C_{C2} = 4.2 μF; Alternatively, C_{C2} = 16.2 μF, C_{C1} = 3.12 μF.

7.30 – 6.79 V/V, 6.6 MHz.

7.31 Poles at 4.40 Hz, 8.42 Hz, 28.2 Hz, with a zero at 1.86 Hz; 29.6 Hz.

7.32 20.7 V/V, 113 MHz.

7.33 Poles at 71.4 Hz, 8.42 Hz, 8.70 Hz and a zero at 3.98 Hz; 72.2 Hz.

7.34 – 13.6 V/V, 25.6 MHz.

7.35 0.963 V/V, 50.0 MHz, 15.6 Hz.

7.36 0.909 V/V, 1.46 MHz.

7.37 – 24.4 V/V, 3.96 MHz, 86 Hz.

7.38 a) – 18.7 V/V, 5.26×10^6 Hz, 68.5 Hz; b) – 15.0 V/V, 1.24 MHz, 51.2 Hz.

7.39 11.05 V/V, 7.75 MHz.

7.40 For R_L connected to the collector of the input transistor: – 7.12 V/V, 6 MHz; For R_L connected to the collector of the grounded-base amplifier: + 7.12 V/V, 4.54 MHz.

7.41 7.12 V/V, 29.3 MHz.

7.42 6.02 V/V, 13.3 MHz.

7.43 14.9 mV, 20 MHz, 88 MHz.

7.44 167 Ω, 3.66 V/V, 37.8 MHz.

Chapter 8

Feedback

8.1 0.01 V, 300 V/V, 0.33 V/V, 300 V/V, 3.0 V/V, 300 V; The output limits.

8.2 7.696, 0.115, 22 dB, 1 V, 0.115 V, 0.01 V, 4.12%.

8.3 As designed, 11; as fabricated, 6; 16.7%, 83.3.

8.4 47.9%.

8.5 10^6 Hz, 10^8 Hz, 10^8 Hz.

8.6 2×10^5 Hz, 0.707×10^4 V/V; The desensitivity factor maintains the gain for frequencies above cutoff.

8.7 100 V/V, 37 dB.

8.8 90.9 V/V, 1.1 mV, 0.1 V, 50 V/V, 20 mV, 1.0 V, 0 V/V.

8.9 a) Shunt-Shunt, $-1/R_2$; b) Shunt-Series, $-r/R_2$ or $-r/(+R_2)$; c) Series-Shunt, $R_1/(R+R_2)$; d) Series-Series, $rR_1/(R_1+R_2)$ or $rR_1/(r+R_1+R_2)$.

8.10 a) Series-Series, r; b) Shunt-Shunt, $-1/R_F$; c) Series-Shunt, 1; d) Series-Series, r.

8.11 2.22 S, 50 mΩ.

8.12 10.6 V/V, 11.2 kΩ, 1.06 Ω, 0.963 Vrms.

8.13 227 V/V, 0.05 V/V, 9.5 kΩ, 200 kΩ, 18.4 V/V, 478 kΩ, 47.1 Ω, 18.4 V/V.

8.14 10.9 V/V, 26 MΩ, 96.3 Ω, 26 MΩ, 96.3 Ω.

8.15 0.951 V/V, ∞ Ω, 166 Ω, 0.816 V/V, 29.8 mV.

8.16 0.974 V/V, 249 Ω, 0.998 V/V, 13.8 Ω.

8.17 26.0 V/V, 2.5 kΩ, 10 kΩ, 1.856 V/V, 362 Ω.

8.18 204.5 mA/V, 18.3 mA/V, 438 kΩ, 23.6 kΩ.

8.19 20.7 A/V, 11.9 Ω, 83.7 mA/V, 3.38 MΩ, 183 kΩ.

8.20 10 Ω, 10 Ω, 10 Ω, 260 mA/V, 72.2 mA/V, 244 kΩ, 36.8 kΩ.

8.21 18 MΩ, -10^{-5} A/V, 100 kΩ, 100 kΩ, – 2.8 V/μA, – 96.6 V/mA, 221 Ω, 17.5 Ω, – 93.3 V/mA, 9.65 kΩ.

8.22 Series-shunt at low frequencies; Shunt-series at high frequencies; – 0.167 A/A, 6 kΩ, 0.83 kΩ, – 183 A/A, 0.58 mA/V, 50 Ω, ∞ Ω.

8.23 – 1.0, 10 kΩ, ∞ Ω, – 0.98 V/V, 168 Hz, 3.98 Hz.

8.24 – 0.298 A/A, 4.45 kΩ, 1.24 kΩ, – 115.4 A/A, 0.335 mA/V.

8.25 1.77 μF, 1.14 μF.

8.26 $2 \times (10/11 + R_L)$ mA/V, 12.9 kΩ, 31 kΩ; For loads from 0 kΩ to 12.9 kΩ the transconductance varies by only 3 dB, and by another 3 dB for R_L up to 31 kΩ.

8.27 0.65, 61.3 Ω, 26 Hz; 15.5 μF, 15.5%.

8.28 See page 249: 63.5 V/V, 1550 V/V, 5 Hz, 244 Hz, 3100 V/V, 48 V/V, 650 Ω.

8.29 1550 V/V, 63.3 V/V, 32.5 μF, 0.10 Hz, 2.45 Hz, 6.8 kΩ, 150 kΩ.

8.30 120 V/V, 1.21×10^4 V/V.

8.31 100.5 V/V, 0.0995 V/V.

8.32 a) 19.3 V/V, 0.951 V/V; b) 4.36 V/V, 0.813 V/V.

8.33 10^5 rad/s, 5 V/V, $\beta =$ 0.02 or less.

8.34 See page 250.

8.35 174.3 °, Oscillation does not occur; 454 Ω.

8.36 7.98 V/V, 2.5 MHz, 20 MHz; The pole is shifted by the amount of the feedback factor, namely 500.

8.37 200, 199, 0.05, 19.9 V/V.

8.38 1.11, 1.11×10^4 V/V, with poles at 10^5 Hz and 9×10^5 Hz, 1.78×10^{-4}, 4×10^3 V/V.

8.39 $10.5 \times 10^6 (1 \pm j)$ Hz, 0.707, 14.9 MHz, 14.9 MHz, 0.043 V/V, 22.7 V/V.

8.40 See page 252: $\beta =$ 0.024; poles at -1.41×10^5 rad/s and at $(-0.298 \pm j\,0.298) \times 10^5$ rad/s, 143°.

8.41 60°, 29°, 5.7°.

8.42 0.01, 0.001, 0.0032, 909 V/V, 306.5 V/V, – 15 °.

8.43 $-\tan^{-1} f/10^6 - 2\tan^{-1} f/10^8$; Margins are zero at $f = 1.01 \times 10^8$ Hz; Margins are 45° at $f = 4.3 \times ^7$ Hz, with a gain of 196.2 V/V and $\beta =$ 0.0051; Margins are 78 ° at $f = 1.4 \times 10^7$ Hz with a gain of 698 V/V and $\beta =$ 0.00143.

8.44 $K =$ 1.11, 900 V/V, 0.9 MHz, 0.9 MHz; $K =$ 0.10, 10 V/V, 10 MHz, 10 MHz.

8.45 10^3 Hz, 10^2 Hz; 10^6 Hz, 10^6 Hz.

8.46 4.14×10^4 Hz, 4.14×10^3 Hz; 41 MHz, 41 MHz.

8.47 1.59 pF, 0.146 pF, 77.6 MHz, 10 MHz, 10 MHz; Factor of 2.4, 0.066 pF; Poles are at 24 kHz, 10 MHz, 39.8 MHz; 10 MHz, 4.1 MHz.

Chapter 9
Output Stages and Power Amplifiers

9.1 For a 1 kΩ load, peak load currents are 1.4 mA, 14 mA, 141 mA, with operation in modes A, A, AB, respectively; For a 0.25 kΩ load, peak load currents are 5.7 mA, 57 mA, 566 mA, with operation in modes A, AB, AB, respectively; Class B operation is possible, for large signals and low bias currents.

9.2 1.53 Vp, 2.7 Vp, ≥ 1.76 kΩ, ≥ 0.88 kΩ.

9.3 4.1 Vp, 4.56 mA.

9.4 a) 43.5 mW, 180 mW, 24.2%; b) 21.8 mW, 180 mW, 12.1%; c) 43.5 mA, 270 mW, 16.1%; d) 21.75 mW, 270 mW, 8.06%.

9.5 0 V, – 7 V for – 7.9 V input; 6.43 V for +7 V input; For v_O = 6.43 V: 41.3 mW, 238 mW, 17.4%; For v_O = – 7.0 V: 49 mW, 117 mW, 41.9%; For largest sine wave output of 6.43 Vp: 20.7 mW, 198.5 mW, 10.4%.

9.6 v_O is 9 Vp or 18 Vpp, with input of 20 Vpp for no load and 22 Vpp for a 10 kΩ load. Gain is 0.9 V/V or 0.82 V/V. Supply power is 0 mW or 6.36 mW. Load power is 0 mW or 4.05 mW. Efficiency is 100% or 63.7%.

9.7 +10.2 mV, +102 mV, +1.02 V.

9.8 For 6 Vp or 4.2 Vrms; 1.125 W, 1.43 W, 78.7%, 0.305 W. For 4 Vp: 0.5 W, 0.95 W, 52.6%, 0.45 W. For a +14.5 V supply and 6 Vp: 1.125 W, 1.73 W, 65%, 0.6 W.

9.9 2.5 mA, 1.35 V, 5.07 V, 0.982 V/V, 0.910 V/V, 0.995 V/V.

9.10 4.48 V.

9.11 21.9 mA, 1.00 V.

9.12 R_1 = 10 kΩ, R_2 = 8.7 kΩ; At the peak, 0.55 V/V; At 0 V, 0.84 V/V.

9.13 $r_{eq} = (r_e(R_1+R_2)+R_1R_2/(\beta+1))/(R_1+r_e) = (r_e(kR_1)+R_1^2(k-1)/(\beta+1))/(R_1+r_e)$, 114 Ω.

9.14 123.8 °C, 2.08 °C/W, 16I, 400 mV.

9.15 86.4 W, 102.5 °C, 0.29 °C/W, 1.68 °C/W, 10.4 cm, 1.79 cm, 52.3 cm.

9.16 42.3 W, 43.1 W; The problem lies in the transistor itself with its dominating thermal resistance.

9.17 0.595 Ω, 0.80 Ω.

9.18 2.76 mA, 15 Ω, 140 Ω.

9.19 a) 1.93 V; Current increases by more than 5 times; b) 2.04 V; Current increases by a few tens of %.

9.20 27 Ω, 100 mA.

9.21 For β from 50 to 150, g_m ranges from 204 mA/V to 207 mA/V, gain ranges from – 977 V/V to 981 V/V, R_n ranges from 946 Ω to 1010 Ω, R_{out} ranges from 4.75 kΩ to 4.79 kΩ; Rhat is, there is little effect.

9.22 95 kΩ, 1 μA, 5.05 kΩ, 0.0405 μA, 10.34 V, 10.56 V.

9.23 Raise R_1 to 115 kΩ with 57.5 kΩ in each half. Raise R_2 and R_3 to 2.3 kΩ and 57.5 kΩ, respectively. I_{12} reduces by 2.3 times while I_9 reduces by less than 2.3 times.

9.24 – 1818 V/V, 87.5 Hz.

9.25 28 Ω, Factor of 2.76, $R_5 = R_6$ = 35 Ω.

9.26 Use R_4 = 10 MΩ, R_3 = 100 kΩ, R_2 = 1 MΩ, R_1 = 101 kΩ, for an input resistance of 100 kΩ.

9.27 a) Drive A_1 as shown, but with R_3 connected to the output of A_1: $R_1 =$ 10 kΩ, $R_2 =$ 90 kΩ, $R_3 = R_4 =$ 100 kΩ; b) Merge R_3 and R_1 into $R_{13} =$ 10 kΩ with $R_2 =$ 90 kΩ, $R_4 =$ 100 kΩ, using 3 resistors overall.

9.28 7 V, 7.5 A, 6.0675 A, 18 A, 0.9 A/V, 3 A/V, 1.5 A/V.

9.29 7.12 V, 863 Ω; 0.429 of V_{14} must appear across Q_6; a) 0.927 V/V; b) 0.988 V/V; On average, the gain is 0.960 V/V.

9.30 0.1 mA; a) 0.899 V/V, 0 V, 1.22 V, – 1.22 V, 0 V; b) 0.952 V/V, 10.78 V, 12 V, 9.40 V, 10 V; On average, the gain is 0.928 V/V.

Chapter 10

Analog Integrated Circuits

10.1 0.733 mA, 0.2205 mA, 12 kΩ.

10.2 18.3 μA, 13.9 μA, 3.40 kΩ.

10.3 Use a transistor Q_{25} whose collector is connected to the emitter Q_{16}, emitter to a resistor R_{12} connected to $-V_{EE}$, and base to either the base of Q_{11} or the emitter of Q_7; 7.4 kΩ, 1 kΩ; The latter is best.

10.4 a) OK; b) From 8.4 V above the positive supply to – 20.7 V below the negative supply.

10.5 $V_T \ln(kI_{REF}/I_{C10}) = I_{C10}R_4$; $I_{C10} =$ 15.7 μA, 4 kΩ.

10.6 – 1.034 V.

10.7 $I_{C6}/I_{C3} = 1/(1+1/(2\beta^2)+1.19/\beta)$, 0.994.

10.8 For R_1 shorted, 0.727 A/A; for R_2 shorted, 1.375 A/A.

10.9 ± 16 mV.

10.10 38.1 kΩ, 56 kΩ.

10.11 For $R_6 = R_7 =$ 0 Ω, $R_{10} =$ 13.46 kΩ; For $R_6 = R_7 =$ 27 Ω, a design is not possible.

10.12 5.03 μA, 0.1 mA/V.

10.13 2.65 kΩ, 5.25 MΩ; Now 1000 V/V versus 815 V/V then.

10.14 0.497 mV, 67% larger.

10.15 ± 3.4%.

10.16 Add resistors R_E in series with the emitters of both Q_8 and Q_9, 18.2 kΩ, 15.6 MΩ, 0.0088 μA/V, 86.7 dB.

10.17 5.81 MΩ versus 4.0 MΩ previously.

10.18 55.4 Ω, 10.0 mA/V, 0.59 MΩ, 78.8 kΩ, – 788 V/V versus – 526.5 V/V previously, a gain increase of 50%.

10.19 4.38, – 421 V/V, 330 Ω.

10.20 21.7 mA, 21.6 mA, 17.9 mA.

10.21 111.2 dB, 2.77 Hz, 1.00 MHz, the same as before.

10.22 a) For 45°: 0.123 pF, 244 kHz; 0.0123 pF, 253 kHz; b) For 60°: 0.212 pF, 142 kHz; 0.0212 pF, 146 kHz.

10.23 154 V/μs, 1540 V/μs; 2.45 MHz, 24.5 MHz.

10.24 Class AB, 2.5 mA, 11.8×10^3 V/V, 13.5 pF.

10.25 For Q_1 through Q_8: I_D are 12.5, 12.5, 12.5, 12.5, 25, 25, 25, 25 μA; $|V_{gs}|$ are 1.35, 1.35, 1.35, 1.35, 1.50, 1.35, 1.50, 1.50 V; g_m are 70.8, 70.8, 70.8, 70.8, 100, 141.6, 100, 100 μA/V; r_o are 2, 2, 2, 2, 1, 1, 1, 1 MΩ; 35.4 V/V, 70.8 V/V, 2506 V/V; from 3.15 V to – 4.64 V on the input; from 4.5 V to – 4.65 V on the output.

10.26 For Q_1 through Q_8: I_D are 6, 6, 6, 6, 12, 12, 12, 12 μA; $|V_{gs}|$ are 1.28, 1.28, 1.35, 1.35, 1.40, 1.35, 1.40, 1.40 V; g_m are 42.5, 42.5, 34.6, 34.6, 60, 69.2, 60, 60 μA/V; r_o are 4.2, 4.2, 4.2, 4.2, 2.1, 2.1 2.1, 2.1 MΩ; – 89.3 V/V, – 72.7 V/V, 6488 V/V, from 3.32 V to – 6.5 V on the input; from 4.60 V to – 4.65 V on the output.

10.27 1.707 V, 2212 V/V, 3.3 mV.

10.28 6.76 pF, 14.5 kΩ, 1.10 MHz, 42.3°, 58.3 pF, 0.116 MHz; 1.78 V/μs, 0.206 V/μs.

10.29 Minimum voltage is $2V_{GS}-V_t$, $R_O = g_m r_o^2$; Thus, as measured by the output resistance and output-voltage overhead, the cascode and (modified) Wilson are the same.

10.30 484 MΩ, 18.7×10^3 V/V.

10.31 $R_O = (g_{m4CC}\, r_{O4CC}\, g_{m4C}\, r_{O4C}\, r_{O3}) || (g_{m2C}\, r_{O2C}\, r_{O2})$, 243 MΩ, 14.4×10^3 V/V, 3.23 V.

10.32 3.75 V, – 2.75 V, – 3 V, 242 MΩ, 9.36×10^3 V/V.

10.33 62.8 μA/V, 6170 V/V, 162 Hz, 2 V/μs.

10.34 55.15 MHz, 550 MHz.

10.35 See page 273: 50 kHz, 21.7 Ω.

10.36 7 bits, 9.77 mV.

10.37 128 kΩ or 256 kΩ, 3.91 Ω or 7.81 Ω, 1.95 Ω or 3.91 Ω, 0.2%.

10.38 10 kΩ, 78.4 Ω, 78 Ω to 156 Ω.

10.39 5 bits.

10.40 See page 275: Use 3 comparators with references at – ½ V, 0 V, +½ V.

10.41 a) 0 V; b) 0 V; c) Saturated at ± 10 V; d) Stays saturated; Specifically, for i) $V_A > V_{REF}$, $V_O =$ 0 V, 0 V, – 10 V, – 10 V, for a), b), c), d), respectively; for ii) $V_A < V_{REF}$, $V_O =$ 0 V, 0 V, +10 V, +10 V for a), b), c), d), respectively.

Chapter 11

Filters and Tuned Amplifiers

11.1 See Table, page 277.

11.2 0 dB, – 1 dB, – 50 dB.

11.3 0.915 dB, 26 dB, 1.25.

11.4 2.86×10^3 rad/s, 100.5 rad/s, 28.5; 0.455×10^{-3} s, 0.35 kHz, 11.2 dB.

11.5 $T(s) = (s^3+0.1s^2)/(s^3+2s^2+1.89s+0.89)$.

11.6 See page 278.

11.7 For $N =$ 13, 43.9 dB, 0.21 dB, 44.8 dB; twelfth order would suffice.

11.8 Use $N = 8$, $\omega_o = 136.7$ krad/s; poles at $\omega_o(-0.1951 \pm j0.9808)$, $\omega_o(-0.5556 \pm j0.8315)$, $\omega_o(-0.8315 \pm j0.5556)$, $\omega_o(-0.9808 \pm j0.1951)$; $T(s) = \omega_o^8/[(s^2+0.3902\omega_o s+\omega_o^2)(s^2+1.111\omega_o s+\omega_o^2)(s^2+1.663\omega_o s+\omega_o^2)(s^2+1.962\omega_o s+\omega_o^2)]$, 10.1 dB, 42.3 dB.

11.9 For Butterworth, 12.45 dB; For 4 Chebyshev, 22.5 dB.

11.10 $N = 7$, 48.5 dB, 0.074 dB, 43.8 dB.

11.11 For Butterworth, $N = 13$, $\omega_o = 1.084 \times 10^3$ rad/s; Poles at $10^3(-0.131 \pm j1.076)$ rad/s, $10^3(0.384 \pm j1.013)$ rad/s, $10^3(-0.616 \pm j0.892)$ rad/s, $10^3(-0.811 \pm j0.719)$ rad/s, $10^3(-0.969 \pm j0.505)$ rad/s, $10^3(-1.052 \pm j0.259)$ rad/s, 1.08×10^3 rad/s; For Chebyshev $N = 7$, $\omega_p = 10^3$ rad/s; Poles at $10^3(-0.057 \pm j1.006)$ rad/s, $10^3(-0.160 \pm j0.807)$ rad/s, $10^3(-0.231 \pm j0.448)$ rad/s, $10^3(-0.256)$ rad/s.

11.12 10 kΩ, 100 kΩ, 159.2 pF, 100 kHz, 1 V/V; More precisely: $f_z = 100$ kHz, 100 V/V, with $R_1 = 10$ kΩ and $R_2 = 990$ kΩ.

11.13 0.1 μF, 0.01 μF, 15.9 kΩ, 15.9 kΩ, with input resistance of 15.9 kΩ.

11.14 $T(s) = (R_{2a}/R_{1a})((s+1/(R_{1b}C_1))/(s+1/(R_{2b}C_2)))(s+1/(R_{2a} \| R_{2b}C_2))/(s+1/(R_{1a} \| R_{1b}C_1))$; $C_1 = 0.2$ μF with $R_{1b} = R_{2b} = 79.58$ kΩ and $R_{1a} = R_{2a} = 8.84$ kΩ, $C_2 = 0.002$ μF, for poles at 100 Hz and 1 kHz, and zeros at 10 Hz and 10 kHz.

11.15 $C_1 = 0.001$ μF, $R_{1a} = 17.7$ kΩ, $R_{1b} = 159$ kΩ, $R_{2a} = 177$ kΩ, $R_{2b} = 1.591$ MΩ, $C_2 = 10$ nF, with zeros at 100 Hz and 1 kHz, and poles at 10 Hz and 10 kHz.

11.16 $T(s) = (s-1/RC)/(s+1/RC)$, $-90°$; 524 Ω, 1051 Ω, 2.68 kΩ, 5.77 kΩ, 10 kΩ, 17.3 kΩ, 37.3 kΩ, 95.1 kΩ, 190.8 kΩ.

11.17 $T(s) = 200s/(s^2+200s+10^6)$, 414 rad/s, and 2414 rad/s.

11.18 a) $T(s) = s^2/(s^2+1.414s+1)$, with $Q = 0.707$ and $\omega_o = \omega_{3db} = 1$ rad/s; $A_{min} = 12.3$ dB; b) $Q = 1.707$ with $\omega_o = 1.287$; $T(s) = 0.707s^2/(s^2+0.754s+1.657)$ where $\omega_{3db} = 1$ rad/s and $A_{min} = 18.3$ dB, a 6 dB improvement.

11.19 $T(s) = (s^2+377^2)/(s^2+s\,377/Q+377^2)$, with $Q = 6.136$ and a 3 dB bandwidth of 9.78 Hz; 3dB frequencies at 55.3 Hz and 65.1 Hz; 1 dB frequencies at 49.0 Hz and 70.4 Hz; 1% frequencies at 25.9 Hz and 94.1 Hz.

11.20 796 pF, 264 μH, 10 Vrms.

11.21 530.6 pF, 0.0048 μH, 26 dB.

11.22 112.5 pF, 22.5 mH, $r_{series} = 283$ Ω, $R_{parallel} = 707$ kΩ.

11.23 $R_1 = R_2 = R_3 = R_5 = 10$ kΩ, $C_4 = 0.1$ μF for $L = 10$ H and 1 nF for $L = 0.1$ H; $R_1 = R_2 = R_3 = 10$ kΩ, $C_4 = 10$ nF, $R_5 = 100$ kΩ or 1 kΩ.

11.24 a) $R_1 = R_2 = R_3 = 10^4$Ω, $R_5 = 10^3$Ω, $C_4 = 10$ nF, $C = 254$ nF, for $Q = 31.9$; b) $R_1 = R_2 = R_3 = 10^4$Ω, $R_5 = 10^5$Ω, $C_4 = 10$ nF, $C = 254$ nF, for $Q = 0.32$; c) $R_1 = R_2 = R_3 = 10^4$Ω, $C_4 = C = 0.05$ μF, $R_5 = 1.013$ kΩ, for $Q = 6.28$; with $C_4 = C = 0.1$ μF and $R_5 = 253$ Ω, $Q = 12.6$.

11.25 For the first-order section: $\omega_o = 1/(2\pi 10^4)$, $Q = 0.5$, $R_1 = 33$ kΩ‖33 kΩ, $R_2 = 33$ kΩ, $C = 400$ pF‖80 pF; For one second-order section: $\omega_o = 1/(2\pi 10^4)$, $Q = 3.953$, $C_4 = C_6 = C = 400$ pF ‖ 80 pF, $R_1 = R_2 = R_3 = R_5 = 33$ kΩ, $R_6 = 120$ kΩ; For the other second-order section: $\omega_o = 1/(2\pi 10^4)$, $Q = 0.7198$, $C_4 = C_6 = C = 400$ pF‖80 pF, $R_1 = R_2 = R_3 = R_5 = 33$ kΩ, $R_6 = 24$ kΩ.

11.26 Use $C_4 = C_6 = C$, with R_6 to control Q (and ω_o), and $R_1 = R_2 = R_3 = R$, with R_5 to control ω_o; $\omega_o = 1/(CR^{1/2}R_5^{1/2})$, $Q = R_6/(RR_5)^{1/2}$; R_6 must vary by a factor of 200.

11.27 $R_x = 2.7$ kΩ, $R_1 = R_2 = R_{3a} = R_{3b} = R_f = 100$ kΩ, -39 V/V.

11.28 $R_L =$ 10 kΩ, $R_H =$ 22.5 kΩ, $R_B = \infty$ Ω, $R_F =$ 15.79 kΩ, $C =$ 1 nF, $R =$ 31.83 kΩ, $R_1 = R_2 = R_f =$ 10 kΩ, $R_3 =$ 190 kΩ.

11.29 $R_1 =$ 102.6 kΩ, $R_2 = R_3 = \infty$ Ω, $r =$ 100 kΩ, $R =$ 200 kΩ, $QR =$ 4 MΩ, $C_1 =$ 0, $C =$ 769 pF, $R_2 = R_3 = \infty$ Ω.

11.30 $C_1 = C_2 =$ 20 pF, $R_3 =$ 0.707 MΩ, $R_4 =$ 0.354 MΩ; Alternatively, use $C_1 =$ 10 pF, $C_2 =$ 20 pF, $R_3 =$ 1 MΩ and 62 kΩ in series, $R_4 =$ 470 kΩ; Alternatively, use $C_1 =$ 10 pF, $C_2 =$ 50 pF, $R_3 =$ 0.849 MΩ, $R_4 =$ 0.236 MΩ; Of the 3 solutions, the first with equal 20 pF capacitors is the most straightforward.

11.31 $R_3 =$ 637 Ω, $R_4 =$ 398 Ω, $R_4/\alpha =$ 318 kΩ, $R_4/(1-\alpha) =$ 398.5 Ω; $R_{in} =$ 318 kΩ at both low and high frequencies.

11.32 $Z_i = (R_4/\alpha)(s^2+s(1/R_3C_1+1/R_3C_2)+1/(C_1C_2R_3R_4))/(s^2+s(1/R_3C_1+1/R_3C_2)+(1-\alpha)/(C_1C_2R_3R_4))$ $= (R_4/\alpha)(s^2+\omega_o s/Q+\omega_o^2)/(s^2+\omega_o s/Q+(1-\alpha)\omega_o^2)$; $Z_i = R_4/(\alpha(1-\alpha))$, as s approaches 0; $Z_i = R_4/\alpha$ as s approaches ∞; at the centre frequency, $Z_i = R_4(1+j\alpha Q)/(\alpha(1+\alpha^2Q^2))$.

11.33 $C =$ 3.3 nF; For the first-order section, $R =$ 0.965 kΩ; For the lowest-Q Sallen and Key section, $R_1 =$ 0.998 kΩ, $R_2 =$ 0.933 kΩ; For the second Sallen and Key section, $R_1 =$ 1.281 kΩ, $R_2 =$ 0.727 kΩ; For the third Sallen and Key section, $R_1 =$ 2.68 kΩ, $R_2 =$ 0.348 kΩ.

11.34 $S^{\omega_o}_{C_{3a}} = -1/(2(1+k_3))$, $S^{\omega_o}_{C_{3b}} = -k_3/(2(1+k_3))$, $S^{\omega_o}_{C_{4a}} = -1/(2(1+k_4))$, $S^{\omega_o}_{C_{4b}} = -k_4/(2(1+k_4))$; All become – ¼; $S^{Q}_{C_{3a}} = -1/(2(1+k_3)$, $S^{Q}_{C_{3b}} = -k_3/(2(1+k_3))$; Both become – ¼; $S^{Q}_{C_{4a}} = 1/(2(1+k_4))$, $S^{Q}_{C_{4b}} = k_4/(2(1+k_4))$; Both become ¼.

11.35 $S^{\omega_o}_k = 0$, $S^{\omega_o}_T = aT_0$ around $T = T_0$; $S^Q_k = -k^2$, $S^Q_T = 0$.

11.36 0.1 pC, 0.1 μA, 10 MΩ, 50 mV in the negative direction, 400 cycles, – 50,000 V/s, 5,000 V/s.

11.37 $C_1 = C_2 =$ 2.000 pF, $C_3 = C_4 =$ 1.257 pF, $C_5 = C_6 =$ 1.777 pF, bandpass output, 10^5 Hz, 1.41×10^5 Hz, 1 V/V.

11.38 a) 57.2×10^6 rad/s, 87.7, 652 krad/s, – 51.1 V/V; b) 67.1×10^6 rad/s, 124, 540 krad/s, – 16.7 V/V.

11.39 a) 62.9×10^6 rad/s, 106, 592 krad/s, – 66.8 V/V; b) 70.3×10^6 rad/s, 139.2, 505 krad/s, – 19.6 V/V; Thus the tapped coil gives the best results in general; see page 294 for a comparison.

11.40 0.63 Ω, 25.1 kΩ, 126.7 pF, 8.38 kΩ.

11.41 10; Use three stages: For one stage, bandwidth is 3.16 MHz and selectivity is 31.6; For the cascade, bandwidth is 0.294 MHz and selectivity is 5.88.

11.42 $C_1 =$ 73.27 pF, $C_2 =$ 74.23 pF, $R_1 =$ 30.616 kΩ, $R_2 =$ 30.416 kΩ; Relative gain is 1.007.

Chapter 12

Signal Generators and Waveform-Shaping Circuits

12.1 $\omega_0 = 1/(LC)^{1/2}$, $G_m = n/R_o + nRC_{ls}/L + n/R_{cp} + 1/(nR_{in})$.

12.2 See page 297: Note that amplifier polarity change can be accounted for by coil-tapped connection reversal.

12.3 $G_m = 1/(nR_o + n/R_{cp} + n/R_{lp} + n/R_{in})$.

12.4 $R_1 =$ 10 kΩ, $R_2 =$ 31.0 kΩ, $R_3 =$ 5.6 kΩ, $R_f =$ 50 kΩ, 5.83 kHz.

12.5 For zeners which are normal {or better}, gain: for υ_O=0 is {– 1.33 V/V} (– 1.33 V/V), For υ_O> 0.7 V is – 1.18 V/V {– 1.32 V/V}, For υ_O> 7.5 V is – 0.67 V/V {– 0.67 V/V}, with the input threshold at 6.36 V {5.68 V}.

12.6 $C = 10$ nF, $R = 1.59$ kΩ, $R_1 = 3.3$ kΩ, $R_2 = 8.2$ kΩ.

12.7 0.280 MHz; The closed-loop op-amp gain must be 3.02/(1 + 0.755j).

12.8 $L(\omega) = (1+R_2/R_1)/\left[1+j\omega/(\omega_t(1+R_2/R_1))-j/(\omega RC)+1/(RC\omega_t(1+R_2/R_1))\right]$, $\omega = \left[\omega_t(1+R_2/R_1)/(RC)\right]^{1/2}$, $R_2/R_1 = 1/(RC\omega_t(1+R_2/R_1))$, $R_2 = 0.0625R_1$; Excess phase shift implies operation at a frequency lower than in the simple case; see page 300.

12.9 $L(s) = -R_fCs/\left[1+(R_1+2R_2+3R_3)Cs+(R_1R_2+2R_2R_3+2R_1R_3)C^2s^2+R_1R_2R_3C^3s^3\right]$; $\omega_o = 1/(C(R_1R_2+2R_2R_3+2R_1R_3)^{1/2})$, which for $R_1 = R_2 = R_3 = R$ is $\omega_o = 1/(CR\sqrt{5})$ with $R_f = 6.26R$; $S^{\omega_o}_{R_1,R_2} = -0.3$, $S^{\omega_o}_{R_3} = -0.5$, $S^{R_f}_{R_1} = 0.174$, $S^{R_f}_{R_2} = 0.335$, $S^{R_f}_{R_3} = 0.497$.

12.10 $L(s) = -(R_f/R)\left[3+s(4RC+3R_fC)+s^2(R^2C^2+4RR_fC^2)+R^2R_fC^3s^3\right]^{-1}$.

12.11 Five sections needed, $\omega_o = 0.191/(RC)$, $|L| = 5804$ V/V.

12.12 $L(s) = -R_fC\left[1/(R^3C^3s^4)+6/(R^2C^2s^3)+10/(RCs^2)+4/s\right]^{-1}$, $\omega = 1.224/(RC)$, $R_f = 6.22$ R.

12.13 $R = 1.59$ kΩ, $C = 10$ nF, $Q = 12.5$, 1.78 Vpeak, $R_1 = 1.1$ kΩ, $R_6 = 19.9$ kΩ.

12.14 $C_2/C_1 = 10.83$, $C_1 = 2.767$ nF, $C_2 = 29.97$ nF; For high supply voltages, the peak voltage can be estimated as 6.66 V, 7.1 V or 8.1 V; For 3 volt supplies a peak signal of about 1 V results.

12.15 0.523 H, 0.0119 pF, 1325 Ω; $C_2/C_1 = 6.4$ for which $C_1 = 1.56$ pF with $R_1 = 0$, or $C_1 = 1$ pF with $R_1 =$ 4.1 kΩ.

12.16 $R_1 = 1.07$ kΩ, $R_2 = 5.13$ kΩ; V_{IL} ranges from 1.04 V to 2.94 V, and V_{IH} ranges from 2.06 V to 3.96 V; Estimates of some critical delay times include 0.4 ns, 0.64 ns, 10 ns, 75.3 ns, 3 μs, (see page 307).

12.17 $R_1 = 2.87$ kΩ with $R_2 = 7.8$ kΩ, or $R_1 = \infty$ Ω with $R_2 = 10$ kΩ and $R_3 = 5.6$ kΩ; $V_{TL} = +0.7$ V, $V_{TH} = +1.4$ V.

12.18 See page 309: $V_{IH} = 3.83$ V, $V_{IL} = 1.17$ V.

12.19 $R = 200$ kΩ, $R_1 = 33$ kΩ, $R_2 = 240$ kΩ, $R_3 = 3.9$ kΩ; The slope is 0.036 V/μs at $t = 0$, 0.028 V/μs at $t = 50$ μs, and 0.032 V/μs on average, $R = 424$ kΩ; Use $R_1' = 10$ kΩ and R_2' as 10 kΩ in series with 680 Ω.

12.20 $R_A = 1$ kΩ, $R_B = 7.2$ kΩ, wih thresholds of ± 1 V; $R_1 = 10$ kΩ, $R_2 = 82$ kΩ; $C = 1000$ pF, $R = 410$ kΩ, $R_3 = 1.5$ kΩ.

12.21 – 0.7 V, – 1.67 V, – 0.833 V, – 10 V, 1.07 V step, 0.25 ms pulse, > 118 V/s; R_4 controls recovery; Retriggering is possible 10 μs (plus the amplifier recovery time) after output falls.

12.22 54.2 μs; For very long input pulses, the loop is open, and the output fall time is an amplified version of that at node *A*; 124 ns.

12.23 $T = 0.11$ ms; Input less than 0.11 ms; For longer inputs, set and reset are applied together.

12.24 – 1.8%.

12.25 4.83 kHz, 66.7%; 12.1 kHz, 91.7%; 14.5 kΩ; There is no combination of resistors which will produce 10% duty cycle: Use 90% duty cycle and an inverter!

12.26 400 Ω; See the Table on page 313: The maximum error is about 8% at about 45°.

12.27 Progressively: $R_1 = 5$ kΩ, $V_2 = 2.4$ V, $V_3 = 6.4$ V, $R_2 = 1.15$ kΩ, $R_3 = 1.15$ kΩ; Errors: – 0.3 mA at 3 V, + 0.1 mA at 5 V, – 0.3 mA at 7 V, 0 mA at 10 V; With 1 mA diodes: $V_2 = 2.4$ V, $V_3 = 6.4$ V, $R_1 = 5$ kΩ, $R_2 = 1.139$ kΩ, $R_3 = 1.659$ kΩ; Errors: – 0.25 mA at 3 V, + 0.14 mA at 5 V, – 0.13 mA at 7 V, – 0.28 mA at 10 V.

12.28 Requires 7 amplifiers and 4 diodes in total, with extra inverters from input υ_3 to the lower amplifier, and on the output.

12.29 This is an absolute-value circuit.

12.30 See page 316: $R_3 = R_1 \| R_2 = R/2$ with $R \gg R_0$.

12.31 See page 316: R_1 is 56 kΩ in series with 620 Ω, $R_2 =$ 10 kΩ, R_5 is to 12 kΩ resistors in parallel.

12.32 + 10 V, 0 V, + 10 V; Input resistance is infinite; The circuit is called a full-wave doubler rectifier.

12.33 Up to $\upsilon_I =$ 14 mV, $\upsilon_O/\upsilon_I =$ 101 V/V, with υ_O reading 1.414 V; From $\upsilon_I =$ 14 mV to $\upsilon_I =$ 150 mV, $\upsilon_O/\upsilon_I =$ 51 V/V, with υ_O reading 8.35 V; Above $\upsilon_I =$ 150 mV, $\upsilon_O/\upsilon_I =$ 1 V/V.

12.34 The output is a dc level of 200 mV which remains when the input is lowered; $R_1 =$ 3.3 / (fC) to ground at the output; Output rises to represent the maximum peak-to-peak value of the combined signal at $f/100$; add $R_2 =$ 0.48 / (fC) from node B to ground.

Chapter 13
MOS Digital Circuits

13.1 3 V, 0.542 V.

13.2 $NM_H =$ 1.3 V, $NM_L =$ 0.7 V, 1.0 V, – 3 V/V.

13.3 $t_{PHL} =$ 3.92 ns, $t_{PLH} =$ 12.35 ns, 29.7 MHz, 20.8 mW, 256 pJ.

13.4 $V_{OH} =$ 1.083 V, $V_{OL} =$ 0.614 V, 0.287 mA from, 1.28 mA to.

13.5 30 MHz, 35 pJ.

13.6 5 MHz; See page 320.

13.7 0.83 $V^{1/2}$.

13.8 a) $V_{OH} =$ 4 V, $V_{IL} =$ 1 V, $V_{OL} =$ 0.204 V, $V_{IH} =$ 2.03 V, $NM_H =$ 1.97 V, $NM_L =$ 0.796 V; b) $V_{OH} =$ 3.32 V, $V_{IH} =$ 1.98 V; c) $V_{OL} =$ 0.197 V, $V_{OH} =$ 3.40 V, $V_{IH} =$ 1.99 V, $NM_L =$ 1.41 V, $NM_L =$ 0.803 V.

13.9 a) 1.67 V; b) 1.59 V.

13.10 6.8 ns, 0.64 ns.

13.11 $(W/L)_1 =$ 1.79, $(W/L)_2 =$ 0.286, $V_{OL} =$ 0.375 V; $(W/L)_1 =$ 3.5, $(W/L)_2 =$ 0.572, $V_{OL} =$ 0.449 V.

13.12 5 V, 0.093 V, 1.99 V, 1.18 V, $NM_H =$ 3.01 V, $NH_L =$ 1.09 V.

13.13 3.57 kΩ, 79.5 kΩ.

13.14 1.99 V, 0.495 V; 2.86 V; 1.94 V, 0.472 V.

13.15 0.092 V; 2 times.

13.16 0.50 ns, 9.17 ns; 4.84 ns; 0.33 ns, 13.1 ns, 0.43 pJ.

13.17 a) 4 units 4/2 with 1 unit 2/4, 32 μm^2, 40 μm^2; b) 4 units 16/2 with 1 unit 2/4, 128 μm^2, 136 μm^2; c) Use 4 units 4/2 with 1 unit 2/4 and 5 units 4/2 with 5 units 2/4, 32 μm^2, 120 μm^2.

13.18 72 μm^2, 104 μm^2.

13.19 4 V, 0.317 V, 0.00 V, 35.9 μA, 3.68 V; 8.68 V, 5 V, 4 V, 6.4 mA; 1.8 mA, 4.0 V, 0.32 V, 0.0 V.

13.20 3.125 kΩ, 3.125 kΩ.

13.21 Both 640 μA; both 550 μA; both 150 μA.

13.22 V_{IL} from 1.59 V to 2.66 V, V_{IH} from 2.16 V to 3.59 V, V_{OL} always 0 V, V_{OH} from 3.75 V to 6.25 V; NM_H from 4.09 V to 0.16 V, NH_L from 2.66 V to 1.59 V.

13.23 90 μA, 13.5 μA, 50 μA, 63.5 μA total, 317.5 μW, 67.5 μW with no capacitor.

13.24 $t_{PHL} = t_{PLH} = t_p =$ 2.10 ns, 49.8 MHz, 1.25 pJ.

13.25 a) 0.8 to 1.3 V, 0 to 0.5 V, neither conducts; b) 0 to 1.3 V, 1.3 V, 0.0 V; c) 0.8 V, 0.5 V, no current flows; d) See page 326; e) 0.39 μs, 0.64 μs; f) 156 kHz.

13.26 $V_{OH} =$ 5 V, $V_{OL} =$ 0 V, $V_{IL} =$ 1.275 V, $V_{IH} =$ 2.02 V, $V_{th} =$ 1.75V, $t_{PLH} =$ 5.57 ns, $t_{PHL} =$ 2.89 ns.

13.27 $V_{OH} =$ 5 V, $V_{OL} =$ 0 V, $V_{IL} =$ 1.78 V, $V_{IH} =$ 2.68 V, $V_{th} =$ 2.10 V, $t_{PLH} =$ 2.30 ns, $t_{PHL} =$ 2.35 ns.

13.28 See page 327: Uses $\overline{Y} = \overline{A}B + \overline{C} + D$.

13.29 $(W/L)_{n,D,\overline{C}} =$ 2, $(W/L)_{n,\overline{A},\overline{B}} =$ 4 $(W/L)_p =$ 12.

13.30 1.36 kΩ, threshold is 2.10 V.

13.31 $F = CXY\overline{Z}$; Use an additional p-channel MOSFET at the left connected to the supply and $\overline{X}$; $(W/L)_4 =$ 10/5, $(W/L)_{1,2,3} =$ 60/5.

13.32 2 mA, 1.64 mA, 10 mW, 8.2 mW.

13.33 Total areas of 39 and 48 units, with input capacitances of 9 units versus 3 units for an inverter; Area is 111 units with 3 units of input capacitance.

13.34 See page 329: equal at (0 V, 0 V), (5 V, 5 V), (2.64 V, 2.64 V); gain is 74 V/V in the middle, and 0 V/V at the ends of the range.

13.35 See page 330.

13.36 W_5 values of 0.56W, 0.25W, 1.0W; use the same values for W_6.

13.37 30.2 μs, 4.96 V, 2.7 V, 0.024 V, 0.225 mA, 12.8 mA.

13.38 3.67 pF.

13.39 22 kΩ with 220 kΩ at the gate.

13.40 9 bits; The 9 bits on the right are 0 0 1 0 0 1 1 0 0 ≡ 76_{10}.

13.41 See page 332: 10 transistors.

13.42 2.16 V, 62.3 μA to the cell, 31.3 μA from the cell, 3.6 μm, $i_5 =$ 27.6 μA, $i_1 =$ 105.1 μA: Writing cannot occur by Q_5 overpowering Q_1; $i_6 =$ 240 μA, $i_4 =$ 53.4 μA: Writing can occur by Q_6 overpowering Q_4, and the design is viable.

13.43 0.027 fF.

13.44 See page 332: 25 in the array, 10 for inverters, 35 total.

13.45 $V_{OH} =$ 0.7 V, $V_{OL} =$ 0.114 V, $V_{IL} =$ 0.51 V, $V_{IH} =$ 0.497 V, $NM_H =$ 0.203 V, $NM_L =$ 0.396 V.

13.46 0.544 mA, 0.816 mW, 13.4 ps, 35.2 μW, 0.011 pJ.

13.47 $V_{OH} =$ 0.7 V, $V_{OL} =$ – 1.27 V, $V_{IL} =$ – 0.26 V, $V_{IH} =$ – 0.16 V, $NM_H =$ 0.86 V, $NM_L =$ 1 V; a) $V_{OH} =$ 0.7 V, $V_{OL} =$ – 1.29 V, $V_{IH} =$ – 0.234 V, $V_{IH} =$ – 0.146 V, $NM_H =$ 0.85 V, $NM_L =$ 1.06 V; b) $V_{OH} =$ 0.7 V, $V_{OL} =$ – 1.09 V, $V_{IL} =$ – 0.293 V, $V_{IH} =$ – 0.184 V, $NM_H =$ 0.88 V, $NM_L =$ 0.80 V; c) $V_{OH} =$ 0.7 V, $V_{OL} =$ 0.032 V, $V_{IL} =$ – 0.582 V, $V_{OL} =$ – 1.37 V, $V_{IH} =$ – 0.532 V, $NH_H =$ 1.23 V, $NH_L =$ 0.79 V.

13.48 $W_S = W_L =$ 20 μm, $W_{PD} =$ 10 μm, $V_{OH} =$ 2.1 V, $V_{OL} =$ 0.13 V, $V_{IL} =$ 1.14 V, $V_{IH} =$ 1.24 V, $NM_H =$ 0.86 V, $NM_L =$ 1.01 V; For a design for which $W_{PD} =$ 2.5 μm, these results are an approximation for fanout of 4.

Chapter 14

Bipolar Digital Circuits

14.1 a) 1 mA, 0.11 μA, 645 mV; b) 0.999 mA, 1.1 μA, 691 mV.

14.2 $\beta_{forced} =$ 0, 58.0 mV to 57.7 mV.

14.3 49.3 mV, 48.7 mV, 12 Ω.

14.4 $\beta_{forced} =$ 0.1; V_{ECsat} (mV): ∞, 58.1, 17.8, 6.1, 3.1, 0.75, 0.50.

14.5 199; V_{ECsat} (mV): ∞, 223, 168, 133, 114, 64, 40; In normal mode, 42.2 mV; In inverted mode, 12.9 mV.

14.6 a) $i_{DC} =$ 2.80 mA, $i_{DE} =$ 12.87 mA; b) 697 mV, 649 mV, 48 mV; c) 45.6 mV.

14.7 380 ns, 18.1 ns, 1.89 ns.

14.8 15.2 ns, 35.8, 12.8 pF, 3 ns, 99.9 ns.

14.9 0.643 V, 0.625 V.

14.10 0.820 V, 0.972 V.

14.11 a) 16.7 μA; b) 1.376 V; c) 0.409 mA; d) 1.64 mA; e) Add another diode in series with D_3, D_4; $R_1 =$ 1.26 kΩ, 2.60 mA, fan out of 2.

14.12 0.41 mA, 2.31 mA.

14.13 1.46 mA, 64, 95.6 mV.

14.14 42.7 ns, 15.5 ns, 47.0 ns, 26.2 ns.

14.15 505 Ω, 15.5 ns, 5.16 ns.

14.16 $V =$ 2.0 V, $R =$ 600 Ω; For $V =$ 2.5 V, use $R =$ 1.0 kΩ for which $N \le 21$; For $V =$ 2.0 V, use $R =$ 0.6 kΩ for which $N \le 27$.

14.17 See page 344.

14.18 $V_{B3} =$ 0.7 V, $I_{1K} =$ 0.7 mA, $V_{C2} =$ 0.9 V, $I_{C2} =$ 2.56 mA, $V_{B2} =$ 1.4 V, $V_{B1} =$ 2.1 V, $I_{B1} =$ 0.725 mA, $I_{E1} =$ 35 μA, $I_{E2} =$ 3.33 mA, $I_{B3} =$ 2.59 mA, $V_{CEsat2} =$ 0.124 V, $V_{CEsat3} =$ 0.107 V, $N \le 23$.

14.19 $V_{E1} =$ 0.3 V, $V_{B1} =$ 1.0 V, $V_{C1} =$ 0.5 V, $I_{B4} =$ 0.3 mA, $V_{B4} =$ 4.5 V, $V_{E4} =$ 3.8 V, $V_{out} =$ 3.1 V, $I_{B4} =$ 0.13 mA, $V_{B4} =$ 4.5 V, $V_{C4} =$ 4.64 V.

14.20 $V_{B3} =$ 0.7 V, $V_{B2} =$ 1.4 V, $V_{B1} =$ 2.1 V, $V_{E1} =$ 1.4 V, $V_{E4} =$ 1.96 V, $V_{B4} =$ 2.66 V, $V_O =$ 1.29V with $I_{C3} =$ 4.4 mA, $I_{B3} =$ 0.49 mA, $I_{E3} =$ 4.9 mA, $I_{E2} =$ 1.19 mA, $I_{E1} =$ 0.61 mA; $V_{CEsat1} =$ 72 mV.

14.21 a) $I_{C3} =$ 1.1 mA, $I_{E4} =$ 7.0 mA, $I_{B4} =$ 0.70 mA, $I_{C2} =$ 0.74 mA, $I_{E1} =$ 0.82 mA, $I_{B3} =$ 0.12 mA, $I_{C3} =$ 1.1 mA, with $V_O =$ 1.29 V and $I_L =$ 6.5 mA; b) $I_{C3} =$ 18 mA, $I_{B3} =$ 2 mA, $I_{E2} =$ 2.7 mA, $I_{C2} =$ 2.43 mA, $V_{C2} =$ 1.1 V, $I_{B2} =$ 0.27 mA, $I_{E1} =$ 0.46 mA, $V_O =$ 1.3 V, with Q_4 cut off.

14.22 $R_2 =$ 1.37 kΩ; Gross slope = – 1 V/V, $R_1 =$ 1.17 kΩ; Increase is 0.94 mA or 36%.

14.23 For source at – 55 °C and load at 125 °C: $NM_L =$ 0.70 V, $NM_H =$ 1.16 V; For source at 125 °C and load at – 55 °C: $NM_L =$ 1.42 V, $NM_H =$ 1.74 V.

14.24 4.76 mA, 17.8 ns.

14.25 1.34 V.

14.26 1.305 mA.

14.27 a) 0.75 V, 0.25 V, 19.6 μA, 0.98 mA; b) 0.767 V, 0.266 V, 0.04 mA, 0.96 mA; c) 0.810 V, 0.316 V, 0.216 mA, 0.784 mA; d) 0.825 V, 0.268 V, 0.392 mA, 9.61 mA; e) 0.810 V, 0.256 V, 0.216 mA, 9.78

mA.

14.28 a) 0.675 mA, 0.675 mA; 3.73 mW, 3.73 mW; b) 0.213 mA, 33.5 mA; 1.07 mW, 167.5 mW; $DP =$ 54.9 pJ.

14.29 a) $V_{OH} = -0.75$ V, $V_{OL} = -(RI + 0.75)$V; b) $V_{th}A = -(RI/2 + 0.75)$V, $V_{BE} =$ 682.7 mV; c) 748.7 mV; d) 675.1 mV, 73.6 mV; e) $V_{IL} = -(RI/2 + 0.824)$V, $V_{IH} = -(RI/2 + 0.676)$V; f) $NM_H = (RI/2 - 0.074)$V, $NM_L = (RI/2 - 0.074)$V; g) $IR =$ 0.442 V; h) $V_{OH} = -0.750$ V, $V_{OL} = -1.192$ V, $V_{IH} = -0.897$ V, $V_{IL} = -1.045$ V, $V_R = -0.971$ V.

14.30 285 Ω, – 0.750 V, – 1.89 V, – 1.493 V, – 1.147 V, 0.397 V, 0.397 V.

14.31 285 Ω; Use 2 diodes for $V_{OH} =$ 2.75 V; Use $R_1 =$ 570 Ω for $V_{OL} =$ 0.47 V.

14.32 $R_2 =$ 148 Ω, $I_{R2} =$ 0.95 mA, $V_{OH} = -0.75$ V or slightly more, and $V_{OL} = -1.04$ V to – 1.14 V.

14.33 0.205 V.

14.34 2 cm.

14.35 a) 2.24 V, 5.0 V, 0.0 V; b) 2.24 V, 4.17 V, 0.96 V.

14.36 $(W/L)_p =$ 2W/L, $(W/L)_n =$ W/L, $V_{th} =$ 2.06 V.

STANDARD COMPONENT VALUES

• Standard 5% Values with Marked $\overline{10\%}$ and $\underline{20\%}$ Values

$\underline{\overline{1.0}}$, 1.1, $\overline{1.2}$, 1.3, $\underline{\overline{1.5}}$, 1.6, $\overline{1.8}$, 2.0, $\underline{\overline{2.2}}$, 2.4, $\overline{2.7}$, 3.0, $\underline{\overline{3.3}}$,

3.6, $\overline{3.9}$, 4.3, $\underline{\overline{4.7}}$, 5.1, $\overline{5.6}$, 6.2, $\underline{\overline{6.8}}$, 7.5, $\overline{8.2}$, 9.1, $\underline{\overline{(10)}}$.

• Standard 1% Values with Marked "$\underline{\textit{Unit}}$" Values

$\underline{1.00}$, 1.02, 1.05, 1.07, 1.10, 1.13, 1.15, 1.18, 1.21, 1.24, 1.27, 1.30, 1.33, 1.37,

1.40, 1.43, 1.47, 1.50, 1.54, 1.58, 1.62, 1.65, 1.69, 1.74, 1.78, 1.82, 1.87, 1.91,

1.96, $\underline{2.00}$, 2.05, 2.10, 2.15, 2.21, 2.26, 2.32, 2.37, 2.43, 2.49, 2.55, 2.61, 2.67,

2.74, 2.80, 2.87, 2.94, $\underline{3.01}$, 3.09, 3.16, 3.24, 3.32, 3.40, 3.48, 3.57, 3.65, 3.74,

3.83, 3.92, $\underline{4.02}$, 4.12, 4.22, 4.32, 4.42, 4.53, 4.64, 4.75, 4.87, $\underline{4.99}$, 5.11, 5.23,

5.36, 5.49, 5.62, 5.76, 5.90, $\overline{6.04}$, 6.19, 6.34, 6.49, 6.65, 6.81, $\underline{6.98}$, 7.15, 7.32,

7.50, 7.68, 7.87, $\underline{8.06}$, 8.25, 8.45, 8.66, 8.87, $\underline{9.09}$, 9.31, 9.53, 9.76, (10.0).

• "Unit" Values

1, 2, 3, 4, 5, 6, 7, 8, 9, (10).

• Tens Values (with {Geometric} and [Arithmetic] "Means")

1, {3}, [5], (10).